Office of Population Censuses and Surveys
Health and Safety Executive

Series DS no. 10

Occupational Health

Decennial Supplement

The Registrar General's decennial supplement for England and Wales

Edited by Frances Drever

London: HMSO

First published 1995

ISBN 0 11 691618 4

Printed in the United Kingdom for HMSO
Dd300387 3/95 C18 G3397 10170

Foreword

For over 150 years, data on occupational mortality have been made available in the Registrar Generals' Decennial Supplements.

The present volume - Occupational Health - is a joint venture of the Office of Population Censuses and Surveys and the Health and Safety Executive. Not only is occupational mortality commented on, but health related issues are discussed. There are chapters on injuries at work, asbestos-related diseases, sickness absence and smoking and drinking. Authors outside government have been involved in writing several chapters, noticeably on mortality and cancer registration.

This is the first in a new series of decennial supplements. *The Health of Adult Britain* and *The Health of Our Children - a review in the 1990s* will be available later in 1995. OPCS is planning a volume on socio-economic differences in health for publication in late 1997.

This book is only possible because of the hard work of many people. The authors of individual chapters are to be commended in managing to co-operate with each other. Mr Tim Devis and Dr Morris Greenberg did much of the preliminary work including ensuring authors had the data they required. Mrs Frances Drever joined the team late in the day but edited the volume and saw the volume through the production process. Ms Karen Dunnell and Dr John Osman oversaw the project from beginning to end. Without these and all the people in the background this volume would never have been produced.

Dr John Fox
CHIEF MEDICAL STATISTICIAN
OPCS

Main contributors

Mrs Frances Drever
Ms Ann Bethune
Dr Penny Babb
Dr Morris Greenberg
OPCS
St Catherine's House
10 Kingsway
London
WC2B 6JP

Dr John Osman
Dr Richard Elliott
Ms Jacky Jones
Health and Safety Executive
Magdalen House
Stanley Precinct
Bootle
Merseyside
L20 3QZ

Ms Jacqui Cooper
Health and Safety Executive
Daniel House
Trinity Road
Bootle
Merseyside
L20 7HE

Mr Philip McCrea
Health and Safety Executive
Baynards House
1 Chepstow Place
Westbourne Grove
London
W2 4TF

Dr David Coggon
Ms Heather Inskip
Mr Paul Winter
MRC
Environmental Epidemiology Unit
University of Southampton
Southampton General Hospital
Southampton
SO16 6YD

Dr Eve Roman
Dr Lucy Carpenter
Imperial Cancer Research Fund
Cancer Epidemiology Unit
Gibson Building
The Radcliffe Infirmary
Oxford
OX2 6HE

Contents

Page

Appendices

List of tables

Chapter 8

Chapter 9

Chapter 13

Page

Chapter 14

Chapter 16

List of figures

Summary

The background (Chapter 1)

Traditionally, the Occupational Mortality Decennial Supplement has been written and collated by the Office of Population Censuses and Surveys (OPCS). This decennial supplement has been the collaboration of both OPCS and the Health and Safety Executive (HSE). The title has also been changed to Occupational Health as aspects of health as well as mortality are examined.

Sources and methods (Chapter 2)

More sources than ever before have been used to put together this decennial supplement. Not only have deaths been analysed, but aspects of occupational health have also been analysed, for example injuries at work and smoking consumption. This has meant that different techniques have been used in different chapters. This chapter brings together a brief description of the sources and methods used.

Demographic and employment trends (Chapter 3)

Movements in demographic, employment and occupational patterns to changes in workplace health and safety risks over the past twenty years are described and the implications for future occupational health and safety standards are considered in this chapter.

A combination of supply-side and demand-side factors have altered employment and occupational patterns since the early 1970s. The increasing importance of females in the labour force has been accommodated by the increased demand for higher-skilled, non-manual and part-time jobs. Technological change has increased the importance of such occupations within industries. However, the main force has been the shift in industrial emphasis away from the traditionally male-dominated primary and manufacturing sectors and towards the service sector. As a result, between 1973 and 1993, employment in service-sector related occupations increased by nearly 4 million, offset by falls in the numbers employed in the primary and manufacturing sectors.

These trends are expected to continue into the next decade as factors such as technological change, changes in patterns of consumer demand and growing competitive pressures continue to impose themselves on the labour market. The scale of such changes are, however, expected to be less dramatic.

Although difficult to isolate the impact of any single factor affecting the incidence of workplace injuries and illnesses, the employment and occupational trends outlined above are seen to have generally positive implications for workplace health and safety. There are reasons, however, why we should not be complacent.

As the nature of risks faced by employees in the economy alter, the emphasis of health and safety management may be required to change as new or previously neglected forms of workplace injuries and illnesses emerge in the place of those associated with the declining traditional sectors of the economy. Additionally, the expected decline in the incidence of 'traditional' illnesses are unlikely to become immediately apparent owing to the often long latency periods between first exposure and diagnosis. We can also expect the general increased tendency to report occupational illnesses as a result of improvements in knowledge, compensation information, and health and safety expectations to continue into the next decade, implying that further progress in occupational health and safety may therefore increasingly lie in the control of risks (such as slips, trips and falls on the same level) more common to all sectors of the economy, and in further programmes aimed at improving general health and safety management and work procedures.

Occupational mortality of men (Chapter 4)

Occupational mortality among men aged 20-74 in England and Wales during 1979-80 and 1982-90 was analysed for the 194 job groups defined in Appendix 2. The most important occupational causes of mortality apparent in the analysis were accidental injury, asbestos and coal mine dust. Coal miners had elevated mortality from both pneumonoconiosis and chronic bronchitis and emphysema, but it was notable that whereas their risk of coalworkers' pneumonoconiosis varied markedly between coal fields, that for chronic bronchitis and emphysema was much more uniform geographically.

Clues emerged to several possible occupational hazards previously unrecognised. In particular, a marked excess of deaths from pneumonococcal and unspecified lobar pneumonia was observed among welders and several other occupations entailing exposure to metal fume. This is consistent with previous decennial supplements, and suggests that inhalation of metal fume or an associated exposure depresses the immune response of the lung to infection.

Other findings of note included elevated mortality from suicide and certain types of accidental injury among farmers, particularly in those who were self-employed; and a high proportion of deaths from oesophageal cancer in farmers from Hereford and Worcester and Kent, possibly related to consumption of cider.

Occupational mortality of women (Chapter 5)

Mortality in women was analysed by their own occupation and, for those who have married, by their husband's occupation. Data were available for women dying at ages 20-74 in

England and Wales during the years 1979 to 1990 excluding 1981, thus providing the largest dataset ever available for such analyses.

Many well recognised associations between mortality and occupation in men were also found in women. Thus, suicide was high for those in the health professions and in farming, while those in entertainment and catering had excesses of alcohol-related and accidental and violent deaths. Interestingly, wives of those in health-related professions and in literary and artistic occupations also had elevated mortality from suicide.

Various hypotheses were examined using these data. For example, the numerous associations documented elsewhere between textile work and various malignancies were not found here, though excesses of cardiovascular and respiratory diseases were apparent. Wives of those working with carcinogenic substances were not found to be at increased risk of dying from cervical cancer, but there was a tendency for the wives of men whose work entailed periods away from home to have higher mortality from this malignancy. There was also limited support for the hypothesis that delayed exposure to common infections could be a factor in the development of certain chronic diseases in teachers.

These data have revealed many new associations which can be examined when the next body of data of comparable size becomes available. The majority are likely to be isolated chance findings, but data for the next decade can be used to assess them.

Occupational mortality by cause of death (Chapter 6)

This chapter reports patterns of occupational mortality among men and women aged 20-74 in England and Wales during 1979-80 and 1982-90 according to cause of death. For many diseases, differences in mortality between job groups appeared to be determined mainly by non-occupational influences. For example, deaths from cancer of the colon, melanoma of the skin, cancer of the kidney, brain cancer and aortic aneurysm were generally more common in professional and administrative occupations, suggesting a role of factors related to lifestyle. The pattern for aortic aneurysm was especially striking since deaths from other types of vascular disease occurred more frequently in manual occupations. Mortality from viral hepatitis, immunodeficiency and alcohol-related disease varied substantially between jobs, but again this variation is likely to have resulted largely from non-occupational influences.

Diseases caused by asbestos showed the expected occupational associations, but it was notable that the rankings of jobs according to their mortality from pleural and peritoneal cancer were different. This may reflect different exposure-response relationships for the two diseases.

Mortality from tuberculosis was highest in occupations with exposure to silica, a well recognised hazard. Other findings which may have been attributable to known occupational hazards included a high proportion of deaths from urothelial cancer in rubber workers and from asthma in electrical assemblers.

Leukaemia, lymphoma, aplastic anaemia and agranulocytosis all occurred with above average frequency in teachers, suggesting a possible hazard from exposure to childhood infections.

The high risk of death from occupational injury in certain jobs was highlighted, especially in aircraft flight deck officers, fishing workers, rail construction and maintenance workers, scaffolders and roofers. Suicide was most frequent in the health care professions and in farmers and foresters, perhaps in part because they have more ready access to means of suicide such as drugs, pesticides and guns.

Cancer incidence in England 1981-87 (Chapter 7)

This chapter summarises the main findings of an analysis of cancers registered in men and women aged 20-74 years in England during the seven-year period 1981-87. Results are presented according to occupation recorded at cancer registration using the 194 job groups listed in Appendix 2. In the commentary, particular attention is given to findings for twelve cancers with a known or suspected occupational aetiology, including cancers of the respiratory and genito-urinary systems, skin, brain, and leukaemia.

The results demonstrate several previously well-established links between occupations and cancer. Two such examples are the increased risks of bladder cancer seen among individuals engaged in rubber manufacture, and of cancers of the nose and nasal sinuses in those employed in the furniture and cabinet making industry. The analyses also provide some support for the previously suggested increased risks of leukaemia in farmers, leather and shoe workers and electrical workers.

Several occupational associations were also noted for which the prior evidence for a cancer link is either weak or lacking. Increased risks of bladder cancer among plastics goods makers, prostatic cancer among dental technicians, and acute myeloid leukaemia among biological scientists are noted. Because these findings were identified in the course of examining a very large number of potential associations, they should be interpreted cautiously.

This is the first time that national cancer registration data for England have been examined systematically for occupational associations in both men and women. The analyses demonstrate that occupational information collected at cancer registration is capable of identifying occupations which may be at increased risk of cancer. Further, more exhaustive analyses, which will include data for 1971-80, are planned. These will allow several of the findings noted here to be examined in greater detail.

Mortality of Longitudinal Study 1971 and 1981 census cohorts (Chapter 8)

With longer term follow-up and occupations as stated at 1971 Census, s two occupation orders showed statistically significant high mortality for both men and women — labourers not elsewhere classified [order XVII] and textile workers [X]. For men the other high mortality orders were miners and quarrymen [II], warehousemen, storekeepers, packers and bottlers [XX], painters and decorators [XVI], furnace, forge, foundry and rolling mill workers [V], drivers of stationary engines, cranes, etc [XVII] and service, sport and recreation workers [XXIII]. Women in the order of service, sport and recreation workers did not have raised mortality although the component units of barmaids [155] and charwomen, office cleaners, etc [166] showed significant excesses.

For the 1981 Census occupations,there was significantly high mortality for textile workers [groups 070, 071], both men and women. Female launderers and dry cleaners [051] also showed a significant excess, as did the following male occupations — other coalminers [088], face trained coalminers [175], other glass and ceramics [091], plastic goods makers [093] and packers and sorters [164].

Asbestos related diseases (Chapter 9)

Deaths due to mesothelioma have increased strongly since the national mesothelioma register was set up in 1968. From 1968 to 1991 the numbers of deaths have increased seven-fold for men and four-fold for women. There were 1,009 deaths in 1991.

Comparison of age specific rates in successive birth cohorts show rates increasing steeply and continuously from the earliest cohorts, born at the end of the last century, up to the cohort born in the early 1940s. Rates for more recent birth cohorts are lower, by 30 per cent for births around 1950, and by 40 per cent for births around 1955. Despite these decreases, the numbers of deaths in these latest cohorts are still substantially above the levels seen for the earliest cohorts.

The highest occupational PMRs are seen for occupations with a clear potential for substantial asbestos exposure, notably shipyard workers, laggers, railway coach builders, plumbers and gas fitters and carpenters.

The typically long delay between exposure to asbestos and the occurrence of mesothelioma means that the occupation recorded on the death certificate will often not indicate the true source of risk. However the wide range of occupations appearing on mesothelioma death certificates suggests — though it does not prove — that asbestos exposure sufficient to lead to mesothelioma is not confined to those occupational situations where asbestos exposure is routine and obvious.

Mortality follow-up of some 55,000 asbestos workers covered by the HSE asbestos worker survey shows an overall excess of 200 deaths from lung cancer, 183 from mesothelioma and 90 deaths involving asbestosis (in the absence of lung cancer or mesothelioma). These excesses are significantly more marked among those reporting their first exposure before 1970. The lung cancer SMR falls from 141 for those exposed before 1970 to 124 for those first exposed in 1970 or later. For the same groups the excess mortality from mesothelioma falls from 5 per cent to 2 per cent and the percentage of deaths involving asbestosis falls from 2 per cent to less than ½ per cent.

When broken down by sector, the highest excess mortality is seen in insulation workers who had 33 per cent excess mortality from lung cancer, mesothelioma and asbestosis combined. No other sector approached this level of excess deaths, showing combined excess mortality for the same three diseases of between 13 per cent (manufacture of dry mixes) and 4 per cent (stripping).

Monitoring occupational diseases (Chapter 10)

There is no single, comprehensive source of information for occupational disease statistics. The statistical picture must be pieced together from different sources. This chapter describes those sources currently used by HSE to produce this picture and summarises the available data relevant to the period covered by this supplement.

The sources used range from self reports of individuals' perceptions of the relationship between their work and their health obtained by means of a 'trailer' questionnaire to the 1990 Labour Force Survey (LFS), to assessments made by doctors with special expertise in the diagnosis of specific occupational diseases. They include data from the statutory reporting system for reportable occupational diseases (the Reporting of Injuries, Diseases and Dangerous Occurrences Regulations 1986 - RIDDOR); information on Prescribed Diseases derived from the Industrial Injuries scheme administered by the Department of Social Security; death certificates referring to specific occupational diseases; reporting schemes involving consultants working in the National Health Service and industry; ad-hoc surveys; and data on exposures to lead and ionising radiation.

The information described has contributed to a fundamental review by HSE of occupational health risks in the UK.

Injuries at work (Chapter 11)

Over 350 people were killed at work every year in the 10 years ending 1991/92. The fatal injury rate in 1991/92 was less than 1½ per 100,000 employees compared with over 2 per 100,000 ten years earlier. The rate in 1991/92 was less than half of that in the early 1970s and less than a quarter of that in the early 1960s.

There were around 20,000 major injuries to employees in each of the 5 years 1986\87 to 1990\91 and a lower number (17,000) in 1991/92. The major injury rate dropped from 99 per 100,000 employees in 1986/87 to 82 per 100,000 in 1991/92. This decrease was attributable to the change in employment patterns, rather than to improvements in safety standards.

Three major injuries out of every 10 were caused by a slip, trip or fall on the same level and a further 2 in 10 by a fall from a height. Nearly a third of all fatal injuries were caused by a fall from a height; a similar proportion resulted from being struck by a moving vehicle or a moving object. Over a third of over 3-day injuries were caused by handling, lifting or carrying with slips, trips and falls accounting for a further 20 per cent.

On average, there were 160,000 injuries to employees reported each year which caused an absence from work of more than 3 days. The over 3-day injury rate dropped from 760 per

100,000 employees in 1986/87 to 710 per 100,000 in 1991/92. As with major injuries, this improvement was attributable to change in the pattern of employment.

Smoking, drinking and occupation (Chapter 12)

This chapter contains an analysis by occupation and age of smoking and drinking data derived from the General Household Surveys (GHS) of 1988 and 1990. It is the most comprehensive set of reference data on these lifestyle factors published to date in the decennial supplement series. The data are intended to be used for reference as an aid to interpreting statistics on occupational morbidity and mortality from conditions where smoking or drinking may be a contributory cause. Chapters 4 to 6 mention several examples of such diseases and the reader should refer to Chapter 12 for detailed supporting information.

The chapter describes the overall patterns in the data with comments and cautions on their interpretation and goes on to describe the findings in each occupational order in more detail. The accompanying tables present the analyses by occupational order and details for 83 male and 46 female occupational units, representing all the occupations for which the GHS sample was large enough to generate meaningful statistics.

Many of the findings are not new. For example the inverse relationship between cigarette smoking and social class is apparent, and the high alcohol consumption of publicans and bar staff is once again demonstrated. Construction workers are also shown to have very high rates of both smoking and heavy drinking. The detailed analyses by age and sex, however, show some surprising results which are masked in previous overall breakdowns of smoking and drinking by occupation. For example, there is evidence of very high smoking and heavy drinking rates in the younger members of some of the health related professions, even though the overall rates for these professions are below average.

It is hoped that the information will be of value not only in the context of occupational health but also to those involved in planning general health education strategies.

Occupation and sickness absence (Chapter 13)

This chapter presents data on sickness absence and work limiting health conditions for the years 1987-1991. The data, derived from the LFS, are presented as values scaled up to be representative of the UK population for the years 1987-1991 and are broken down by sex, age and occupational order. Caution must be exercised in interpretation of some of the smaller scaled up values as these represent a small sample size. However, several general trends are evident. A clear difference between sexes is evident with higher sickness absence rates and a greater proportion of short-term sickness among female workers. Increasing age is related to an increase in the length of sickness absence spells, whereas the rate of sickness absence appears high in both the youngest and oldest members of the work-force. General trends by occupational order appear to reflect the socio-economic standing of the orders. High rates of sickness absence, a greater proportion of long-term sickness and a greater percentage of work limiting health conditions all appear more common in orders with, on the whole, lower socio-economic status. Other underlying fluctuations by occupational order may be related to specific facets of the occupations or the populations involved. Some of these factors that may influence sickness absence are presented.

Occupation and fertility (Chapter 14)

Analyses of fertility by parents' occupations were carried out on birth registrations from the 1980s. A number of father's occupations with potentially hazardous exposures were associated with low fertility (as compared with England and Wales), for example, laboratory technicians and workers in the printing industry.

The mean age of mothers at childbirth and proportion of (jointly registered) births outside marriage were also calculated. The rise in mean age which has occurred nationally was also seen across the occupations, although differences in mean age at childbirth persisted between occupations. Professional groups such as lawyers and teachers showed the highest average ages at childbirth, while the lowest mean ages occurred in manual occupations such as painters and decorators, packers and sorters. Similar social class differences were found in the proportion of births occurring outside marriage, with the highest proportions found where the fathers had manual occupations.

Ad hoc occupational mortality studies (Chapter 15)

From early on the Registrars General took an active interest in the variations in mortality associated with occupation and with residence, progressively developing the sophistication of their analyses, and publishing more of their findings.

After the Second World War, researchers frustrated by loss of data and overwhelmed by the burden of work involved in pursuing personal initiatives, turned to OPCS for assistance with their studies. OPCS developed services to meet the developing demand, initially providing data on death registration and latterly including cancer registration, subject to ethical controls.

Changes in information technology have made data handling easier and promoted linkage between data registers. Occupational and environmental studies benefit from the basic provisions and stand to benefit even further from record linkage and electronic data transfer.

OPCS has facilitated some 130 hypothesis creations and testing studies in the fields of occupational and environmental health. The occupations studies have ranged from small groups to total industries in the UK and have contributed to international collaborative studies. The findings in a substantial proportion of the studies have been of national and international interest and have contributed to occupational and environmental standards in the UK and abroad.

It is foreseeable that OPCS assisted studies will increase, both to investigate the long term health effects of occupational and environmental exposures to specific groups of agents and to monitor the benefits of intervention.

International comparisons (Chapter 16)

The comparison of occupational disease, illness, accident and mortality statistics between countries could serve several important purposes. The 'best' scores would indicate the standard of control that was economically and technically achievable, and serve to act as an interim target. Insofar as health and safety provisions constitute a charge on manufacturing costs, these statistics could be used to monitor that the financial burdens have been agreed to be equitably shared between signatories to an economic agreement. Where environmental and genetic variables contribute differently between countries to the susceptibility to disease, international health studies could contribute to investigations of the aetiology of occupational diseases and subsequently to the strategy for their amelioration.

Analyses of selected data show extreme variations in the quality and quantity of data currently available internationally. Applying the necessary caveats, there are indications of significant differences in mortality and disease experience in developed countries, that may not entirely be due to data quality and selection, and merit further investigations.

Before maximum benefit can be derived from international comparative studies, the data of interest require to be standardised. A major potential contribution to this, initiated by William Farr, is the International Statistical Classification of Diseases and related health problems (ICD), now in its 10th Revision by WHO An international classification has been agreed for occupational injuries, and data have been assembled under the aegis of ILO for a number of years now, prior to this standardisation. Standards for classifying occupations have proliferated but an agreed international version remains to be adopted. Standards for the diagnosis of industrial diseases remain to be established.

The different sectors of the work force that fail to be enumerated for occupational illness, disease and accidents, as a result of different national social security provisions, and the bias produced by failure of notification among certain employment groups, severely impair the interpretation of the data generally available at present internationally.

The need to produce better data for international comparative purposes would also produce better national data. The time scale for advances to be made has so far been measured over decades. From Farr's initiative to the publication of the first international classification of diseases took some 50 years and its widespread adoption even longer. If there were a will, a scheme for the production of useful international occupational health and safety statistics could be in place within ten years.

Classifications, definitions and reference statistics

Eight appendices contain details of classifications and definitions, and a series of tables presenting detailed statistics.

Appendix 1 Classification of causes of death

This appendix lists the causes of deaths used in the analyses in chapters 4 to 6 with their ICD(9) codes.

Appendix 2 Definition of the Southampton classification of job groups

A special classification of job groups was used by many of the authors of this volume. this classification was devised at the MRC environmental epidemiology unit at the University of Southampton. This appendix gives the definition of the job groups with the codes from the 1980 OPCS classification (CO80) used to form the groups.

Appendix 3 Description of job groups of the Southampton classification

The title of a job group cannot always give a full idea of the nature of the occupation. This appendix gives fuller descriptions and includes social class and some of the known or suspected hazards associated with the job group. Again this was drawn up by the MRC group in Southampton.

Appendix 4 Mortality of men and women aged 20 to 74

This is a listing of all the statistically significant (5 per cent) PMRs for men and women for each of the 194 job groups of the Southampton classification.

Appendix 5 Cancer registrations for men

This listing gives the significant PRRs for cancer registrations for men.

Appendix 6 Cancer registrations for women

This listing gives the significant PRRs for cancer registrations for women.

Appendix 7 Mortality of men aged 20 to 64

The PMR for men aged 20 to 74 in each job group is given for each of the fourteen chapters of ICD(9).

Appendix 8 Mortality of women aged 20 to 59

The PMR for women aged 20 to 59 in each job group is given for each of the fourteen chapters of ICD(9).

Chapter 1 The background to this decennial supplement

John Osman

Epidemiology & Medical Statistics Unit, HSE

1.1 Introduction

Work is an essential and generally beneficial element of modern society and there is a general expectation that individuals will, as far as is possible, contribute to society through employment, and provide for themselves and their families by working through the major part of their adult lives. But work may be associated with risks to individuals' health and safety, and society expects those risks to be reduced to the minimum that is reasonably practicable. The aims of the UK Health and Safety Commission (HSC) and Executive (HSE), whose existence and functions derive from the Health and Safety at Work etc Act 1974, are to protect the health, safety and welfare of employees (and to safeguard others, principally the public) who may be exposed to risks from economic activity. These aims are pursued in a variety of ways among which are carrying out and promoting research, publishing information and encouraging well informed public discussion of the nature, scale and tolerability of risk.[1]

The long-standing series of Registrar Generals' decennial supplements on occupational mortality have been a major source of data on occupational disease in England and Wales, presenting information for each decade on occupational differences in the distribution of overall mortality and morbidity from specific diseases. These data provide a valuable means of generating hypotheses about work related risks to health as well as some insight into the effectiveness of preventive measures. A number of similar reports have been produced for Scotland. The supplements on occupational mortality use information recorded on death certificates, one of a number of national datasets which are the responsibility of the Office of Population Censuses and Surveys (OPCS) in England and Wales and the General Registration Office in Scotland (GRO(S)).

A wide range of other relevant data, e.g. data from the national census, from cancer registries and from national surveys such as the Employment Department's Labour Force Survey and the General Household Survey are collected, collated and published by OPCS and GRO(S). Data on occupational illness from yet other sources are collated and published by HSE[2] which recognises that no single source of information will provide a comprehensive picture of the scale and nature of occupational ill-health. HSE has a strategy of basing its policy development and enforcement strategy on data derived from all relevant sources.

Against this background HSE welcomed the opportunity to join with OPCS in bringing together as many as possible of the available sources of statistical information relevant to the last decade in this latest volume in the series of decennial supplements. To emphasise the broad nature of occupational illness the title has been changed from one referring only to occupational mortality to another — *Occupational Health: Decennial Supplement* — which encompasses the many non-fatal work-related diseases as well.

1.2 History of the Decennial Supplements

A series of Registrar Generals' reports based on the recently instituted registers of births, marriages and deaths began in 1837. As far as occupational data were concerned the earliest of these reports were restricted to studying the relationship between occupation and violent death. By the 14th annual report analyses were being presented for deaths from other causes of males aged from 20 upwards in 10-year age groups, in certain occupations and occupational classes, using recent census figures as denominators to derive mortality rates. For 150 years now decennial supplements dealing with occupational mortality have appeared regularly, and routinely collected death registration data have been combined with other information and subjected to increasingly detailed analysis to better monitor the health of people employed in different occupations.

The Registrars General recognised early on the limitations of their data but considered them worth publishing in the hope that they would lead to reductions in what was a considerable burden of occupational mortality in many occupations in the 19th and early 20th centuries. Meanwhile they were active in improving the quality of the data. A standard nomenclature of diseases was proposed by William Farr shortly after he joined the General Register Office in 1837. Although there was considerable delay in it gaining international acceptance, it formed the basis for the International Classification of Diseases (ICD) which is now into its tenth revision under the sponsorship of the World Health Organisation (WHO).

Classifications of occupation by industrial grouping (an industrial classification) and by individual job (an occupational classification) were clearly required and standards were set and periodically amended. Despite the improvement in the quality of data arising from such standardisation, the Registrars General recognised significant residual problems affecting the accuracy of numerators of cases and of denominators of persons at risk. These arose from disparities between a person's major occupation, that person's occupation as stated at the census, and that reported at death registration. Furthermore the Standard Industrial Classification and Standard Occupational Classification were primarily derived for economic and employment purposes, and although of considerable use, have not allowed analyses of mortality (or any other data) by direct reference to specific hazards, which may be present across a range of industries and occupations. Nevertheless, the use of the current classifications in analyses of vital statistics continues to provide information of considerable value for the monitoring of occupational groups, justifying and perpetuating the policy of pragmatism of the early Registrars General in exploiting available if imperfect data, rather than waiting on the perfect.

A number of examples illustrate the potential value of decennial supplement data in the elucidation of possible associations between work and ill-health. One such example relates to the decision by a working group brought together under the auspices of the International Agency for Research on Cancer to classify occupational exposure as a painter as carcinogenic. This classification was based on an assessment of data from a wide range of sources including the decennial supplement series in which excesses of 'all malignant neoplasms' and lung cancer in those employed in such work were recorded in each of the last four supplements.[3] The 1951 supplement[4] noted that welders had a striking excess of mortality from pneumonia and a similar finding has been noted in each successive supplement. It persists in more recent mortality data[5] (see also Chapter 4) which shows that the increase is attributable mainly to an excess of pneumococcal and unspecified lobar pneumonia.

While the mortality data published in decennial supplements do not prove associations such as these are causal, they do provide a valuable source of hypotheses which can, to some extent, be tested in subsequent mortality data and can be followed up by other research. Other associations which are currently being followed up in further research and for which decennial supplement mortality data have been relevant include those between lung cancer and work as a butcher, and leukaemia and work in electrical trades. The latter association shows the wider relevance of these data in view of the interest shown in possible risks to members of the public as well as occupational risks.

1.3 Broadening the scope of the Decennial Supplement

Although the data presented in previous decennial supplements have been, and will continue to be, of considerable relevance, mortality data alone cannot describe the nature and scale of all occupational diseases since many of them are non-fatal.

Even where diseases have in the past given rise to considerable mortality, e.g. the pneumoconioses, other systems for monitoring their occurrence have often been put in place. Thus data on coal miners' pneumoconiosis is available from the prevention programme introduced by the industry and from the Industrial Injuries (II) scheme for no-fault compensation administered by the Department of Social Security (DSS). Where a disease is usually fatal and most cases can be ascribed to a specific occupational cause then there can be advantages in establishing special registers. Additional data on death certificates can be incorporated into such registers, e.g. geographic data and dates of birth, which increase the scope for analysis. The register can be used to document all fatal cases of the disease including those where it is an associated cause of death rather than just the underlying cause. A mesothelioma register for Great Britain has been developed on this basis.

The prognosis for many cancers, e.g. leukaemia and lymphoma has improved with advances in treatment. The relevance of mortality data on cancer will, to some extent, decrease in parallel with these advances in treatment. However, the existence of a network of cancer registries throughout Great Britain means that data on cancer incidence, where these include occupational information, can provide the basis for analyses along the same lines as those so successfully presented for mortality.

Fatalities from some previously important occupational diseases should never now occur. For example, there have been virtually no fatal cases of lead poisoning since the 1940s. When controls were first put in place, cases of lead poisoning continued to occur but as the controls improved these reduced considerably in number (Figure 1.1). Today risks from exposure to lead at work are monitored not by recording cases of illness, but by measurements of lead in air and biological monitoring of exposed workers. Similarly risks from ionising radiation are now principally monitored by measurement of radiation exposure since cases of disease related to this hazard are infrequent.

Figure 1.1 Reported occupational lead poisonings, in Great Britain 1900 to 1990/91

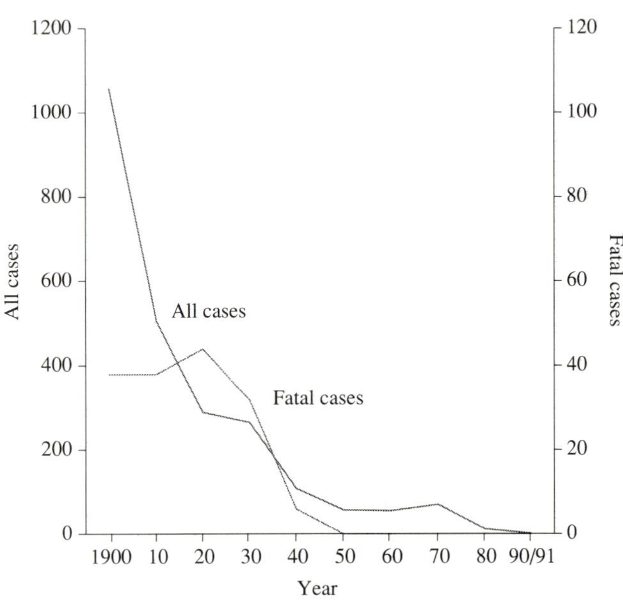

Cases of a number of 'prescribed diseases' are eligible for no fault compensation through the II scheme mentioned above provided (for most diseases) they have worked in a specified occupation. For many years the II scheme generated the bulk of the occupational ill-health statistical data used by HSE because the scheme provided benefit for spells of industrial sickness absence (which had a greater value than normal sickness benefit) and industrial death benefit, as well as compensation for disablement arising from prescribed diseases. The general entitlement to Industrial Sickness Benefit was abolished in 1983, and Industrial Death Benefit was abolished in 1987. There have been other changes in the scheme which have reduced its usefulness as a means of monitoring occupational disease. Despite this, it remains a valuable source of data because all cases are individually examined and validated and the figures derived from the scheme represent an absolute lower limit to the number of cases occurring.

As the more 'traditional' occupational diseases have come under better control, attention has focused on others which have become relatively more important, either because of the numbers apparently affected, or because potential risks associated with new technologies and industrial processes have emerged. There has been an increasing recognition of the economic importance of work-related illness some of which may be of relatively minor severity but costly to industry because of decreased efficiency of those at work, or lost time arising through sickness absence.

Some illnesses of concern today, such as musculoskeletal disorders, dermatitis and asthma, commonly have non-occupational as well as occupational causes, and this presents a particular challenge to those who wish to document the contribution made by work to the overall burden of illness. In the case of lung disease and dermatitis, specialists with particular expertise in identifying occupationally related cases, now report, on a voluntary basis, those they identify to central monitoring schemes. Such specialists may see only the more serious end of a spectrum of severity. Some data on occupational disease, notably on dermatitis, are available from surveys of the work of general practitioners who can be expected to see milder cases as well. Although diagnosis by medical practitioners, particularly those with specialist knowledge, may be thought to provide the most valid information on occupational ill-health, individual sufferers may be well placed to judge the work-relatedness of their symptoms and illnesses and information of this nature is available from surveys of the general population. A particular advantage of population surveys based on self reports of ill-health is that currently they are the only means of obtaining data across a full range of health conditions and occupational groups (see Figures 1.2 and 1.3).

Finally, although this supplement is mainly concerned with occupational illness, many people continue to suffer accidents at work and there is a considerable body of data about them.

The scope of the decennial supplement series has therefore been broadened to cover the range of relevant data described above. It has been placed in the context of information on the changing structure of UK industry, and other relevant data such as that on sickness absence and factors such as smoking and alcohol consumption which may vary across occupations and be relevant to the interpretation of findings relating

Figure 1.2 Estimated 12-month prevalence of self-reported work-related illness, by disease category, in England and Wales, Spring 1990

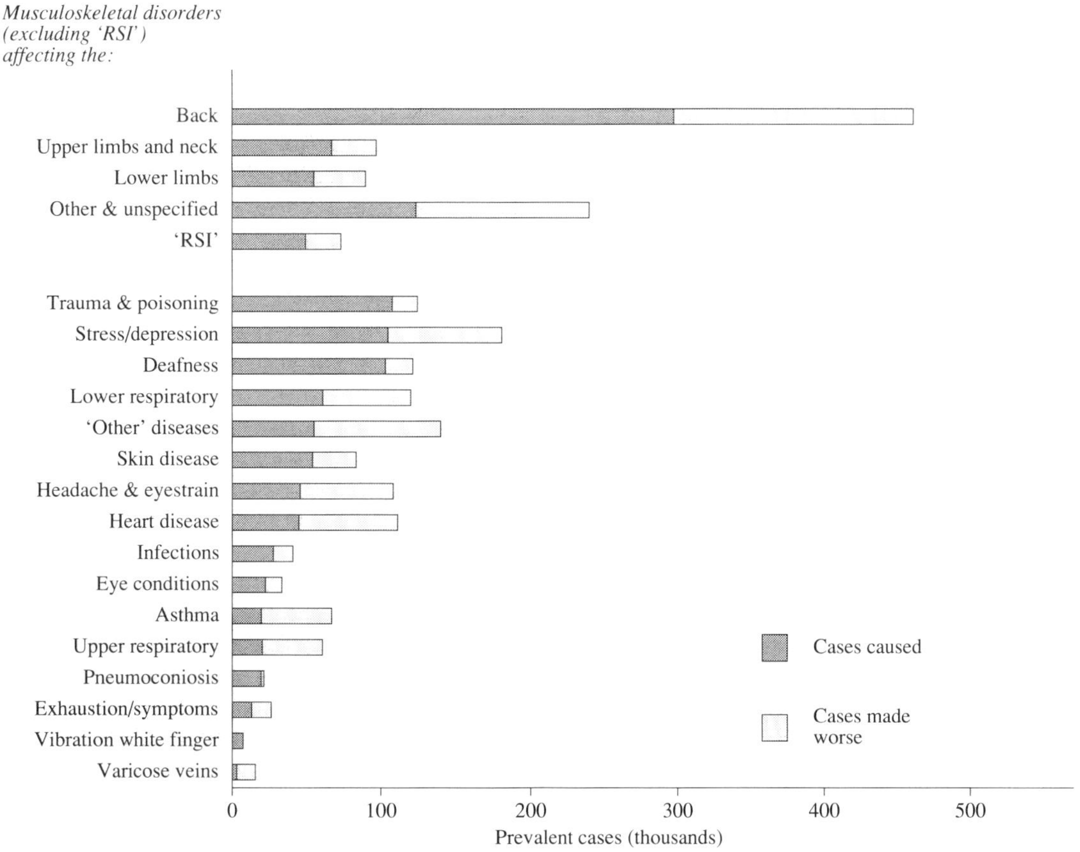

Musculoskeletal disorders (excluding 'RSI') affecting the:
Back
Upper limbs and neck
Lower limbs
Other & unspecified
'RSI'
Trauma & poisoning
Stress/depression
Deafness
Lower respiratory
'Other' diseases
Skin disease
Headache & eyestrain
Heart disease
Infections
Eye conditions
Asthma
Upper respiratory
Pneumoconiosis
Exhaustion/symptoms
Vibration white finger
Varicose veins
Cases caused
Cases made worse
0 100 200 300 400 500
Prevalent cases (thousands)

Note: Diseases ordered by number of cases 'caused', musculoskeletal conditions treated as a group.
Data derived from 1990 Labour Force Survey: see also Chapter 10.

Figure 1.3 Relative risk and estimated numbers of prevalent cases of self-reported work-related illness by occupation category, in England and Wales, Spring 1990

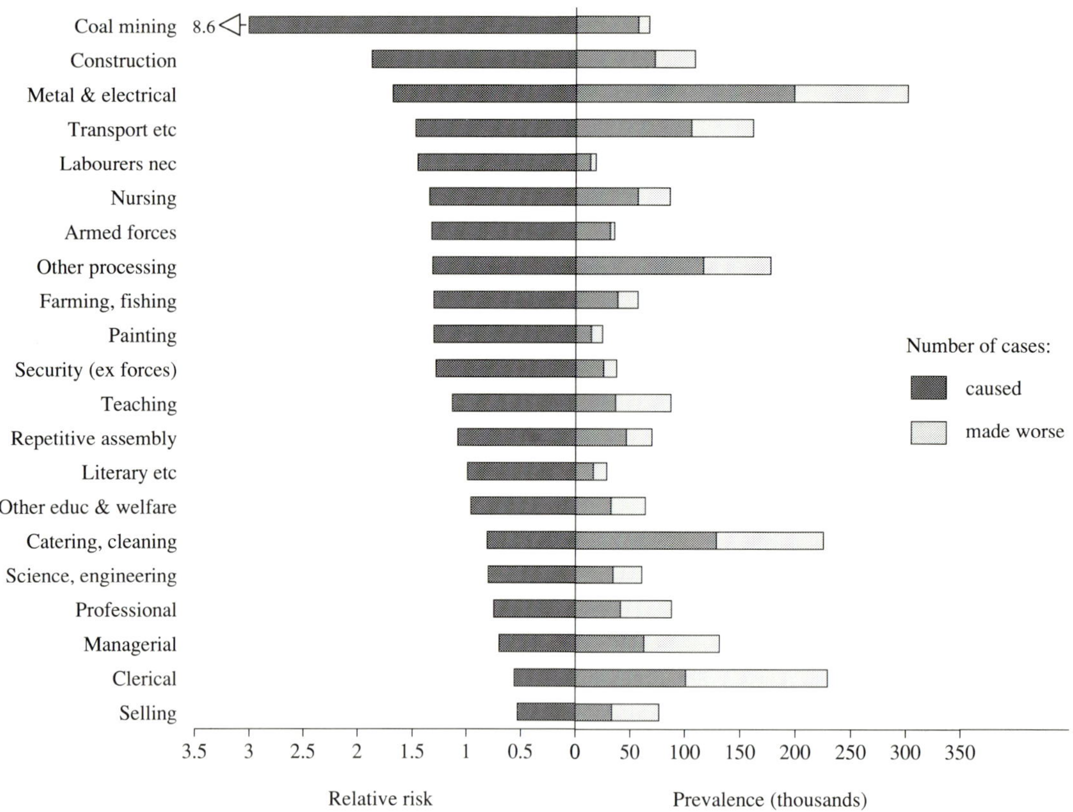

Data derived from 1990 Labour Force Survey: see also Chapter 10

to them. Furthermore, the UK national mortality and cancer registration data have long been used as measures of the health experience of specific occupational cohorts, or as the source of cases for other epidemiological approaches to the elucidation of concern about possible hazards presented by the working environment, and a chapter has been devoted to this facility and the studies that have been based on it.

1.4 Authorship and sponsorship of this volume

In the past, the production of a decennial supplement, from the planning of analyses, to their interpretation and writing up in the form of a commentary, has essentially been a task undertaken by OPCS. As a consequence of the expanded content of this supplement, it has been necessary to involve additional specialists from academic units and HSE. OPCS and HSE are jointly responsible for coordinating production of the supplement, but the responsibility for the contents of individual chapters is that of the authors.

The contents of the supplement should not be taken to represent government policy, but should be regarded as a source of information which can contribute to well informed discussion of work-related illness and its prevention.

1.5 Presentation of the data

In keeping with the intention of making information widely available this volume contains a considerable amount of data in the form of summary figures and tables. These have themselves been derived from more comprehensive databases and in order that this basic information can be made more easily accessible much of it will be made available in the form of electronic media (computer disks) in a standardised format. Using these media, individuals will be able not only to access the additional data but also will be able to re-work it and derive their own analyses.

References

1 *Health and Safety Commission Plan of Work for 1994/95.* Sudbury: HSE Books, 1994.

2 *Health and Safety Commission Annual Report 1993/94: Statistical Supplement.* Sudbury: HSE Books, 1994.

3 *Monographs on the Evaluation of Carcinogenic Risks to Humans, Vol. 47, Some organic solvents, resin monomers and related compounds, pigments and occupational exposures in paint manufacture and painting.* Lyon: IARC, 1989.

4 Registrar General. *Decennial supplement. England and Wales, 1951 Occupational Mortality* Part II Vol. 1. London: HMSO, 1958.

5 Coggon D, Inskip H, Winter P, Pannett B. Lobar pneumonia: an occupational disease in welders. *Lancet* 1994; 344: 41-43.

Chapter 2 Sources and methods

Barbara Noble
formerly Health Statistics, OPCS

Frances Drever
Nita Shah
Health Statistics, OPCS

2.1 Introduction

More data sources have been exploited in this decennial supplement than were used previously. The main source of data on occupational health is still the routine collection of the occupation of a deceased person and the cause of death via the registration systems of England and Wales, and of Scotland. Every ten years this is supplemented by a wealth of census data. Since 1971, these two sources have been linked by the OPCS Longitudinal Study (LS). The LS is able to provide more accurate information on retired study members linking their deaths with their occupations when working.

Major surveys such as the General Household Survey (GHS) and the Labour Force Survey (LFS) have become increasingly important sources of data. These sources are described in more detail below, together with some comparisons and pitfalls associated with their use.

2.2 Sources of data

2.2.1 Death registration data

Under existing legislation, all deaths must be registered with a Registrar of Births and Deaths, usually within five days in England and Wales, or eight days in Scotland.

In order to register a death, a medical certificate of cause of death is needed. This may be signed by a registered medical practitioner who has attended the deceased in his last illness. In the case of sudden death, or where there is the possibility of an unnatural death, or when an industrial disease is suspected, the case will be referred to a Coroner in England and Wales, or to the Procurator Fiscal in Scotland. He will then certify the death, possibly after holding a post-mortem and/or an inquest.

The medical certificate of death, shown in Figure 2.1, includes a detailed listing of causes of death as follows:

1a Cause or condition directly leading to death
1b Other disease or condition, if any, leading to 1a
1c Other disease or condition, if any, leading to 1b

2 Other significant conditions CONTRIBUTING TO THE DEATH but not relating to the disease or condition causing it.

Coding follows rules laid down by the World Health Organisation (WHO).[1] The general rule is that the condition entered alone on the lowest used line of part 1 is the underlying cause of death, unless it is highly improbable that this condition could have given rise to all the conditions entered above it. All cases where an industrial disease is mentioned, even when it is not the underlying cause of death, should be referred to a Coroner in England and Wales and the Procurator Fiscal in Scotland. The back of the death certificate lists some categories of death which may be of industrial origin, see Figure 2.2. The book of medical certificates[2] contains a more detailed list.

A special code used by OPCS picks out cases which indicate that the death might have been due to, or contributed to by, the employment followed at some time by the deceased. These data are used only for statistical purposes.

When a death is registered, the registrar fills in a draft entry form (Form 310) shown in Figure 2.3. This draft entry is the source of statistical data collected by OPCS and GRO(S). It includes details of cause of death, copied from the death certificate, and of the occupation of the deceased. In addition, the registrar will try to obtain sufficiently full details of the last occupation of those aged 16 and over for it to be coded. There is no statutory obligation to provide this information which includes the exact nature of the work, the industry and employment status (see boxes G and H of Figure 2.3).

Full employment details may be difficult to obtain, particularly for those who have retired some time ago, or if the death is registered via a coroner, or by an informant (such as the matron of a nursing home) who may not have had personal knowledge of the deceased. The last occupation recorded on a death certificate may not be that followed for the greater part of the deceased's life. In later working years, the occupation may change from a more strenuous or dangerous job to one more suited to a person nearing retirement. This is particularly significant if a person's health has been adversely affected by their employment. These possible changes of employment have important consequences when looking at the links between occupation and cause of death.

Occupation data are not available for many women since the occupation of a woman not working when she died is only registered if she has been in paid employment for most of her

Figure 2.1 Medical certificate of cause of death - front page

BIRTHS AND DEATHS REGISTRATION ACT 1953

(Form prescribed by Registration of Births and Deaths Regulations 1987)

Registrar to enter No. of Death Entry ....

MEDICAL CERTIFICATE OF CAUSE OF DEATH

For use only by a Registered Medical Practitioner WHO HAS BEEN IN ATTENDANCE during the deceased's last illness, and to be delivered by him forthwith to the Registrar of Births and Deaths.

Name of deceased ....

Date of death as stated to me .... day of .... Age as stated to me ....

Place of death ....

Last seen alive by me .... day of ....

1 The certified cause of death takes account of information obtained from post-mortem.
2 Information from post-mortem may be available later
3 Post mortem not being held.
4 I have reported this death to the Coroner for further action. *(See overleaf)*

Please ring appropriate digit(s) and letter

a Seen after death by me.
b Seen after death by another medical practitioner but not by me
c Not seen after death by a medical practitioner.

CAUSE OF DEATH

The condition thought to be the 'Underlying Cause of Death' should appear in the lowest completed line of Part I.

I (a)Disease or condition directly leading to death† ....

(b)Other disease or condition, if any, leading to: I(a) ....

(c) Other disease or condition, if any, leading to: I(b) ....

II Other significant conditions CONTRIBUTING TO THE DEATH but not related to the disease or condition causing it ....

SPECIMEN

These particulars not to be entered in death register

Approximate interval between onset and death

....
....
....
....

The death might have been due to or contributed to by the employment followed at some time by the deceased ☐ Please tick where applicable

† *This does not mean the mode of dying, such as heart failure, asphyxia, asthenia, etc: it means the disease, injury, or complication which caused death.*

I hereby certify that I was in medical attendance during the above named deceased's last illness, and that the particulars and cause of death above written are true to the best of my knowledge and belief.

Signature .... Qualifications as registered by General Medical Council ....

Residence .... Date ....

For deaths in hospital: Please give the name of the consultant responsible for the above- named as a patient ....

Figure 2.2 Medical certificate of death - back page: list of some of the categories of death which may be of industrial origin

LIST OF SOME OF THE CATEGORIES OF DEATH WHICH MAY BE OF INDUSTRIAL ORIGIN

MALIGNANT DISEASES		Causes include
(a)	Skin	- radiation and sunlight - pitch or tar - mineral oils
(b)	Nasal	- wood or leather work - nickel
(c)	Lung	- asbestos - nickel - radiation
(d)	Pleura	- asbestos
(e)	Urinary Tract	- benzidine - dyestuff - chemicals in rubbers
(f)	Liver	- PVC manufacture
(g)	Bone	- radiation
(h)	Lymphatics and haematopoietic	- radiation - benzene
	POISONING	
(a)	Metals	e.g. arsenics, cadmium, lead
(b)	Chemicals	e.g chlorine, benzene
(c)	Solvents	e.g. trichlorethylene

INFECTIOUS DISEASES		Causes include
(a)	Anthrax	- imported bone, bonemeal - hide or fur
(b)	Brucellosis	- farming or veterinary
(c)	Tuberculosis	- contact at work
(d)	Leptospirosis	- farming, sewer or under ground workers
(e)	Tetanus	- farming or gardening
(f)	Rabies	- animal handling
(g)	Viral hepatitis	- contact at work
	BRONCHIAL ASTHMA AND PNEUMONITIS	
(a)	Occupational asthma	- sensitising agent at work
(b)	Allergic Alveoitis	- farming
	PNEUMOCONIOSIS	- mining and quarrying - potteries - asbestos

life. In England and Wales, all regular occupations should be noted, whether full or part time, but only full time (30 or more hours a week) occupations are recorded in Scotland. Occupation details for women aged 75 and over are not coded.

2.2.2 Birth registration data

Information on the occupation of the father has always been collected when a birth is registered. It has been coded for a 10 per cent sample of births since 1970. It was not coded prior to 1970. This information is only available when a birth is within marriage or registered jointly by both parents. During the 1980s, the proportion of sole registrations was fairly constant at about 8 per cent.

The occupation of the mother could be registered for births outside marriage from 1905 but this information was not coded. From 1986, registrars have been expected to ask if the mother had been in employment at any time before the baby was born. This information is coded, when it was provided, for all stillbirths and for a 10 per cent sample of live births. However, the mother's occupation was stated for only 31 per cent of live births in 1986. This increased to 48 per cent in 1990.[3] These proportions were even lower for stillbirths. The proportions varied with the mother's age and with whether the child was the first or subsequent child in a family. Higher status and full-time occupations are more likely to be recorded.

2.2.3 Census data

In the 1981 and 1991 Censuses, questions were asked on the current main job with details required for the last week. If the person had not worked in the last week, questions referred to the previous job if it was within the last 10 years.

In 1991, a new question on long-term illness was asked (see question 12, Figure 2.4). This will provide useful data on occupational health.

The 1991 Census questions 13, 15 and 16 on employment, are shown in Figure 2.4. The same questions were asked in England, Wales and Scotland. The wording has changed slightly since 1981 and the employment status information is asked for in question 13. However, the same employment information is available. Although census questions are detailed, self-completion without extra probing from an interviewer may mean that there is insufficient detail to code all occupation details correctly.

Figure 2.3 Draft death entry (Form 310)

Reg Dist. | District & SD Nos. | Entry No.
Sub Dist. | Date of registration

DRAFT OF PARTICULARS OF DEATH TO BE REGISTERED

1. Date and place of death
(date) ....

2. Name and surname | 3. Sex
4. Maiden surname of woman who has married

5. Date and place of birth
(date) ....

6. Occupation and usual address

8. Cause of death
Ia ....
b ....
c ....
II ....
Certified by

7. (a) Name and surname of informant | (b) Qualification
(c) Usual address

O National Health Service medical card collected?
* YES NO
If NO, NHS No. ....
Signature of registrar

SP(T)162 5/85

DEATH | District & SD Nos. | Entry No.
D
E * 6 months or over | Under 6 months | (i) | Date of registration
D & SD No. | (ii) | Z

CONFIDENTIAL PARTICULARS

These particulars which are required under the Population (Statistics) Acts will not be entered in the register. This information will be confidential and used only for the preparation of statistics by the Registrar General.

At date of death the deceased was *
Single 1
Married 2 → (If married insert date of birth of spouse) Day Month Year
Widowed 3
Divorced 4
Not known 5

Q | (iii)
G(a) Deceased or †Mother | POSTCODE
H(a)* 1 2 3 4 5
See cover for employment status codes
G(b) Husband or †Father
(viia) | (viib)
H(b)* 1 2 3 4 5
See cover for employment status codes
J | (iv)
J
J
J
R Last seen alive | Day Month Year
S * Seen or Not Seen after death
a b c
T * Referred to Coroner by
1 Doctor 2 Registrar
B * SD Enq YES NO | (viii) 1 ME 2
U 1 | (v) | (vi) | 3 4 5 6
N * Post Mortem YES NO | (ix) a b c z e
M | W Employment 1 | (x) | Edit control

*Ring as appropriate
†If deceased is under 16 years of age
FORM 310

Figure 2.4 1991 Census questions on long-term illness and employment

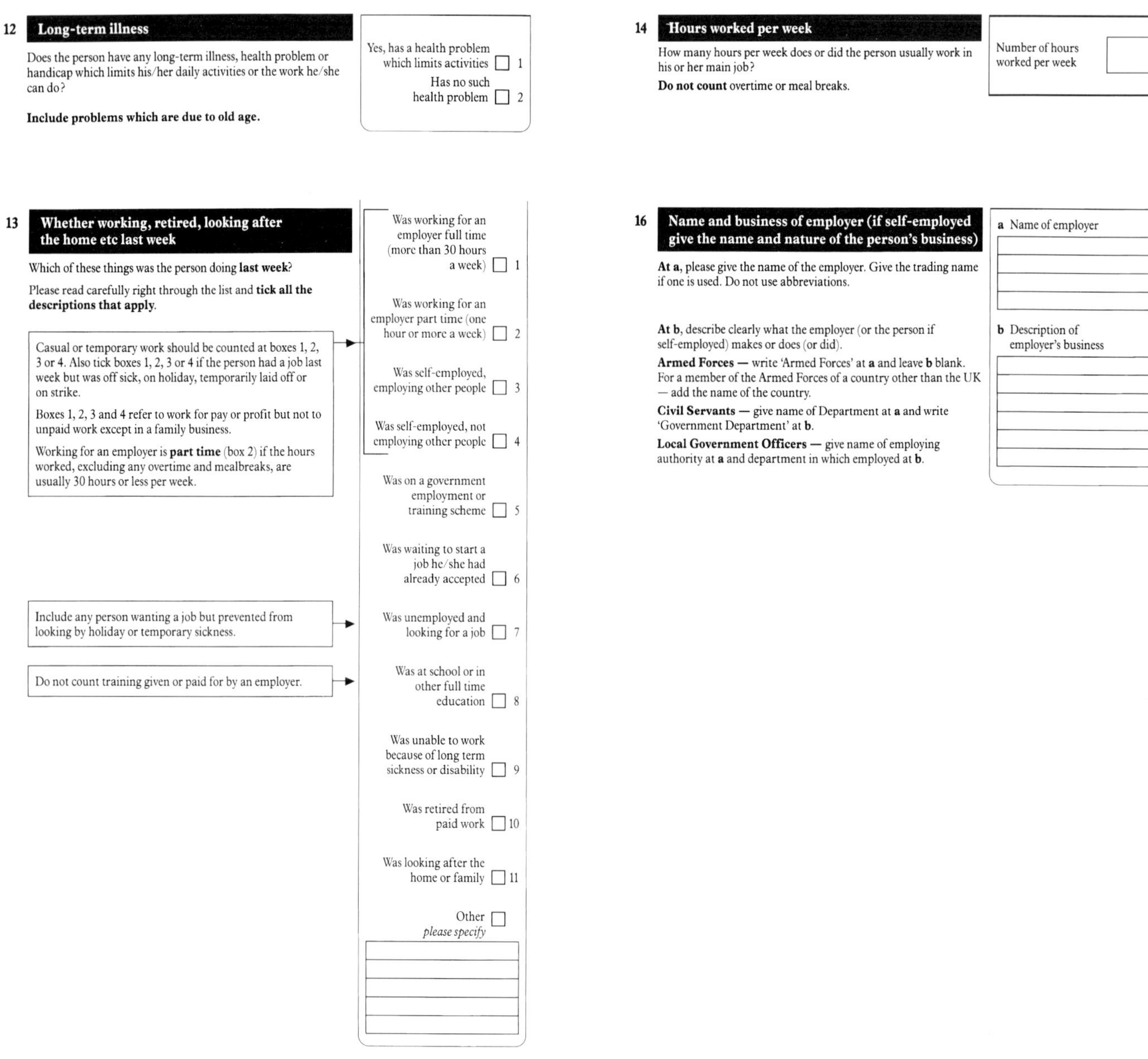

12 Long-term illness

Does the person have any long-term illness, health problem or handicap which limits his/her daily activities or the work he/she can do?

Include problems which are due to old age.

Yes, has a health problem which limits activities ☐ 1
Has no such health problem ☐ 2

13 Whether working, retired, looking after the home etc last week

Which of these things was the person doing **last week**?

Please read carefully right through the list and **tick all the descriptions that apply**.

Casual or temporary work should be counted at boxes 1, 2, 3 or 4. Also tick boxes 1, 2, 3 or 4 if the person had a job last week but was off sick, on holiday, temporarily laid off or on strike.

Boxes 1, 2, 3 and 4 refer to work for pay or profit but not to unpaid work except in a family business.

Working for an employer is **part time** (box 2) if the hours worked, excluding any overtime and mealbreaks, are usually 30 hours or less per week.

Include any person wanting a job but prevented from looking by holiday or temporary sickness.

Do not count training given or paid for by an employer.

Was working for an employer full time (more than 30 hours a week) ☐ 1
Was working for an employer part time (one hour or more a week) ☐ 2
Was self-employed, employing other people ☐ 3
Was self-employed, not employing other people ☐ 4
Was on a government employment or training scheme ☐ 5
Was waiting to start a job he/she had already accepted ☐ 6
Was unemployed and looking for a job ☐ 7
Was at school or in other full time education ☐ 8
Was unable to work because of long term sickness or disability ☐ 9
Was retired from paid work ☐ 10
Was looking after the home or family ☐ 11
Other ☐ *please specify*

14 Hours worked per week

How many hours per week does or did the person usually work in his or her main job?

Do not count overtime or meal breaks.

Number of hours worked per week ☐

16 Name and business of employer (if self-employed give the name and nature of the person's business)

At a, please give the name of the employer. Give the trading name if one is used. Do not use abbreviations.

At b, describe clearly what the employer (or the person if self-employed) makes or does (or did).

Armed Forces — write 'Armed Forces' at **a** and leave **b** blank. For a member of the Armed Forces of a country other than the UK — add the name of the country.

Civil Servants — give name of Department at **a** and write 'Government Department' at **b**.

Local Government Officers — give name of employing authority at **a** and department in which employed at **b**.

a Name of employer

b Description of employer's business

2.2.4 NHSCR flagging

The National Health Service Central Register (NHSCR) is a register held by OPCS, on behalf of the NHS, of all registered NHS patients in England and Wales. There is a similar register in Scotland, the General Register Office (Scotland) Central Register. The registers hold minimal patient details of NHS number, name, date of birth and current Family Health Services Authority (FHSA) of the patient's GP. These details are sufficient to allow the transfer of NHS records from one FHSA to another when a patient moves. Other data held by FHSAs, such as address, are not available to OPCS.

It is possible to mark, or 'flag', an individual's record at NHSCR. This allows linkage of death registration with more detailed records via the person's name and date of birth. This method is used by the LS and for cancer registrations. It is also used for medical research. A recent review was made of this use of NHSCR[4] which included a list of studies that have been made. Many of these research studies have flagged workers in particular industries where there has been concern about a particular hazard. Chapter 15 describes OPCS involvment in these studies.

2.2.5 OPCS Longitudinal Study

The LS was set up following the 1971 Census. A sample of about 1 per cent of the population was drawn from Census records by choosing everybody with four particular birth dates. Since 1971, the members of the LS have been followed, using flagging on the NHSCR, by linking data from birth and death registrations, from the NHSCR, from cancer registrations and by linking data from the 1981 and 1991 Censuses.

A birth to a mother included in the LS or the death of an LS member or the spouse of an LS member will be noted. The NHSCR provides information on LS members entering or leaving England and Wales.

The LS sample is topped up with any birth registered in England and Wales occurring on one of the four LS sample birthdates, and with migrants entering the NHSCR born on one of these days. Extra members are identified at each Census.

The LS is a very rich source of occupational health data because occupational details at one or more Censuses can be linked to death registrations giving more detail than is available from registration alone. This is particularly important for women whose occupations are often not recorded at death, and for men who may have changed occupation.

Because the LS is only a 1 per cent sample, it is often necessary to aggregate data from several years. This is particularly true when using occupational data where there may be a very small number of people in the sample following a particular occupation.

2.2.6 Cancer registration system

Data on cancers are collected by 12 Regional Cancer Registries in England and Wales. The registries cover the same areas as the pre-1995 Regional Health Authorities (RHAs), except that a single registry now covers all Thames RHAs. Further changes in the NHS are occurring, but the cancer registries' coverage remains unchanged at present. Data are collected from the NHS on cancer incidence, but occupation is an optional item.

Records from the registries are used to flag, at NHSCR, people who have had cancer. The death of a person who has had cancer, who may have died from other causes, is linked via NHSCR flagging to data on the cancer collected by the registries. This system has potential to study the connections between occupation and cancer but it cannot be fully exploited unless full occupational details are routinely collected.

2.2.7 General Household Survey

The General Household Survey (GHS)[5] is a large annual survey conducted by the Social Survey Division of OPCS since 1971. In 1991, interviews were carried out in nearly 10,000 private households with more than 19,000 people aged 16 and over.

Each year, some standard questions on employment and health are asked. The employment questions refer to the last week for those working, or to the last job at any time. Since 1979, questions on long-standing illness or disability have been included. The current wording is:

> "Do you have any long-standing illness, disability or infirmity? "
> IF YES "Does this illness or disability limit your activities in any way?"

Tabulations include prevalence of long-standing illness, by age, sex and 'collapsed' socio-economic group. There are differences in the wording of this question and the 1991 Census question on long-term illness.

2.2.8 Labour Force Survey

Data from the Labour Force Survey (LFS)[6] provide valuable background information, especially in non-census years, on the age, sex and occupation structure of the workforce. The first LFS took place in 1973. Since then, it has expanded in frequency and sample size. It is now a quarterly panel survey of about 60,000 households in Great Britain, with one fifth of households being replaced each quarter. Each member, aged 16 and over, of a household is interviewed in the first quarter. These interviews are followed up for a further four quarters before the household leaves the panel.

The employment questions refer to the last week for those working, or else to the previous job in the last three years. In 1990 extra questions on accidents at work and time off work were asked as a 'trailer' to the LFS.[7] These questions provided useful checks on reporting of accidents to the HSE, described in Chapter 11.

2.3 Classification systems

2.3.1 Definitional differences

There are differences among the data sources on the definition of 'occupation'. These are outlined in the paragraphs above describing the source. The quality will vary according to who fills in the form, and whether an occupation is that presently followed, or was the last job which may have been followed some years before.

Skilled interviewers on the GHS and LFS are likely to get good quality occupation data. Workers who are suffering from an occupational illness may have changed job as a result so the occupation causing the illness may be missed.

In calculating death rates for different diseases, the numerator, number of deaths, can be obtained each year from death registrations although this has limitations. It is more difficult to obtain data for the denominator of the rate, the population at risk. The decennial census is usually the only source of data. This creates obvious problems in non-census years and problems even in census years where registration and census statistics are not always comparable.

Some of these problems are removed by using LS data. However, the size of the sample usually means that aggregation of data from several years is necessary. This issue in relation to mortality differences by social class is explored in a recent article by Devis.[8]

2.3.2 Classification systems based on occupation

It is necessary to update occupational classifications regularly, as some titles become obsolete and new occupations develop. A careful balance is needed to provide an up-to-date classification which can also allow sufficient comparability over time with other classifications.

There are several different occupation coding systems that have been in use in the 1980s. Firstly, the 1980 OPCS classification (CO80)[9] consisted of 549 Occupational Unit Groups (OUG) arranged in 161 Condensed Key List groups, and also in 17 Occupation Orders. This was used for the coding of occupation in mortality data from 1980 to 1990. CO80 was based on the much more detailed Classification of Occupations and Directory of Occupational Titles (CODOT)[10] developed and used by the Employment Department (ED). Categories within CODOT were aggregated to groups to form the Key Occupations for Statistics (KOS) and further aggregated to 'Condensed KOS'. This was also an aggregation of CO80. Thus condensed KOS acted as a bridge between CO80 and CODOT.

The International Standard of Classification of Occupations (ISCO 88)[11] is not used directly in Great Britain. It may be used in the future, particularly for EC statistics. It is difficult to use an international classification for national statistics as types of job done, and job titles, vary considerably from country to country.

The most recent classification in use is the 1990 Standard Occupational Classification (SOC)[12], which is replacing both CO80 and CODOT. Its development is described by Thomas and Elias[13]. The hierarchical structure of the SOC consists of 371 unit categories grouped into 77 Minor, 22 sub-Major and 9 Major groups. Information on this can be obtained from Table 2 of SOC volume 1.[12] The SOC is based on skill level and the nature of work activities and does not depend on employment status (manager, foreman, etc), or distinguish between employed or self-employed workers. A particular effort was made to avoid the coding of many jobs to 'not elsewhere classified'.

SOC has been used since 1991 for published classifications based on occupation for the Census, the fourth Morbidity Study of General Practice (MSGP4), LFS and GHS. However, a special 'bridging' coding system (component codes) was used by OPCS to code 1991 Census data so that it is also possible to produce tabulations using CO80 for direct comparison with 1981 Census data. This bridging system has also been used since January 1991 to code occupations from birth and death registrations. Figure 2 of SOC volume 3[12] illustrates the links between these classifications.

The LS will contain occupation component codes for 1991 Census data to allow use of either CO81 or SOC. In 1991, birth and death registrations affecting LS members from January to census day were specially coded to CO80 so that data for the complete intercensal period are available on a consistent basis.

Since most of the data in this volume relate to mortality in the 1980s, it was necessary to use a coding system based on CO80. However, CO80 placed supervisors and employees from the same job in different employment categories. This distinction is important from the socio-economic viewpoint but is less relevant to studies of occupational health where supervisors and employees tend to share the same hazards. The numbers of deaths among supervisors are relatively few, limiting the statistical power of analyses which look at them separately. From the CO80, occupational units have been aggregated into 194 larger 'job groups' by the MRC in Southampton by type of risk. This Southampton classification has been used where possible in this volume. It is described in detail in Appendix 3. As well as combining supervisors and employees in the same job groups, this revised classification amalgamated other units in which any occupational hazards that were likely to affect mortality were similar.

Readers who wish to conduct alternative or more detailed analyses will find data for individual CO80 occupational units on the disks which will be available later.

2.3.3 Socio-economic group with brief definitions

Socio-economic group (SEG) is derived from occupational coding in order to bring together people with jobs of similar social and economic status. This system was first used in 1951. The current system is based on SOC,[12] but the 17 broad groupings mean that there is little discontinuity between this and the groupings used in the 1980s. Table 2.1 shows the groupings. Recoding a sample of occupations from the 1981 Census resulted in 2 per cent of cases being allocated to a different SEG under SOC. The greatest discontinuity was in the treatment of 'domestics' and 'cleaners' which had previously been allocated to different SEGs. This particularly affected women.

The GHS and LFS use a 'collapsed' version of SEG for some tables, into the following seven groups:

Professional
Employers and managers
Intermediate non-manual
Junior non-manual
Skilled manual (including foremen and supervisors) and own account non-professional
Semi-skilled manual and personal service
Unskilled manual

2.3.4 Social class based on occupation

The concept of social class has been used for classification since 1911, though its use has been overtaken by other classifications for many uses in recent years. The social classes currently used are

I Professional etc. occupations
II Managerial and technical occupations (previously 'Intermediate')
III Skilled occupations
(N) non-manual
(M) manual
IV Partly skilled occupations
V Unskilled occupations

Social class is assigned by considering both SOC unit and employment status, using Table A1 of Volume 3 of SOC.[12] The use of SOC to define social class produces little discontinuity.

Table 2.1 Social economic groups

(1) Employers and mangers in central and local government, industry, commerce, etc. - large establishments
- 1.1 Employers in industry, commerce, etc.
- 1.2 Managers in central and local government, industry, commerce, etc. Persons who generally plan and supervise in non-agricultural enterprises employing 25 or more persons.

(2) Employers and managers in industry, commerce, etc - small establishments.
- 2.1 Employers in industry, commerce, etc - small establishments. As in 1.1 but in establishments employing fewer than 25 persons.
- 2.2 Managers in industry, commerce, etc - small establishments. As in 1.2 but in establishments employing fewer than 25 persons.

(3) Professional workers - self employed
Self-employed persons engaged in work normally requiring qualifications of university degree standard.

(4) Professional workers - employees
Employees engaged in work normally requiring qualifications of university degree standard.

(5) Intermediate non-manual workers
- 5.1 Ancillary workers and artists
  Employees engaged in non-manual occupations ancillary to the professions, not normally requiring qualifications of university degree standard; persons engaged in artistic work and not employing others therein. Self-employed nurses, medical auxiliaries, teachers, work study engineers and technicians are included.
- 5.2 Foremen and supervisors non-manual
  Employees (other than managers) engaged in occupations included in group 6, who formally and immediately supervise others engaged in such occupations.

(6) Junior non-manual workers
Employees, not exercising general planning or supervisory powers, engaged in clerical, sales and non-manual communications occupations, excluding those who have additional and formal supervisory functions (these are included in group 5.2).

(7) Personal service workers
Employees engaged in service occupations caring for food, drink, clothing and other personal needs.

(8) Foremen and supervisors - manual
Employees (other than managers) who formally and immediately supervise others engaged in manual occupations, whether or not themselves engaged in such occupations.

(9) Skilled manual workers
Employees engaged in manual occupations which require considerable and specific skills.

(10) Semi-skilled manual workers
Employees engaged in manual occupations which require slight but specific skills.

(11) Unskilled manual workers
Other employees engaged in manual occupations.

(12) Own account workers (other than professional)
Self-employed persons engaged in any trade, personal service or manual occupation not normally requiring training of university degrees standard and having no employees other than family workers.

(13) Farmers - employers and managers
Persons who own, rent or manage farms, market gardens or forests, employing people other than family workers in the work of the enterprise.

(14) Farmers - own account
Persons who own or rent farms, market gardens or forests and having no employees other than family workers.

(15) Agricultural workers
Persons engaged in tending crops, animals, game or forests, or operating agricultural or forestry machinery.

(16) Members of armed forces

(17) Inadequately described and not stated occupations.

This is Figure 4 of SOC volume 3[12]

2.3.5 Industry

The occupational classifications described above are based on the nature of a person's work. However, the health risks associated with a particular job may vary according to the industry.

The current codes used are those from the Standard Industrial Classification (SIC(80)),[14] revised in 1980. Industries are grouped into 10 Divisions, which are subdivided in a hierarchical manner into 60 Classes, 222 Groups and 334 Activities. The 1992 revision, SIC(92),[15] has 17 sections, 14 subsections, 60 divisions, 222 groups, 503 classes and 142 sub-classes. Details of these are lengthy and can be found in the appropriate publications.

2.3.6 Cause of death

Since 1979, OPCS has coded underlying cause of death according to the 9th revision of the International Statistical Classification of Diseases, Injuries and Causes of Death[1] (ICD9). In 1985 and 1986, all causes of death as well as underlying cause were coded. Appendix 1 gives details of these codes.

2.4 Statistical methods

2.4.1 Summary measures of mortality rates

The purpose of this volume is to highlight health differences between occupations, in particular in mortality. It is evident that different numbers of people follow different occupations and so consideration of mortality rates rather than total deaths is necessary. The simplest measure of mortality, the crude death rate, is the number of deaths divided by the population at risk. Decennial supplements rely on occupational data from the decennial census to provide the necessary data on the population at risk although occupational details from the census and from death certificates are not strictly comparable.

In order to compare mortality in different occupations, it is necessary to allow for different age structures. An occupation generally followed by older people (such as judges) would be expected to have a higher crude mortality rate than one followed mainly by younger people (such as policemen). Comparability is achieved by the use of age-standardisation.

2.4.2 SMRs

A common method of age-standardisation used in the study of occupational health is a summary measure known as the standardised mortality ratio (SMR). This is an indirect method of age-standardisation. Age-specific death rates in a standard population are applied to the age structure of the occupation group being considered to calculate an expected number of deaths in the occupation group. The ratio of the observed number of deaths to the expected number of deaths in the group is calculated and multiplied by 100 to give the SMR. The SMR of the standard population is 100. An SMR over or under 100 indicates a higher or lower than expected mortality rate in a specific group.

Table 2.2 below illustrates the calculation of an SMR for men in occupation group A. The 1981 Census population is used as the standard population, column 1. Column 3, column 2 divided by column 1, gives age-specific death rates in the standard population. These rates are applied to the population in occupation A, given in column 4, to give expected numbers of deaths in this occupation, in column 6. The total observed number of deaths, shown in column 5, (301) divided by the expected number of deaths (295), multiplied by 100 gives an SMR of 102.

The advantage of a single summary measure, such as the SMR, is that comparison of mortality rates between the standard population and a particular group is very easy. However, much valuable detail is lost. Accurate population data by age are needed both for the standard population, which is not usually a problem, and for the particular occupation group which can be a problem in non-census years. Major problems of bias, known as numerator/denominator bias, can arise because of the different data sources used to calculate the observed and expected numbers of deaths. The problems associated with these data sources are described above.

It should be noted that an SMR is useful in comparing mortality in a particular occupation with that of the population in general. It should not be used to compare different occupation groups with each other.

The problems associated with recording womens' occupations on death certificates and on census forms greatly increase the likelihood of bias in the calculation of SMRs.

2.4.3 PMRs

Proportional mortality ratios (PMR) are widely used in this volume to compare mortality from particular causes among particular occupation groups with that of the general population. It should be noted that PMRs are not a measure of overall mortality in a particular group.

To calculate a PMR, the proportion of deaths in the general population from a particular cause is needed. This proportion is applied to the number of deaths in the occupation group being considered to produce an expected number of deaths from the particular cause. The ratio of the actual number of deaths to the expected number is multiplied by 100 to give the PMR. If the observed number of deaths is greater, or less, than expected, the PMR will be greater, or less, than 100.

The main advantage of using PMRs is that they are free of the numerator/denominator bias that affects SMRs. All data used are from the same source, occupation and cause of death from death certificates. Because census data are not required, analyses can be performed at any time, not just in census years.

It is necessary to interpret PMRs with caution. If mortality in general is low in a particular group, indicated by a low overall SMR, a high PMR may arise from a particular cause even though the true death rate from that cause is not actually high.

Table 2.2 Data to illustrate calculation of an SMR

Age group	All men			Men in occupation *A*		
	Population 1981 (thousands)	Deaths from all causes	Death rate per 1000	Population 1981 (thousands)	Actual deaths from all causes	Expected deaths
	(1)	(2)	(3)= (2)/(1)	(4)	(5)	(6)= (3)*(4)
20-24	2,044	1,821	0.89	6.95	4	6.2
25-34	3,854	3,560	0.92	13.42	8	12.4
35-44	3,246	6,344	1.95	10.54	77	20.6
45-54	3,005	19,178	6.38	10.28	77	65.6
55-64	2,877	53,371	18.55	10.27	192	190.5
All ages 20-64	15,026	84,274		51.46	301	295.3

Table 2.3 Data to illustrate calculation of a PMR

Age group	Deaths				
	All men			Men in occupation *A*	
	All causes	Cause *X*	Proportion from cause *X*	All causes (deaths)	Expected deaths cause *X*
	(1)	(2)	(3)= (2)/(1)	(4)	(5)= (3)*(4)
20-24	21,732	984	0.0453	3	0.136
25-29	18,072	1,104	0.0611	10	0.611
30-34	20,544	2,976	0.1449	21	3.042
35-39	27,300	6,768	0.2479	30	7.437
40-44	42,576	16,224	0.3811	78	29.723
45-49	61,236	27,396	0.4474	121	54.133
50-54	102,900	51,636	0.5018	150	75.271
55-59	187,416	94,236	0.5028	245	123.190
60-64	308,988	152,148	0.4924	452	222.568
65-69	433,956	217,572	0.5014	698	349.955
70-74	550,296	275,136	0.5000	879	439.481
All ages 20-74	1,775,016	846,180		2,687	1,305

Table 2.3 illustrates the calculation of a PMR for men in occupation group *A* from cause *X*. Column 3 gives the proportion of all deaths by age from cause *X*, column 2 divided by column 1. This proportion is applied to the number of deaths from all causes by age in occupation *A*, given in column 4, to give expected numbers of deaths in this occupation from cause *X* in column 5. The total observed number of deaths (1,544 - not shown in Table), divided by the total expected number of deaths (1,305) shown in column 5, multiplied by 100 gives an PMR of 118.

2.4.4 Statistical significance

When considering SMRs or PMRs in different occupation groups, random variation will lead to some values being greater than 100 and some less than 100. Tests of statistical significance are used to determine the likelihood of a particular high (or low) value being due to random variation or to a genuine difference in mortality in the occupation group.

The standard statistical technique is to state a 'null hypothesis' against which results can be tested. For occupational mortality, the null hypothesis might be 'that the mortality of occupational group *A* does not differ from that of the population as a whole'. This null hypothesis would be rejected only if the mortality of group *A* differed 'significantly' from that of the whole population.

Usually, such hypotheses would be rejected only if the mortality rate could have been found by chance in less than 1 in 20 such groups. More formally, a result is said to be 'statistically significant at the 95 per cent confidence level' if there is only a 5 per cent (1 in 20) chance that the result could occur as a result of random variation. A stiffer test is to look for significance at the 99 per cent level, 1 in 100 chance.

In order to test a hypothesis, a statistical model of the data is necessary. In any interval of time, the probability that an individual dies is small, so the Poisson distribution provides a useful model for mortality data. The Poisson model has been used to test statistical significance in groups with less than 20 deaths.

For 20 or more deaths, the 'chi-squared' statistic has been used, where

$$X^2 = \sum \frac{(\text{observed-expected})^2}{\text{expected}}$$

The expected deaths are calculated as described in sections 2.4.2 and 2.4.3 above. This is distributed as a chi-square with one degree of freedom.

Table 2.4 Formulae for calculating upper and lower confidence limits

	If d is between 100 and 900	If d is > 900
L	$0.96 + d - 1.96\sqrt{(d + 0.11)}$	$0.962 + d - 1.9602\sqrt{d}$
U	$1.94 + d + 1.96\sqrt{(d + 0.96)}$	$1.94 + d + 1.96\sqrt{(d + 0.96)}$

2.4.5 Confidence intervals

In general, a 95 per cent confidence interval (CI) for any number of events, *D*, is a range of the form (*L,U*). The confidence interval is such that, if we do the calculation over and over again with different samples of the events, then 95 per cent of the time, the interval between *L* and *U* will contain the true value of *D*. Different samples usually produce different intervals. The advantage of the CI over a test of significance is that it gives some idea of variability.

In order to calculate *L* and *U*, we need to know the distribution of the number of events, *D*. In the case of the Poisson distribution, *L* is such that the probability of *D* or more events from a Poisson distribution with mean *L* is 2½ per cent, 0.025 so

$$1 - \Sigma(L^r/r!)\exp(-L) = 0.025$$

Similarly, *U* is such that the probability of *D* or fewer events from a Poisson with mean *U* is 2½ per cent. So,

$$\Sigma(U^r/r!)\exp(-U) = 0.025.$$

Table 2.5 95 per cent confidence limits where observed numbers are less than 100

Obsd	L(ower)	U(pper)	Obsd	L(ower)	U(pper)
0	0.0000	3.6889	50	37.1110	65.9188
1	0.0253	5.5716	51	37.9278	67.0556
2	0.2422	7.2247	52	38.8361	68.1911
3	0.6187	8.7673	53	39.7006	69.3253
4	1.0899	10.2416	54	40.5665	70.4583
5	1.6235	11.6683	55	41.4335	71.5901
6	2.2019	13.0595	56	42.3018	72.7207
7	2.8144	14.4227	57	43.1712	73.8501
8	3.4538	15.7632	58	44.0418	74.9784
9	4.1154	17.0848	59	44.9135	79.1057
10	4.7954	18.3904	60	45.7863	77.2319
11	5.4912	19.6820	61	46.6602	78.3571
12	6.2006	20.9616	62	47.5350	79.4812
13	6.9220	22.2304	63	48.4109	80.6044
14	7.6539	23.4896	64	49.2878	81.7266
15	8.3954	24.7402	65	50.1656	82.8478
16	9.1454	25.9830	66	51.0444	83.9682
17	9.9031	27.2186	67	51.9241	85.0876
18	10.6679	28.4478	68	52.8047	86.2062
19	11.4392	29.6709	69	53.6861	87.3239
20	12.2165	30.8884	70	54.5684	88.4408
21	12.9993	32.1007	71	55.4516	89.5568
22	13.7873	33.3083	72	56.3356	90.6721
23	14.5800	34.5113	73	57.2203	91.7865
24	15.3773	35.7101	74	58.1059	92.9002
25	16.1787	36.9049	75	58.9923	94.0131
26	16.9841	38.0960	76	59.8794	95.1253
27	17.7932	39.2836	77	60.7672	96.2368
28	18.6058	40.4678	78	61.6558	97.3475
29	19.4218	41.6488	79	62.5450	98.4576
30	20.2409	42.8269	80	63.4350	99.5669
31	21.0630	44.0020	81	64.3257	100.6756
32	21.8880	45.1745	82	65.2170	101.7836
33	22.7157	46.3443	83	66.1090	102.8910
34	23.5460	47.5116	84	67.0017	103.9977
35	24.3788	48.6765	85	67.8950	105.1038
36	25.2140	49.8392	86	68.7889	106.2093
37	26.0514	50.9996	87	69.6834	107.3142
38	26.8911	52.1580	88	70.5786	108.4185
39	27.7328	53.3143	89	71.4743	109.5222
40	28.5766	54.4686	90	72.3706	110.6253
41	29.4223	55.6211	91	73.2675	111.7278
42	30.2699	56.7718	92	74.1650	112.8298
43	31.1193	57.9207	93	75.0630	113.9313
44	31.9705	59.0679	94	75.9616	115.0322
45	32.8233	60.2135	95	76.8607	116.1326
46	33.6778	61.3576	96	77.7603	117.2324
47	34.5338	62.5000	97	78.6605	118.3318
48	35.3914	63.6410	98	79.5611	119.4306
49	36.2505	64.7806	99	80.4623	120.5289

There are no explicit algebraic solutions to these equations for L and U. For known values of D, numerical solutions can be obtained.[16,17] For d events, L and U are calculated as shown in Table 2.4.

When d is less than 100, then L and U can be found using Table 2.5.

Both SMRs and PMRs are constructed as the ratio of a number of events, observed deaths, to an expected number of events. A 95 per cent CI for an SMR or PMR is constructed by relating the values of L and U to the expected number of events, e by

$$l = 100L/e \text{ and } u = 100U/e.$$

Example 1,
using the SMR calculated in Table 2.2,

$$d = 301 \quad e = 295.3 \quad \text{SMR is 102}$$

using the formula above,

$$l = 100(0.96 + 301 - 1.96\sqrt{(301 + 0.11)})/295.3$$
$$l = 90.7$$

and
$$u = 100(1.94 + 301 + 1.96\sqrt{(301 + 0.96)})/295.3$$
$$u = 114.1$$

So, the 95 per cent CI for the SMR of 102 is 91 to 114.

Example 2,
For a particular cause of death, the SMR is 150 when there were 75 deaths,

$$d = 75 \quad e = 50.1 \quad \text{SMR is } 150$$

using the values in Table 2.5,

$$l = 100 \times 58.9923 / 50.1$$
$$l = 117.7$$
and $$u = 100 \times 94.0131 / 50.1$$
$$u = 187.7$$

So, the 95 per cent CI for the SMR of 150 is 118, 188.

The standard value for an SMR or PMR is 100. If the 95 per cent CI does not contain 100, then the SMR or PMR is statistically significant at the 95 per cent level. So the SMR in example 1 is not statistically significant but that in example 2 is statistically significant.

References

1 World Health Organisation. *Manual of the International Statistical Classification of Diseases, Injuries and Causes of Death.* Vol 1 Geneva: WHO, 1977; Vol 2 (Index) Geneva: WHO, 1978.

2 This book is issued to medical practitioners. It contains the certificates to be filled out and notes on their completion.

3 Cooper J and Botting B. Analysing fertility and infant mortality by mother's social class as defined by occupation. *Population Trends* 1992; 70: 15-21.

4 Office of Population Censuses and Surveys. *Uses of OPCS records for medical research.* OPCS Occasional Paper 41, London: OPCS, 1993.

5 A Bridgwood and D Savage. *General Household Survey 1991.* London: HMSO, 1993.

6 Office of Population Censuses and Surveys. *Labour Force Survey 1990 and 1991.* London: HMSO, 1992.

7 *Employment Gazette* Dec 92.

8 Devis T. Measuring mortality differences by cause of death and social class defined by occupation *Population Trends* 1993; 73:00-00.

9 Office of Population Censuses and Surveys. *Classification of Occupations 1980.* London: HMSO, 1980.

10 Department of Employment. *Classification of Occupations and Directory of Occupational Titles. London:* HMSO, 1972.

11 International Labour Office. *International Standard Classification of Occupations 1968.* ILO, 1969.

12 Office of Population Censuses and Surveys. *Standard Occupational Classification 1990.* London: HMSO,1991.

13 Thomas R and Elias P. Development of the Standard Occupational Classification. *Population Trends* 1989; 55: 16-21.

14 Central Statistical Office. *Standard Industrial Classification 1980 revision.* London: HMSO, 1980.

15 Central Statistical Office. *Standard Industrial Classification 1992 revision.* London: HMSO, 1992.

16 Breslow NE and Day NE. The standard mortality ratio. In Sen PK(ed). *Biostatistics: statistics in biomedical public health and environmental science.* New York: Elsevier, 1985.

17 Goldblatt P and Jones D. Chapter 3 Methods. In Goldblatt P (ed). *Longitudinal Study — Mortality and social organisation.* London: HMSO, 1990.

Chapter 3 Demographic and employment trends

Stewart Taylor
Neil Davies
Philip McCrea

Health and Safety Executive

This chapter outlines the major changes in demography and employment over the past twenty years and considers how they have contributed to changes in occupational exposure to injury and ill-health risks. The future prospects for these risks are then assessed using Employment Department (ED) labour force projections and expectations of future employment changes produced by the Institute for Employment Research (IER) at the University of Warwick.

3.1 Demographic changes

In 1993 the population of working age (ages 16 to 64 for men and 16 to 59 for women) in Britain was 34.6 million, having shown a steady increase over the past two decades - from 31.7 million in 1973 and 33.3 million in 1983. The rate of population growth has been similar for both males and females. In 1993 just over half (52 per cent) of the population of working age was male, virtually unchanged since 1973.

More significant changes are observed when we look at the labour force. The labour force is the total number of people aged 16 or over in the economy in work or available for work. It excludes those who are in full-time education or are not available for work for other reasons. In 1993 the labour force stood at 27.1 million.

Changes in the size of the labour force can come from changes in the population of working age or from changes in the proportion of the population of working age that is actively seeking work. This latter proportion is known as the activity rate.

Figure 3.1 shows changes in the labour force over the last twenty years. Since 1973 the labour force has increased by around 3.2 million (13 per cent). The large majority of this increase (2.8 million) was accounted for by women. This increase was far in excess of the rise in the female population of working age. In contrast, the rise in the male labour force was well below the increase in the male population of working age.

These patterns are the outcome of changes in activity rates, which worked in opposite ways for males and females. While male activity rates fell from 90 per cent to 85 per cent between 1973 and 1993, female activity rates over this period increased from 59 per cent to 71 per cent.

There are a number of factors behind these changes in activity rates. That more women have decided to enter the labour market, particularly after the birth of children, would seem to reflect the increased availability of part-time work and changes in social attitudes towards women working. Evidence on the latter is available from the British Social Attitudes Survey,[1] although this covers only a relatively short time period. The survey found that the proportion of respondents who disagreed with the view that 'a husband's job is to earn the money; a wife's job is to look after the home and family' rose from 37 per cent to 45 per cent between 1984 and 1991.

Figure 3.1 The labour force, 1973-93, Great Britain - ages 16+ (millions)

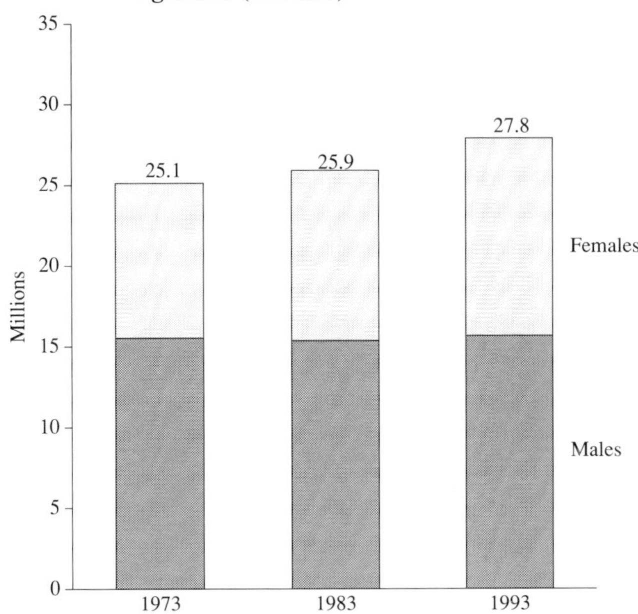

Source: Data taken from Employment Gazette (April 1994)

The fall in male activity rates has been sharpest for those aged 55 and over, influenced by a trend towards early retirement. There has also been an accompanying fall in activity rates of prime age males (defined here as those aged between 25 and 44) during the 1980s as demand shifted away from traditionally male-dominated industries (such as coal extraction and solid fuels; metal goods; and engineering). Fewer job opportunities were produced in these sectors, acting to discourage older workers from remaining in the labour market.

These changes are reflected in the age composition of the labour force. Figure 3.2 shows a fall in the number aged 55 and over since 1973. Another significant change has been in the number of new young labour market entrants (aged 16 to

19), the principal cause being demographic. The number of 16-19 year olds rose up to the early 1980s, reflecting the baby boom of the 1960s. Since then the lower birth rates of the 1970s have depressed the number of new young labour market entrants.

Another factor has been the increasing proportion of young people staying on into further or higher education. Between 1980/81 and 1991/92, the proportion of 16-18 year olds in full-time education increased from 29 per cent to 48 per cent. Figure 3.2 also illustrates the rise in the number of prime age workers, from just over 40 per cent of the labour force in 1973 to 50 per cent in 1993, reflecting the rise in female activity rates and the baby-boomers moving up the age bands.

Figure 3.2 Age composition of the labour force, 1973-93, Great Britain - males and females, ages 16+

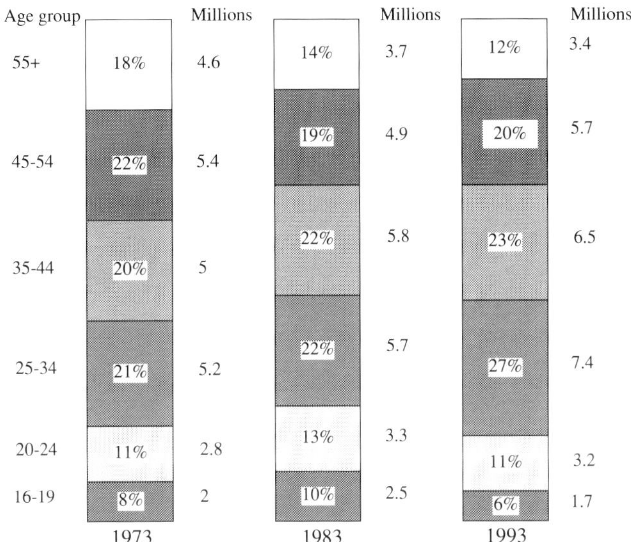

Source: Data taken from Employment Gazette (April 1994)

3.2 Patterns of employment

The changes in labour supply outlined above have combined with changes in the demand for labour to alter the pattern of employment in the economy.

3.2.1 Industry

The major industrial shift in the structure of employment over the past few decades has been the broad movement away from what are known as the primary (agriculture and energy) and manufacturing (e.g. engineering and metal goods) sectors towards the service sector (e.g. retail distribution and business services).

Overall, employment in 1993 stood below its 1973 level. Total employment fell from 24.3 million in 1973 to 22.8 million in 1983, before rising to 23.9 million a decade later.

From Figure 3.3 we can see that manufacturing's share of employment has fallen the most sharply, from around a third in 1973 to under a fifth by 1993. More than 3 million less people were employed in manufacturing in 1993 compared to 1973. Figure 3.4 shows that employment fell throughout the manufacturing sector, with the sharpest falls in metal manufacture, textiles and clothing, and vehicles.

Agriculture and energy are relatively small sectors in terms of their share of total employment, although both have experienced a decline in jobs. Within the energy sector, coal extraction and solid fuels has experienced a particularly sharp fall in numbers - from over 300,000 in 1973 to less than 50,000 in 1993, with the majority of this decline occurring in more recent years. The construction industry accounts for around 6 per cent of employment. Although lower than in 1973, employment in the construction industry is similar to its 1983 level.

Figure 3.3 Employment by industrial sector, 1973-93, Great Britain (millions)

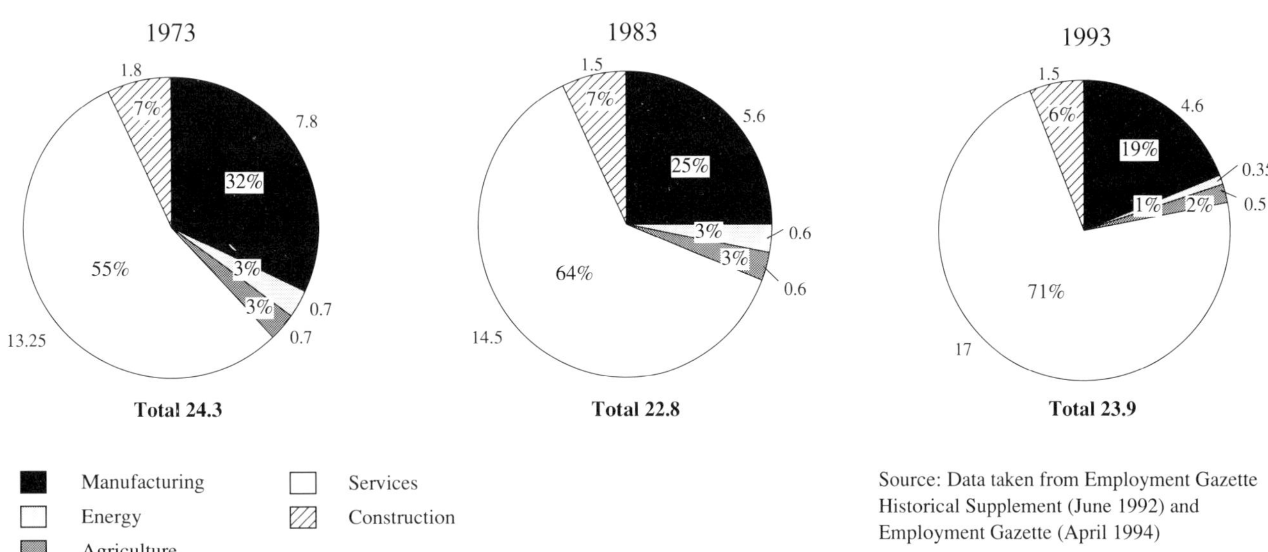

Source: Data taken from Employment Gazette Historical Supplement (June 1992) and Employment Gazette (April 1994)

Figure 3.4 Employment changes within manufacturing, 1973-1993, Great Britain (per cent) employees only

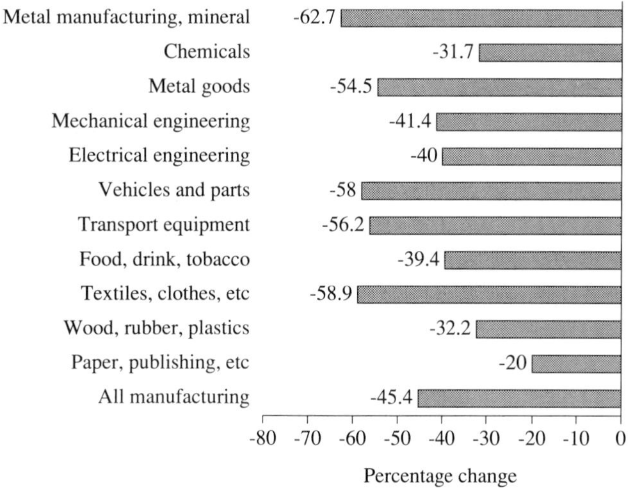

Source: Data taken from Employment Gazette Historical Supplement (June 1992) and Employment Gazette (April 1994)

In contrast, service sector employment has increased substantially since the early 1970s. In 1973 just over half of total employment was in this sector. By 1993 this had increased to over 70 per cent - an additional 3.7 million jobs. Figure 3.5 shows a breakdown in employment within the service sector. The most notable increases have been in banking, finance and insurance, experiencing a 90 per cent increase between 1973 and 1993. Distribution and Other services, including the public sector, have also shown increases.

Figure 3.5 Employment changes within the service sector, 1973-93, Great Britain (per cent)

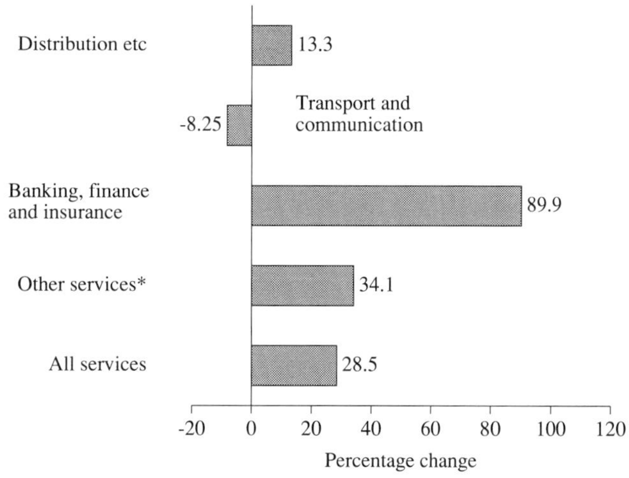

* Includes public administration, education and health services

Source: Data taken from Employment Gazette Historical Supplement (June 1992) and Employment Gazette (April 1994)

3.2.2 Occupation

The changing pattern of the economy's industrial structure has acted to benefit some occupations more than others. Associated with the move towards the service sector has been a broad shift away from blue-collar, manual jobs, towards more skilled, white-collar non-manual jobs. This is illustrated in Figure 3.6.

This shows that the number employed in service sector-related occupations has risen since the early 1970s. In particular, demand for higher-skilled, non-manual labour has increased, with the number of managerial, administrative and professional jobs rising by nearly 3 million between 1971 and 1993. This increase was accompanied by an increase of 1 million personal and protective service jobs, which includes occupations such as police officers, traffic wardens, waiters, ambulance staff, hairdressers, caretakers and bookmakers.

In contrast, the two occupations of fastest decline have been those associated with the manufacturing sector. The share of craft and related manual occupations in total employment has fallen from 20 per cent to 13 per cent between 1971 and 1993. In addition, the number of plant operatives declined by more than a million over this period.

However, the trend towards higher skilled non-manual jobs has not solely been due to movements in the employment shares of different industrial sectors. A variety of factors, including the impact of new technology, has served to alter the composition of employment within industries. For example, managers and professionals increased their share of total employment within manufacturing from 14 per cent to 26 per cent between 1971 and 1993.

3.2.3 Employment status

Over the last couple of decades, especially during the 1980s, two types of employment have increased in their importance to the UK labour market.

Figure 3.7 compares the growth in part-time employment and self-employment to the growth in total employment since 1973. Twenty years ago, 16 per cent of jobs were part-time. By 1993 this had increased to 24 per cent. Four-fifths of those employed on a part-time basis are women and the majority work in a relatively small number of occupations, notably clerical and secretarial, personal services, teaching, health and retail distribution.

Meanwhile, the number of people who are self-employed has increased from just over 2 million in 1973 to over 3 million in 1993. The large majority of this growth occurred during the last ten years. Self-employment appears to be particularly important amongst men, who, in 1993, accounted for around 75 per cent of self-employment, compared with 25 per cent for females, and within the construction and agriculture sectors. Since the late 1970s, however, self-employment has also shown steady growth in the service sector, with just under 60 per cent of all self-employed workers operating in the service sector in 1993. Within this, growth has been particularly strong in banking, finance and insurance, whose share of the economy's self-employed has increased from 13 per cent in 1973 to 22 per cent two decades later.

Figure 3.6 Employment by broad occupational groups, 1971-93, Great Britain (millions)

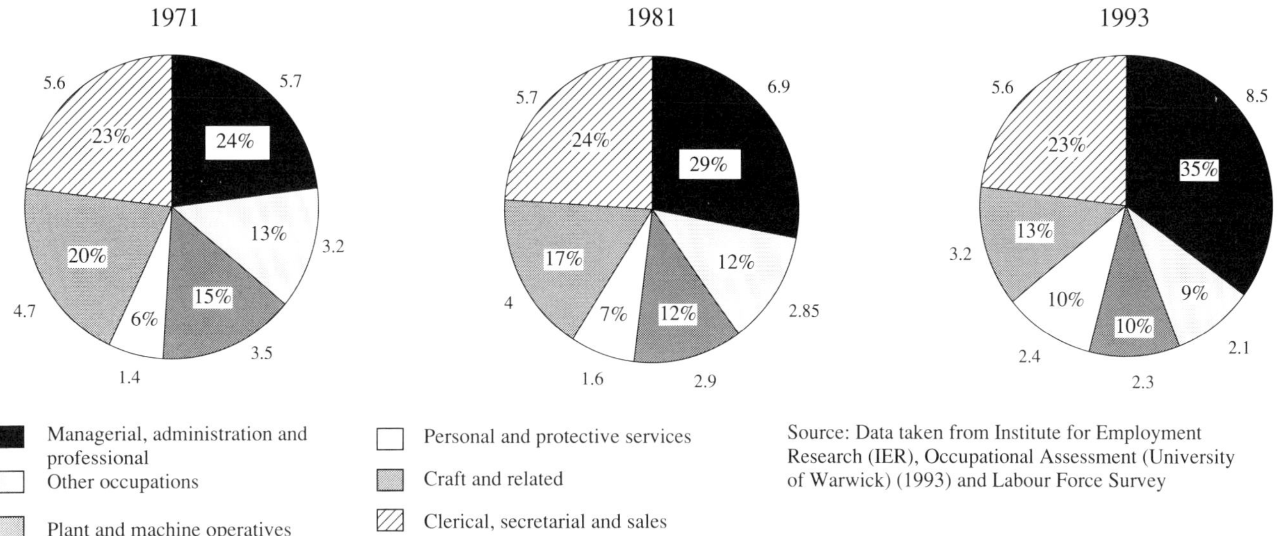

Source: Data taken from Institute for Employment Research (IER), Occupational Assessment (University of Warwick) (1993) and Labour Force Survey

Figure 3.7 Changes in employment by status, 1973-93, Great Britain (millions)

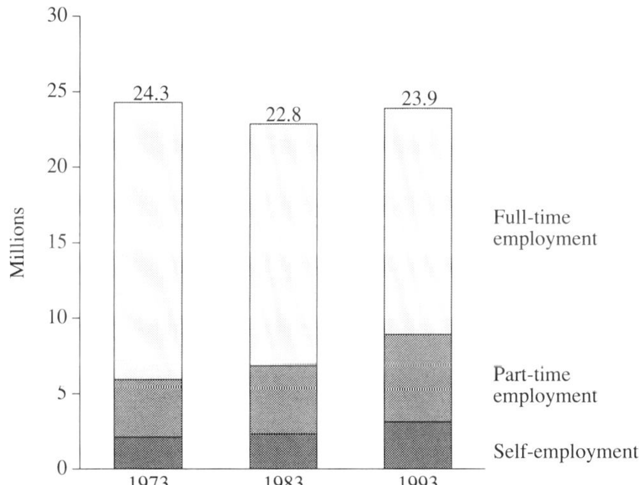

Source: Data taken from Employment Gazette Historical Supplement (June 1992) and Employment Gazette (April 1994)

Figure 3.8 The labour force, 1993-2006, Great Britain - ages 16+ (millions)

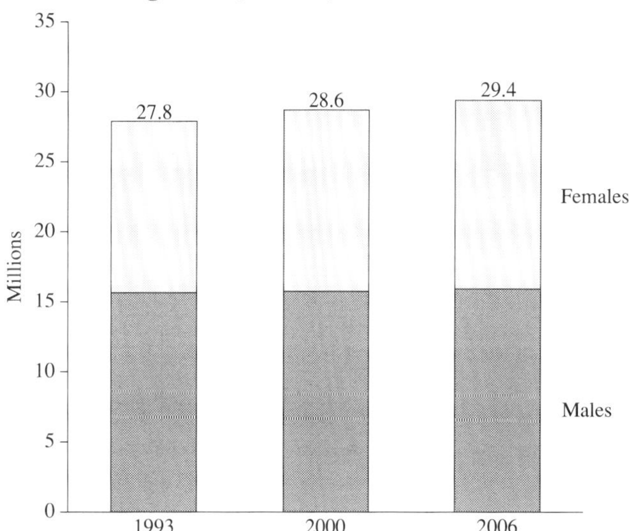

Source: Data taken from Employment Gazette (April 1994)

3.3 Prospects to the year 2000

3.3.1 Labour supply

Each year the Employment Department publishes its labour force projections. As can be seen from Figure 3.8, the labour force is expected to grow by over 800,000 between 1993 and 2000. Approximately 87 per cent of this will be accounted for by women, as pre-2000 female activity rates broadly continue. However, the female contribution to the labour force increment of over 714,000 between 2000 and 2006 will decline to about 73 per cent, with the male labour force experiencing faster rates of growth for this latter period.

Figure 3.9 shows that the overall labour force growth of 1.5 million between 1993 and 2006 will be concentrated amongst those aged 35 and over, accounting for 55 per cent of the labour force in 1993 and 64 per cent in 2006. The 35-44 age group shows particularly strong growth, reflecting the upward movement of the 1960s 'baby-boomers' through the age bands. Similarly, the diminishing contribution of the 16-34 age groups reflects the lower fertility rates of the 1970s.

3.3.2 Employment

This section outlines the expectations of the Institute for Employment Research (IER), as detailed in their Occupa-

Figure 3.9 Age composition of the labour force, 1993-2006, Great Britain - males and females, ages 16+

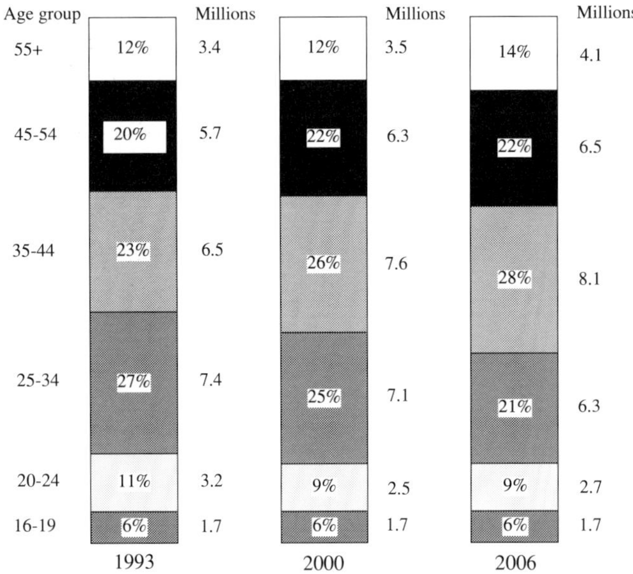

Source: Data taken from Employment Gazette (April 1994)

tional Assessment, 1992-93.[2] The IER's projections as summarised here were based upon actual data only up to 1991. They are set alongside the now available 1993 data. However, it should be noted that it is possible that these projections would have been significantly different if they had been based on the more recent data.

Trends in employment are expected to be broadly similar to those seen during the last few decades as many of the forces behind the changes seen recently are expected to continue. For example, technological change, changing patterns of consumer demand and growing competitive pressures are all expected to remain key influences. However, the scale of future changes are unlikely to be dramatic and should be less than those experienced over the past decade or so.

3.3.3 Industrial change

Figure 3.10 presents IER's expectations for employment by industry in the year 2000.

Both agriculture and energy are expected to experience some further declines in employment. Agricultural employment falls because of slower output growth and productivity improvements, while the main reasons for the fall in employment in the energy sector are greater competition and rationalisation. Subsequently, both sectors experience reductions in their shares of total employment.

The number of manufacturing jobs is expected to continue its long-term decline, falling to around 17 per cent of total employment in the year 2000. Even though manufacturing output is projected to increase, potential employment benefits are expected to be cancelled out by productivity gains needed to maintain international competitiveness.

Construction has been especially hard hit by the recent recession. Recovery is likely to come slowly as the demand for infrastructure and confidence in the housing market both begin to rise. Overall employment is expected to increase between 1993 and 2000, although at a modest rate.

The service sector is expected to experience positive employment growth, although slower than that experienced during the 1980s. Within the sector, the strongest growth is expected to be in services such as research and development, and recreational and cultural services, health and education, and banking, insurance and business services. However, some sectors, such as transport and communications, are expected to experience small job losses.

3.3.4 Occupational change

A substantial growth in employment in managerial, administrative and professional occupations is anticipated, with the effect that their share of total employment should grow to above 40 per cent by the year 2000. This is the largest increase

Figure 3.10 Employment by industrial sector, 1993 and 2000, Great Britain (millions)

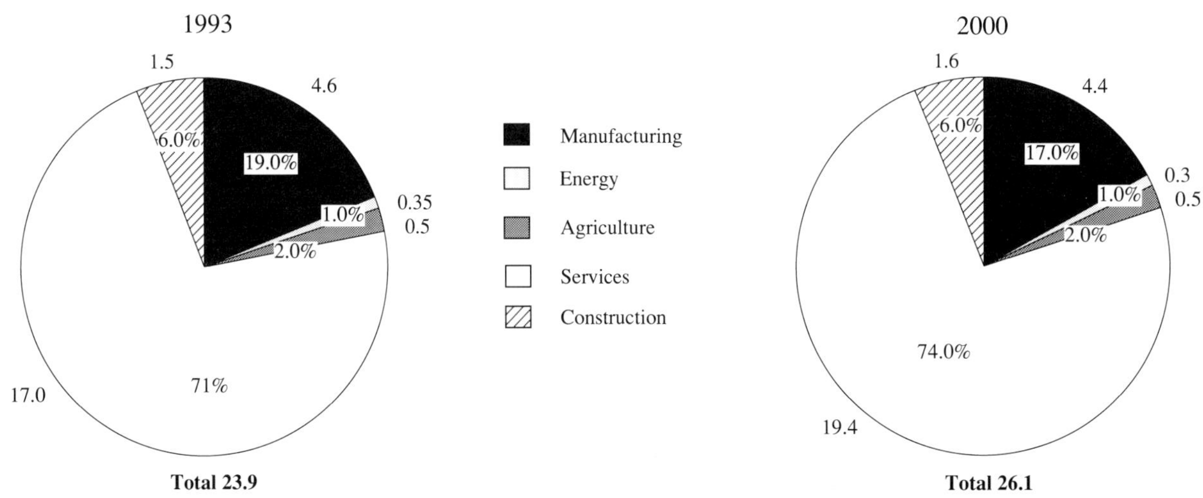

Source: 1993 data taken from Employment Gazette (April 1993), 2000 data taken from Institute for Employment Research (IER), Occupational Assessment (University of Warwick) (1993)

Note: IER projections are based on 1991 data

Figure 3.11 Employment by broad occupational groups, 1993 and 2000, Great Britain (millions)

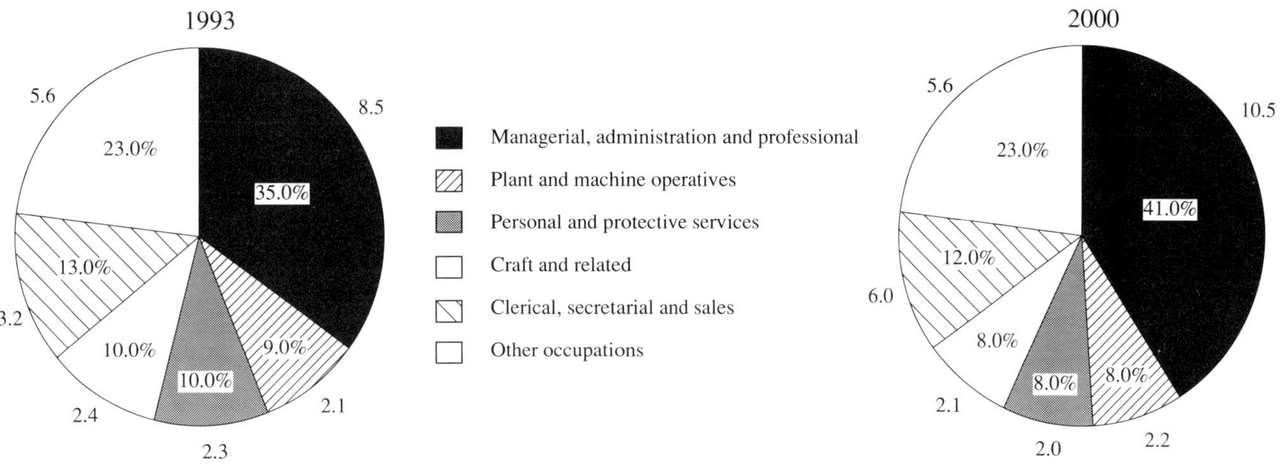

Source: 1993 data taken from Labour Force Survey, 2000 data taken from Institute for Employment Research (IER), Occupational Assessment (University of Warwick) (1993)

Note: IER projections are based on 1991 data

experienced by the occupational groups illustrated in Figure 3.11. Within this, corporate managers and administrators are expected to experience the fastest annual rates of growth.

Drops in employment in clerical and secretarial occupations, expected because of the labour-saving effects of new technology, are projected to be largely offset by more positive trends displayed by those working in sales occupations as buyers, brokers and sales representatives, for example.

Figure 3.11 shows that manual employment is expected to continue to decline as a proportion of total employment. Craft and skilled, plant and machine operatives and unskilled labourers are expected to experience absolute falls in the number of people employed, especially for those working as industrial plant and machine operators.

3.3.5 Changes in employment status

As Figure 3.12 illustrates, both part-time employment and self-employment are expected to continue to grow faster than employment in general. However, the rate of growth - particularly for the self-employed - is expected to be more modest than during the 1980s.

Figure 3.12 Changes in employment by status, 1973-2000, Great Britain (millions)

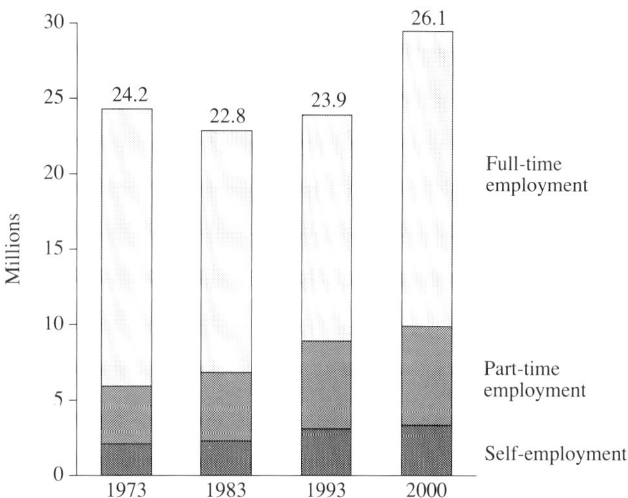

Source: 1993 data taken from Employment Gazette (April 1994) 2000 data taken from IER Occupational Assessment (1993)

3.4 Implications for health and safety

As can be seen in Chapter 11, injury rates in more traditional sectors of the economy such as energy, manufacturing and construction tend to be higher than those in the service sector. While there are difficulties in isolating the impact of changes in employment from other factors, a continuation of the trend towards service sector and non-manual employment, although expected to be at a much slower pace than in the recent past, should therefore generally have positive implications for workplace health and safety. However, there are several reasons not to be complacent.

Firstly, although we may expect the move away from heavy industry to produce a decline in the incidence of illnesses commonly reported in such sectors (such as pneumoconiosis, occupational asthma and occupational deafness), such a trend may not become immediately evident.

Drawing conclusions about changes in the incidence of many areas of ill-health and changes in the conditions that influence them is made somewhat difficult by the existence of often long and variable delay periods between first exposure and onset of disability. Some disease, such as occupational asthma, set in relatively quickly and, therefore, incidence figures can be expected to reflect employment patterns within a relatively short time-scale. Other illnesses, however, develop over much longer time periods. An example is pneumoconiosis arising from mining, quarrying, asbestos, foundries and potteries industries, which takes on average between 15 and 30 years to develop. This means that many sufferers have passed retirement age before the disease is diagnosed.

In this respect, current and future patterns of work-related ill-health will reflect past patterns of employment to some extent. It will be, therefore, some time still before many areas of ill-health that are associated with exposure in declining heavy industries show the substantial falls we may expect.

The expected decline in the incidence of these 'traditional' illnesses may be offset to an extent by increased awareness and tendency to report occupational illnesses in general. This may be a result of improved knowledge of diseases and arrangements for compensation, influenced by improved health and safety management.

Secondly, the shift towards new industries and occupations and the growing proportion of female workers in the economy will change the nature of risks that employees face. Much progress in reducing injury rates in individual industries in recent years has been made through reducing specific and well-recognised risks such as machinery accidents. In the coming years, further progress may require increased emphasis on controlling more widespread risks, such as slips, trips and falls on the same level — accountable for 32% of all major injuries in 1991/92. These risks are faced in many cases to a similar extent by workers in services as well as those in the more traditional sectors and we may expect their incidence to become of increasing importance in the future.

The change in the nature of risks faced by Britain's workforce will have implications for the type of injury likely to occur. In accordance with this, more emphasis may need to be placed on the already increasing number of accidents involving fractures and strains/sprains, the former accounting for three-quarters of non-fatal major injuries to employees and the self-employed in 1991/92, and the latter for two-fifths of over 3-day injuries to employees.

With females accounting for an increasing proportion of non-fatal and over 3-day injuries, the expected continued increase of their relative importance in the economy, along with the expected continued ageing of the working population, may act to emphasis these injury trends.

In many cases, these risks can only be tackled through better management and work procedures rather than specific regulation or control. The increased use of risk assessments, such as those required under the 1992 Management of Health and Safety at Work Regulations, will help bring about better management and work procedures.

Thirdly, the further shift towards non-manual occupations, together with the impact of new technology, is likely to highlight new or previously neglected forms of ill-health. These may include the range of generally minor illnesses caused by factors common to the indoor environment (i.e. air-conditioning and humidification systems, poor lighting and air pollution) collectively known as 'sick building syndrome'. Also, illnesses such as stress and depression (which the 1990 LFS reported as the second most common self-reported condition), migraines and headaches, and musculoskeletal disorders might be expected to follow an upward trend.

Lastly, people's concerns and expectations over health and safety standards can be expected to rise over time as incomes and living standards increase. As people become better off, health and safety may become an increasingly important component of their overall welfare.

References

1. The Institute for Employment Research (University of Warwick) *Review of the economy and employment: occupational assessment.* Warwick: IER, 1993.
2. *Employment Gazette Historical Supplement.* June 1992.
3. Labour force projections: 1993-2006. *Employment Gazette* April 1993, 139-147.
4. *Employment Gazette* April 1994.
5. *Education Statistics for the United Kingdom - 1993 Edition.*
6. Serial and Community Planning Research. *British Social Attitudes - the 8th report.* London: 1991.

Chapter 4 Occupational mortality of men

David Coggon
Hazel Inskip
Paul Winter
Brian Pannett

MRC Environmental Epidemiology Unit, University of Southampton

4.1 Introduction

This chapter and the two that follow report an analysis of occupational mortality for England and Wales during the period 1979-80 and 1982-90. Data for 1981 were omitted because industrial action involving Registrars of Births, Deaths and Marriages produced occupational details of uncertain quality for registrations in that year. The analysis had three main aims:

- to monitor known occupational causes of mortality
- to provide information about suspected occupational hazards
- to generate clues to previously unsuspected occupational hazards.

It was based on some 1.8 million deaths and as such was the largest survey of occupational mortality ever carried out in Britain. This allowed examination of rarer occupations and causes of death than has been possible previously.

The main outcome measure used in these chapters is the proportional mortality ratio (PMR) for each combination of job and diagnostic group. PMRs were calculated with stratification for age in five-year bands, and with the distribution of deaths among all people with adequately described occupations as the standard. The results presented in this chapter and in Chapter 5 are standardised also for social class. More detailed information about PMRs is contained in Chapter 2.

The job and diagnostic groups analysed were defined as described in Appendices 2 and 3. For some combinations of job and cause of death that were of special interest, PMRs were also derived for subsets of the job group, defined either by employment status (self-employed/manager/employee) or place of residence at the time of death.

The sections that follow describe some of the more important findings from the analysis for men.

4.2 Overall burden of occupational disease

The two most important occupational causes of mortality to emerge from the analysis of deaths in men were accidental injury and asbestos. Crude calculations based on the excess proportional mortality in occupations exposed to specific hazards of injury at work suggest that over the eleven years of the study more than 2,500 deaths were attributable to such injuries (see section 6.13.2). This method of calculation gives a conservative estimate, and the data presented in Chapter 11, which are based on notifications of occupational accidents, indicate that the true figure was approximately 5,000 deaths. During the same period, some 150 deaths from asbestosis, and more than 850 deaths from pleural and peritoneal cancer, could confidently be attributed to occupational exposure to asbestos. Again the true figure is likely to be higher than this because the calculation is based only on deaths in occupations with an excess of pleural or peritoneal cancer. It does not take into account cases who were exposed to asbestos at work but then moved to another occupation which did not carry an increased risk. The number of lung cancer deaths caused by asbestos is more difficult to estimate because the confounding effects of smoking must be taken into account. However, it is possible that the excess of lung cancer from occupational exposure to asbestos is more than twice that of mesothelioma.[1,2]

Several other respiratory diseases also have occupational causes that accounted for a substantial number of deaths, in particular, pneumoconioses other than asbestosis (more than 1,000 deaths), farmers' lung disease and other allergic pneumonitis (63 deaths) and chronic bronchitis and emphysema in coal miners (an excess of some 1,900 deaths).

The occupational associations with accidental injuries, mesothelioma, pneumoconiosis and farmers' lung disease stand out because relative risks are high. It is possible that substantial numbers of occupationally related deaths occur from other more common causes of death, but are not clearly discernible because relative risks are lower.

4.3 Metal workers (job groups 112-120, 145-146, 149)

4.3.1 Pneumococcal and unspecified lobar pneumonia

One of the aims of the analysis was to identify new occupational hazards. The strongest pointer to a previously unrecognised occupational disease concerned pneumococcal and unspecified lobar pneumonia in welders. An excess of pneumonia in welders has been reported in previous decennial supplements.[3-5] In 1949-53 the SMR was 226, (70 deaths), in 1959-63 184, (101 deaths) and in 1970-72 157 (66 deaths).

Table 4.1 Mortality from pneumococcal and unspecified lobar pneumonia (ICD 481) in occupations entailing exposure to metal fume - men England and Wales, 1979-80 and 1982-90

Job group		Ages 20-64			Ages 65-74		
		Deaths	PMR	95% CI	Deaths	PMR	95% CI
112	Furnace operatives (metal)	6	154	57-336	12	171	88-298
116	Moulders and coremakers (metal)	18	292	173-461	14	144	79-241
119	Galvanisers and tin platers	1	142	4-790	3	333	69-972
145	Sheet metal workers	21	190	117-291	17	116	67-185
149	Welders	55	255	192-333	20	107	65-165

These findings have drawn little comment. The current analysis distinguished different types of pneumonia, and a clear excess was found in mortality from pneumococcal and unspecified lobar pneumonia (PMR 186, 75 deaths). This excess was confined to men below age 65 years (PMR 255, 55 deaths). Furthermore, a similar pattern was observed in several other occupations that entail exposure to metal fume (Table 4.1). Taken together, these findings point strongly to an occupational hazard.[5]

4.3.2 Chronic bronchitis and emphysema

Similar pathogenetic mechanisms may underlie increased mortality from chronic bronchitis and emphysema in various metal manufacturing occupations (Table 4.2). A hazard of chronic bronchitis in the metal production industry has been suspected for some time.[6] However, when assessing mortality from this disease, it is important to take into account the potential confounding effects of smoking. Many of the metal making occupations also had high proportional mortality from lung cancer (see section 6.10.1), suggesting that part, at least, of their excess of chronic bronchitis and emphysema is attributable to smoking.

Table 4.2 Mortality from chronic bronchitis and emphysema (ICD 491, 492, 496) in metal production workers - men aged 20-74, England and Wales, 1979-80 and 1982-90

Job group		Deaths	PMR	95% CI
112	Furnace operatives (metal)	248	126	111-143
113	Rollers (metal)	50	140	104-184
114	Smiths and forge workers	147	89	75-104
115	Metal drawers	57	117	88-151
116	Moulders and coremakers (metal)	409	146	132-161
117	Electroplaters	68	144	112-182
118	Annealers, hardeners, temperers (metal)	46	94	69-125
119	Galvanisers and tin platers	36	138	96-191
120	Other metal manufacturers	524	110	101-120

Much of the metal production industry in England and Wales is geographically localised, and Table 4.3 shows PMRs for chronic bronchitis and emphysema in furnacemen and moulders and coremakers from the main industrial centres. For comparison, PMRs are also given for all occupations combined in the same places. For almost all of the areas, the mortality of moulders and coremakers was above that expected from national death rates, and also above that of other occupations in the same place. The exceptions were Rotherham and Sheffield with PMRs of 98 and 88 respectively.

4.4 Coal miners (job groups 088, 175)

4.4.1 Pneumoconiosis

Altogether, 942 deaths were ascribed to coal workers' pneumoconiosis during the eleven-year study period. Figure 4.1 shows mortality from coal workers' pneumoconiosis in face-trained and other coal miners according to their county of residence. PMRs tended to be highest in North-west England and Wales and exceeded 2,000 in Lancashire, Merseyside, South Glamorgan, Mid Glamorgan and Dyfed. In contrast, PMRs in South Yorkshire, Staffordshire, Derbyshire, Nottinghamshire, Leicestershire, Tyne and Wear, and Northumberland were all below 500. This accords with a previous analysis of mortality for the period 1968-78[7], and may reflect differences both in coal rank and in dust concentrations in mines.

Table 4.3 Mortality from chronic bronchitis and emphysema (ICD 491, 492, 496) in furnace operatives and moulders and coremakers by area of residence - men aged 20-74, England and Wales, 1979-80 and 1982-90

Area	Furnacemen, and moulders and coremakers			All occupations		
	Deaths	PMR	95% CI	Deaths	PMR	95% CI
Manchester	11	288	144-515	1326	133	126-140
Derby	18	201	119-318	439	107	97-118
Dudley	37	184	129-254	649	116	107-125
Sandwell	67	182	141-232	762	113	105-122
Middlesbrough	10	177	85-326	346	113	101-126
Walsall	28	163	108-236	539	107	98-116
Bradford	12	154	80-269	925	110	103-117
Wolverhampton	22	144	90-218	511	103	94-112
Birmingham	19	128	77-200	1994	101	96-105
Langbaurgh	9	126	58-239	295	105	93-117
Leeds	14	125	68-210	1434	108	103-114
The Wrekin	7	124	50-255	234	114	100-130
Rotherham	10	98	47-181	560	114	105-124
Sheffield	31	88	60-126	1133	100	94-106

Figure 4.1 Mortality from coal workers' pneumoconiosis (ICD 500) in face-trained (175) and other coal miners (088) by county: Men aged 20-74, England and Wales, 1979-80 and 1982-90

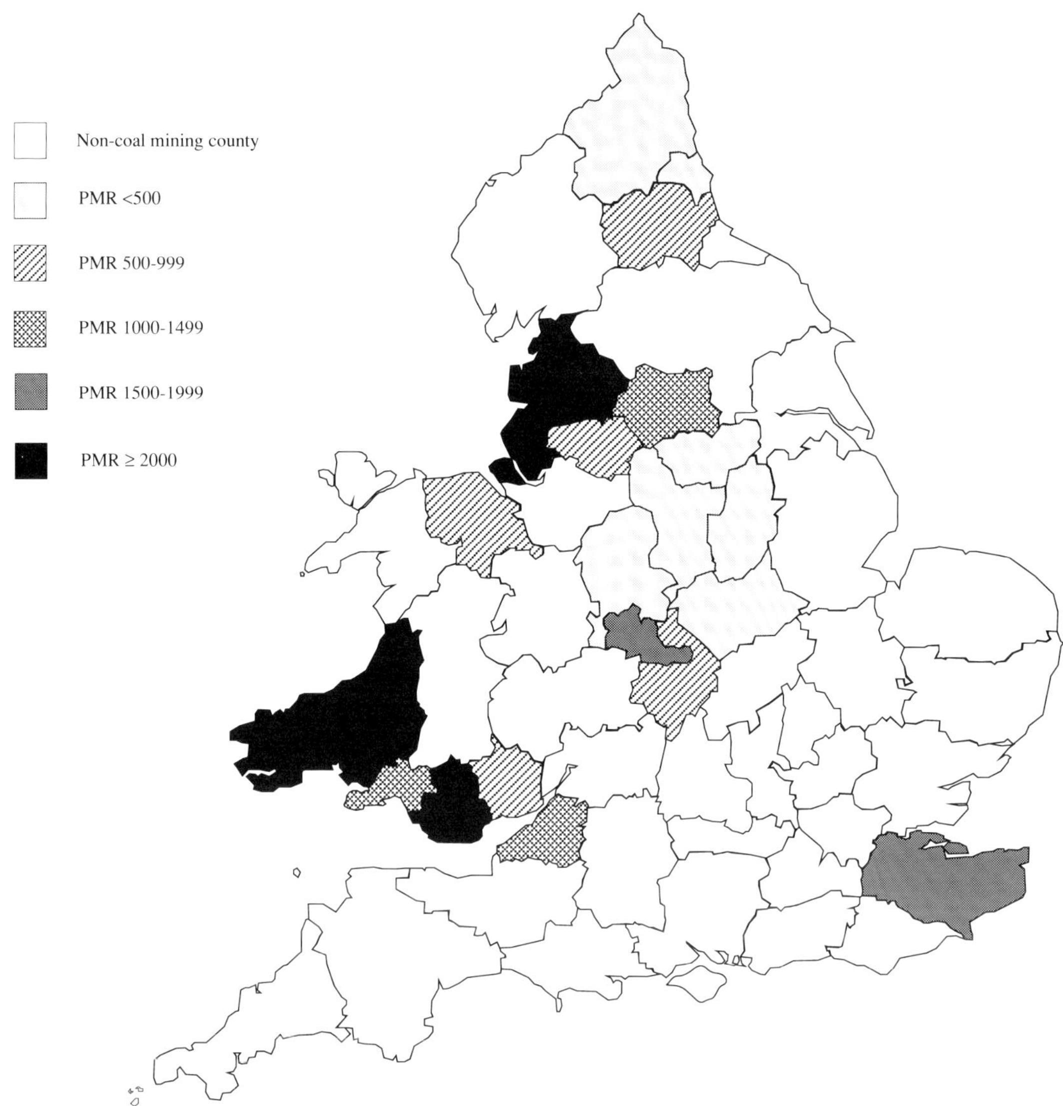

Only 26 deaths from silicosis were recorded in coal miners, but it is notable that 12 of these were from Llanelli and five from Carmarthen, both in the county of Dyfed. This again is consistent with earlier studies,[7] and probably reflects the presence of silica in the bedrock in which coal seams in these areas lie.

4.4.2 Chronic bronchitis and emphysema

Chronic bronchitis and emphysema is another occupational disease of coal miners. Figure 4.2 shows PMRs from chronic bronchitis and emphysema in coal miners by county, and in Figure 4.3 these PMRs are plotted against the PMRs for all occupations combined in the same counties. With the exception of Avon (PMR = 223) and Leicestershire (PMR = 96), PMRs were consistently in the range 115-170, and showed no relation to mortality from chronic bronchitis and emphysema in other occupations. Nor was the geographical distribution of mortality from chronic bronchitis and emphysema in miners related to that from coal workers' pneumoconiosis (Figure 4.4). This suggests that the features of dust exposure which cause bronchitis and emphysema are different from those that produce pneumoconiosis.

4.4.3 Chronic and unspecified myocarditis

Another remarkable finding was a marked excess mortality from chronic and unspecified myocarditis both in coal miners and in other miners (Table 4.4). Myocarditis is an unusual diagnosis as a cause of death and therefore further information was sought from the death certificates of these men. Certificates were available for those who had died during

Figure 4.2 Mortality from chronic bronchitis and emphysema (ICD 491, 492, 496) in face-trained (175) and other coal miners (088) by county: Men aged 20-74, England and Wales, 1979-80 and 1982-90

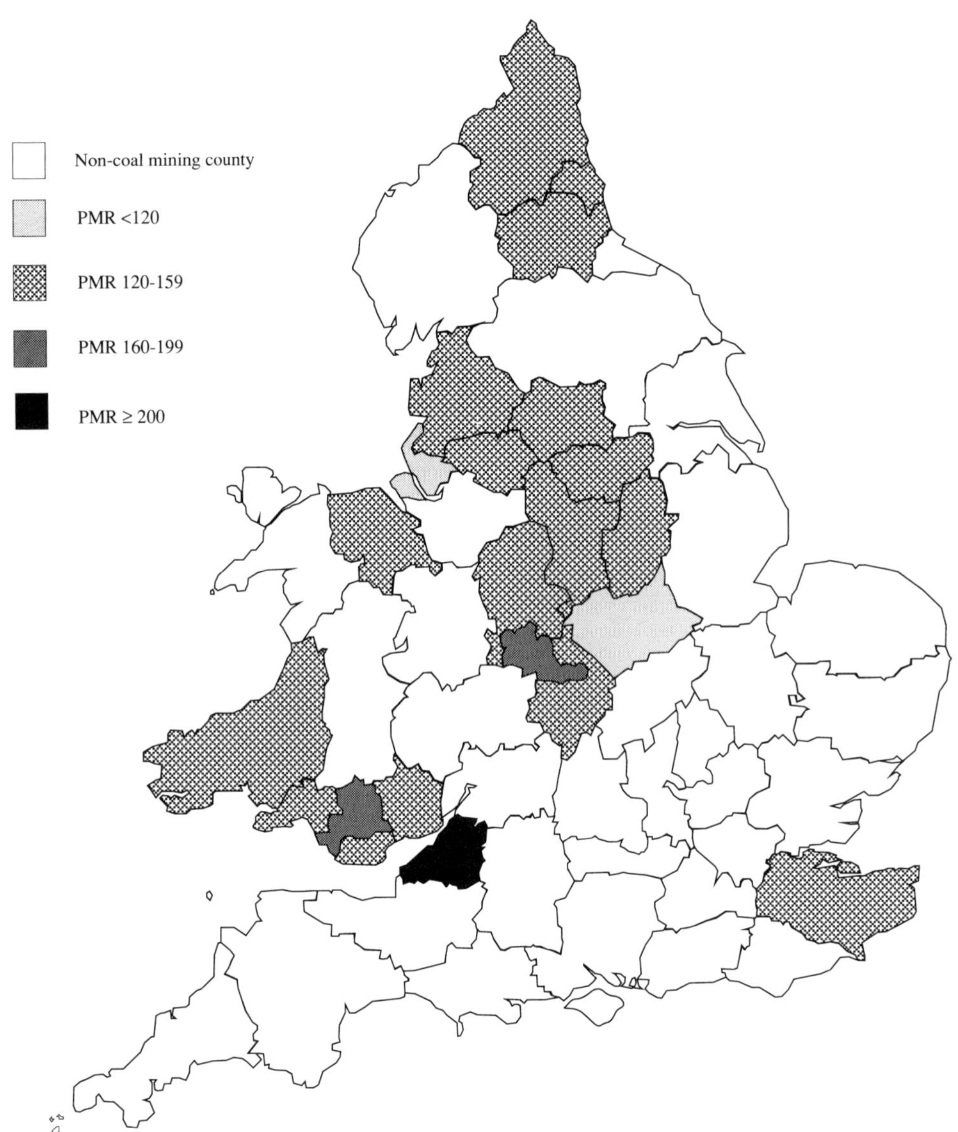

Figure 4.3 Mortality from chronic bronchitis and emphysema (ICD 491,492,496) in face-trained (175) and other coal miners (088) and in all occupations combined by county: Men aged 20-74, England and Wales, 1979-80 and 1982-90

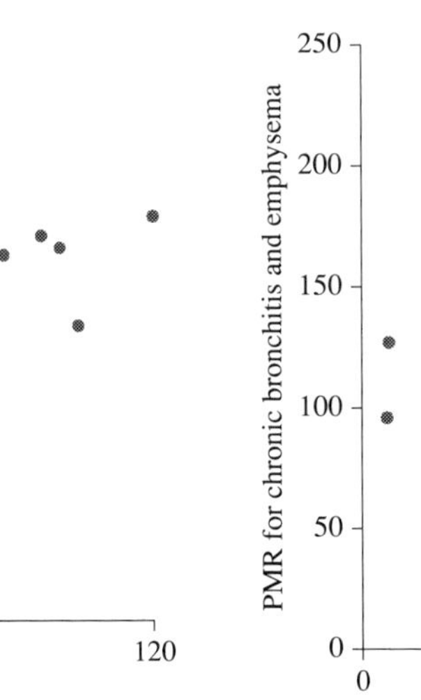

Figure 4.4 Mortality from chronic bronchitis and emphysema (ICD 491,492,496) and from coal workers' pneumoconiosis (ICD 500) in face-trained (175) and other coal miners (088) by county: Men aged 20-74, England and Wales, 1979-80 and 1982-90

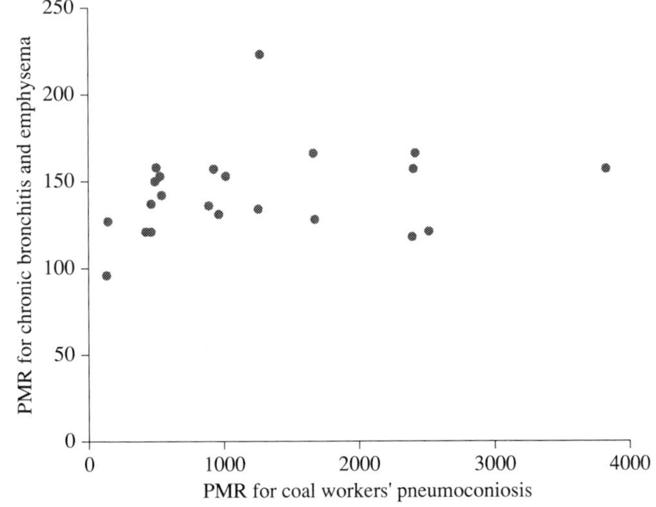

Table 4.4 Mortality from chronic and unspecified myocarditis (ICD 429.0) in miners - men aged 20-74, England and Wales, 1979-80 and 1982-90

Job group		Deaths	PMR	95% CI
088	Other coal miners	157	226	192-264
175	Face trained coalminers	41	288	207-391
176	Miners (not coal) and quarry workers	9	191	88-363

1982-90, and 135 of 159 deaths had been certified by the coroner or deputy coroner for South Yorkshire (East) District. Almost all of these deaths were ascribed to 'myocardial fibrosis' attributed to arteriosclerosis. This excess probably reflects unusual diagnostic practice in one area rather than an unrecognised hazard.

4.5 Urothelial cancer in rubber workers (job groups 085, 092)

During the 1950s, rubber manufacturers were found to have elevated mortality from bladder cancer.[8] This was shown to result from the use of 2-naphthylamine as an antioxidant, and the chemical was banned. Subsequent studies suggested that rubber workers who entered employment after 2-naphthylamine was withdrawn had no increased risk of bladder cancer.[9]

The current analysis indicated increased mortality from urothelial cancer both in rubber process workers and in men making and repairing rubber goods, and the excess was apparent even in those below age 65 years (Table 4.5). A persistent risk from exposure to 2-naphthylamine in the 1950s is still a possible explanation if the cases were exposed very early in their careers. Future analyses, however, should support or refute the contention.

4.6 Mortality in farmers (job group 047)

The health of farmers has been investigated less than that of many occupational groups, partly because they work mainly in small businesses, making it more difficult to identify populations for study. They are, however, exposed to a wide range of occupational hazards, and this is reflected in their mortality. Farmers had high death rates from various accidental injuries, from several occupationally related respiratory disorders, from hernias (which may result from the extremely heavy lifting in agriculture, especially in the past), and from suicide (Table 4.6). More attention is needed to the occupational health problems of agricultural workers.

Table 4.6 Mortality from selected causes in farmers - men aged 20-74, England and Wales, 1979-80 and 1982-90

047 Farmers

Cause of death (ICD)	Deaths	PMR	95% CI
Farmers' lung disease (495.0)	56	1089	823-1416
Other and unspecified allergic pneumonitis (495.1,495.3-495.9)	7	787	317-1622
Inguinal hernia (550)	41	191	137-259
Other hernia (551-553)	41	149	107-202
Infections of skin, joints and bone (680-686,711,730)	36	181	127-251
Off-road motor vehicle accidents (E820-E825)	38	255	180-350
Animal transport accidents (E827-E828)	15	468	262-773
Pesticide poisoning (E863)	4	1455	396-3724
Poisoning by other gases (E869)	7	417	168-859
Slipping and tripping (E885)	17	193	112-309
Injury by animals and plants (E905-E906)	21	775	479-1186
Injury by falling object (E916)	35	156	109-217
Injury by machinery (E919)	147	457	386-538
Injury by firearms (E922)	23	670	424-1006
Injury by electric current (E925)	29	213	143-307
Suicide (E950-E959)	1215	156	147-165

4.6.1 Mortality from accidents and suicide in farmers according to employment status

Many farmers are self-employed, and it is of interest to compare their mortality from accidental injury and suicide with that of employees. Table 4.7 shows PMRs at ages 20-64 for injuries likely to be work-related, according to employment status. The overall proportion of deaths from occupationally related injury and poisoning was markedly higher in self-employed farmers (PMR 624, 149 deaths) than in those who worked as employees (PMR 194, 112 deaths), while mortality in managers and foremen was intermediate (PMR 333, 23 deaths). The increased mortality in self-employed farmers was attributable to higher PMRs for most of the specific types of injury making up the total, and particularly for injury by machinery, for which the PMR was almost five times that in employees. This pattern suggests that self-employed farmers are exposed to greater risks at work, and it may have implications for targeting preventive strategies.

The proportion of deaths from suicide at ages 20-64 was also higher in self-employed farmers (PMR 193, 415 deaths) than in employees (PMR 136, 498 deaths) or managers and foremen (PMR 134, 70 deaths). This differential would be consistent with an influence of financial pressures in farmers running their own businesses, but the fact that PMRs were also elevated in other categories of employment status suggests that even if such pressures are relevant, they are not the only explanation for the excess. Another factor may have been ready access to means of successful suicide such as guns and poisons.

Table 4.5 Mortality from urothelial cancer (ICD 188, 189.1-189.8) in rubber workers - men aged 20-74, England and Wales, 1979-80 and 1982-90

Job group		Ages 20-74			Ages 20-64		
		Deaths	PMR	95% CI	Deaths	PMR	95% CI
085	Rubber manufacturers	31	136	93-194	12	160	83-279
092	Rubber goods makers	24	134	85-199	10	148	71-272

Table 4.7 Mortality from injury and poisoning in farmers according to employment status - men aged 20-64, England and Wales, 1979-80 and 1982-90

047 Farmers

Cause of death (ICD)	Employees			Managers & foremen			Self-employed		
	Deaths	PMR	95% CI	Deaths	PMR	95% CI	Deaths	PMR	95% CI
Off-road motor vehicle accidents (E820-E825)	17	186	108 -298	3	274	56 -800	12	359	186 -628
Animal transport accidents (E827-E828)	2	430	52 -1554	2	685	83 -2474	8	526	227 -1036
Pesticide poisoning (E863)	1	1429	36 -7959	1	3333	84 -18572	1	1923	49 -10715
Poisoning by other gases (E869)	2	273	33 -986	0	0	0 -2072	5	741	241 -1729
Slipping and tripping (E885)	6	370	136 -806	0	0	0 -1272	2	170	21 -613
Injury by animals and plants (E905-E906)	9	826	378 -1569	0	0	0 -2493	8	896	387 -1765
Injury by falling object (E916)	16	104	60 -169	3	239	49 -698	11	272	136 -487
Injury by machinery (E919)	43	211	153 -285	10	465	223 -855	74	1079	847 -1355
Injury by firearms (E922)	3	282	58 -823	3	836	172 -2442	14	880	481 -1476
Injury by electric current (E925)	13	167	89 -285	1	91	2 -508	14	375	205 -628
Total	112	194	160 -234	23	333	211 -501	149	624	528 -733

Figure 4.5 Mortality of farmers (047) from melanoma of skin (ICD 172) by region: Men aged 20-74, England and Wales, 1979-80 and 1982-90

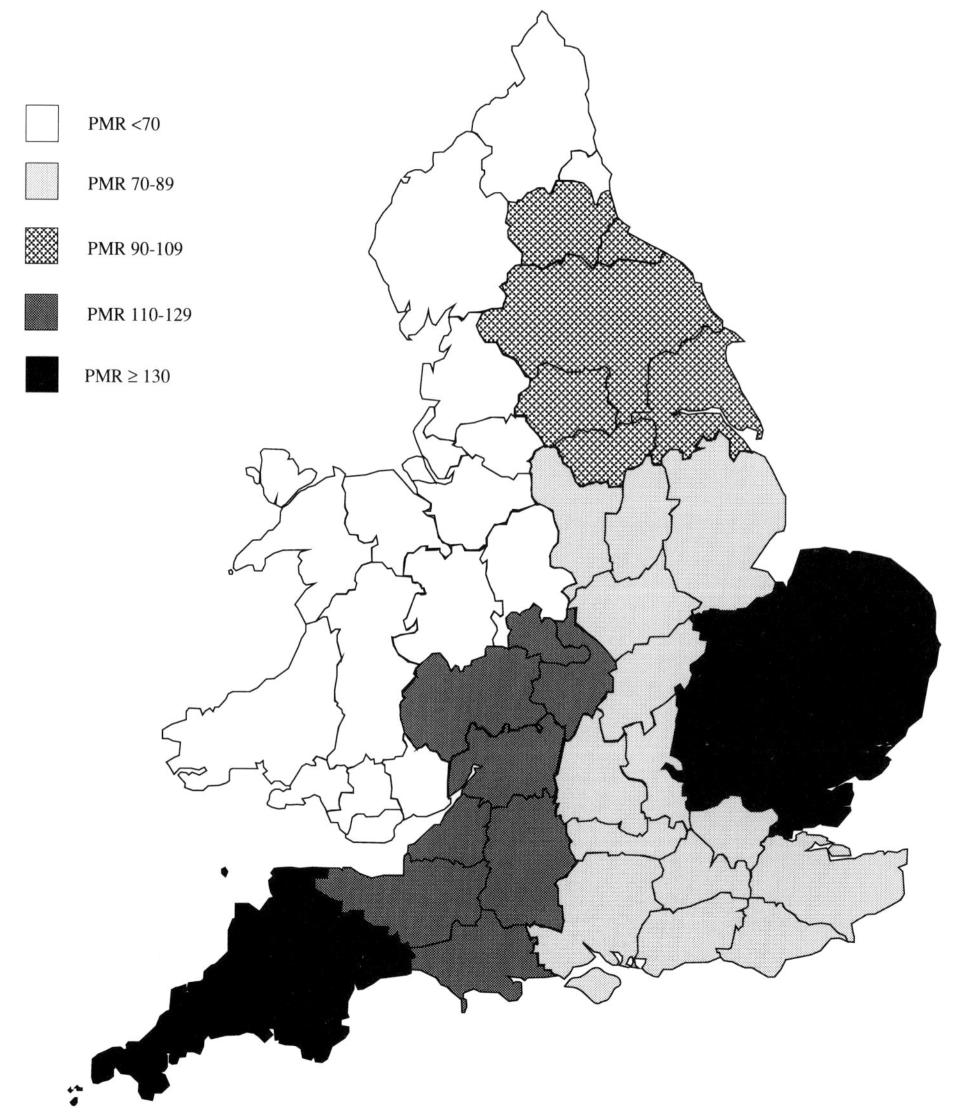

4.6.2 Geographical differences in mortality among farmers

Agricultural activities vary from one part of the country to another, depending on climate, terrain and soil characteristics. If these differences lead to different occupational hazards, they may be reflected in geographical differences in mortality. In addition, farmers may exhibit regional variation in mortality as a result of non-occupational influences that apply to the population more generally. The findings for several causes of death merit comment.

4.6.3 Oesophageal cancer

Mortality from oesophageal cancer was significantly elevated in farmers from two counties - Somerset (PMR 156, 23 deaths) and Hereford and Worcester (PMR 175, 36 deaths). These are both areas where cider apples are grown.[10] Alcohol is an established cause of oesophageal cancer, and another alcoholic drink made from apples, calvados, has been linked with a high incidence of the disease in the Normandy region of France.[11] It is possible that consumption of rough cider made on farms contributed to the high mortality from oesophageal cancer in farmers from Somerset and Hereford and Worcester. In two other counties where cider is made on farms, Devon and Dorset, PMRs were also above average (127 and 128 respectively).

4.6.4 Melanoma of skin

Cutaneous melanoma is related to solar radiation and is more common in the sunnier south of England than in the north.[12] However, farmers and other outdoor occupations do not have particularly high mortality from the disease (see section 6.2.5), suggesting that intermittent exposure to sunlight may be more harmful than regular exposure. The numbers of deaths from melanoma in farmers were too few for analysis by county, but Figure 4.5 shows PMRs for the disease in the agricultural regions of the Ministry of Agriculture, Fisheries and Food (MAFF). The counties which make up these regions are listed in Annex 4.1. As in the general population,

Figure 4.6 Mortality of farmers (047) cancer of the prostate (ICD 185) by region: Men aged 20-74, England and Wales, 1979-80 and 1982-90

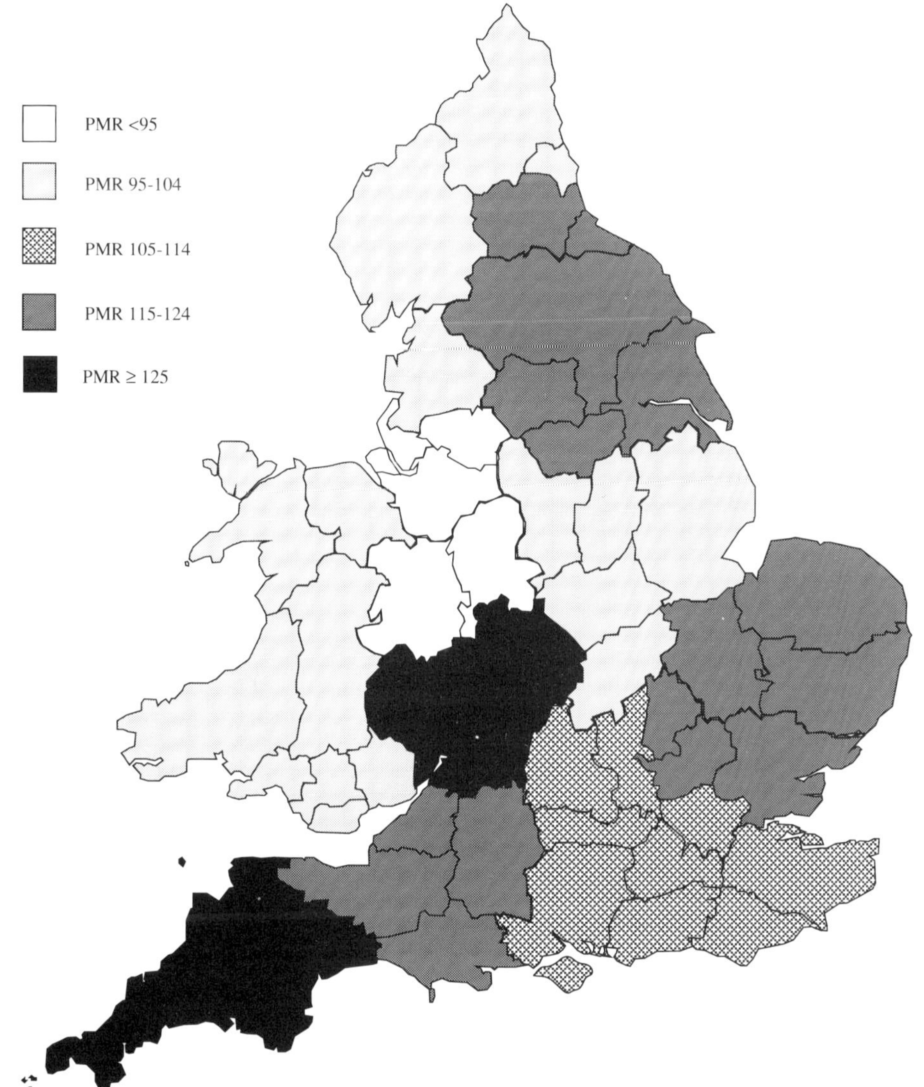

PMRs from melanoma in farmers show a north/south gradient with values ranging from 49 in the Northern region to 164 in the South-west. Thus, even if regular sun exposure is less dangerous than intermittent exposure, it nevertheless appears to carry a higher risk when more intense. Alternatively, the geographical differences in mortality from melanoma could be largely a late manifestation of exposure to sunlight in childhood.

4.6.5 Cancer of the prostate

High rates of cancer of the prostate have been reported in farmers,[13] but in this analysis the proportion of deaths from prostatic cancer in farmers was only slightly above average (PMR 112, 1361 deaths). To explore whether the disease might be more common in relation to particular types of farming (eg arable, dairy), PMRs were derived for each of the MAFF regions (Figure 4.6). If anything, mortality tended to be higher in the South, but the pattern did not suggest an association with specific agricultural activities.

4.6.6 Non-Hodgkin's lymphoma

Non-Hodgkin's lymphoma is another tumour which has been reported as more common in farmers.[14] In particular, it has been associated in some studies with exposure to phenoxy herbicides such as 2,4-D and 2,4,5-T.[15, 16, 17] In England and Wales the phenoxy herbicides used most widely in agriculture in recent years have been MCPA, mecoprop and mecoprop-P. Figure 4.7 shows PMRs from non-Hodgkin's lymphoma in farmers by MAFF region plotted against rates of phenoxy herbicide usage. The relation between the two variables was weak (correlation coefficient = 0.29), and provides little support for the hypothesised association.

Figure 4.7 Mortality from non-Hodgkin's lymphoma (ICD 200,202) in farmers (047) and rates of phenoxy herbicide usage by region: Men aged 20-74, England and Wales, 1979-80 and 1982-90

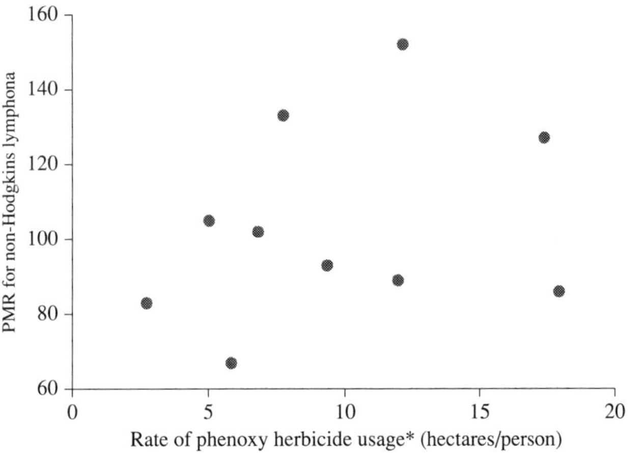

* Calculated as a ratio of the estimated area (in hectares) of arable[14] and grassland[15] crops in the region treated by phenoxy herbicides annually to the number of whole-time farmers, partners and directors, salaried farm managers and male family and hired agricultural workers in the region as assessed in the 1992 Agricultural and Horticultural Census of England and Wales[11]

4.6.7 Farmers' lung disease

Farmers' lung disease is an occupational hazard specific to work in agriculture. Table 4.8 shows PMRs from farmers' lung disease and other and unspecified allergic pneumonitis by region. There were more than 20-fold variations with particularly high proportions in farmers from Wales (PMR 2868, 22 deaths) and the North-east of England (PMR 2063, 15 deaths). These are areas with a lot of hill farming and a relatively damp climate predisposing to formation of the moulds in hay which cause the disease.

Table 4.8 Mortality from farmers' lung disease (ICD 495.0) and other and unspecified allergic pneumonitis (ICD 495.1, 495.3-495.9) by region, men aged 20-74, England and Wales, 1979-80 and 1982-90

047 Farmers

Region	Deaths	PMR	95% CI
Northern	5	956	310-2231
North East	15	2063	1155-3403
North Mercia	4	707	193-1809
East Midlands	2	332	40-1198
South Mercia	3	746	154-2181
Anglia	1	141	4-787
Wessex	1	247	6-1376
South West	6	1220	448-2654
South East	4	509	139-1303
Wales	22	2868	1795-4349

4.7 Lung cancer in butchers (job group 042)

Previous analyses of occupational mortality and cancer incidence, including several Decennial Supplements, have suggested high rates of lung cancer in butchers and slaughtermen.[18] This has led to the hypothesis of an occupational hazard, possibly from fumes generated when meat is wrapped in plastic[19] or from a carcinogenic papillomavirus.[20] (Meat workers are known to have a high prevalence of cutaneous warts caused by an unusual papillomavirus, HPV 7.) Alternatively, the high rates may have resulted from a confounding effect of smoking.

In the current analysis, the excess of lung cancer in butchers was less marked than in previous studies (PMR 108, 1268 deaths), making a major hazard less likely.

4.8 Summary for each job group

This section highlights some of the more interesting findings for each of the 194 job groups analysed. A list of significantly high and low PMRs is given in Appendix 4, but inevitably when many combinations of occupation and cause of death are examined, some will show unusually high or low PMRs just by chance. It is therefore important to evaluate findings in the context of what is known from elsewhere. An observation with high statistical significance may quite reasonably be attributed to chance if it is inconsistent with the results of other studies. At the same time, some results are important even though they are not statistically significant.

001 - LAWYERS
High rates of liver cancer (PMR 212, 32 deaths), cirrhosis (PMR 141, 24 deaths) and 'other alcohol-related diseases' (PMR 173, 35 deaths) suggest high alcohol consumption within the legal profession. Lawyers also had elevated mortality from immunodeficiency (PMR 216, 12 deaths).

002 - ACCOUNTANTS
Elevated mortality from melanoma of skin (PMR 142, 86 deaths) is unlikely to reflect a hazard in the workplace [see section 6.2.5].

003 - PERSONNEL MANAGERS ETC.
Mortality from fibrosing alveolitis was increased (PMR 210, 14 deaths), but there is no obvious occupational cause for this.

004 - ECONOMISTS AND STATISTICIANS
High mortality from melanoma of skin (PMR 291, 8 deaths) is unlikely to reflect a hazard in the workplace [see section 6.2.5].

005 - COMPUTER PROGRAMMERS
Three cancers showed elevated rates - cancer of bone (PMR 247, 11 deaths), cancer of soft tissue (PMR 203, 11 deaths) and brain cancer (PMR 145, 43 deaths).

006 - SALES MANAGERS ETC.
A high rate of skin melanoma (PMR 132, 88 deaths) is unlikely to reflect a hazard in the workplace [see section 6.2.5]. Elevated mortality was also recorded for retroperitoneal cancer (PMR 199, 12 deaths) and acute lymphatic leukaemia (PMR 161, 21 deaths), but there is no obvious occupational cause for these findings.

007 - GOVERNMENT INSPECTORS
Mortality patterns were unremarkable.

008 - GOVERNMENT ADMINISTRATORS
Proportional mortality was marginally elevated for cancer of the colon (PMR 120, 151 deaths), cancer of the prostate (PMR 118, 138 deaths), cancer of the brain (PMR 128, 70 deaths) and aortic aneurysm (PMR 133, 143 deaths), but these findings are more likely to reflect aspects of lifestyle than occupational hazards.

009 - OTHER ADMINISTRATORS
A high rate of glomerulonephritis (PMR 205, 16 deaths) was observed, but this seems unlikely to reflect an occupational hazard. Mortality was also high for immunodeficiency (PMR 162, 23 deaths) indicating high rates of HIV infection in these occupations.

010 - TEACHERS IN HIGHER EDUCATION
Mortality was high for cancer of the prostate (PMR 167, 150 deaths), brain cancer (PMR 185, 116 deaths), non-Hodgkin's lymphoma (PMR 146, 69 deaths), myeloma (PMR 148, 38 deaths), chronic lymphatic leukaemia (PMR 184, 16 deaths), immunodeficiency (PMR 177, 16 deaths), Parkinson's disease (PMR 170, 21 deaths) and motor neurone disease (PMR 170, 30 deaths). The high rates of lymphatic and haematological cancers and of degenerative neurological diseases might be caused by exposure to infections. There is no obvious occupational cause for the other findings.

011 - TEACHERS NEC
As in many professional occupations, the rate of skin melanoma was elevated (PMR 154, 105 deaths) [see section 6.2.5]. High mortality from aplastic anaemia (PMR 266, 18 deaths), agranulocytosis (PMR 296, 4 deaths), Parkinson's disease (PMR 186, 90 deaths) and multiple sclerosis (PMR 152, 51 deaths) could be the result of exposure to infections. In particular, multiple sclerosis, rates of which were also increased in female teachers, has been linked with infection by Epstein-Barr virus. There is no obvious occupational cause for the elevated mortality that was observed from thyroid cancer (PMR 199, 21 deaths). A high rate of deaths from accidental falls (PMR 357, 19 deaths) may have resulted from recreational activities such as rock climbing. Mortality was also high from immunodeficiency (PMR 138, 53 deaths).

012 - VOCATIONAL TRAINERS, SOCIAL SCIENTISTS, ETC.
As in teachers in higher education, mortality was increased from brain cancer (PMR 128, 66 deaths) and non-Hodgkin's lymphoma (PMR 134, 54 deaths).

013 - WELFARE WORKERS
A high rate of bone cancer (PMR 225, 10 deaths) has no obvious occupational cause.

014 - CLERGY
An elevated rate of diabetes (PMR 148, 59 deaths) has no obvious occupational cause. Mortality was unusually high from motor vehicle accidents (PMR 140, 50 deaths) and a similar excess was also observed in female clergy [see section 6.13.1].

015 - DOCTORS
A high death rate from viral hepatitis (PMR 435, 14 deaths) is probably attributable in part to persistence of infection acquired early in life among doctors born in countries where hepatitis B is more prevalent, but there is also a well recognised hazard of hepatitis B virus transmission through 'sharps' injuries. [see section 6.4] As in previous decennial supplements, doctors had elevated mortality from liver cancer (PMR 190, 40 deaths), cirrhosis (PMR 203, 49 deaths) and 'other alcohol-related diseases' (PMR 133, 37 deaths) which reflects high alcohol consumption within the profession. A high rate of suicide (PMR 162, 141 deaths) has also been found before. Occupational stresses may have contributed, but the medical knowledge to ensure the success of a suicide attempt and access to drugs are probably more important factors. The excess was distributed more or less uniformly across the age range from 20 to 74.

016 - DENTISTS
Like doctors, dentists had a high rate of suicide (PMR 194, 38 deaths). Medical knowledge to ensure the success of suicide attempts and access to drugs are probably major factors. There is no obvious occupational cause for dentists' elevated mortality from prostatic cancer (PMR 159, 37 deaths).

017 - NURSES

The rate of suicide was high in male nurses (PMR 127, 135 deaths) as it was also in female nurses [see section 6.13.3]. Medical knowledge to ensure the success of suicide attempts and access to drugs are likely to be important factors. Occupational stress may also contribute.

018 - PHARMACISTS

Mortality from suicide was elevated (PMR 170, 46 deaths). Medical knowledge to ensure the success of suicide attempts and access to drugs are probably important factors.

019 - MEDICAL RADIOGRAPHERS

Mortality patterns were unremarkable.

020 - PHYSIOTHERAPISTS

Mortality patterns were unremarkable.

021 - HEALTH PROFESSIONS NEC

Mortality patterns were unremarkable.

022 - VETERINARIANS

Mortality from suicide was particularly high (PMR 361, 35 deaths). As in other health professions this may partly reflect medical knowledge to ensure the success of suicide attempts and access to drugs, but occupational stress may also be a factor. Thirty two of the 35 suicide deaths were in men less than 65 years of age.

023 - DRIVING INSTRUCTORS

Mortality from asthma was elevated (PMR 186, 14 deaths), but only nine of the asthma deaths occurred before age 65, suggesting that this was not an acute effect of exposure to traffic fumes. Suicide rates were high (PMR 142, 54 deaths).

024 - LITERARY AND ARTISTIC OCCUPATIONS

High death rates were observed for a number of infectious diseases including tuberculosis (PMR 165, 29 deaths), septicaemia (PMR 154, 16 deaths), viral hepatitis (PMR 242, 15 deaths) and unspecified pneumonia (PMR 172, 28 deaths). It is likely that many of these deaths were related to HIV infection, and mortality from immunodeficiency was also high (PMR 341, 140 deaths). Deaths ascribed directly to drug dependence were increased (PMR 352, 24 deaths) as were deaths from accidental poisoning by drugs (PMR 263, 44 deaths). High consumption of alcohol in this occupational group is indicated by elevated death rates from liver cancer (PMR 134, 60 deaths), cirrhosis (PMR 151, 82 deaths), cancer of the oral cavity (PMR 154, 42 deaths), cancer of the pharynx (PMR 166, 33 deaths) and 'other alcohol-related diseases' (PMR 175, 158 deaths). Elevated mortality from non-melanomatous skin cancer (PMR 182, 19 deaths) probably reflects deaths from Kaposi's sarcoma, a common complication of HIV infection. These occupations also had high mortality from homicide (PMR 229, 29 deaths).

025 - PERSONS INVOLVED IN SPORT

The risk of accidental injury in certain sports was reflected in high mortality from off-road motor vehicle accidents (PMR 3005, 11 deaths) and animal transport accidents (PMR 2667, 2 deaths).

026 - BIOLOGICAL SCIENTISTS

Mortality patterns were unremarkable.

027 - CHEMICAL ENGINEERS AND SCIENTISTS

High mortality was found from pleural cancer (PMR 238, 13 deaths). This is probably the result of past occupational exposure to asbestos in chemical works, but the association may have been exaggerated by misclassification of workers with manual occupations in the chemical industry (e.g. a mechanical engineer being recorded as a chemical engineer). Also some chemical engineers may have worked previously in manual occupations in the chemical industry where exposure to asbestos has been more common.

028 - PHYSICAL SCIENTISTS AND MATHEMATICIANS

High mortality was observed for cancer of the mediastinum (PMR 787, 4 deaths) and brain cancer (PMR 169, 65 deaths).

029 - ELECTRICAL AND ELECTRONIC ENGINEERS (PROFESSIONAL)

A high rate of urothelial cancer (PMR 152, 58 deaths) has no obvious occupational cause. Mortality from brain cancer (PMR 96, 65 deaths) and leukaemia was unremarkable.

030 - PROFESSIONAL ENGINEERS NEC

Mortality from pleural cancer was moderately elevated (PMR 139, 53 deaths) reflecting past exposure to asbestos in some of these occupations.

031 - DRAUGHTSPERSONS

High mortality was observed from brain cancer (PMR 133, 89 deaths) and multiple sclerosis (PMR 190, 37 deaths). There is little evidence to suggest that either of these findings indicates an occupational hazard.

032 - LABORATORY TECHNICIANS

Mortality was elevated from dementia (PMR 163, 29 deaths) and non-alcoholic gastritis and duodenitis (PMR 663, 6 deaths), but direct occupational causes for these findings seem unlikely.

033 - ARCHITECTS AND SURVEYORS

Increased mortality from pleural cancer (PMR 148, 21 deaths) suggests that exposure to asbestos has occurred in some of these occupations. High death rates from septicaemia (PMR 204, 12 deaths) and multiple sclerosis (PMR 149, 31 deaths) are unlikely to result from occupational hazards.

034 - AIRCRAFT FLIGHT DECK OFFICERS

The major occupational hazard is from air traffic accidents (PMR 7760, 44 deaths) which caused 11 per cent of deaths below age 65. A high death rate from 'other alcohol-related diseases' (PMR 184, 10 deaths) is of special concern in this occupational group, and it is possible that alcohol contributed to some of the accidental deaths. There is no obvious occupational cause for the finding of increased mortality from cancer of the colon (PMR 162, 26 deaths), cancer of the prostate (PMR 208, 25 deaths) and dementia (PMR 440, 7 deaths).

035 - AIR TRAFFIC CONTROLLERS

High mortality from liver cancer (PMR 412, 3 deaths) and 'other alcohol-related diseases' (PMR 597, 8 deaths) is a special concern in this occupation where impairment of performance by alcohol can have disastrous consequences. It is possible that occupational stress contributes to the apparently high alcohol consumption in some members of this group.

036 - SEAFARERS
Mortality rates were high for cancer of the oral cavity (PMR 273, 56 deaths), cancer of the pharynx (PMR 290, 45 deaths), cancer of the liver (PMR 154, 46 deaths) cancer of the larynx (PMR 242, 67 deaths), cirrhosis (PMR 256, 81 deaths), pancreatitis (PMR 170, 27 deaths) and 'other alcohol-related diseases' (PMR 309, 154 deaths). These findings are consistent with those in previous decennial supplements and reflect unusually high consumption of alcohol among seamen. High rates of tuberculosis (PMR 185, 32 deaths), pneumococcal and unspecified lobar pneumonia (PMR 150, 46 deaths), bronchopneumonia (PMR 144, 222 deaths), unspecified pneumonia (PMR 175, 22 deaths) and gastric ulcer (PMR 184, 37 deaths) may be attributable to a combination of high alcohol consumption and poor nutrition. The risk of accidental injury is highlighted by the large excess of deaths from water transport accidents (PMR 2088, 104 deaths), and indeterminate injury (PMR 184, 93 deaths), almost all of which occurred in men below age 65. If anything the mortality ratios for accidental injury are likely to be an underestimate because deaths at sea may not always be registered in Britain.

037 - TECHNICIANS NEC
Mortality patterns were unremarkable.

038 - PRODUCTION AND MAINTENANCE MANAGERS
A high rate of cancer of the gall bladder (PMR 137, 64 deaths) seems unlikely to be due to an occupational hazard.

039 - MANAGERS IN CONSTRUCTION
High rates of cancer of the peritoneum (PMR 426, 6 deaths) and pleura (PMR 319, 32 deaths) are attributable to asbestos exposure. Increased mortality from fibrosing alveolitis (PMR 195, 23 deaths) may in part reflect misdiagnosis of asbestosis.

040 - MANAGERS IN TRANSPORT, UTILITIES AND MINING
Mortality was increased from testicular cancer (PMR 174, 21 deaths) and cancer of the penis (PMR 193, 13 deaths), but current evidence does not suggest an occupational cause for these findings.

041 - OFFICE MANAGERS
A high mortality from cancer of the gall bladder (PMR 140, 36 deaths) is unlikely to be due to an occupational hazard.

042 - BUTCHERS
An excess of accidents caused by cutting and piercing instruments or objects (PMR 735, 4 deaths) is probably due to injuries at work. Death rates from lung cancer were only moderately elevated (PMR 108, 1268 deaths).

043 - FISHMONGERS AND POULTRY DRESSERS
High mortality from cancer of the larynx (PMR 242, 10 deaths) and bronchus (PMR 139, 255 deaths) could indicate an occupational hazard (high rates of lung cancer have been reported in several studies of meat workers), but they may simply reflect patterns of smoking in these occupations.

044 - RETAILERS AND DEALERS
Mortality was high from homicide (PMR 181, 78 deaths), the excess being concentrated in proprietors and managers of garages and shops, scrap dealers and market and street traders. No increase in homicide was observed for shop assistants.

045 - PUBLICANS AND BAR STAFF
Mortality was high from cancer of the oral cavity (PMR 275, 117 deaths), cancer of the pharynx (PMR 230, 71 deaths), cancer of the liver (PMR 162, 112 deaths), cancer of the larynx (PMR 262, 119 deaths), oesophageal varices (PMR 271, 9 deaths), gastric ulcer (PMR 188, 64 deaths), cirrhosis (PMR 301, 243 deaths), pancreatitis (PMR 134, 45 deaths) and 'other alcohol-related diseases' (PMR 365, 458 deaths). This is likely to reflect not only ready access to alcohol in these occupations, but also the selection of people who habitually consume alcohol into such work. High death rates from falling on stairs or steps (PMR 229, 43 deaths), unspecified falls (PMR 193, 37 deaths) and fire (PMR 202, 33 deaths) may also be attributable to alcohol.

046 - CATERERS
High mortality was observed from cancer of the pharynx (PMR 243, 36 deaths), cancer of the liver (PMR 193, 56 deaths), cirrhosis (PMR 157, 52 deaths) and 'other alcohol-related diseases' (PMR 177, 97 deaths). This may reflect both ease of access to alcohol at work and selective recruitment of habitual alcohol drinkers into these occupations. Elevated mortality from viral hepatitis (PMR 352, 14 deaths) and immunodeficiency (PMR 404, 62 deaths) were observed. The high rate of immunodeficiency may in turn have contributed to high mortality from tuberculosis (PMR 141, 23 deaths). High death rates were also observed for duodenal ulcer (PMR 181, 54 deaths) and non-alcoholic gastritis and duodenitis (PMR 627, 7 deaths).

047 - FARMERS
The hazards of exposure to allergenic dusts are reflected in high mortality from farmers' lung disease (PMR 1089, 56 deaths) and other and unspecified allergic pneumonitis (PMR 787, 7 deaths). They may also have contributed to a high death rate from influenza (PMR 163, 46 deaths). High mortality from inguinal hernia (PMR 191, 41 deaths) and other hernia (PMR 149, 41 deaths) is probably attributable to the heavy lifting that is carried out in many of these jobs. Striking excesses of mortality were observed for several categories of accidental injury, including off-road motor vehicle accidents (PMR 255, 38 deaths), accidents involving animal transport (PMR 468, 15 deaths), poisoning by pesticides (PMR 1455, 4 deaths), poisoning by gases (PMR 417, 7 deaths), slipping and tripping (PMR 193, 17 deaths), injuries caused by animals and plants (PMR 775, 21 deaths),

injuries causing by falling objects (PMR 156, 35 deaths), accidents with machinery (PMR 457, 147 deaths), accidents with firearms (PMR 670, 23 deaths) and injury by electric current (PMR 213, 29 deaths). The risk of acute injury in many agricultural activities may also account for a high death rate from epilepsy (PMR 176, 148 deaths). A high mortality from suicide (PMR 156, 1215 deaths) may have been influenced both by access to means of suicide (firearms and pesticides) and also by occupational stresses, especially in the self-employed. Other excesses of mortality which might be due to occupational hazards include haemolytic anaemia (PMR 268, 10 deaths) and infections of skin, joints and bone (PMR 181, 36 deaths). Death rates from cancer were unremarkable.

048 - ARMED FORCES
Mortality was high from cancer of the oral cavity (PMR 207, 45 deaths), cancer of the pharynx (PMR 173, 29 deaths), cirrhosis (PMR 182, 59 deaths) and 'other alcohol-related diseases' (PMR 221, 132 deaths). Alcohol may have contributed also to high death rates from falling on stairs and steps (PMR 183, 25 deaths), other falls (PMR 190, 13 deaths) and unspecified falls (PMR 147, 22 deaths). Many excesses of accidental deaths are likely to be attributable to injuries at work, including those from off-road motor vehicle accidents (PMR 220, 19 deaths), air transport accidents (PMR 1128, 72 deaths), non-recreational drowning (PMR 681, 4 deaths), injury by cutting and piercing instruments or objects (PMR 425, 6 deaths), firearm accidents (PMR 417, 7 deaths) accidental explosions (PMR 339, 7 deaths) and other accidents (PMR 629, 26 deaths). Eleven deaths were ascribed to war (PMR 2973) and some of the excess mortality from homicide (PMR 205, 42 deaths) may also have been directly related to work. Military personnel had unusually high mortality from multiple sclerosis (PMR 255, 48 deaths) which might be related to patterns of infection associated with life in barracks.

049 - POLICE
Much of the excess mortality from motor vehicle accidents (PMR 165, 226 deaths), non-recreational drowning (PMR 1008, 5 deaths) and homicide (PMR 263, 16 deaths) is likely to have been caused by work. There is no obvious occupational cause for high rates of motor neurone disease (PMR 186, 44 deaths) and of nasal cancer (PMR 235, 9 deaths).

050 - FIRE SERVICE PERSONNEL
Eight deaths were ascribed to fire (PMR 311). An excess of deaths from accidental falls (PMR 455, 5 deaths) is probably also attributable to work. Of the cancers, oesophagus (PMR 135, 46 deaths), gall bladder (PMR 224, 10 deaths) and brain (PMR 136, 41 deaths) showed the most marked elevations of mortality.

051 - LAUNDERERS AND DRY CLEANERS
Increased mortality was observed from cancer of the stomach (PMR 130, 57 deaths) and cancer of soft tissue (PMR 415, 7 deaths), but there is no strong evidence from elsewhere to link these cancers with laundry or dry cleaning work. Mortality from renal failure was also elevated (PMR 211, 12 deaths).

052 - HAIRDRESSERS
Excess mortality from immunodeficiency (PMR 931, 31 deaths) and viral hepatitis (PMR 388, 4 deaths) were observed.

053 - OFFICE WORKERS AND CASHIERS
Mortality patterns were unremarkable.

054 - POSTAL WORKERS
Excesses of fibrosing alveolitis (PMR 157, 42 deaths) and Crohn's disease (PMR 247, 10 deaths) are unlikely to be caused directly by occupation.

055 - PETROL PUMP ATTENDANTS
Mortality patterns were unremarkable.

056 - VAN SALES PERSONS
Mortality patterns were unremarkable.

057 - SALES REPRESENTATIVES
Mortality patterns were unremarkable.

058 - SECURITY WORKERS
Four deaths from poisoning by liquified petroleum gas (PMR 660) may have been related to work. Otherwise mortality patterns were unremarkable.

059 - COOKS AND KITCHEN PORTERS
Mortality was high from cancer of the oral cavity (PMR 152, 31 deaths), cancer of the pharynx (PMR 334, 54 deaths), cancer of the liver (PMR 273, 70 deaths), gastric ulcer (PMR 155, 32 deaths), cirrhosis (PMR 168, 41 deaths) and 'other alcohol-related diseases' (PMR 220, 114 deaths). Alcohol may also have contributed to elevated death rates from falls on stairs and steps (PMR 157, 19 deaths) and unspecified falls (PMR 167, 21 deaths). Increased mortality from immunodeficiency (PMR 496, 31 deaths) and viral hepatitis (PMR 492, 15 deaths) were observed. Infection by HIV may in turn explain the high death rate observed from tuberculosis (PMR 206, 34 deaths). Increased mortality was also observed from duodenal ulcer (PMR 170, 56 deaths), non-alcoholic gastritis and duodenitis (PMR 431, 5 deaths) and homicide (PMR 183, 27 deaths). Mortality from lung cancer was close to expectation (PMR 98, 971 deaths).

060 - OTHER SERVICE OCCUPATIONS
High mortality was observed from falls on or from ladders and scaffolding (PMR 259, 51 deaths) and from falls from buildings or other structures (PMR 162, 42 deaths), the excess being confined to the subgroup of 'cleaners, window cleaners, chimney sweeps, road sweepers' (occupational unit 071.02). Increased mortality from immunodeficiency (PMR 450, 41 deaths) was concentrated particularly in travel stewards.

061 - HOSPITAL PORTERS AND WARD ORDERLIES
Mortality was high from accidental poisoning by drugs (PMR 218, 12 deaths) and from viral hepatitis (PMR 334, 7 deaths). The death rate from immunodeficiency (PMR 199, 5 deaths) was also elevated. Three of the deaths from accidental poisoning by drugs were attributed to opiates and related narcotics.

062 - AMBULANCE WORKERS
Mortality was elevated for cancer of the pancreas (PMR 143, 37 deaths) and cancer of the brain (PMR 154, 24 deaths), but there is no obvious occupational cause for these findings.

063 - RAILWAY STATION WORKERS
High mortality was observed from railway accidents (PMR 1248, 25 deaths).

064 - UNDERTAKERS
Mortality patterns were unremarkable.

065 - FORESTERS
Increased mortality from off-road motor vehicle accidents (PMR 545, 3 deaths), pesticide poisoning (PMR 11111, 1 death), injury by falling objects (PMR 1566, 14 deaths) and injury by machinery (PMR 585, 7 deaths) probably results from accidents at work. Rates of suicide (PMR 159, 38 deaths) were also raised, perhaps reflecting access to toxic chemicals and firearms. Cancer mortality was unremarkable.

066 - FISHING AND RELATED WORKERS
Mortality was high from cancer of the oral cavity (PMR 265, 9 deaths), cancer of the larynx (PMR 279, 14 deaths) and cancer of the bronchus (PMR 127, 232 deaths), all of which are diseases caused by smoking. A high risk of occupational injury is indicated by the mortality from water transport accidents (PMR 3376, 48 deaths). If anything, this figure is likely to be an underestimate because not all deaths at sea are registered in Britain. There was also elevated mortality from injuries where it could not be determined if the cause was accidental or a purposeful act (PMR 220, 29 deaths).

067 - TANNERY WORKERS
Mortality patterns were unremarkable.

068 - LEATHER AND SHOE WORKERS
A high death rate was observed from nasal cancer (PMR 237, 7 deaths), reflecting the known hazard of this disease in shoe manufacturers. Mortality was also elevated from renal failure (PMR 176, 35 deaths), but there is no obvious occupational cause for this finding.

069 - PREPARATORY FIBRE PROCESSORS
Three deaths were ascribed to byssinosis (PMR 10345), two to asbestosis (PMR 743) and three to accidents with machinery (PMR 822). As in 1971, mortality from cancer of the stomach was elevated (PMR 125, 29 deaths).

070 - SPINNERS AND WINDERS
Four deaths were ascribed to byssinosis (PMR 8000), and this disease may also have contributed to elevated mortality from cor pulmonale (PMR 277; 7 deaths). An excess of tuberculosis (PMR 255, 7 deaths) was also observed.

071 - WARP PREPARERS AND WEAVERS
Mortality patterns were unremarkable.

072 - KNITTERS
A high mortality from bronchopneumonia (PMR 152, 33 deaths) is unlikely to be directly attributable to an occupational hazard.

073 - BLEACHERS, DYERS AND FINISHERS
Two deaths from injuries caused by hot substances (PMR 1961) are likely to have been occupational. Mortality was also high from gastric ulcer (PMR 262, 12 deaths), bronchopneumonia (PMR 150, 57 deaths), diabetes (PMR 153, 28 deaths) and tuberculosis (PMR 271, 9 deaths), but likely occupational causes for these findings are not apparent. Mortality from bladder cancer was slightly elevated (PMR 117, 29 deaths).

074 - OTHER TEXTILE WORKERS
Fourteen deaths were ascribed to byssinosis (PMR 2414). High mortality from chronic rheumatic heart disease (PMR 175, 49 deaths) may reflect recruitment to these jobs of people with poor social circumstances in childhood. A high death rate was also observed from viral hepatitis (PMR 359, 8 deaths).

075 - CHEMICAL WORKERS
A high death rate from pleural cancer (PMR 159, 39 deaths) is attributable to asbestos exposure. The risk of occupational injury in these occupations is highlighted by their increased mortality from accidents involving hot substances (PMR 559, 5 deaths), explosive material (PMR 797, 12 deaths), machinery (PMR 199, 22 deaths) and gases and vapours (PMR 1093, 4 deaths).

076 - BAKERS
High mortality from asthma (PMR 151, 26 deaths) is consistent with the known hazard of the disease in bakers. Mortality from influenza was also elevated (PMR 390, 10 deaths), but all but one of the deaths occurred after the normal retirement age of 65.

077 - BREWERY WORKERS
An increased death rate from rectal cancer (PMR 131, 31 deaths) is consistent with earlier reports that this tumour is associated with beer drinking, a habit which is likely to be more common in brewery workers. Mortality was also high from 'other alcohol-related diseases' (PMR 232, 14 deaths).

078 - FOOD PROCESSORS
Mortality patterns were unremarkable.

079 - PAPER MANUFACTURERS
Three deaths were ascribed to accidents with machinery (PMR 616). Mortality was high from non-Hodgkin's lymphoma (PMR 203, 14 deaths), but this cancer has not been linked with work in the paper industry in other studies.

080 - BOOKBINDERS
An excess of deaths from epilepsy (PMR 955, 8 deaths), all of which occurred below age 65, may reflect patterns of recruitment to these occupations or may be a chance finding.

081 - PAPER CUTTERS
Increased mortality from aplastic anaemia (PMR 2667, 4 deaths) and agranulocytosis (PMR 5882, 1 death) could indicate an occupational hazard. There was also an excess of nasal cancer (PMR 830, 2 deaths) which raises the possibility that paper dust, like wood dust, might be a cause of this tumour.

082 - GLASS AND CERAMICS FURNACE WORKERS
As in 1971, mortality was increased from chronic bronchitis and emphysema (PMR 135, 93 deaths).

083 - GLASS FORMERS AND DECORATORS
Mortality was elevated from carcinoma of the colon (PMR 141, 34 deaths), but there is no obvious occupational cause for this finding.

084 - CERAMICS CASTERS
Four deaths were ascribed to pneumoconiosis (PMR 3846).

085 - RUBBER MANUFACTURERS
Mortality was increased from motor neurone disease (PMR 244, 12 deaths). An excess of deaths from urothelial cancer (PMR 136, 31 deaths) extended to men below age 65 (PMR 160, 12 deaths).

086 - PLASTICS WORKERS
Mortality from cancer of the bronchus was somewhat elevated (PMR 128, 190 deaths).

087 - MAN-MADE FIBRE MAKERS
Mortality patterns were unremarkable. The rate of lung cancer was close to expectation (PMR 108, 60 deaths).

088 - OTHER COAL MINERS
Causes of death with particularly high death rates included coal workers' pneumoconiosis (PMR 770, 693 deaths), silicosis (PMR 179, 21 deaths), pulmonary fibrosis (PMR 143, 66 deaths), chronic bronchitis and emphysema (PMR 142, 3799 deaths), acute bronchitis (PMR 161, 53 deaths) and cor pulmonale (PMR 146, 110 deaths), all of which are recognised hazards of work in coal or other mines. The other major occupational cause of mortality was from accidents — in particular, accidents involving non-road vehicles (PMR 1261, 32 deaths), falling objects (PMR 275, 33 deaths) and machinery (PMR 223, 34 deaths). The deaths from these accidents were all in men below age 65. In contrast, excess mortality from injuries caused by falling on stairs (PMR 175, 73 deaths) and by fire (PMR 165, 57 deaths) was equally apparent above and below retirement age, suggesting that these accidents occurred mainly outside work. Mortality from chronic and unspecified myocarditis (PMR 226, 157 deaths) was also unusually high [see section 4.4.3].

089 - TOBACCO WORKERS
Mortality patterns were unremarkable. Deaths from lung cancer were close to expectation (PMR 99, 62 deaths).

090 - OTHER WOOD AND PAPER PROCESSORS
Mortality was high from Parkinson's disease (PMR 280, 11 deaths) but there is no obvious occupational cause for this finding.

091 - OTHER OCCUPATIONS — GLASS AND CERAMICS
Three deaths were attributed to silicosis (PMR 645) and six to other and unspecified pneumoconiosis (PMR 593). In addition, excess mortality was observed from chronic rheumatic heart disease (PMR 150, 24 deaths), chronic and unspecified myocarditis (PMR 237, 17 deaths), chronic bronchitis and emphysema (PMR 125, 375 deaths) and gastric ulcer (PMR 154, 21 deaths). The high rate of rheumatic heart disease is likely to reflect the social background of people recruited to these jobs.

092 - RUBBER GOODS MAKERS
Mortality from urothelial cancer was moderately elevated (PMR 134, 24 deaths). A high death rate from bronchopneumonia (PMR 142, 37 deaths) is unlikely to have a direct occupational cause.

093 - PLASTIC GOODS MAKERS
Increased mortality was observed from acute lymphatic leukaemia (PMR 574, 6 deaths) and diabetes (PMR 158, 27 deaths).

094 - COMPOSITORS
Mortality patterns were unremarkable.

095 - PRINTING PLATE PREPARERS
Increased mortality was observed from acute myeloid leukaemia (PMR 426, 7 deaths), multiple sclerosis (PMR 626, 5 deaths) and fibrosing alveolitis (PMR 523, 4 deaths).

096 - PRINTING MACHINE MINDERS
Mortality was elevated from motor neurone disease (PMR 208, 15 deaths).

097 - PRINTERS (SO DESCRIBED)
Increased mortality from dementia (PMR 144, 38 deaths) might be related to solvent exposure - an association which has been suggested by several studies. Death rates were also high for multiple sclerosis (PMR 189, 23 deaths) and hypertensive disease (PMR 148, 91 deaths), but a direct occupational cause for these findings seems less likely.

098 - TAILORS AND DRESSMAKERS
Mortality was high from immunodeficiency (PMR 1057, 5 deaths), and HIV infection may have contributed to an excess of tuberculosis (PMR 402, 16 deaths). A high rate of deaths from diabetes (PMR 215, 49 deaths) is unlikely to have arisen directly from an occupational hazard.

099 - CLOTHING CUTTERS
As in tailors, there was high mortality from tuberculosis (PMR 308, 6 deaths).

100 - SEWERS AND EMBROIDERERS
Mortality patterns were unremarkable.

101 - UPHOLSTERERS
Elevated mortality from cancer of the pleura (PMR 264, 16 deaths) suggests that there has been exposure to asbestos in these occupations. Mortality was also increased from cancer of the prostate (PMR 149, 76 deaths) and from meningeal tumours (PMR 563, 5 deaths) but there is no strong evidence to suggest that these excesses are attributable to occupational hazards.

102 - CARPET FITTERS
An excess of rectal cancer (PMR 188, 13 deaths) is unlikely to have a direct occupational cause.

103 - OTHER WORKERS WITH FABRICS
Mortality patterns were unremarkable.

104 - CARPENTERS
A high death rate from cancer of the pleura (PMR 262, 167 deaths) indicates that exposure to asbestos has been quite common in these occupations. Mortality from nasal cancer was only slightly above expectation (PMR 128, 19 deaths). Death rates from falls, both from ladders and scaffolding (PMR 182, 31 deaths) and also from buildings (PMR 227, 40 deaths) were high, but surprisingly, mortality from accidents with machinery was low (PMR 14, 3 deaths). Other diseases with increased death rates included arteritis (PMR 234, 9 deaths), endocarditis (PMR 179, 27 deaths), cancer of the eye (PMR 208, 19 deaths) and tuberculosis (PMR 157, 69 deaths).

105 - CABINET MAKERS
There was a clear excess of deaths from nasal cancer (PMR 522, 8 deaths). As in carpenters and joiners, mortality was high from cancer of the eye (PM 416, 4 deaths). An association with myeloma (PMR 204, 23 deaths) has not previously been suspected in this occupational group, and may be a chance finding.

106 - CASE AND BOX MAKERS
Two deaths were ascribed to nasal cancer (PMR 512). Increased mortality from stomach cancer (PMR 152, 30 deaths) is unlikely to represent an occupational hazard of these occupations.

107 - PATTERN MAKERS
Mortality was increased from cancer of the colon (PMR 144, 32 deaths) and Parkinson's disease (PMR 269, 10 deaths).

108 - WOODWORKING MACHINISTS
Eight deaths were ascribed to nasal cancer (PMR 357). Otherwise, mortality patterns were unremarkable.

109 - OTHER WOODWORKERS
Two deaths were recorded from nasal cancer (PMR 305). Otherwise mortality patterns were unremarkable.

110 - DENTAL TECHNICIANS
Increased mortality from cancer of the prostate (PMR 204, 24 deaths) has not been reported previously. Mortality from lung cancer was below expectation (PMR 92, 77 deaths).

111 - OTHER MAKERS OF PAPER GOODS
Mortality patterns were unremarkable.

112 - FURNACE OPERATIVES (METAL)
Mortality from chronic bronchitis and emphysema was moderately increased (PMR 126, 248 deaths). As in several other occupations with exposure to metal fumes, mortality from pneumococcal and unspecified lobar pneumonia was increased (PMR 165, 18 deaths) [see section 4.3.1].

113 - ROLLERS (METAL)
Mortality from chronic bronchitis and emphysema was elevated (PMR 140, 50 deaths). High mortality was also observed from bronchopneumonia (PMR 189, 21 deaths), some cases of which may have been related to chronic lung disease. Cancer mortality was unremarkable.

114 - SMITHS AND FORGE WORKERS
Three deaths from peritoneal cancer (PMR 607) suggest that there has been significant exposure to asbestos in these occupations. Mortality was also increased from duodenal ulcer (PMR 202, 21 deaths), but there is no obvious occupational cause for this finding.

115 - METAL DRAWERS
Mortality was high from gastric ulcer (PMR 433, 8 deaths), but there is no obvious occupational cause for this finding.

116 - MOULDERS AND COREMAKERS (METAL)
Three deaths were ascribed to silicosis (PMR 872) and nine to other and unspecified pneumoconiosis (PMR 1240). Mortality from chronic bronchitis and emphysema was also elevated (PMR 146, 409 deaths) as was that from pneumococcal and unspecified lobar pneumonia (PMR 201, 32 deaths) and cancer of the larynx (PMR 149, 22 deaths). An excess of pneumococcal and unspecified lobar pneumonia was also observed in other occupations with exposure to metal fume, suggesting that this is an occupational hazard [see section 4.3.1].

117 - ELECTROPLATERS
Mortality was increased from both lung cancer (PMR 126, 134 deaths) and from chronic bronchitis and emphysema (PMR 144, 68 deaths).

118 - ANNEALERS, HARDENERS, TEMPERERS (METAL)
An excess of pneumococcal and unspecified lobar pneumonia (PMR 288, 8 deaths) was most marked in men below 65 years of age (PMR 392, 4 deaths).

119 - GALVANISERS AND TIN PLATERS
Mortality from chronic bronchitis and emphysema was somewhat increased (PMR 138, 36 deaths).

120 - OTHER METAL MANUFACTURERS
Six deaths were ascribed to silicosis (PMR 740) and 12 to other and unspecified pneumoconiosis (PMR 677). Mortality from several other respiratory diseases was also elevated — in particular, cancer of the larynx (PMR 153, 40 deaths) and chronic bronchitis and emphysema (PMR 110, 524 deaths). Other diseases with high death rates included gastric ulcer (PMR 144, 33 deaths), cirrhosis (PMR 156, 31 deaths) and glomerulonephritis (PMR 198, 13 deaths), but direct occupational causes for these excesses are perhaps less likely. Nine deaths resulted from accidents with machinery (PMR 230).

121 - PRESS AND MACHINE TOOL SETTERS
Mortality from asthma was increased (PMR 149, 28 deaths) but most of the excess occurred in men above the normal retirement age of 65.

122 - CENTRE LATHE TURNERS
Mortality was high from peripheral vascular disease (PMR 167, 33 deaths), but a direct occupational cause for this finding

seems unlikely. Deaths from bladder cancer were below expectation (PMR 85, 51 deaths).

123 - MACHINE TOOL SETTER OPERATORS
Mortality patterns were unremarkable.

124 - MACHINE TOOL OPERATORS
Increased mortality from pleural cancer (PMR 143, 116 deaths) suggests that some exposure to asbestos has occurred in these occupations. Mortality was also high from viral hepatitis (PMR 175, 28 deaths).

125 - PRESS AND AUTOMATIC MACHINE OPERATORS
Mortality was increased from lung cancer (PMR 122, 354 deaths), although this risk estimate may be somewhat inflated because of below average overall mortality in these occupations.

126 - METAL POLISHERS
High mortality was observed from cancer of the stomach (PMR 131, 75 deaths), lung cancer (PMR 122, 332 deaths) and chronic bronchitis and emphysema (PMR 136, 168 deaths).

127 - FETTLERS AND DRESSERS (METAL)
Four deaths were recorded from silicosis (PMR 911) and 15 from other and unspecified pneumoconiosis (PMR 3650). Mortality was also increased from tuberculosis (PMR 319, 11 deaths) — a known complication of silicosis — and from lung cancer (PMR 127, 259 deaths) and chronic bronchitis and emphysema (PMR 128, 131 deaths).

128 - SHOT BLASTERS
Three deaths from coal workers' pneumoconiosis (PMR 307) may have arisen from misclassification of shot firers as shot blasters. An excess of tuberculosis (PMR 315, 4 deaths) could be related to silica exposure. Two deaths resulted from off-road motor vehicle accidents (PMR 885) which may have occurred at work.

129 - TOOLMAKERS
Mortality was increased from cancer of the small intestine (PMR 221, 12 deaths), cancer of the testis (PMR 203, 12 deaths), chronic myeloid leukaemia (PMR 181, 20 deaths), multiple sclerosis (PMR 182, 24 deaths) and chronic rheumatic heart disease (PMR 148, 46 deaths), but there are no obvious occupational causes for these excesses and they have not been a consistent finding in other studies.

130 - PRECISION INSTRUMENT MAKERS
Mortality was increased from cancer of the rectum (PMR 132, 52 deaths) and multiple sclerosis (PMR 248, 11 deaths), but a direct occupational cause for these findings seems unlikely.

131 - WATCH AND CLOCK MAKERS
Mortality was high from tuberculosis (PMR 445, 8 deaths).

132 - PRODUCTION FITTERS
Increased mortality from pleural cancer (PMR 151, 192 deaths) suggests that exposure to asbestos has occurred in these occupations. There was also a marked excess of deaths from accidents — in particular, accidents with machinery (PMR 271, 107 deaths), explosion of a pressure vessel (PMR 424, 5 deaths), accidents involving electricity (PMR 126, 25 deaths) and injury caused by hot and corrosive substances (PMR 310, 10 deaths). High death rates from cancer of the salivary glands (PMR 153, 31 deaths) and meningeal tumours (PMR 147, 26 deaths) are less likely to be attributable directly to work.

133 - MOTOR MECHANICS
Only 12 deaths were recorded from cancer of the pleura (PMR 33) and three from cancer of the peritoneum (PMR 97). Excess mortality from off-road motor vehicle accidents (PMR 201, 17 deaths), accidental poisoning by motor vehicle exhaust (PMR 245, 11 deaths), injury by falling objects (PMR 202, 21 deaths) and injury caused by explosive materials (PMR 253, 6 deaths) indicates a risk of various types of occupational injury.

134 - AIRCRAFT ENGINE FITTERS
Mortality patterns were unremarkable.

135 - OFFICE MACHINERY MECHANICS
Mortality patterns were unremarkable.

136 - ELECTRICAL AND ELECTRONIC PRODUCTION FITTERS
Mortality patterns were unremarkable.

137 - ELECTRICIANS
A high rate of pleural cancer (PMR 254, 127 deaths) reflects the exposure to asbestos that has occurred in these occupations. Their risk of occupational injury is apparent in high death rates from accidents involving falls from ladders or scaffolding (PMR 153, 23 deaths), explosive material (PMR 317, 10 deaths) and electricity (PMR 385, 42 deaths). There was a moderate excess of deaths from brain cancer (PMR 117, 204 deaths). Mortality was also increased from cancer of the bone (PMR 167, 25 deaths), melanoma of skin (PMR 162, 79 deaths), cancer of the thyroid (PMR 183, 18 deaths), amyloidosis (PMR 259, 14 deaths), multiple sclerosis (PMR 197, 70 deaths) and subarachnoid haemorrhage (PMR 126, 184 deaths). Rates of leukaemia were close to expectation.

138 - ELECTRICAL PLANT OPERATORS
Fourteen deaths were from cancer of the pleura (PMR 219), indicating exposure to asbestos in these occupations. Mortality was also increased from cancer of the oral cavity (PMR 243, 13 deaths) and cancer of the oesophagus (PMR 161, 52 deaths), but these excesses are more likely to be due to non-occupational causes.

139 - TELEPHONE FITTERS
Mortality was increased from brain cancer (PMR 149, 86 deaths) and also from non-Hodgkin's lymphoma (PMR 145, 73 deaths) and multiple sclerosis (PMR 222, 26 deaths). The last two associations have not been consistently reported in previous studies and may be a chance finding.

140 - ELECTRIC CABLE AND LINE WORKERS
Seven deaths involved accidental falls from buildings (PMR 693) and 12 injury by electric current (PMR 2178).

141 - RADIO AND TV MECHANICS
Five deaths resulted from accidental injury by electric current (PMR 387). Mortality was also increased from motor neurone disease (PMR 240, 14 deaths), multiple sclerosis (PMR 290, 12 deaths) and suicide (PMR 142, 62 deaths). There was a slight excess of deaths from brain cancer (PMR 117, 23 deaths) and acute myeloid leukaemia (PMR 149, 11 deaths).

142 - OTHER ELECTRONIC MAINTENANCE ENGINEERS
Mortality was increased from brain cancer (PMR 156, 39 deaths), acute myeloid leukaemia (PMR 179, 17 deaths) and chronic myeloid leukaemia (PMR 242, 10 deaths). High death rates were also observed from cancer of the colon (PMR 161, 71 deaths), cancer of the prostate (PMR 159, 54 deaths), and non-Hodgkin's lymphoma (PMR 161, 32 deaths), but these associations were not suspected from previous studies.

143 - ELECTRICAL ENGINEERS (SO DESCRIBED)
High mortality from cancer of the pleura (PMR 186, 31 deaths) is attributable to asbestos exposure. Six deaths were caused by accidents with electricity (PMR 262) and four by accidents with explosive material (PMR 536). Mortality was also increased from cancer of the colon (PMR 133, 208 deaths), cancer of the thyroid (PMR 273, 11 deaths), myeloma (PMR 141, 47 deaths) motor neurone disease (PMR 161, 36 deaths) and multiple sclerosis (PMR 185, 23 deaths).

144 - PLUMBERS AND GAS FITTERS
High death rates from cancer of the peritoneum (PMR 309, 11 deaths), cancer of the pleura (PMR 327, 134 deaths) and asbestosis (PMR 411, 13 deaths) are the result of exposure to asbestos in these occupations. A risk of occupational injuries is indicated by elevated mortality from accidents involving electricity (PMR 167, 12 deaths), falls from buildings (PMR 179, 21 deaths), falls from ladders and scaffolding (PMR 184, 21 deaths), accidents involving gases (PMR 600, 4 deaths) and water transport accidents (PMR 307, 13 deaths).

145 - SHEET METAL WORKERS
Elevated mortality was observed from cancer of the pleura (PMR 135, 25 deaths) suggesting that some exposure to asbestos has occurred in these occupations. High death rates also occurred from cancer of the pharynx (PMR 167, 22 deaths) and pneumococcal and unspecified lobar pneumonia (PMR 147, 38 deaths). The excess of pneumococcal and unspecified lobar pneumonia was confined to men of working age. A similar increase in risk was found in welders, and it may indicate a previously unrecognised occupational hazard. [See section 4.3.1].

146 - METAL PLATE WORKERS
A marked excess of pleural cancer (PMR 515, 73 deaths) reflects the exposure to asbestos in these occupations. Asbestos may also have contributed to elevated mortality from lung cancer (PMR 120, 894 deaths). Rates were also increased for cancer of the liver (PMR 159, 29 deaths), cancer of the larynx (PMR 205, 37 deaths) and pancreatitis (PMR 182, 17 deaths), possibly as a consequence of high alcohol consumption.

147 - STEEL ERECTORS
A high risk of occupational injuries is indicated by elevated mortality from accidental falls from ladders and scaffolding (PMR 754, 20 deaths), falls from buildings (PMR 1136, 30 deaths), other falls (PMR 563, 8 deaths), injury by falling objects (PMR 409, 9 deaths) and accidents with machinery (PMR 243, 8 deaths). A high rate of lung cancer (PMR 123, 567 deaths) is probably attributable to smoking. Increased mortality from cancer of the oral cavity (PMR 247, 20 deaths), cancer of the larynx (PMR 182, 21 deaths) and other alcohol-related disease (PMR 165, 29 deaths) suggests above average alcohol consumption in this occupational group.

148 - SCAFFOLDERS
High mortality was observed from falls from ladders and scaffolding (PMR 2,175, 38 deaths), falls from buildings (PMR 228, 9 deaths), other falls (PMR 370, 4 deaths), injury by falling objects (PMR 267, 6 deaths) and accidents with machinery (PMR 241, 7 deaths). An increased death rate from drug dependence (PMR 340, 7 deaths) probably reflects patterns of recruitment to these jobs. High mortality from stomach cancer (PMR 143, 59 deaths) and chronic bronchitis and emphysema (PMR 123, 100 deaths) is likely to result from non-occupational causes.

149 - WELDERS
Mortality from cancer of the pleura was increased (PMR 179, 56 deaths) as would be expected from the exposure of welders to asbestos. A high death rate from pneumococcal and unspecified lobar pneumonia (PMR 186, 75 deaths) is consistent with earlier analyses of occupational mortality and with similar findings in other occupations exposed to metal fume. The excess was confined to men of working age, and it may well represent a previously unrecognised occupational hazard. [See section 4.3.1].

150 - RIGGERS
Four deaths resulted from falls from ladders or scaffolding (PMR 369) and ten from falls from buildings (PMR 1025). Excess mortality from cancer of the oesophagus (PMR 177, 38 deaths) and cancer of the larynx (PMR 219, 12 deaths) may reflect above-average smoking and/or alcohol consumption in this occupational group.

151 - JEWELLERY WORKERS
High mortality from bronchopneumonia (PMR 245, 21 deaths) is unlikely to be directly occupational.

152 - ENGRAVERS AND ETCHERS (PRINTING)
Mortality patterns were unremarkable.

153 - VEHICLE BODY BUILDERS
Increased mortality from cancer of the peritoneum (PMR 957, 4 deaths), cancer of the pleura (PMR 470, 24 deaths) and asbestosis (PMR 1139, 5 deaths) is the result of asbestos exposure in these occupations.

154 - OILERS AND GREASERS
Mortality patterns were unremarkable.

155 - ELECTRONICS LINE WORKERS
Mortality patterns were unremarkable.

156 - COIL WINDERS
Mortality was increased from cancer of the brain (PMR 364, 10 deaths).

157 - POTTERY DECORATORS
One death was ascribed to other or unspecified pneumoconiosis (PMR 4762). It is possible that this individual had also worked in other jobs in the pottery industry where exposure to pneumoconiotic dusts is known to occur.

158 - COACH PAINTERS
Mortality patterns were unremarkable. Deaths from lung cancer were below expectation (PMR 87, 69 deaths).

159 - OTHER SPRAY PAINTERS
As has been reported in previous studies of painters, mortality from lung cancer was elevated (PMR 126, 557 deaths). Increased death rates were also observed for cancer of the testis (PMR 315, 10 deaths) and subarachnoid haemorrhage (PMR 146, 33 deaths).

160 - PAINTERS AND DECORATORS NEC
The most clear occupational cause of mortality was from accidental injury, with increased death rates from falls on stairs (PMR 158, 44 deaths), falls from ladders or scaffolding (PMR 350, 64 deaths), falls from buildings (PMR 134, 25 deaths), and other falls (PMR 175, 18 deaths). High mortality from drug dependence (PMR 297, 38 deaths), accidental poisoning by drugs (PMR 251, 76 deaths), homicide (PMR 203, 42 deaths) and indeterminate injury (PMR 163, 214 deaths) probably reflects patterns of recruitment to these occupations. The death rate from lung cancer was moderately elevated (PMR 112, 4110 deaths).

161 - ELECTRICAL, ELECTRONIC ASSEMBLERS
Increased mortality from asthma (PMR 239, 10 deaths) could be related to the presence of respiratory sensitisers in soldering fluxes. Eight of the ten asthma deaths occurred before normal retirement age. [see section 6.10.3] A high death rate was also found for subarachnoid haemorrhage (PMR 221, 15 deaths), but there is no obvious occupational cause for this finding.

162 - INSTRUMENT ASSEMBLERS
Mortality patterns were unremarkable.

163 - ASSEMBLERS (VEHICLES AND OTHER METAL GOODS)
Mortality from liver cancer was elevated (PMR 157, 25 deaths), but the death rate from cirrhosis (PMR 58, 9 deaths) was low, and this may have been a chance finding.

164 - PACKERS AND SORTERS
High mortality from epilepsy (PMR 230, 33 deaths) may reflect patterns of recruitment to these occupations.

165 - BRICKLAYERS AND TILESETTERS
A high death rate from falls from ladders and scaffolding (PMR 208, 19 deaths) is likely to be due largely to accidents at work. In contrast, although there was excess mortality from falls on stairs (PMR 175, 25 deaths), most of these deaths occurred after normal retirement age and were probably not occupational.

166 - MASONS AND STONECUTTERS
Eleven deaths were from silicosis (PMR 6471) and five from other and unspecified pneumoconiosis (PMR 1351). Mortality was also increased from oesophageal cancer (PMR 145, 40 deaths), but there is no obvious occupational cause for this finding.

167 - PLASTERERS
Eighteen deaths were ascribed to cancer of the pleura (PMR 149) suggesting that some exposure to asbestos has occurred in these occupations. Asbestos may also have contributed to elevated mortality from lung cancer (PMR 127, 805 deaths), a finding that was also apparent in the 1971 Decennial Supplement. High death rates from accidental poisoning by drugs (PMR 196, 12 deaths) and indeterminate injury (PMR 159, 41 deaths) may be the result of patterns of recruitment to these jobs. There was also an excess of nasal cancer (PMR 249, 7 deaths), but the reasons for this are unclear.

168 - ROOFERS AND GLAZIERS
The risk of occupational accidents is reflected in high death rates from falls, both from ladders and scaffolding (PMR 725, 27 deaths) and from buildings (PMR 1238, 116 deaths). Elevated mortality from drug dependence (PMR 238, 13 deaths) and ‘other alcohol-related diseases’ (PMR 154, 38 deaths) probably reflects recruitment patterns in a casual labour force. Rates of lung cancer were also increased (PMR 121, 439 deaths).

169 - BUILDERS ETC.
Mortality from pleural cancer was elevated (PMR 136, 65 deaths) as a result of the asbestos exposure in these occupations, but there were no deaths from asbestosis (4.96 expected). The major occupational cause of mortality appeared to be accidental injury - in particular from electric current (PMR 163, 14 deaths), falling objects (PMR 143, 18 deaths), falling into holes (PMR 332, 6 deaths), falling from buildings (PMR 211, 35 deaths) and falling from ladders or scaffolding (PMR 378, 53 deaths). High death rates from cancer of the gall bladder (PMR 141, 43 deaths) and cancer of the breast (PMR 197, 16 deaths) have no obvious occupational cause.

170 - RAIL TRACK WORKERS
55 deaths resulted from railway accidents (PMR 4754). Otherwise, mortality patterns were unremarkable.

171 - ROAD CONSTRUCTION WORKERS AND PAVIORS
High death rates from injury by electric current (PMR 494, 6 deaths) and motor vehicle accidents (PMR 140, 89 deaths) are likely in part to reflect occupational accidents. There is no obvious occupational cause for an elevated death rate from Parkinson’s disease (PMR 155, 25 deaths).

172 - SEWAGE PLANT ATTENDANTS
Mortality patterns were unremarkable.

173 - MAINS AND SERVICE LAYERS
Seven deaths were caused by injury from falling objects (PMR 537) and two from falling into holes (PMR 1000).

High death rates from cancer of the rectum (PMR 187, 53 deaths), cancer of the pancreas (PMR 141, 39 deaths), cancer of the larynx (PMR 224, 16 deaths) and cancer of the prostate (PMR 155, 50 deaths) have no obvious occupational cause.

174 - CONSTRUCTION WORKERS NEC
The exposure to asbestos in these occupations resulted in high mortality from cancer of the peritoneum (PMR 956, 64 deaths), cancer of the pleura (PMR 191, 77 deaths) and asbestosis (PMR 1274, 71 deaths). Elevated death rates were also observed from falls from ladders or scaffolding (PMR 184, 37 deaths), falls from buildings (PMR 259, 87 deaths), falling into holes (PMR 379, 13 deaths), injury by falling objects (PMR 289, 56 deaths) and accidents with machinery (PMR 170, 42 deaths).

175 - FACE TRAINED COALMINERS
Excess mortality was observed from various respiratory diseases — in particular, coal workers' pneumoconiosis (PMR 3771, 131 deaths), silicosis (PMR 695, 5 deaths), chronic bronchitis and emphysema (PMR 156, 920 deaths), acute bronchitis (PMR 246, 17 deaths) and pneumothorax (PMR 336, 5 deaths). These findings are consistent with previous studies. The other major occupational cause of mortality was accidental injury from 'other vehicle accidents' (PMR 4749, 17 deaths), falling objects (PMR 816, 21 deaths), accidents with machinery (PMR 458, 16 deaths) and accidents with explosive material (PMR 1102, 7 deaths). An increased rate of stomach cancer (PMR 123, 326 deaths) has been reported before in coal miners, but may not be directly occupational. Mortality from chronic and unspecified myocarditis (PMR 288, 41 deaths) was also unusually high [see section 4.4.3].

176 - MINERS (NOT COAL) AND QUARRY WORKERS
58 deaths were ascribed to silicosis (PMR 8123), 7 to other and unspecified pneumoconiosis (PMR 992) and 11 to pulmonary fibrosis (PMR 362). In addition, there was excess mortality from tuberculosis (PMR 431, 24 deaths) — a known complication of silicosis. Increased mortality from 'other vehicle accidents' (PMR 800, 2 deaths), injury by falling objects (PMR 646, 8 deaths) and accidents with machinery (PMR 445, 7 deaths) reflects the risk of occupational accidents in these jobs.

177 - RAILWAY GUARDS
17 deaths were from railway accidents (PMR 2751). Increased mortality from diabetes (PMR 158, 28 deaths) is unlikely to be directly occupational.

178 - RAILWAY SIGNAL WORKERS
Increased mortality from diabetes (PMR 150, 29 deaths) is unlikely to be directly occupational.

179 - SHUNTERS AND POINT OPERATORS
13 deaths were from railway accidents (PMR 4887). Otherwise, mortality patterns were unremarkable.

180 - RAILWAY ENGINE DRIVERS
28 deaths were from railway accidents (PMR 2244). Otherwise, mortality patterns were unremarkable.

181 - ROAD TRANSPORT INSPECTORS
Mortality patterns were unremarkable.

182 - BUS AND COACH DRIVERS
Mortality patterns were unremarkable. Deaths from motor vehicle accidents were below expectation (PMR 97, 129 deaths).

183 - LORRY DRIVERS
Mortality was increased from motor vehicle accidents (PMR 155, 1409 deaths), off-road motor vehicle accidents (PMR 291, 75 deaths), accidents involving liquified petroleum gas (PMR 253, 10 deaths) and injury by falling objects (PMR 181, 65 deaths). These excesses are likely to be attributable mainly to accidents at work. High death rates were also observed from cancer of the larynx (PMR 132, 249 deaths) and acute monocytic leukaemia (PMR 203, 15 deaths), but these have not been consistent findings in earlier studies. Deaths from urothelial cancer were close to expectation (PMR 102, 687 deaths).

184 - OTHER MOTOR DRIVERS
Mortality patterns were unremarkable.

185 - BUS CONDUCTORS AND DRIVERS' MATES
High mortality was observed from cancer of the kidney (PMR 172, 30 deaths), myeloma (PMR 201, 24 deaths), diabetes (PMR 154, 51 deaths) and pneumococcal and unspecified lobar pneumonia (PMR 157, 22 deaths), but there are no strong reasons to expect direct occupational causes for these findings. As in 1971, mortality from chronic bronchitis and emphysema was elevated (PMR 119, 291 deaths).

186 - MECHANICAL PLANT DRIVERS
22 deaths were from accidents with machinery (PMR 768). Otherwise, mortality patterns were unremarkable.

187 - CRANE DRIVERS
24 deaths were from accidents with machinery (PMR 549) and seven from off-road motor vehicle accidents (PMR 332). Otherwise, mortality patterns were unremarkable.

188 - FORK LIFT TRUCK DRIVERS
27 deaths were from accidents with machinery (PMR 351). There was also increased mortality from laryngeal cancer (PMR 145, 30 deaths), but a direct occupational cause for this excess seems unlikely.

189 - SLINGERS
Six deaths were from accidents with machinery (PMR 836). Increased mortality from hypertensive disease (PMR 185, 21 deaths) and cor pulmonale (PMR 441, 10 deaths) is unlikely to be directly occupational.

190 - STOREKEEPERS
Mortality patterns were unremarkable.

191 - DOCKERS AND GOODS PORTERS
Elevated mortality from cancer of the pleura (PMR 214, 36 deaths) reflects the exposure to asbestos that has occurred in these occupations. A risk of occupational accidents is indicated by high death rates from water transport accidents

(PMR 683, 12 deaths), accidents with machinery (PMR 179, 12 deaths) and accidental crushing (PMR 402, 3 deaths). Mortality was also increased from cancer of the oral cavity (PMR 136, 53 deaths), cancer of the pharynx (PMR 123, 34 deaths), cancer of the larynx (PMR 127, 66 deaths), cirrhosis (PMR 163, 61 deaths) and 'other alcohol-related diseases' (PMR 138, 102 deaths), suggesting above average alcohol consumption among these men.

192 - REFUSE COLLECTORS
Mortality patterns were unremarkable.

193 - LABOURERS IN COKE OVENS
One death was from accidental poisoning by solvents (PMR 7143). An excess of bronchopneumonia (PMR 140, 44 deaths) is unlikely to be directly occupational. High mortality from lung cancer (PMR 118, 156 deaths) may have been related in part to exposure to polycyclic aromatic hydrocarbons.

194 - BOILER OPERATORS
Asbestos exposure underlies an increased mortality from cancer of the pleura (PMR 269, 24 deaths). Elevated death rates were also observed from cancer of the larynx (PMR 166, 34 deaths) and urothelial cancer (PMR 136, 99 deaths)

References

1 Doll R, Peto R. *The causes of cancer.* Oxford: OUP, 1981.

2 De Vos Irvine H, Lamont DW, Hole DJ, Gillis CR. Asbestos and lung cancer in Glasgow and the West of Scotland. *Br Med J* 1993;306:1503-6.

3 Office of Population Censuses and Surveys. *Occupational mortality, decennial supplement 1970-72.* London: HMSO, 1978.

4 Office of Population Censuses and Surveys. *Registrar General's Decenial Supplement 1961 :Occupational Mortality Tables.* London: HMSO, 1971.

5 Coggon D, Inskip H, Winter P, Pannett B. Lobar pneumonia: an occupational disease in welders. *Lancet* 1994;344:41-3.

6 Silverstein M, Maizlish N, Park R, Silverstein B, Brodsky L, Mirer F. Mortality among ferrous foundry workers. *Am J Ind Med* 1986;10:27-43.

7 Gardner MJ, Winter PD, Pannett B, Powell CA. The geographical distribution of mortality ascribed to certain pneumoconioses during 1968-78 and related industries in England and Wales. *Ann Occup Hyg* 1988 2;32 Suppl 2: 515-22.

8 Case RAM, Hosker ME. Tumour of the urinary bladder as an occupational disease in the rubber industry in England and Wales. *Br J Prev Soc Med* 1954;8:39-50.

9 Parkes HG, Veys CA, Waterhouse JAH, Peters A. Cancer mortality in the British rubber industry. *Br J Ind Med* 1982;39:209-20.

10 Anon. *Agricultural statistics in England and Wales 1992.* London: HMSO, 1993.

11 Tuyns AJ, Péquignot G, Abbatucci JS. Oesophageal cancer and alcohol consumption: importance of type of beverage. *Int J Cancer* 1979;23:443-7.

12 Swerdlow AJ. Incidence of malignant melanoma of the skin in England and Wales and its relationship to sunshine. *Br Med J* 1979;2:1324-7.

13 Nomura AMY, Kolonel LN. Prostate cancer: a current perspective. *Epidemiol Rev* 1991;13:200-27.

14 Blair A, Zahm SH, Pearce NE, Heineman EF, Fraumeni JF. Clues to cancer etiology from studies of farmers. *Scand J Work Environ Health* 1992;18:209-15.

15 Hoar SK, Blair A, Holmes FF. Agricultural herbicide use and risk of lymphoma and soft tissue sarcoma. *JAMA* 1986;256:1141-7.

16 Davies RP, Thomas MR, Garthwaite DG, Bowen HM. Pesticide usage survey report 108. *Arable farm crops in Great Britain 1992.* London: MAFF, 1993.

17 Thomas MR, Garthwaite DG. Pesticide usage survey report 119. *Grassland and fodder crops in Great Britain 1993.* London: MAFF, 1994.

18 Fox AJ, Lynge E, Malker H. Lung cancer in butchers. *Lancet* 1982;1:165-6.

19 Johnson ES, Fischman HR, Matanoski GM, Diamond E. Occurrence of cancer in women in the meat industry. *Br J Ind Med* 1986;43:597-604.

20 Pegum JS. Lung cancer in butchers. *Lancet* 1982;1:561.

Annex 4.1 Counties making up MAFF regions

Region	Counties
Northern	Cumbria, Lancashire, Tyne and Wear, Northumberland
North East	Cleveland, Durham, South Yorkshire, North Yorkshire, West Yorkshire, Humberside
North Mercia	Cheshire, Merseyside, Shropshire, Staffordshire, Greater Manchester
East Midlands	Derbyshire, Leicestershire, Lincolnshire, Northamptonshire, Nottinghamshire
South Mercia	Gloucestershire, Hereford and Worcester, Warwickshire, West Midlands
Anglia	Bedfordshire, Cambridgeshire, Essex, Hertfordshire, Norfolk, Suffolk
Wessex	Dorset, Avon, Somerset, Wiltshire
South West	Devon, Cornwall, Isles of Scilly
South East	Berkshire, Buckinghamshire, Hampshire, Isle of Wight, Kent, Greater London, Oxfordshire, Surrey, East Sussex, West Sussex
Wales	Clwyd, Dyfed, Gwent, Gwynedd, Mid Glamorgan, Powys, South Glamorgan, West Glamorgan

Chapter 5 Occupational mortality of women

Hazel Inskip
David Coggon
Paul Winter
Brian Pannett

MRC Environmental Epidemiology Unit, University of Southampton

5.1 Introduction

Data on mortality of women by occupation are available in two forms. Most interest focuses on the analysis of women by their own occupation, but for certain causes of death and occupational groups it is of interest to consider married women classified by their husbands' occupation. Both are discussed in this chapter.

The data on women, both classified by their own and by their husbands' occupations, were analysed in a similar way to the data for men. Certain features of the analysis and interpretation differed however, due to the nature of the data available. In this chapter the details of the methods are described, followed by a discussion of women in a variety of occupational groups. Section 5.8 is devoted to the analysis of married women in relation to their husbands' occupation, but all other sections consider women by their own occupation. Section 5.10 gives a summary of the interesting findings for women in each occupational group.

As mentioned in Chapter 2, the majority of married women do not have an occupation recorded on their death certificate because they have not been employed for most of their working life. This has the advantage that any occupational effects are less likely to be diluted out by the inclusion of women who only worked for a few years or were employed part-time. The disadvantage is that only 330,678 (30 per cent) of the 1,110,074 deaths which occurred at ages 20-74 in women in England and Wales during the eleven-year period could be used in the analysis. This, however, was an increase on the proportion of data available for the 1970-72 decennial supplement for which only 21 per cent of deaths to women had an occupation recorded.

With fewer deaths available for analysis for women there were many job groups which contained small numbers of deaths. This is illustrated in Figure 5.1 from which it can be seen that for 92 of the 194 occupational groups (47 per cent) there were fewer than 100 deaths from all causes at ages 20-74. At the other extreme, the largest job group of all, namely office workers, comprised 42,482 deaths which was 24 per cent of the total deaths in the analysis.

As described in Chapter 2, proportional mortality ratios (PMRs) were used to assess the relationship between mortality and occupation. The age range 20-74 years was used

Figure 5.1 Number of job groups having particular numbers of all cause deaths in females aged 20-74, England and Wales, 1979-80 and 1982-90

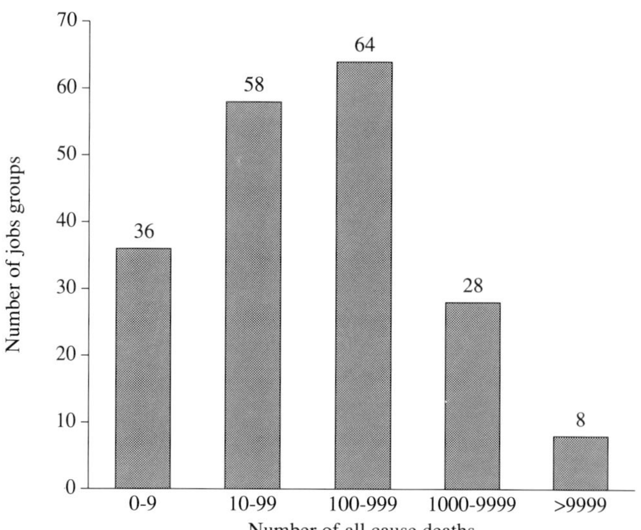

except where interest focused on deaths during working life, such as those from accidents or violence, in which case the age range 20-59 was used. But unless specified, all PMRs are quoted for the age range 20-74. All PMRs were adjusted for age and social class, and are statistically different from 100 at the 5 per cent level unless otherwise stated.

Given that occupation was only recorded for women who were employed for most of their lives, there is likely to be a higher proportion of women in this dataset with no, or small numbers of, children than in the general population.[1] However, since the standard set of proportions used was all deaths for which the woman's occupation was recorded, any bias should be reduced, though not eliminated. Some occupations, such as teaching, comprise a higher percentage of single women with no children than other occupations.[1] Thus it was difficult to assess many of the female reproductive cancers by occupation because of possible biases. This is unfortunate, as for such cancers there is obviously no information on occupational hazards from the data on men.

Appendix 4 gives a summary of all PMRs of interest in each of the 194 occupational groups as defined in Appendices 2 and 3. All PMRs which differed from 100 at the 5 per cent level of statistical significance are listed unless they were based on fewer than three observed or expected deaths. In section 5.10 a commentary on these PMRs is given.

5.2 Teachers (job group 011)

Deaths to teachers comprised nearly five per cent of deaths in women for whom an occupation was recorded. There are few hazardous materials to which they are exposed though they do have an increased exposure to childhood infections. As a group, however, their proportionate mortality pattern differed quite considerably from that of all occupied women. Table 5.1 shows the PMRs, with 95 per cent confidence intervals, discussed for this group.

Table 5.1 PMRs for female teachers (job group 011), aged 20-74 for selected causes of death in England and Wales 1979-80, 1982-90

Disease (ICD)	Deaths	PMR*	95% CI
Breast cancer (174)	2,404	128	123 - 133
Cancer of the body of uterus (182)	140	128	108 - 152
Cancer of the ovary (183)	740	130	121 - 140
Cervical cancer (180)	125	53	44 - 63
Cancer of the colon (153)	572	118	109 - 131
Multiple sclerosis (340)	105	152	125 - 185
Non-Hodgkin's lymphoma (200,202)	180	118	101 - 137
Hodgkins disease (201)	45	142	104 - 191
Parkinson's disease (332)	47	124	91 - 165
Motor neurone disease (335.2)	70	125	97 - 158

* Standardised for age and social class

Certain reproductive cancers which are related to low levels of childbearing[2] were raised in this group, namely breast (PMR 128, 2404 deaths), body of uterus (PMR 128, 140 deaths) and ovary (PMR 130, 740 deaths). Not surprisingly, cervical cancer for which the association with childbearing is reversed showed a deficit (PMR 53, 125 deaths). Although only deaths to women employed for most of their working lives were used as the standard against which each group was compared, the proportion of single women among teachers has tended to be higher than in many other occupations[1] and so, in general, they have fewer children. Similar findings were obtained from the occupational mortality data for 1970-72[3] and breast cancer was found to be particularly high for teachers in a similar analysis in the United States for 1979-87.[4]

Cancer of the colon in teachers was discussed extensively in the previous decennial supplement for 1979-80 and 1982-3[5]. In that analysis, the data for men referred to Great Britain as a whole but those for women were for England and Wales, as were analysed here. Thus the dataset for women in the previous decennial supplement is a subset of the data analysed here. For the years 1979-80 and 1982-83, the PMR for colon cancer in teachers aged 20-59 was 139 based on 59 deaths but not adjusted for social class. Adding the later years provided an extra 160 deaths, which gave a PMR for the later period of 126 and a PMR for the whole period of 129 as shown in Table 5.2. Social class adjustment reduced the overall PMR for the eleven year period to 118 (as shown in Table 5.1)and thus the excess does not appear to be as marked as was first thought.

There has been considerable interest in recent years in delayed exposure to infectious agents giving rise to serious illness later. The specific agents have not been identified in most cases, though late exposure to poliomyelitis may influence the risk of motor neurone disease,[6] and Epstein Barr virus is thought to be a risk factor for multiple sclerosis.[7] The PMR for this latter disease was raised in teachers (PMR 152, 105 deaths) with comparable figures for those aged 20-59 (PMR 151, 63 deaths). Raised PMRs for non-Hodgkin's lymphoma (PMR 118, 180 deaths) and Hodgkin's disease (PMR 142, 45 deaths) were also seen. This was also the case with PMRs for Parkinson's disease (PMR 124, 47 deaths), and motor neurone disease (PMR 125, 70 deaths), though the latter two did not reach the standard 5 per cent level of statistical significance. There have been suggestions that all of these diseases may be due to infectious agents,[6-8] particularly if first exposure to the infection occurs later than normal. The finding that these PMRs were raised in teachers, who may be first exposed to certain infections when entering the classroom as adults, lends some support to these theories. This is discussed further in Chapter 6 where these data are examined alongside those for men.

Table 5.2 PMRs for cancer of the colon (ICD 153) in female teachers (job group 011) aged 20-59, England and Wales in various time periods

Time period	Deaths	PMR*	95% CI
1979-80, 1982-83	59	139	106-179
1984-1990	160	126	107-147
1979-80, 1982-90	219	129	113-148

* Standardised for age but *not* for social class

5.3 Health-related professions (job groups 015-021, 061, 062)

Deaths to those employed in the health-related professions comprised 10 per cent of all deaths to women aged 20-59 and 8 per cent of those aged 20-74 years. There were 20,734 deaths to nurses aged 20-74, and 4,211 to hospital porters and ward orderlies, but there were fewer than 1,000 deaths in each of the other specific occupational groups in the medical field. This is illustrated in Figure 5.2.

Figure 5.2 Number of all cause deaths in various health professions for women aged 20-74, England and Wales, 1979-80 and 1982-90

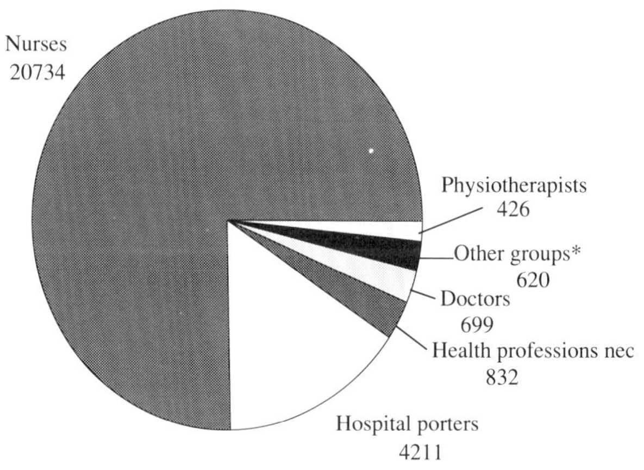

* Dentists, pharmicists, medical radiographers and ambulance staff

Many of the health-related professions had elevated PMRs for suicide as shown in Table 5.3. This included a profession allied to health, namely veterinarians, who had a high PMR at ages 20-59 (PMR 459, 7 deaths). Deaths at ages 20-59 from accidental poisoning by drugs were also raised for nurses (PMR 161, 57 deaths), and for medical practitioners the PMR was of borderline significance (PMR 300, 5 deaths).

Table 5.3 Significantly raised PMRs in women in health and health-related professions for suicide (ICD E950-E959) at ages 20-59 in England and Wales, 1979-80, 1982-90

Job group		Deaths	PMR*	95% CI
015	Doctors	41	191	137-259
017	Nurses	372	135	121-149
020	Physiotherapists	12	209	108-366
021	Health professions nec	25	165	106-243
022	Veterinarians	7	459	184-945
062	Ambulance workers	5	581	189-1357

* Standardised for age and social class

There is nothing new in the finding that suicide is raised in the health-related professions.[9, 10] Such workers know how to commit suicide if they chose to do so, and they have much easier access to the means than do those employed elsewhere. Prevention of suicide will be hard in this group as it is not easy to remove the means from the workplace. Greater concern for the workers in such professions and the stresses placed upon them should be encouraged.

Few other PMRs from accidental and violent deaths reached significance in these job groups at ages 20-59 with the exception of nurses who had a raised PMRs for injury undetermined as accidental or purposeful (PMR 143, 176 deaths) and hospital porters and ward orderlies for whom the PMR for slipping and tripping accidents was elevated (PMR 538, 4 deaths). Although the number of deaths involved was small in the latter case, it is possible that these deaths were related to the occupation.

Alcohol-related diseases, such as cirrhosis of the liver, tended to be elevated in men in the health-related professions, but this was not the case for women. There was even a deficit of cancer of the oral cavity in female nurses (PMR 54, 21 deaths). These patterns indicate that alcohol consumption may differ between the males and females of the profession.

5.4 Farmers (job group 047)

The pattern of mortality in female farmers was similar to that observed in men. Notably there were excesses of a number of accidental and violent deaths, most of which occurred under 60 years of age, and these are shown in Table 5.4. Many, such as animal transport accidents (PMR 943, 3 deaths), injury by animals and plants (PMR 2632, 4 deaths) and injury by machinery (PMR 2632, 4 deaths) were likely to have to occurred at work. Although the numbers of deaths were small, the conclusion that female farmers are at risk from accidents of many types has to be drawn. This is not surprising given the nature of the work.

Table 5.4 Significantly raised PMRs for accidental and violent deaths in female farmers (job group 047) at ages 20-59, England and Wales, 1979-80, 1982-90

Cause of death (ICD)	Deaths	PMR*	95% CI
Motor vehicle traffic accidents (E810-E819)	31	154	105-219
Animal transport accidents (E827-E828)	3	943	193-2740
Fall from building (E882)	3	644	132-1866
Injury by animals or plants (E905-E906)	2	6452	810-24083
Injury by machinery (E919)	4	2632	727-6828
Suicide (E950-E959)	29	151	101-217

* Standardised for age and social class

High mortality from suicide in men has been reported before[9,10] but previous decennial supplements have not identified it as a particular problem in women, although it was found to be significantly raised here with a PMR of 151 (29 deaths) at ages 20-59. The explanation for elevated mortality from suicide in this occupation may be access to means and the stress of the job, particularly in the self-employed. But since it is also recorded in vets, it has been noted that both professions are used to seeing ill or distressed animals being killed.[10] This may lead them to consider suicide more readily for themselves when they are suffering. In general it appears that female farmers face similar risks to men in the occupation with regard to accidental and violent deaths.

The PMRs for certain respiratory disorders are presented in Table 5.5. Not surprisingly there was an excess of farmers' lung disease (PMR 7,692, 2 deaths) which although well known in male farmers[9] is less well documented in women, presumably because the number of deaths from this cause is very small. Fibrosing alveolitis was also high (PMR 261, 7 deaths), based on slightly larger numbers of deaths and there is a possibility that some of these could have been misdiagnosed cases of farmers' lung disease. Indeed the cause of death for one of these seven women was noted on the death certificate as being due to occupational exposure in farming.

Table 5.5 PMRs for certain respiratory disorders in female farmers (job group 047) aged 20-74 in England and Wales, 1979-80, 1982-90

Cause of death (ICD)	Deaths	PMR*	95% CI
Lung cancer (162)	167	92	78-107
Acute bronchitis (466)	7	288	116-593
Bronchopneumonia (485)	41	84	60-114
Chronic bronchitis and emphysema (491,492,496)	54	72	54-93
Asthma (493)	18	111	66-176
Farmers' lung disease (495.0)	2	7692	932-27787
Fibrosing alveolitis (516.3)	7	261	105-538

* Adjusted for age and social class

Mortality from acute bronchitis based on the same number of deaths was also high (PMR 288, 7 deaths) and the PMR from asthma was marginally raised (PMR 111, 18 deaths), though the latter did not reach the standard 5 per cent level of statistical significance. In contrast, deficits of chronic lung disorders such as lung cancer (PMR 92, 167 deaths, not significant), and chronic bronchitis and emphysema (PMR 72, 54 deaths), were observed which indicate that smoking is not a major health problem in this group.

5.5 Textile workers (job groups 069-074)

Textile workers have usually received specific attention in the decennial supplements,[5, 9] as it has been suggested that they are at increased risk of a wide range of diseases.

Table 5.6 gives the PMRs for textile workers, combining job groups 069-074, for a variety of cancers for which associations with textile work have been noted elsewhere.[3, 9, 11, 12] There was no significant excess for any of these cancers. Notable is the combined PMR for laryngeal cancer which was only 63, although for preparatory fibre processors alone, the PMR was 556 based on 3 deaths. A non-significant excess of cervical cancer was also observed in preparatory fibre processors (PMR 178, 11 deaths), but for all textile production workers combined the PMR was only 103 on 174 deaths. Although non-Hodgkin's lymphoma has been linked to textile work in the past,[11] none of the textile groups showed an excess, while four groups exhibited non-significantly low PMRs, namely warp preparers and weavers (PMR 80, 16 deaths), other textile workers (PMR 75, 25 deaths), spinners and winders (PMR 60, 10 deaths), and those in preparatory fibre processes (no deaths, 3.36 expected). The overall PMR for this cause in textile workers was 74 (61 deaths).

Table 5.6 PMRs in female textile workers (job groups 069-074) at ages 20-74 in England and Wales, 1979-80, 1982-90 for cancers that have been associated with textile work in other studies

Cause of death (ICD)	Deaths	PMR*	95% CI
Nasal cancer (160)	4	83	23 -213
Laryngeal cancer (161)	10	63	30 -116
Soft tissue cancer (171)	11	74	37 -132
Melanoma of skin (172)	28	89	59 -129
Cervical cancer (180)	174	103	89 -120
Urothelial cancer (188,189.1-189.8)	96	107	87 -131
Thyroid cancer (193)	15	99	55 -163
Non-Hodgkin's lymphoma (200, 202)	61	74	57 -95

* Standardised for age and social class

The recent monograph from the International Agency for Research in Cancer on the carcinogenic risk of exposures in textile manufacturing concluded that there was limited evidence for such a risk in this work.[12] Their evaluation was based on findings of increased bladder cancer among dyers and weavers and nasal cancer in weavers and other textile workers. In the occupational mortality data for women, there were non-significant excesses of bladder cancer in warp preparers and weavers (PMR 125, 27 deaths) and amongst bleachers, dyers and finishers (PMR 233, 4 deaths). There was only one death from nasal cancer in weavers giving a PMR of 71 and among other textile workers, (job group 074), there was a PMR of 177 based on 3 deaths. In general it appears that the occupational mortality data provide little or no support for an association between textile production and any of the cancers given in the table.

Support for elevated mortality from respiratory diseases amongst textile workers[9] was, however, much more apparent in these data. PMRs for certain cardiovascular and respiratory disorders are given in Table 5.7. Not surprisingly, the only job groups for women with significantly raised PMRs for byssinosis were among textile workers, but no deaths

Table 5.7 PMRs in female textile workers (job groups 069-074) at ages 20-74 in England and Wales, 1979-80, 1982-90 for certain cardiovascular and respiratory disorders

Cause of death (ICD)	Deaths	PMR*	95% CI
Ischaemic heart disease (410-414)	3643	114	11 -118
Other cerebrovascular disease (431-438)	1338	109	103 -115
Lung cancer (162)	750	79	74 -85
Pneumonia (480-486)	473	118	108 -129
Chronic bronchitis and emphysema (491, 492,496)	567	119	110 -129
Asthma (493)	55	83	63 -108
Byssinosis (504)	39	1140	811 -1560

* Standardised for age and social class

from this cause occurred in two of the textile work groups (job groups 072 knitters, and 073 bleachers, dyers and finishers). The four groups with high PMRs are given in Table 5.8. The overall PMR for all textile workers (job groups 069-074) was 1,140 based on 39 deaths as seen in Table 5.7.

Table 5.8 PMRs for female textile workers aged 20-74 for byssinosis (ICD 504), England and Wales 1979-80, 1982-90

Job group		Deaths	PMRs*	95% CI
069	Preparatory fibre processors	5	7576	2460 -17679
070	Spinners and winders	5	1511	490 -3525
071	Warp preparers and weavers	4	1338	365 -3425
074	Other textile workers	25	955	617 -1411

* Standardised for age and social class

Chronic bronchitis and emphysema was high in textile workers overall and in specific textile groups. It should be noted however, that smoking is unlikely to be the explanation for this excess as textile workers had low PMRs for lung cancer. Asthma was also somewhat low in such workers but there was an excess of deaths from pneumonia of all types. Raised PMRs for respiratory diseases, notably chronic obstructive airways disease, byssinosis and pneumonia were observed in the female occupational mortality data from 1970-72[9]. A deficit of lung cancer was also apparent and asthma mortality was generally low.

Ischaemic heart disease and the broad group of other cardiovascular diseases were raised in textile workers. An excess of ischaemic heart disease for such workers was also noted in the previous decennial supplement.

In general, the mortality pattern for textile workers was similar to that seen in 1970-72,[9] with the dominant excesses occurring in the respiratory and circulatory disease groups.

5.6 Entertainment and catering (job groups 024, 045, 046)

Alcohol-related deaths and injury and poisoning featured prominently in these occupations. The PMRs for selected causes for those in literary and artistic occupations (job group 024) are given in Table 5.9. Significantly raised PMRs were found for many causes of death for which alcohol is known or thought to play a role, namely cirrhosis of the liver (PMR 177, 16 deaths), other alcohol-related diseases (PMR 217, 40 deaths) and cancers of the oesophagus (PMR 146, 31 deaths) and oral cavity (PMR 336, 14 deaths). A range of accidental

Table 5.9 PMRs for women aged 20-74 in literary and artistic occupations (job group 024) for selected causes of death, England and Wales, 1979-80, 1982-90

Cause of death (ICD)	Deaths	PMR*	95% CI
Cancer of oral cavity (141,143,144,145)	14	336	184-564
Oesophageal cancer (150)	31	146	99-208
Drug dependence (304)	4	448	112-1148
Anorexia nervosa (307.1)	5	383	124-893
Cirrhosis of liver (571.5)	16	177	101-288
Other alcohol-related diseases (303,305.0,425.5,535.3,571.0-571.3, E860.0,E860.1)	40	217	155-296
Accidental poisoning by drugs (E850-E858)	15	210	117-346
Injured by fire (E890-E899)	11	317	158-568
Suicide (E950-E959)	90	165	132-202
Injury undetermined as accidental or purposeful (E980-E989)	35	143	99-199

* Standardised for age and social class

and violent deaths were high in this group too, namely suicide (PMR 165, 90 deaths), poisoning by drugs (PMR 210, 15 deaths), injury undetermined as accident or purposeful (PMR 143, 35 deaths), and death by fire (PMR 317, 11 deaths). An excess of suicide among authors, journalists and related workers was noted in the 1970-72 decennial supplement.[9] This group is also the only one for which significantly raised PMRs for anorexia nervosa (PMR 383, 5 deaths) and drug dependence (PMR 448, 4 deaths) were observed (see Table 5.9).

None of these observations is particularly surprising. The stress caused to those in entertainment due to lack of job security and in face of failure may lead to suicide or excessive risk-taking or to alcohol drinking or drug taking. Women in certain sections of this occupational group, such as actresses and models, face considerable pressure to remain slim in order to continue in employment and so may increase their risk of anorexia nervosa.

Table 5.10 PMRs for female publicans and bar workers (job group 045) aged 20-74 for selected causes of death, England and Wales, 1979-80, 1982-90

Cause of death (ICD)	Deaths	PMR*	95% CI
Cancer of oral cavity (141,143,144,145)	30	271	183-387
Cancer of the pharynx (specified) (146-148)	22	251	157-380
Cirrhosis of liver (571.5)	73	336	264-423
Other alcohol-related diseases (303,305.0,425.5,535.3,571.0-571.3, E860.0,E860.1)	101	291	237-354
Accidental poisoning by drugs (E850-E858)	19	181	109-283
Suicide (E950-E959)	48	65	48-86
Homicide (E960-E969)	28	315	209-456

* Standardised for age and social class

The mortality pattern for hotel, bar and club workers (job group 045), was broadly similar to those in the Arts as can be seen in Table 5.10 in which PMRs for selected causes of death are given. Thus deaths from cirrhosis of the liver (PMR 336, 73 deaths), other alcohol-related diseases (PMR 291, 101 deaths), cancers of the oral cavity (PMR 271, 30 deaths) and pharynx (PMR 251, 25 deaths) and accidental poisoning by drugs (PMR 181, 19 deaths) were all high in this group. The PMR for drug dependence was also high (PMR 359, 4 deaths) but this was not statistically significant. In contrast, mortality from suicide was low (PMR 65, 48 deaths), but homicide was high (PMR 315, 28 deaths). Such workers are possibly at risk from customers who consume too much alcohol. Suicide was also low for restaurateurs and caterers, (job group 046), with a PMR of 86 based on 126 deaths, and homicide was high (PMR 138, 27 deaths), possibly for the same reason as for those in hotels, bars and clubs, but neither of these PMRs differed significantly from 100. Deaths from 'other alcohol-related diseases' were also elevated in this group but the excess was not great and failed to reach statistical significance (PMR 113, 74 deaths).

5.7 Other raised PMRs of note

There are a variety of additional findings which are of interest. These are listed in Table 5.11 and are discussed in this section.

Exposure to solder flux fumes is a recognised occupational hazard leading to an increase in asthma.[13] That there was an excess of mortality from asthma (PMR 169, 23 deaths) in electrical and electronic assemblers (job group 161), is of note, especially as this also occurred in men in this occupation.

Excesses of bladder cancer in rubber workers have been well documented.[14] They are due to exposure to 2-naphthylamine, legislation against which was introduced in the 1950s. In this analysis, there was still an excess of this cancer observed both in male and female rubber goods makers (job group 092). In women the PMR was 457 (9 deaths), but for women under 60 years of age it was 955 (3 deaths). The younger deaths are of particular concern in that their exposure to 2-naphthylamine could only have taken place over a short time before its use was phased out. If these bladder cancers were due to this antioxidant then a latent period about 30 years occurred before development of the tumour. By the time of publication of the next decennial supplement this excess should have disappeared, unless other chemicals in rubber manufacturing and processing give rise to this tumour.

Table 5.11 Miscellaneous PMRs of interest in selected job groups for women aged 20-74, England and Wales, 1979-80, 1982-90

Job Group	Cause of death (ICD)	Deaths	PMR*	95% CI
161 Electrical and electronic assemblers	Asthma (493)	23	169	107-254
092 Rubber goods makers	Urothelial cancer (188,189.1-189.8)	9	457	209-867
117 Electroplaters	Lung cancer (162)	6	273	100-594
051 Launderers and dry cleaners	Renal failure (584-586)	23	165	104-247

* Standardised for age and social class

Chromium platers, who form part of occupational category 117, have been found in other studies to have elevated rates of lung cancer.[15] It is notable that the only significantly raised PMR in women in this group was for this malignancy (PMR 273, 6 deaths). The number of deaths was however very small and there is no information on the smoking habits of these workers.

A raised PMR of 165 based on 23 deaths from renal failure was seen in women working in laundries and dry cleaners (job group 051). That the PMR was even higher in men is of note (PMR 211, 12 deaths) though based on fewer deaths. The broader group of diseases of the genito-urinary system gave rise to a significantly raised PMR of 239 (based on 16 deaths) in women in this occupation in the 1970-72 decennial supplement.[9] Without further information on the specific types of disease and more details of the occupational exposures of the sufferers, it is difficult to take this observation further.

5.8 Mortality of women by husband's occupation

Traditionally, the mortality of married women was analysed by their husbands' occupation and only for single women was the woman's own occupation used. Starting with the 1970-72 decennial supplement,[9] and as in the other sections of this chapter, all women have been analysed by their own occupation, if recorded on the death certificate, regardless of marital status. In certain areas however, an analysis of married women by their husbands' occupations can be informative. Three areas are discussed here. Cervical cancer and suicide were examined to see if there was any relationship between these causes of death and the husband's work; and the pattern of mortality in farmers' wives was assessed to see if it was similar to that in farmers.

5.8.1 Cervical cancer (ICD 180)

Two hypotheses about cervical cancer that relate to husbands' occupations have been suggested and have been examined using data from previous decennial supplements.[9,16] Both are discussed here.

There is emerging evidence that sexually transmitted papilloma viruses may be a cause of cervical cancer.[17] An early examination of the hypothesis that this cancer might be caused by a sexually transmitted infection utilised the 1959-63 decennial supplement[16] and noted that high rates of cervical cancer were observed in wives of men whose work involved prolonged absences from home.[18] Couples who are frequently separated because the husband is away from home for work are likely to have more frequent opportunities for extra-marital sexual relationships which would encourage the spread of the relevant viruses. The association between male occupations entailing time away from home and cervical cancer in the wives has been examined here. The statistically significant PMRs for cervical cancer by husband's occupation for 1979-80 and 1982-90 are presented in Table 5.12. These give some support to the findings of Beral[18] in that many of the occupations listed involve periods away from home. Of the ten occupations given in the table, those working as merchant seamen, fishermen, lorry drivers, and bus and coach drivers all may have extended periods away from home, although most bus drivers may only work on short journeys.

Table 5.12 Significantly raised PMRs for cervical cancer (ICD 180) in women aged 20-74 by husband's occupation, England and Wales, 1979-80, 1982-90

Husband's job group		Deaths	PMR*	95% CI
045	Publicans and bar staff	235	180	158-205
065	Foresters	21	179	110-274
036	Sea farers	106	162	133-196
066	Fishing and related workers	21	158	98-242
186	Mechanical plant drivers	43	152	110-205
183	Lorry drivers	854	141	132-151
046	Caterers	87	141	113-175
147	Steel erectors	49	139	103-184
182	Bus and coach drivers	168	138	118-160
168	Roofers and glaziers	48	136	100-181
044	Retailers and dealers	590	126	116-137
174	Construction workers nec	280	123	109-138
058	Security workers	212	123	107-140
169	Builders etc.	257	117	103-132
048	Armed forces**	159	121	103-142

* Standardised for age and social class
** Adjustment for social class is not possible for this job category

Previous analyses of mortality by husband's occupation have used SMRs and Robinson[19] noted that SMRs for wives of fishermen may have been artificially inflated because many fishermen were at sea at the time of the census and thus omitted from the denominator of the rates used in the calculation of the SMR. Use of proportional mortality analyses obviates this problem, however, and with a PMR of 158 (21 deaths) the excess cannot be due to denominator biases.

Another occupation which entails time away from home also had a significantly raised PMR, namely the armed forces with a PMR (not adjusted for social class) of 121 based on 159 deaths. An excess amongst the wives of military personnel has been noted before with an SMR of 201 in 1970-72.[20] Like fishermen, the armed forces may be under-represented in censuses, and this may explain why the SMR in 1970-72 was much higher than the PMR in the current analysis.

In contrast, as has been found before,[20] there were certain occupations for which the husband is frequently away from home but for which the wives did not have elevated mortality from cervical cancer, namely sales representatives (PMR 100, 190 deaths), railway engine drivers (PMR 96, 39 deaths) and railway guards (PMR 74, 13 deaths). Many in these occupations however, may work in areas close to home and so they are less likely to be away for long periods. This may be a reason for the mortality pattern differing from the groups mentioned above although, if this is so, the PMR of 138 for wives of bus drivers is surprisingly high.

It appears therefore that there is some evidence for the hypothesis that wives of men whose work involves extended periods away from home are at greater risk of cervical cancer, although there are exceptions to this which weaken the association.

The second hypothesis, proposed by Robinson,[19, 20] is that aspects of the work performed by certain men may also increase the risk of cervical cancer in his wife. In particular she suggested that those working with chemical and other carcinogens may bring home contaminants to which the wife would then be exposed. In her analyses she examined all the occupations which gave rise to SMRs over 120 in the 1970-72 decennial supplement, but these SMRs were not adjusted for social class. After the wives of the military, the highest SMR was for wives of miners and quarrymen with an SMR of 185. In the 1979-80 and 1982-90 analysis, this group was split into two categories and the PMRs for wives of these men are given in Table 5.13. The results do not indicate an association between mining and cervical cancer as the PMRs are not significantly raised. Furthermore, in the past it has been the wives of coal face workers who have appeared to be most at risk of cervical cancer whereas, if anything, the opposite was observed in this analysis.

Table 5.13 PMRs for cervical cancer (ICD 180) in women aged 20-74 whose husbands worked in coalmining, England and Wales, 1979-80, 1982-90

Husband's job group		Deaths	PMR*	95% CI
175	Face-trained miners	76	92	72 - 115
088	Other coal miners	394	108	98 - 119
088 & 175	All coal miners	470	105	96 - 115

* Adjusted for age and social class

Wives of textile workers have been noted in the past to have elevated mortality from cervical cancer.[9, 19] Again the current analysis does not support those findings. PMRs for wives of textile workers are presented in Table 5.14, from which it can be seen that none of the groups had particularly high mortality from this cause. None of the PMRs listed was significantly greater than 100.

Table 5.14 PMRs for cervical cancer (ICD 180) at ages 20-74 in wives of textile workers, England and Wales, 1979-80, 1982-90

Husband's job group		Deaths	PMR*	95% C.I.
069	Preparatory fibre processors	6	88	32 - 192
070	Spinners and winders	10	78	38 - 144
071	Warp preparers and weavers	23	121	77 - 181
072	Knitters	12	115	59 - 200
073	Bleachers, dyers and finishers	14	83	45 - 139
074	Other textile workers	58	81	62 - 105
069-074	All textile workers	123	90	74 - 107

* Standardised for age and social class.

Of the occupational groups listed in Table 5.12 with significantly raised PMRs for cervical cancer not discussed above, few are likely to entail exposure to carcinogens at work which could then be brought home. Possible exceptions are foresters exposed to pesticides and construction workers exposed to asbestos and various dusts. The raised PMRs for wives of publicans and barmen and caterers were more likely to be due to lifestyle factors, such as smoking and sexual activity, than the occupation *per se*.

The findings from this analysis therefore do not give much support to Robinson's findings from 1970-72.[19, 20] It appears unlikely that exposures at work to substances which are then brought into the home are having a major effect on the risk of cervical cancer in wives. This analysis is in closer agreement with findings from an alternative method of examining the 1970-72 data using occupational groupings defined by exposure to a variety of agents which might be carcinogenic to the wives.[21] Using this method there appeared to be no association in the 1970-72 data between exposure to any of the various substances and the risk of cervical cancer in wives.

It is possible, of course, that hazards of this nature were present in the past but, with improved hygiene practices in workplaces, such risks have now disappeared. Thus an analysis of occupational mortality can no longer test the theory that carcinogenic substances could be transferred from husband to wife. However, if there ever was a risk from occupational exposures, it does not appear to constitute a problem nowadays.

5.8.2 Suicide (ICD E950-E959)

Significantly-raised PMRs for suicide by husband's occupation are given in Table 5.15. The age range 20-59 has been used in an attempt to coincide with the time during which the husbands are most likely to have been in employment. The wives of medical practitioners head the list, and of the 51 suicides in this group, 12 were in women who themselves were doctors or nurses. However, this is not sufficient to explain the excess. Doctors' wives may be able to obtain lethal drugs more easily because of their husbands' profession. Similarly, six of the 25 wives of male nurses who committed suicide were nurses themselves which might explain part of the excess, but ready access to means of successful suicide may have been a function in the others.

Table 5.15 Significantly raised PMRs for suicide (ICD E950-E959) in women aged 20-59 by husband's occupation, England and Wales, 1979-80, 1982-90

Husband's job group		Deaths	PMR*	95% CI
015	Doctors	51	226	168 - 297
122	Centre lathe turners	15	188	105 - 310
061	Hospital porters and ward orderlies	18	174	103 - 276
017	Nurses	25	151	98 - 224
024	Literary and artistic occupations	62	143	110 - 183
137	Electricians	73	136	107 - 171
169	Builders etc.	76	129	101 - 161
011	Teachers nec	92	126	102 - 155
124	Machine tool operators	141	123	103 - 145

* Standardised for age and social class

Three out of the top four PMRs for suicide in women by husband's occupation occurred to those whose husbands were in the health professions. This suggests that it is easier for such women to commit suicide than for those married to other workers, but it is not clear how the wives of hospital porters and ward orderlies have more ready access to drugs than other women. It should be noted that, while both male and female doctors and nurses had significantly raised PMRs for suicide during working life, male and female hospital porters and ward orderlies did not, though males in this occupation were at greater risk of accidental poisoning by drugs (PMR 225, 11 deaths). Explanations for the excess of suicide in wives in terms of lifestyle or pressures at work do not seem plausible, as doctors and hospital porters are very different in these respects.

Wives of those in literary and artistic occupations were also at higher risk of suicide. Only 10 per cent of women who themselves were in literary and artistic occupations were recorded as the wife of a man in the same occupation. It is of note that whether in the profession or married to someone in it, women were at increased risk of committing suicide.

Interestingly, for all but one of the other occupations listed in Table 5.12, concordance was observed with the findings in men by their own occupation during their working years of 20-64. The PMRs for suicide in male electricians, builders and handymen, teachers and machine tool operators were all significantly raised. Only for centre lathe turners, for whom the PMR was significantly low at 56, was a contrary pattern observed.

5.8.3 Wives of farmers (job group 047)

Many farmers' wives also work on the farm and are exposed to the same risks as their husbands or other women working in agriculture. Some 30 per cent of the women farmers who died were also recorded as being the wife of a farmer, and so this analysis is not independent of that for female farmers. However, the coding rules for occupations at death mean that many women do not have an occupation recorded for themselves and so there were 14 times as many deaths recorded amongst farmers' wives (35,222 deaths) as there were in female farmers (2,512 deaths). If it is correct to assume that many farmers' wives actually work in farming, then an analysis of such women will have more power than examining women who are recorded as farmers in their own right.

The significantly raised PMRs for farmers' wives are given in Table 5.16. For some causes the findings were similar to those for male and female farmers. Notably there was an excess of deaths from farmers' lung disease which indicates that at least some of the farmers' wives in this analysis were working in farming. The excess of deaths from injury by animals and plants in this group is also not surprising.

Table 5.16 Significantly raised PMRs at ages 20-74 for wives of farmers (job group 047), England and Wales 1979-80, 1982-90

Cause of death (ICD)	Deaths	PMR	95% C.I.
Cancer of gall bladder (156)	134	120	101 -142
Skin cancer excluding melanoma (173)	32	147	100 -207
Cancer of uterus (part unspecified) (179)	93	128	103 -157
Cancer of the body of uterus (182)	281	126	111 -141
Thyroid cancer (193)	70	148	115 -187
Diabetes (250)	685	144	133 -155
Hypertensive disease (401-405)	351	116	104 -129
Pulmonary embolism/phlebitis (415.1, 451, 453)	458	116	106 -127
Aortic valve disorders (424.1)	162	133	113 -155
Atrial fibrillation (427.3)	83	128	102 -158
Other cerebrovascular disease (431-438)	3853	115	111 -118
Other bacterial pneumonia (482)	18	176	104 -278
Influenza (487)	36	155	109 -215
Farmers' lung disease (495.0)	6	862	316 -1876
Pulmonary fibrosis (515)	50	156	116 -206
Other hernia (551-553)	53	138	104 -181
Glomerulonephritis (580-583)	49	141	104 -186
Motor vehicle traffic accidents (E810-E819)	210	123	107 -140
Injury by animals and plants (E905-E906)	4	506	138 -1296

Excesses were also seen for skin cancer, motor vehicle accidents, influenza and other hernia, the first two being high in female and the last two in male farmers. It must be noted however, that many of the excesses recorded in Table 5.16 were not apparent in the data for men or women by their own occupation. The excess of ovarian and the deficit of breast cancer in female farmers was not seen in farmers' wives, though high PMRs for uterine cancer were found in both.

The causes of death for which there was concordance with the male and/or female data are not surprising but where there was no agreement it is likely that spurious associations have been observed. It is of interest that the PMR for suicide (PMR 111, 134 deaths) in farmers' wives was much lower than for male and female farmers.

Farmers' wives therefore shared some of the risks that farmers (male or female) face but their mortality pattern was not identical. This may be because only a proportion (of unknown magnitude) of farmers' wives actually worked in farming. Also, it is possible that some of the farmers' wives only worked on the farm for short periods or that they chose to participate in selected areas of farming.

5.9 Conclusion

In a few situations an analysis of women by their husbands' occupations can be informative and some points of interest have emerged. Cervical cancer showed some association with occupations which lead the husbands away from home a great deal, while similar excesses of suicide were seen for men in certain occupations and for their wives. The mortality pattern of farmers' wives was similar in part to that for male and female farmers but associations with farming may have been diluted by wives who do not work in this occupation.

Despite there being limited data on mortality by occupation for women by their own occupation, interesting patterns have emerged. Such an analysis cannot conclusively identify new hazards but raised PMRs can provide indications as to where problems may be occurring. The significantly raised PMRs based on three or more observed or expected deaths are documented in Appendix 4 and a commentary on each occupational group follows in section 5.10.

Many of the excesses for women by their own occupation that have been described in detail in this chapter are not surprising, but it is reassuring that the PMR method applied to only 30 per cent of female deaths does identify some well-known associations with occupation. That these excesses are still occurring even though the hazards were identified many years ago is not so reassuring, however, in terms of prevention of disease. Admittedly some women may have been exposed to substances before they were recognised as hazards, and, with long latent periods, the excess of mortality is still evident. However there are other causes of death for which action should take effect over a shorter period. An example is the excess of suicide in health-related professions and farmers. Reducing the suicide rate is one of the aims stated in the Health of the Nation strategy[22] and so these occupational groups may need to be targeted. The Health of the Nation also tackles the problem of excessive alcohol drinking. In terms of mortality caused by alcohol, certain occupations are at particular risk and so extra emphasis may need to be put on advising such groups.

5.10 Summaries of results for each occupational group

This section gives details of the mortality of women in each of the 194 job groups for which notable associations were apparent. Many occupational groups are omitted because there were no PMRs of interest, usually because the number of deaths from all causes was small. Appendix 4 lists all the PMRs which differed significantly at the 5 per cent level from 100, for ages 20-74, standardised for age and social class, and based on three or more observed or expected deaths. The commentary which follows discusses these findings but also comments on certain PMRs of interest which are based on fewer deaths or relate to a different age group.

Although many of the findings are likely to be spurious associations, they are given here so that should similar patterns be observed in other datasets, or in subsequent decennial supplements, then more detailed studies could be pursued. Despite being based on small numbers, many PMRs are highlighted for two reasons. Firstly, no dataset of this size on mortality by occupation has ever been available before and thus associations between occupation and causes of death in women may have been missed in the past because the numbers were too small. Secondly, if there really are associations between certain causes of death in women and their occupations, then raised PMRs based on small numbers of deaths may indicate a greater underlying burden of disease. Married women who do not work for all of their potential working life do not have an occupation recorded on the death certificate, and so links with the occupation may be missed. Nonetheless, the PMRs here should be viewed with caution as only a small minority are likely to relate to real hazards at work.

001 - LAWYERS
There was little of note for this occupational group except a raised PMR for unspecified pneumonia (PMR 1064, 3 deaths), which is not readily explained.

002 - ACCOUNTANTS
This group had raised PMRs for sarcoidosis (PMR 845, 3 deaths) and myeloma (318, 9 deaths) but low mortality from suicide (PMR 56, 12 deaths).

003 - PERSONNEL MANAGERS ETC.
A raised PMR for ovarian cancer was observed (PMR 142, 38 deaths) along with a rare reproductive cancer, namely cancer of the placenta (PMR 2500, 2 deaths) although based on only two deaths. An elevated ratio was also observed for renal failure (PMR 318, 6 deaths) but there were deficits from suicide (PMR 44, 6 deaths) and the miscellaneous group of other cerebrovascular diseases (ICD 431-438, PMR 68, 27 deaths).

005 - COMPUTER PROGRAMMERS
A raised PMR for non-Hodgkin's lymphoma (PMR 298, 8 deaths) was seen in this group.

006 - SALES MANAGERS ETC.
A raised PMR of 146 was seen for cancer of the ovary based on 73 deaths. Mortality from sub-arachnoid haemorrhage was also raised (PMR 147, 38 deaths) in contrast to a deficit of ischaemic heart disease (PMR 75, 159 deaths). The PMR for motor vehicle traffic accidents (PMR 158, 46 deaths) was also high with the excess being confined to those under the age of 60 (PMR 166, 43 deaths) which would indicate an association with working life.

007 - GOVERNMENT INSPECTORS
There were seven deaths from suicide in this group, all of which occurred at under 60 years of age, giving rise to a PMR of 495 for ages 20-59. The only other high PMR of note was a PMR of 14286 based on only one death, the cause being due to injury by animals and plants. This however occurred in a woman over 60 years of age, making it less likely that the cause was related to her occupation.

008 - GOVERNMENT ADMINISTRATORS
An excess of breast cancer (PMR 129, 117 deaths) occurred equally in those less than 60 years of age (PMR 124, 58 deaths) and those over 60 (PMR 134, 59 deaths). Other raised PMRs were based on only two deaths, but there was a deficit of suicide (PMR 20, 2 deaths).

009 - OTHER ADMINISTRATORS
The well-documented excess of melanoma of the skin in office workers was apparent in this group (PMR 155, 30 deaths). Small excesses of ovarian (PMR 120, 143 deaths) and breast cancer (PMR 111, 436 deaths) were observed but a high proportion of women with no or few children in this group might explain these findings. Dementia was also high (PMR 186, 26 deaths) whereas deficits of diabetes (PMR 40, 10 deaths) and chronic rheumatic heart disease (PMR 52, 11 deaths), ischaemic heart disease (PMR 86, 425 deaths) and suicide (PMR 70, 34 deaths) were seen in this group. There was an elevated PMR for chronic bronchitis and emphysema (PMR 129, 76 deaths) but before adjustment for social class this was only 101.

010 - TEACHERS IN HIGHER EDUCATION
This group had marked deficits from the smoking-related disorders of lung cancer (PMR 62, 30 deaths) and chronic bronchitis and emphysema (PMR 36, 5 deaths). Two other causes had deficits, the first also associated with smoking, namely ischaemic heart disease (PMR 71, 91 deaths) and pulmonary embolism and phlebitis (PMR 27, 2 deaths). Excesses of breast cancer (PMR 133, 167 deaths) which may be explained by the childbearing pattern in this group, and rectal cancer (PMR 171, 18 deaths) were seen.

011 - TEACHERS NEC
Excesses of a number of malignancies were seen in this group. Cancers of the breast (PMR 128, 2404 deaths) body of uterus (PMR 128, 140 deaths), and ovary (PMR 130, 740 deaths) are possibly related to low levels of childbearing, the opposite being seen for cancer of the cervix (PMR 53, 125 deaths) which tends to decrease with increasing numbers of children. Other cancers which were high are colon (PMR 118, 572 deaths), gall bladder (PMR 137, 57 deaths), brain (PMR 123, 252 deaths), non-Hodgkin's lymphoma

(PMR 118, 180 deaths) and Hodgkin's disease (PMR 142, 45 deaths). Melanoma of the skin was also high (PMR 143, 134 deaths) which may be explained by intermittent exposure to the sun during holidays as for office workers.

Another excess of note was that of multiple sclerosis (PMR 152, 105 deaths) which may be due to late exposure to Epstein Barr virus. Non-significantly raised PMRs were also seen for Parkinson's disease (PMR 124, 47 deaths) and motor neurone disease (PMR 125, 70 deaths) which are diseases for which delayed exposure to viruses may be an aetiological factor. Myasthenia, another disorder of the nervous system was also high (PMR 369, 4 deaths). Hodgkin's disease mentioned above may also have a infectious component. Additionally, all leukaemias combined (ICD codes 204-208) gave rise to a significantly-raised PMR of 117 based on 186 deaths, and infections may play a role here too.

The PMR for motor vehicle traffic accidents was raised (PMR 118, 218 deaths) but the excess occurred mainly in those over 60 years of age (PMR 157, 58 deaths). In contrast there was a deficit of deaths from 'other alcohol-related diseases' (PMR 69, 75 deaths).

Deficits were seen for certain smoking related disorders, namely cancers of the lung (PMR 62, 557 deaths) and larynx (PMR 35, 5 deaths), chronic bronchitis and emphysema (PMR 56, 164 deaths) and ischaemic heart disease (PMR 88, 2174 deaths). There were also deficits of deaths from diabetes (PMR 79, 100 deaths) and tuberculosis (PMR 39, 5 deaths).

012 - VOCATIONAL TRAINERS, SOCIAL SCIENTISTS, ETC

The only excesses of note in this group were chronic myeloid leukaemia (PMR 350, 6 deaths), peripheral vascular disease (PMR 366, 5 deaths) and fibrosing alveolitis (PMR 622, 3 deaths).

013 - WELFARE WORKERS

There were excesses of cancer of the lung (PMR 119, 335 deaths), gall bladder (PMR 177, 23 deaths) and uterus part unspecified (PMR 169, 19 deaths), but a deficit of deaths from brain cancer (PMR 73, 48 deaths). There were also deficits of deaths from 'other alcohol-related diseases' (PMR 32, 11 deaths), accidental poisoning by drugs (PMR 11, 1 death) and multiple sclerosis (PMR 41, 9 deaths). Ischaemic heart disease was slightly elevated (PMR 114, 828 deaths).

014 - CLERGY

Excesses of cancer of the oesophagus (PMR 187, 16 deaths) and motor vehicle accidents (PMR 171, 21 deaths) occurred, the latter possibly being due to many hours spent driving to perform pastoral visits. Deficits were seen for smoking related disorders, namely lung cancer (PMR 39, 13 deaths) and chronic bronchitis and emphysema (PMR 38, 4 deaths). Cancer of the pancreas (PMR 43, 8 deaths) and suicide (PMR 35, 5 deaths) were also low.

015 - DOCTORS

The well-known excess of deaths from suicide amongst doctors was observed here (PMR 193, 50 deaths), the excess being also apparent during the working ages of 20-59 years (PMR 191, 41 deaths). A raised PMR from accidental poisoning by drugs was also seen during working life (PMR 300, 5 deaths) but this was not statistically significant at the 5 per cent level. The only other raised PMR of note was that for cancer of the pancreas (PMR 160, 23 deaths).

016 - DENTISTS

The only significantly raised PMR in this group was for cancer of the pharynx (PMR 1143, 2 deaths) but the number of deaths is particularly small. Both deaths occurred at under 60 years of age which gave a PMR in this age group of 1802.

017 - NURSES

Deaths from accidents and violence featured prominently in this group with raised PMRs during the working ages of 20-59 for accidental poisoning by drugs (PMR 161, 57 deaths), suicide (PMR 135, 372 deaths), and injury undetermined as accidental or purposeful (PMR 143, 176 deaths). There was an excess of unspecified falls in the full age group of 20-74 (PMR 150, 49 deaths) but the excess was confined to those over 60 and so these are less likely to be related to the occupation.

No marked excess was noted for any malignancy but deficits existed for breast cancer (PMR 85, 2245 deaths), ovarian cancer (PMR 83, 674 deaths), colon cancer (PMR 90, 635 deaths), cancer of the oral cavity (PMR 54, 21 deaths), melanoma (PMR 68, 89 deaths) and other skin cancer (PMR 43, 6 deaths).

In contrast, and possibly partially explaining the deficits in the cancers, two large disease categories showed marginally raised PMRs, namely ischaemic heart disease (PMR 105, 3881 deaths) and chronic bronchitis and emphysema (PMR 110, 482 deaths). Both of these were significantly below 100 prior to social class adjustment.

Amongst other diseases a few had raised PMRs, namely diabetes (PMR 119, 224 deaths), sarcoidosis (PMR 171, 19 deaths), Parkinson's disease (PMR 128, 72 deaths) and 'other' (i.e. not aortic) aneurysm (PMR 238, 15 deaths), but deaths from meningococcal infection were particularly low (PMR 15, 1 death).

018 - PHARMACISTS

Cervical cancer was high (PMR 244, 8 deaths) but this was the only PMR which showed a statistically significant difference from 100. Although not significant, the PMR for suicide was however raised (PMR 164, 17 deaths) and this may be associated with the availability of drugs.

019 - MEDICAL RADIOGRAPHERS

A PMR of 1754 based on only two deaths was seen for aplastic anaemia in this group, but no other PMRs were noteworthy.

020 - PHYSIOTHERAPISTS

An excess of ovarian cancer was seen in this group (PMR 161, 28 deaths). Raised PMRs were also seen for chronic myeloid leukaemia (PMR 399, 4 deaths), chronic and unspecified myocarditis (PMR 673, 3 deaths), and falls on stairs (PMR 488, 3 deaths) but these were based on small numbers of deaths and there are no obvious explanations for them.

021 - HEALTH PROFESSIONS NEC
Excesses were seen for melanoma of the skin (PMR 207, 12 deaths), possibly due to intermittent sun exposure, and suicide (PMR 158, 28 deaths) possibly due to easy access to drugs. The PMR for suicide at ages 20-59 was also raised (PMR 165, 25 deaths).

022 - VETERINARIANS
The only excess in this small occupational category was for suicide (PMR 414, 7 deaths) which is likely to be due to the availability of the means. All these deaths occurred in women under 60 years of age (PMR 459).

024 - LITERARY AND ARTISTIC OCCUPATIONS
Alcohol-related disorders featured highly in this group with PMRs of 177 (16 deaths) for cirrhosis of the liver and 217 (40 deaths) for 'other alcohol-related diseases'. Alcohol is also thought to be a causal factor in oesophageal cancer (PMR 146, 31 deaths) and cancer of the oral cavity (PMR 336, 14 deaths). The only other raised malignancy was connective and soft tissue neoplasms (PMR 218, 13 deaths).

A number of PMRs for accidental deaths were raised, some of which may be related to the pressures of the job and the lifestyle of those employed in these professions. They were accidental poisoning by drugs (PMR 210, 15 deaths), suicide (PMR 165, 90 deaths), injury undetermined as accidental or purposeful (PMR 143, 35 deaths) and death by fire (PMR 317, 11 deaths), but deaths caused by motor vehicle traffic accidents were low (PMR 72, 41 deaths). Also possibly related to the lifestyle of those in this occupational group were the raised PMRs for anorexia nervosa (PMR 383, 5 deaths) and drug dependence (PMR 448, 4 deaths).

Also raised were the PMRs for meningococcal infection (PMR 370, 4 deaths) and bronchopneumonia (PMR 143, 43 deaths), but there were deficits of deaths from ischaemic heart disease (PMR 75, 289 deaths) and sub-arachnoid haemorrhage (PMR 54, 26 deaths).

025 - PERSONS INVOLVED IN SPORT
The greatest PMR in this group was for animal transport accidents (PMR 8000, 6 deaths). All the deaths from this cause occurred at under 60 years of age and thus it would appear to be associated with the type of work, particularly for horse-riders. Based on small numbers, there were also excesses of Hodgkin's disease (PMR 633, 3 deaths) and of melanoma of skin, but the latter only reached statistical significance in those under 60 years of age (PMR 549, 3 deaths) since all three deaths from this cause occurred in that age group.

026 - BIOLOGICAL SCIENTISTS
The only raised PMR in this group was for motor neurone disease (PMR 524, 3 deaths), but the number of deaths was very small and an association with the occupation is not obvious.

028 - PHYSICAL SCIENTISTS AND MATHEMATICIANS
Some leukaemias appeared to be raised in this group. Chronic lymphatic leukaemia had a PMR of 905 (2 deaths) and other leukaemias (ICD codes 207 and 208) had a PMR of 2532 also based on two deaths. The number of deaths from each cause is however very small and it is unlikely that exposures at work could be responsible.

031 - DRAUGHTSPERSONS
The two PMRs of note in this group were those for the main cancers associated with low levels of childbearing, namely breast cancer (PMR 147, 44 deaths) and ovarian cancer (PMR 198, 18 deaths).

032 - LABORATORY TECHNICIANS
Excesses of cancer of the kidney except pelvis (PMR 215, 14 deaths), bronchiectasis (PMR 346, 5 deaths) and complications of puerperium (PMR 836, 3 deaths) were seen in this group. There was a deficit of hypertensive disease (PMR 25, 2 deaths).

033 - ARCHITECTS AND SURVEYORS
The only excess in this group was for pulmonary embolism and phlebitis (PMR 382, 4 deaths).

034 - AIRCRAFT FLIGHT DECK OFFICERS
A PMR of 11111 based on one death for air transport accidents was not a surprise in this group.

037 - TECHNICIANS NEC
Raised PMRs for cancers of the rectum (PMR 415, 12 deaths), body of uterus (PMR 311, 5 deaths) and kidney, except pelvis (PMR 357, 5 deaths) were seen in this group, but a deficit of other cerebrovascular disease (ICD codes 431-438) was also observed (PMR 37, 4 deaths). Pneumococcal pneumonia was also elevated in this group (PMR 562, 3 deaths).

038 - PRODUCTION AND MAINTENANCE MANAGERS
Lung cancer (PMR 128, 96 deaths) and stomach cancer (PMR 156, 28 deaths) were elevated in this group, but there were fewer deaths than expected from suicide (PMR 26, 4 deaths) and other cerebrovascular disease (ICD 431-438, PMR 73, 59 deaths).

040 - MANAGERS IN TRANSPORT, UTILITIES AND MINING
Mortality from ischaemic heart disease (PMR 137, 107 deaths) and gastric ulcer (PMR 389, 4 deaths) was elevated. There was also one death from gastrojejunal ulcer leading to a PMR of 5000. Deaths from motor vehicle traffic accidents were surprisingly few (PMR 28, 2 deaths).

041 - OFFICE MANAGERS
Bladder cancer was high in this group (PMR 178, 21 deaths) as too were arteritis (PMR 565, 3 deaths) and the miscellaneous group of 'other accidents' (PMR 569, 3 deaths), but deaths from diabetes (PMR 51, 9 deaths) and suicide (PMR 56, 22 deaths) were fewer than expected.

042 - BUTCHERS
There was a marked deficit of deaths from ovarian cancer

(PMR 27, 2 deaths), but two deaths from appendicitis gave a PMR of 2632.

043 - FISHMONGERS AND POULTRY DRESSERS
The only raised PMRs in this group were all based on small numbers of deaths. They were cancer of the body of uterus (PMR 401, 4 deaths), diabetes (PMR 349, 5 deaths) and 'other alcohol-related diseases' (PMR 683, 4 deaths).

044 - RETAILERS AND DEALERS
There was an excess of acute myocarditis (PMR 181, 17 deaths) in this group and deficits of glomerulonephritis (PMR 62, 18 deaths), cirrhosis of the liver (PMR 77, 82 deaths), systemic lupus erythematosus (PMR 61, 17 deaths), multiple sclerosis (PMR 65, 88 deaths) and acute lymphatic leukaemia (PMR 59, 16 deaths). There was also an excess of retroperitoneal cancer in those under 60 years of age (PMR 267, 13 deaths).

This is a large occupational category and there were many PMRs which were significantly different from 100 at the 5% level of significance even though they did not appear to be very high or low as they lay in the range 80-120. They are listed below for completeness.

A number of malignancies differed significantly from 100, namely cancers of the colon (PMR 89, 923 deaths), lung (PMR 105, 2220 deaths), brain (PMR 88, 317 deaths) and kidney except pelvis (PMR 83, 143 deaths). The PMR for brain cancer was however 100 prior to adjustment for social class. Amongst reproductive cancers, cervical cancer was elevated (PMR 117, 500 deaths), but breast (PMR 84, 2995 deaths) and ovarian cancer (PMR 92, 1059 deaths) were low, possibly indicating an association with parity. The PMR for ovarian cancer was however elevated to 105 prior to adjustment for social class.

PMRs were elevated for a number of circulatory and respiratory diseases, namely ischaemic heart disease (PMR 110, 6997 deaths), chronic rheumatic heart disease (PMR 117, 330 deaths), hypertensive disease (PMR 114, 258 deaths), subarachnoid haemorrhage (PMR 110, 607 deaths), other cerebrovascular disease (PMR 106, 2625 deaths), chronic bronchitis and emphysema (PMR 111, 862 deaths) and bronchopneumonia (PMR 112, 573 deaths). The PMRs for hypertensive disease and bronchopneumonia were however, 100 and 95 without social class adjustment. In contrast the PMR for other aneurism (i.e. not aortic) (PMR 33, 3 deaths) was low.

The only other PMR to mention is that for motor vehicle traffic accidents (PMR 86, 279 deaths) which was low in this occupational group.

045 - PUBLICANS AND BAR STAFF
An almost three-fold excess of 'other alcohol-related diseases' (PMR 291, 101 deaths) was seen in this group and an even higher PMR for cirrhosis of the liver (PMR 336, 73 deaths). Cancer of the oral cavity (PMR 271, 30 deaths) and pharynx (PMR 251, 22 deaths) were high which may also have been due to alcohol drinking. Accidental poisoning by drugs (PMR 181, 19 deaths) was also high but suicide was low (PMR 65, 48 deaths) in contrast to the elevated PMR for homicide (PMR 315, 28 deaths). Oesophageal varices were also high (PMR 526, 5 deaths).

Respiratory disorders appeared to be high in this group, with specific excesses being seen for chronic bronchitis and emphysema (PMR 152, 239 deaths) and asthma (PMR 138, 56 deaths). The PMR for lung cancer was also raised (PMR 128, 544 deaths) which presumably was due to the smoking habits of members of this job group.

Cervical cancer was raised (PMR 158, 159 deaths), in contrast to breast cancer (PMR 73, 481 deaths) and ovarian cancer (PMR 70, 149 deaths) so the explanation for these may lie in the childbearing pattern of the workers concerned. Deficits were seen for brain cancer (PMR 70, 49 deaths), melanoma (PMR 63, 19 deaths), non-Hodgkin's lymphoma (PMR 69, 40 deaths), Hodgkin's disease (PMR 28, 3 deaths), myeloma (PMR 63, 22 deaths), acute lymphatic leukaemia (0 deaths, 4.79 expected) and chronic myeloid leukaemia (PMR 27, 3 deaths). Multiple sclerosis deaths were also low in this group (PMR 56, 13 deaths).

046 - CATERERS
Given the similarity between this job group and group 045 it is of interest that the PMR for homicide (PMR 138, 27 deaths) was raised and that for melanoma of the skin was low (PMR 79, 49 deaths) though these were not statistically significant. 'Other alcohol-related diseases' were not significantly raised in this group.

A deficit of multiple sclerosis was also observed in this group (PMR 60, 26 deaths), but Guillain Barre syndrome (PMR 363, 5 deaths) was high.

Two malignancies were raised, namely lung cancer (PMR 114, 1564 deaths) and stomach cancer (PMR 116, 376 deaths), but there was a deficit of deaths from breast cancer (PMR 93, 1389 deaths).

There was a slight excess of deaths from aortic aneurysm (PMR 117, 178 deaths), in contrast to deficits from chronic rheumatic heart disease (PMR 75, 122 deaths) and bronchopneumonia (PMR 88, 315 deaths). Deaths from motor vehicle traffic accidents were somewhat higher than expected (PMR 118, 186 deaths).

047 - FARMERS
The well-documented excess of suicide in farmers was observed here (PMR 145, 38 deaths) along with a number of accidental causes, namely motor vehicle traffic accidents (PMR 146, 39 deaths), animal transport accidents (PMR 855, 3 deaths), falls from or out of buildings (PMR 543, 3 deaths), deaths caused by animals or plants (PMR 1770, 2 deaths) and deaths caused by machinery (PMR 2632, 4 deaths). Most of these causes are likely to be related to the occupation of the individuals concerned although it is possible that some of the falls were in fact suicides. All of the accidental deaths (excluding suicide), occurred in those under 60 years of age, except for eight of the motor vehicle accident deaths.

Not surprisingly there was an excess of farmers' lung disease (PMR 7692, 2 deaths), and the PMR for fibrosing alveolitis was also raised (PMR 261, 7 deaths). Acute bronchitis was also elevated (PMR 288, 7 deaths) but deficits of chronic bronchitis and emphysema (PMR 72, 54 deaths) and sub-arachnoid haemorrhage were seen (PMR 62, 26 deaths).

There was an excess of ovarian cancer (PMR 123, 94 deaths) but a deficit of breast cancer (PMR 85, 194 deaths) and so the excess of the former cannot readily be explained by the childbearing patterns of the farmers. Excesses were also observed for three less common malignancies, namely other mediastinum (ICD 164.1-164.9, PMR 761, 3 deaths), skin cancer excluding melanoma (PMR 419, 7 deaths) and other endocrine glands (not thyroid) (PMR 1227, 2 deaths) though the numbers of deaths in all cases are small.

048 - ARMED FORCES
Motor vehicle traffic accidents (PMR 195, 15 deaths), air transport accidents (PMR 1905, 2 deaths) and other accidents (ICD code E928) (PMR 2941, 2 deaths) were all raised, as too were deaths from poisoning by gas and other domestic fuels (PMR 1042, 2 deaths). The only non-accidental causes of death which were raised were multiple sclerosis (PMR 772, 7 deaths) and infections of the kidney and urinary tract (PMR 540, 3 deaths), although all three deaths from the latter cause occurred at ages over 60 years.

It should be noted that adjustment for social class could not be performed for this occupational group.

049 - POLICE
Although the numbers of deaths were small, excesses were seen in the police for soft tissue malignancies (PMR 526, 3 deaths), melanoma (PMR 421, 6 deaths), pneumococcal pneumonia (PMR 704, 3 deaths) and one death from non-recreational drowning which gave a PMR of 14286. The latter death occurred in a woman under 60 years of age and may have occurred in the course of work as she died attempting to rescue a holiday maker.

051 - LAUNDERERS AND DRY CLEANERS
Excesses were seen for bronchopneumonia (PMR 127, 110 deaths), cholelithiasis and cholecystitis (PMR 209, 10 deaths), polyarteritis nodosa (PMR 462, 4 deaths) and renal failure (PMR 165, 23 deaths). The latter was also raised in men in this occupational group with a PMR of 211 based on 12 deaths.

052 - HAIRDRESSERS
Cirrhosis of the liver (PMR 210, 14 deaths) and 'other alcohol-related diseases' (PMR 211, 26 deaths) were high in this group. There were also excesses of cancer of the oral cavity (PMR 293, 8 deaths) and duodenal ulcers (PMR 202, 12 deaths) both of which may have alcohol as a contributory factor. The only other raised PMRs were for breast cancer (PMR 114, 249 deaths) and multiple sclerosis (PMR 200, 21 deaths). Deficits were seen for cancer of the gall bladder (PMR 17, 1 death), diabetes (PMR 48, 11 deaths) and aortic aneurysm (PMR 50, 9 deaths).

053 - OFFICE WORKERS AND CASHIERS
This occupational category is the largest of all and it is of interest that the only significantly raised PMR over 110 was for multiple sclerosis (PMR 111, 463 deaths). It was also significantly raised for men in this occupational group (PMR 113, 289 deaths). A possible explanation for this is that sufferers from multiple sclerosis are drawn to occupations within this group because they are largely sedentary.

Other marginally-raised PMRs which reached statistical significance were excesses of three cancers, namely breast (PMR 106, 10522 deaths), colon (PMR 105, 2780 deaths) and brain (PMR 107, 1069 deaths). In contrast, ischaemic heart disease was lower than expected (PMR 96, 14535 deaths).

054 - POSTAL WORKERS
Breast cancer was elevated in this group (PMR 122, 110 deaths), but there was a deficit of deaths from aortic valve disorders (no deaths, 3.73 expected) and other cerebrovascular disease (ICD 431-438, PMR 74, 76 deaths).

055 - PETROL PUMP ATTENDANTS
There were excesses of peripheral vascular disease (PMR 828, 4 deaths), cancer of the meninges and meningioma (PMR 1010, 2 deaths) and poisoning by gas and other domestic fuels (PMR 3509, 2 deaths) but the numbers of deaths in each category were small. Mortality from chronic bronchitis and emphysema was also high but based on slightly larger numbers (PMR 240, 8 deaths).

057 - SALES REPRESENTATIVES
Motor vehicle traffic accidents were high in this group (PMR 206, 42 deaths) with the excess being confined to those under 60 years of age. Travelling by car is often a requirement for many of the occupations within this group and so the excess is not surprising. There was also evidence of an excess in men. The only other significantly raised PMRs were for amyloidosis (PMR 748, 3 deaths) and pneumococcal pneumonia (PMR 234, 8 deaths).

058 - SECURITY WORKERS
An excess of carcinoma of the colon (PMR 167, 35 deaths) contrasts in this group with a deficit of stomach cancer (PMR 22, 3 deaths). Motor vehicle traffic accidents were higher than expected (PMR 204, 14 deaths) and, based on only two deaths, there was an excess of poisoning by gas and other domestic fuels (PMR 1042).

059 - COOKS AND KITCHEN PORTERS
Excesses of pancreatic cancer (PMR 118, 203 deaths), stomach cancer (PMR 115, 249 deaths), the miscellaneous group of other cardiomyopathies (ICD 425.0-425.4, 425.6-425.9, PMR 168, 39 deaths) and diabetes (PMR 126, 176 deaths) were evident in this group. Suicide was low (PMR 73, 55 deaths) as too was mortality from cor pulmonale (PMR 40, 7 deaths).

060 - OTHER SERVICE PERSONNEL
There were deficits of multiple sclerosis (PMR 73, 54 deaths), chronic rheumatic heart disease (PMR 84, 332 deaths), ischaemic heart disease (PMR 97, 9805 deaths) and chronic bronchitis and emphysema (PMR 87, 1351 deaths) although the latter two cause groups had PMRs of 106 and 105 respectively before adjustment for social class, though the PMR for chronic bronchitis and emphysema was not signifi-

cantly raised. It is worth noting that this is the second largest occupational group and all these causes of death exhibited excesses (though not necessarily significant) in clerical workers which is the largest group.

061 - HOSPITAL PORTERS AND WARD ORDERLIES
Excesses were seen in this group for lung cancer (PMR 114, 399 deaths), diabetes (PMR 136, 65 deaths), suicide (PMR 132, 62 deaths) and deaths from slipping and tripping (PMR 448, 6 deaths). The latter excess was most marked in those under 60 years of age (PMR 538, 4 deaths) and these deaths may have occurred at work, given the nature of the occupation. There was also an elevated PMR for hepatitis (PMR 261, 8 deaths) which may be due to infections acquired at work.

Deficits were seen for deaths from chronic rheumatic heart disease (PMR 66, 26 deaths) and bronchopneumonia (PMR 74, 58 deaths).

062 - AMBULANCE WORKERS
As for other groups with access to the means, suicide was high (PMR 520, 5 deaths), but in addition lung cancer was also raised (PMR 227, 10 deaths).

068 - LEATHER AND SHOE WORKERS
The PMR for acute bronchitis (PMR 277, 7 deaths) was raised, but there were deficits of deaths from infections of the kidney and urinary tract (PMR 16, 1 death) and motor vehicle traffic accidents (PMR 34, 5 deaths).

069 - PREPARATORY FIBRE PROCESSORS
Not surprisingly there was an excess of byssinosis (PMR 7576, 5 deaths) in this group. Other excesses were cancer of the larynx (PMR 556, 3 deaths) and renal failure (PMR 275, 6 deaths). A deficit of breast cancer (PMR 42, 13 deaths) was observed.

070 - SPINNERS AND WINDERS
An excess of byssinosis (PMR 1511, 5 deaths) was obviously related to the occupation, as too may be the excess of chronic bronchitis and emphysema (PMR 123, 127 deaths). In contrast the PMR for lung cancer was low (PMR 78, 158 deaths).

Amongst other malignancies, an excess of retro-peritoneal cancer was evident (PMR 567, 5 deaths), but there was a deficit of breast cancer (PMR 64, 102 deaths) and the low PMR for ovarian cancer (PMR 74, 43 deaths) was of borderline significance.

Ischaemic heart disease (PMR 124, 805 deaths) and other cerebrovascular disease (ICD 431-438, PMR 113, 289 deaths) were elevated, but the PMR for diabetes (PMR 61, 21 deaths) was low.

071 - WARP PREPARERS AND WEAVERS
Byssinosis was also high in this group (PMR 1338, 4 deaths) as too were chronic bronchitis and emphysema (PMR 135, 152 deaths) and another respiratory disease, bronchopneumonia (PMR 140, 117 deaths). But, as for job group 070, there was also a deficit of lung cancer (PMR 79, 164 deaths).

Also similar to job group 070, there were excesses of ischaemic heart disease (PMR 110, 888 deaths) and other cerebrovascular diseases (ICD 431-438, PMR 116, 374 deaths) and a deficit of breast cancer (PMR 68, 136 deaths).

There was also an excess of acute monocytic leukaemia (PMR 860, 3 deaths), but deficits of myeloma (PMR 47, 7 deaths), dementias (PMR 56, 12 deaths) and other cardiomyopathies (ICD 425.0-425.4, 425.6-425.9, PMR 18, 1 death).

072 - KNITTERS
There was a raised PMR for aortic aneurysm (PMR 181, 15 deaths) in this group but the only other excesses were based on small numbers of deaths. They were for cancer of the skin excluding melanoma (PMR 500, 3 deaths), cancer of the eye (PMR 682, 3 deaths) and bacterial and unspecified meningitis (PMR 743, 4 deaths).

073 - BLEACHERS, DYERS AND FINISHERS
A marked deficit of cancer of the colon was seen in this group (PMR 13, 1 death).

074 - OTHER TEXTILE WORKERS
Byssinosis was particularly high in this group (PMR 955, 25 deaths) and there was also a raised PMR for chronic bronchitis and emphysema (PMR 115, 227 deaths). As in other textile groups, ischaemic heart disease was slightly raised (PMR 113, 1481 deaths), in contrast to a deficit of lung cancer (PMR 74, 300 deaths).

Cor pulmonale (PMR 184, 16 deaths) and systemic lupus erythematosus (PMR 312, 7 deaths) were also high, but there were deficits of mitral valve disorders (no deaths, 6.63 expected) oesophageal cancer (PMR 68, 31 deaths) and 'other alcohol-related diseases' (PMR 44, 6 deaths).

075 - CHEMICAL WORKERS
Cancer of the pleura was high in this group (PMR 498, 4 deaths), probably due to asbestos exposure. Also high were haemolytic anaemia (PMR 1215, 3 deaths), oesophageal varices (PMR 1504, 2 deaths) and deaths due to explosive material (PMR 2353, 2 deaths). The numbers of deaths were small but it is possible that the two deaths from explosions occurred at work as both women were under 60 years of age at death. Deficits were seen for bladder cancer (no deaths, 8.04 expected) and aortic aneurysm (PMR 32, 3 deaths).

076 - BAKERS
Deaths from ischaemic heart disease (PMR 117, 209 deaths) were somewhat higher than expected in this group.

078 - FOOD PROCESSORS
Excesses of gastric ulcer deaths (PMR 179, 17 deaths), non-inguinal hernia (PMR 292, 10 deaths), ischaemic heart disease (PMR 110, 790 deaths) and other cerebrovascular disease (ICD 431-438, PMR 114, 306 deaths) were seen in this group, but deficits of chronic rheumatic heart disease (PMR 57, 17 deaths), motor vehicle traffic accidents (PMR 55, 13 deaths) and suicide (PMR 46, 11 deaths) were evident.

079 - PAPER MANUFACTURERS
The only excess in this group was for breast cancer (PMR 238, 13 deaths).

080 - BOOKBINDERS
Breast cancer deaths (PMR 126, 82 deaths) were higher than expected, but the only other excess in this group was based on three deaths and was for cancer of the meninges and meningioma with a PMR of 563.

081 - PAPER CUTTERS
Myeloma was raised in this group (PMR 1205, 2 deaths) but the number of deaths was very small.

083 - GLASS FORMERS AND DECORATORS
The only excess of note in this group was for sub-arachnoid haemorrhage (PMR 302, 8 deaths).

084 - CERAMICS CASTERS
Unspecified falls were high in this group (PMR 1185, 5 deaths) and one death from silicosis gave a PMR of 6250, which is likely to be due to the occupation. There was also an excess of ischaemic heart disease (PMR 127, 74 deaths)

085 - RUBBER MANUFACTURERS
The miscellaneous group of other cerebrovascular disease (ICD 431-438) was high in this group (PMR 161, 21 deaths).

089 - TOBACCO WORKERS
Two deaths from appendicitis give rise to a PMR of 2020 and there was a deficit of deaths from bronchopneumonia (PMR 32, 3 deaths).

090 - OTHER WOOD AND PAPER PROCESSORS
Dementias (PMR 285, 7 deaths) and biliary cirrhosis (PMR 621, 3 deaths) were high in this occupational group.

091 - OTHER OCCUPATIONS - GLASS AND CERAMIC
There was one death from silicosis in this group giving a PMR of 1667 which was not statistically significantly raised, but there were significant excesses of chronic bronchitis and emphysema (PMR 174, 61 deaths), oesophageal cancer (PMR 182, 15 deaths) and unspecified pneumonia (PMR 408, 5 deaths). There were also excesses of peripheral vascular disease (PMR 244, 12 deaths), other cardiomyopathies (ICD 425.0-425.4, 425.6-425.9, PMR 298, 6 deaths), tuberculosis (PMR 368, 4 deaths) and Addison's disease (PMR 1026, 2 deaths) though the number of deaths for the latter was very small. There was a deficit of ovarian cancer (PMR 62, 17 deaths).

092 - RUBBER GOODS MAKERS
Excesses of bladder cancer in those working with rubber have been documented and this was apparent here (PMR 457, 9 deaths). It is of concern that there was still an excess, particularly as it also occurred in those under 60 years old (PMR 955, 3 deaths) since legislation to control the hazardous exposure was introduced in the 1950s.

093 - PLASTIC GOODS MAKERS
Four deaths from chronic and unspecified myocarditis gave rise to a PMR of 560, but a deficit of suicide was observed (no deaths, 4.45 expected).

094 - COMPOSITORS
There was an excess of breast cancer in this group (PMR 214, 12 deaths).

096 - PRINTING MACHINE MINDERS
In this group there was an excess of cancer of the small intestine (PMR 781, 3 deaths) but a deficit of rectal cancer (PMR 14, 1 death).

097 - PRINTERS (SO DESCRIBED)
Excesses of lung cancer (PMR 161, 47 deaths) and unspecified pneumonia (PMR 680, 3 deaths) occurred in this group.

098 - TAILORS AND DRESSMAKERS
The only raised PMRs in this group were for malignant neoplasms of the breast (PMR 120, 326 deaths) and pleura (PMR 420, 12 deaths). The majority of the latter deaths occurred to women resident in Leeds and this is being examined further. In contrast there was a deficit of lung cancer (PMR 81, 214 deaths).

099 - CLOTHING CUTTERS
An excess of gall bladder cancer occurred in this group (PMR 343, 6 deaths).

100 - SEWERS AND EMBROIDERERS
Excesses of chronic rheumatic heart disease (PMR 145, 138 deaths), pulmonary hypertension (PMR 228, 9 deaths), rheumatoid arthritis (PMR 138, 43 deaths), amyloidosis (PMR 252, 7 deaths) and suicide (PMR 122, 109 deaths) were seen in this group. The PMRs for breast cancer (PMR 109, 835 deaths) and lung cancer (PMR 92, 738 deaths) exhibited the opposite pattern prior to social class adjustment, being 85 and 103 respectively (though only that for breast cancer was significantly low). PMRs of 50 were seen for cancer of the oral cavity (9 deaths) and homicide (8 deaths).

101 - UPHOLSTERERS
Cancer of the thyroid was raised in this group (PMR 645, 3 deaths).

105 - CABINET MAKERS
An excess of chronic bronchitis and emphysema was seen in this group (PMR 371, 5 deaths).

108 - WOODWORKING MACHINISTS
Myeloma (PMR 489, 3 deaths) and cor pulmonale (PMR 1020, 2 deaths) were elevated but based on small numbers of deaths.

109 - OTHER WOODWORKERS
Lung cancer was raised in this group (PMR 209, 10 deaths).

111 - OTHER MAKERS OF PAPER AND GOODS
In this group there were excesses of rheumatoid arthritis (PMR 243, 9 deaths) and brain cancer (PMR 180, 16 deaths), but a deficit of colon cancer (PMR 61, 18 deaths).

116 - MOULDERS AND COREMAKERS (METAL)
There was one death from silicosis in this group giving rise to a PMR of 10000 which was likely to be occupationally related.

117 - ELECTROPLATERS
There was an excess of lung cancer in this group (PMR 273, 6 deaths) which may be related to exposures at work.

120 - OTHER METAL MANUFACTURERS
There were excesses of some circulatory diseases in this group, notably other cerebrovascular disease as defined by ICD codes 431-438 (PMR 181, 20 deaths) and peripheral vascular disease (PMR 488, 3 deaths). An excess of cancer of the larynx was seen (PMR 1124, 2 deaths) though the number of deaths was very small, but there was a deficit of breast cancer (PMR 31, 3 deaths).

122 - CENTRE LATHE TURNERS
Two deaths from cancer of the larynx in this group gave rise to a PMR of 2899.

124 - MACHINE TOOL OPERATORS
Lung cancer (PMR 122, 363 deaths), chronic bronchitis and emphysema (PMR 140, 183 deaths) and peripheral vascular disease (PMR 167, 26 deaths) were high. Also raised was the PMR for acute monocytic leukaemia though the number of deaths was small (PMR 566, 3 deaths). In contrast there were deficits of cancer of the gall bladder (PMR 34, 4 deaths), breast cancer (PMR 83, 223 deaths), motor neurone disease (PMR 37, 4 deaths) and motor vehicle traffic accidents (PMR 53, 12 deaths).

125 - PRESS AND AUTOMATIC MACHINE OPERATORS
There were excesses of two respiratory diseases, namely chronic bronchitis and emphysema (PMR 145, 76 deaths) and viral pneumonia (PMR 486, 3 deaths), as well as raised levels of cancer of the bone (PMR 594, 4 deaths) and hydronephrosis (PMR 1005, 2 deaths) though the numbers were small. A deficit of deaths from aortic aneurysm (PMR 25, 3 deaths) was apparent.

132 - PRODUCTION FITTERS
Two deaths from pleural cancer gave rise to a PMR of 2151, which is likely to be due to asbestos exposure at work, and the same number of deaths gave a PMR of 1299 for non-inguinal hernia.

142 - OTHER ELECTRONIC MAINTENANCE ENGINEERS
Two deaths from cancer of the gall bladder gave a PMR of 1190, but there was a deficit of deaths from ischaemic heart disease (PMR 43, 5 deaths).

143 - ELECTRICAL ENGINEERS (SO DESCRIBED)
Two relatively rare circulatory deaths occurred in this group. One was from mitral valve disorders giving a PMR of 5000 and the other from chronic and unspecified myocarditis with a PMR of 4762.

145 - SHEET METAL WORKERS
Three deaths from cervical cancer in this group gave rise to a PMR of 557.

146 - METAL PLATE WORKERS
Two pleural cancer deaths led to a PMR of 8000 in this group, and these presumably were due to the working environment. There was also a raised PMR for injury undetermined as accidental or purposeful (PMR 1639, 2 deaths), but again there were only two deaths on which this was based.

149 - WELDERS
A raised PMR for stomach cancer was observed (PMR 203, 14 deaths) in this group.

151 - JEWELLERY WORKERS
Diabetes was raised in this group (PMR 388, 4 deaths).

155 - ELECTRONICS WIRE WORKERS
Three deaths from Hodgkin's disease gave rise to a PMR of 1250.

157 - POTTERY DECORATORS
Suicide (PMR 338, 10 deaths) and unspecified falls (PMR 579, 5 deaths) were high in this group, as too were deaths from gastric ulcer (PMR 331, 5 deaths). Deaths from bronchopneumonia were fewer than expected (PMR 35, 4 deaths).

160 - PAINTERS AND DECORATORS NEC
Raised PMRs were seen for motor neurone disease (PMR 709, 3 deaths) and infections of the kidney and urinary tract (PMR 1166, 5 deaths). Two deaths from railway accidents (PMR 7407) occurred, both of them in women under 60 years of age.

161 - ELECTRICAL, ELECTRONIC ASSEMBLERS
There were excesses of chronic myeloid leukaemia (PMR 269, 7 deaths), diabetes (PMR 162, 40 deaths) and asthma (PMR 169, 23 deaths) in this group along with two other disorders with small numbers of deaths, namely Cushing's disease (PMR 1923, 2 deaths) and acute myocarditis (PMR 676, 3 deaths). Deficits of pneumococcal pneumonia (PMR 23, 2 deaths), bronchopneumonia (PMR 64, 27 deaths) and pulmonary embolism and phlebitis (PMR 48, 11 deaths) were observed, and certain violent causes of death also had low PMRs, namely motor vehicle traffic accidents (PMR 44, 8 deaths), suicide (PMR 50, 9 deaths) and injury undetermined as accidental or purposeful (PMR 17, 2 deaths).

162 - INSTRUMENT ASSEMBLERS
Chronic rheumatic heart disease (PMR 729, 5 deaths) and tuberculosis (PMR 3846, 3 deaths) were raised. It is interesting to note that both have an infectious cause but it is difficult to relate this to the occupation and the numbers of deaths involved are very small.

163 - ASSEMBLERS (VEHICLES AND OTHER METAL GOODS)
An excess of other leukaemia (ICD codes 207 and 208) was seen in this group (PMR 680, 3 deaths) in contrast to deficits from aortic aneurysm (PMR 20, 2 deaths) and motor vehicle traffic accidents (no deaths, 8.13 expected).

164 - PACKERS AND SORTERS
Epilepsy (PMR 153, 24 deaths), chronic rheumatic heart disease (PMR 126, 86 deaths), mitral valve disorders (PMR 182, 16 deaths) and other cardiomyopathies (ICD

codes 425.0-425.4 and 425.6-425.9, PMR 168, 31 deaths) were all raised in this group, but they may be a reflection of the type of person who can be recruited into the occupation rather than those who suffer as a result of their work. Excesses were also seen for appendicitis (PMR 317, 6 deaths) and cholelithiasis and cholecystitis (PMR 177, 16 deaths). Deficits of glomerulonephritis (PMR 17, 1 death), peripheral vascular disease (PMR 49, 15 deaths) and motor vehicle traffic accidents (PMR 71, 46 deaths) and homicide (PMR 37, 4 deaths) were seen.

169 - BUILDERS ETC.
Three deaths from bronchopneumonia gave rise to a PMR of 800 in this group.

174 - CONSTRUCTION WORKERS NEC
Colon cancer was raised with a PMR of 601 based on 3 deaths in this group and there were two deaths from pneumococcal pneumonia giving a PMR of 3030.

177 - RAILWAY GUARDS
Two deaths from oesophageal cancer gave rise to a PMR of 4762 in this group.

178 - RAILWAY SIGNAL WORKERS
There was an excess of deaths from ischaemic heart disease (PMR 164, 21 deaths) in this group. One death from a slipping or tripping accident occurred in a woman under 60 years of age, which gave a PMR for the 20-59 age group of 50000. This may have occurred at work.

182 - BUS AND COACH DRIVERS
There was an excess of cancer of the kidney, excluding pelvis, in this group (PMR 706, 3 deaths) although the number of deaths was small.

183 - LORRY DRIVERS
Not surprisingly in this group, deaths from motor vehicle traffic accidents were high (PMR 271, 19 deaths) as too were other accidents (ICD E928, PMR 1460, 2 deaths). There was also an excess of oesophageal disease (PMR 1515, 3 deaths) though the number of deaths was small.

184 - OTHER MOTOR DRIVERS
Excesses arose in this group for lung cancer (PMR 197, 20 deaths), melanoma of the skin (PMR 488, 4 deaths) and rheumatoid arthritis (PMR 733, 3 deaths). There was a deficit of deaths from other cerebrovascular disease (ICD 431-438, PMR 39, 4 deaths)

185 - BUS CONDUCTORS AND DRIVERS' MATES
Certain respiratory and circulatory disorders were high in this group. They were lung cancer (PMR 131, 103 deaths), chronic bronchitis and emphysema (PMR 145, 51 deaths), cor pulmonale (PMR 337, 5 deaths) and acute myocarditis (PMR 1493, 2 deaths).

187 - CRANE DRIVERS
Chronic bronchitis and emphysema was high in this group (PMR 273, 7 deaths).

190 - STOREKEEPERS
Five malignancies were raised in this group, namely breast cancer (PMR 117, 195 deaths), gall bladder cancer (PMR 206, 14 deaths), cancer of the body of the uterus (PMR 173, 23 deaths), Hodgkin's disease (PMR 317, 7 deaths) and chronic myeloid leukaemia (PMR 268, 7 deaths). Two circulatory disorders were high, these being pulmonary embolism and phlebitis (PMR 148, 33 deaths) and portal vein thrombosis (PMR 2247, 2 deaths). Peritonitis was also high (PMR 475, 5 deaths) but dementias were low (PMR 39, 4 deaths).

191 - DOCKERS AND GOODS PORTERS
Based on only two deaths there was a raised PMR for cancer of the small intestine (PMR 4762).

References

1. Martin J, Roberts C. *Women and employment: a lifetime perspective. The report of the 1980 DE/OPCS women and employment survey*. London: HMSO, 1984.
2. Beral V. Long term effects of childbearing on health. *J Epidemiol Community Health* 1985;39:343-346.
3. Roman E, Beral V, Inskip H. Occupational mortality among women in England and Wales. *BMJ* 1985;291:194-6.
4. Hogfoss Rubin C, Burnett CA, Halperin WE, Seligman PJ. Occupation as a risk identifier for breast cancer. *Amer J public Health* 1993;83:1311-5.
5. Office of Population Censuses and Surveys. *Occupational mortality, decenial supplement 1979-80, 1982-83. Part I commentary*. London: HMSO, 1986.
6. Martyn CN, Barker DJP, Osmond C. Motoneuron disease and past poliomyelitis in England and Wales. *Lancet* 1988;i:1319-1322.
7. Martyn CN, Cruddas M, Compston DAS. Symptomatic Epstein-Barr virus infection and multiple sclerosis. *Journal of Neurology, Neurosurgery and Psychiatry* 1993; 56:167-168.
8. Mims CA. Viral aetiology of diseases of obscure origin. *Br Med Bull* 1985;41:63-69.
9. Office of Population Censuses and Surveys. *Occupational mortality, decennial supplement 1970-72*. London: HMSO, 1978.
10. Charlton J, Kelly S, Dunnell K, Evans B, Jenkins R. Suicide deaths in England and Wales: trends in factors associated with suicide deaths. *Population Trends* 1993;71:34-42.
11. Delzell E, Grufferman S. Cancer and other causes of death among female textile workers 1976-78. *JNCI* 1983;71:735-740.
12. International Agency for Research on Cancer. IARC monographs on the evaluation of carcinogenic risks to humans. *Some flame retardants and textile chemicals, and exposures in the textile manufacturing industry*. Volume 48. Lyon: IARC, 1990.
13. Meridith S. Reported incidence of occupational asthma in the United Kingdom, 1989-90. *J Epidemiol Comm Health* 1993;47:459-63.
14. International Agency for Research on Cancer. IARC monographs on the evaluation of the carcinogenic risk of chemicals to humans. *The rubber industry*. Volume 28. Lyon: IARC, 1982.
15. International Agency for Research on Cancer. IARC monographs on the evaluation of carcinogenic risks to humans. *Chromium, nickel and welding*. Volume 49. Lyon: IARC, 1990.
16. The Registrar General. *Decennial supplement for England and Wales 1961. Occupational mortality tables*. London: HMSO, 1971.
17. Munoz N, Bosch FX, Jensen OM (eds). *Human papillomavirus and cervical cancer*. IARC Scientific Publications No 94, Lyon: IARC, 1989.

18 Beral V. Cancer of the cervix: a sexually transmitted infection? *Lancet* 1974;i:1037-1040.

19 Robinson J. Cervical cancer; occupational risks. *Lancet* 1983;ii:1496-7.

20 Robinson J. Occupational risks of husbands and wives and possible preventive strategies. In: Jordan JA, Sharp F, Singer A (eds). *Pre-clinical neoplasia of the cervix: Proceedings of the ninth study group of the Royal College of Obstetricians and Gynaecologists* (October 1981) London: RCOG, 1982:11-27.

21 Zakelj MP, Fraser P, Inskip H. Cervical cancer and husband's occupation. *Lancet* 1984;i:510.

22 *The Health of the Nation. A strategy for health in England.* ISBN: 0101198620. London: HMSO, 1992.

Chapter 6 Occupational mortality by cause of death

David Coggon
Hazel Inskip
Paul Winter
Brian Pannett

MRC Environmental Epidemiology Unit, University of Southampton

6.1 Introduction

As well as looking at patterns of mortality in each job separately as in Chapters 4 and 5, it is also helpful to place the findings for different jobs alongside each other. Often hazards are not specific to a single occupation, and their existence may only become apparent when several jobs which share an exposure are all found to have high mortality from the same cause of death. For hazards that are already recognised, examination of death rates across the range of occupations allows the full scale of resultant mortality to be estimated, and helps to define which occupations have the highest risk. Also, comparison of mortality in different jobs can give clues to non-occupational causes of disease, related to aspects of lifestyle. For example, occupations with high mortality from alcohol-related diseases also tend to have high mortality from falls on stairs, suggesting that alcohol may play a part in causing such falls.

This chapter examines mortality across different job groups for selected causes of death. Because the interest is in both occupational and non-occupational causes of disease, the PMRs presented are adjusted only for age and not for social class. Unless otherwise stated they are for men aged 20-74. Where relevant, PMRs are also given for women (again aged 20-74), but in many cases the numbers of deaths among women in relevant occupations were too few for meaningful analysis.

6.2 Cancers

6.2.1 Cancer of the stomach (ICD 151)

In the past cancer of the stomach has been linked with jobs that entail exposure to inorganic dusts, in particular coal mining and rubber manufacture.[1-3] Face-trained coal miners (PMR 132, 326 deaths) and rubber manufacturers (PMR 119, 57 deaths) did have above average mortality from stomach cancer, but much more striking was the relation of the disease to social class. Occupations at the top of the PMR ranking were all manual jobs, while those at the bottom were predominantly white collar (Tables 6.1 and 6.2). The steep social class gradient in stomach cancer incidence has been recognised for many years. The causes of the disease include infection by the bacterium, H pylori,[4] and low consumption of fresh fruit and vegetables.[5] Both of these risk factors are likely to be more common in the manual classes.

Table 6.1 Job groups with highest and lowest PMRs for stomach cancer (ICD 151) - men aged 20-74, England and Wales, 1979-80, 1982-90

	Job group	Deaths	PMR	95% CI
106	Case and box makers	30	164	110-234
148	Scaffolders	59	154	118-199
126	Metal polishers	75	142	111-178
111	Other makers of paper goods	49	141	104-186
051	Launderers and dry cleaners	57	139	105-180
127	Fettlers and dressers (metal)	55	137	103-178
175	Face trained coalminers	326	132	118-147
024	Literary and artistic occupations	164	60	51-70
015	Doctors	59	57	43-73
016	Dentists	13	56	30-95
005	Computer programmers	13	56	30-95
131	Watch and clock makers	15	54	30-90
001	Lawyers	36	49	34-68
018	Pharmacists	21	44	27-68

Notes 1 PMRs are standardised for age but not for social class.
2 Only job groups with PMRs significantly ($p<0.05$) different from 100 are listed.

Table 6.2 Job groups with highest and lowest PMRs for stomach cancer (ICD 151) - women aged 20-74, England and Wales, 1979-80, 1982-90

	Job group	Deaths	PMR	95% CI
102	Carpet fitters	2	847	103-3061
077	Brewery workers	6	290	106-631
073	Bleachers, dyers and finishers	10	224	108-412
149	Welders	14	222	121-373
059	Cooks and kitchen porters	249	126	111-143
046	Caterers	376	124	112-137
060	Other service occupations	874	118	110-126
044	Retailers and dealers	495	91	83-100
053	Office workers and cashiers	1175	87	82-92
011	Teachers nec	194	82	71-94
024	Literary and artistic occupations	24	64	41-95
058	Security workers	3	24	5-70

Notes 1 PMRs are standardised for age but not for social class.
2 Only job groups with PMRs significantly ($p<0.05$) different from 100 are listed.

6.2.2 Cancers of the colon and pancreas (ICD 153, 157)

Both of these diseases showed an association with social class in the opposite direction to stomach cancer. Professional and administrative occupations featured high in the PMR rankings while the jobs with the lowest PMRs were manual. The reasons for this are unclear, but nutritional differences may have an influence. Smoking is a known cause of pancreatic cancer,[6] but rates of smoking tend to be lower in the professions.

6.2.3 Cancer of the nose (ICD 160)

Cancer of the nose and nasal sinuses is known to be caused by wood dust, especially from hardwoods, and by dust from vegetable-tanned leather.[1,7] Table 6.3 shows that mortality in occupations which entail exposure to wood and leather dusts was elevated, but it is unclear whether this reflects recent exposure or a latent effect of exposures in the past before controls were introduced to reduce dust levels.

Table 6.3 Mortality from cancer of the nose and nasal sinuses (ICD 160) in occupations which entail exposure to wood or leather dust - men aged 20-74, England and Wales 1979-80, 1982-90

	Job group	Deaths	PMR	95% CI
068	Leather and shoe workers	7	257	103 -530
104	Carpenters	19	138	83 -216
105	Cabinet makers	8	568	245 -1119
106	Case and box makers	2	554	67 -2001
107	Pattern makers	0	0	0 -608
108	Woodworking machinists	8	386	167 -761
109	Other woodworkers	2	330	40 -1192

Note PMRs are standardised for age but not for social class.

It is noteworthy that several other occupations with high rates of nasal cancer involve exposure to other biological dusts, in particular, paper cutters (PMR 897, 2 deaths), brewery workers (PMR 343, 3 deaths) and rubber manufacturers (PMR 312, 3 deaths).

6.2.4 Cancer of the trachea, bronchus and lung (ICD 162)

Several of the occupations high in the PMR rankings for cancer of the bronchus entail exposure to known lung carcinogens - in particular, electroplaters (PMR 137, 134 deaths for men and PMR 273, 6 deaths for women) who are exposed to chromates, and labourers in coke ovens (PMR 128, 156 deaths for men) who have high exposure to polycyclic aromatic hydrocarbons. In general, however, the effects of occupational hazards are difficult to discern because of the strong confounding effects of smoking and because many hazards, e.g. bischloromethyl ether and chromates in chemical workers, affect only a small subset of the relevant job group. Any effects are therefore diluted. Most occupations with high rates of asbestos-related disease (see section 6.10.4) also had elevated mortality from cancer of the bronchus, but they were by no means the highest jobs in the rankings.

6.2.5 Melanoma of skin (ICD 172)

Melanoma is of interest because it is one of the few cancers whose incidence is currently rising.[4] This increase has been attributed to changing patterns of exposure to sunlight. A causal role for solar ultraviolet radiation is supported by the geographical distribution of the tumour with higher rates in white people living at lower latitudes, and by the increased risk that has been demonstrated in individuals with fair skin and a history of sunburn.[8] However, it has been observed that occupations with the highest exposure to sunlight, e.g. agricultural workers and fishermen, do not have unusually high mortality from melanoma. This has led to the hypothesis that risk arises from intermittent exposure to sun while prolonged exposure produces a protective tan.

In the current analysis, the occupations with highest PMRs for melanoma were predominantly indoor jobs, but it is striking that they were also mainly professions (Table 6.4). In contrast, jobs such as face-trained coal miner and other coal miner, which if anything involve even less exposure to sun, had low mortality from the disease (PMRs 49 and 52 respectively). It may be that the professional classes take more frequent holidays in the sun, but the explanation could lie in other risk factors, perhaps nutritional. Another possibility is that exposure to sunlight in childhood is more important than that occurring later in life.

Table 6.4 Job groups with the highest PMRs for melanoma of skin (ICD 172) - men and women aged 20-74, England and Wales, 1979-80, 1982-90

	Job group	Deaths	PMR	95% CI
Men				
035	Air traffic controllers	4	780	212 -1996
004	Economists and statisticians	8	589	254 -1161
034	Aircraft flight deck officers	8	384	166 -757
028	Physical scientists and mathematicians	18	255	151 -403
001	Lawyers	22	254	159 -385
002	Accountants	86	244	195 -302
011	Teachers nec	105	243	198 -294
033	Architects and surveyors	49	239	177 -317
003	Personnel managers etc.	24	230	147 -343
010	Teachers in higher education	29	221	148 -318
Women				
025	Persons involved in sport	3	545	112 -1594
049	Police	6	505	185 -1098
184	Other motor drivers	4	391	107 -1002
021	Health professions nec	12	240	124 -419
009	Other administrators	30	183	123 -261
011	Teachers nec	134	170	143 -202

Notes 1 PMRs are standardised for age but not for social class.
2 Only job groups with PMRs significantly ($p<0.05$) different from 100 are listed.

6.2.6 Cancers of the female reproductive system (ICD 174, 180, 182, 183)

Table 6.5 shows mortality for job groups in which the PMRs for at least two of the three female reproductive cancers - breast, ovary and cervix - were significantly different from 100. The most striking feature of the table is that PMRs for breast and ovarian cancer tended to be high or low together, while PMRs for cervical cancer showed an opposite trend. The occupations with high rates of breast and ovarian cancer tended to be professional and administrative, but adjustment for social class did not entirely eliminate the relationship.

The explanation may lie partly in differences in sexual behaviour. Both breast and ovarian cancer are less common in women who bear children at an early age,[9,10] while early sexual activity is associated with an increased incidence of cervical cancer.[11] Nutritional differences could also have an influence.

As in the 1971 Decennial Supplement, cancer of the body of the uterus followed a similar pattern to cancers of the breast and ovary. Eight job groups had PMRs for cancer of the body of the uterus significantly different from 100, including four listed in Table 6.5 - teachers nec (PMR 147, 140 deaths), clergy (PMR 210, 12 deaths), other service occupations

Table 6.5 Mortality from cancers of the breast (ICD 174), ovary (ICD 183) and cervix (ICD 180) - women aged 20-74, England and Wales, 1979-80, 1982-90

	Job group	Cancer of breast			Cancer of the ovaries			Cancer of the cervix		
		Deaths	PMR	95% CI	Deaths	PMR	95% CI	Deaths	PMR	95% CI
003	Personnel managers etc.	95	125	102-153	38	161	114-221	11	77	39-138
006	Sales managers etc	173	123	105-142	73	166	130-209	29	107	72-154
009	Other administrators	436	132	120-145	143	137	116-162	43	76	55-103
010	Teachers in higher education	167	160	137-186	47	148	108-197	12	65	33-113
011	Teachers nec	2404	151	145-157	740	148	138-159	125	47	39-56
014	Clergy	90	127	102-156	47	197	144-262	6	58	21-126
015	Doctors	98	125	102-153	24	96	62-144	3	21	4-62
020	Physiotherapists	67	140	108-178	28	185	123-267	6	75	27-162
028	Physical scientists and mathematicians	25	155	100-230	11	227	113-407	4	101	28-260
031	Draughtspersons	44	167	121-224	18	224	133-354	2	37	4-133
045	Publicans and bar staff	481	77	70-84	149	74	62-87	159	156	132-182
046	Caterers	1389	81	77-86	478	84	76-92	311	120	107-134
051	Launderers and dry cleaners	237	71	63-81	90	80	64-98	49	101	74-133
053	Office workers and cashiers	10522	120	118-123	3286	116	112-120	1252	85	81-90
059	Cooks and kitchen porters	896	84	78-89	303	84	75-94	175	111	95-128
060	Other service personnel	3178	76	74-79	1139	82	77-87	717	116	108-125
070	Spinners and winders	102	50	41-61	43	60	44-82	25	88	57-130
074	Other textile workers	318	71	64-80	114	75	62-90	74	114	89-143
100	Sewers and embroiderers	835	85	79-91	294	91	81-102	185	118	101-136
124	Machine tool operators	223	65	57-74	95	82	66-100	71	142	111-179
125	Press and automatic machine operators	100	74	60-90	25	55	36-81	29	145	97-208
161	Electrical, electronic assemblers	168	78	67-91	72	101	79-127	48	146	107-193
164	Packers and sorters	525	73	67-80	198	83	72-95	132	118	99-140
185	Bus conductors and drivers' mates	65	75	58-95	17	57	33-91	19	155	93-242

Notes 1 PMRs are standardised for age but not for social class.
2 The job groups listed are those for which PMRs were significantly ($p<0.05$) different from 100 for two or more of these cancers.

(PMR 84, 252 deaths) and machine tool operators (PMR 59, 15 deaths).

Because of the potential confounding effects of sexual behaviour and nutrition, it is difficult to identify potential occupational causes for these diseases. However, clues may arise where the normal pattern is broken. Thus, for example, the PMR for breast cancer in farmers was significantly low (PMR 82, 194 deaths) while that for ovarian cancer was elevated (PMR 120, 94 deaths). The PMR for cervical cancer was raised (PMR 117, 44 deaths), but not significantly so. This excess of ovarian cancer where a deficit would have been expected is suspicious, and would be worth pursuing if found also in other studies.

6.2.7 Cancer of the prostate (ICD 185)

Little is known about the causes of prostatic cancer, but previous studies have suggested an increased risk in farmers[12] and in men exposed to radiation in the nuclear industry.[13] In the current analysis, agricultural workers had moderately elevated rates (PMR 117, 1361 deaths), and professional occupations tended to feature high in the PMR rankings.

6.2.8 Urothelial cancer (ICD 188, 189.1-189.8)

Certain aromatic amines, which used to be handled in the dyestuffs and rubber industries, are potent causes of bladder and other urothelial cancer.[1,14] The most potent carcinogens of this group were withdrawn from use in the 1950s, but because they act with long latency, disease has continued to occur in excess in workers exposed before the ban came into force.

In the current analysis male rubber manufacturers (PMR 139, 31 deaths) and rubber goods makers (PMR 137, 24 deaths for men and PMR 434, 9 deaths for women) featured high in the ranking of PMRs, with excesses that were apparent even in men and women below age 65. These excesses could represent a delayed effect of exposure in the 1950s, but will require continued monitoring to check that they do not persist.

6.2.9 Cancer of the renal parenchyma (ICD 189.0)

There are no established occupational causes of renal cancer, although a link with hydrocarbon compounds has been proposed.[15] PMRs tended to be higher in professional occupations, but the reasons for this are unclear. Smoking, which is a probable cause of the disease,[9,16] is less common in the professions than in manual occupations.

6.2.10 Cancer of the eye (ICD 190)

Cancer of the eye is a rare cause of death and has been little studied. It is striking that two of the jobs with the highest PMRs were woodworking occupations - cabinet makers (PMR 398, 4 deaths) and carpenters (PMR 193, 19 deaths). This may be a chance finding, but should be borne in mind when future analyses are carried out.

6.2.11 Cancer of the brain (ICD 191)

Like melanoma, the incidence of brain cancer is rising.[4] The reasons for this are unknown. The best established cause of brain tumours is ionising radiation,[17,18] but this would not explain the trend. Many different occupations have been linked with an increased risk of brain cancer in isolated studies, but the most consistent association has been with 'electrical' occupations.[19] It has been proposed that low frequency magnetic fields are a cause of brain cancer and might explain this association, but it is not clear that 'electrical' occupations necessarily have the highest exposure to such fields.

Table 6.6 shows the PMRs from brain cancer that were found in electrical jobs. Several were above average, but much more striking was the high rate of brain cancer in professional and administrative occupations. (See Table 6.7). This suggests a cause related in some way to affluence, perhaps nutritional or infectious.

Table 6.6 Mortality from brain cancer (ICD 191) in electrical occupations — men aged 20-74, England and Wales, 1979-80, 1982-90

	Job group	Deaths	PMR	95% CI
029	Electrical and electronic engineers (professional)	65	165	127-210
136	Electrical and electronic production fitters	12	102	53-178
137	Electricians	204	111	96-127
138	Electrical plant operators	16	80	46-129
139	Telephone fitters	86	141	113-174
140	Electric cable and line workers	11	76	38-135
141	Radio and TV mechanics	23	111	70-166
142	Other electronic maintenance engineers	39	149	106-203
143	Electrical engineers (so described)	67	108	84-138

Note PMRs are standardised for age but not for social class

Table 6.7 Job groups with highest PMRs for brain cancer (ICD 191) - men and women aged 20-74, England and Wales, 1979-80 and 1982-90

Job group	Deaths	PMR	95% CI
Men			
156 Coil winders	10	344	165-633
028 Physical scientists and mathematicians	65	292	225-372
010 Teachers in higher education	116	269	222-322
016 Dentists	17	206	120-330
027 Chemical engineers and scientists	46	197	144-263
001 Lawyers	53	197	147-258
033 Architects and surveyors	127	193	161-230
011 Teachers nec	257	186	164-210
005 Computer programmers	43	183	132-247
012 Vocational trainers, social scientists, etc.	66	181	140-231
008 Government administrators	70	179	140-226
Women			
006 Sales managers etc	29	181	121-260
010 Teachers in higher education	19	175	105-273
011 Teachers nec	252	155	136-175
041 Office managers	38	144	102-198

Notes 1 PMRs are standardised for age but not for social class.
2 Only job groups with significantly ($p<0.05$) elevated PMRs are listed.

6.2.12 Lymphoma, myeloma, leukaemia, aplastic anaemia, agranulocytosis (ICD 200-208, 284, 288.0)

Several diseases within this group share similar causes. Thus, ionising radiation causes acute leukaemia and chronic myeloid leukaemia;[20] benzene causes aplastic anaemia, acute myeloid leukaemia, and possibly other leukaemias and myeloma;[21] viral infections have been implicated in the aetiology of Hodgkin's disease, non-Hodgkin's lymphoma and the leukaemias;[22] and low frequency magnetic fields are suspected of causing various leukaemias, especially in 'electrical' occupations.[19]

The occupational hazards from benzene and ionising radiation have been recognised and controlled for many years. Moreover, job title as classified in this analysis, is not a specific index of exposure to these agents. Thus, the analysis would not be expected to show an effect from them.

Table 6.8 gives the mortality from different leukaemias and lymphomas in electrical occupations. PMRs were most consistently elevated for Hodgkin's disease, myeloma and acute myeloid leukaemia.

Also of note is the high mortality from lymphatic and haematopoietic cancer among teachers, both in schools and in higher education (see Table 6.9). In addition, male teachers had increased rates of aplastic anaemia and agranulocytosis and in females aplastic anaemia was raised although not significantly at the 5 per cent level (PMR 128, 10 deaths). Social class adjustment reduced these excesses in men but only partially, whilst in women the PMR for aplastic anaemia rose to 151. Such excesses are consistent with an aetiological role of infections acquired as an adult through frequent contact with large numbers of young people.

6.3 Diabetes (ICD 250)

Diabetes is only recorded as the underlying cause of death in a small proportion of diabetic people, and mortality is not a reliable index of incidence. Furthermore, there are several types of diabetes with distinct causes. Nevertheless it is of interest that among men, three job groups concerned with the manufacture and repair of clothing appear near the top of the mortality ranking (see Table 6.10). One possible explanation for this finding is that the three job groups, tailors and dressmakers, sewers and embroiderers, and other workers with fabrics, include a high proportion of immigrants from the Indian sub-continent, among whom the prevalence of diabetes is known to be unusually high.[23] Death certificates were reviewed for men from these occupations who died from diabetes during 1982-90, and of 57 men, 27 had been born in the UK, 16 in India or Pakistan, and 14 in other countries. No parallel excess of diabetes was observed in women from the same jobs.

6.4 Viral hepatitis and immunodeficiency (ICD 070, 279.1)

Viral hepatitis and immunodeficiency are considered together because they share risk factors. Both are more prevalent in homosexual men and both are transmitted by sharing of needles among intravenous drug abusers.[24,25] These links may explain why certain job groups were high in the PMR rankings for both of the disorders. (See Table 6.11).

The occupation with highest mortality from hepatitis in men was doctors (PMR 916, 14 deaths). Death certificates were reviewed for doctors who died of hepatitis during 1982-90, and only five out of 12 had been born in the UK. In particular, four had been born in Africa. This suggests that the excess may be attributable more to persistence of infection acquired early in life than to injury at work by sharp instruments or tissues contaminated by infected blood.

The numbers of women dying from these causes were too few for meaningful analysis.

Table 6.8 Mortality from lymphoma and leukaemia in electrical occupations - men aged 20-74, England and Wales, 1979-80, 1982-90

Job group	Non-Hodgkin's lymphoma (ICD 200,202)			Hodgkin's disease (ICD 201)			Myeloma (ICD 203)			Acute lymphatic leukaemia (ICD 204.0)		
	Deaths	PMR	95% CI	Deaths	PMR	95% CI	Deaths	PMR	95% CI	Deaths	PMR	95% CI
029 Electrical and electronic engineers (professional)	59	190	145- 245	17	169	98- 270	24	167	107- 249	5	125	41- 292
136 Electrical and electronic production fitters	7	69	28- 141	4	187	51- 478	7	121	48- 248	3	374	77- 1092
137 Electricians	137	89	75- 105	51	128	95- 168	97	120	97- 146	14	89	49- 149
138 Electrical plant operators	14	77	42- 129	3	93	19- 271	17	155	90- 248	0	0	0- 316
139 Telephone fitters	73	138	108- 174	18	147	87- 233	33	113	78- 159	6	132	48- 287
140 Electric cable and line workers	10	77	37- 142	3	115	24- 337	10	134	64- 247	0	0	0- 382
141 Radio and TV mechanics	19	113	68- 176	5	107	35- 250	5	59	19- 137	1	59	1- 328
142 Other electronic maintenance engineers	32	155	106- 219	12	184	95- 321	13	134	71- 228	4	156	43- 401
143 Electrical engineers (so described)	68	123	96- 156	18	163	96- 257	47	146	108- 195	6	148	54- 322

Job group	Chronic lymphatic leukaemia (ICD 204.1)			Acute myeloid leukaemia (ICD 205.0)			Chronic myeloid leukaemia (ICD 205.1)		
	Deaths	PMR	95% CI	Deaths	PMR	95% CI	Deaths	PMR	95% CI
029 Electrical and electronic engineers (professional)	4	76	21- 195	11	75	38- 135	13	198	105- 338
136 Electrical and electronic production fitters	4	179	49- 459	9	206	94- 391	1	55	1- 306
137 Electricians	28	91	61- 132	83	121	96- 149	25	84	54- 124
138 Electrical plant operators	1	23	1- 129	9	118	54- 224	3	96	20- 280
139 Telephone fitters	9	80	37- 152	24	104	66- 154	13	131	70- 224
140 Electric cable and line workers	1	35	1- 193	7	127	51- 261	2	86	10- 310
141 Radio and TV mechanics	4	126	34- 323	11	145	72- 259	2	59	7- 214
142 Other electronic maintenance engineers	4	112	31- 287	17	177	103- 284	10	232	111- 426
143 Electrical engineers (so described)	11	88	44- 157	31	131	89- 187	12	121	63- 212

Note PMRs are standardised for age but not for social class.

Table 6.9 Mortality from lymphatic and haematopoietic cancer, aplastic anaemia and agranulocytosis in teachers - men and women aged 20-74, England and Wales, 1979-80, 1982-90

Cause of death (ICD)	Teachers in higher education (job group 010)						Teachers nec (job group 011)					
	Men			Women			Men			Women		
	Deaths	PMR	95% CI	Deaths	PMR	95% CI	Deaths	PMR	95% CI	Deaths	PMR	95% CI
Non-Hodgkin's lymphoma (200,202)	69	198	154 -250	9	111	51- 211	201	172	149- 197	180	138	118- 160
Hodgkin's disease (201)	10	131	63 -242	1	63	2- 351	44	157	114- 212	45	185	135- 248
Myeloma (203)	38	205	145 -281	7	161	65- 332	107	169	139- 205	95	125	101- 153
Acute lymphatic leukaemia (204.0)	6	242	89 -526	0	0	0- 416	17	175	102- 280	17	131	76- 209
Chronic lymphatic leukaemia (204.1)	16	233	133 -379	3	405	83- 1183	35	145	101- 202	21	151	93- 231
Acute myeloid leukaemia (205.0)	26	177	116 -260	9	186	85- 354	85	165	132- 204	94	128	104- 157
Chronic myeloid leukaemia (205.1)	14	220	120 -368	4	226	62- 578	40	177	127- 242	38	140	99- 193
Acute monocytic leukaemia (206.0)	0	0	0 -718	0	0	0- 2493	2	110	13- 398	6	238	87- 519
Other leukaemia (207,208)	4	188	51 -480	1	239	6- 1333	5	65	21- 152	10	142	68- 262
Aplastic anaemia (284)	2	129	16 -467	1	221	6- 1230	18	322	191- 508	10	128	62- 236
Agranulocytosis (288.0)	0	0	0 -1716	0	0	0- 3617	4	556	152- 1424	1	61	2- 341

Note PMRs are standardised for age but not for social class.

Table 6.10 **Mortality from diabetes (ICD 250) in occupations concerned with the manufacture and repair of clothing - men aged 20-74, England and Wales, 1979-80, 1982-90**

	Job group	Deaths	PMR	95% CI
098	Tailors and dressmakers	49	205	152 -271
100	Sewers and embroiderers	12	190	98 -331
103	Other workers with fabrics	14	175	96 -294

Note PMRs are standardised for age but not for social class.

Table 6.11 **Job groups with highest PMRs for viral hepatitis and immunodeficiency - men aged 20-74, England and Wales 1979-80, 1982-90**

	Job group	Viral hepatitis (ICD 070)			Immunodeficiency (ICD 279.1)		
		Deaths	PMR	95% CI	Deaths	PMR	95% CI
001	Lawyers	4	347	94- 887	12	313	162- 547
006	Sales managers etc	5	87	28- 203	39	244	173- 333
008	Government administrators	1	67	2- 372	9	306	140- 582
009	Other administrators	4	145	39- 370	23	353	224- 531
010	Teachers in higher education	0	0	0- 210	16	363	207- 589
011	Teachers nec	9	156	71- 295	53	308	231- 403
012	Vocational trainers, social scientists etc.	2	135	16- 489	8	233	101- 459
014	Clergy	4	376	102- 962	7	330	133- 679
015	Doctors	14	916	501- 1537	5	115	37- 269
017	Nurses	4	213	58- 546	20	312	191- 482
018	Pharmacists	3	610	126- 1782	1	84	2- 468
024	Literary and artistic occupations	15	284	159- 468	140	725	610- 855
041	Office managers	4	75	21- 193	30	236	159- 338
045	Publicans and bar staff	12	176	91- 308	41	223	160- 303
046	Caterers	14	392	214- 658	62	559	429- 718
052	Hairdressers	4	405	110- 1037	31	992	674- 1410
053	Office workers and cashiers	38	118	84- 162	157	224	191- 262
059	Cooks and kitchen porters	15	394	221- 650	31	245	167- 349
061	Hospital porters and ward orderlies	7	338	136- 695	5	118	38- 274
074	Other textile workers	8	323	140- 637	2	41	5- 147
098	Tailors and dressmakers	2	331	40- 1194	5	524	170- 1222

Notes 1 PMRs are standardised for age but not for social class.
2 The job groups listed are those with significantly ($p<0.05$) elevated PMRs >200 for either disease.

6.5 Tuberculosis (ICD 010-018, 137)

Tuberculosis can be a complication of immunodeficiency,[26] and several of the occupations with high death rates from immunodeficiency also had high rates of tuberculosis - in particular, tailors and dressmakers, and cooks and kitchen porters (see Table 6.12). The jobs with the highest mortality from tuberculosis, however, were those which entail exposure to silica. This reflects the known tendency for tuberculosis to complicate silicosis.[27]

Table 6.12 **Job groups with the highest PMRs for tuberculosis (ICD 010-018,137) - men aged 20-74, England and Wales, 1979-80, 1982-90**

	Job group	Deaths	PMR	95% CI
176	Miners (not coal) and quarry workers	24	495	317-738
131	Watch and clock makers	8	396	171-781
128	Shot blasters	4	376	102-962
127	Fettlers and dressers (metal)	11	375	187-671
098	Tailors and dressmakers	16	357	204-580
070	Spinners and winders	7	287	115-591
099	Clothing cutters	6	274	100-596
073	Bleachers, dyers and finishers	9	242	111-459
059	Cooks and kitchen porters	34	216	149-301
036	Seafarers	32	189	129-268

Notes 1 PMRs are standardised for age but not for social class.
2 Only job groups with significantly ($p<0.05$) elevated PMRs are listed.

6.6 Drug dependence (ICD 304)

The jobs with highest mortality from drug dependence were mainly from the construction industry (see Table 6.13). This may reflect selection of drug abusers into a casual labour market, but the possibility of drug abuse spreading through social pressures at work should not be ruled out.

Table 6.13 **Job groups with the highest PMRs for drug dependence (ICD 304) - men aged 20-74, England and Wales, 1979-80, 1982-90**

	Job group	Deaths	PMR	95% CI
148	Scaffolders	7	355	143-731
160	Painters and decorators nec	38	299	212-411
168	Roofers and glaziers	13	242	129-415
024	Literary and artistic occupations	24	230	147-343
174	Construction workers nec	34	203	141-284

Notes 1 PMRs are standardised for age but not for social class.
2 Only job groups with significantly ($p<0.05$) elevated PMRs are listed.

6.7 Parkinson's disease, motor neurone disease and multiple sclerosis (ICD 332, 335.2, 340)

These degenerative diseases of the nervous system all have been postulated to occur as a late effect of earlier infection. For example, it has been proposed that motor neurone disease

Table 6.14 Mortality from Parkinson's disease, motor neurone disease and multiple sclerosis - men and women aged 20-74, England and Wales, 1979-80, 1982-90

Cause of death (ICD)	Teachers in higher education (job group 010)						Teachers nec (job group 011)					
	Men			Women			Men			Women		
	Deaths	PMR	95% CI	Deaths	PMR	95% CI	Deaths	PMR	95% CI	Deaths	PMR	95% CI
Parkinson's disease (332)	21	200	124-306	0	0	0-222	90	199	160-244	47	136	100-181
Motor neurone disease (335.2)	30	226	152-323	3	104	21-304	69	159	123-201	70	145	113-183
Multiple sclerosis (340)	17	178	104-285	7	174	70-358	51	169	126-222	105	180	148-218

Note PMRs are standardised for age but not for social class.

might be a delayed manifestation of subclinical polio infection,[28] and multiple sclerosis has been linked with delayed, into adolescence or early adult life, infection by Epstein-Barr virus.[29] In addition, it is known that Parkinson's disease can be induced by the chemical, MPTP (methyl phenyl tetrahydropyridine),[30] and it is possible that other toxins also have a causal role.

Mortality patterns from Parkinson's disease did not point to an obvious toxic cause, but it was notable that teachers both in schools and in higher education, had high death rates from all of these neurological disorders (see Table 6.14). This would be compatible with an infectious aetiology.

6.8 Epilepsy (ICD 345)

Deaths from epilepsy usually occur through injury resulting from loss of consciousness during fits, and a history of seizures may lead to exclusion from jobs where the risks of such injury, both to the worker and others, are particularly high. Selection of this kind may explain the low mortality from epilepsy in occupations such as roofers and glaziers (PMR 45, 6 deaths), the armed forces (PMR 43, 15 deaths) and police (PMR 42, 7 deaths). Of the jobs with high mortality, farmers (PMR 190, 148 deaths) give most cause for concern. The excess of deaths was confined to men of working age and may well have resulted from injury through fits at work. If so, exclusion of known epileptics from the most dangerous farming activities might reduce mortality.

6.9 Diseases of the circulatory system

6.9.1 Ischaemic heart disease (ICD 410-414)

Ischaemic heart disease was the most common of the causes of death examined, accounting for more than a third of deaths in men and some 23 per cent in women. Chance variation in mortality between occupations was therefore relatively low, and the highest PMR recorded in men was only 120 (in clergy). The ranking of PMRs did not point to any obvious occupational hazards and the jobs at the top of the ranking were not those that would be considered unusually stressful (see Table 6.15). Nor were they all sedentary occupations.

6.9.2 Aortic aneurysm and peripheral vascular disease (ICD 441, 443, 557)

These disorders result from arterial atheroma and might be expected to have similar causes. Smoking, for example, increases the risk of both.[31,32] It is therefore remarkable that

Table 6.15 Job groups with highest PMRs for ischaemic heart disease (ICD 410-414) - men and women aged 20-74, England and Wales, 1979-80, 1982-90

	Job group	Deaths	PMR	95% CI
Men				
014	Clergy	1696	120	114-126
107	Pattern makers	516	119	109-130
181	Road transport inspectors	1121	115	109-122
094	Compositors	817	113	106-121
031	Draughtspersons	2502	113	108-117
040	Managers in transport, utilities and mining	6853	113	110-115
023	Driving instructors	787	112	104-120
Women				
178	Railway signalworkers	21	174	108-267
084	Ceramics casters	74	136	107-170
137	Electricians	54	133	100-173
070	Spinners and winders	805	130	121-139
076	Bakers	209	124	108-142
074	Other textile workers	1481	123	116-129
069	Preparatory fibre processors	157	121	103-142

Notes 1 PMRs are standardised for age but not for social class.
2 Only job groups with significantly ($p<0.05$) elevated PMRs are listed.

in men there was an inverse relation between occupational mortality from the two diseases (Figure 6.1). Aortic aneurysm was relatively more common in professional and administrative occupations while peripheral vascular disease mainly afflicted men in manual jobs. The explanations for this are unclear and it should be a spur to further investigation.

Figure 6.1 Proportional mortality ratios for aortic aneurysm and peripheral vascular disease by occupation - men aged 20-74, England and Wales, 1979-80 and 1982-90

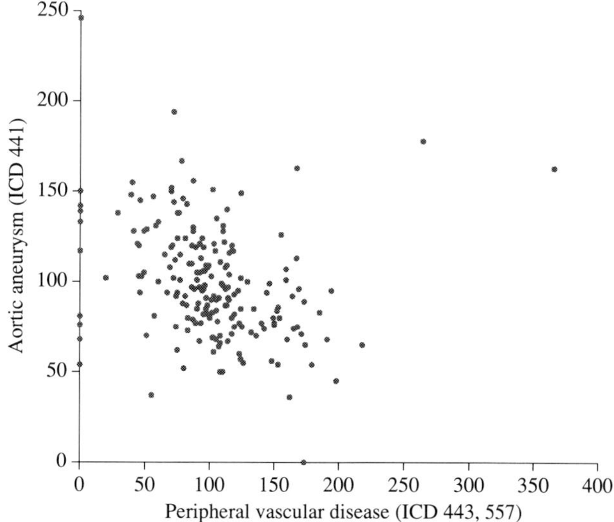

6.9.3 Chronic and unspecified myocarditis (ICD 429.0)

The top five jobs in the PMR ranking included face-trained coal miners (PMR 297, 41 deaths), other coal miners (PMR 280, 157 deaths) and other miners and quarry workers (PMR 232, 9 deaths). This may have arisen from unusual diagnostic practice in a district of South Yorkshire (see section 4.4).

6.10 Respiratory and related problems

6.10.1 Acute bronchitis, chronic bronchitis and emphysema (ICD 466, 491, 492, 496)

Important causes of chronic bronchitis include respiratory infection in infancy[33] and smoking.[34] In addition the disease is known to be an occupational hazard in coal miners[35] and is a suspected hazard in metal manufacture.[36] These occupations featured high in the ranking of PMRs (Table 6.16), and coal miners also had elevated mortality from acute bronchitis. Many of the metal making occupations also had high death rates from lung cancer (Table 6.16), and a confounding effect of smoking may have contributed to their excess of bronchitis.

6.10.2 Pneumococcal and unspecified lobar pneumonia (ICD 481)

As described in Chapter 4, there was unusually high mortality from pneumococcal and unspecified lobar pneumonia in welders. The excess extended to other occupations involving exposure to metal fume (Table 4.4) and, as in welders, was most marked in men of working age. Further investigation is needed to define the nature of this apparent hazard more precisely. In particular, we need to know whether it is related to specific metals and by what biological mechanism it operates.

6.10.3 Asthma (ICD 493)

Asthma is one of the more common occupational diseases, the most frequent occupational causes being exposure to isocyanates, wood dust, dust from grains, work with laboratory animals and fumes from solder flux.[37] However, these occupational causes account for only a small proportion of the asthma that is prevalent in the community. Moreover, relatively few asthmatics die from the disease. Despite this, electrical assemblers emerged as the occupation with highest mortality from asthma, both in men (PMR 241, 10 deaths) and in women (PMR 164, 23 deaths). It is unclear whether this excess reflects occupationally induced asthma or selection of people with respiratory impairment into jobs that are physically undemanding.

6.10.4 Asbestos-related disease (ICD 163, 158.8, 158.9, 501)

Three causes of death - cancer of the pleura, cancer of the peritoneum and asbestosis - are strongly related to asbestos exposure.[27] In addition, asbestos causes many deaths from lung cancer, but the relation to lung cancer does not stand out so clearly because the disease has many other causes. Table 6.17 shows the occupations with high death rates from asbestos-related diseases ranked according to their PMRs from cancer of the pleura. It illustrates the continued high mortality from this hazard despite the stringent controls on exposure that are now in place. This is partly because of the long latency with which asbestos causes death, especially from pleural and peritoneal cancer.

Identifying the occupations with asbestos-related deaths demonstrates where significant exposure has occurred. At first sight some of the jobs listed in Table 6.17 may seem surprising. For example, there were 167 deaths from pleural cancer in carpenters, giving them a higher PMR than most occupations in the construction industry. This probably reflects work with asbestos board. Notable for an absence of excess risk are garage mechanics, about whom concerns have been raised because of the asbestos in brake linings. If this does pose a hazard then the risk appears to be small.

Also remarkable is the difference in ranking for the different diseases. In particular, 'other construction workers,' who include laggers, have by far the highest mortality from cancer of peritoneum and asbestosis, but appear well down the ranking for cancer of the pleura. It is possible that this reflects differences in the types of asbestos fibres to which different occupations are exposed or different dose-response relations for the various asbestos-related diseases.

Job groups with high mortality from pleural cancer in women are shown in Table 6.18 (there were too few deaths from peritoneal cancer or asbestosis in women for meaningful analysis). All of these occupations, except tailors and dressmakers, also had high rates in men. A review of death

Table 6.16 Job groups with the highest PMRs for chronic bronchitis and emphysema - men aged 20-74, England and Wales, 1979-80, 1982-90

	Job group	Chronic bronchitis and emphysema (ICD 491,492,496)			Cancer of the trachea, bronchus and lung (ICD 162)		
		Deaths	PMR	95% CI	Deaths	PMR	95% CI
088	Other coal miners	3799	167	162-172	4610	98	95-101
175	Face trained coalminers	920	166	155-177	1137	98	92-104
128	Shot blasters	42	155	112-210	84	132	105-163
116	Moulders and coremakers (metal)	409	155	141-171	736	129	120-139
117	Electroplaters	68	152	118-193	134	137	115-163
127	Fettlers and dressers (metal)	131	151	126-179	259	136	120-154
113	Rollers (metal)	50	148	110-196	92	129	104-159
119	Galvanisers and tin platers	36	146	102-202	67	121	94-154

Notes 1 PMRs are standardised for age but not for social class.
2 Only job groups with significantly ($p<0.05$) elevated PMRs for chronic bronchitis and emphysema are listed.

Table 6.17 Job groups with high mortality from cancer of the pleura, cancer of the peritoneum and asbestosis - men aged 20-74, England and Wales, 1979-80, 1982-90

	Job group	Cancer of the pleura (ICD 163)			Cancer of the peritoneum (ICD 158.8,158.9)			Asbestosis (ICD 501)		
		Deaths	PMR	95% CI	Deaths	PMR	95% CI	Deaths	PMR	95% CI
146	Metal plate workers	73	709	556-892	1	78	2-432	3	292	60-854
153	Vehicle body builders	24	649	416-968	4	877	239-2246	5	1302	423-3039
144	Plumbers and gas fitters	134	450	377-533	11	283	141-506	13	457	243-782
101	Upholsterers	16	366	209-594	0	0	0-688	0	0	0-802
104	Carpenters	167	362	309-421	6	102	37-221	9	196	90-372
137	Electricians	127	349	291-415	4	83	23-212	1	29	1-164
138	Electrical plant operators	14	301	165-506	0	0	0-643	1	212	5-1180
027	Chemical engineers and scientists	13	274	146-468	0	0	0-597	0	0	0-813
149	Welders	56	247	186-320	1	33	1-186	3	142	29-415
039	Managers in construction	32	240	164-340	6	349	128-759	1	77	2-429
194	Boiler operators	24	240	153-357	2	171	21-616	4	366	100-937
143	Electrical engineers (so described)	31	227	154-323	1	58	1-325	1	74	2-410
132	Production fitters	192	208	180-240	12	103	53-179	17	189	110-303
167	Plasterers	18	207	122-327	3	265	55-773	0	0	0-431
145	Sheet metal workers	25	186	120-274	4	235	64-602	4	300	82-768
169	Builders etc	65	166	128-212	5	98	32-229	0	0	0-98
030	Professional engineers nec	53	162	121-212	2	46	6-168	0	0	0-120
174	Construction workers nec	77	160	126-200	64	990	762-1265	71	1592	1243-2009
033	Architects and surveyors	21	160	99-245	1	58	1-324	0	0	0-297
191	Dockers and goods porters	36	156	109-216	7	242	97-500	4	170	46-435
075	Chemical workers	39	144	103-198	5	146	47-341	5	189	61-441
069	Preparatory fibre processors	2	139	17-501	0	0	0-2049	2	1316	159-4753
114	Smiths and forge workers	6	133	49-289	3	561	116-1639	0	0	0-765
124	Machine tool operators	116	132	109-158	15	136	76-224	11	126	63-226

Notes 1 PMRs are standardised for age but not for social class.
2 Only job groups with significantly ($p<0.05$) elevated PMRs for at least one of the diseases are listed.

Table 6.18 Job groups with high mortality from cancer of the pleura (ICD 163) - women aged 20-74, England and Wales, 1979-80, 1982-90

	Job group	Deaths	PMR	95% CI
146	Metal plate workers	2	10000	1211-36123
132	Production fitters	2	2740	332-9897
101	Upholsterers	2	870	105-3141
098	Tailors and dressmakers	12	555	287-969
075	Chemical workers	4	554	151-1419

Notes 1 PMRs are standardised for age but not for social class.
2 Only job groups with significantly ($p<0.05$) elevated PMRs are listed.

certificates indicated that of 11 female tailors and dressmakers who died of pleural cancer during 1982-90, eight lived in Leeds and a further two elsewhere in West Yorkshire. In only one case was the disease attributed to occupation. This suggests a previously unrecognised occupational cluster and further investigation is being carried out to establish exactly where the women had worked.

In some cases the asbestos exposure giving rise to the excess will have been a general feature of the occupation concerned, but in others it may have been limited to those carrying out the job in specific industries where asbestos has been handled. Such industries include shipbuilding, manufacture and repair of railway engines and rolling stock, manufacture of asbestos textiles and production of construction materials containing asbestos, and tend to be geographically localised.[38] This is apparent in the heterogeneity of mortality from asbestos-related disease by place. Table 6.19 shows local authority areas with significantly elevated PMRs for asbestos-related disease in all occupations combined. Most are centres of shipbuilding, railway engineering or other asbestos industries.

Table 6.19 Local authority areas with high PMRs for asbestos related disease (ICD 158.8, 158.9, 163, 501)

Area	Deaths	PMR	95% CI
Category (i) - shipbuilding			
Barrow-in-Furness	48	744	548-987
Plymouth	94	558	451-683
Gillingham	19	353	212-551
Rochester	30	327	220-467
Southampton	48	322	237-427
Stockton-on-Tees	40	305	218-415
Newcastle upon Tyne	73	303	237-381
North Tyneside	52	301	225-395
South Tyneside	45	289	211-387
Hartlepool	21	252	156-386
Sunderland	61	249	191-321
Fareham	12	243	126-425
Gateshead	47	238	175-316
Portsmouth	30	227	153-325
Caradon	10	220	106-405
Medina	11	217	108-388
Swale	15	215	120-355
Havant	15	211	118-348
Wirral	54	203	152-264
Langbaurgh	21	179	111-274
Sefton	33	145	100-204
Category (ii) - railway engineering			
Eastleigh	17	313	182-500
Thamesdown	29	294	197-422
Crewe & Nantwich	17	223	130-357
Doncaster	41	183	131-248
Derby	30	177	119-253
Leeds	95	174	141-213
Category (iii) - other areas with high PMRs			
Barking & Dagenham	62	484	371-621
Newham	52	336	251-441
Havering	47	303	223-404
Stevenage	11	254	127-454
Thurrock	21	247	152-377
Castle Point	12	240	124-419
Basildon	20	221	135-342
Aylesbury Vale	16	220	126-358
Dartford	11	206	103-369
Sutton	19	196	118-307
Redbridge	26	179	117-263
Hillingdon	26	178	116-262
Calderdale	26	167	109-245

To explore the nature of the asbestos exposure giving rise to high occupational mortality from cancers of the pleura and peritoneum and asbestosis, PMRs by job were calculated separately for four categories of local authority area:

(i) areas with significantly high PMRs attributable to shipbuilding
(ii) areas with significantly high PMRs attributable to railway engineering
(iii) other areas with significantly high PMRs
(iv) areas in which PMRs were not significantly elevated.

The areas making up categories (i) to (iii) are indicated in Table 6.19.

Table 6.20 shows the outcome of this analysis. For some jobs (professional engineers nec., architects and surveyors, chemical workers, smiths and forge workers, machine tool operators, production fitters, sheet metal workers, welders, plasterers, builders etc. and dockers and goods porters) the excess mortality was confined to places with significantly increased PMRs for all occupations. This suggests industry specific exposure. For other jobs (chemical engineers and scientists, managers in construction, preparatory fibre processors, upholsterers, carpenters, electricians, electrical plant operators, electrical engineers (so described), plumbers and gas fitters, metal plate workers, vehicle body builders, construction workers nec, and boiler operators) PMRs tended to be highest in places with asbestos industries, but were also elevated in areas with no significant excess of asbestos related disease overall. This would be compatible with a more general hazard of the occupation. For example, plumbers and electricians may encounter asbestos lagging outside the main asbestos industries. An exception to this may be preparatory fibre processors among whom exposure to asbestos is likely to be restricted to the asbestos textiles industry. The four cases in this job group were from Newham, Southwark, Manchester and Bradford.

6.11 Alcohol-related diseases (ICD see Table 6.21)

Table 6.21 shows occupations with high mortality from eight causes of death related to alcohol. All of these occupations had elevated PMRs for at least three of the causes, although the pattern varied from one job to another. Thus, for example, doctors and lawyers had no excess of laryngeal cancer, presumably because they smoke less than the average and smoking is also a powerful cause of this disease.[39] The high PMRs in many of these occupations for falls on stairs suggests a link between alcohol and such deaths.

6.12 Gastric ulcer, duodenal ulcer (ICD 531, 532)

Several of the occupations with high rates of alcohol-related diseases in men also had high mortality from gastric ulcer - in particular, merchant sea farers (PMR 178, 37 deaths), cooks and kitchen workers (PMR 173, 32 deaths) and men working in hotels, pubs and bars (PMR 154, 64 deaths). The last two groups also had markedly elevated death rates from duodenal ulcer with PMRs of 187 (56 deaths) and 169 (54 deaths) respectively. In women, the only occupation with a

Table 6.20 Mortality from asbestos related disease in four categories of local authority area - men aged 20-74, England and Wales 1979-80, 1982-90

Job group	(i) Shipbuilding			(ii) Railway engineering			(iii) Other high PMRs			(iv) Other areas		
	Deaths	PMR	95% CI	Deaths	PMR	95% CI	Deaths	PMR	95% CI	Deaths	PMR	95% CI
027 Chemical engineers and scientists	2	410	50 -1480	0	0	0 -1563	0	0	0 -2476	11	222	111 -398
030 Professional engineers nec	13	471	251 -806	5	308	100 -719	2	159	19 -576	35	101	71 -141
033 Architects and surveyors	5	478	155 -1116	3	530	109 -1549	1	311	8 -1730	13	92	49 -157
039 Managers in construction	10	737	353 -1355	1	133	3 -742	3	592	122 -1729	25	182	118 -269
069 Preparatory fibre processors	0	0	0 -24593	1	769	19 -4286	0	0	0 -7528	3	190	39 -555
075 Chemical workers	18	391	232 -619	3	227	47 -662	0	0	0 -466	28	106	71 -154
101 Upholsterers	2	513	62 -1852	1	351	9 -1955	6	2844	1044 -6189	7	156	63 -322
104 Carpenters	44	749	544 -1006	17	648	377 -1037	12	675	349 -1179	109	235	193 -283
114 Smiths and forge workers	5	711	231 -1660	2	1439	174 -5198	0	0	0 -1597	2	45	5 -162
124 Machine tool operators	32	442	302 -624	13	303	161 -518	9	209	96 -397	88	96	77 -118
132 Production fitters	55	496	374 -647	17	388	226 -622	36	671	470 -930	113	123	101 -148
137 Electricians	37	802	564 -1106	9	450	206 -854	7	504	202 -1038	79	216	171 -269
138 Electrical plant operators	3	694	143 -2029	2	580	70 -2094	0	0	0 -1346	10	215	103 -396
143 Electrical engineers (so described)	4	317	86 -811	5	663	215 -1548	2	403	49 -1457	22	155	97 -235
144 Plumbers and gas fitters	29	736	492 -1058	11	638	318 -1141	17	1472	857 -2357	101	340	277 -413
145 Sheet metal workers	16	937	536 -1522	4	513	140 -1315	1	139	4 -773	12	90	47 -158
146 Metal plate workers	41	737	529 -1001	5	1064	345 -2483	4	928	253 -2376	27	439	289 -640
149 Welders	25	584	378 -864	4	407	111 -1041	3	308	64 -900	28	130	86 -188
153 Vehicle body builders	0	0	0 -1430	0	0	0 -2509	12	2727	1409 -4764	21	569	352 -871
167 Plasterers	5	525	171 -1226	1	211	5 -1175	3	840	173 -2456	12	135	70 -235
169 Builders etc	4	163	45 -418	6	280	103 -609	4	345	94 -883	56	133	100 -173
174 Construction workers nec	55	900	678 -1172	26	1179	769 -1729	8	387	167 -762	123	253	210 -302
191 Dockers and goods porters	18	459	272 -726	13	516	275 -882	1	222	6 -1235	15	70	39 -115
194 Boiler operators	6	505	185 -1099	4	739	201 -1893	1	195	5 -1086	19	189	114 -295

Notes 1 PMRs are for cancer of the pleura (ICD 163), cancer of the peritoneum (ICD 158.8,158,9) and asbestosis (ICD 501) combined
2 PMRs are standardised for age but not for social class
3 Definition of the areas in the text

Table 6.21 Job groups with high mortality from alcohol-related diseases - men and women aged 20-74, England and Wales, 1979-80, 1982-90

Job group	Cancer of the oral cavity (ICD 141, 143-145)			Cancer of the pharynx (ICD 146-148)			Cancer of the oesophagus (ICD 150)			Cancer of the liver (ICD 155)		
	Deaths	PMR	95% CI	Deaths	PMR	95% CI	Deaths	PMR	95% CI	Deaths	PMR	95% CI
Men												
001 Lawyers	12	172	89-301	8	151	65-297	38	108	76-148	32	324	222-458
015 Doctors	18	184	109-290	13	173	92-296	59	117	89-151	40	286	204-389
024 Literary and artistic occupations	42	152	110-206	33	155	107-218	126	93	78-111	60	155	119-200
036 Sea farers	56	267	202-347	45	279	203-373	118	109	90-131	46	154	113-205
045 Publicans and barstaff	117	271	224-325	71	214	167-270	261	119	105-135	112	184	152-222
046 Caterers	26	126	83-186	36	227	159-315	99	97	79-118	56	194	147-253
048 Armed forces	45	207	151-277	29	173	116-248	150	134	113-157	38	118	84-163
059 Cooks and kitchen porters	31	157	107-223	54	353	265-461	76	79	62-98	70	254	198-320
066 Fishing and related workers	9	255	117-484	4	147	40-376	22	124	77-187	6	120	44-260
147 Steel erectors	20	223	136-345	10	144	69-265	49	110	81-145	17	136	79-218
191 Dockers and goods porters	53	178	133-233	34	149	103-209	177	110	95-128	25	57	37-85
Women												
024 Literary and artistic occupations	14	378	207-635	5	147	48-344	31	161	109-228	10	129	62-237
045 Publicans and bar staff	30	297	200-424	22	252	157-381	65	118	91-150	18	94	56-149
052 Hairdressers	8	238	103-469	2	66	8-238	11	62	31-112	6	85	31-184

Job group	Cancer of the larynx (ICD 161)			Other alcohol related diseases (ICD see below)			Cirrhosis (ICD 571.5)			Fall on stairs (ICD 880)		
	Deaths	PMR	95% CI	Deaths	PMR	95% CI	Deaths	PMR	95% CI	Deaths	PMR	95% CI
Men												
001 Lawyers	11	126	63-226	35	196	136-272	24	233	149-347	3	79	16-231
015 Doctors	6	48	18-104	37	157	111-217	49	341	252-452	10	197	95-363
024 Literary and artistic occupations	32	96	66-135	158	201	171-235	82	198	157-246	19	118	71-185
036 Sea farers	67	249	193-316	154	312	265-365	81	265	211-330	14	132	72-222
045 Publicans and barstaff	119	218	181-261	458	431	393-473	243	383	336-434	43	194	140-261
046 Caterers	31	122	83-173	97	180	146-220	52	171	127-224	14	125	68-210
048 Armed forces	34	124	86-174	132	221	185-263	59	182	138-234	25	183	119-271
059 Cooks and kitchen porters	35	146	102-203	114	209	172-251	41	140	100-190	19	169	102-263
066 Fishing and related workers	14	318	174-533	13	141	75-240	9	172	79-327	3	153	32-448
147 Steel erectors	21	187	116-286	29	132	88-189	12	92	47-160	4	89	24-228
191 Dockers and goods porters	66	165	127-210	102	170	138-206	61	144	110-185	20	143	87-221
Women												
024 Literary and artistic occupations	2	89	11-320	40	276	197-376	16	215	123-349	5	166	54-387
045 Publicans and bar staff	12	182	94-317	101	341	278-415	73	378	296-475	12	173	89-302
052 Hairdressers	1	49	1-271	26	210	137-307	14	211	116-355	4	145	39-370

Notes 1 PMRs are standardised for age but not for social class.
2 Other alcohol related diseases (ICD 303,305.0,425.5,535.3,571.0-571.3,860.0,860.1)

significantly raised PMR for this cause was hairdressers (PMR 203, 12 deaths), a group which also had high mortality from alcohol-related disease. Whether these results reflect a causal role of alcohol is unclear.

6.13 Injury and poisoning

6.13.1 Motor vehicle accidents (ICD E810-E819)

The highest mortality from motor vehicle accidents among men was in clergy (PMR 159, 50 deaths), while in women clergy ranked second (PMR 426, 21 deaths). These excesses were not attributable to a large number of deaths in a single accident, and the explanation is unclear.

Several occupations that entail professional driving appeared high in the PMR rankings, in particular lorry drivers, and police. Elevated mortality in road surfacers is also likely to reflect a hazard at work.

6.13.2 Occupational accidents (ICD see Tables 6.22 and 6.23)

Tables 6.22 and 6.23 show the frequency with which deaths could be attributed with reasonable confidence to accidents at work. Among men, the largest excess of deaths, some 1,200 in total, was from transport accidents, and especially road transport accidents in lorry drivers (more than 500 excess deaths), water transport accidents in sea farers and fishermen (approximately 150 excess deaths), rail accidents in railway staff (some 130 excess deaths) and air transport accidents in aircraft flight deck officers and members of the armed forces (approximately 100 excess deaths). The greatest risks to individual workers were from air transport accidents in aircraft flight deck officers (more than one in ten deaths below age 65), water transport accidents in fishermen (some 6 per cent of deaths below age 65) and rail transport accidents in rail track workers (almost 5 per cent of deaths below age 65). The figure for fishermen may be an underestimate if not all deaths at sea are registered in England and Wales.

Table 6.22 Mortality from accidents related to work - men aged 20-64, England and Wales 1979-80, 1982-90

Cause of death	Job group		Deaths	PMR	95% CI	Estimated excess of deaths from accident	Excess as a proportion of all deaths in job group (per 1000)
Railway accidents (ICD E800-E807)	063	Railway station workers	21	1642	1015 -2514	19.7	6.9
	170	Rail track workers	55	6790	5114 -8843	54.2	47.6
	177	Railway guards	17	3142	1831 -5031	16.5	21.5
	179	Shunters and points operators	13	5804	3090 -9924	12.8	34.8
	180	Railway engine drivers	27	2634	1734 -3837	26.0	11.6
Motor vehicle traffic accidents (ICD E810-E819)	049	Police	214	156	136 -179	77.1	21.0
	171	Road construction workers and paviors	81	154	122 -191	28.3	13.6
	183	Lorry drivers	1327	164	155 -173	516.8	16.2
Off-road motor vehicle accidents (ICD E820-E825)	025	Persons involved in sport	11	2311	1154 -4135	10.5	46.2
	047	Farmers	32	222	151 -313	17.6	0.8
	048	Armed forces	19	225	135 -351	10.5	2.2
	065	Foresters	3	592	122 -1729	2.5	4.2
	133	Motor mechanics	17	246	143 -394	10.1	1.4
	183	Lorry drivers	75	380	299 -476	55.2	1.7
	187	Crane drivers	7	453	182 -934	5.5	1.5
Animal transport accidents (ICD E827-E828)	025	Persons involved in sport	2	6061	734 -21893	2.0	8.6
	047	Farmers	12	786	406 -1373	10.5	0.5
Water transport accidents (ICD E830-E838)	036	Sea farers	104	3449	2818 -4181	101.0	23.1
	066	Fishing and related workers	48	6138	4524 -8143	47.2	60.5
	191	Dockers and goods porters	12	421	218 -736	9.2	1.7
Air transport accidents (ICD E840-E845)	034	Aircraft flight deck officers	44	16858	12245 -22647	43.7	106.7
	048	Armed forces	72	1141	892 -1437	65.7	13.9
Other vehicle accidents (ICD E846-E848)	088	Other coal miners	32	2817	1925 -3981	30.9	2.3
	175	Face trained coal miners	17	6137	3575 -9826	16.7	4.9
	176	Miners (not coal) and quarry workers	2	1667	202 -6021	1.9	1.9
	179	Shunters and points operators	2	5263	637 -19012	2.0	5.3
	180	Railway engine drivers	2	1070	130 -3863	1.8	0.8
Pesticide poisoning (ICD E863)	047	Farmers	3	1010	208 -2952	2.7	0.1
Poisoning by motor vehicle exhaust (ICD E868.2)	133	Motor mechanics	10	326	156 -599	6.9	1.0
Fall from ladder or scaffolding (ICD E881)	060	Other service personnel	38	307	217 -422	25.6	1.5
	104	Carpenters	27	268	176 -390	16.9	1.5
	144	Plumbers and gas fitters	18	256	152 -405	11.0	1.4
	147	Steel erectors	20	1172	716 -1810	18.3	8.9
	148	Scaffolders	38	3169	2242 -4354	36.8	33.3
	150	Riggers	3	455	94 -1328	2.3	2.7
	160	Painters and decorators nec	53	491	368 -643	42.2	3.4
	165	Bricklayers and tilesetters	17	328	191 -525	11.8	1.9
	168	Roofers and glaziers	26	1045	682 -1533	23.5	11.1
	169	Builders etc	47	520	382 -692	38.0	3.7
	174	Construction workers nec	31	241	163 -342	18.1	1.3
Fall from building (ICD E882)	060	Other service personnel	37	230	162 -317	20.9	1.2
	104	Carpenters	35	221	154 -307	19.1	1.7
	140	Electric cable and line workers	6	708	260 -1540	5.2	6.7
	144	Plumbers and gas fitters	20	189	116 -292	9.4	1.2
	147	Steel erectors	30	1263	852 -1805	27.6	13.4
	148	Scaffolders	9	409	187 -776	6.8	6.1
	150	Riggers	8	971	419 -1913	7.2	8.1
	160	Painters and decorators nec	23	137	87 -207	6.3	0.5
	165	Bricklayers and tilesetters	15	193	108 -318	7.2	1.2
	168	Roofers and glaziers	116	2202	1819 -2641	110.7	52.3
	169	Builders etc	29	216	145 -311	15.6	1.5
	174	Construction workers nec	80	379	301 -472	58.9	4.2
Fall into hole (ICD E883)	169	Builders etc	6	418	153 -909	4.6	0.4
	173	Mains and service layers	2	1587	192 -5734	1.9	1.8
	174	Construction workers nec	13	583	311 -998	10.8	0.8
Other fall (ICD E884)	050	Fire service personnel	4	450	123 -1152	3.1	2.4
	147	Steel erectors	8	702	303 -1383	6.9	3.3
	148	Scaffolders	4	412	112 -1055	3.0	2.7
	160	Painters and decorators	16	205	117 -332	8.2	0.7
Injured by fire (ICD E890-E899)	050	Fire service personnel	8	321	138 -632	5.5	4.3
Heat injury (ICD E900)	048	Armed forces	4	1646	449 -4215	3.8	0.8
Injury by animals or plants (ICD E905-E906)	047	Farmers	17	1493	869 -2390	15.9	0.8

Note PMRs are standardised for age but not for social class.

Table 6.22 - *continued*

Cause of death	Job group		Deaths	PMR	95% CI	Estimated excess of deaths from accident	Excess as a proportion of all deaths in job group (per 1000)
Non-recreational drowning (ICD E910.3)	048	Armed forces	4	681	186 -1745	3.4	0.7
	049	Police	5	1923	624 -4488	4.7	1.3
Injury by falling object (ICD E916)	047	Farmers	30	164	111 -234	11.7	0.6
	065	Foresters	14	2233	1221 -3746	13.4	22.7
	088	Other coal miners	33	417	287 -586	25.1	1.9
	133	Motor mechanics	17	200	116 -320	8.5	1.2
	147	Steel erectors	9	513	234 -973	7.2	3.5
	148	Scaffolders	6	391	143 -850	4.5	4.0
	169	Builders etc	18	183	109 -290	8.2	0.8
	173	Mains and service layers	7	779	313 -1604	6.1	5.8
	174	Construction workers nec	56	377	285 -490	41.1	2.9
	175	Face trained coal miners	21	1102	681 -1688	19.1	5.7
	176	Miners (not coal) and quarry workers	8	954	412 -1879	7.2	7.1
	183	Lorry drivers	62	215	165 -276	33.2	1.0
Injury by being caught between objects (ICD E918)	183	Lorry drivers	7	273	110 -562	4.4	0.1
	191	Dockers and goods porters	3	870	179 -2541	2.7	0.5
Injury by machinery (ICD E919)	047	Farmers	127	492	410 -585	101.2	4.8
	065	Foresters	6	678	249 -1476	5.1	8.7
	069	Preparatory fibre processors	3	1111	229 -3247	2.7	8.5
	075	Chemical workers	22	267	167 -404	13.7	2.0
	079	Paper manufacturers	3	817	169 -2389	2.6	6.6
	088	Other coal miners	34	309	214 -432	23.0	1.7
	120	Other metal manufacturers	9	292	133 -554	5.9	2.0
	132	Production fitters	107	364	298 -440	77.6	3.2
	147	Steel erectors	8	324	140 -638	5.5	2.7
	148	Scaffolders	7	324	130 -668	4.8	4.4
	174	Construction workers nec	42	200	144 -270	21.0	1.5
	175	Face trained coal miners	16	609	348 -989	13.4	4.0
	176	Miners (not coal) and quarry workers	7	606	244 -1249	5.8	5.8
	186	Mechanical plant drivers	22	1022	639 -1549	19.8	12.0
	187	Crane drivers	24	720	461 -1073	20.7	5.6
	188	Fork lift truck drivers	27	473	311 -689	21.3	5.2
	189	Slingers	6	1085	398 -2362	5.4	8.9
	191	Dockers and goods porters	12	220	113 -383	6.5	1.2
Injury by cutting or piercing instruments or objects (ICD E920)	042	Butchers	4	810	221 -2073	3.5	1.0
	048	Armed forces	5	384	125 -895	3.7	0.8
Injury by explosion of pressure vessel (ICD E921)	132	Production fitters	4	404	110 -1036	3.0	0.1
Injury by firearms (ICD E922)	047	Farmers	20	716	437 -1106	17.2	0.8
	048	Armed forces	7	427	172 -879	5.4	1.1
Injury by explosive material (ICD E923)	048	Armed forces	7	351	141 -724	5.0	1.1
	075	Chemical workers	12	886	458 -1548	10.6	1.5
	133	Motor mechanics	6	309	113 -672	4.1	0.6
	137	Electricians	10	393	189 -723	7.5	0.7
	143	Electrical engineers (so described)	4	668	182 -1710	3.4	1.0
	175	Face trained coal miners	7	1580	635 -3256	6.6	1.9
Injury by hot substances (ICD E924)	073	Bleachers, dyers and finishers	2	4167	505 -15051	2.0	2.6
	075	Chemical workers	3	598	123 -1746	2.5	0.4
	112	Furnace operatives (metal)	2	2667	323 -9633	1.9	1.6
	132	Production fitters	8	451	195 -890	6.2	0.3
Injury by electric current (ICD E925)	047	Farmers	28	210	139 -303	14.6	0.7
	132	Production fitters	24	164	105 -244	9.3	0.4
	137	Electricians	42	513	370 -694	33.8	3.2
	140	Electric cable and line workers	12	2970	1535 -5189	11.6	15.0
	141	Radio and TV mechanics	5	530	172 -1237	4.1	3.4
	143	Electrical engineers (so described)	6	353	130 -768	4.3	1.3
	144	Plumbers and gas fitters	12	226	117 -395	6.7	0.8
	169	Builders etc	14	207	113 -347	7.2	0.7
	171	Road construction workers and paviors	6	514	189 -1118	4.8	2.3
	187	Crane drivers	5	347	113 -809	3.6	1.0

Note PMRs are standardised for age but not for social class.

Table 6.23 Mortality from accidents related to work - women aged 20-59, England and Wales, 1979-80, 1982-90

Cause of death (ICD)	Job group		Deaths	PMR	95% CI	Estimated excess of deaths from accident	Excess as a proportion of all deaths in job group (per 1000)
Motor vehicle traffic accidents (ICD E810-E819)	057	Sales representatives	41	211	151-286	21.6	41.7
	183	Lorry drivers	18	318	188-503	12.3	74.3
Animal transport accidents (ICD E827-E828)	025	Persons involved in sport	6	8955	3286-19492	5.9	148.3
	047	Farmers	3	744	154-2176	2.6	4.0
Air transport accidents (ICD E840-E845)	048	Armed forces	2	1905	231-6881	1.9	24.9
Fall from building (ICD E882)	047	Farmers	3	968	200-2828	2.7	4.2
Injury by animals or plants (ICD E905-E906)	047	Farmers	2	6061	734-21893	2	3.1
Injury by machinery (ICD E919)	047	Farmers	4	10256	2795-26260	4	6.2

Note PMRs are standardised for age but not for social class.

Table 6.24 Job groups with highest PMRs from suicide (ICD E950-959) - men aged 20-64 and women aged 20-59, England and Wales 1979-80, 1982-90

	Job group	Deaths	PMR	95% CI
Men				
022	Veterinarians	32	431	294-609
157	Pottery decorators	8	340	147-671
016	Dentists	32	227	155-321
018	Pharmacists	40	227	162-310
015	Doctors	124	198	165-236
047	Farmers	983	161	151-172
065	Foresters	33	159	109-224
023	Driving instructors	50	158	117-209
027	Chemical engineers and scientists	64	155	119-198
021	Health professions nec	34	151	105-212
Women				
022	Veterinarians	7	820	330-1689
007	Government inspectors	7	574	231-1183
062	Ambulance workers	5	466	151-1087
015	Doctors	41	421	302-572
018	Pharmacists	14	338	185-568
020	Physiotherapists	12	249	129-435
021	Health professions nec	25	202	131-299
010	Teachers in higher education	21	179	111-274
024	Literary and artistic occupations	70	168	131-212
017	Nurses	372	161	145-178

Notes 1 PMRs are standardised for age but not for social class.
2 Only job groups with significantly (p<0.05) elevated PMRs are listed.

Table 6.25 Job groups with highest PMRs from homicide (ICD E960-E969) - men aged 20-64 and women aged 20-59, England and Wales, 1979-80, 1982-90

	Job group	Deaths	PMR	95% CI
Men				
103	Other workers with fabrics	3	515	106-1504
148	Scaffolders	9	290	132-550
167	Plasterers	10	216	104-397
059	Cooks and kitchen porters	27	208	137-304
048	Armed forces	40	201	144-274
049	Police	16	187	107-303
160	Painters and decorators nec	41	177	127-241
024	Literary and artistic personnel	28	170	113-246
060	Other service occupations	35	163	113-227
044	Retailers and dealers	74	145	114-182
174	Construction workers nec	42	143	103-194
Women				
045	Publicans and bar staff	26	395	258-580
060	Other service personnel	45	169	123-226
046	Caterers	22	157	98-238

Notes 1 PMRs are standardised for age but not for social class.
2 Only job groups with significantly (p<0.05) delevated PMRs are listed.

Another major cause of accidental deaths was falls of various types in occupations related to the construction industry (more than 500 excess deaths). Of the individual occupations, those most at risk were roofers and glaziers (approximately 6 per cent of deaths below age 65) and scaffolders (4 per cent of deaths below age 65).

An estimated 350 deaths were attributable to accidents at work with machinery, the highest risk being in drivers of mechanical plant (just over 1 per cent of deaths below age 65).

Relatively few deaths from occupational accidents occurred in women, reflecting the smaller numbers of women in hazardous occupations.

6.13.3 Suicide (ICD E950-E959)

The occupations with highest mortality from suicide are listed in Table 6.24. Medical and related professions figure prominently, probably reflecting their knowledge of how to achieve successful suicide and possibly their access to pharmaceuticals. Ready access to means of suicide (guns and poisons) may also contribute to the high mortality of farmers and foresters from suicide.

6.13.4 Homicide (ICD E960-E969)

The occupations with highest mortality from homicide (Table 6.25) include several with obvious occupational hazards of criminal injury, in particular the armed forces and police. Most of the others also had high rates of drug dependence (scaffolders, literary and artistic occupations, painters and decorators) or alcohol-related diseases (cooks and kitchen porters, publicans and barmen and caterers).

References

1. Alderson MR. *Occupational cancer.* London: Butterworths, 1986.
2. Dubrow R, Wegman DH. Setting priorities for occupational cancer research and control: synthesis of the results of occupational disease surveillance studies. *J Natl Cancer Inst* 1983;71:1123-42.

3 Wright WE, Bernstein L, Peters JM, Garabrant DH, Mack TM. Adenocarcinoma of the stomach and exposure to occupational dust. *Am J Epidemiol* 1988;128:64-73.

4 Coggon D, Inskip H. Is there an epidemic of cancer? *Br Med J* 1994;308:705-8.

5 Howson CP, Hiyama T, Wynder EL. The decline in gastric cancer: epidemiology of an unplanned triumph. *Epidemiol Rev* 1986;8:1-27.

6 Mack TM. Pancreas. In Schottenfeld D, Fraumeni JF (eds). *Cancer epidemiology and prevention.* Philadelphia: Saunders, 1982: 638-67.

7 International Agency for Research on Cancer. IARC monographs on the evaluation of the carcinogenic risk of chemicals to humans. Vol 25. *Wood, leather and some associated industries.* Lyon: IARC, 1981.

8 Lee JAH. Melanoma and exposure to sunlight. *Epidemiol Rev* 1982;4:110-36.

9 Doll R, Peto R. *The causes of cancer.* Oxford: OUP, 1981.

10 Kelsey JL, Gammon MD, John EM. Reproductive factors and breast cancer. *Epidemiol Rev* 1993;15:36-47.

11 Cramer DW. Uterine cervix. In Schottenfeld D, Fraumeni JF (eds). *Cancer epidemiology and prevention.* Philadelphia: Saunders, 1982: 881-900.

12 Nomura AMY, Kolonel LN. Prostate cancer: a current perspective. *Epidemiol Rev* 1991;13:200-27.

13 Fraser P, Carpenter L, Maconochie N, Higgins C, Booth M, Beral V. Cancer mortality and morbidity in employees of the United Kingdom Atomic Energy Authority 1946-1986. *Br J Cancer* 1993;67:615-24.

14 International Agency for Research on Cancer. IARC monographs on the evaluation of the carcinogenic risk of chemicals to man. Vol 4. *Some aromatic amines, hydrazine and related substances, N-nitroso compounds and miscellaneous alkylating agents.* Lyon: IARC, 1974.

15 Partanen T, Heikkilä P, Hernberg S, Kauppinen T, Moneta G, Ojajärvi A. Renal cell cancer and occupational exposure to chemical agents. *Scand J Work Environ Health* 1991;17:231-9.

16 Morrison AS, Cole P. Urinary tract. In Schottenfeld D, Fraumeni JF (eds). *Cancer epidemiology and prevention.* Philadelphia: Saunders, 1982: 925-37.

17 Modan B, Biadatz D, Mart H, Steinitz R, Levin SG. Radiation-induced head and neck tumours. *Lancet* 1974;1:277-9.

18 Shore RE, Albert RE, Pasternack BS. Follow-up study of patients treated by x-ray epilation for tinea capitis: resurvey of post-treatment illness and mortality experience. *Arch Env Health* 1976;31:21-8.

19 National Radiological Protection Board. *Electromagnetic fields and the risk of cancer.* Documents of the NRPB 1992;3:1-138.

20 Boice JD, Land CE. Ionising radiation. In Schottenfeld D, Fraumeni JF (eds). *Cancer epidemiology and prevention.* Philadelphia: Saunders, 1982: 231-53.

21 Goldstein BD. Benzene toxicity. *Occupational Medicine: State of the Art Reviews* 1988;3:541-54.

22 Mims CA. Viral aetiology of diseases of obscure origin. *Br Med Bull* 1985;41:63-9.

23 Simmons D, Williams DRR, Powell MJ. Prevalence of diabetes in a predominantly Asian community: preliminary findings of the Coventry diabetes study. *Br Med J* 1989;298:18-21.

24 Shapiro CN, Margolis HS. Hepatitis B epidemiology and prevention. *Epidemiol Rev* 1990;12:221-7.

25 Peterman TA, Drotman P, Curran JW. Epidemiology of the acquired immunodeficiency syndrome (AIDS). *Epidemiol Rev* 1985;7:1-21.

26 Rieder HL, Cauthen GM, Comstock GW, Snider DE. Epidemiology of tuberculosis in the United States. *Epidemiol Rev* 1989;11:79-98.

27 Elmes PC. Inorganic dusts. In Raffle PAB, Adams PH, Baxter PJ, Lee WR (eds). *Hunter's diseases of occupations.* London: Edward Arnold, 1994: 410-57.

28 Martyn CN, Barker DJP, Osmond C. Motoneuron disease and past poliomyelitis in England and Wales. *Lancet* 1988;1:1319-22.

29 Martyn CN, Cruddas M, Compston DAS. Symptomatic Epstein-Barr virus infection and multiple sclerosis. *J. Neurol Neurosurg Psychiat* 1993;56:167-8.

30 Langston JW. MPTP and Parkinson's disease. *Trends in Neurosciences* 1985;8:79-83.

31 Reed D, Reed C, Stemmermann G, Hayashi T. Are aortic aneurysms caused by atherosclerosis? *Circulation* 1992;85:205-11.

32 Powell JT. Smoking. In Fowkes FGR (ed). *Epidemiology of peripheral vascular disease.* London: Springer Verlag, 1991: 141-53.

33 Barker DJP, Godfrey KM, Fall C, Osmond C, Winter PD, Shaheen SO. Relation of birthweight and childhood respiratory infection to adult lung function and death from chronic obstructive airways disease. *Br Med J* 1991;303:671-5.

34 Fletcher CM, Peto R, Tinker CM, Speizer FE. The natural history of chronic airflow obstruction. *Br Med J* 1977;1:1645-8.

35 Marine WM, Gurr D, Jacobsen M. Clinically important respiratory effects of dust exposure and smoking in British coal miners. *Am Rev Respir Dis* 1988;137:106-12.

36 Maizlish N, Park R, Silverstein B, Brodsky L, Mirer F. Mortality among ferrous foundry workers. *Am J Ind Med* 1986;10:27-43.

37 Meredith SK, Taylor VM, McDonald JC. Occupational respiratory disease in the United Kingdom 1989: a report to the British Thoracic Society and the Society of Occupational Medicine by the SWORD project group. *Br J Ind Med* 1991;48:292-8.

38 Gardner MJ, Acheson ED, Winter PD. Mortality from mesothelioma of the pleura during 1968-78 in England and Wales. *Br J Cancer* 1982; 46: 81-8.

39 Austin DF. Larynx. In Schottenfeld D, Fraumeni JF (eds). *Cancer epidemiology and prevention.* Philadelphia: Saunders, 1982: 554-63.

Chapter 7 Cancer incidence in England 1981-1987

Eve Roman
Cancer Epidemiology Unit, Imperial Cancer Research Fund

Lucy Carpenter
Department of Public Health and Primary Care, University of Oxford

7.1 Introduction

This chapter summarises the main findings of an analysis of cancers registered in England during the seven-year period 1981-87. Cancers registered in men during the four years 1966-69 were examined in relation to occupation in the 1970-72 Decennial Supplement.[1] The present report is based on substantially larger numbers of events and incorporates data on women as well as men. The inclusion of women is particularly important in view of the sustained growth in the number of women participating in work outside the home.[2]

Mortality during the period 1979-90, including mortality from specific cancers, is discussed in relation to occupation in Chapters 4-6. Some degree of overlap between data collected at death certification and cancer registration is to be expected for malignancies with poor survival, such as cancers of the lung and stomach. For others, such as cancers of the skin and bladder, where the five-year survivals are over 50 per cent, the degree of overlap will not be so great. It is for cancers such as these that cancer registration data are particularly valuable. Moreover, unlike at death certification, the occupation recorded at cancer registration has the advantage of not having to rely solely on information supplied by the next of kin.

7.2 Overview of the data

During the 1980s, information on cancers diagnosed in England and Wales was supplied to OPCS by fifteen cancer registries. Cancer registries vary both with respect to the methods they employ to ascertain information about individuals diagnosed with cancer in their area, and with respect to the amount of data they collect and subsequently provide to OPCS.[3] During the 1980s, no information on occupation was collected by the Welsh cancer registry and the data analysed here consequently refer to England alone.

Individual anonymised records were supplied for 1,034,759 cancers registered in England among people aged 20-74 years, of which 48 per cent were in men and 52 per cent were in women (Table 7.1). The analyses presented here are restricted to the 371,890 patients (36 per cent) for whom OPCS received a valid occupational code, this figure being more than twice as high for men (51 per cent) as for women (22 per cent). Of the remainder, occupational codes were not provided for 35 per cent of men and 63 per cent of women. A detailed examination of the data revealed that certain registries had not supplied valid occupational codes for all years, resulting in the removal of a further 14 per cent. In addition, 3,613 individual notifications could not be used because the registry of origin, which was needed for the analysis, was missing from the computerized registration record.

To facilitate comparison with the findings for cancer mortality, results are presented here using the same occupational and disease groupings as those employed in the analysis of mortality (see Chapters 4-6, Appendix 4). Figures 7.1 and 7.2 show the numbers of cancer registrations for the fifteen commonest job groups in men and women. Among women with an occupation recorded, the commonest job group was office workers and cashiers, which accounted for 27 per cent of the total. For men, no single job group dominated although, as for women, office workers and cashiers was the most common, accounting for 7 per cent of all registrations. The fifteen commonest cancers in men and women with an

Table 7.1 Number of cancer registrations in men and women aged 20-74 years, England, 1981-87

	Men		Women		Total	
	Number	%	Number	%	Number	%
All registrations	496,152	100	538,607	100	1,034,759	100
of which						
Total included	252,663	51	119,227	22	371,890	36
Total excluded	243,489	49	419,380	78	662,869	64
of which						
Code not given	174,323	35	340,268	63	514,591	50
Incorrect/incomplete codes	67,357	14	77,308	14	144,665	14
Registry unknown	1,809	<1	1,804	<1	3,613	<1

Figure 7.1 Fifteen most common occupations reported at cancer registration, men aged 20-74, England 1981-87

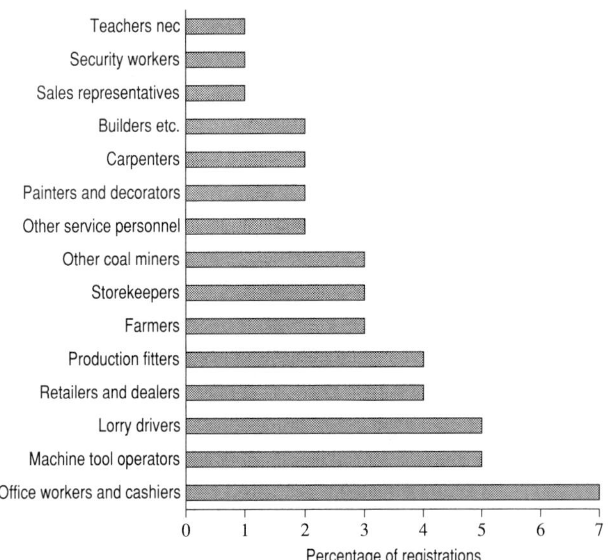

Figure 7.2 Fifteen most common occupations reported at cancer registration, women aged 20-74, England 1981-87

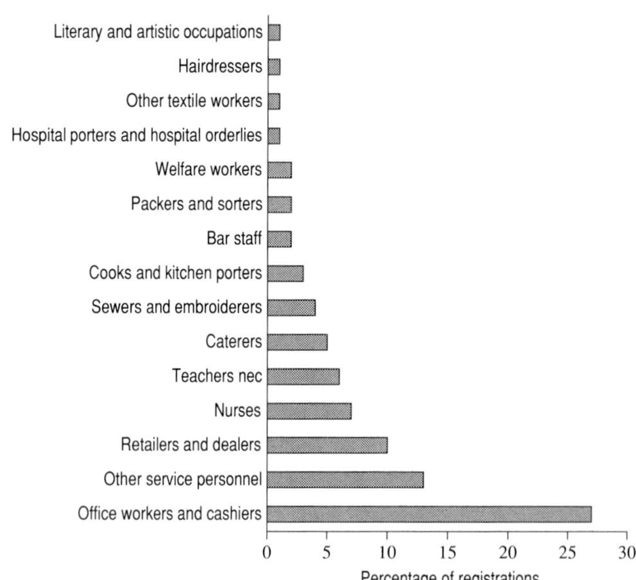

occupation recorded are presented in Figures 7.3 and 7.4. For men, cancer of the lung was by far the most frequent, accounting for 32 per cent of the total. For women, cancer of the breast was commonest (24 per cent) although cancers of in-situ cervix (15 per cent) and lung (10 per cent) also contributed substantial numbers of registrations.

7.3 Issues for the interpretation of the results

The measure of effect used in this chapter is the proportional registration ratio (PRR) which is calculated in the same way as the proportional mortality ratio (PMR), all registrations with an adequately described occupation forming the standard for comparison (see Chapter 2 for details). The reasons for basing analyses of cancer registration on proportional indices are similar to those put forward for mortality. Several issues need to be borne in mind when interpreting the results of proportional analyses. Some of these are addressed elsewhere in this volume. Considerations more specific to the interpretation of cancer registration data are discussed briefly below.

An advantage of the occupational information collected at cancer registration is that details are usually provided by the individual concerned and not by their next of kin. In practice, this means that occupational information is likely to be more accurate than that collected at death certification. The completeness of recording of occupation at cancer registration is,

Figure 7.3 Fifteen most common cancers in men aged 20-74, England 1981-87

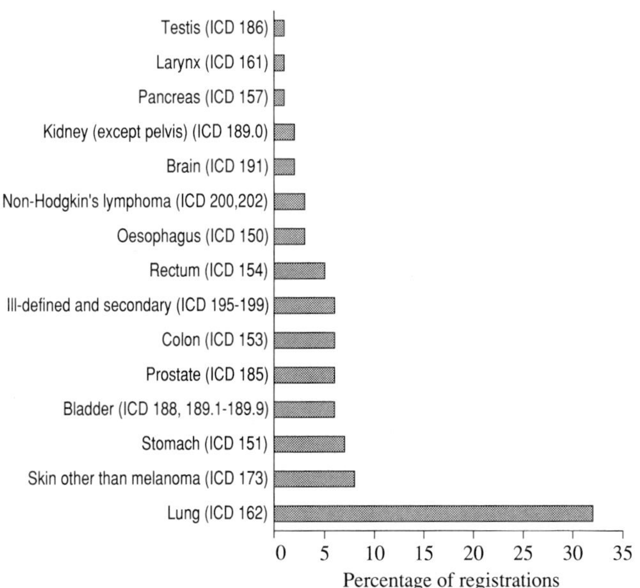

Figure 7.4 Fifteen most common cancers in women aged 20-74, England 1981-87

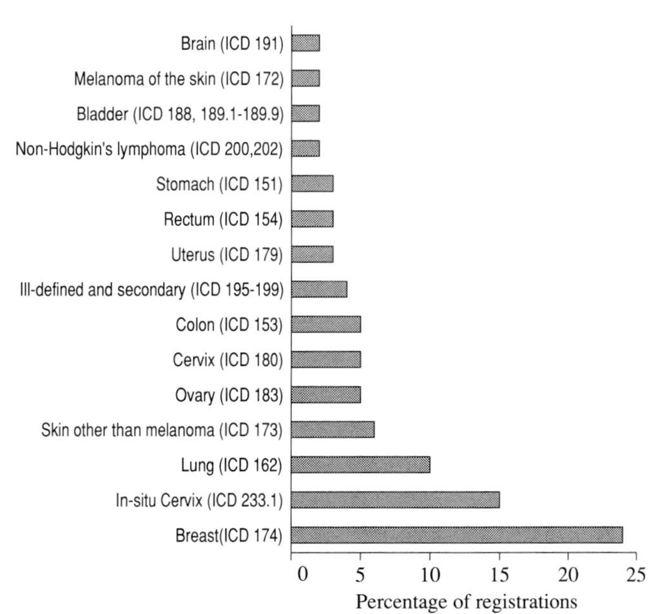

however, much lower than at death: more than half of all cancers registered in those aged 20-74 years could not be included in the analysis presented in this chapter (Table 7.1). As at death, women are far less likely to have an occupation recorded than men. One reason for the poor reporting of occupation at cancer registration is that information is gathered from a variety of sources.[3,4] Whether or not incomplete recording results in bias is difficult to assess. It is possible, however, that issues of industrial compensation influence the likelihood of occupational information being recorded. For example, the fact that pleural mesothelioma is a compensatible disease may result in occupations involving asbestos exposure being more likely to be recorded at cancer registration (see Chapter 9).

The effect of selection into or out of a particular occupation on the basis of health was considered as a potential source of bias for mortality (Chapter 4). Similar issues may apply here, although the effect of health related selection on cancer morbidity is less well understood. It is possible that individuals may change job, or retire from work completely, during the pre-diagnosis symptomatic stages of certain cancers. This could introduce bias if the most recent, rather than longest held, occupation is recorded at cancer registration.

A related issue of particular importance to cancer analyses is latency, since carcinogenic effects of occupational exposure may take years — perhaps even decades — to develop. The occupation recorded at cancer registration may therefore be less relevant than the occupation held several years previously. In so far as an occupational title can be taken to indicate exposure to a potentially carcinogenic substance, this could lead to misclassification of exposure and to an underestimation of the strength of associations.

The PRRs presented here are adjusted for age (five-year age groups), social class (six classes) and registry of origin. The possibility of confounding by other factors needs, however, to be considered. An obvious example is smoking, as this is a known risk factor for several cancers including lung, oesophagus, larynx, oral cavity and pharynx. Other factors known to affect cancer risk include alcohol consumption and parity. As several of these exposures are also related to social class (e.g. smoking is more common in social classes IV and V than in I and II) and, to a lesser extent, region of residence, the adjustment procedure might be expected to remove some of their confounding effects. At least some residual confounding by lifestyle factors is likely to remain and should be considered when interpreting the results (see Chapter 12).

Results have been examined for 60 specific cancers across 194 different occupational groups. When such a large number of associations are examined, some statistically significant findings are likely to occur by chance alone. Where similar results have been described before, attention is drawn to PRRs which were significant at the two-sided 5 per cent level of statistical significance. Associations which have not been reported previously should be interpreted with care.

7.4 Results by job group

The main findings for men and women according to the occupational groups listed in Appendix 2 are tabulated in Appendices 5 and 6. PRRs are given if they are significantly greater than 100 at the two-sided 5 per cent level of significance (based on three or more registrations), or significantly less than 100.

The main results for all 194 job groups are summarized in section 7.8, with reference to the findings for cancer mortality (Chapters 4 and 5).

7.5 Results by cancer site: cancers with known occupational associations

Results for six cancers with a known occupational aetiology are presented here; namely nose and nasal sinuses, lung, pleura, skin other than melanoma, bladder and leukaemia. For the purpose of this summary, PRRs have been tabulated if they were significantly greater than 100, based on three or more registrations at the two-sided 5 per cent level of significance. To aid interpretation, PRRs have been ordered according to the magnitude of the lower 95 per cent confidence interval. This method of ordering has the advantage of taking account of both the magnitude of the PRR and the level of statistical variability associated with the PRR.

7.5.1 Cancer of the nose and nasal sinuses (ICD 160)

Cancers of the nose and nasal sinuses are very rare, accounting for less than half of one per cent of all malignancies in these data (508 registrations in men and 120 in women). That exposures in the furniture and cabinet making industry are important determinants of these cancers is clearly demonstrated in Table 7.2. As documented elsewhere,[1,5-8] the risk is greatest for furniture makers and woodworking machinists. Among men, the highest PRR was for cabinet makers who had an eight-fold risk (9 registrations) and the second highest was for woodworking machinists who had a seven-fold risk (10 registrations).

Table 7.2 Job groups with significantly high PRRs[a] for cancers of the nose and nasal sinuses (ICD=160), ordered by the lower 95% confidence interval (CI). Men and women aged 20-74 years, England, 1981-87

	Job group	Number	PRR	(95% CI)
Men				
105	Cabinet makers	9	803	(367 -1525)
108	Woodworking machinists	10	710	(341 -1307)
109	Other woodworkers	5	676	(220 -1580)
164	Packers and sorters	6	312	(115 -681)
Women				
060	Other service personnel	29	153	(103 -220)

[a] P<0.05, based on at least three registrations. Adjusted for age, social class and region of registration.

A further occupational group worthy of mention are leather and shoe workers because of the previously reported link between exposure to leather dust and nasal cancer.[7] A two-fold relative risk was observed in leather and shoe workers but this was based on only four cases and was not statistically significant. No cases of this malignancy were reported in women in this occupational group.

7.5.2 Cancer of the lung (ICD 162)

Cancer of the lung accounted for the largest percentage of registrations in men and was the third commonest cancer in women (Figures 7.3 and 7.4). Although around 90 per cent of all lung cancers can be attributed to smoking, many occupational associations have also been suggested.[6-8] In the data examined here, several occupations had statistically significant PRRs (Table 7. 3). With few exceptions, these represent only moderately raised risks based on very large numbers of cases. Most PRRs are less than 125 and excesses of this magnitude may well be explained by variations in smoking habits (see Chapter 12).

Table 7.3 Job groups with significantly high PRRs[a] for cancer of the lung (ICD=162), ordered by the lower 95% confidence interval (CI). Men and women aged 20-74 years, England, 1981-87

	Job group	Number	PRR	(95% CI)
Men				
030	Professional engineers nec	549	125	(115-136)
045	Publicans and bar staff	816	122	(115-131)
116	Moulders and coremakers (metal)	329	123	(111-138)
040	Managers in transport, utilities and mining	560	120	(111-131)
044	Retailers and dealers	2,536	114	(110-119)
179	Shunters and pointsmen	65	141	(109-181)
174	Construction workers nec	1,144	114	(109-122)
194	Boiler operators	387	114	(104-127)
183	Lorry drivers	3,666	107	(104-111)
167	Plasterers	306	115	(103-129)
160	Painters and decorators	1,664	108	(103-114)
012	Vocational trainers, social scientists etc	135	121	(102-144)
126	Metal polishers	157	120	(102-141)
120	Other metal manufacturers	507	111	(102-122)
064	Undertakers	49	136	(101-180)
127	Fettlers and dressers (metal)	102	124	(101-151)
056	Van sales persons	233	115	(101-131)
164	Packers and sorters	421	111	(101-123)
038	Production and maintenance managers	632	108	(101-118)
165	Bricklayers and tilesetters	1,059	107	(101-114)
115	Metal drawers	64	130	(100-166)
029	Electrical and electronic engineers (professional)	86	124	(100-154)
187	Crane drivers	526	109	(100-119)
088	Other coalminers	2,337	104	(100-109)
Women				
174	Construction workers nec	16	269	(154-437)
057	Sales representatives	54	165	(124-216)
045	Publicans and bar staff	293	138	(124-156)
086	Plastic workers	7	306	(123-631)
126	Metal polishers	16	191	(109-310)
044	Retailers and dealers	1,036	112	(106-120)
104	Carpenters	20	171	(105-266)
124	Machine tool operators	94	129	(105-159)
012	Vocational trainers, social scientists etc	15	180	(101-297)
125	Press and automative machine operators	69	127	(100-162)

[a] P≤0.05, based on at least three registrations. Adjusted for age, social class and region of registration.

For a number of occupational groups excesses of lung cancer are evident in both sexes. Whilst these excesses could reflect the smoking habits of individuals employed in the occupations concerned, the occupational environment may also be important. The increased risk of lung cancer among publicans and bar staff, for example, may reflect heavier smoking among people working in these occupations and/or exposure to environmental tobacco smoke in the workplace.

It is also of interest to note that several of the occupations listed involve work with metals, for example, moulders and coremakers, metal polishers, other metal manufacturers, fettlers and dressers and metal drawers. Although it is possible that some of the excesses represent true occupational hazards,[7] it is difficult to rule out the possibility that they might have resulted from the confounding effects of smoking.

7.5.3 Pleural cancer (ICD 163)

The majority of pleural cancers are mesotheliomas, which are strongly related with exposure to asbestos. Many of the jobs listed in Table 7.4 are likely to have involved contact with this mineral. Pleural cancer is a highly fatal disease and, with very few exceptions, the occupations listed are also those with increased mortality (see Table 6.7). Among women, only two occupational groups had a significant excess of pleural cancer based on three or more registrations. For one of these, namely carpenters, there was also a significant excess in men.

Table 7.4 Job groups with significantly high PRRs[a] for cancer of the pleura (ICD=163), ordered by the lower 95% confidence interval (CI). Men aged 20-74 years, England, 1981-87

	Job group	Number	PRR	(95% CI)
Men				
146	Metal plate workers	41	378	(271-513)
138	Electrical plant operators	16	382	(219-622)
137	Electricians	61	250	(192-322)
153	Vehicle body builders	10	385	(185-709)
174	Construction workers nec	42	245	(177-331)
145	Sheet metal workers	25	266	(173-394)
144	Plumbers and gas fitters	50	220	(164-291)
104	Carpenters	70	206	(161-261)
132	Production fitters	121	181	(150-217)
003	Personnel managers etc	8	336	(146-664)
038	Production fitters	23	227	(144-341)
149	Welders	34	196	(136-274)
143	Electrical engineers (so described)	18	207	(123-328)
194	Boiler operators	11	231	(115-414)
075	Chemical workers	30	169	(114-242)
031	Draughtspersons	9	234	(107-445)
101	Upholsterers	7	257	(104-531)
Women				
104	Carpenters	3	1596	(329-4665)
185	Bus conductors and drivers' mates	3	753	(156-2203)

[a] P≤0.05, based on at least three observations. Adjusted for age, social class and region of registration.

Like pleural cancer, most peritoneal cancers (ICD 158.8 - 158.9) are mesotheliomas. These are extremely rare, but it is interesting to note that among men the only two groups with significantly raised risks were electrical plant operators (PRR 990, 3 registrations) and construction workers (PRR 449, 14 registrations) (Appendix 5). Both these groups had significant excesses of pleural cancer (Table 7.4).

7.5.4 Skin cancer other than melanoma (ICD 173)

The results presented in Table 7.5 support the suggestion that outdoor work and consequent exposure to ultra violet radiation is associated with cancers of the skin other than melanoma. Obvious examples are seafarers (PRR 150, 142 registra-

Table 7.5 Job groups with significantly high PRRs[a] for cancers of the skin other than melanoma (ICD=173), ordered by the lower 95% confidence interval (CI). Men and women aged 20-74 years, England, 1981-87

	Job group	Number	PRR	(95% CI)
Men				
011	Teachers nec	512	163	(150-178)
034	Aircraft flight deck officers	25	207	(134-306)
036	Seafarers	142	150	(127-178)
015	Doctors	161	148	(126-173)
028	Physical scientists and mathematicians	60	153	(117-198)
165	Bricklayers and tilesetters	256	126	(112-144)
047	Farmers	745	118	(110-127)
014	Clergy	101	133	(109-163)
169	Builders etc.	346	119	(108-133)
062	Ambulance workers	43	147	(107-199)
139	Telephone fitters	107	128	(106-156)
124	Machine tool operators	905	111	(105-119)
013	Welfare workers	86	129	(104-160)
063	Railway station workers	108	126	(104-152)
006	Sales managers etc	253	118	(104-134)
097	Printers	146	121	(103-143)
066	Fishing and related workers	28	151	(101-219)
039	Managers in construction	81	127	(101-158)
104	Carpenters	357	112	(101-125)
Women				
180	Railway engine drivers	5	354	(115-828)
011	Teachers nec	400	117	(107-130)
015	Doctors	43	144	(104-194)
069	Preparatory fibre processors	17	172	(101-277)
017	Nurses	460	110	(101-122)

[a] P≤0.05, based on at least three registrations. Adjusted for age, social class and region of registration.

tions), farmers (PRR 118, 745 registrations) and fishing and related workers (PRR 151, 28 registrations). For many of the other groups listed in Table 7.5, it is likely that a large proportion of the working day is spent outside, for example bricklayers and tile setters, (PRR 126, 256 registrations) and builders etc. (PRR 119, 346 registrations). The raised risk among aircraft flight deck officers (PRR 207, 25 registrations) could also be related to ultra violet or, possibly, other forms of radiation.

Several categories of health care professionals also appear to be at increased risk. The PRRs for male and female doctors, male ambulance workers and female nurses are 148 (16 registrations), 144 (43 registrations), 147 (43 registrations) and 110 (460 registrations) respectively.

7.5.5 Bladder cancer (ICD 188, 189.1-189.9)

Bladder cancer is more common in men than women. In the data examined here bladder cancer accounted for six per cent of registrations (14,598) in men and two percent of registrations (2,068) in women. The well established link between employment in the rubber industry and bladder cancer[6,7] is reflected in the results presented in Table 7.6: the PRRs for men and women engaged in rubber manufacture are 226 (58 registrations) and 350 (7 registrations) respectively. The data further suggest that other occupations potentially exposed to chemical compounds, such as aromatic amines (used in the manufacture of dyes, pigments, rubber, etc.) may also be at increased risk. The risk for men described as plastic goods makers, which is nearly twice that expected at 187 (19 registrations), is particularly noteworthy.

Table 7.6 Job groups with significantly high PRRs[a] for bladder cancer (ICD=188, 189.1-189.9), ordered by the lower 95% confidence interval (CI). Men and women aged 20-74 years, England, 1981-87

	Job group	Number	PRR	(95%CI)
Men				
085	Rubber manufacturers	58	226	(172-293)
093	Plastic goods makers	19	187	(113-293)
023	Driving instructors	36	161	(113-224)
124	Machine tool operators	735	112	(105-121)
166	Masons and stonecutters	23	164	(104-247)
158	Coach painters	12	198	(103-347)
184	Other motor drivers	112	125	(103-151)
018	Pharmacists	35	146	(102-203)
063	Railway station workers	90	126	(102-156)
097	Printers	115	123	(102-148)
006	Sales managers	168	117	(100-137)
054	Postal workers	222	114	(100-131)
Women				
085	Rubber manufacturers	7	350	(141-723)
077	Brewery workers	3	589	(122-1723)
072	Knitters	14	193	(106-324)
017	Nurses	138	120	(101-142)

[a] P≤0.05, based on at least three registrations. Adjusted for age, social class and region of registration.

7.5.6 Leukaemia (ICD 204-208)

The leukaemias contributed around two per cent of all registrations in this age group (20-74 years). Acute myeloid leukaemia was the commonest, accounting for 35 per cent of all leukaemias in men (1,800 registrations) and 45 per cent in women (691 registrations). Several chemical and physical exposures have been implicated in the aetiology of leukaemia, including ionizing radiation and benzene. Prior evidence for excess risks exist for farmers, leather and shoe workers and, more recently, for electrical workers.[6,7] Several of these occupational associations are evident in the present analyses (Table 7.7), where PRRs are presented for occupations in which the risk was significantly raised either for all leukaemias combined or for one of the main subtypes.

Among other occupational groups listed, the most notable finding is for women employed as biological scientists, where the PRR for acute myeloid leukaemia was 1,244. Whilst impressive, it should be noted that this PRR is based on only four cases.

7.6 Results by cancer site: cancers with suspected occupational associations

Results for six cancers for which an occupational link is suspected are presented here; namely mouth and pharynx, colon, larynx, melanoma, prostate and brain. As in section 7.5 PRRs have been tabulated if they were significantly greater than 100, based on three or more registrations, and ordered according to the magnitude of the lower 95 per cent confidence limit.

7.6.1 Cancers of the mouth and pharynx (ICD 140-149)

Cancers of the mouth and pharynx are a heterogeneous group, which together accounted for around 2 per cent of malignancies in men and one per cent in women. Occupational groups with raised PRRs for all cancers of the mouth

Table 7.7 Job groups with significantly high PRRs[a] for all leukaemias (or one of the listed subtypes) ordered by the lower 95% confidence interval (CI) for all leukaemia. Men and women aged 20-74 years, England, 1981-87

	Job group	All leukaemias (ICD=204-208)			Acute lymphatic (ICD=204.0)			Chronic lymphatic (ICD=204.1)		
		Number	PRR	(95%CI)	Number	PRR	(95% CI)	Number	PRR	(95% CI)
Men										
142	Other electronic maintenance engineers	19	277	(167 -434)	2	246	(30 -890)	5	408	(133 -954)
068	Leather and shoe workers	28	159	(106 -230)	1	95	(3 -531)	6	114	(42 -250)
047	Farmers	213	117	(102 -134)	13	128	(68 -219)	74	127	(100 -160)
130	Precision instrument makers	13	190	(101 -325)	1	206	(5 -1152)	5	280	(91 -656)
180	Railway engine drivers	20	158	(97 -245)		0	[b]	6	153	(56 -334)
007	Government inspectors	17	141	(83 -227)	1	136	(4 -759)	9	234	(107 -445)
173	Mains and service layers	10	164	(79 -303)	2	735	(89 -2658)	1	53	(1 -301)
011	Teachers nec	89	95	(77 -118)	4	55	(15 -142)	22	90	(57 -137)
042	Butchers	37	101	(71 -140)	1	39	(1 -218)	18	175	(104 -278)
027	Chemical engineers (so described)	15	117	(66 -193)		0	[b]	7	270	(109 -557)
059	Cooks and kitchen porters	26	97	(64 -143)	3	131	(27 -386)	13	203	(108 -348)
150	Riggers	7	138	(56 -286)		0	[b]		0	[b]
177	Railway guards	6	108	(40 -235)		0	[b]	5	319	(104 -745)
Women										
075	Chemical workers	18	305	(181 -483)	1	221	(6 -1236)	3	262	(54 -768)
161	Electrical, electronic assemblers	8	320	(138 -632)		0	[b]	1	231	(6 -1291)
038	Production and maintenance managers	6	290	(107 -632)		0	[b]		0	[b]
026	Biological scientists	5	307	(100 -718)	1	190	(5 -1061)		0	[b]
182	Bus and coach drivers	4	323	(88 -828)		0	[b]		0	[b]

	Job group	Acute myeloid (ICD=205.0)			Chronic myeloid (ICD=205.1)		
		Number	PRR	(95%CI)	Number	PRR	(95% CI)
Men							
142	Other electronic and maintenance engineers	8	293	(127 -579)	1	73	(2 -411)
068	Leather and shoe workers	13	222	(119 -381)	5	219	(71 -513)
047	Farmers	66	114	(89 -146)	32	112	(77 -159)
130	Precision instrument makers	5	204	(67 -478)	1	90	(2 -502)
180	Railway engine drivers	9	223	(102 -425)	2	113	(14 -409)
007	Government inspectors	6	158	(58 -346)		0	[b]
173	Mains and service layers	2	103	(13 -375)	5	450	(146 -1052)
011	Teachers nec	29	92	(62 -133)	25	155	(100 -229)
042	Butchers	10	80	(39 -149)	5	91	(30 -214)
027	Chemical engineers (so described)	5	98	(32 -230)	2	82	(10 -299)
059	Cooks and kitchen porters	5	51	(17 -121)	3	66	(14 -194)
150	Riggers	2	125	(15 -456)	4	484	(132 -1239)
177	Railway guards	1	54	(1 -305)		0	[b]
Women							
075	Chemical workers	10	382	(183 -703)	1	129	(3 -721)
161	Electrical, electronic assemblers	3	247	(51 -722)	1	276	(7 -1541)
038	Production and maintenance managers	5	500	(162 -1168)	1	317	(8 -1769)
026	Biological scientists	4	1244	(339 -3188)		0	[b]
182	Bus and coach drivers	4	584	(159 -1497)		0	[b]

[a] P≤0.05, based on at least three registrations. Adjusted for age, social class and region of registration.
[b] no registrations observed, and P>0.05.

and pharynx combined, or for cancers of the oral cavity and pharynx alone, are listed in Table 7.8.

That cancers of the mouth and pharynx are related to smoking and alcohol consumption is borne out by several of the associations observed. The most striking finding is for publicans and bar staff. The PRRs for men and women employed in this group are 300 (138 registrations) and 181 (36 registrations) respectively. As well as active smoking, these elevated risks could reflect the effects of environmental tobacco smoke in the workplace (see Chapter 12).

Several other factors have been implicated in the aetiology of specific cancers of the mouth and pharynx.[6,7] Nasopharyngeal cancer (ICD 147), for example, has been related to the preparation and consumption of salted fish and preserved food. In this context, it is worth noting that the risk of nasopharyngeal cancer was markedly increased among men employed as cooks and kitchen porters (PRR 847, 18 registrations), and caterers (PRR 520, 9 registrations) (see Appendix 5). Another well known association is the excess of lip cancer (ICD 140) among outdoor workers.[1,5] In the data examined here the evidence was particularly strong for farmers (PRR 288, 47 registrations) (see Appendix 5).

7.6.2 Cancer of the colon (ICD 153)

Cancer of the colon is one of the commoner cancers (Figures 7.3, 7.4). The evidence for an occupational aetiology is relatively weak, but several occupational associations have been noted in previous Decennial Supplements.[1,5] Table 7.9 contains a mix of occupational groups, and no striking occupational relationships are evident. Contrary to expectation, sedentary or professional occupations are not especially prominent. It should be noted, however, that these results are adjusted for social class, which may have removed some of

Table 7.8 Job groups with significantly high PRRs[a] for all cancers of the mouth and pharynx (or one of the listed groups), ordered by the lower 95% confidence interval (CI) for all sites combined. Men and women aged 20-74 years, England, 1981-87

	Job group	All mouth & pharynx (ICD=140-149)			Oral cavity (ICD= 141, 143-145)			Pharynx (ICD=146-148)		
		No.	PRR	(95% CI)	No.	PRR	(95% CI)	No.	PRR	(95% CI)
Men										
045	Publicans and bar staff	138	300	(253 -355)	70	316	(247 -400)	48	334	(247 -444)
059	Cooks and kitchen porters	67	256	(199 -325)	22	185	(116 -281)	37	397	(280 -548)
036	Seafarers	58	240	(183 -311)	34	300	(208 -420)	18	214	(127 -339)
046	Caterers	40	212	(152 -289)	20	234	(143 -361)	15	235	(132 -389)
051	Launderers and dry cleaners	11	250	(125 -448)	4	195	(53 -500)	4	277	(76 -711)
015	Doctors	37	178	(125 -246)	24	258	(166 -385)	9	137	(63 -262)
066	Fishing and related workers	12	227	(118 -397)	5	193	(63 -451)	3	190	(39 -558)
024	Literary and artistic occupations	45	139	(102 -186)	21	148	(92 -227)	13	115	(62 -198)
085	Rubber manufacturers	14	184	(101 -310)	7	206	(83 -426)	6	190	(70 -415)
044	Retailers and dealers	156	106	(91 -125)	64	93	(72 -120)	62	131	(101 -169)
058	Security workers	77	111	(88 -140)	45	140	(103 -189)	14	61	(34 -103)
147	Steel erectors	15	147	(82 -243)	11	237	(118 -424)	1	26	(1 -146)
077	Brewery workers	9	176	(81 -336)	3	133	(28 -389)	6	311	(114 -678)
161	Electrical, electronic assemblers	4	251	(69 -645)	1	144	(4 -805)	3	556	(115 -1628)
Women										
040	Managers in transport utilities and mining	5	544	(177 -1270)		0	[b]	2	759	(92 -2745)
045	Publicans and bar staff	36	181	(127 -252)	17	192	(112 -309)	12	201	(104 -351)
143	Electrical engineers (so described)	3	609	(126 -1780)	2	744	(90 -2689)	1	1049	(27 -5848)

[a] P≤0.05 based on at least three registrations. Adjusted for social class and region of registration
[b] No registrations observed and P>0.05

Table 7.9 Job groups with significantly high PRRs[a] for cancer of the colon (ICD=153), ordered by the lower 95% confidence interval (CI). Men and women aged 70-74 years, England, 1981-87

Job group	Number	PRR	(95% CI)
Men			
103 Other workers with fabrics	15	231	(129-382)
089 Tobacco workers	7	305	(123-630)
008 Government administrators	35	155	(108-216)
001 Lawyers	51	143	(107-189)
071 Warp preparers and weavers	32	155	(106-220)
129 Toolmakers	94	129	(104-158)
182 Bus and coach drivers	114	125	(103-150)
124 Machine tool operators	654	110	(102-119)
011 Teachers nec	255	114	(101-130)
Women			
039 Managers in construction	3	629	(130-1840)
011 Teachers nec	315	116	(104-130)

[a] P≤0.05, based on at least three registrations. Adjusted for age, social class and region of registration.

the potential for identifying sedentary occupations at increased risk.

PRRs were raised to a similar degree for both male and female teachers. An excess of colon cancer mortality was noted for men in this group in the previous Decennial Supplement[5] and an excess of a similar magnitude to that seen here is evident in women's mortality (Chapter 5). The magnitude of the excess risks are modest, however, and the fact that they are statistically significant largely reflects the substantial numbers of cases on which they are based.

7.6.3 Cancer of the larynx (ICD 161)

Laryngeal cancer is a relatively rare malignancy which is about seven times more common in men than in women. The major known risk factors are alcohol and tobacco consumption, which together possibly explain up to 90 percent of all these cancers. Occupations with significantly high risks for laryngeal cancer are listed in Table 7.10. The high risk among men employed as publicans or bar staff, and possibly also seafarers, may reflect drinking and smoking habits of individuals in these occupations. The PRRs for petrol pump attendants and coach painters may also be worth noting.

Table 7.10 PRRs[a] for cancer of the larynx (ICD=161), ordered by the lower 95% confidence interval (CI). Men and women aged 20-74 years, England, 1981-87

Job group	Number	PRR	(95% CI)
Men			
045 Publicans and bar staff	75	194	(153-244)
055 Petrol pump attendants	6	368	(135-801)
036 Seafarers	33	173	(119-243)
158 Coach painters	5	365	(119-853)
120 Other metal manufacturers	27	154	(102-225)
051 Launderers and dry cleaners	8	233	(101-461)
043 Fishmongers and poultry dressers	9	219	(100-417)
Women			
124 Machine tool operators	5	325	(204-6080)

[a] P≤0.05, based on at least three registrations. Adjusted for age, social class and region of registration.

7.6.4 Melanoma (ICD 172)

In these data melanoma was more common in women than men, accounting for 2 per cent of registrations in women and less than one per cent in men. Occupations found to be at increased risk of melanoma of the skin are listed in Table 7.11. In contrast with the mortality analyses, relatively few occupational groups demonstrated a statistically significant excess risk of melanoma. Although exposure to sunlight is a known risk factor, the lack of representation of outdoor occupations is notable, and contrasts with findings for other cancers of the skin (Table 7.5).

Table 7.11 Job groups with significantly high PRRs[a] for melanoma of the skin (ICD=172), ordered by the lower 95% confidence interval (CI). Men and women aged 20-74 years, England, 1981-87

Job group	Number	PRR	(95% CI)
Men			
011 Teachers nec	93	151	(123-186)
124 Machine tool operators	102	139	(114-170)
001 Lawyers	22	168	(105-255)
098 Tailors and dressmakers	6	275	(101-600)
Women			
130 Precision instrument makers	3	765	(158-2237)
019 Medical radiographers	10	272	(131-501)
011 Teachers nec	223	145	(128-166)

a P≤0.05, based on at least three registrations. Adjusted for age, social class and region of registration.

Increased risks were evident for teachers and male lawyers in the cancer registration data, but not for several other professional occupations noted for their high melanoma mortality (see Table 6.4). This discrepancy may reflect the fact that the PRRs presented in Table 7.11 are adjusted for social class. An almost three-fold risk of melanoma was observed among women employed as medical radiographers (PRR 272, 10 registrations) but this may be a chance finding.

7.6.5 Prostate cancer (ICD 185)

Prostate cancer is predominately a cancer of older men, and was the fifth commonest cancer in these data (Figure 7.3). There is little consistent evidence linking prostate cancer with exposures in the workplace, although several occupational analyses have been conducted. Eleven occupations had statistically significant PRRs for cancer of the prostate (Table 7.12). The highest risk was for dental technicians (PRR 249, 15 registrations). In the mortality data, excess prostate cancer is also evident among dental technicians and dentists (see Appendix 4).

Table 7.12 Job groups with significantly high PRRs[a] for cancer of the prostate (ICD=185), ordered by the lower 95% confidence interval (CI). Men aged 20-74 years, England, 1981-87

Job group	Number	PRR	(95% CI)
110 Dental technicians	15	249	(140-412)
011 Teachers nec	255	129	(114-146)
124 Machine tool operators	682	117	(109-127)
101 Upholsterers	39	151	(108-207)
047 Farmers	641	117	(108-127)
052 Hairdressers	45	146	(107-196)
025 Persons involved in sport	16	185	(106-302)
143 Electrical engineers (so described)	82	132	(106-165)
033 Architects and surveyors	101	128	(104-156)
097 Printers (so described)	107	122	(101-149)
104 Carpenters	266	112	(100-127)

a P≤0.05, based on at least three registrations. Adjusted for age, social class and region of registration.

A marginally increased risk of prostatic cancer among farmers was found (PRR 117, 641 registrations). This association has been reported before.[1] At the time, it was speculated that it might have been an artefact caused by the relatively low lung cancer rates in this occupation. This seems a less likely explanation for the current analysis, where lung cancer risk in farmers was close to expectation (PRR 92, 2258 registrations).

7.6.6 Cancer of the brain (ICD 191)

Brain cancer accounted for two per cent of all registrations (Figures 7.3, 7.4). Little is known about the aetiology of brain tumours, but many factors have been implicated. The suspected role played by electromagnetic radiation, which is receiving increasing attention, is of particular importance to electrical occupations.[1,5,8] Female electrical engineers and electrical and electronic assemblers demonstrated more than three-fold excess risks (3 and 7 registrations respectively) (Table 7.13). Among men, none of the PRRs for individual electrical occupational groups were statistically significant (Table 7.13). An excess of brain cancer among draughtsmen, noted in a previous decennial supplement[1] is apparent here (PRR 168, 50 registrations).

Table 7.13 Job groups with significantly high PRRs[a] for cancer of the brain (ICD=191), ordered by the lower 95% confidence interval (CI). Men and women aged 20-74 years, England, 1981-87

Job group	Number	PRR	(95% CI)
Men			
031 Draughtspersons	50	168	(125-222)
011 Teachers nec	144	126	(107-149)
008 Government administrators	16	181	(104-295)
001 Lawyers	27	152	(100-222)
169 Builders etc.	94	123	(100-151)
Women			
143 Electrical engineers (so described)	3	630	(130-1843)
161 Electrical and electronic assemblers	7	311	(125-642)

a P≤0.05, based on at least three registrations. Adjusted for age, social class and region of registration.

7.7 Comment

The analyses presented in this chapter clearly demonstrate that occupational information collected at cancer registration can be used to identify groups which may be at increased risk of cancer. Many of the associations observed have been noted previously, but others have not. These occupational data are extremely valuable, and efforts should be made to improve the quality and completeness of the information collected.

This is the first time that national cancer registration data for England have been systematically examined for occupational associations in both men and women. Only selected findings could be summarized here. Further, more exhaustive analyses, which will include data for 1971-80, are planned.

7.8 Results by occupational group: commentary

The main findings for all 194 occupational groups are summarized below for men and women. PRRs are presented if they are significantly different from 100 based on three or more registrations. Results are also given for specific cancers where an association was noted in the mortality analyses (Chapters 4 and 5). In occupational groups where no associations were observed, only the total numbers of registrations are given.

001 - LAWYERS

Men Number of registrations: 544

There was some suggestion of an excess of colon cancer (PRR 143, 51 registrations), melanoma (PRR 168, 22 registrations) and brain cancer (PRR 152, 27 registrations). A small excess of lip cancer was also observed (PRR 763, 3 registrations). The increased risk of primary liver cancer (PRR 169, 9 registrations) was weaker than that observed for mortality, and was not statistically significant. Lung cancer was significantly below expectation (PRR 75, 67 registrations).

Women Number of registrations: 96

002 - ACCOUNTANTS

Men Number of registrations: 1567

There was a marginal deficit of primary liver cancer in this group (PRR 39, 5 registrations). The increased risk of melanoma, noted among men at death, was not evident at registration (PRR 102, 29 registrations).

Women Number of registrations: 185

In accord with the findings for womens' mortality, the risk of multiple myeloma was raised at cancer registration although not significantly so (PRR 251, 3 registrations).

003 - PERSONNEL MANAGERS ETC.

Men Number of registrations: 563

There was some evidence of an excess of pleural cancer (PRR 336, 8 registrations).

Women Number of registrations: 197

There was some evidence of an excess of multiple myeloma (PRR 473, 5 registrations). In contrast with mortality, the level of ovarian cancer was unexceptional (PRR 103, 9 registrations).

004 - ECONOMISTS AND STATISTICIANS

Men Number of registrations: 70

Melanoma was increased at death, but there were only two cancer registrations, lending little support to the mortality findings.

Women Number of registrations: 32

005 - COMPUTER PROGRAMMERS

Men Number of registrations: 235

There was some evidence of an excess of primary liver cancer (PRR 387, 5 registrations). As for mortality, cancers of the bone (PRR 255, 5 registrations), soft tissue (PRR 156, 5 registrations) and brain (PRR 105, 13 registrations) were raised, but not significantly so.

Women Number of registrations: 119

006 - SALES MANAGERS ETC

Men Number of registrations: 2354

A high level of skin cancer other than melanoma was observed (PRR 118, 253 registrations) and the risk of bladder cancer was also marginally increased (PRR 117, 168 registrations). Three cancer sites were high at death, but there was little suggestion of a corresponding increase at registration: the PRRs for melanoma, retroperitoneal cancer and acute lymphatic leukaemia being 99 (36 registrations), 108 (3 registrations) and 76 (3 registrations) respectively. In addition, the incidence of chronic lymphatic leukaemia was lower than expected (PRR 52, 9 registrations)

Women Number of registrations: 572

There were fewer registrations for non-Hodgkin's lymphoma than expected (PRR 37, 4 registrations). The increased mortality for ovarian cancer noted at death, was not matched by an increase at cancer registration (PRR 102, 26 registrations).

007 - GOVERNMENT INSPECTORS

Men Number of registrations: 495

There was some evidence that chronic lymphatic leukaemia was above expectation in this group (PRR 234, 9 registrations).

Women Number of registrations: 39

There was some evidence of a reduced incidence of cancer in-situ of the cervix: there were no registrations whereas four were expected.

008 - GOVERNMENT ADMINISTRATORS

Men Number of registrations: 341

In accord with mortality, increased levels of colon cancer (PRR 155, 35 registrations) and brain cancer (PRR 181, 16 registrations) were observed. There were fewer lung cancer registrations than expected (PRR 76, 63 registrations); and the excess risk of prostate cancer evident at death was not apparent in the cancer data (PRR 87, 22 registrations).

Women Number of registrations: 121

The increased risk of breast cancer noted at death, was less conspicuous at cancer registration (PRR 114, 34 registrations).

009 - OTHER ADMINISTRATORS

Men Number of registrations: 1109

There was some evidence of an excess of non-Hodgkin's lymphoma (PRR 142, 48 registrations) and cancer of the eye (PRR 281, 6 registrations).

Women Number of registrations: 777

An excess ovarian cancer (PRR 135, 54 registrations) accords with a similar finding for mortality. In contrast with mortality, the risks for melanoma (PRR 92, 17 registrations) and breast cancer (PRR 101, 209 registrations) are unremarkable.

010 - TEACHERS IN HIGHER EDUCATION

Men Number of registrations: 548

None of the excess cancer risks found at death were significantly elevated at cancer registration: prostate cancer (PRR 131, 42 registrations), brain cancer (PRR 148, 26 registrations), non-Hodgkin's lymphoma (PRR 135, 28 registrations), myeloma (PRR 58, 4 registrations), and chronic lymphatic leukaemia (PRR 136, 5 registrations). Lung cancer risk was reduced (PRR 78, 83 registrations), concurring with the findings for women's mortality and women's cancer registration.

Women Number of registrations: 188

In correspondence with the findings for male mortality and

cancer registration, the level of lung cancer was low (PRR 43, 6 registrations). The excesses of breast cancer and rectal cancer noted at death were not apparent at cancer registration: the PRRs being 100 (52 registrations) and 138 (7 registrations) respectively.

011 - TEACHERS NEC

Men Number of registrations: 3451

As for mortality, an excess of melanoma was observed (PRR 151, 93 registrations). The risk of skin cancers other than melanoma was also raised (PRR 163, 512 registrations). Significant excesses were also observed for six other cancers — colon (PRR 114, 255 registrations), prostate (PRR 129, 255 registrations), testis (PRR 128, 117 registrations), multiple myeloma (PRR 144, 60 registrations), brain (PRR 126, 44 registrations) and chronic myeloid leukaemia (PRR 155, 25 registrations). The excess mortality from thyroid cancer was reflected in the cancer registration data, but the association was weaker and not statistically significant (PRR 148, 16 registrations). There were fewer registrations than expected for cancers of the mouth and pharynx (PRR 70, 43 registrations), lung (PRR 56, 425 registrations), pancreas (PRR 67, 33 registrations), and ill-defined and secondary cancers (PRR 75, 125 registrations).

Women Number of registrations: 6321

As for men, raised risks were suggested for both melanoma (PRR 145, 223 registrations) and, to a lesser extent, other skin cancers (PRR 117, 400 registrations). Excesses of cancer of the breast (PRR 118, 1942 registrations), uterus (PRR 114, 244 registrations) and colon (PRR 116, 315 registrations), and deficits of cancer of the cervix (PRR 64, 177 registrations) and in-situ cervix (PRR 81, 714 registrations) also accord with the findings for mortality. By contrast, cancer of the brain (PRR 117, 121 registrations), non-Hodgkin's lymphoma (PRR 111, 132 registrations), Hodgkin's disease (PRR 115, 44 registrations) and cancer of the gallbladder (PRR 105, 23 registrations), which were high at death, were not significantly raised at registration. As for mortality, there were fewer than expected cancers of the lung (PRR 51, 231 registrations) and larynx (PRR 42, 6 registrations). Deficits were also suggested for cancers of the stomach (PRR 68, 72 registrations) and pharynx (PRR 46, 7 registrations).

012 - VOCATIONAL TRAINERS, SOCIAL SCIENTISTS, ETC.

Men Number of registrations: 510

An increased risk of lung cancer (PRR 121, 135 registrations), and a decreased risk of skin cancer other than melanoma (PRR 66, 33 registrations) were suggested. An excess of lung cancer is also apparent among women employed in this group. Retroperitoneal cancer risk was raised (PRR 577), although this observation is based on only three cases. In accord with the data on male mortality and female cancer incidence, the risk of non-Hodgkin's lymphoma was increased, although not significantly so (PRR 123, 21 registrations). Also in agreement with male mortality, there were more registrations from brain cancer than expected, although again the excess is not statistically significant (PRR 126, 20 registrations).

Women Number of registrations: 160

As for male mortality, the risk of non-Hodgkin's lymphoma in women employed in this group was above expectation (PRR 271, 8 registrations). In accord with data on male cancer registration, the risk of lung cancer was marginally increased (PRR 180, 15 registrations). Ill-defined and secondary cancers were also raised (PRR 274, 12 registrations). There was only one registration for chronic myeloid leukaemia, which was increased in women at death.

013 - WELFARE WORKERS

Men Number of registrations: 745

An excess risk of skin cancer other than melanoma was suggested (PRR 129, 86 registrations). There was only one case of bone cancer reported, contrasting with an apparently raised risk at death.

Women Number of registrations: 1689

There was a weak suggestion of an excess of meningeal cancer (PRR 183, 15 registrations) but — as for mortality — brain cancer was below expectation (PRR 61, 17 registrations). In contrast with mortality, there was little evidence of increased risks for cancers of the lung (PRR 102, 121 registrations), gallbladder (PRR 124, 7 registrations) or uterus (PRR 102, 60 registrations).

014 - CLERGY

Men Number of registrations: 677

High levels of skin cancer other than melanoma were apparent in this group (PRR 133, 101 registrations). In accord with the data on female cancer incidence and male mortality, the risk of lung cancer was very low (PRR 54, 69 registrations).

Women Number of registrations: 294

Uterine cancer risk was elevated (PRR 165, 23 registrations) and, as with male cancer incidence and female mortality, lung cancer risks were low (PRR 39, 7 registrations). In contrast to the data on female mortality there was little evidence of an increased risk of oesophageal cancer (PRR 88, 5 registrations).

015 - DOCTORS

Men Number of registrations: 989

Evidence of raised risks were strongest for cancers of the oral cavity (PRR 258, 24 registrations) and skin other than melanoma (PRR 148, 161 registrations). Non-Hodgkin's lymphoma also appeared to be in excess but only weakly so (PRR 139, 50 registrations). As for mortality, liver cancer was above expectation (PRR 178, 18 registrations). Deficits were suggested for cancers of the lung (PRR 73, 127 registrations), stomach (PRR 70, 34 registrations) and ill-defined and secondary cancers (PRR 56, 26 registrations). In addition, there were no registrations for pleural cancer whereas five were expected.

Women Number of registrations: 418

There was some, albeit fairly weak, suggestion of an excess of skin cancer other then melanoma (PRR 144, 43 registrations). Risks of cancers of the ovary (PRR 52, 11 registrations) and cervix (PRR 36) were reduced, although the latter finding was based on only four cases.

016 - DENTISTS

Men Number of registrations: 205

There was little evidence of an increased risk of prostatic cancer (PRR 104, 16 registrations), which was elevated at death.

Women Number of registrations: 52
There was only one pharyngeal cancer registration, providing little support for the association at death.

017 - NURSES
Men Number of registrations: 781
There was some evidence of an excess of cancer of the rectum (PRR 141, 52 registrations) and a deficit of cancer of the oesophagus (PRR 53, 10 registrations).

Women Number of registrations: 8037
Nasopharyngeal cancer was raised in this group (PRR 204, 13 registrations), and there was weak evidence of an excess of skin cancers other than melanoma (PRR 110, 460 registrations), which contrasts with the deficit noted for mortality. Bladder cancers were also slightly higher than expected (PRR 120, 138 registrations). There was little evidence of reduced risks for any specific cancer, including those cancers suggested by mortality although each was below expectation: breast (PRR 96, 1862 registrations), ovary (PRR 93, 359 registrations), colon (PRR 93, 322 registrations), oral cavity (PRR 78, 22 registrations) and melanoma (PRR 89, 176 registrations).

018 - PHARMACISTS
Men Number of registrations: 404
There were more registrations from bladder cancer than expected (PRR 146, 35 registrations).

Women Number of registrations: 151
The risk of cervix cancer, although increased at death, was not significantly raised (PRR 131, 6 registrations)

019 - MEDICAL RADIOGRAPHERS
Men Number of registrations: 28

Women Number of registrations: 122
There was some suggestion of an excess of melanoma (PRR 272, 10 registrations).

020 - PHYSIOTHERAPISTS
Men Number of registrations: 33
The observation that pancreatic cancer was increased in this group, was based on only three registrations (PRR 767).

Women Number of registrations: 219
Breast cancer was marginally raised (PRR 128, 71 registrations). Ovarian cancer was increased at death, but not at registration (PRR 112, 12 registrations). Chronic myeloid leukaemia was also increased at death, but there were no registrations for this cancer. Some evidence for a marginal deficit of stomach cancer was also apparent: no registrations were observed whereas four were expected.

021 - HEALTH PROFESSIONS NEC
Men Number of registrations: 292

Women Number of registrations: 356
There was weak evidence of an excess of meningeal cancer (PRR 322, 5 registrations). As for mortality, melanoma incidence was above average but of only borderline significance (PRR 174, 15 registrations).

022 - VETERINARIANS
Men Number of registrations: 67
A reduced risk of lung cancer was evident in this group (PRR 33, 4 registrations)

Women Number of registrations: 18

023 - DRIVING INSTRUCTORS (not HGV)
Men Number of registrations: 346
The above average number of bladder cancers was statistically significant (PRR 161, 36 registrations).

Women Number of registrations: 42

024 - LITERARY AND ARTISTIC OCCUPATIONS
Men Number of registrations: 1835
Risk of soft tissue sarcoma was increased at cancer registration (PRR 170, 20 registrations). In accord with the data on male mortality, the PRRs for cancers of the liver, oral cavity, and pharynx were also raised, although not significantly so, being 120 (15 registrations), 148 (21 registrations) and 115 (13 registrations) respectively. Skin cancers other than melanoma were also high at death, but there was little indication of a raised risk at registration (PRR 98, 174 registrations).

Women Number of registrations: 924
In the mortality data several alcohol-related cancers were found to be high: the PRRs for cancers of the liver, oral cavity, and oesophagus were similarly increased at registration, although not significantly so, being 185(3 registrations), 137(4 registrations), and 173(14 registrations) respectively.

025 - PERSONS INVOLVED IN SPORT
Men Number of registrations: 131
There was some evidence of an excess of cancer of the prostate (PRR 185, 16 registrations).

Women Number of registrations: 67
No notable associations were apparent, although there were three cases of Hodgkin's disease — for which mortality appeared raised — (PRR 386, 3 registrations). Only two cases of melanoma were observed and, in contrast with mortality, there was little evidence of a raised risk (PRR 124).

026 - BIOLOGICAL SCIENTISTS
Men Number of registrations: 163

Women Number of registrations: 71
There were four registrations for acute myeloid leukaemia in this group, yielding a PRR of 1244.

027 - CHEMICAL ENGINEERS AND SCIENTISTS
Men Number of registrations: 428
Weak evidence of a raised risk of chronic lymphatic leukaemia was observed (PRR 270, 7 registrations). Only three cases of pleural cancer provide little support for the excess risk noted for mortality (PRR 134).

Women Number of registrations: 28

028 - PHYSICAL SCIENTISTS AND MATHEMATICIANS
Men Number of registrations: 343
There were more registrations than expected for skin cancers other than melanoma (PRR 153, 60 registrations). In accord

with male mortality, the PRR for brain cancer was raised although not significantly so (PRR 147,18 registrations). There were no registrations for cancer of the mediastinum, which was raised at death.

Women Number of registrations: 67
Although levels of leukaemia were increased at death, only two leukaemias were registered in women in this group.

029 - ELECTRICAL AND ELECTRONIC ENGINEERS (PROFESSIONAL)
Men Number of registrations: 450
There were more cases than expected for cancers of the stomach (PRR 169, 35 registrations) and, to a lesser extent, lung (PRR 124, 86 registrations). As for mortality, bladder cancer was raised (PRR 128, 33 registrations).

Women Number of registrations: 29

030 - PROFESSIONAL ENGINEERS NEC
Men Number of registrations: 2561
A moderately increased risk of lung cancer (PRR 125, 549 registrations) and a non-significantly increased risk of pleural cancer (PRR 131, 18 registrations) was observed. Deficits of melanoma (PRR 68, 36 registrations) and of other skin cancers (PRR 77, 221 registrations) were also apparent.

Women Number of registrations: 100
There were more registrations than expected for in-situ cervix cancer (PRR 187, 19 registrations), but only one registration for invasive cervix cancer.

031 - DRAUGHTSPERSONS
Men Number of registrations: 962
As for mortality, an excess risk of brain cancer was suggested (PRR 168, 50 registrations). In addition, there was weak evidence of an excess of cancer of the pleura (PRR 234, 9 registrations).

Women Number of registrations: 117
For two cancers where an excess mortality was suggested, namely breast and ovary, risks were raised, but not significantly so (PRR 111, 32 registrations and PRR 124, 7 registrations respectively).

032 - LABORATORY TECHNICIANS
Men Number of registrations: 848
An excess of stomach cancer was found in this group (PRR 133, 68 registrations). Deficits of colon cancer (PRR 62, 32 registrations) and melanoma of the skin (PRR 40, 5 registrations) were also observed.

Women Number of registrations: 481
In accord with the data on women's mortality, an excess of kidney cancer was observed (PRR 255, 9 registrations). A deficit of uterine cancer was also noted (PRR 46, 7 registrations).

033 - ARCHITECTS AND SURVEYORS
Men Number of registrations: 1195
An increased risk was seen for prostate cancer (PRR 128, 101 registrations) and, to a lesser extent, cancer of the small intestine (PRR 278, 7 registrations). There was weak evidence of a reduced risk of stomach cancer (PRR 72, 40 registrations) and of mouth and pharyngeal cancers (PRR 53, 13 registrations). As for mortality, risk of pleural cancer was above expectation, but not significantly so (PRR 173, 9 registrations).

Women Number of registrations: 73

034 - AIRCRAFT FLIGHT DECK OFFICERS
Men Number of registrations: 115
An excess of skin cancers other than melanoma (PRR 207, 25 registrations), was apparent. Mortality from colonic cancer and prostatic cancer was high, but registrations were only marginally elevated; the PRRs being 115 (8 registrations) and 117 (8 registrations) respectively.

Women Number of registrations: 0

035 - AIR TRAFFIC CONTROLLERS
Men Number of registrations: 30
There was only one registration for primary liver cancer, lending little support to the excess noted for mortality.

Women Number of registrations: 2

036 - SEAFARERS
Men Number of registrations: 1361
As for mortality, risks were high for a number of sites including cancer of the oral cavity (PRR 300, 34 registrations), cancer of the pharynx (PRR 214, 18 registrations), cancer of the larynx (PRR 173, 33 registrations), and skin cancers other than melanoma (PRR 150, 142 registrations). There was, however, no suggestion of an excess of primary liver cancer (PRR 102, 11 registrations). There were fewer registrations than expected for four sites; rectal cancer (PRR 67, 41 registrations), bladder cancer (PRR 75, 58 registrations), Hodgkin's disease (PRR 37, 4 registrations), and multiple myeloma (PRR 29, 4 registrations).

Women Number of registrations: 18

037 - TECHNICIANS NEC
Men Number of registrations: 778

Women Number of registrations: 140
Four cases of Hodgkin's disease were observed (PRR 428). As for mortality, risks for cancers of the rectum and uterus were raised, but not significantly so: the PRRs being 215 (6 registrations) and 155 (8 registrations) respectively. There were no registrations for kidney cancer in this group, so the excess reported for mortality was not supported.

038 - PRODUCTION AND MAINTENANCE MANAGERS
Men Number of registrations: 2374
Registrations for three sites were higher than expected: cancer of the lung (PRR 108, 632 registrations), cancer of the pleura (PRR 227, 23 registrations), and cancer of the soft tissues (PRR 173, 20 registrations). There were fewer registrations than expected for two sites: cancer of the oral cavity (PRR 50, 10 registrations) and cancer of the prostate (PRR 79, 118 registrations). In contrast to the mortality data, there was no indication of an increased risk of cancer of the gall bladder (PRR 97, 11 registrations).

Women Number of registrations: 132
In accord with the findings for mortality, stomach cancer risk

was elevated (PRR 353, 9 registrations). There were also more registrations for acute myeloid leukaemia than expected (PRR 500, 5 registrations).

039 - MANAGERS IN CONSTRUCTION

Men Number of registrations: 679

There was an excess of cancer of the skin other than melanoma (PRR 127, 81 registrations). As for mortality, pleural cancer risk was above expectation but this was based on only four cases and was not statistically significant (PRR 129, 4 registrations). There were no registrations for cancer of the peritoneum, hence the excess found at death was not supported.

Women Number of registrations: 10

There was some evidence of an increased risk of colon cancer (PRR 629), but this was based on only three registrations.

040 - MANAGERS IN TRANSPORT, UTILITIES AND MINING

Men Number of registrations: 1883

An excess of lung cancer was evident in this group (PRR 120, 560 registrations). There were also more registrations than expected for ill-defined and secondary cancers (PRR 124, 120 registrations). The excess for cancers of other endocrine organs was based on only three registrations (PRR 541). In contrast to mortality, no excesses of testicular cancer (PRR 85, 19 registrations) or penile cancer (PRR 96, 4 registrations) were noted. Three sites had significantly fewer registrations than expected: prostatic cancer (PRR 78, 99 registrations), skin cancer other than melanoma (PRR 83, 135 registrations), and multiple myeloma (PRR 54, 13 registrations).

Women Number of registrations: 113

There was some evidence of a raised risk for cancers of the mouth and pharynx (PRR 544, 5 registrations).

041 - OFFICE MANAGERS

Men Number of registrations: 2145

A marginal excess of cancer of the hypopharynx was observed (PRR 229, 9 registrations). The slight excess of gallbladder cancer suggested at death was evident at registration, but was not statistically significant (PRR 135, 13 registrations).

Women Number of registrations: 686

No notable associations were apparent. The excess risk of bladder cancer found at death, was not supported at cancer registration (PRR 94, 9 registrations).

042 - BUTCHERS

Men Number of registrations: 1661

The risk of chronic lymphatic leukaemia was higher than expected (PRR 175, 18 registrations), but risks of colon cancer (PRR 75, 65 registrations) and pleural cancer were lower (PRR 27, 3 registrations).

Women Number of registrations: 86

There were more registrations for oesophageal cancer than expected (PRR 402, 5 registrations). There were only two registrations for ovarian cancer hence the excess found at death was not supported.

043 - FISHMONGERS AND POULTRY DRESSERS

Men Number of registrations: 300

There was weak evidence of an excess risk for laryngeal cancer (PRR 219, 9 registrations) which accords with the finding for mortality. In contrast, the risk of lung cancer, which was high at death, appeared unremarkable (PRR 110, 108 registrations).

Women Number of registrations: 45

There were only three cases of cancer of the uterus (PRR 178, 3 registrations) and none of chronic lymphoid leukaemia, lending little support for the excess noted for mortality.

044 - RETAILERS AND DEALERS

Men Number of registrations: 8988

In this group there were excesses of pharyngeal cancer (PRR 131, 62 registrations) and lung cancer (PRR 114, 2536 registrations); and deficits of colon cancer (PRR 90, 531 registrations), skin cancers other than melanoma (PRR 80, 624 registrations) and non-Hodgkin's lymphoma (PRR 86, 219 registrations).

Women Number of registrations: 10910

In accord with the data on women's mortality and men's cancer registration, the risk of lung cancer was marginally raised (PRR 112, 1036 registrations). Registrations for two other sites were also high: invasive cervix (PRR 125, 547 registrations) and in-situ cervix (PRR 107, 1536 registrations). As for mortality, there were marginally fewer registrations than expected for breast cancer (PRR 95, 2471 registrations) and melanoma (PRR 75, 162 registrations). Ovarian cancer risk, also reduced at death, was close to expectation at registration (PRR 97, 548 registrations).

045 - PUBLICANS AND BAR STAFF

Men Number of registrations: 2664

There were several risks which differed from expectation in this occupational group. As for mortality, clear excesses were observed for cancers of the oral cavity (PRR 316, 70 registrations), pharynx (PRR 334, 48 registrations), larynx (PRR 194, 75 registrations), liver (PRR 202, 38 registrations) and lung (PRR 122, 816 registrations). The marked excess of mouth and pharyngeal cancers were apparent in several individual sites, notably floor of mouth (PRR 316, 70 registrations), oropharynx (PRR 431, 25 registrations) and hypopharynx (PRR 356, 19 registrations). Ill-defined and secondary cancers were also above expectation but the association was substantially weaker than the excesses already noted (PRR 118, 160 registrations). Reduced risks were observed for several cancer sites: brain (PRR 63, 45 registrations), colon (PRR 78, 136 registrations), melanoma (PRR 55, 17 registrations), other skin (PRR 70, 166 registrations), bladder (PRR 84, 140 registrations), prostate (PRR 72, 122 registrations), testis (PRR 60, 18 registrations), non-Hodgkin's lymphoma (PRR 67, 50 registrations), multiple myeloma (PRR 51, 17 registrations), all leukaemias combined (PRR 58, 37 registrations), and acute myeloid leukaemia alone (PRR 35, 7 registrations).

Women Number of registrations: 2590

Risks were increased for several of the cancers noted above for men: oral cavity (PRR 192, 17 registrations), pharynx (PRR 201, 12 registrations), and lung (PRR 138, 293 registrations). Excess risks were also evident for cancers of the cervix (PRR 140, 198 registrations for invasive cancer and PRR 116, 640 registrations for in-situ cancer). In addition, there were seven registrations for cancer of the mediastinum

(PRR 691, 7 registrations). A similar pattern of deficits to that already described for mortality was evident: breast cancer (PRR 80, 416 registrations), ovarian cancer (PRR 81, 90 registrations), uterine cancer (PRR 65, 45 registrations), melanoma (PRR 59, 26 registrations), other skin cancers (PRR 62, 80 registrations), kidney cancer (PRR 46, 9 registrations), and all leukaemias combined (PRR 55, 17 registrations).

046 - CATERERS

Men Number of registrations: 1025

In agreement with the mortality data, excesses were observed for cancer of the pharynx (PRR 235, 15 registrations) and primary cancer of the liver (PRR 227, 17 registrations). Cancer of the oral cavity was also raised (PRR 234, 20 registrations). Deficits were noted for two sites; colon (PRR 72, 45 registrations) and skin cancers other than melanoma (PRR 61, 55 registrations).

Women Number of registrations: 5015

There were fewer registrations than expected from multiple myeloma (PRR 54, 19 registrations). In accord with the data on women's mortality, the risks of lung cancer (PRR 105, 637 registrations) and stomach cancer (PRR 114, 178 registrations) were raised, although not significantly so.

047 - FARMERS

Men Number of registrations: 8245

Risks were significantly raised for cancers of the lip (PRR 288, 47 registrations), skin cancer other than melanoma (PRR 118, 745 registrations), prostate (PRR 117, 641 registrations), multiple myeloma (PRR 126, 120 registrations), all leukaemias combined (PRR 117, 213 registrations), and chronic lymphatic leukaemia (PRR 127, 74 registrations). Reduced risks were suggested for cancers of the oral cavity (PRR 68, 41 registrations) and gallbladder (PRR 69, 29 registrations) and for several cancers of the respiratory system: lung (PRR 92, 2258 registrations), pleura (PRR 28, 10 registrations) and larynx (PRR 80, 88 registrations).

Women Number of registrations: 838

Non-Hodgkin's lymphoma was the only malignancy for which there was suggestion of an excess (PRR 166, 25 registrations). As for mortality, the risk of ovarian cancer was above expectation but not significantly so (PRR 110, 43 registrations). In contrast with mortality, there was little evidence of a deficit of breast cancer (PRR 99, 171 registrations) or skin cancers other than melanoma (PRR 80, 39 registrations). There was only one registration for cancer of the mediastinum, and none for cancer of the endocrine glands other than thyroid.

048 - ARMED FORCES

Men Number of registrations: 1222

In contrast with mortality, there were no indications of increases in cancer of the oral cavity (PRR 105, 20 registrations) or cancer of the pharynx (PRR 103, 15 registrations).

Women Number of registrations: 77

049 - POLICE

Men Number of registrations: 1455

As for male mortality, nasal cancers were above average but based on only four cases, the excess was not statistically significant (PRR 140).

Women Number of registrations: 122

No notable associations were apparent. Only four cases of melanoma (PRR 174), and one case of a soft tissue malignancy, were reported lending only weak support to the findings for cancer mortality.

050 - FIRE SERVICE PERSONNEL

Men Number of registrations: 554

There was some evidence of an increased risk of oropharyngeal cancer in this group (PRR 469, 4 registrations), and a deficit of primary liver cancer: no cases were observed whereas four were expected.

Women Number of registrations: 2

051 - LAUNDERERS AND DRY CLEANERS

Men Number of registrations: 249

Mouth and pharyngeal cancers in general, and of the hypopharynx in particular, were increased in this group, the PRRs being 250 (11 registrations) and 580 (3 registrations) respectively. There was some suggestion of an excess of laryngeal cancers (PRR 233, 8 registrations) and of other leukaemias (PRR 675, 3 registrations). An above average risk of death from stomach cancer was not supported by the cancer registration data (PRR 101, 18 registrations), and only two soft tissue cancers were observed (PRR 206, 2 registrations).

Women Number of registrations: 751

There was some suggestion of an excess of ovarian cancer (PRR 139, 48 registrations). Brain cancer was below expectation (PRR 34) but this was based on only three cases.

052 - HAIRDRESSERS

Men Number of registrations: 486

Increased levels of prostate cancer (PRR 146, 45 registrations) and decreased levels of lung cancer (PRR 83, 133 registrations) and of oesophageal cancer (PRR 25, 3 registrations) were noted.

Women Number of registrations: 1061

A low risk of stomach cancer was evident (PRR 57, 13 registrations). There were only two registrations for cancer of the oral cavity (PRR 64), lending little support to the findings from mortality.

053 - OFFICE WORKERS AND CASHIERS

Men Number of registrations: 16265

Women Number of registrations: 29843

There was little suggestion of an excess of any specific cancer, including those for which mortality was statistically significantly raised: breast (PRR 101, 7265 registrations), colon (PRR 102, 1296 registrations) and brain (PRR 97, 420 registrations). Invasive cancer of the cervix was marginally below expectation (PRR 92, 1194 registrations).

054 - POSTAL WORKERS

Men Number of registrations: 3301

This group was characterized by a marginally increased risk

of bladder cancer (PRR 114, 222 registrations), and decreased risks of oesophageal cancer (PRR 72, 61 registrations), lung cancer (PRR 92, 1010 registrations) and pleural cancer (PRR 52, 12 registrations).

Women Number of registrations: 417
Breast cancer was high in the mortality data, but the risk at cancer registration was only marginally increased (PRR 109, 96 registrations).

055 - PETROL PUMP ATTENDANTS
Men Number of registrations: 125
Raised risks were suggested for cancers of the larynx (PRR 368, 6 registrations) and, to a lesser extent, lung (PRR 135, 47 registrations). The number of prostatic cancers was below expectation (PRR 21, 2 registrations).

Women Number of registrations: 59

056 - VAN SALES PERSONS
Men Number of registrations: 666
The only observations of note were a slight increase in lung cancer (PRR 115, 233 registrations), and a deficit of cancers of the mouth and pharynx (PRR 41, 5 registrations).

Women Number of registrations: 43

057 - SALES REPRESENTATIVES
Men Number of registrations: 3587
No excess risks were apparent. A lower than expected proportion of lung cancers was observed (PRR 89, 794 registrations).

Women Number of registrations: 528
In contrast with men, lung cancer was above expectation in this group (PRR 165, 54 registrations). In addition, raised risks were also suggested for in-situ cervix cancer (PRR 123, 140 registrations), liver cancer (PRR 347, 4 registrations) and oesophageal cancer (PRR 228, 9 registrations). A below average proportion of breast cancers was also observed (PRR 71, 86 registrations).

058 - SECURITY WORKERS
Men Number of registrations: 3553
There were more registrations than expected for two cancer sites: oral cavity (PRR 140, 45 registrations) and endocrine organs excluding thyroid (PRR 433, 5 registrations).

Women Number of registrations: 413
A marginally increased risk of uterine cancer was evident in this group (PRR 165, 20 registrations). In agreement with the findings for women's mortality there were fewer registrations than expected for stomach cancer (PRR 28, 3 registrations) and more for colon cancer (PRR 145, 25 registrations) although the latter was not statistically significant.

059 - COOKS AND KITCHEN PORTERS
Men Number of registrations: 1239
Excesses were apparent for the same three cancers noted for mortality, namely oral cavity (PRR 185, 22 registrations), pharynx (PRR 397, 37 registrations) and primary liver (PRR 215, 17 registrations). In addition, excesses of cancers of the penis (PRR 264, 57 registrations) and chronic lymphatic leukaemia (PRR 203, 13 registrations) were suggested. Skin cancer other than melanoma (PRR 74, 69 registrations) and colon cancer (PRR 71, 43 registrations) were below expectation.

Women Number of registrations: 3522
Pancreatic cancer, for which an excess was noted at death, was marginally increased although not significantly so (PRR 129, 51 registrations). In contrast to mortality, however, stomach cancer risk was not above average in this group (PRR 98, 130 registrations).

060 - OTHER SERVICE PERSONNEL
Men Number of registrations: 5608

Women Number of registrations: 14621
Marginal departures from expectation were noted for five sites. Excesses were seen for cancers of the nose and nasal sinuses (PRR 153, 29 registrations) and breast (PRR 103, 3003 registrations); and deficits were seen for cancers of the lung (PRR 94, 1782 registrations), pleura (PRR 57, 15 registrations), and ill-defined and secondary sites (PRR 92, 577 registrations).

061 - HOSPITAL PORTERS AND WARD ORDERLIES
Men Number of registrations: 911
There was a deficit of pleural cancer: no cases were observed whereas five were expected.

Women Number of registrations: 1322
An excess risk of thyroid cancer was suggested (PRR 195, 13 registrations). In contrast with mortality, lung cancer risk did not appear anomalous (PRR 102, 147 registrations).

062 - AMBULANCE WORKERS
Men Number of registrations: 395
Skin cancer other than melanoma was slightly increased in this group (PRR 147, 43 registrations). In contrast to mortality, there were no significant findings for brain cancer (PRR 97, 8 registrations) or pancreatic cancer (PRR 145, 8 registrations).

Women Number of registrations: 38
There were no statistically significant associations noted within this group. In contrast with mortality, the PRR for lung cancer was 95, based on three registrations.

063 - RAILWAY STATION WORKERS
Men Number of registrations: 1333
Excess risks were suggested for cancers of the skin other than melanoma (PRR 126, 108 registrations), eye (PRR 441, 6 registrations) and bladder (PRR 126, 90 registrations). Cancers of the lung (PRR 86, 416 registrations) and pharynx (PRR 32, 3 registrations) were below expectation.

Women Number of registrations: 59
As for men, an excess of skin cancer other than melanoma was observed (PRR 211, 9 registrations), although this finding was not statistically significant.

064 - UNDERTAKERS
Men Number of registrations: 148
There were slightly more registrations observed than expected for lung cancer (PRR 136, 49 registrations).

Women Number of registrations: 7

065 - FORESTERS

Men Number of registrations: 194

There was some suggestion of an increased risk of multiple myeloma (PRR 322, 7 registrations).

Women Number of registrations: 6

066 - FISHING AND OTHER RELATED WORKERS

Men Number of registrations: 292

There were more registrations than expected for skin cancers other than melanoma (PRR 151, 28 registrations), lip cancer (PRR 504, 3 registrations), and cancers of the mouth and pharynx combined (PRR 227, 12 registrations); and fewer than expected for bladder cancer (PRR 36, 6 registrations).

Women Number of registrations: 3

067 - TANNERY WORKERS

Men Number of registrations: 94

Women Number of registrations: 14

068 - LEATHER AND SHOE WORKERS

Men Number of registrations: 796

Risks were raised for all leukaemias combined (PRR 159, 28 registrations), acute myeloid leukaemia (PRR 222, 13 registrations), cancer of the soft tissues (PRR 273, 7 registrations), cancer of the rectum (PRR 136, 48 registrations) and cancer of the gum (PRR 714, 3 registrations). In accord with the mortality findings, nasal cancer risks were increased although not significantly so (PRR 211, 4 registrations).

Women Number of registrations: 488

069 - PREPARATORY FIBRE PROCESSORS

Men Number of registrations: 155

As for mortality, stomach cancer was above expectation, but not significantly so (PRR 136, 16 registrations). There were fewer registrations for skin cancers other than melanoma than expected (PRR 30, 3 registrations)

Women Number of registrations: 137

Raised risks were suggested for ill-defined and secondary cancers (PRR 209, 17 registrations), skin cancers other than melanoma (PRR 172, 17 registrations) and gallbladder cancer (PRR 625, 4 registrations). As for mortality, cancer of the larynx was above average (PRR 383, 3 registrations) and cancer of the breast below (PRR 58, 12 registrations), although neither association was statistically significant.

070 - SPINNERS AND WINDERS

Men Number of registrations: 235

Apart from an excess of ill-defined and secondary cancers (PRR 167, 21 registrations), no significant associations were apparent.

Women Number of registrations: 611

An excess of stomach cancer was evident in this group (PRR 154, 46 registrations). In accord with the mortality findings, breast cancer risks were low (PRR 76, 73 registrations), but lung cancer risks were unremarkable (PRR 110, 121 registrations). There was only one registration for retroperitoneal cancer lending little support to the findings for mortality.

071 - WARP PREPARERS AND WEAVERS

Men Number of registrations: 403

An excess of colon cancer was observed (PRR 155, 32 registrations). As for mortality in women in this group, lung cancer was below expectation (PRR 81, 111 registrations).

Women Number of registrations: 798

In accord with mortality, there were fewer than expected lung cancers (PRR 88, 116 registrations), breast cancers (PRR 96, 125 registrations) and multiple myelomas (PRR 85, 6 registrations), but none of these PRRs was statistically significant. Only two cases of acute monocytic leukaemia were reported, lending little support to the findings for mortality.

072 - KNITTERS

Men Number of registrations: 220

An excess of cancer of the gallbladder occurred in this group (PRR 426, 5 registrations).

Women Number of registrations: 306

Bladder cancer risk was increased (PRR 193, 14 registrations) and lung cancer risk was decreased (PRR 58, 20 registrations). In contrast with mortality, there was little indication of an increased risk of skin cancer other than melanoma (PRR 105, 17 registrations).

073 - BLEACHERS, DYERS AND FINISHERS

Men Number of registrations: 356

There was some suggestion of an excess of cancer of the gallbladder (PRR 326, 6 registrations). As for mortality, bladder cancer risk was slightly elevated but not significantly so (PRR 121, 27 registrations).

Women Number of registrations: 51

Only two cases of colon cancer were reported, adding little further data with which to assess the deficit noted for mortality.

074 - OTHER TEXTILE WORKERS

Men Number of registrations: 1156

There was some evidence of an excess of thyroid cancer (PRR 397, 7 registrations), and a deficit of skin cancers other than melanoma (PRR 69, 52 registrations).

Women Number of registrations: 1157

075 - CHEMICAL WORKERS

Men Number of registrations: 2953

The only cancer for which an excess was suggested was pleura (PRR 169, 30 registrations), supporting the findings for mortality.

Women Number of registrations: 534

Excess risks were suggested for all leukaemias combined (PRR 305, 18 registrations), and for acute myeloid leukaemia in particular (PRR 382, 10 registrations). Stomach cancer risk was also increased (PRR 185, 31 registrations). As for mortality in women, and mortality and cancer incidence in men, pleural cancer was above average (PRR 328, 3 registrations), but not significantly so. A deficit of breast cancer was notable (PRR 73, 69 registrations) but, in contrast with mortality, the incidence of bladder cancer was not unusual (PRR 90, 8 registrations).

076 - BAKERS

Men Number of registrations: 791

There was slight evidence of a raised risk of nasopharyngeal cancer in this group (PRR 366, 4 registrations).

Women Number of registrations: 155

077 - BREWERY WORKERS

Men Number of registrations: 300

As for mortality, an excess of cancer of the rectum was evident (PRR 165, 24 registrations). The risks of pharyngeal cancer (PRR 311, 6 registrations) and thyroid cancer (PRR 562, 3 registrations) were also higher than expected, and colon cancer was lower (PRR 44, 7 registrations).

Women Number of registrations: 19

078 - FOOD PROCESSORS

Men Number of registrations: 867

Women Number of registrations: 429

079 - PAPER MANUFACTURERS

Men Number of registrations: 158

The excess of non-Hodgkin's lymphoma noted for mortality was only weakly supported in these data (PRR 187, 6 registrations).

Women Number of registrations: 21

In contrast with mortality, breast cancer was below average in this group (PRR 95, 4 registrations).

080 - BOOKBINDERS

Men Number of registrations: 99

Women Number of registrations: 164

In accord with the mortality data, the risk of breast cancer was greater than 100 (PRR 124, 42 registrations), but there were no registrations for meningeal cancer.

081 - PAPER CUTTERS

Men Number of registrations: 42

No cases of nasal cancer were reported, so the excess noted for mortality could not be investigated further.

Women Number of registrations: 2

082 - GLASS AND CERAMICS FURNACE WORKERS

Men Number of registrations: 145

Women Number of registrations: 4

083 - GLASS FORMERS AND DECORATORS

Men Number of registrations: 188

There were slightly more registrations from primary liver cancer than expected (PRR 372, 4 registrations). In contrast with mortality, colon cancer was below expectation (PRR 95, 9 registrations).

Women Number of registrations: 27

084 - CERAMICS CASTERS

Men Number of registrations: 63

Women Number of registrations: 41

085 - RUBBER MANUFACTURERS

Men Number of registrations: 457

A two-fold risk of bladder cancer was suggested (PRR 226, 56 registrations), which is somewhat stronger than the finding noted for mortality. The risk for all cancers of the mouth and pharynx combined was also slightly higher than expected (PRR 184, 14 registrations), although there was little suggestion of an increase in any individual site.

Women Number of registrations: 72

As for men, a marked excess of bladder cancer was evident (PRR 350, 7 registrations).

086 - PLASTICS WORKERS

Men Number of registrations: 129

Although mortality from lung cancer was high, the number of registrations was only slightly greater than expected (PRR 108, 48 registrations).

Women Number of registrations: 23

There was an increased risk of lung cancer in this group (PRR 306, 7 registrations).

087 - MAN-MADE FIBRE MAKERS

Men Number of registrations: 24

Excesses of non-Hodgkin's lymphoma (PRR 616, 3 registrations) and of ill-defined and secondary cancers were noted in this group (PRR 386, 5 registrations).

Women Number of registrations: 1

088 - OTHER COAL MINERS

Men Number of registrations: 6447

Marginally increased risks of rectal cancer (PRR 116, 288 registrations) and lung cancer (PRR 104, 2337 registrations) and decreased risks of oesophageal cancer (PRR 77, 118 registrations) and melanoma (PRR 62, 19 registrations) were evident.

Women Number of registrations: 1

089 - TOBACCO WORKERS

Men Number of registrations: 43

An excess of colon cancer was suggested (PRR 305, 7 registrations). In contrast with mortality, an above average number of lung cancer was observed, but the excess was not statistically significant (PRR 128, 17 registrations).

Women Number of registrations: 48

There was weak suggestion of an excess of invasive cervical cancer (PRR 281, 6 registrations).

090 - OTHER WOOD AND PAPER PROCESSORS

Men Number of registrations: 99

Women Number of registrations: 50

091 - OTHER OCCUPATIONS - GLASS AND CERAMICS

Men Number of registrations: 708

There was some suggestion of a deficit of acute myeloid leukaemia: no cases were observed whereas four were expected.

Women Number of registrations: 285
Skin cancer other than melanoma was slightly below expectation (PRR 23, 3 registrations). As for mortality, ovarian cancer was lower than average but not significantly so (PRR 59, 9 registrations), and there was little evidence of an excess of cancer of the oesophagus (PRR 103, 6 registrations)

092 - RUBBER GOODS MAKERS
Men Number of registrations: 140

Women Number of registrations: 34
There was some evidence of an increased risk for ill-defined and secondary cancers (PRR 362, 6 registrations).

093 - PLASTIC GOODS MAKERS
Men Number of registrations: 175
An excess of bladder cancer was evident (PRR 187, 19 registrations). Only one case of acute lymphatic leukaemia was reported, which lends only weak support to the excess noted for mortality.

Women Number of registrations: 66

094 - COMPOSITORS
Men Number of registrations: 277
There was some evidence of an excess of kidney cancer in this group (PRR 238, 11 registrations).

Women Number of registrations: 30
Mortality from breast cancer was high in this group, but there was no evidence of a corresponding increase in registrations (PRR 96, 6 registrations).

095 - PRINTING PLATE PREPARERS
Men Number of registrations: 89
Only one case of acute myeloid leukaemia was reported, providing no additional support for the excess noted for mortality.

Women Number of registrations: 6

096 - PRINTING MACHINE MINDERS
Men Number of registrations: 348

Women Number of registrations: 81
Ovarian cancer was high in this group (PRR 233, 10 registrations). There were no cases of cancer of the small intestine and only three cases of rectal cancer (PRR 119), providing little support for the associations found for mortality.

097 - PRINTERS (SO DESCRIBED)
Men Number of registrations: 1533
There was some suggestion of an excess of skin cancer other than melanoma (PRR 121, 146 registrations), bladder cancer (PRR 123, 115 registrations) and prostate cancer (PRR 122, 107 registrations). Risk of multiple myeloma appeared below average in this group (PRR 38, 6 registrations).

Women Number of registrations: 215
An excess of lung cancer noted for mortality was less evident at registration (PRR 137, 32 registrations).

098 - TAILORS AND DRESSMAKERS
Men Number of registrations: 380
There were significantly more registrations than expected for two sites: stomach cancer (PRR 163, 46 registrations), and melanoma of the skin (PRR 275, 6 registrations). Lung cancer risk was below expectation (PRR 74, 94 registrations).

Women Number of registrations: 834
There was a slight excess of invasive cervical cancer in this group (PRR 142, 42 registrations) but not of cancer in-situ of the cervix (PRR 87, 38 registrations). As for mortality, cancers of the breast and pleura were increased but neither was statistically significant (PRR 107, 175 registrations and PRR 189, 4 registrations respectively). As for men, risk of lung cancer was reduced but the deficit was not statistically significant (PRR 84, 105 registrations).

099 - CLOTHING CUTTERS
Men Number of registrations: 211

Women Number of registrations: 101
Only one case of gallbladder cancer was observed, providing little support for the excess noted at mortality.

100 - SEWERS AND EMBROIDERERS
Men Number of registrations: 140
There was a deficit of skin cancers other than melanoma (PRR 31, 3 registrations).

Women Number of registrations: 4115
There was an excess of meningeal cancer (PRR 162, 22 registrations). In contrast with mortality, PRRs for lung cancer and breast cancer were close to expectation being 100 (503 registrations) and 99 (728 registrations) respectively. In accord with mortality, a below average number of cancers of the oral cavity were observed, but the deficit was not statistically significant (PRR 85, 14 registrations). A deficit of other leukaemias was nominally significant based on zero cases observed.

101 - UPHOLSTERERS
Men Number of registrations: 436
Excesses of cancers of the pleura (PRR 257, 7 registrations) and prostate (PRR 151, 39 registrations) mirror those found for mortality. An excess of meningeal tumours, also suggested by the mortality data, was evident in the registrations but was not significant (PRR 294, 3 registrations). There was some suggestion of a deficit of oesophageal cancer (PRR 37, 4 registrations).

Women Number of registrations: 76
As for mortality, there was an excess of thyroid cancer in this group, but this was based on only two cases (PRR 856, 2 registrations).

102 - CARPET FITTERS
Men Number of registrations: 111
The PRR for rectal cancer (high in the mortality data) was above 100, but not significantly so (PRR 126, 6 registrations).

Women Number of registrations: 5

103 - OTHER WORKERS WITH FABRICS
Men Number of registrations: 125
There was some evidence of a raised risk of colon cancer in this group (PRR 231, 15 registrations).

Women Number of registrations: 85

104 - CARPENTERS
Men Number of registrations: 4462
In accord with mortality, there was an excess of pleural cancer (PRR 206, 70 registrations), and a slight, but non-significant, excess of nasal cancer (PRR 120, 12 registrations). There was also a marginal excess of skin cancer other than melanoma (PRR 112, 357 registrations). Deficits were seen for three sites; lung (PRR 91, 1314 registrations), stomach (PRR 87, 278 registrations), and chronic myeloid leukaemia (PRR 48, 7 registrations). Cancer of the eye was below expectation, in contrast with the excess noted for mortality (PRR 81, 5 registrations).

Women Number of registrations: 90
As for men employed in this group, there was evidence of an excess of pleural cancer (PRR 1596, 3 registrations). In contrast to the significant deficit of lung cancer among men, however, the risk of lung cancer among women was increased (PRR 171, 20 registrations).

105 - CABINET MAKERS
Men Number of registrations: 453
As for mortality, there was strong evidence of an excess of nasal cancer in this group (PRR 803, 9 registrations). In addition, an above average risk was suggested for testicular cancer (PRR 237, 10 registrations). In contrast with mortality, there was little suggestion of an excess of multiple myeloma (PRR 102, 5 registrations) and, although the PRR was above 100 for cancer of the eye, this was based on only two cases and was not statistically significant (PRR 346).

Women Number of registrations: 31

106 - CASE AND BOX MAKERS
Men Number of registrations: 112
In contrast with mortality, there was little suggestion of a raised risk of stomach cancer (PRR 103, 9 registrations) and there were no registrations for nasal cancer.

Women Number of registrations: 15

107 - PATTERN MAKERS
Men Number of registrations: 161
Although based on only four cases, there was some suggestion of an excess of Hodgkin's disease (PRR 399). In contrast with mortality, colon cancer rates appeared unremarkable (PRR 141, 12 registrations).

Women Number of registrations: 13

108 - WOODWORKING MACHINISTS
Men Number of registrations: 594
The risk of nasal cancer was markedly increased (PRR 710, 10 registrations). A smaller, but statistically significant, excess of oesophageal cancer was also observed (PRR 165, 24 registrations).

Women Number of registrations: 16
No cases of multiple myeloma were registered, in contrast with the excess noted for mortality.

109 - OTHER WOODWORKERS
Men Number of registrations: 289
An almost seven-fold excess of nasal cancer was found in this group (PRR 676, 5 registrations).

Women Number of registrations: 24
The excess of lung cancer at death was seen at registration, but this was based on only five cases and was not statistically significant (PRR 197, 5 registrations).

110 - DENTAL TECHNICIANS
Men Number of registrations: 105
As in the mortality data, the risk of the prostatic cancer was raised (PRR 249, 15 registrations).

Women Number of registrations: 7

111 - OTHER MAKERS OF PAPER GOODS
Men Number of registrations: 192

Women Number of registrations: 195
Unlike mortality, risks of cancers of the brain and colon appeared relatively unremarkable (PRR 207, 7 registrations and PRR 51, 5 registrations; respectively)

112 - FURNACE OPERATIVES (METAL)
Men Number of registrations: 460
An excess of ill-defined and secondary cancers was of borderline statistical significance (PRR 143, 40 registrations).

Women Number of registrations: 0

113 - ROLLERS (METAL)
Men Number of registrations: 80

Women Number of registrations: 0

114 - SMITHS AND FORGE WORKERS
Men Number of registrations: 446
In accord with the mortality findings, two registrations for peritoneal cancer constituted an excess of borderline statistical significance (PRR 975)

Women Number of registrations: 0

115 - METAL DRAWERS
Men Number of registrations: 147
There was weak evidence of an excess of lung cancer (PRR 130, 64 registrations).

Women Number of registrations: 9

116 - MOULDERS AND COREMAKERS (METAL)
Men Number of registrations: 762
There was an excess of lung cancer in this group (PRR 123, 329 registrations). The excess of laryngeal cancer noted for mortality was reflected in the registration data, although it was not statistically significant (PRR 139, 13 registrations). PRRs were significantly low for two sites — prostate (PRR 54, 24 registrations) and skin cancers other than melanoma (PRR 57, 27 registrations).

Women Number of registrations: 42

117 - ELECTROPLATERS

Men Number of registrations: 177

Lung cancer, for which mortality was raised, was unexceptional (PRR 114, 67 registrations). A statistically significant deficit of cancers of the oesophagus was observed (based on 0 registrations)

Women Number of registrations: 12

Although lung cancer was raised (PRR 170, 3 registrations) unlike mortality, this was not statistically significant.

118 - ANNEALERS, HARDENERS, TEMPERERS (METAL)

Men Number of registrations: 83

Three cases of liver cancer constituted a significant excess (PRR 649).

Women Number of registrations: 2

119 - GALVANISERS AND TIN PLATERS

Men Number of registrations: 75

A deficit of prostatic cancer was of borderline statistical significance (PRR 0, 0 registrations).

Women Number of registrations: 6

120 - OTHER METAL MANUFACTURERS

Men Number of registrations: 1223

There were more registrations than expected for two sites: laryngeal cancer (PRR 154, 27 registrations) and lung cancer (PRR 111, 507 registrations). There were fewer registrations than expected from skin cancers other than melanoma (PRR 71, 48 registrations).

Women Number of registrations: 30

In contrast with mortality, there were no registrations for laryngeal cancer. The five registrations for breast cancer lend little support to the deficit seen for mortality (PRR 110).

121 - PRESS AND MACHINE TOOL SETTERS

Men Number of registrations: 1073

There is a suggestion of an excess of scrotal skin cancer (PRR 676), but this is based on only three cases. Brain cancer appeared below expectation but the strength of the association was weak (PRR 56, 12 registrations).

Women Number of registrations: 68

122 - CENTRE LATHE TURNERS

Men Number of registrations: 807

Excesses were observed for scrotal cancer (PRR 1132, 4 registrations), cancer of the salivary glands (PRR 395, 4 registrations), and ill-defined cancers of the mouth and pharynx (PRR 777, 5 registrations).

Women Number of registrations: 11

There were no registrations for cancer of the larynx, in contrast with the excess noted for mortality.

123 - MACHINE TOOL SETTER OPERATORS

Men Number of registrations: 76

Women Number of registrations: 1

124 - MACHINE TOOL OPERATORS

Men Number of registrations: 11199

Nine sites had more registrations than expected: scrotal skin cancer (PRR 261, 18 registrations), melanoma (PRR 139, 102 registrations), skin cancers other than melanoma (PRR 111, 905 registrations), prostate cancer (PRR 117, 682 registrations), bladder cancer (PRR 112, 735 registrations), testicular cancer (PRR 119, 141 registrations), kidney cancer (PRR 115, 210 registrations), cancer of the soft tissues (PRR 134, 60 registrations), and colon cancer (PRR 110, 654 registrations). Interestingly, pleural cancer, the only cancer which was increased at death, was not significantly elevated at registration (PRR 113, 59 registrations). Four sites had fewer registrations than expected: lung cancer (PRR 90, 3319 registrations), stomach cancer (PRR 88, 717 registrations), pharyngeal cancer (PRR 67, 48 registrations), and primary liver cancer (PRR 72, 52 registrations).

Women Number of registrations: 540

As for mortality, lung cancer risk was increased at registration (PRR 129, 94 registrations). Laryngeal cancer was also high (PRR 325, 5 registrations). No cases of acute monocytic leukaemia were registered, in contrast with the excess noted for mortality. Breast cancer was low, which agrees with the mortality data (PRR 75, 81 registrations) but a deficit of deaths from gallbladder cancer was not reflected in these data (PRR 126, 4 registrations).

125 - PRESS AND AUTOMATIC MACHINE OPERATORS

Men Number of registrations: 402

Risk of lung cancer, for which an excess was noted for mortality, appeared unremarkable (PRR 110, 147 registrations).

Women Number of registrations: 432

An excess of lung cancer was suggested although the evidence is weak (PRR 127, 69 registrations). No cases of bone cancer were reported in this group, in contrast with the excess noted for mortality. Breast cancer risk was below expectation (PRR 76, 64 registrations).

126 - METAL POLISHERS

Men Number of registrations: 380

Lung cancer risk was increased (PRR 120, 157 registrations), in agreement with the findings for womens' cancer registration and mens' mortality. There was, however, little indication of an increased risk of stomach cancer (PRR 101, 33 registrations). Two sites had significantly reduced risks: skin cancers other than melanoma (PRR 50, 12 registrations), and colon cancer (PRR 54, 11 registrations).

Women Number of registrations: 55

In accord with the registration and mortality findings among men, lung cancer risks were increased (PRR 191, 16 registrations).

127 - FETTLERS AND DRESSERS (METAL)

Men Number of registrations: 232

As for mortality, an excess of lung cancer was suggested (PRR 124, 102 registrations).

Women Number of registrations: 2

128 - SHOT BLASTERS

Men Number of registrations: 98

There was some evidence of a deficit of prostatic cancer where no registrations were observed and five were expected.

Women Number of registrations: 2

129 - TOOLMAKERS

Men Number of registrations: 1383

There was some suggestion of an excess of colon cancer (PRR 129, 94 registrations) but cancer incidence was unremarkable for three cancers with above average mortality — small intestine, testis and chronic myeloid leukaemia: the PRRs being 111 (3 registrations), 68 (9 registrations), and 124, (5 registrations) respectively. Statistically significant deficits were observed for cancers of the lung (PRR 86, 397 registrations) and pleura (based on 0 registrations).

Women Number of registrations: 19

130 - PRECISION INSTRUMENT MAKERS

Men Number of registrations: 322

There were more registrations than expected from myeloma (PRR 284, 9 registrations) and all leukaemias combined (PRR 190, 13 registrations). In accord with mortality, rectal cancer risk was increased although not significantly so (PRR 140, 20 registrations).

Women Number of registrations: 25

More registrations were observed for melanoma than were expected (PRR 765, 3 registrations).

131 - WATCH AND CLOCK MAKERS

Men Number of registrations: 140

Women Number of registrations: 12

132 - PRODUCTION FITTERS

Men Number of registrations: 8908

Pleural cancer was high in this group, in agreement with the findings for mortality (PRR 181, 121 registrations). In addition, more eye cancers were registered than expected (PRR 175, 21 registrations) and more for cancers of other parts of the mouth (PRR 169, 24 registrations). The risk of nasal cancer was low (PRR 48, 9 registrations). In contrast with mortality, risks of cancer of the salivary glands and meningeal tumours appeared unremarkable (PRR 109, 13 registrations and PRR 79, 15 registrations respectively).

Women Number of registrations: 157

Five gallbladder cancers were registered in this group, resulting in a PRR of 619. There were two pleural cancer registrations which, whilst not statistically significant, accords with the excess noted for mortality (PRR 403). Breast cancer risk was below expectation (PRR 58, 18 registrations).

133 - MOTOR MECHANICS

Men Number of registrations: 1826

There was a suggestion of an excess of gallbladder cancer (PRR 187, 17 registrations) and deficits of skin cancers other than melanoma (PRR 74, 95 registrations) and ill-defined and secondary cancers (PRR 73, 70 registrations). As for mortality, there was a deficit of pleural cancer (PRR 23) but this was based on only three cases.

Women Number of registrations: 5

134 - AIRCRAFT ENGINE FITTERS

Men Number of registrations: 17

Women Number of registrations: 0

135 - OFFICE MACHINERY MECHANICS

Men Number of registrations: 34

Women Number of registrations: 3

136 - ELECTRICAL AND ELECTRONIC PRODUCTION FITTERS

Men Number of registrations: 172

Women Number of registrations: 6

137 - ELECTRICIANS

Men Number of registrations: 3141

As for mortality, there was a clear excess of pleural cancer (PRR 250, 61 registrations). Risks also appeared raised for meningeal cancers (PRR 214, 17 registrations) and for multiple myeloma (PRR 143, 43 registrations). Apart from pleural cancer, PRRs for other cancers with notable mortality were relatively unremarkable: (melanoma (PRR 126, 34 registrations) bone (PRR 102, 8 registrations), thyroid (PRR 147, 11 registrations) and brain (PRR 121, 84 registrations)). There were suggestions of deficits of cancer of the lung (PRR 85, 822 registrations) and liver (PRR 43, 8 registrations).

Women Number of registrations: 8

A higher than average risk of invasive cervical cancer was observed in this group (PRR 625) but this was based on only three cases.

138 - ELECTRICAL PLANT OPERATORS

Men Number of registrations: 516

Pleural cancer and peritoneal cancer were above expectation which agrees with the findings for mortality (PRR 382, 16 registrations and PRR 990, 3 registrations, respectively). Five registrations for cancer of the small intestine resulted in a PRR of 566. Lung cancer was below expectation (PRR 83, 145 registrations). An excess of cancer of the oesophagus, noted for mortality, was evident in the registration data, but was of only borderline statistical significance (PRR 158, 21 registrations). In contrast with mortality, there was no suggestion of an excess of cancer of the oral cavity (PRR 102, 4 registrations).

Women Number of registrations: 9

139 - TELEPHONE FITTERS

Men Number of registrations: 1089

Raised risks were suggested for skin cancers other than melanoma (PRR 128, 107 registrations) and kidney cancer (PRR 153, 29 registrations). In contrast with mortality, there was little suggestion of an excess of brain cancer (PRR 101, 23 registrations) or non-Hodgkin's lymphoma (PRR 99, 29 registrations). A lower than expected proportion of lung cancer was observed (PRR 83, 283 registrations).

Women Number of registrations: 26

140 - ELECTRIC CABLE AND LINE WORKERS

Men Number of registrations: 233

Cancers of the testis (PRR 385, 8 registrations) and stomach

(PRR 157, 28 registrations) were elevated in this group. A deficit of cancer of the colon was observed, but this was based on only four cases (PRR 33).

Women Number of registrations: 5

141 - RADIO AND TV MECHANICS

Men Number of registrations: 326

The proportion of Hodgkin's disease was above expectation (PRR 240, 9 registrations). As for mortality, above average proportions of brain cancer and acute myeloid leukaemia were reported, but neither of these excesses were statistically significant (PRR 118, 9 registrations and PRR 224, 6 registrations respectively).

Women Number of registrations: 9

142 - OTHER ELECTRONIC MAINTENANCE ENGINEERS

Men Number of registrations: 259

As for mortality, there was evidence of increased leukaemia in this group: the PRRs for all leukaemias combined, for acute myeloid leukaemia and for chronic lymphatic leukaemia being 277 (19 registrations), 293 (8 registrations) and 408 (5 registrations) respectively. In contrast with mortality, significant excesses were not evident for cancers of the colon (PRR 109, 14 registrations), prostate (PRR 110, 10 registrations) and non-Hodgkin's lymphoma (PRR 122, 10 registrations). There was a deficit of lung cancer in this group (PRR 72, 51 registrations).

Women Number of registrations: 9

No cases of gallbladder cancer were registered, in contrast with the excess noted for mortality.

143 - ELECTRICAL ENGINEERS (SO DESCRIBED)

Men Number of registrations: 1152

As for mortality, there was a clear excess of pleural cancer (PRR 207, 18 registrations). In addition, an increased risk of prostatic cancer was suggested (PRR 132, 82 registrations). Above average proportions were observed for three other cancers in excess at death, namely colon, thyroid and multiple myeloma, but none of these were significantly raised at registration, the PRRs being 120 (70 registrations), 118 (3 registrations), and 151 (17 registrations) respectively. Below average proportions of cancers of the lung (PRR 75, 277 registrations) and larynx (PRR 43, 7 registrations) were observed.

Women Number of registrations: 31

Above average proportions of cancers of the mouth and pharynx (PRR 609, 3 registrations) and cancers of the brain (PRR 630, 3 registrations) were noted in this group.

144 - PLUMBERS AND GAS FITTERS

Men Number of registrations: 2824

As for mortality, pleural cancer risk was increased (PRR 220, 50 registrations), and peritoneal cancer was high but this was not statistically significant (PRR 230, 4 registrations). There were marginally fewer registrations than expected from multiple myeloma (PRR 61, 17 registrations).

Women Number of registrations: 4

145 - SHEET METAL WORKERS

Men Number of registrations: 1282

As for mortality, an excess of pleural cancer was observed (PRR 266, 25 registrations) but there was less support for the association noted for pharyngeal cancer (PRR 145, 12 registrations).

Women Number of registrations: 33

Four cases of cervical cancer were reported, of which three were in-situ cancers (PRR 81), lending little support to the finding noted for cervical cancer mortality in this group.

146 - METAL PLATE WORKERS

Men Number of registrations: 1036

As for mortality, risks were raised for cancers of the pleura (PRR 378, 41 registrations) and liver (PRR 196, 11 registrations) but there was less support for excesses of cancers of the lung (PRR 107, 398 registrations) or larynx (PRR 120, 15 registrations). Decreased risks were seen for multiple myeloma (PRR 30, 3 registrations) and non-Hodgkin's lymphoma (PRR 55, 11 registrations).

Women Number of registrations: 28

An excess of lung cancer was of borderline statistical significance (PRR 230, 8 registrations). Only one case of pleural cancer was registered which contrasts with the excess noted for mortality.

147 - STEEL ERECTORS

Men Number of registrations: 579

Of the cancers for which mortality was above average (lung, larynx and oral cavity), only for oral cavity was an excess clearly evident at cancer registration (PRR 237, 11 registrations). For cancers of the larynx (PRR 156, 13 registrations) and lung (PRR 109, 207 registrations) PRRs were above 100, but not statistically significant.

Women Number of registrations: 5

148 - SCAFFOLDERS

Men Number of registrations: 298

Oesophageal cancer risk was raised in this group (PRR 187, 14 registrations). Stomach cancer was elevated at death, but there was no corresponding excess of cancer registrations (PRR 98, 20 registrations).

Women Number of registrations: 4

149 - WELDERS

Men Number of registrations: 2142

As for mortality, there was clear evidence of an excess of pleural cancer (PRR 196, 34 registrations).

Women Number of registrations: 110

An excess of cancer of the stomach, noted for mortality, was less evident in these data (PRR 139, 6 registrations).

150 - RIGGERS

Men Number of registrations: 270

An increased risk of chronic myeloid leukaemia (PRR 484), and a decreased risk of prostatic cancer (PRR 34), was seen in this group but both were based on only four registrations. Risks of oesophageal cancer and laryngeal cancer, which were increased at death, were raised but not significantly so

(PRR 133, 10 registrations and PRR 158, 6 registrations, respectively).

Women Number of registrations: 6

151 - JEWELLERY WORKERS
Men Number of registrations: 59
The occurrence of lung cancer was lower than expected (PRR 51, 10 registrations).

Women Number of registrations: 25

152 - ENGRAVERS AND ETCHERS (PRINTING)
Men Number of registrations: 74

Women Number of registrations: 18

153 - VEHICLE BODY BUILDERS
Men Number of registrations: 376
As for mortality, a clear excess of pleural cancer was seen in these data (PRR 385, 10 registrations). The excess of cancer of the peritoneum, noted for mortality, was also evident but this was based on only two cases (PRR 1012).

Women Number of registrations: 10

154 - OILERS AND GREASERS
Men Number of registrations: 80

Women Number of registrations: 0

155 - ELECTRONICS WIRE WORKERS
Men Number of registrations: 72

Women Number of registrations: 49
No cases of Hodgkin's disease were reported in this group, in contrast with the excess noted for mortality.

156 - COIL WINDERS
Men Number of registrations: 36
For brain cancer, which had increased mortality, there was only one registration.

Women Number of registrations: 58

157 - POTTERY DECORATORS
Men Number of registrations: 23

Women Number of registrations: 99

158 - COACH PAINTERS
Men Number of registrations: 100
Laryngeal cancer (PRR 365, 5 registrations) and bladder cancer (PRR 198, 12 registrations) were increased in this group.

Women Number of registrations: 2

159 - OTHER SPRAY PAINTERS
Men Number of registrations: 585
There was some suggestion of an excess of other leukaemias, based on four cases (PRR 404). Excesses of cancers of the lung and testis noted at death were less evident at registration (PRR 111, 213 registrations and PRR 133, 9 registrations, respectively). The risk of prostatic cancer was lower than expected (PRR 58, 17 registrations).

Women Number of registrations: 36
There was some suggestion of an excess of stomach cancer (PRR 371, 5 registrations).

160 - PAINTERS AND DECORATORS NEC
Men Number of registrations: 4725
There was some evidence of an excess of hypopharyngeal cancer in this group (PRR 174, 18 registrations). As at death, the risk of lung cancer was marginally elevated (PRR 108, 1664 registrations). The risks at three sites were reduced: prostate (PRR 81, 213 registrations), pleura (PRR 57, 20 registrations), and melanoma (PRR 62, 20 registrations).

Women Number of registrations: 118

161 - ELECTRICAL, ELECTRONIC ASSEMBLERS
Men Number of registrations: 85
With the possible exception of pharyngeal cancer, where three cases were registered (PRR 556), no notable associations were apparent.

Women Number of registrations: 211
Excesses were suggested for brain cancer (PRR 311, 7 registrations) and all leukaemias combined (PRR 320, 8 registrations). Only one case of chronic myeloid leukaemia was reported providing little support to the excess noted for mortality.

162 - INSTRUMENT ASSEMBLERS
Men Number of registrations: 9

Women Number of registrations: 7

163 - ASSEMBLERS (VEHICLES AND OTHER METAL GOODS)
Men Number of registrations: 505
Three cases of acute lymphatic leukaemia produced a PRR of borderline statistical significance (PRR 520). The evidence for an excess of primary liver cancer was weaker than that noted for mortality (PRR 172, 6 registrations).

Women Number of registrations: 61
No cases of other leukaemia were registered, in contrast with the excess noted for mortality.

164 - PACKERS AND SORTERS
Men Number of registrations: 1111
Registrations were elevated for two cancers; nasal cancer (PRR 312, 6 registrations) and lung cancer (PRR 111, 421 registrations). Registrations for two sites were reduced skin cancers other than melanoma (PRR 71, 55 registrations) and colon cancer (PRR 62, 37 registrations).

Women Number of registrations: 1937

165 - BRICKLAYERS AND TILESETTERS
Men Number of registrations: 2957
Excesses were suggested for skin cancers other than melanoma (PRR 126, 256 registrations), lung cancer (PRR 107, 1059 registrations) and testicular cancer (PRR 142, 37 registrations). Deficits were suggested for bladder cancer (PRR 74, 125 registrations), colon cancer (PRR 72, 111 registrations) and pharyngeal cancer (PRR 47, 8 registrations).

Women Number of registrations: 0

166 - MASONS AND STONECUTTERS
Men Number of registrations: 240
Bladder cancer was increased in this group (PRR 164, 23 registrations). Although increased at death, only five cancers of the oesophagus were registered (PRR 83, 5 registrations).

Women Number of registrations: 4

167 - PLASTERERS
Men Number of registrations: 835
A suggestion of an excess of cancers of other parts of the mouth was apparent (PRR 396, 5 registrations). As noted for mortality, there was some suggestion of an excess of lung cancer (PRR 115, 306 registrations). Three cases of nasal cancer (PRR 173) provide weak support for the excess noted at death, and an excess of pleural cancer was not reflected in the registration data (PRR 74, 5 registrations). Incidence of cancer of the gallbladder was significantly below expectation (PRR 0, 0 registrations).

Women Number of registrations: 15

168 - ROOFERS AND GLAZIERS
Men Number of registrations: 556
Pancreatic cancer was increased in this group (PRR 208, 16 registrations). An excess of lung cancer deaths was not reflected in the registration data (PRR 99, 170 registrations).

Women Number of registrations: 16

169 - BUILDERS ETC.
Men Number of registrations: 3760
There was evidence of an excess of skin cancers other than melanoma (PRR 119, 346 registrations) and, to a lesser extent, brain cancer (PRR 123, 94 registrations). Three cancers for which mortality appeared raised (pleura, gallbladder and breast) were less remarkable in these data (PRR 119, 30 registrations, PRR 92, 17 registrations and PRR 154, 10 registrations, respectively).

Women Number of registrations: 79

170 - RAILWAY TRACK WORKERS
Men Number of registrations: 358

Women Number of registrations: 2

171 - ROAD CONSTRUCTION WORKERS
Men Number of registrations: 662
The PRR for ill-defined and secondary cancers was significantly below 100 (PRR 64, 24 registrations).

Women Number of registrations: 9

172 - SEWAGE PLANT ATTENDANTS
Men Number of registrations: 257

Women Number of registrations: 9

173 - MAINS AND SERVICE LAYERS
Men Number of registrations: 314
There was evidence of an excess for several cancers: oesophagus (PRR 194, 16 registrations), pancreas (PRR 222, 10 registrations), primary liver (PRR 289, 6 registrations) and chronic myeloid leukaemia (PRR 450, 5 registrations). With the exception of the finding for pancreas, cancers noted as raised in the mortality analyses were unremarkable (rectum PRR 130, 19 registrations, larynx PRR 60, 3 registrations and prostate PRR 87, 12 registrations).

Women Number of registrations: 3

174 - CONSTRUCTION WORKERS NEC
Men Number of registrations: 3023
In accord with the mortality findings, peritoneal cancer (PRR 449, 14 registrations) and pleural cancer (PRR 245, 42 registrations) were raised. Lung cancer was also increased (PRR 114, 1144 registrations). The PRRs for three cancer sites were decreased: bladder (PRR 78, 134 registrations), prostate (PRR 77, 111 registrations), and multiple myeloma (PRR 57, 17 registrations).

Women Number of registrations: 41
In accord with the findings for male cancer registration, lung cancer was high in this group (PRR 269, 16 registrations). Although there was evidence of increased mortality from colon cancer, only one registration was observed.

175 - FACE TRAINED COALMINERS
Men Number of registrations: 2471
An excess of kidney cancer was suggested (PRR 146, 54 registrations) and, to a lesser degree, ill-defined and secondary cancers (PRR 116, 203 registrations). An excess of stomach cancer noted in the mortality analyses was less marked in these data (PRR 108, 219 registrations). Deficits were evident for pleural cancer (PRR 22, 6 registrations) and skin cancers other than melanoma (PRR 72, 83 registrations).

Women Number of registrations: 0

176 - MINERS (NOT COAL) AND QUARRYMEN
Men Number of registrations: 258

Women Number of registrations: 0

177 - RAILWAY GUARDS
Men Number of registrations: 263
There was a weak suggestion of an excess of chronic lymphatic leukaemia, based on five cases (PRR 319).

Women Number of registrations: 5
No cases of oesophageal cancer were reported for this group, in contrast with the excess noted for mortality.

178 - RAILWAY SIGNAL WORKERS
Men Number of registrations: 307

Women Number of registrations: 13

179 - SHUNTERS AND POINTS OPERATORS
Men Number of registrations: 136
Risks appeared raised for cancers of the pancreas (PRR 383, 8 registrations) and lung (PRR 141, 65 registrations).

Women Number of registrations: 1

180 - RAILWAY ENGINE DRIVERS
Men Number of registrations: 640
Risks were raised for two sites: thyroid cancer (PRR 382, 5

registrations) and acute myeloid leukaemia (PRR 223, 9 registrations). A deficit of pharyngeal cancer was of borderline significance based on zero registrations.

Women Number of registrations: 17
There were slightly more registrations than expected for skin cancers other then melanoma than but this was based on only five cases (PRR 354).

181 - ROAD TRANSPORT INSPECTORS
Men Number of registrations: 306
Three cases of cancer of the small intestine resulted in a PRR of borderline statistical significance (PRR 553).

Women Number of registrations: 8

182 - BUS AND COACH DRIVERS
Men Number of registrations: 1785
Three sites had significantly elevated risks:- oesophagus (PRR 132, 59 registrations), colon (PRR 125, 114 registrations) and retroperitoneum (PRR 336, 5 registrations). The risk of pharyngeal cancer was slightly reduced (PRR 35), but this is based on only four registrations. A significant deficit of pleural cancer was based on only three registrations (PRR 21).

Women Number of registrations: 69
Acute myeloid leukaemia was increased in this group, but this is based on only four registrations (PRR 584). No cases of kidney cancer were registered, which contrasts with the excess noted for mortality.

183 - LORRY DRIVERS
Men Number of registrations: 10834
There were suggestions of modest excesses of lung cancer (PRR 107, 3666 registrations), and ill-defined cancers of the mouth and pharynx (PRR 176, 17 registrations). In contrast with mortality, PRRs for cancer of the larynx and acute monocytic leukaemia were relatively unremarkable (PRR 109, 177 registrations and PRR 111, 4 registrations). Reduced risks were suggested for four cancers, of which pleura was the most notable (PRR 30, 26 registrations). The other three were cancers of the bone (PRR 48, 10 registrations), colon (PRR 91, 505 registrations) and breast (PRR 48, 7 registrations).

Women Number of registrations: 335

184 - OTHER MOTOR DRIVERS
Men Number of registrations: 1478
Bladder cancer was marginally increased in this group (PRR 125, 112 registrations), and cancers of the mouth and pharynx marginally decreased (PRR 60, 17 registrations).

Women Number of registrations: 80
In accord with mortality, lung cancer risk was increased, but not significantly so (PRR 149, 10 registrations). Only three cases were registered with melanoma of the skin, lending weak support to the excess noted for mortality (PRR 251, 3 registrations).

185 - BUS CONDUCTORS AND DRIVERS' MATES
Men Number of registrations: 463
Excesses were suggested for cancers of the stomach (PRR 144, 50 registrations) and gallbladder (PRR 315, 7 registrations). As for mortality, an excess of multiple myeloma was also suggested but this was of borderline statistical significance (PRR 220, 10 registrations). An excess of kidney cancer noted in the mortality data was mirrored in these data but was not statistically significant (PRR 160, 11 registrations).

Women Number of registrations: 182
A raised risk was suggested for cancer of the pleura (PRR 753) although this was based on only three cases. Breast cancer risk was below expectation (PRR 62, 20 registrations). Unlike mortality, lung cancer risk appeared unremarkable in this group (PRR 110, 34 registrations).

186 - MECHANICAL PLANT DRIVERS
Men Number of registrations: 359
There were four registrations for male breast cancer, resulting in a PRR of 829.

Women Number of registrations: 0

187 - CRANE DRIVERS
Men Number of registrations: 1402
A marginally increased proportion of lung cancer was observed in this group (PRR 109, 526 registrations). Five cases of breast cancer were reported, giving rise to a significant excess (PRR 336). A reduced risk of prostate cancer was also suggested (PRR 73, 53 registrations).

Women Number of registrations: 0

188 - FORK LIFT TRUCK DRIVERS
Men Number of registrations: 1032
The PRR for laryngeal cancer — which was high at death — was 144 (22 registrations).

Women Number of registrations: 18

189 - SLINGERS
Men Number of registrations: 226

Women Number of registrations: 1

190 - STOREKEEPERS
Men Number of registrations: 6934
Pancreatic cancer was marginally increased in this group (PRR 125, 129 registrations). Five sites had reduced risks: other parts of the mouth (PRR 33, 4 registrations) pleura (PRR 52, 18 registrations), soft tissue (PRR 49, 12 registrations), meningeal (PRR 36, 4 registrations), and chronic myeloid leukaemia (PRR 55, 11 registrations).

Women Number of registrations: 688
An excess of invasive cervical cancer was of borderline statistical significance (PRR 138, 47 registrations). None of the five malignancies with raised risks at death had significantly raised risks at registration. PRRs for three of these — breast, gallbladder and uterine body — were 107 (147 registrations), 201 (8 registrations) and 103 (22 registrations) respectively. For the remaining two malignancies — Hodgkin's disease and chronic myeloid leukaemia — the numbers of cases were one and two, respectively.

191 - DOCKERS AND GOODS PORTERS

Men Number of registrations: 2568

In contrast with mortality, there was little suggestion of an excess for cancers of the pleura (PRR 96, 16 registrations), oral cavity (PRR 103, 32 registrations), pharynx (PRR 90, 20 registrations) or larynx (PRR 112, 47 registrations). A deficit of non-Hodgkin's lymphoma was suggested by these data (PRR 65, 29 registrations).

Women Number of registrations: 30

No cases of cancer of the small intestine were reported so the excess noted for mortality was not supported in these data.

192 - REFUSE COLLECTORS

Men Number of registrations: 664

Women Number of registrations: 11

193 - LABOURERS IN COKE OVENS

Men Number of registrations: 108

An excess of lung cancer of similar magnitude to that noted for mortality was seen in these data but was not statistically significant (PRR 117, 49 registrations).

Women Number of registrations: 2

194 - BOILER OPERATORS

Men Number of registrations: 953

In accord with mortality, pleural cancer risk was increased (PRR 231, 11 registrations). Lung cancer was also raised (PRR 114, 387 registrations). In contrast to the mortality data, there was little evidence of an increased risk of bladder cancer (PRR 97, 54 registrations) or laryngeal cancer (PRR 105, 14 registrations).

Women Number of registrations: 10

Acknowledgements

We would particularly like to thank Diana Bull, who carried out the majority of the analyses, and Nicola Fear, who assisted with the statistical summaries. We would also like to thank Frances Drever whose support and encouragement made this chapter possible. The work was supported by the Imperial Cancer Research Fund.

References

1. Office of Population Censuses and Surveys. *Occupational mortality, decennial supplement 1970-72*. London: HMSO, 1978.
2. Office of Population Censuses and Surveys. *Labour Force Survey 1990 and 1991*. Series LFS no.9. London: HMSO, 1992.
3. Office of Population Censuses and Surveys. *Review of the national cancer registration system*. Series MBI no 17. London: HMSO, 1990.
4. Swerdlow AJ. Cancer registration in England and Wales: some aspects relevant to interpretation of the data. *Journal of the Royal Statistical Society* 1986; vol 149, part 2: 146-60.
5. Office of Population Censuses and Surveys. *Occupational mortality, decennial supplement 1979-80, 1982-83*. London: HMSO, 1986.
6. Doll R, Peto R. *The causes of cancer*. Oxford: Oxford University Press, 1981.
7. International Agency for Research on Cancer (ed L Tomatis). *Cancer causes occurrence and control*. IARC scientific publications no. 100, Lyon: IARC,1990.
8. Office of Population Censuses and Surveys. *1988 Cancer statistics registrations*. Series MBI no.21. London: HMSO,1994.

Chapter 8 Mortality of Longitudinal Study 1971 and 1981 Census cohorts

Ann Bethune
Seeromanie Harding
Anne Scott
Haroulla Filakti

Longitudinal Study - Medical Analysis Section, OPCS

8.1 Introduction

One of the leading objectives of the Longitudinal Study (LS) is to shed light on occupational differentials in mortality. With data from two decades now available, the 1970s and the 1980s, this chapter presents prospective occupational mortality analyses for two cohorts. Mortality of the 1971 census cohort of men and women up to 1981[1,2,3] has been presented previously and findings are updated here with follow-up extended to the end of 1989. The mortality of the 1981 census cohort is examined for the first time.

By directly linking death registration data with occupation data from censuses, the LS overcomes the numerator-denominator biases arising from the use of unlinked data[2,3] where occupation is that stated on the death certificate. The ability to explore the mortality of women[4] and of persons over retirement age in relation to occupation, and the mortality of the unoccupied is also an advantage. This chapter deals with overall mortality, although with the longer follow-up period of the 1971 cohort mortality is examined by cause for some occupations. The cause-specific findings complement results presented in Chapters 4 and 5. The analyses in this chapter differ from the others in this volume where occupation is that recorded at death.

Changes in the structure of British employment throughout the 1970s and 1980s, with the decline in production industries and the expansion of the services sector, have led to a redistribution of the workforce into new or altered occupations (see Chapter 3). The economic activity and occupations of the 1981 Census cohort will reflect the changes in the 1970s.

8.2 Method of analysis

The analyses presented in this chapter are derived using the person-years at risk approach, as outlined previously,[5] and standardised mortality ratios (SMRs) are used to investigate the relationships between occupation and mortality. The death rates used for standardisation are those for all men or all women in the LS cohorts, by five-year age groups and by single calendar years, and are cause specific where relevant.

Confidence intervals are calculated at the 95 per cent level and an SMR is considered statistically significant when the confidence interval (CI) excludes the population standard of 100.

The sample distributions of men and women in the 1971 cohort by occupation order and number of observed deaths are shown in Annex 8.1. A corresponding table for the 1981 Census cohort is shown in Annex 8.2.

8.3 Occupation data

In the 1971 census, people 15 years and over were asked whether or not they had a job in the week before census and the type of employment. Those out of work or retired were asked to record their most recent occupation. Only people employed, temporarily sick, seeking work or retired were classified to an occupation. Anyone whose occupation was vaguely described or not stated was classified as 'inadequately described'. Those who declared they were permanently sick or otherwise economically inactive were categorised as 'unoccupied'. The units of occupation are aggregated into the 26 occupation orders based on the 1970 Classification of Occupations.

The 1981 census economic activity question requested all people not in a job in the previous week to provide details of their most recent full-time job. Thus for the 1981 census cohort, unlike the 1971 cohort, men or women who declared themselves permanently sick or disabled could potentially be classified to an occupation. This has implications for interpreting the effects of health selection[6] on the mortality of the 1981 census cohort and will be explored in further work.

It has been noted previously that LS mortality estimates were lower than decennial supplement figures as the LS was able to measure accurately the high mortality of the unoccupied.[7] However, despite methodological differences, the LS confirmed the patterns of high and low mortality occupations found in cross-sectional studies.

For the 1981 census cohort the units of the 1980 classification of occupations are aggregated to correspond with the University of Southampton 'job groups' described in Appendices 2 and 3 of this volume. Briefly, this brings together units exposed to similar occupational hazards likely to affect mortality. Some of these 194 groups are fairly similar to the occupation orders used for the 1971 census cohort. As the LS represents only one per cent of the population of England and Wales some of the reclassified groups are too small for meaningful analysis and SMRs were only calculated for groups with 25 or more deaths.

Figure 8.1 LS women 1971 census cohort, percentage distribution by age and economic activity in 1971

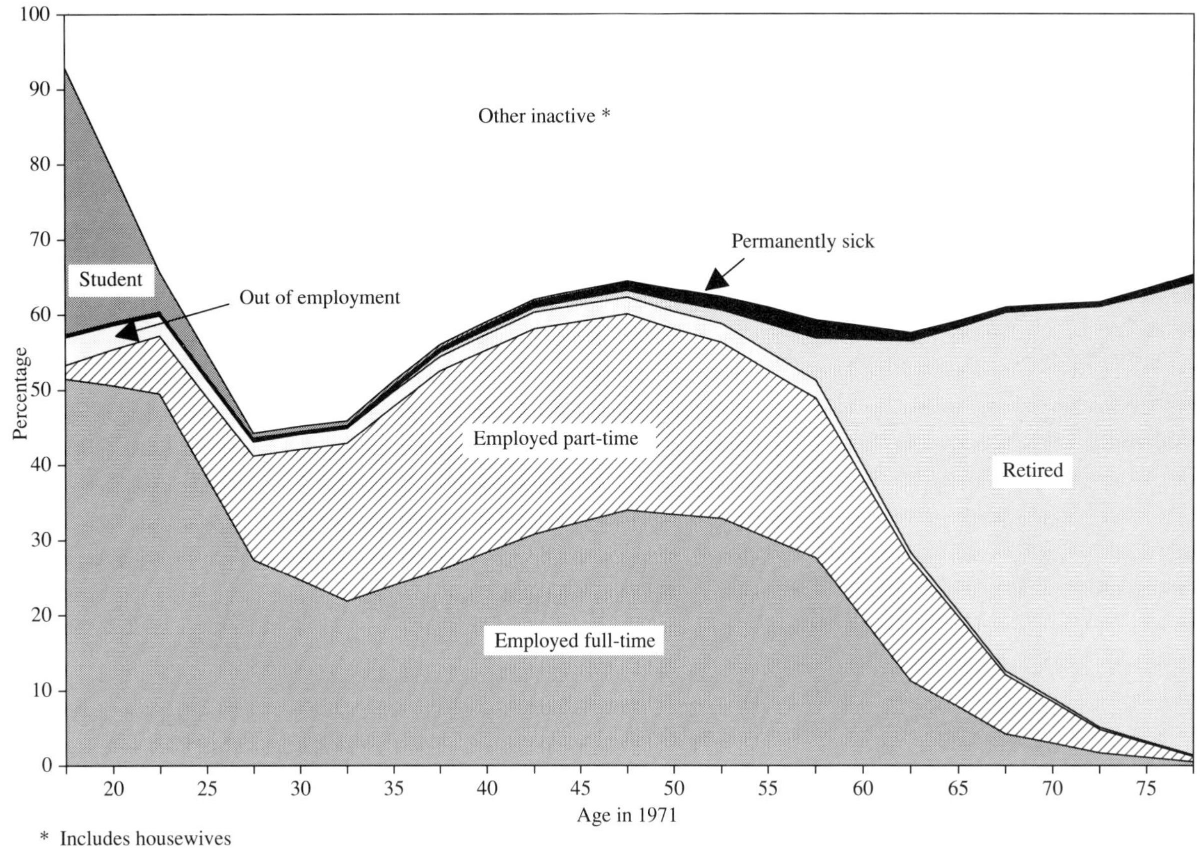

* Includes housewives

8.3.1 Women in occupations

It is widely recognised that the occupational information recorded at census, based on economic activity in the previous week, is unrepresentative of the lifetime work experiences of large numbers of women. Their employment rates are affected by age, by marital status, number and ages of children, and by full- or part-time working.[8] Single women are more likely than married women to have a paid job, but the association between child care responsibilities and work outside the home is even stronger.[9] Retired women are also less likely than retired men to state their last occupation. This is largely due to lack of clarity in the preamble to the 1971 census questions on occupation.[8] Housewives who were working in the week before both censuses were classified to an occupation, as were housewives who stated their last full-time job at the 1981 census. However, by asking for details of their most recent full-time job, the 1981 census question discriminated against women whose most recent jobs would have been part-time.

At the 1971 census, 45 per cent of women were classified to an occupation. This rose to 48 per cent at the 1981 census. Figures 8.1 and 8.2 provide an overview of the patterns of women's economic activity by age at entry to the study.

8.4 Mortality data

This chapter focuses mainly on overall mortality which is tabulated by age at death with standardisation based on five-year age groups. Where health selection[3,7,10] into or out of an occupation varies by age, overall SMRs may obscure mortality differentials. Findings are therefore also presented for working ages and for early and later retirement ages.

To help interpret the SMRs and to provide a link with the proportional mortality ratios (PMRs) in Chapters 4, 5 and 6, cause-specific mortality is presented for some occupations of the 1971 cohort. Cause of death is classified according to the eighth and ninth revisions of the International Classification of Diseases (ICD), as appropriate to year of death. The ICD codes used are shown in Table 8.1.

Both cohorts are followed to the end of 1989. For the 1971 census cohort, data from the first five years of follow-up have been ignored to allow for the effects of health selection. Although preliminary examination suggests there is evidence of a health selection effect in the 1980s this requires further investigation and the earlier years have not been excluded from analyses of the 1981 census cohort.

Table 8.1 Major causes of death analysed in this chapter

Cause group	ICD8 codes	ICD9 codes
All malignant neoplasms	140 - 209	140 - 208
Lung cancer	162	162
Breast cancer	180	180
Circulatory diseases	390 - 458	390 - 459
Ischaemic heart disease	410 - 414	410 - 414
Cerebrovascular disease	430 - 438	430 - 438
Respiratory diseases	460 - 519	460 - 519
Bronchitis, emphysema, asthma	490 - 493	490 - 493
Accidents, poisonings and violence	E800 - E999	E800 - E999

Figure 8.2 LS women 1981 census cohort, percentage distribution by age and economic activity in 1981

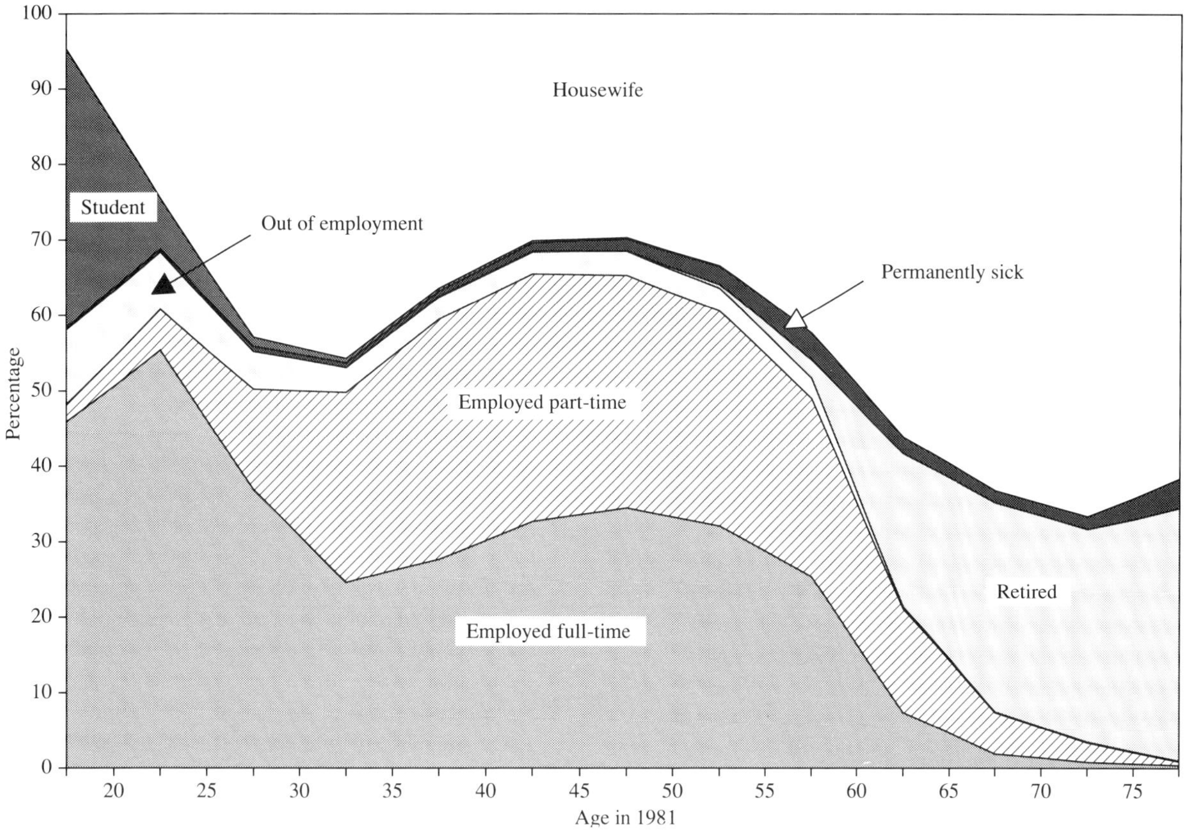

Housewife
Student
Out of employment
Permanently sick
Employed part-time
Retired
Employed full-time
Percentage
Age in 1981

8.5 Long-term follow-up from 1971 Census

This section covers the mortality between 1976 and 1989 of men and women aged 15 and above at the 1971 census, by occupation order. The analysis of disease patterns associated with an occupation is limited to the examination of four broad cause of death groups (malignant neoplasms, circulatory diseases, respiratory diseases, and accidents, poisonings and violence) for the orders showing significant excesses or deficits in overall mortality. All cause mortality for men and women at all ages is shown in Annex 8.1.

8.5.1 Mortality 1976-89, men

The male 1971 census cohort comprises 187,871 individuals of whom 40,368 died between 1976 and the end of 1989. The findings here support those of Fox and Goldblatt[7] for the shorter follow-up period of 1976-81. When the SMRs are ranked from low to high (Figure 8.3), men classified as labourers [order XVIII] experienced the highest mortality (SMR 120) and professional and technical workers [XXV] the lowest (SMR 72). Other occupations which show significant excesses in mortality are textile workers [X] (SMR 118), miners and quarrymen [II] (SMR 117), furnace, forge, foundry workers and rolling mill workers [V] (SMR 112), warehousemen, storekeepers, packers and bottlers [XX] (SMR 112), painters and decorators [XVI] (SMR 112), drivers of stationary engines, cranes, etc [XVII] (SMR 111) and service, sport and recreation workers [XXIII] (SMR 105). Two orders now show significant excesses compared to previous work based on the shorter follow-up period, 1976-81. Drivers of stationary engines, cranes, etc previously showed a non-significant excess in overall mortality, and painters and decorators did not show an excess.[7]

In addition to professional workers, the other occupations which show significant deficits in mortality are administrators and managers [XXIV] (SMR 80), clothing workers [XI] (SMR 81), electrical and electronic workers [VI] (SMR 86), sales workers [XXII] (SMR 90), clerical workers [XXI] (SMR 93) and farmers, foresters and fishermen [I] (SMR 93).

As with previous findings, the unoccupied and inadequately described show substantial excesses in mortality which attenuate with age. The impact of the mortality of the permanently sick on the SMRs of men classified to an occupation have been discussed elsewhere.[11] Allowance is made for these health selection effects by the exclusion of the first five years of follow-up. The rest of this section will focus on the occupational orders with significant excesses and deficits in mortality.

8.5.2 Age pattern of mortality and length of follow-up, men

The age patterns of the orders which show significant excesses and deficits suggest generally (see Annex 8.3) that these relative disadvantages and advantages persisted at all ages. The pattern is largely replicated in the units also shown in the table. Labourers [XVIII] and warehousemen, storekeepers, packers and bottlers [XX] had significant excesses in all three age groups.

While the LS has advantages over cross-sectional analysis in terms of the numerator-denominator biases, there are discrepancies with the results of other studies which may be associated with health selection, sample size and length of follow-up, as shown for the following occupations.

Figure 8.3 LS men 1971 Census cohort, mortality 1976-89 SMRs of occupation orders ranked from low to high

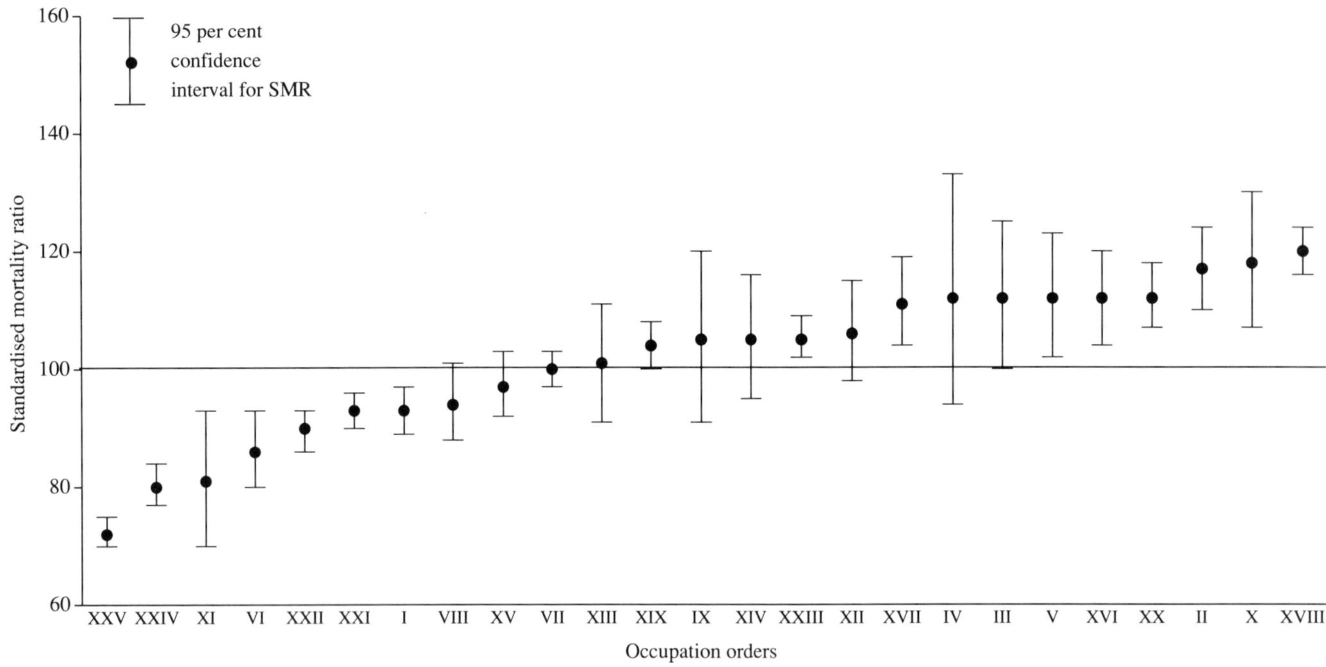

Contrary to previous decennial supplement[1] findings, only a small excess is evident for miners and quarrymen [II] at working ages. This is shown in Annex 8.3. The table also shows that above ground miners [unit 008] had a non-significant excess of 14 per cent, while underground miners [007] had mortality similar to that of all men. There is some evidence to suggest that the lack of excess mortality of underground miners may be explained by health selection out of the occupation.[3] When the SMRs are disaggregated by follow-up periods (1976-81, 1982-85, 1986-89) an excess (non-significant) is seen only for the last period. Deaths of underground miners from ischaemic heart disease, respiratory diseases and lung cancer also show the same pattern of excess mortality in the latter years of follow-up. This may be because men dying from smoking related diseases, for example, were no longer employed as underground miners and therefore did not give underground mining as their occupation at the 1971 census.

The all age SMR for transport and communication workers [XIX] is not significantly different from that of all men (SMR 104). However of the 22 units, 13 show raised SMRs, of which 2 are significantly higher than expected; deck and engineroom ratings, barge and boatmen [116] experienced a 34 per cent excess and porters, ticket collectors, railway [132], a 25 per cent excess.

In contrast to previous work using LS data from 1971-81, the longer follow-up supports the excess mortality of painters and decorators [XVI] at working ages shown by the last decennial supplement.[2,7] When disaggregated by follow-up periods, the SMRs for the period 1982-85 are significantly higher in all three age groups. This occupation may have a constant movement of men in and out, particularly at older ages. Further work will look at men who were classified to this occupation at both 1971 and 1981 censuses.

8.5.3 Cause of death, men

Associations between selected occupations and causes of death for the orders with significant deficits or excesses in overall mortality are shown in Tables 8.2, 8.3, 8.4 and 8.5.

For the orders with significantly low all cause mortality, the pattern is repeated for the cause specific SMRs, with the exception of farmers, foresters and fishermen [II] and clerical workers [XXI]. Each of these orders showed significantly raised mortality for one cause group.

The mortality of farmers, foresters and fishermen from accidents, poisonings and violence was significantly greater by 39 per cent (Table 8.2). This pattern of mortality corresponds with findings elsewhere[2,12] and in this volume (Chapter 4). All units of the order show higher than expected mortality from this cause although the SMRs are only significant for gardeners and groundsmen [005] (SMR 236) aged 75 and over at death.

Clerical workers had significantly greater mortality (SMR 114) from circulatory diseases at ages 20-64 (Table 8.3). The excess is significant for ischaemic heart disease but not for cerebrovascular disease. These findings agree with other studies which have shown, for example, that lower grade male civil servants are at greater risk of heart disease than their higher grade colleagues.[13]

Three orders with significant excesses in all cause mortality show significantly raised SMRs from several causes. These are labourers [XVIII], miners and quarrymen [II], and warehousemen, storekeepers, packers and bottlers [XX] (Table 8.4).

The mortality of labourers [XVIII] is significantly raised for all four cause groups. The risks of dying from respiratory

Figure 8.4 LS women 1971 Census cohort, mortality 1976-89 SMRs of occupation orders (50+ deaths only) ranked from low to high

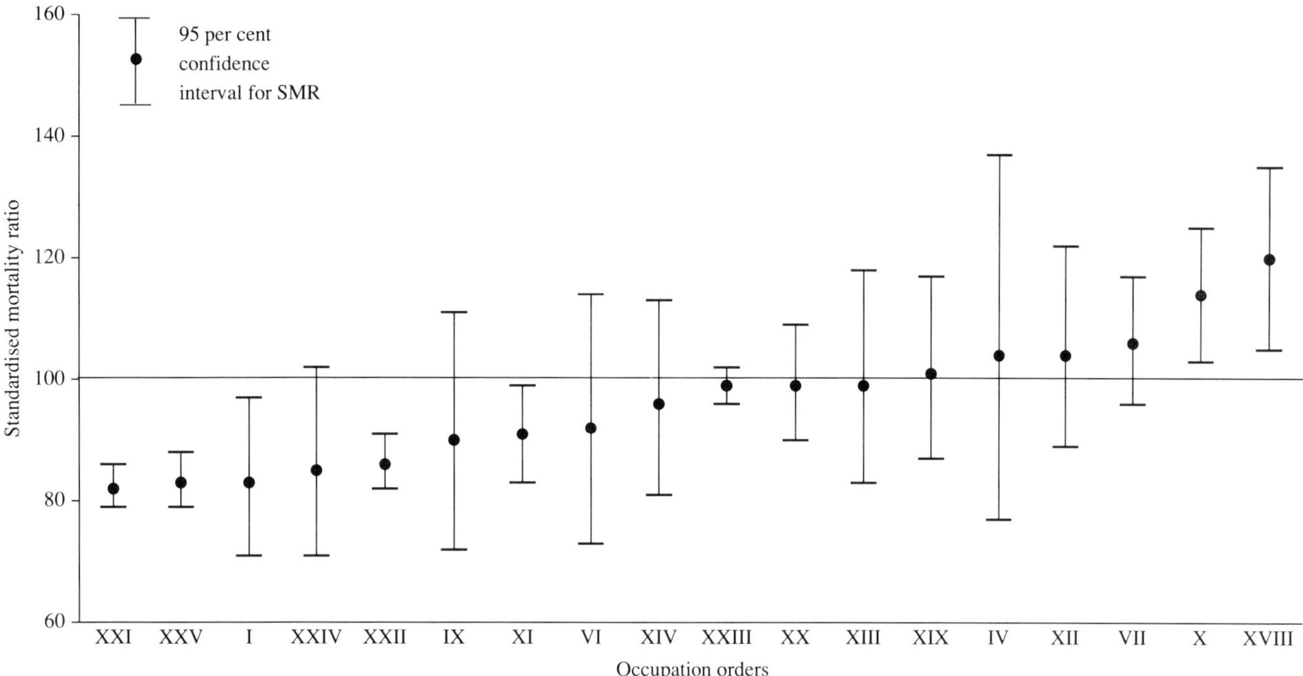

diseases (SMR 150) and accidents, poisonings and violence (SMR 151) were greater relative to the other causes. At ages 20-64, more than twice the number of expected deaths from respiratory diseases were recorded (SMR 220). The small excesses from circulatory diseases at older ages are the result of lower mortality from ischaemic heart disease at ages 65 and over, and lower mortality from strokes at 75 and over. As labouring may be a new, post-retirement occupation for some men, health selection may have depressed the SMRs at older ages. Further work will look at the mortality of men recorded as labourers at both 1971 and 1981 censuses and explore the hypothesis of a protective mechanism of physical activity at older ages.

Miners and quarrymen also had a greater risk of dying from respiratory diseases compared to the other causes. Deaths from bronchitis, emphysema and asthma are significantly higher than expected at ages 65-74 (SMR 234) and 75 and over (SMR 172).

Drivers of stationary engines and cranes [XVII] had excesses from circulatory and respiratory diseases and lung cancer. The significantly increased risk of dying from ischaemic heart disease at working ages (SMR 134) agrees with results from the Turin longitudinal study.[14] Higher than expected lung cancer ratios have also been found previously.[1]

Table 8.2 LS men 1971 census cohort, mortality 1976-89: mortality of farmers, foresters and fishermen (I) from accidents, poisonings and violence

	Age at death							
	20-64		65-74		75 and over		All ages	
	SMR	CI†	SMR	CI†	SMR	CI†	SMR	CI†
Accidents, poisonings and violence	130	(87 - 186)	142	(76 - 243)	149	(95 - 221)	139	(107 - 176)*
All causes	85	(76 - 95)*	90	(82 - 97)*	97	(92 - 103)	93	(89 - 97)*

† 95 per cent confidence interval for SMR
* 95 per cent confidence interval for SMR excludes 100

Table 8.3 LS men 1971 census cohort, mortality 1976-89: mortality of clerical workers (XXI) from circulatory diseases

	Age at death							
	20-64		65-74		75 and over		All ages	
	SMR	CI†	SMR	CI†	SMR	CI†	SMR	CI†
Circulatory diseases	114	(102 - 127)*	96	(88 - 105)	96	(89 - 104)	100	(95 -105)
Ischaemic heart disease	114	(101 - 129)*	97	(87 - 108)	102	(92 - 113)	103	(97 - 110)
Cerebrovascular disease	112	(80 - 151)	90	(72 - 111)	95	(81 - 110)	95	(85 - 107)
All causes	101	(93 - 109)	90	(85 - 96)*	91	(86 - 96)*	93	(90 - 96)*

† 95 per cent confidence interval for SMR
* 95 per cent confidence interval for SMR excludes 100

Table 8.4 LS men 1971 census cohort, mortality 1976-89: SMRs of high mortality orders by cause and age at death

Occupation order	Age at death							
	20-64		65-74		75 and over		All ages	
	SMR	CI†	SMR	CI†	SMR	CI†	SMR	CI†
XVIII labourers								
All causes	138	(129 - 148)*	119	(112 - 126)*	112	(106 - 118)*	120	(116 - 124)*
All malignant neoplasms	135	(119 - 153)*	114	(102 - 127)*	120	(107 - 134)*	121	(113 - 130)*
Lung cancer	147	(120 - 178)*	121	(102 - 144)*	133	(109 - 160)*	132	(118 - 146)*
Circulatory diseases	128	(115 - 142)*	109	(100 - 119)*	103	(96 - 111)	110	(105 - 116)*
Respiratory diseases	220	(176 - 272)*	162	(140 - 187)*	133	(118 - 148)*	150	(138 - 163)*
Accidents, poisonings and violence	151	(115 - 195)*	148	(93 - 224)	154	(102 - 223)*	151	(124 - 182)*
II Miners and quarrymen								
All causes	102	(88 - 117)*	130	(118 - 143)*	113	(104 - 123)*	117	(110 - 124)*
All malignant neoplasms	115	(90 - 146)	121	(100 - 145)	110	(90 - 134)	116	(103 - 130)*
Lung cancer	111	(73 - 162)	101	(72 - 139)	119	(83 - 166)	110	(89 - 133)
Circulatory diseases	92	(73 - 113)	123	(107 - 141)*	113	(100 - 127)	112	(103 - 122)*
Respiratory diseases	138	(82 - 218)	183	(143 - 232)*	135	(112 - 162)*	149	(129 - 171)*
Bronchitis, emphysema, asthma	153	(74 - 282)	234	(166 - 322)*	172	(125 - 231)*	191	(154 - 234)*
XX Warehousemen, storekeepers, packers, bottlers								
All causes	113	(102 - 126)*	110	(102 - 119)*	113	(105 - 122)*	112	(107 - 118)*
All malignant neoplasms	102	(83 - 125)	116	(100 - 135)	112	(95 - 133)	111	(101 - 123)*
Lung cancer	118	(86 - 158)	129	(102 - 160)*	116	(85 - 155)	122	(105 - 142)*
Circulatory diseases	119	(103 - 138)*	113	(101 - 126)*	107	(96 - 119)	112	(104 - 120)*
Respiratory diseases	116	(74 - 172)	101	(78 - 130)*	127	(107 - 150)*	117	(103 - 134)*
XVII Drivers of stationary engines, cranes, etc								
All causes	116	(102 - 131)*	110	(98 - 123)	110	(98 - 123)	111	(104 - 119)*
All malignant neoplasms	104	(81 - 132)	95	(75 - 119)	101	(76 - 131)	100	(86 - 114)
Lung cancer	128	(88 - 179)	122	(87 - 167)	153	(101 - 223)*	131	(107 - 160)*
Circulatory diseases	131	(109 - 155)*	113	(96 - 132)	108	(91 - 127)	116	(106 - 127)*
Respiratory diseases	108	(60 - 178)*	146	(107 - 195)*	126	(95 - 163)	130	(108 - 156)*

† 95 per cent confidence interval for SMR
* 95 per cent confidence interval for SMR excludes 100

Table 8.5 LS men 1971 census cohort, mortality 1976-89: mortality of painters and decorators (XVI), service, sport and recreation workers (XXIII), furnace, forge, foundry, rolling mill (V) and service, sport and recreation workers (XXIII) from malignant neoplasms by age at death

	Age at death							
	20-64		65-74		75 and over		All ages	
	SMR	CI†	SMR	CI†	SMR	CI†	SMR	CI†
XVI Painters and decorators								
All malignant neoplasms	136	(106 - 172)*	137	(111 - 169)*	118	(90 - 153)	132	(115 - 150)*
Lung cancer	155	(105 - 220)*	142	(100 - 195)	164	(106 - 242)*	151	(122 - 185)*
XXIII Service, sport and recreation workers								
All malignant neoplasms	112	(97 - 130)	119	(106 - 133)*	119	(107 - 133)*	117	(110 - 126)*
Lung cancer	104	(81 - 132)	111	(91 - 133)	129	(106 - 155)*	115	(102 - 129)*
V Furnace, forge, foundry, rolling mill workers								
All malignant neoplasms	131	(95 - 176)	133	(98 - 177)	91	(60 - 133)	120	(99 - 144)
Lung cancer	130	(76 - 208)	149	(92 - 228)	130	(67 - 227)	137	(102 - 181)*
XIX Transport and communication workers								
All malignant neoplasms	110	(98 - 124)	123	(110 - 136)*	105	(92 - 120)	114	(106 - 121)*
Lung cancer	136	(114 - 161)*	135	(114 - 158)*	118	(94 - 146)	131	(118 - 145)*

† 95 per cent confidence interval for SMR
* 95 per cent confidence interval for SMR excludes 100

Painters and decorators [XVI] and workers in service, sport and recreation [XXIII] show excess mortality from all cancers (Table 8.5). The all age SMRs for both orders are also significantly raised for lung cancer. All cause mortality of transport and communication workers [XIX] shows a small excess (Annex 8.1) but this order also experienced significantly higher than expected deaths from malignant neoplasms (Table 8.5). The lung cancer ratios are as high at working ages as at ages 65-74. Furnace, forge, foundry and rolling mill workers [V] also had a higher risk of dying from lung cancer although only the all age SMR is significant. Similar findings for deaths from all cancers for these three orders were reported in the last decennial supplement.

Textile workers [X] show excesses for all causes but the SMR is only significant for stroke mortality at ages 75 and over (SMR 162). This accounts for the older age pattern of mortality found in Table 8.4 and agrees with previous findings.[2]

8.5.4 Mortality 1976-89, women

There are 202,789 individuals in the 1971 female Census cohort. Of these 40,774 died between 1976 and the end of 1989. From the information they provided at the census, 45 per cent of women aged 15 and above could be classified to an occupation. A further 12 per cent were assigned to the 'inadequately described' category. The overwhelming majority of the remaining group described themselves as housewives and no attempt was made to collect details of any previous occupations of these women.

Women's employment is concentrated in a relatively small number of occupations — of women aged 15-59 at the 1971 census who could be classified, 75 per cent were in four occupation orders. Twenty-eight per cent were clerical workers [order XXI], 24 per cent service, sport and recreation workers [XXIII], 12 per cent professional, technical workers and artists [XXV] and 12 per cent sales workers [XXII].

Half of the women who could be classified were in just six units. Seventeen per cent were clerks and cashiers [unit 139]; 9 per cent were typists, shorthand writers, secretaries [141]; 8 per cent were shop saleswomen and assistants [144]; 5 per cent were charwomen, office cleaners, window cleaners and chimney sweeps [166]; and 5 per cent were maids, valets and related service workers and 4 per cent were nurses [183].

Some of the demographic events which affect female employment rates are indicated by the cross-sectional data for the 1971 population which are summarised in Figure 8.1. Detailed information from women's work histories[9] reveal a pattern which can only be inferred from this age breakdown of the 1971 data; this is that many women move out of the labour market around the time they begin childbearing but return when domestic responsibilities are less demanding.

The occupational mortality of all women in the period 1976-89 is shown in Annex 8.1. The SMR of 92 for women who were classified to an occupation indicates that, compared with all women in the study, those who were classified experienced significantly reduced mortality.

There are 18 orders with 50 or more observed deaths and the SMRs for these orders are illustrated in Figure 8.4. Women in 12 of these 18 orders experienced lower mortality levels than the general population. The advantages are greatest and statistically significant for the large orders of clerical workers [XXI] (SMR 82), professional, technical workers and artists [XXV] (SMR 83) and sales workers [XXII] (SMR 86), and the smaller orders of farmers, foresters and fishermen [I] (SMR 83) and clothing workers [XI] (SMR 91). Only two orders have significantly raised mortality — textile workers [X] (SMR 114) and labourers not elsewhere classified [XVIII] (SMR 120). Men in these occupations also have elevated mortality.

8.5.5 Age pattern of mortality, women

SMRs by age at death are presented for occupation orders in Table 8.6 and for units in Table 8.7. At ages 20-59, 60-74 and 75 and above, women who could be classified to an occupation show mortality levels significantly lower than those of all women. There is very little difference in the SMRs for the three broad age ranges but a slight attenuation of the advantage is discernible after age 74.

In contrast, there is considerable variation in the excess mortality experienced by women in inadequately described occupations. At ages 20-59 their mortality was 27 per cent higher than that of all women, at 60-74 the excess was 17 per cent and at older ages it was only 2 per cent. However, even the latter figure is significantly different from that for all women of the same age. Another important feature which is revealed by the information in Table 8.6 is the large number of retired women whose occupations were inadequately described. In the oldest age group the number of retired women who could not be classified to an occupation far exceeded the number who could.

The relative level of mortality of the unoccupied also varies by age. The excess of deaths is greatest at ages 20-59 (SMR 109), lower in the 60-74 age group (SMR 107) and almost the same as all women at ages 75 and above (SMR 101).

A comparison of the SMRs for the three age groups for each occupation shows general uniformity in the direction of the SMRs but no consistent age-related patterns or trends. In the four units with the largest numbers of women (clerks and cashiers [139], typists, shorthand writers and secretaries [141], shop saleswomen and assistants [144] and nurses [183]) mortality was low at every age, two-thirds of these SMRs being significantly low (Table 8.7). Of these units, typists, shorthand writers and secretaries [141] fared best in survival terms with effectively the same significant advantage at all ages — SMRs of 77, 76 and 78 respectively. The other three units show wider variations by age but in each of these units the relative level was lowest at ages 60-74.

Women in several other occupations experienced a significant deficit of deaths at ages 60-74. These are primary and secondary school teachers [193], cooks [162] and canteen assistants and counter hands [161]. At other ages the patterns for these units are rather different (Table 8.7).

The age pattern of mortality in the large clerical [XXI], sales [XXII] and professional [XXV] orders reflect those described for their component units. Of the other low mortality orders, clothing workers [XI] and administrators and

Table 8.6 LS women 1971 census cohort, mortality 1976-89: all cause mortality by age at death, selected occupation orders

		Age at death								
		20-59			60-74			75 and over		
		Deaths	SMR	(CI)†	Deaths	SMR	(CI)†	Deaths	SMR	(CI)†
I	Farmers, foresters, fishermen	16	65	(37 - 106)	57	83	(63 - 108)	88	87	(70 - 107)
IV	Glass and ceramic makers	7	90	(36 - 185)	24	96	(62 - 144)	19	123	(74 - 193)
VI	Electrical and electronic workers	22	88	(55 - 134)	44	89	(65 - 120)	19	104	(62 - 162)
VII	Engineering and allied trade workers nec	88	116	(93 - 142)	197	99	(86 - 114)	136	110	(92 - 130)
IX	Leather workers	12	94	(48 - 163)	27	72	(47 - 104)	49	103	(77 - 138)
X	Textile workers	41	122	(88 - 166)	136	110	(92 - 130)	256	115	(101 - 130)*
XI	Clothing workers	53	82	(61 - 107)	163	84	(72 - 99)*	272	97	(86 - 110)
XII	Food, drink and tobacco workers	19	82	(49 - 128)	79	114	(91 - 143)	58	102	(77 - 132)
XIII	Paper and printing workers	16	78	(45 - 127)	64	115	(88 - 147)	51	92	(68 - 120)
XIV	Makers of other products	29	104	(70 - 149)	70	101	(79 - 127)	43	84	(61 - 113)
XVIII	Labourers nec	29	113	(76 - 162)	113	125	(103 - 151)*	111	116	(96 - 140)
XIX	Transport and communication workers	33	90	(62 - 127)	79	105	(83 - 131)	64	103	(80 - 132)
XX	Warehousemen, storekeepers, packers, bottlers	74	107	(84 - 135)	231	110	(96 - 125)	135	82	(69 - 97)*
XXI	Clerical workers	434	82	(74 - 90)*	828	77	(72 - 82)*	845	88	(82 - 94)*
XXII	Sales workers	222	94	(82 - 108)	626	84	(78 - 91)*	659	86	(79 - 93)*
XXIII	Service, sport and recreation workers	479	104	(95 - 114)	1744	98	(94 - 103)	2024	98	(94 - 102)
XXIV	Administrators and managers	18	89	(53 - 140)	41	68	(49 - 93)*	62	101	(77 - 129)*
XXV	Professional, technical workers, artists	181	77	(66 - 89)*	414	77	(70 - 85)*	732	89	(82 - 95)*
XXVII	Inadequately described occupations	113	127	(105 - 153)*	1336	117	(111 - 123)*	10166	102	(101 - 105)*
All classified occupations		1787	92	(88 - 96)*	4969	91	(88 - 93)*	5646	94	(92 - 97)*
Unoccupied		1614	109	(104 - 114)*	4902	107	(104 - 110)*	10421	101	(99 - 103)

Only orders contributing at least 50 observed deaths are included in this table.

† 95 per cent confidence interval for SMR
* 95 per cent confidence interval for SMR excludes 100

Table 8.7 LS women 1971 census cohort, mortality 1976-89: all cause mortality by age at death, selected occupation orders and units

		Age at death								
		20-59			60-74			75 and over		
		Deaths	SMR	(CI)†	Deaths	SMR	(CI)†	Deaths	SMR	(CI)†
VII	Engineering and allied trades workers nec	88	116	(93 - 142)	197	99	(86 - 114)	136	110	(92 - 130)
	039 Machine tool operators	17	98	(57 - 158)	39	103	(73 - 141)	30	146	(98 - 208)
	054 Other metal making, working; jewellery and electrical production process workers	19	133	(80 - 208)	36	103	(72 - 142)	22	94	(59 - 142)
XXI	Clerical workers	434	82	(74 - 90)*	828	77	(72 - 82)*	845	88	(82 - 94)*
	139 Clerks, cashiers	273	82	(73 - 92)*	596	78	(72 - 85)*	601	93	(85 - 100)
	141 Typists, shorthand writers, secretaries	126	77	(65 - 92)*	198	76	(66 - 87)*	218	78	(68 - 89)*
XXII	Sales workers	222	94	(82 - 108)	626	84	(78 - 91)*	659	86	(79 - 93)*
	143 Proprietors and managers, sales	57	103	(78 - 134)	194	89	(77 - 103)	273	87	(77 - 97)*
	144 Shop salesmen and assistants	148	89	(76 - 105)	408	83	(75 - 91)*	353	85	(76 - 95)*
XXIII	Service, sport and recreation workers	479	104	(95 - 114)	1744	98	(94 - 103)	2024	98	(94 - 102)
	154 Publicans, innkeepers	6	93	(34 - 203)	24	106	(68 - 157)	24	83	(54 - 124)
	155 Barmen, barmaids	21	115	(71 - 175)	67	141	(109 - 179)*	54	143	(107 - 187)*
	161 Canteen assistants, counter hands	87	104	(83 - 128)	245	87	(76 - 98)*	193	82	(71 - 95)*
	162 Cooks	28	105	(70 - 152)	98	80	(65 - 98)*	135	99	(83 - 117)
	164 Maids, valets, related service workers nec	94	103	(83 - 126)	403	96	(87 - 106)	531	97	(89 - 106)
	166 Charwomen, office/window cleaners, chimney sweeps	113	117	(97 - 141)	473	109	(99 - 119)	518	105	(96 - 114)
	168 Launderers, dry cleaners and pressers	21	151	(93 - 230)	61	110	(84 - 141)	60	104	(79 - 134)
XXV	Professional, technical workers, artists	181	77	(66 - 89)*	414	77	(70 - 85)*	732	89	(82 - 95)*
	183 Nurses	71	82	(64 - 104)	155	77	(65 - 90)*	206	99	(86 - 114)
	193 Primary and secondary school teachers	56	75	(57 - 98)*	113	71	(59 - 86)*	253	91	(80 - 103)

† 95 per cent confidence interval for SMR
* 95 per cent confidence interval for SMR excludes 100

managers [XXIV] experienced significantly low mortality in the early years of retirement but none of the SMRs for farmers [I] is significant.

At the other end of the scale are the orders and units with elevated risks of dying. In the two orders with significant excesses of deaths at all ages — textile workers [X] and labourers nec [XVIII] — mortality was raised in every age group but no clear age pattern emerges. Among textile workers the level of mortality was significantly higher than that for all women aged 75 and over only, and lowest at ages 60-74 (Table 8.6). In contrast, among labourers the level was highest and statistically significant in the early years of retirement and lower at other ages.

Barmaids [155] had the highest mortality ratios. Their level of mortality was over 40 per cent higher than that of all women aged 60 and above (Table 8.7). At working ages their risks are raised but, unlike at older ages, the excess of deaths of 15 per cent could have arisen by chance. Women in two units show a mortality gradient in the opposite direction from that of barmaids, their greatest excesses being at working ages. These are charwomen, office cleaners etc [166] and launderers, dry cleaners and pressers [168]. None of the age specific ratios for these women is significant but the level for charwomen, office cleaners, etc at all ages (SMR 108) is significantly higher than that of all women.

The findings for mortality at ages 20-59 in the period 1976-89 can be compared with the published data for the years 1976-81.[15] There are very limited differences between the two sets of results. In both the shorter and longer periods of follow-up, women in only two orders — clerical workers [XXI] and professionals, technical workers and artists [XXV] — experienced levels of mortality significantly different from those of the corresponding standard populations. More generally, the direction of the estimates — relative high or low mortality — are unchanged.

8.5.6 Cause of death, women

Cause-specific SMRs for women of all ages in selected occupations are presented in Tables 8.8, 8.9 and 8.10.

Table 8.8 LS women 1971 census cohort, mortality 1976-89: low mortality orders and units, cause of death, all ages

	Deaths	SMR	(CI)†
Order XXI Clerical workers			
All causes	2107	82	(79 - 86)*
All malignant neoplasms	704	90	(84 - 97)*
Breast cancer	196	111	(96 - 128)
Lung cancer	103	83	(68 - 101)
Circulatory diseases	911	79	(74 - 85)*
Respiratory diseases	176	78	(67 - 90)*
Accidents, poisonings and violence	67	80	(62 - 102)
Order XXII Sales workers			
All causes	1507	86	(82 - 91)*
All malignant neoplasms	445	92	(84 - 101)
Breast cancer	108	106	(87 - 128)
Lung cancer	66	84	(65 - 107)
Circulatory diseases	710	85	(79 - 92)*
Respiratory diseases	117	71	(59 - 86)*
Accidents, poisonings and violence	31	68	(47 - 97)*
Order XXV Professional, technical workers, artists			
All causes	1327	83	(79 - 88)*
All malignant neoplasms	396	96	(86 - 105)
Breast cancer	88	98	(78 - 120)
Lung cancer	58	91	(69 - 118)
Circulatory diseases	630	82	(76 - 89)*
Respiratory diseases	113	68	(56 - 82)*
Accidents, poisonings and violence	50	115	(85 - 151)
Unit 183 Nurses			
All causes	432	87	(79 - 96)*
All malignant neoplasms	133	94	(79 - 111)
Breast cancer	26	82	(54 - 121)
Lung cancer	27	121	(79 - 175)
Circulatory diseases	200	87	(75 - 100)
Respiratory diseases	33	70	(48 - 99)*
Accidents, poisonings and violence	24	164	(105 - 245)*
Unit 193 Primary and secondary school teachers			
All causes	422	82	(75 - 91)*
All malignant neoplasms	116	90	(75 - 108)
Breast cancer	34	121	(84 - 169)
Lung cancer	11	56	(28 - 101)
Circulatory diseases	217	87	(76 - 100)
Respiratory diseases	36	65	(45 - 90)*
Accidents, poisonings and violence	9	66	(30 - 126)

† 95 per cent confidence interval for SMR
* 95 per cent confidence interval for SMR excludes 100

In all three large, low mortality occupation orders — clerical workers [XXI], sales workers [XXII] and professional, technical workers and artists [XXV] — the death ratios for both respiratory and circulatory diseases are significantly lower than in the standard population (Table 8.8). There are larger variations in mortality from all malignant neoplasms and accidents, poisonings and violence in these same three occupation orders. Only clerical workers had significantly low SMRs from all cancers, and only sales workers had significantly low SMRs from accidents, poisonings and violence. An exception to these reduced risks of dying is the 15 per cent excess deaths from accidents, poisonings and violence for women classified as professional, technical workers and artists.

The patterns of death by cause for all the large clerical and sales occupation units are very similar to those for their corresponding orders. In contrast, there are distinct differences between the cause-specific SMRs for the component units of professional, technical workers and artists [XXV]. The findings for two of the units — nurses [183] and primary and secondary school teachers [193] — are included in Table 8.8. The most striking difference between the two professions is the variation in the number of deaths from accidents, poisonings and violence. Mortality from these causes amongst nurses was significantly higher than that of all women by 64 per cent, whereas teachers had 34 per cent fewer than expected deaths, although this could have been due to chance. This excess of deaths from external causes amongst nurses therefore seems largely responsible for the 15 per cent excess of the whole order [XXV].

The findings presented in Table 8.8 are consistent with other results. The evidence for excess risks of mortality from

Table 8.9 LS women 1971 census cohort, mortality 1976-89: textile workers, cause of death, all ages

	Deaths	SMR	(CI)†
Order X Textile workers			
All causes	433	114	(103 - 125)*
All malignant neoplasms	91	103	(83 - 127)
Breast cancer	9	51	(23 - 97)*
Lung cancer	12	89	(46 - 156)
Circulatory diseases	210	109	(95 - 125)
Respiratory diseases	61	142	(109 - 183)*
Accidents, poisonings and violence	14	155	(85 - 260)
Units 067, 068 Warpers, sizers, drawers-in, weavers			
All causes	135	116	(97 - 137)
All malignant neoplasms	16	72	(41 - 117)
Breast cancer	3	72	(15 - 210)
Lung cancer	-	-	
Circulatory diseases	73	118	(92 - 148)
Respiratory diseases	24	160	(103 - 239)*
Accidents, poisonings and violence	5	206	(67 - 481)
Unit 065 Spinners, doublers, twisters			
All causes	70	178	(139 - 225)*
All malignant neoplasms	17	198	(116 - 318)*
Breast cancer	2	118	(14 - 426)
Lung cancer	3	234	(48 - 685)
Circulatory diseases	30	148	(100 - 212)
Respiratory diseases	8	168	(73 - 331)
Accidents, poisonings and violence	2	217	(26 - 785)
Unit 070 Bleachers and finishers of textiles			
All causes	18	176	(104 - 278)*
All malignant neoplasms	4	137	(37 - 351)
Breast cancer	-	-	
Lung cancer	-	-	
Circulatory diseases	9	185	(85 - 351)
Respiratory diseases	1	108	(3 - 604)
Accidents, poisonings and violence	1	372	(9 - 2073)

† 95 per cent confidence interval for SMR
* 95 per cent confidence interval for this SMR excludes 100

Table 8.13 LS women 1971 cohort, mortality 1976-89: orders and units, cause of death, all ages

	Deaths	SMR	(CI)†
Order XVIII Labourers nec			
All causes	253	120	(105 - 135)*
All malignant neoplasms	70	123	(96 - 155)
Breast cancer	9	76	(35 - 145)
Lung cancer	18	195	(116 - 309)*
Circulatory diseases	130	128	(107 - 152)*
Respiratory diseases	21	102	(63 - 155)
Accidents, poisonings and violence	3	57	(12 - 167)
Order XXIII Service, sport and recreation workers			
All causes	4247	99	(96 - 102)
All malignant neoplasms	1163	103	(97 - 109)
Breast cancer	221	96	(84 - 109)
Lung cancer	211	116	(101 - 133)*
Circulatory diseases	2042	97	(93 - 101)
Respiratory diseases	436	104	(94 - 114)
Accidents, poisonings and violence	82	80	(64 - 99)*
Unit 155 Barmen, barmaids			
All causes	142	137	(115 - 162)*
All malignant neoplasms	46	147	(108 - 197)*
Breast cancer	7	101	(41 - 208)
Lung cancer	11	217	(109 - 389)*
Circulatory diseases	62	131	(100 - 168)
Respiratory diseases	15	161	(90 - 266)
Accidents, poisonings and violence	4	138	(38 - 353)
Unit 154 Publicans, innkeepers			
All causes	54	93	(70 - 122)
All malignant neoplasms	19	126	(76 - 197)
Breast cancer	4	130	(36 - 334)
Lung cancer	4	165	(45 - 423)
Circulatory diseases	19	67	(40 - 105)
Respiratory diseases	8	139	(60 - 273)
Accidents, poisonings and violence	-	-	
Unit 166 Charwomen, office cleaners, window cleaners, chimney sweeps			
All causes	1104	108	(102 - 114)*
All malignant neoplasms	310	116	(104 - 130)*
Breast cancer	49	92	(68 - 121)
Lung cancer	59	136	(104 - 176)*
Circulatory diseases	529	105	(96 - 114)
Respiratory diseases	109	110	(91 - 133)
Accidents, poisonings and violence	18	77	(46 - 122)

† 95 per cent confidence interval for SMR
* 95 per cent confidence interval for SMR excludes 100

accidents, poisonings and violence amongst nurses continues to accumulate (see Chapter 5 in this volume). There is remarkably close agreement between these cause-specific, all age SMRs over the extended period 1976-89 for the three large orders, and the published SMRs for women aged 15-59 at death in the period 1976-81.[8]

Research suggests that both men and women with a history of employment in the textile industry have raised risks of dying.[1,2] The significant excesses of deaths for all causes for women in the 1971 census cohort (Annex 8.1 and Table 8.6) are consistent with this and are supported by the more detailed findings presented in Table 8.9. For textile workers [X], the SMRs for all the four broad cause groups are raised. However, only the 42 per cent excess of deaths from respiratory diseases is statistically significant. This is in contrast to the findings for the shorter follow-up period 1971-81, in which there were no female deaths from respiratory diseases.[8]

The SMRs for women in the different textile units within the order are generally high. The 60 per cent excess of deaths from respiratory diseases amongst warpers, sizers and drawers-in [067] and weavers [068] is statistically significant. There is also a significant 98 per cent excess for spinners, doublers and twisters [065] for deaths from all malignant neoplasms. In this case, as only three of the seventeen deaths were from lung cancer, cigarette smoking can be only a small part of the explanation for the unexpectedly large number of deaths from all cancers.

The data in Table 8.10 indicate that the elevated mortality of female labourers [XVIII] is essentially attributable to high rates of circulatory diseases and an excess of deaths from lung cancer. The latter finding mirrors the results for the equivalent cohort of men (Table 8.4) and other studies of male labourers[2] and is probably explained by the smoking habits of these workers.

In the case of circulatory diseases, although the ratios for both men and women show a significant excess of deaths for all ages combined, the age patterns are very different. For women the SMRs for circulatory diseases are 98, 128 and 131 for the age groups 20-59, 60-74 and 75 and above respectively. Only the latter figure is statistically significant. Amongst men, mortality was high at working ages and decreased with age (Table 8.4). There is no obvious explanation for an association between a history of active manual work and the age pattern of deaths from circulatory diseases shown for women. However, the number of women in this occupation is small. Earlier research[1] suggested that female labourers, like their male counterparts, have elevated risks of dying from accidents. This analysis has not found an increased risk for women, although it has for men (see Table 8.4).

Twenty-two per cent of women who could be classified to an occupation at the 1971 census were service, sport and recreation workers [XXIII]. The level of all cause mortality in this order is almost the same as that of all women, as is also the case for deaths from all malignant neoplasms, and circulatory and respiratory diseases (Table 8.10). However, women in this order experienced significantly fewer than expected deaths from accidents, poisonings and violence (SMR 80) and a significant 16 per cent excess of lung cancer deaths. The order comprises a fairly wide range of units and the cause-specific details for three of them — barmen, barmaids [155], publicans and innkeepers [154] and charwomen, office cleaners, window cleaners and chimney sweeps [166] — are presented alongside those for the order.

Barmaids show raised levels of mortality for all four broad causes of death, although the only statistically significant excess is that for all malignant neoplasms (SMR 147). Lung cancer accounts for a significantly high proportion of these deaths (SMR 217). Publicans and innkeepers also experienced high levels of mortality from lung cancer and respiratory diseases but there are too few deaths to reach statistical significance. It is noted elsewhere in this volume that work in bars may entail high levels of passive smoking (see Chapters 4 and 5). However, in the absence of information on the smoking habits of the barmaids themselves, it is difficult to determine whether passive smoking represents a hazard of the occupation.

Charwomen and other cleaners also show significantly high lung cancer (SMR 136) mortality. The same finding has been recorded in earlier studies.[1,2] Again, lifestyle, rather than an occupational hazard, is implicated.

8.6 Follow-up from 1981 census

This section covers all cause mortality by occupation for men and women aged 16 and over at the 1981 census. Occupations are classified as 'groups' corresponding to those used in earlier chapters of this volume, using the Southampton classification. Where component units are discussed these refer to the 1980 classification of occupations (see section 8.3). The SMRs and confidence intervals for groups of men and women with 25 or more deaths are shown in Annex 8.2.

8.6.1 Mortality 1981-89, men

Of the 197,360 men in the 1981 census cohort, 24,875 died between census day 1981 and the end of 1989. Table 8.11 shows the SMRs and confidence intervals for occupation groups with significant deficits and excesses in mortality. Seven groups show significantly raised mortality: other textile workers [group 074] (SMR 125), other coal miners [088] (SMR 121), plastics workers [086] (SMR 180), other occupations - glass and ceramics [091] (SMR 132), plastic goods makers [093] (SMR 150), packers and sorters [164] (SMR 125) and face trained coal miners [175] (SMR 117).

Table 8.11 LS men 1981 census cohort: all cause mortality 1981-89 by occupation groups with statistically significant SMRs≠

Occupation group	Deaths	SMR	(CI)†
Low mortality			
001 Lawyers	36	71	(50 - 99)
002 Accountants	163	74	(63 - 86)
003 Personnel managers etc	36	61	(43 - 85)
006 Sales managers etc	194	80	(69 - 92)
008 Government administrators	96	68	(56 - 85)
009 Other administrators	120	69	(57 - 82)
011 Teachers nec	197	64	(56 - 74)
014 Clergy	62	63	(49 - 81)
015 Doctors	56	67	(51 - 87)
024 Literary and artistic occupations	122	76	(63 - 90)
029 Electrical and electronic engineers (professional)	29	52	(35 - 74)
030 Professional engineers nec	200	70	(60 - 80)
032 Laboratory technicians	89	77	(62 - 94)
033 Architects and surveyers	77	72	(57 - 90)
038 Production and maintenance managers	468	82	(75 - 90)
039 Managers in construction	133	82	(69 - 97)
041 Office managers	230	81	(71 - 92)
044 Retailers and dealers	976	93	(87 - 99)
047 Farmers	965	93	(87 - 99)
053 Office workers and cashiers	1627	91	(87 - 96)
057 Sales representatives	395	85	(77 - 94)
104 Carpenters	281	88	(78 - 99)
130 Precision instrument makers	25	66	(43 - 97)
High mortality			
074 Other textile workers	100	125	(101 - 152)
086 Plastics workers	26	180	(118 - 264)
088 Other coal miners	316	121	(108 - 135)
091 Other occupations — glass and ceramics	60	132	(101 - 170)
093 Plastic goods makers	43	150	(108 - 202)
164 Packers and sorters	176	125	(107 - 145)
175 Face trained coal miners	267	117	(103 - 132)
Inadequately described/not stated	3706	113	(110 - 117)

≠ Only calculated for groups with at least 25 deaths
† 95 per cent confidence interval for SMR

The groups of spinners and winders [070], knitters [072] and bleachers, dyers and finishers [073] together with other textile workers [074] are largely comparable with the 1970 occupation order of textile workers [X]. All four groups experienced excess mortality though this is only significant for other textile workers (see Annex 8.2). This pattern is similar to that of male textile workers in the 1971 census cohort. It would seem therefore, that the risks of working in the textile industry remain.

Similarly, the pattern of excess mortality of the two coal mining groups which are roughly comparable with the 1970 order of miners and quarrymen [II] reflect the continuing hazards of their industry.

Twenty-three groups have significantly low SMRs, largely among the professional and non-manual occupations. An exception is nurses [017] with a 19 per cent excess, although this could have arisen by chance. Among the manual occupation groups, only carpenters [104] (SMR 88) and precision instrument makers [130] (SMR 66) show significant mortality deficits.

Office workers and cashiers [053] is a rather mixed group. Its significantly low SMR of 91 conceals excess mortality in three of the units within the group: stores and despatch clerks [units 045.02, 046.01] (SMR 118), telephone operators [048.03, 051.02] (SMR 107) and radio and telegraph operators [048.04, 051.03] (SMR 111). The men in these units account for only 4 per cent of the whole occupational group and the numbers are too small for the SMRs to reach significance.

8.6.2 Age pattern of mortality, men

All cause SMRs with confidence intervals for men aged 20-64, 65-74 and 75 and over at death by occupation groups are shown in Table 8.15. The occupations with excess overall mortality show consistently raised mortality in each of the age groups though they are significant for only three occupations, and not at working ages. At ages 65-74, other coal miners [088] and face trained coalminers [175] had significant excesses of 31 per cent and 36 per cent respectively. Packers and sorters [164] had significant excesses at both 65-74 and 75 and over, of 29 per cent and 32 per cent respectively.

The mortality pattern of packers and sorters is similar, relative to their standard populations, for men in both the 1971 and 1981 census cohorts. The unit which includes the packers and sorters of the 1971 cohort — packers, labellers and related workers [unit 137] — also had an increased risk of death at working ages and a significant 27 per cent excess at 65-74 (Annex 8.3).

The significantly high SMR of the face trained coalminers is relatively greater than that of underground mine workers [unit 007] in the 1971 census cohort, where the excess mortality compared to that of all men in the age group was 30 per cent[2] (Annex 8.3).

Chemical workers [075] had slightly raised SMRs in the two youngest age groups, with a significant excess of 30 per cent at ages 75 and over. Security workers [058] in both the youngest and oldest age groups had 6 per cent excess mortality with a significant 20 per cent excess at 65-74. Four occupations had significantly raised SMRs at working ages: other service personnel [060] (SMR 139), other metal manufacturers [120] (SMR 189), railway guards [177] (SMR 211) and crane drivers [187] (SMR 158).

Comparing the crane drivers with occupational order XVII, drivers of stationary engines, cranes etc in the 1971 census cohort, shows that the excess mortality for 1981-89 is relatively greater than for the same age group in the period 1976-89 where the excess compared with that of all men in the age group was 16 per cent.[2]

The other service personnel group [060] show a significant 39 per cent excess mortality at working ages. This group includes a range of occupations related to the holiday and travel industries as well as cleaners and domestic workers. Disaggregating the group by units (Table 8.12) shows that the raised mortality at working ages is mainly accounted for by two units: caretakers [071.01, 072.01] (SMR 144) and cleaners, window cleaners, chimney sweeps and roadsweepers [071.02, 072.02] (SMR 132). Caretakers in the 65-74 age group had a significant 26 per cent excess. The numbers of men in these two units account for 82 per cent of the whole group.

As with the all age analysis, mortality advantage is concentrated in the professional and non-manual occupation groups, though only teachers nec [011] and professional engineers nec [030] had significantly low SMRs in all age groups. Nurses [017] are again the exception, with non-significant excess mortality in all age groups. This is not what would be expected according to cross-sectional figures in the last decennial supplement where male nurses showed a mortality deficit.[2]

Four manual occupation groups had significant deficits in mortality in one age group. Postal workers [054] had an SMR of 68 at working ages with a slight excess at the older age groups. Carpenters [104] had a deficit in all age groups though this is significant only at working ages (SMR 79). Motor mechanics [133] had significantly low mortality at working ages (SMR 74), rising with age to equal the mortality of all men at 75 and over. The mortality of toolmakers [129] equalled that of all men at working ages, decreasing with age to a significant deficit at 75 and over.

Men whose occupations were inadequately described or were not stated at the 1981 census had significantly raised SMRs, attenuating with age. The overall excess is relatively lower than that of men in the 1971 census cohort. This is partly explained by men who were 'permanently sick' at the 1981 census being classified to a previous occupation.

Table 8.12 LS men 1981 census cohort: all cause mortality 1981-89 by age and occupation units≠ making up other service personnel group (060)

Occupation unit		Age								
		20-64			65-74			75 and over		
		Deaths	SMR	(CI)†	Deaths	SMR	(CI)†	Deaths	SMR	(CI)†
067.05, 069.03	Hotel porters	10	226	(108 - 415)*	12	116	(60 - 202)	8	178	(77 - 352)
071.01, 072.01	Caretakers	60	144	(110 - 186)*	97	126	(102 - 154)*	82	96	(77 - 120)
071.02, 072.02	Cleaners, window cleaners, chimney sweeps, road sweepers	76	132	(104 - 165)*	133	93	(78 - 110)	167	98	(84 - 114)
071.04, 075.02	Lift and car park attendants	10	198	(95 - 364)	16	148	(84 - 240)	14	77	(42 - 130)

≠ SMRs only calculated for units with at least 25 deaths
† 95 per cent confidence interval for SMR
* 95 per cent confidence interval for SMR excludes 100

8.6.3 Mortality 1981-89, women

The 1981 female census cohort comprises 214,398 individuals, of whom 25,709 died between census day and the end of 1989. Forty-eight per cent of women aged 16 and above were classified to an occupation. A further 41 per cent were classified as housewives. Eighty per cent of deaths were to women without an occupation. This reflects the age profile of the sample, with only 41 per cent of women over retirement age at the 1981 Census classified to an occupation. The age breakdown of the cohort at the 1981 Census is summarised in Figure 8.2.

As with the 1971 census cohort, women's employment is concentrated in a small number of occupations — 53 per cent of women were in three occupation groups. Office workers and cashiers [group 053] account for 31 per cent of women classified to an occupation, with two units, clerks and cashiers (not retail) [units 054.04, 046.03] accounting for 84 per cent of the group. Other service personnel [060] and retailers and dealers [044] account for 11 per cent and 10 per cent respectively.

Seventy-five per cent of the retailers and dealers group [044] were shop saleswomen and assistants [054.01, 055.02]. Cleaners, window cleaners, chimney sweeps and roadsweepers [071.02, 072.02] and other domestic and school helpers [067.02, 068.03] make up 91 per cent of the other service personnel group [060].

The groups and units with significantly low and high overall mortality are shown in Table 8.13. The overall mortality of other groups is in Annex 8.2. As with the 1971 cohort, women classified to an occupation had a significant mortality advantage — women in 11 out of 17 groups with 50 or more observed deaths had lower mortality than all women.

Table 8.13 LS women 1981 census cohort: all cause mortality 1981-89 by occupation groups and units with statistically significant SMRs≠

Occupation group (unit)	Deaths	SMR	(CI)†
Low mortality			
011 Teachers nec	285	80	(71 - 90)
013 Welfare workers	62	77	(59 - 99)
017 Nurses	216	79	(69 - 91)
044 Retailers and dealers	558	88	(81 - 96)
(054.01, 055.01) Shop salespersons and assistants	348	83	(74 - 93)
046 Caterers	226	84	(74 - 96)
053 Office workers and cashiers	1143	83	(78 - 88)
(045.04, 046.03) Clerks and cashiers	690	84	(78 - 91)
(048.01, 049.02) Typists, shorthand writers, secretaries	274	76	(67 - 85)
High mortality			
051 Launderers and dry cleaners	63	132	(101 - 169)
(075.03) Launderers, dry cleaners, pressers	61	137	(105 - 176)
060 Other service personnel			
(071.01, 072.01) Caretakers	30	169	(114 - 241)
070 Spinners and winders	55	137	(103 - 178)

≠ Only calculated for groups with at least 25 deaths
† 95 per cent confidence interval for SMR

Only two groups had significantly higher mortality than expected (Table 8.13). These are launderers and dry cleaners [051] (SMR 132) and spinners and winders [070] (SMR 137). As an occupation group, launderers and dry cleaners are directly comparable with the 1971 cohort [unit 168] whose mortality was raised relative to all women in that cohort, though this was not significant. The spinners and winders group includes units of textile workers over and above the 1971 unit [065]. However, the pattern of excess mortality within the textile industry remains. This is supported by the 25 per cent excess shown by warp preparers and weavers [071] though this was not statistically significant.

A thirty-four per cent mortality excess experienced by barmaids [063.03, 065.02], though not significant, is consistent with findings for the 1971 cohort of women. Caretakers [071.01, 072.01] also had a significant excess (SMR 169), consistent with the findings for men in the 1981 cohort, though it is higher in relative terms.

8.6.4 Age pattern of mortality, women

When the mortality of the 1981 census cohort of women is explored by age at death, the numbers in each age group for some occupations become too small for meaningful comment. However, the overall pattern is similar to that of the 1971 cohort in that women at ages 20-59, 60-74 and 75 and above who could be classified to an occupation, generally had mortality levels lower than those of all women of similar ages.

Table 8.14 shows the occupational groups and units with significantly high or low overall SMRs for women by the three age groups at death. Of the two occupations with increased mortality risks at all ages, only launderers and dry cleaners [051] experienced excesses in each age group, none significant. Spinners and winders [070] had raised mortality after retirement age, significant only at ages 60-74. This pattern is similar to that of the unit of spinners, doublers and twisters [unit 065] in the 1971 cohort of women.

Publicans and bar staff [045], whose overall mortality, though raised (SMR 116), was not significant (Annex 8.2) had a 45 per cent excess at working ages with a mortality deficit at 60-74. The component unit of barmaids [063.03, 065.02] had excess mortality at all ages but this could be a chance finding. It is, however, consistent with the findings for the same unit [155] in the 1971 cohorts.

Of the larger occupation groups, retailers and dealers [044] and office workers and cashiers [053] had significant mortality deficits at working ages and at 60-74. Only the unit of typists, shorthand writers, secretaries [048.01, 049.02] had a significant advantage in every age group, consistent with findings for the 1971 cohort of women.

Interestingly, compared to men who were nurses at the 1981 Census and who experienced raised mortality in every age group, women had a mortality deficit in every age group, significant at ages 60-74. The finding of low overall mortality for female nurses is consistent with that of the 1971 cohort.

Table 8.14 LS women 1981 census cohort: all cause mortality 81-89 by age and occupation groups and units with a statistically significant SMR≠

Occupation group (unit)	Age								
	20-59			60-74			75 and over		
	Deaths	SMR	(CI)†	Deaths	SMR	(CI)†	Deaths	SMR	(CI)†
011 Teachers nec	56	84	(63 - 109)	72	79	(62 - 100)	157	79	(68- 93)*
017 Nurses	59	82	(62 - 105)	66	69	(53 - 88)*	91	87	(70- 107)
044 Retailers and dealers	111	80	(65 - 96)*	217	84	(73 - 96)*	229	98	(86- 112)
(054.01, 055.01) Shop salespersons and assistants	75	74	(58 - 92)*	141	77	(65 - 91)*	151	99	(83- 117)
046 Caterers	39	70	(50 - 96)*	109	97	(80 - 117)	78	78	(62- 98)*
051 Launderers and dry cleaners	10	153	(74 - 282)	21	116	(72 - 178)	31	134	(91- 190)
053 Office workers and cashiers	292	76	(68 - 85)*	452	81	(74 - 89)*	398	92	(83- 102)
(045.04, 046.03) Clerks and cashiers (not retail)	161	75	(64 - 87)*	284	81	(72 - 91)*	245	97	(85- 110)
(048.01, 049.02) Typists, shorthand writers, secretaries	71	67	(53 - 85)*	99	79	(64 - 96)*	104	80	(65- 96)*
070 Spinners and winders	3	97	(20 - 282)	21	201	(125 - 308)*	31	116	(79- 165)
Inadequately described/not stated	849	112	(105 - 120)*	4619	102	(99 - 105)	13738	98	(97- 100)

≠ Only calculated for groups with at least 25 deaths
† 95 per cent confidence interval for SMR
* 95 per cent confidence interval for SMR excludes 100

Women whose occupations were inadequately described or were not stated at the 1981 census had a fairly similar mortality pattern to men in that cohort, and to both men and women in the 1971 cohort, the excesses being greater at working ages. The 12 per cent excess over all women aged 20-59 at death in the 1981 census cohort (Table 8.14) is relatively lower than that of women of the same age in the 1971 cohort.

8.7 Discussion

The aim of this chapter has been, first, to examine overall mortality, and then to identify occupations where relative mortality varied by age at death. The results have confirmed a number of previously reported findings both from longitudinal and cross-sectional studies. Principally, for occupations associated with low mortality, the advantage is maintained over the extended follow-up period, while for occupations associated with higher than expected mortality, the risks are still present. However, death rates in the LS among men and women who stated an occupation at census were generally lower than those of other men and women. This was true for both the 1971 and the 1981 census cohorts.

By classifying the occupations of the 1981 census cohort to the Southampton groups used elsewhere in this volume, many are not directly comparable with those of the 1971 cohort. However, by breaking some occupation orders and groups into their component units, relative patterns of mortality can be shown. The mortality ratios for comparable occupations of the 1971 and 1981 cohorts relative to their respective standard populations are illustrated in Figures 8.5 and 8.6.

For men and women in the low mortality, professional and non-manual occupations, both census cohorts appear equally advantaged, although the advantage of male teachers is relatively greater than that of females. Male nurses are the exception in these occupations but their mortality excesses are not statistically significant. The PMRs listed in Chapter 4 of this volume show a raised suicide rate for male nurses and its authors suggest that occupational stress may be a contributory factor. However, no male nurses in the 1971 cohort died from suicide, or from accidents, poisonings and violence. The PMRs (Chapter 5) and the SMRs (1971 cohort) for female nurses are raised for suicides and for all deaths from accidents, poisonings and violence, although their overall SMR is significantly low.

In the high mortality occupations the pattern for men is virtually identical for both census cohorts and largely conforms to previous findings in the LS and elsewhere[2,7] (Figure 8.5). The one exception is packers, labellers and related workers. Their jobs are likely to be transient occupations, which may explain why previous, cross-sectional findings showed a mortality deficit.[2] The difference in the relative excess overall mortality between the two cohorts is largely explained by the difference between the ratios for men aged 75 and above at death. There is no obvious interpretation of this finding. If it is not due to differences in the reporting of occupations at the two censuses by the same individuals, it may reflect patterns of recruitment into these jobs. For example, by the 1980s such jobs may have attracted older men with poorer than average health.

The numbers of women in the high risk occupations that can be compared are very small (Figure 8.6).

For both male and female textile workers cause-specific mortality of the 1971 census cohort showed excesses for all the broad cause groups. Only deaths from respiratory diseases were statistically significant for women, and only deaths from strokes for men. Mortality was raised in each age group for both men and women but was significant only for deaths at ages 75 and above (see Annex 8.3 and Table 8.6).

Figure 8.5 LS men, 1971 census cohort mortality 1976-89 and 1981 census cohort mortality 1981-89 for comparable occupations

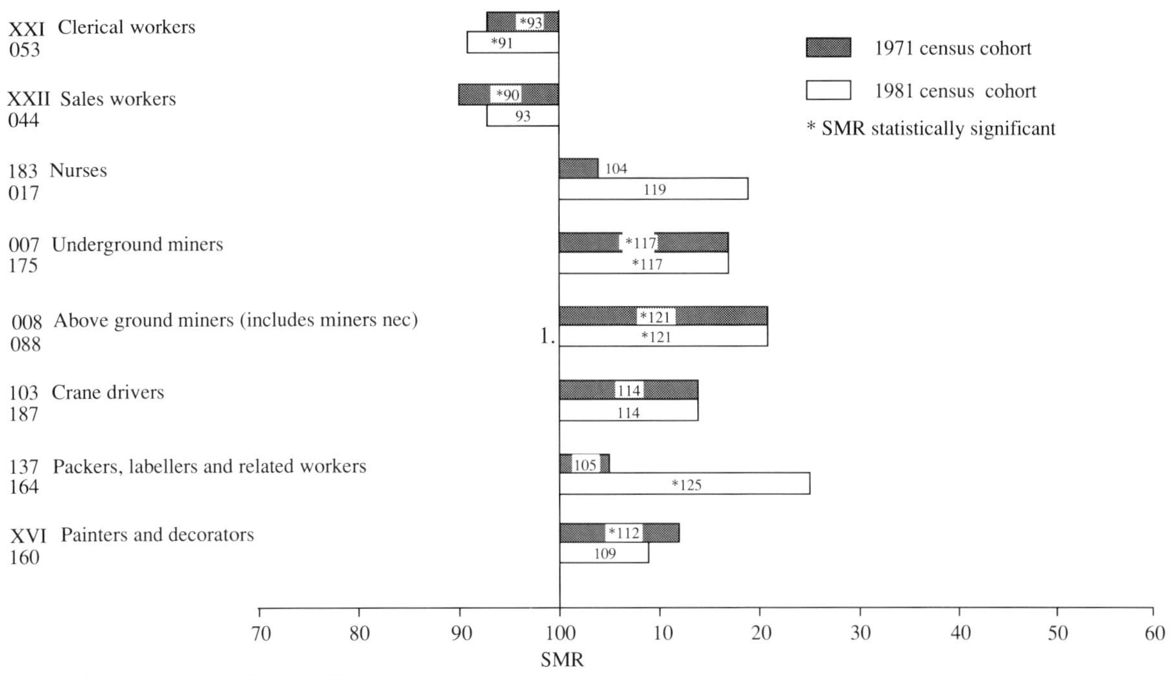

1. Includes miners not elsewhere classified

Figure 8.6 LS women, 1971 census cohort mortality 1976-89 and 1981 census cohort mortality 1981-89 for comparable occupations

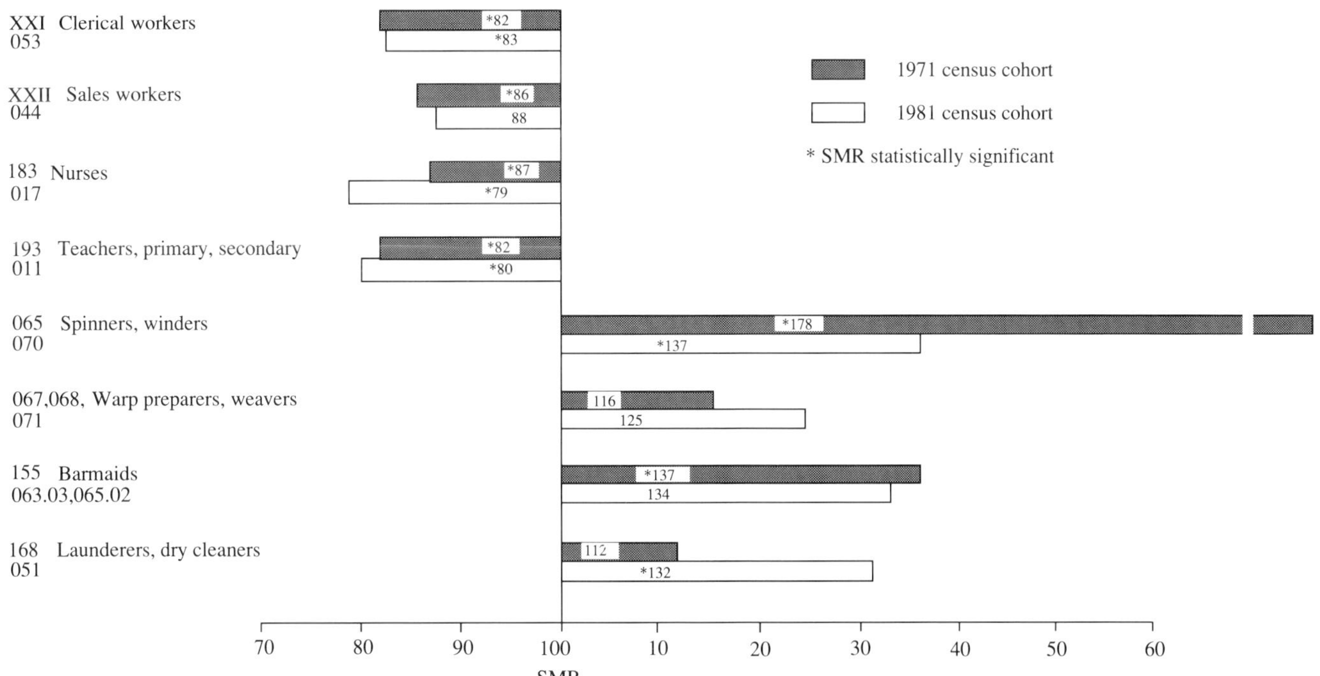

8.8 Conclusion

This chapter is a broad descriptive examination of mortality in relation to occupation. Although no adjustment has been made for social class, the results are consistent with previous findings standardised by social class.[7,17] Unlike cross-sectional studies the LS is not restricted to social class as a measure of socio-economic circumstances and life-styles. Alternative measures such as housing tenure and car access have been shown to be better discriminators of mortality outcomes, particularly for men at older ages and for women.[16] These measures will be used in an attempt to disentangle the influences of extrinsic factors from those associated with working in particular occupations.

Exposure to occupational hazards for women is not the same as for men. Few women work in the heavy manual jobs in which men are at risk. They also tend to have shorter working lives, with breaks for childbearing, and to work part-time. The exploration of women's mortality in relation to their occupations, marital status and the number and ages of any dependent children will hopefully shed more light on the effects of their exposures to the high risk occupations identified in this analysis.

It is anticipated that greater insight into the effects of occupation will be found when the mortality patterns of men and women who were in the same occupations at both censuses, and those who changed occupations between censuses, are explored. There is also scope for studying post-retirement occupations.

References

1. *Occupational mortality: decennial supplement 1970-72.* DS no.1, London: HMSO, 1978.
2. *Occupational mortality: decennial supplement 1979-80, 1982-83.* DS no.6, London: HMSO, 1986.
3. Goldblatt P (ed). *Mortality and social organisation. Longitudinal Study 1971-1981.* LS no.6, London: HMSO, 1990.
4. McDowall M. Measuring women's occupational mortality. *Population Trends* 1983;34:25-29.
5. Goldblatt P, Jones D. Methods. In Goldblatt P (ed). *Mortality and social organisation. Longitudinal Study 1971-1981.* LS no.6, London: HMSO, 1990: 51-66.
6. Goldblatt P, Fox J, Leon D. Mortality of employed men and women. In Goldblatt P (ed). *Mortality and social organisation. Longitudinal Study 1971-1981.* LS no.6, London: HMSO, 1990: 67-80.
7. Goldblatt P, Fox J. Mortality of men by occupation. In Goldblatt P (ed). *Mortality and social organisation. Longitudinal Study 1971-1981.* LS no.6, London: HMSO, 1990: 81-97.
8. Moser K, Goldblatt P, Pugh H. Occupational mortality of women in employment. In Goldblatt P (ed). *Mortality and social organisation. Longitudinal Study 1971-1981.* LS no.6, London: HMSO, 1990: 129-144.
9. Martin J, Roberts C. *Women and employment: a lifetime perspective.* London: HMSO, 1984.
10. Fox AJ, Collier PF. Low mortality rates in industrial cohort studies due to selection for work and survival in the industry. *Brit J Prev Soc Med* 1976; 30: 225-230.
11. Moser K, Goldblatt P, Fox J, Jones D. Unemployment and mortality. In Goldblatt P (ed). *Mortality and social organisation. Longitudinal Study 1971-1981.* LS no.6, London: HMSO, 1990: 81-97.
12. Charlton J, Kelly S, Dunnell K. Suicide deaths in England and Wales: trends in factors associated with suicide deaths. *Population Trends* 1993;71:34-42.
13. Marmot MG, Shipley MJ, Rose G. Inequalities in death-specific explanations of a general pattern? *Lancet* 1984(i): 1003-6.
14. Tuchsen F, Anderson O, Costa G, Filakti H, Marmot M. *Occupation and ischaemic heart disease in the EC. A register based longitudinal study of occupational mortality and morbidity.* Copenhagen: National Institute of Occupational Health, 1993.
15. Moser KA, Goldblatt PO. Occupational mortality of women aged 15-59 years at death in England and Wales. *J of Epid and Community Health* 1991; 45: 117-124.
16. Moser K, Pugh H, Goldblatt P. Mortality and the social classification of women. In Goldblatt P (ed). *Mortality and social organisation. Longitudinal Study 1971-1981.* LS no.6, London: HMSO, 1990: 145-162.
17. Goldblatt P. Mortality and alternative social classifications. In Goldblatt P (ed). *Mortality and social organisation. Longitudinal Study 1971-1981.* LS no.6, London: HMSO, 1990: 145-162

Annex 8.1 LS men and women 1971 census cohort: distribution by 1970 occupation orders and number of deaths and SMRs, 1976-89

Occupation order		Men					Women				
		N	Per cent	Deaths	SMR	(CI)†	N	Per cent	Deaths	SMR	(CI)†
I	Farmers, foresters, fishermen	7139	3.8	1952	93	(89 - 97)*	1065	0.5	161	83	(71 - 97)*
II	Miners and quarrymen	3452	1.8	1159	117	(110 - 124)*	9	<.1	2	83	(10 - 300)
III	Gas, coke and chemical makers	1384	0.7	321	112	(100 - 125)	130	0.1	22	136	(85 - 205)
IV	Glass and ceramic makers	693	0.4	135	112	(94 - 133)	354	0.2	50	104	(77 - 137)
V	Furnace, forge, foundry, rolling mill workers	1777	0.9	433	112	(102 - 123)*	88	<.1	11	107	(53 - 191)
VI	Electrical and electronic workers	5437	2.9	682	86	(80 - 93)*	972	0.5	85	92	(73 - 114)
VII	Engineering and allied trades workers nec	26459	14.1	4739	100	(97 - 103)	3086	1.5	421	106	(96 - 117)
VIII	Woodworkers	4412	2.3	796	94	(88 - 101)	122	0.1	17	96	(56 - 154)
IX	Leather workers	748	0.4	210	105	(91 - 120)	625	0.3	88	90	(72 - 111)
X	Textile workers	1707	0.9	398	118	(107 - 130)*	1951	1.0	433	114	(103 - 125)*
XI	Clothing workers	934	0.5	206	81	(70 - 93)*	3477	1.7	488	91	(83 - 99)*
XII	Food, drink and tobacco workers	2711	1.4	611	106	(98 - 115)	1073	0.5	156	104	(89 - 122)
XIII	Paper and printing workers	2304	1.2	427	101	(91 - 111)	1007	0.5	131	99	(83 - 118)
XIV	Makers of other products	2314	1.2	398	105	(95 - 116)	1179	0.6	142	96	(81 - 113)
XV	Construction workers	5850	3.1	1280	97	(92 - 103)	19	<.1	1	37	(1 - 207)
XVI	Painters and decorators	2993	1.6	726	112	(104 - 120)*	88	<.1	7	84	(34 - 173)
XVII	Drivers of stationary engines, cranes, etc	3332	1.8	842	111	(104 - 119)*	36	<.1	4	88	(24 - 225)
XVIII	Labourers nec	12083	6.4	3459	120	(116 - 124)*	1213	0.6	253	120	(105 - 135)*
XIX	Transport and communication workers	13725	7.3	3073	104	(100 - 108)	1618	0.8	176	101	(87 - 117)
XX	Warehousemen, storekeepers, packers, bottlers	5759	3.1	1647	112	(107 - 118)*	3013	1.5	440	99	(90 - 109)
XXI	Clerical workers	12387	6.6	2824	93	(90 - 96)*	25390	12.5	2107	82	(79 - 86)*
XXII	Sales workers	13079	7.0	2871	90	(86 - 93)*	10740	5.3	1507	86	(82 - 91)*
XXIII	Service, sport and recreation workers	10290	5.5	2958	105	(102 - 109)*	21682	10.7	4247	99	(96 - 102)
XXIV	Administrators and managers	9517	5.1	1997	80	(77 - 84)*	856	0.4	121	85	(71 - 102)
XXV	Professional, technical workers, artists	17513	9.3	2446	72	(70 - 75)*	11161	5.5	1327	83	(79 - 88)*
XXVI	Armed forces (British and foreign)	1986	1.1	128	88	(74 - 105)	107	0.1	5	110	(36 - 258)
XXVII	Inadequately described occupations	6145	3.3	2317	113	(109 - 118)*	24988	12.3	11615	104	(102 - 106)*
	Unoccupied	11741	6.2	1333	175	(166 - 185)*	86740	42.8	16757	103	(102 - 105)*
	Total	187871	100.0	40368	100		202789	100.0	40774	100	

Standard populations - all men for male SMRs, all women for female SMRs
† 95 per cent confidence interval for SMR
* 95 per cent confidence interval for SMR excludes 100

Annex 8.2 LS men and women 1981 census cohorts: distribution by occupation groups (Southampton classification) and number of deaths and all cause SMRs†, 1981-89

Occupation group	Men					Women				
	N	Per cent	Deaths	SMR	(CI)≠	N	Per cent	Deaths	SMR	(CI)≠
001 Lawyers	462	0.2	36	71	(50 - 99)*	70	0.0	1		
002 Accountants	2234	1.1	163	74	(63 - 86)*	222	0.1	3		
003 Personnel managers etc	553	0.3	36	61	(43 - 85)*	211	0.1	8		
004 Economists and statisticians	97	0.1	6			31	0.0	1		
005 Computer programmers	812	0.4	9			191	0.1	1		
006 Sales managers etc	2401	1.2	194	80	(69 - 92)*	456	0.2	26	106	(70 - 156)
007 Government inspectors	300	0.2	32	78	(53 - 110)	17	0.0	-		
008 Government administrators	853	0.4	96	69	(56 - 85)*	222	0.1	10		
009 Other administrators	1064	0.5	120	69	(57 - 82)*	721	0.3	37	79	(56 - 109)
010 Teachers in higher education	959	0.5	68	81	(63 - 103)	338	0.2	15		
011 Teachers nec	2722	1.4	197	64	(56 - 74)*	4770	2.2	285	80	(71 - 90)*
012 Vocational trainers, social scientists, etc	612	0.3	59	80	(61 - 103)	241	0.1	6		
013 Welfare workers	657	0.3	61	90	(69 - 115)	1270	0.6	62	77	(59 - 99)*
014 Clergy	438	0.2	62	63	(49 - 81)*	59	0.0	8		
015 Doctors	669	0.3	56	67	(51 - 87)*	244	0.1	13		
016 Dentists	152	0.1	14			38	0.0	-		
017 Nurses	570	0.3	66	119	(92 - 152)	5866	2.7	216	79	(69 - 91)*
018 Pharmacists	161	0.1	21			88	0.0	4		
019 Medical radiographers	15	0.0	-			98	0.0	2		
020 Physiotherapists	12	0.0	2			144	0.1	1		
021 Health professions nec	184	0.1	22			344	0.2	12		
022 Veterinarians	51	0.0	2			11	0.0	1		
023 Driving instructors	168	0.1	15			23	0.0	1		
024 Literary and artistic occupations	1606	0.8	122	76	(63 - 90)*	692	0.3	31	80	(55 - 114)
025 Persons involved in sport	88	0.0	9			41	0.0	2		
026 Biological scientists	152	0.1	10			48	0.0	-		
027 Chemical engineers and scientists	352	0.2	23			32	0.0	-		
028 Physical scientists and mathematicians	632	0.3	22			152	0.1	2		
029 Electrical and electronic engineers (professional)	755	0.4	29	52	(35 - 74)*	10	0.0	-		
030 Professional engineers nec	3072	1.6	200	70	(60 - 80)*	47	0.0	2		
031 Draughtspersons	1066	0.5	71	79	(62 - 100)	101	0.1	1		
032 Laboratory technicians	1318	0.7	89	77	(62 - 94)*	451	0.0	11		
033 Architects and surveyors	1224	0.6	77	72	(57 - 90)*	38	0.0	-		
034 Aircraft flight deck officers	96	0.1	1			2	0.0	1		
035 Air traffic controllers	28	0.0	2			2	0.0	-		
036 Seafarers	626	0.3	82	121	(97 - 151)	5	0.0	-		
037 Technicians nec	1030	0.5	73	79	(62 - 100)	130	0.1	1		
038 Production and maintenance managers	4331	2.2	468	82	(75 - 90)*	316	0.2	19		
039 Managers in construction	1371	0.7	133	82	(69 - 97)*	45	0.0	1		
040 Managers in transport,utilities and mining	1585	0.8	209	93	(81 - 107)	149	0.1	7		
041 Office managers	2095	1.1	230	81	(71 - 92)*	642	0.3	17		
042 Butchers	1064	0.5	119	96	(80 - 115)	87	0.0	7		
043 Fishmongers and poultry dressers	104	0.1	19	49			0.0	2		
044 Retailers and dealers	8032	4.1	976	93	(87 - 99)*	10688	5	558	88	(81 - 96)*
045 Publicans and bar staff	1424	0.7	185	104	(90 - 120)	1886	0.9	109	116	(95 - 140)
046 Caterers	1238	0.6	118	98	(81 - 117)	3736	1.7	226	84	(74 - 96)*
047 Farmers	5932	3.0	965	93	(87 - 99)*	915	0.4	52	78	(58 - 102)
048 Armed forces	2281	1.2	94			171	0.1	2		
049 Police	1287	0.7	64	91	(70 - 117)	120	0.1	1		
050 Fire service personnel	504	0.3	37	113	(80 - 156)	12	0.0	1		
051 Launderers and dry cleaners	202	0.1	19			535	0.3	63	132	(101 - 169)*
052 Hairdressers	354	0.2	47	111	(82 - 148)	1168	0.5	14		
053 Office workers and cashiers	11554	5.9	1627	91	(87 - 96)*	32034	14.9	1143	83	(78 - 88)*
054 Postal workers	1853	0.9	249	93	(82 - 105)	225	0.1	14		
055 Petrol pump attendants	154	0.1	29	105	(70 - 151)	89	0.0	4		
056 Van salespersons	681	0.3	48	93	(68 - 123)	36	0.0	3		
057 Sales representatives	3969	2.0	395	85	(77 - 94)*	567	0.3	14		
058 Security workers	1605	0.8	328	111	(100 - 124)	244	0.1	9		
059 Cooks and kitchen porters	1145	0.6	93	106	(85 - 129)	2674	1.3	181	88	(76 - 102)
060 Other services personnel	3145	1.6	723	107	(99 - 115)	11752	5.5	883	94	(88 - 101)

Annex 8.2 - *continued*

Occupation group	Men					Women				
	N	Per cent	Deaths	SMR	(CI)≠	N	Per cent	Deaths	SMR	(CI)≠
061 Hospital porters and ward orderlies	469	0.2	70	91	(71 - 115)	1036	0.5	49	103	(76 - 136)
062 Ambulance workers	208	0.1	28	116	(77 - 168)	19	0.0	-		
063 Railway station workers	237	0.1	53	113	(85 - 148)	17	0.0	2		
064 Undertakers	47	0.0	8			5	0.0	-		
065 Foresters	168	0.1	19			5	0.0	-		
066 Fishing and related workers	90	0.1	14			2	0.0	-		
067 Tannery workers	81	0.0	22			32	0.0	5		
068 Leather and shoe workers	430	0.2	78	102	(81 - 128)	402	0.2	20		
069 Preparatory fibre processors	91	0.1	17			49	0.0	10		
070 Spinners and winders	177	0.1	24			279	0.1	55	137	(103 - 178)*
071 Warp preparers and weavers	190	0.1	27	98	(65 - 143)	197	0.1	65	125	(97 - 159)
072 Knitters	141	0.1	22			149	0.1	7		
073 Bleachers, dyers and finishers	178	0.1	37	120	(85 - 166)	58	0.0	5		
074 Other textile workers	576	0.3	100	125	(101 - 152)*	560	0.3	52	116	(86 - 152)
075 Chemical workers	1153	0.6	170	116	(100 - 135)	138	0.1	12		
076 Bakers	486	0.2	74	112	(88 - 140)	145	0.1	11		
077 Brewery workers	112	0.1	12			11	0.0	-		
078 Food processors	998	0.5	102	94	(76 - 114)	702	0.3	22		
079 Paper manufacturers	85	0.0	11			16	0.0	3		
080 Bookbinders	84	0.0	5			154	0.1	8		
081 Paper cutters	87	0.1	14			9	0.0	1		
082 Glass and ceramic furnace workers	100	0.1	13			5	0.0	-		
083 Glass formers and decorators	161	0.1	15			47	0.0	3		
084 Ceramics casters	112	0.1	12			58	0.0	4		
085 Rubber manufacturers	156	0.1	23			20	0.0	4		
086 Plastics workers	176	0.1	26	180	(118 - 264)*	60	0.0	3		
087 Man-made fibre makers	80	0.0	14			16	0.0	1		
088 Other coal miners	1219	0.6	316	121	(108 - 135)*	1	0.0	-		
089 Tobacco workers	39	0.0	3			57	0.0	4		
090 Other wood and paper processors	183	0.1	25	128	(83 - 189)	110	0.1	6		
091 Other occupations - glass and ceramics	419	0.2	60	132	(101 - 170)*	117	0.1	14		
092 Rubber goods makers	221	0.1	23			80	0.0	4		
093 Plastic goods makers	424	0.2	43	150	(108 - 202)*	208	0.1	11		
094 Compositors	257	0.1	31	76	(52 - 108)	34	0.0	1		
095 Printing plate preparers	99	0.1	7			12	0.0	-		
096 Printing machine minders	544	0.3	71	111	(87 - 140)	113	0.1	11		
097 Printers (so described)	501	0.3	56	107	(81 - 139)	138	0.1	7		
098 Tailors and dressmakers	148	0.1	27	76	(50 - 110)	273	0.1	31	61	(87 - 42)
099 Clothing cutters	157	0.1	24			139	0.1	10		
100 Sewers and embroiderers	136	0.1	10			2035	1.0	101	89	(72 - 108)
101 Upholsterers	266	0.1	30	97	(66 - 139)	40	0.0	1		
102 Carpet fitters	123	0.1	-			1	0.0	-		
103 Other workers with fabrics	78	0.0	5			147	0.1	26		
104 Carpenters	3245	1.6	281	88	(78 - 99)*	14	0.0	2		
105 Cabinet makers	260	0.1	23			19	0.0	-		
106 Case and box makers	76	0.0	10			3	0.0	-		
107 Pattern makers	122	0.1	16			3	0.0	-		
108 Woodworking machinists	453	0.2	54	117	(88 - 153)	50	0.0	2		
109 Other woodworkers	140	0.1	14			21	0.0	-		
110 Dental technicians	79	0.0	8			9	0.0	-		
111 Other makers of paper goods	378	0.2	31	98	(66 - 139)	381	0.2	15		
112 Furnace operatives (metal)	213	0.1	36	119	(84 - 165)	-	-	-		
113 Rollers (metal)	72	0.0	15			-	-	-		
114 Smiths and forge workers	174	0.1	37	114	(80 - 157)	-	-	-		
115 Metal drawers	52	0.0	9			2	0.0	2		
116 Moulders and coremakers (metal)	333	0.2	50	107	(80 - 141)	27	0.0	3		
117 Electroplaters	75	0.0	7			10	0.0	1		
118 Annealers, hardeners, temperers (metal)	82	0.0	18			3	0.0	-		
119 Galvanisers and tin platers	52	0.0	2			8	0.0	1		
120 Other metal manufacturers	507	0.3	76	118	(93 - 148)	17	0.0	-		
121 Press and machine tool setters	615	0.3	66	96	(74 - 122)	16	0.0	-		
122 Centre lathe turners	507	0.3	60	107	(82 - 138)	6	0.0	1		
123 Machine tool setter operators	232	0.1	19			3	0.0	2		
124 Machine tool operators	3380	1.7	395	99	(89 - 109)	633	0.3	38	104	(73 - 142)
125 Press and automatic machine operators	318	0.2	32	112	(77 - 159)	229	0.1	17		

Annex 8.2 - *continued*

Occupation group	Men					Women				
	N	Per cent	Deaths	SMR	(CI)≠	N	Per cent	Deaths	SMR	(CI)≠
126 Metal polishers	151	0.1	26	101	(66 - 148)	21	0.0	2		
127 Fettlers and dressers (metal)	156	0.1	20			21	0.0	4		
128 Shot blasters	53	0.0	4			-	-	-		
129 Toolmakers	952	0.5	102	86	(70 - 104)	9	0.0	-		
130 Precision instrument makers	340	0.2	25	66	(43 - 97)*	17	0.0	-		
131 Watch and clock makers	82	0.0	17			4	0.0	-		
132 Production fitters	6039	3.1	604	94	(86 - 102)	60	0.0	4		
133 Motor mechanics	2798	1.4	175	90	(77 - 105)	15	0.0	1		
134 Aircraft engine fitters	83	0.0	6			-	-	-		
135 Office machinery mechanics	28	0.0	3			-	-	-		
136 Electrical and electronic production fitters	212	0.1	19			30	0.0	1		
137 Electricians	3139	1.6	203	88	(76 - 100)	47	0.0	2		
138 Electrical plant operators	250	0.1	42	97	(70 - 132)	6	0.0	-		
139 Telephone fitters	1172	0.6	74	79	(62 - 100)	19	0.0	1		
140 Electric cable and line workers	195	0.1	23			5	0.0	-		
141 Radio and TV mechanics	355	0.2	19			6	0.3	-		
142 Other electronic maintenance engineers	361	0.2	11			6	0.0	-		
143 Electrical engineers (so described)	233	0.1	22			1	0.0	-		
144 Plumbers and gas fitters	2001	1.0	165	98	(83 - 114)	2	0.0	-		
145 Sheet metal workers	932	0.5	80	110	(87 - 137)	-	-	-		
146 Metal plate workers	515	0.3	72	109	(85 - 137)	-	-	-		
147 Steel erectors	270	0.1	22			-	-	-		
148 Scaffolders	206	0.1	12			-	-	-		
149 Welders	1555	0.8	119	103	(86 - 124)	114	0.1	8		
150 Riggers	133	0.1	23			-	-	-		
151 Jewellery workers	92	0.0	9			35	0.0	2		
152 Engravers and etchers (printing)	41	0.0	10			5	0.0	1		
153 Vehicle body builders	287	0.1	32	84	(58 - 119)	-	-	-		
154 Oilers and greasers	89	0.0	15			3	0.0	1		
155 Electronics wire workers	104	0.1	7			69	0.0	4		
156 Coil winders	63	0.0	6			58	0.0	6		
157 Pottery decorators	14	0.0	-			48	0.0	5		
158 Coach painters	28	0.0	8			-	-	-		
159 Other spray painters	463	0.2	49	124	(92 - 165)	32	0.0	3		
160 Painters and decorators nec	2454	1.2	318	109	(98 - 122)	68	0.0	2		
161 Electrical, electronic assemblers	191	0.1	14			711	0.3	28	99	(66 - 143)
162 Instrument assemblers	16	0.0	2			57	0.0	2		
163 Assemblers (vehicles and other metal goods)	555	0.3	44	95	(69 - 127)	216	0.1	18		
164 Packers and sorters	1084	0.5	176	125	(107 - 145)*	2225	1.0	134	105	(88 - 124)
165 Bricklayers and tilesetters	1584	0.8	181	111	(95 - 128)	-	-	-		
166 Masons and stonecutters	150	0.1	26	125	(82 - 184)	-	-	-		
167 Plasterers	522	0.3	51	104	(77 - 137)	-	-	-		
168 Roofers and glaziers	458	0.2	24			5	0.0	-		
169 Builders etc	2005	1.0	266	97	(86 - 110)	16	0.0	-		
170 Rail track workers	331	0.2	63	96	(74 - 122)	4	0.0	1		
171 Road construction workers and paviors	410	0.2	77	126	(100 - 158)	4	0.0	1		
172 Sewage plant attendants	156	0.1	28	113	(75 - 163)	-	-	-		
173 Mains and service layers	241	0.1	36	119	(83 - 165)	-	-	-		
174 Construction workers nec	3728	1.9	387	105	(95 - 116)	19	0.0	2		
175 Face trained coal miners	1230	0.6	267	117	(103 - 132)*	2	0.0	2		
176 Miners (not coal) and quarry workers	152	0.1	16			-	-	-		
177 Railway guards	186	0.1	33	119	(82 - 167)	2	0.0	-		
178 Railway signal workers	133	0.1	35	139	(97 - 193)	5	0.0	1		
179 Shunters and points operators	86	0.0	13	-	-		-			
180 Railway engine drivers	558	0.3	122	106	(88 - 126)	2	0.0	-		
181 Road transport inspectors	157	0.1	28	85	(57 - 123)	3	0.0	-		
182 Bus and coach drivers	1132	0.6	119	92	(77 - 111)	24	0.0	1		
183 Lorry drivers	6108	3.1	640	106	(98 - 114)	221	0.1	7		
184 Other motor drivers	791	0.4	81	94	(74 - 116)	46	0.0	2		
185 Bus conductors and drivers' mates	314	0.2	53	132	(99 - 172)	52	0.0	6		
186 Mechanical plant drivers	479	0.2	36	88	(62 - 122)	4	0.0	-		
187 Crane drivers	579	0.3	101	114	(93 - 138)	10	0.0	-		
188 Fork lift truck drivers	969	0.5	73	106	(83 - 133)	10	0.0	-		
189 Slingers	119	0.1	27	132	(87 - 192)	-	-	-		
190 Storekeepers	4133	2.1	613	103	(95 - 111)	574	0.3	39	112	(80 - 153)

Annex 8.2 - *continued*

Occupation group	Men					Women				
	N	Per cent	Deaths	SMR	(CI)≠	N	Per cent	Deaths	SMR	(CI)≠
191 Dockers and goods porters	1179	0.6	184	111	(95 - 128)	22	0.0	2		
192 Refuse collectors	371	0.2	42	112	(81 - 152)	-	-	-		
193 Labourers in coke ovens	63	0.0	16			-	-	-		
194 Boiler operators	315	0.2	90	112	(90 - 138)	1	0.0	-		
999 Miscellaneous occupations	11803	6.0	1780			3395	1.6	221		
Total stated occupations	185984	94.2				103048	48.1	5261		
Inadequately described, or not stated occupations	11376	5.8	3706	113	(110 - 117)*	111350	51.9	20448	100	(98 - 101)
Total	197360	100.00				214398	100.0			

Standard populations – all men for male SMRs, all women for female SMRs
† Only for groups with at least 25 deaths
≠ 95 per cent confidence interval for SMR
* 95 per cent confidence interval for SMR excludes 100

Annex 8.3 LS men 1971 census cohort, mortality 1976-89: all cause mortality by age at death and selected units

Units of high and low mortality orders		Age at death								
		20-64			65-74			75 and over		
		Deaths	SMR	(CI)†	Deaths	SMR	(CI)†	Deaths	SMR	(CI)†
Units of high mortality orders										
XVIII	Labourers nec	814	138	(129 - 148)*	1173	119	(112 - 126)*	1472	112	(106 - 118)*
	108 Engineering and allied trades	181	135	(116 - 156)*	313	122	(109 - 137)*	355	117	(105 - 130)*
	113 Building and contracting	185	143	(123 - 165)*	175	111	(95 - 129)	196	117	(101 - 135)*
	114 Other labourers nec	322	139	(125 - 155)*	493	117	(107 - 128)*	708	109	(102 - 118)*
X	Textile workers	105	115	(94 - 140)	125	113	(94 - 135)	168	123	(105 - 144)*
	065 Spinners, doublers and twisters	8	87	(38 - 172)	12	125	(65 - 219)	14	129	(71 - 217)
	068 Weavers	13	106	(57 - 182)	22	152	(95 - 229)	30	118	(79 - 168)
	070 Bleachers and finishers of textiles	15	129	(72 - 213)	18	123	(73 - 194)	24	130	(83 - 194)
II	Miners and quarrymen	196	102	(88 - 117)	431	130	(118 - 143)*	532	113	(104 - 123)*
	007 Coal mine - workers underground	159	101	(86 - 118)	343	130	(116 - 144)*	443	114	(104 - 126)*
	008 Coal mine - workers above ground	26	114	(75 - 167)	70	134	(105 - 170)*	66	113	(87 - 143)
V	Furnace, forge, foundry, rolling mill workers	131	120	(100 - 142)	149	121	(102 - 142)*	153	100	(84 - 117)
	020 Moulders and core makers (foundry)	38	116	(82 - 160)	40	138	(99 - 188)	35	90	(63 - 125)
	022 Metal making and treating workers	9	123	(56 - 234)	20	163	(100 - 252)	18	162	(96 - 257)
XVI	Painters and decorators	196	116	(101 - 134)*	256	109	(96 - 123)	274	112	(99 - 127)
	100 Painters and decorators nec	158	111	(94 - 129)	225	108	(94 - 123)	245	115	(101 - 131)*
XX	Warehousemen, storekeepers, packers, bottlers	354	113	(102 - 126)*	598	110	(102 - 119)*	695	113	(105 - 122)*
	136 Warehousemen, storekeepers and assistants	299	111	(99 - 125)	497	107	(98 - 117)	593	116	(107 - 126)*
	137 Packers, labellers and related workers	55	125	(94 - 162)	101	127	(103 - 154)*	102	100	(82 - 122)
XVII	Drivers of stationary engines, cranes, etc.	245	116	(102 - 131)*	308	110	(98 - 123)	289	110	(98 - 123)
	103 Crane and hoist operators, slingers	71	127	(99 - 160)	79	113	(89 - 141)	62	103	(79 - 131)
	105 Stationary engine, materials handling plant operators	105	119	(97 - 144)	138	108	(90 - 127)	130	106	(89 - 126)
XXIII	Service, sport and recreation workers	577	107	(98 - 116)	968	107	(101 - 114)*	1413	103	(98 - 109)
	153 Guards and related workers	74	120	(95 - 151)	185	113	(97 - 131)	307	101	(90 - 113)
	154 Publicans, innkeepers	53	124	(93 - 162)	71	115	(90 - 145)	74	111	(87 - 140)
	155 Barmen, barmaids	24	135	(87 - 201)	34	148	(103 - 207)*	31	127	(86 - 180)
	165 Caretakers and office keepers	56	145	(109 - 188)*	132	114	(96 - 136)	164	109	(93 - 127)
Units of low mortality orders										
XXV	Professional, technical workers, artists	694	67	(62 - 72)*	785	74	(69 - 79)*	967	76	(71 - 81)*
	183 Nurses	13	73	(39 - 126)	32	117	(80 - 166)	36	109	(76 - 150)
	193 Primary and secondary school teachers	76	66	(52 - 83)*	94	65	(53 - 80)*	99	66	(53 - 80)*
	196 Mechanical engineers	51	74	(55 - 97)*	36	72	(50 - 99)*	47	86	(63 - 114)
XXIV	Administrators and managers	527	76	(70 - 83)*	673	78	(72 - 84)*	797	84	(78 - 90)*
	175 Managers in engineering and allied trades	111	86	(71 - 104)	115	79	(65 - 95)*	106	74	(61 - 89)*
	177 Managers in mining and production nec	81	75	(60 - 94)*	111	80	(66 - 96)*	160	78	(66 - 91)*
	180 Managers nec	160	76	(64 - 88)*	224	75	(66 - 86)*	298	93	(83 - 104)
XI	Clothing workers	33	69	(48 - 98)*	69	94	(73 - 119)	104	78	(64 - 95)*
	077 Clothing and related products makers nec	10	61	(29 - 112)	17	77	(45 - 124)	36	64	(45 - 89)*
VI	Electrical and electronic workers	229	79	(69 - 90)*	210	86	(75 - 98)*	243	96	(84 - 109)
	027 Electricians	122	90	(75 - 107)	111	100	(82 - 120)	125	102	(85 - 122)
XXII	Sales workers	721	97	(90 - 104)	860	83	(78 - 89)*	1290	91	(86 - 96)*
	143 Proprietors and managers, sales	313	94	(84 - 105)	450	82	(75 - 90)*	672	92	(85 - 99)*
	144 Shop salesmen and assistants	95	111	(90 - 136)	109	93	(77 - 113)	188	103	(89 - 119)
	150 Salesmen, services	86	81	(65 - 100)	112	78	(65 - 94)*	186	81	(70 - 94)*
XXI	Clerical workers	629	101	(93 - 109)	893	90	(85 - 96)*	1302	91	(86 - 96)*
	139 Clerks and cashiers	564	106	(98 - 115)	789	92	(86 - 98)*	1181	93	(87 - 98)*
I	Farmers, foresters, fishermen	300	85	(76 - 95)*	548	90	(82 - 97)*	1104	97	(92 - 103)
	002 Farmers, farm managers, market gardeners	108	68	(56 - 82)*	222	85	(74 - 97)*	460	91	(83 - 98)*
	005 Gardeners, groundsmen	55	94	(71 - 123)	149	99	(84 - 117)	309	100	(89 - 112)

† 95 per cent confidence interval for SMR
* 95 per cent confidence interval for SMR excludes 100

Annex 8.4 LS men 1981 census cohort: all cause mortality 1981-89 by age and occupation group≠

Occupation group	Age at death 20-64 Deaths	SMR	(CI)†	65-74 Deaths	SMR	(CI)†	75 and over Deaths	SMR	(CI)†
001 Lawyers	10	83	(40 - 153)	8	67	(29 - 132)	18	68	(40 - 108)
002 Accountants	41	63	(45 - 85)*	53	73	(54 - 95)*	69	83	(65 - 105)
003 Personnel managers etc	10	43	(21 - 79)*	15	70	(39 - 115)	11	79	(39 - 141)
006 Sales managers etc	61	68	(52 - 88)*	70	82	(64 - 104)	63	91	(70 - 117)
007 Government inspectors	9	82	(37 - 155)	9	65	(30 - 123)	14	86	(47 - 144)
008 Government administrators	30	76	(51 - 108)	27	51	(34 - 74)*	39	85	(61 - 117)
009 Other administrators	35	81	(56 - 112)	38	64	(45 - 88)*	47	65	(48 - 87)*
010 Teachers in higher education	28	71	(47 - 102)	19	72	(43 - 112)	21	120	(74 - 184)
011 Teachers nec	57	67	(51 - 86)*	61	62	(48 - 80)*	79	65	(51 - 81)*
012 Vocational trainers, social scientists etc	21	85	(53 - 130)	24	90	(58 - 134)	14	62	(34 - 104)
013 Welfare workers	18	72	(43 - 114)	23	90	(57 - 135)	20	115	(70 - 177)
014 Clergy	9	58	(27 - 111)	16	50	(29 - 82)*	37	73	(51 - 101)
015 Doctors	16	70	(40 - 114)	16	66	(38 - 108)	24	65	(42 - 97)*
017 Nurses	19	143	(86 - 224)	19	120	(72 - 187)	28	107	(71 - 155)
024 Literary and artistic occupations	32	70	(48 - 99)*	35	76	(53 - 106)	55	79	(59 - 103)
029 Electrical and electronic engineers (professional)	8	35	(15 - 70)*	11	72	(36 - 128)	10	56	(27 - 102)
030 Professional engineers nec	79	71	(56 - 88)*	65	73	(56 - 93)*	56	65	(49 - 84)*
031 Draughtspersons	21	65	(40 - 99)*	21	82	(51 - 125)	29	91	(61 - 131)
032 Laboratory technicians	35	91	(64 - 127)	25	62	(40 - 92)*	29	77	(52 - 111)
033 Architects and surveyors	26	66	(43 - 97)*	29	86	(58 - 124)	22	65	(41 - 99)*
036 Seafarers	24	125	(80 - 186)	21	122	(76 - 187)	37	119	(84 - 164)
037 Technicians nec	28	85	(57 - 123)	23	73	(46 - 110)	22	80	(50 - 121)
038 Production and maintenance managers	127	71	(59 - 84)*	157	87	(74 - 102)	184	88	(76 - 102)
039 Managers in construction	50	92	(69 - 122)	32	60	(41 - 84)*	51	93	(69 - 122)
040 Managers in transport, utilities and mining	63	97	(74 - 124)	70	92	(72 - 117)	76	91	(72 - 114)
041 Office managers	60	75	(57 - 97)*	69	80	(62 - 101)	101	85	(69 - 103)
042 Butchers	21	79	(49 - 121)	43	105	(76 - 142)	54	96	(72 - 125)
044 Retailers and dealers	247	100	(88 - 113)	279	85	(75 - 95)*	449	95	(87 - 104)
045 Publicans and bar staff	61	118	(90 - 151)	64	113	(87 - 145)	60	86	(66 - 111)
046 Caterers	39	103	(73 - 141)	31	80	(55 - 114)	48	109	(81 - 145)
047 Farmers	145	79	(67 - 93)*	258	89	(78 - 100)	560	99	(91 - 108)
049 Police	15	52	(30 - 85)*	13	103	(55 - 177)	36	126	(89 - 175)
050 Fire service personnel	14	104	(57 - 175)	11	131	(66 - 235)	12	111	(57 - 194)
052 Hairdressers	10	130	(62 - 238)	14	96	(53 - 162)	23	115	(73 - 172)
053 Office workers and cashiers	327	93	(83 - 104)	548	92	(84 - 100)	751	90	(84 - 96)*
054 Postal workers	52	68	(51 - 90)*	102	102	(83 - 123)	95	104	(84 - 127)
055 Petrol pump attendants	4	143	(39 - 366)	7	58	(23 - 119)	18	142	(84 - 224)
056 Van sales persons	11	57	(28 - 104)	13	110	(58 - 188)	24	117	(75 - 174)
057 Sales representatives	106	83	(68 - 100)	127	89	(74 - 106)	162	83	(71 - 97)*
058 Security workers	70	106	(82 - 134)	135	120	(101 - 142)*	123	106	(88 - 126)
059 Cooks and kitchen porters	28	98	(65 - 142)	31	112	(76 - 159)	34	107	(74 - 150)
060 Other service personnel	166	139	(118 - 161)*	273	106	(94 - 120)	283	94	(84 - 106)
061 Hospital porters, ward orderlies	17	98	(57 - 157)	30	98	(66 - 140)	23	79	(50 - 118)
062 Ambulance workers	6	87	(32 - 190)	9	107	(49 - 203)	13	148	(79 - 253)
063 Railway station workers	10	103	(50 - 190)	20	136	(83 - 210)	23	102	(65 - 153)
068 Leather and shoe workers	20	145	(88 - 223)	23	94	(59 - 140)	35	93	(64 - 129)
071 Warp preparers and weavers	8	118	(51 - 232)	8	118	(51 - 232)	11	79	(39 - 141)
073 Bleachers, dyers and finishers	6	93	(34 - 202)	13	151	(80 - 258)	18	115	(68 - 182)
074 Other textile workers	27	142	(94 - 207)	35	127	(89 - 177)	38	112	(80 - 154)
075 Chemical workers	47	109	(80 - 144)	52	108	(80 - 141)	71	130	(102 - 164)*
076 Bakers	8	66	(28 - 129)	18	93	(55 - 147)	48	138	(102 - 184)*
078 Food processors	37	111	(78 - 153)	28	93	(62 - 134)	37	82	(58 - 113)
086 Plastics workers	10	184	(88 - 338)	9	185	(84 - 350)	7	169	(68 - 349)
088 Other coal miners	47	116	(85 - 154)	109	131	(108 - 158)*	160	117	(99 - 136)
090 Other wood and paper processors	10	161	(77 - 295)	8	109	(47 - 216)	7	116	(47 - 240)
091 Other occupations - glass and ceramics	18	138	(82 - 218)	20	136	(83 - 210)	22	125	(79 - 190)
093 Plastic goods makers	20	154	(94 - 238)	13	140	(75 - 239)	10	156	(74 - 286)
094 Compositors	4	49	(13 - 125)	10	84	(40 - 155)	17	83	(48 - 133)
096 Printing machine minders	18	91	(54 - 143)	21	99	(61 - 152)	32	140	(96 - 198)
097 Printers (so described)	14	119	(65 - 201)	11	75	(38 - 135)	30	116	(78 - 165)
098 Tailors and dressmakers	4	112	(30 - 286)	8	89	(38 - 175)	15	65	(36 - 107)
101 Upholsterers	5	72	(23 - 168)	14	136	(75 - 229)	11	81	(40 - 144)
104 Carpenters	76	79	(62 - 99)*	89	88	(71 - 108)	116	96	(79 - 115)
108 Woodworking machinists	15	109	(61 - 180)	21	124	(77 - 190)	18	117	(69 - 185)
111 Other makers of paper goods	10	82	(40 - 151)	11	109	(54 - 195)	10	106	(51 - 195)

≠ SMRs only calculated for groups with at least 25 deaths
† 95 per cent confidence interval for SMR
* 95 per cent confidence interval for this SMR excludes 100

Annex 8.4 - *continued*

Occupation group	Age at death								
	20-64			65-74			75 and over		
	Deaths	SMR	(CI)†	Deaths	SMR	(CI)†	Deaths	SMR	(CI)†
112 Furnace operatives (metal)	12	147	(76 - 256)	9	99	(45 - 188)	15	116	(65 - 192)
114 Smiths and forge workers	11	118	(94 - 336)	6	84	(31 - 183)	20	102	(62 - 158)
116 Moulders and coremakers (metal)	12	114	(59 - 199)	15	113	(63 - 186)	23	101	(64 - 151)
120 Other metal manufacturers	31	189	(129 - 269)*	22	93	(58 - 141)	23	95	(60 - 142)
121 Press and machine tool setters	26	105	(68 - 153)	22	107	(67 - 162)	19	77	(46 - 122)
122 Centre lathe turners	21	120	(75 - 184)	22	108	(67 - 163)	17	95	(55 - 152)
124 Machine tool operators	115	100	(83 - 120)	133	104	(87 - 123)	147	94	(80 - 111)
125 Press and automatic machine operators	14	121	(66 - 202)	8	76	(33 - 151)	10	157	(75 - 288)
126 Metal polishers	9	171	(78 - 325)	11	94	(47 - 168)	6	68	(25 - 149)
129 Toolmakers	31	100	(68 - 143)	35	97	(67 - 135)	36	70	(49 - 97)*
130 Precision instrument makers	11	108	(54 - 193)	5	40	(13 - 94)*	9	59	(27 - 113)
132 Production fitters	181	91	(78 - 105)	191	93	(80 - 107)	231	97	(85 - 110)
133 Motor mechanics	47	74	(55 - 99)*	59	97	(74 - 125)	69	100	(78 - 127)
137 Electricians	70	83	(65 - 105)	52	80	(60 - 105)	80	98	(78 - 122)
138 Electrical plant operators	9	86	(39 - 163)	21	174	(108 - 265)*	12	59	(30 - 102)
139 Telephone fitters	22	66	(41 - 100)	28	98	(65 - 141)	24	77	(49 - 115)
144 Plumbers and gas fitters	47	81	(60 - 108)	50	95	(71 - 126)	68	117	(91 - 148)
145 Sheet metal workers	24	96	(62 - 143)	26	106	(69 - 156)	29	126	(84 - 181)
146 Metal plate workers	16	105	(60 - 171)	22	123	(77 - 187)	34	102	(71 - 143)
149 Welders	55	108	(82 - 141)	37	101	(71 - 139)	27	99	(65 - 144)
153 Vehicle body builders	6	81	(30 - 176)	5	48	(16 - 111)	21	105	(65 - 161)
159 Other spray painters	12	103	(53 - 180)	22	185	(116 - 280)*	15	95	(53 - 157)
160 Painters and decorators nec	82	104	(83 - 129)	112	109	(89 - 131)	123	114	(95 - 136)
163 Assemblers (vehicle and other metal goods)	24	129	(83 - 192)	11	69	(35 - 124)	9	76	(35 - 144)
164 Packers and sorters	41	110	(79 - 150)	70	129	(101 - 163)*	65	132	(102 - 169)*
165 Bricklayers and tilesetters	46	93	(68 - 124)	64	118	(91 - 151)	70	117	(92 - 148)
166 Masons and stonecutters	3	61	(13 - 179)	10	128	(62 - 236)	13	162	(86 - 277)
167 Plasterers	21	138	(85 - 211)	12	89	(46 - 157)	18	88	(52 - 140)
169 Builders etc	65	103	(79 - 131)	85	91	(73 - 112)	116	100	(83 - 120)
170 Rail track workers	13	123	(66 - 211)	21	99	(61 - 151)	29	85	(57 - 122)
171 Road construction workers and paviors	18	141	(83 - 223)	19	120	(72 - 187)	40	123	(88 - 168)
172 Sewage plant attendants	8	139	(60 - 275)	10	146	(70 - 269)	10	82	(39 - 151)
173 Mains and service layers	9	116	(53 - 220)	8	87	(38 - 171)	19	143	(86 - 223)
174 Construction workers nec	111	103	(85 - 124)	119	104	(87 - 125)	157	107	(91 - 125)
175 Face trained coalminers	41	103	(74 - 140)	97	136	(110 - 165)*	129	110	(91 - 130)
177 Railway guards	12	211	(109 - 369)*	4	68	(19 - 174)	17	105	(61 - 168)
178 Railway signal workers	6	114	(42 - 249)	11	131	(65 - 234)	18	155	(92 - 245)
180 Railway engine drivers	14	48	(26 - 81)*	41	133	(96 - 181)	67	120	(93 - 153)
181 Road transport inspectors	3	49	(10 - 142)	7	74	(30 - 153)	18	104	(62 - 164)
182 Bus and coach drivers	39	106	(75 - 144)	35	96	(67 - 134)	45	81	(59 - 108)
183 Lorry drivers	245	112	(98 - 127)	196	99	(86 - 114)	198	105	(91 - 121)
184 Other motor drivers	29	99	(66 - 142)	29	95	(64 - 137)	23	86	(54 - 129)
185 Bus conductors and drivers' mates	11	127	(63 - 227)	19	154	(93 - 241)	23	120	(76 - 180)
186 Mechanical plant drivers	11	73	(36 - 130)	12	103	(53 - 180)	13	93	(50 - 160)
187 Crane drivers	36	158	(111 - 219)*	30	117	(79 - 167)	34	84	(58 - 117)
188 Fork lift truck drivers	33	103	(71 - 145)	22	93	(58 - 140)	18	138	(82 - 218)
189 Slingers	8	154	(66 - 303)	11	153	(77 - 274)	8	98	(43 - 194)
190 Storekeepers	149	100	(85 - 117)	220	102	(89 - 116)	244	106	(93 - 120)
191 Dockers and goods porters	42	110	(80 - 149)	65	113	(87 - 144)	77	109	(86 - 136)
192 Refuse collectors	14	108	(59 - 182)	14	100	(55 - 168)	14	133	(73 - 224)
194 Boiler operators	13	103	(55 - 177)	30	134	(90 - 191)	47	104	(76 - 138)
Inadequately described/not stated	226	155	(135 - 177)*	872	120	(113 - 129)*	2603	108	(104 - 113)*

≠ SMRs only calculated for groups with at least 25 deaths
† 95 per cent confidence interval for SMR
* 95 per cent confidence interval for SMR excludes 100

Chapter 9 Asbestos-related diseases

Sally Hutchings
Jacky Jones
John Hodgson

Epidemiology and Medical Statistics Unit, HSE

9.1 Introduction

The three main asbestos-related diseases — asbestosis, mesothelioma and lung cancer — constitute the most serious single category of occupational disease. This chapter presents analyses based on two large data sources: the national (GB) mesothelioma register, and the Health and Safety Executive's (HSE) mortality survey of British asbestos workers. These two sources give complementary perspectives on the distribution of asbestos-related disease. Analysis of the mesothelioma register data provides an index of the total impact of asbestos exposure on the national population. The 'background' incidence of mesothelioma is generally estimated to be around two per million per year, or about 100 cases a year in Great Britain.[1] The only identified cause of mesothelioma is asbestos exposure, and the majority of cases now occurring can be assumed to be caused by asbestos exposure. That most of this exposure is occupational in nature is suggested by the contrast between rates for men and for women, and by the high proportional mortality ratios (PMRs) recorded among occupations with a clear potential for substantial asbestos exposure. However, the wide range of occupations appearing on mesothelioma death certificates suggests that asbestos exposure sufficient to lead to mesothelioma is not confined to those occupational situations where asbestos exposure is routine and obvious.

The HSE's asbestos worker mortality survey, set up in the early 1970s, has aimed to cover all workers with regular asbestos exposure, but coverage has clearly been more complete in the more easily identifiable areas such as asbestos manufacture and insulation work than in ancillary areas such as general construction and demolition, though all licensed asbestos strippers have been included since 1986. This lack of complete coverage, and the relatively recent start of the survey in terms of the long latency of, for example, mesothelioma, means that one should not expect all mesotheliomas to occur in the asbestos worker survey population. Even so, the total of 183 mesotheliomas in the asbestos survey contrasts markedly with the national total of 10,985 in the same period. Again, this points to the importance of exposures outside the obvious and easily recognised risk situations.

9.2 Mesothelioma register

9.2.1 Sources and methods

Mesothelioma is a rare form of cancer which principally affects the external lining of the lungs (pleura) and stomach (peritoneum). It has a strong association with exposure to asbestos dust, and the long latent period between first exposure to asbestos and the development, and diagnosis of mesothelioma is seldom less than 15 years and can be as long as 60 years. The disease is almost invariably fatal within one year of diagnosis.

Following reports of the association between asbestos exposure and mesothelioma[2,3] the Advisory Panel to Her Majesty's Senior Medical Inspector of Factories recommended the establishment of a national mesothelioma register.[4] This register was duly set up in 1967, and has been maintained since then by the HSE.

Data on mesothelioma cases are received from a number of sources, the main ones being the Office of Population Censuses and Surveys (OPCS) and the General Registrar's Office for Scotland (GRO(S)). Both send copies of death drafts to HSE, with 'mesothelioma' mentioned in part I or part II of the draft (see chapter 2) or in additional information supplied by the certifier. The reason for compiling the data in this way, rather than relying on routine cause of death codes is that the International Classification of Diseases (ICD) is based on tumour site rather than origin, so mesothelioma cannot be uniquely identified by an ICD code (see section 9.4.4). Each case is coded according to sex, area of usual residence at death, last full time occupation and site of mesothelioma (pleura, peritoneum, both or unspecified). Details are entered into the computerised register. Additional sources are used to monitor the completeness of the main data sources. These include cancer registrations received via OPCS from regional cancer registries, details of industrial death benefits awarded by the Department of Social Security(DSS), coroners' and postmortem reports or reports from the HSE's regional employment medical advisers. Only deaths in Great Britain are included in the register.

This chapter presents an analysis of cases in the 24 year period, 1968-91, in which mesothelioma was specifically mentioned on the death certificate. Data for the years 1967 and 1992 are incomplete and have been excluded from this analysis. Results are presented relating deaths to sex, year of death, age at death, date of birth, site of tumour, and last full time occupation, as recorded on the death certificate.

9.2.2 Year of death

The total number of mesothelioma deaths in Great Britain followed a steady upward trend between 1968 and 1984

Table 9.1 Mesothelioma deaths in Great Britain by sex , 1968-91

Year of death	Males	Females	Total
1968	115	39	154
1969	123	36	159
1970	145	49	194
1971	141	40	181
1972	169	43	212
1973	182	42	224
1974	185	58	243
1975	218	52	270
1976	256	56	312
1977	273	60	333
1978	328	63	391
1979	340	93	433
1980	355	102	457
1981	395	73	468
1982	411	91	502
1983	474	95	569
1984	534	84	618
1985	531	83	614
1986	599	101	700
1987	702	106	808
1988	751	111	862
1989	767	132	899
1990	763	117	880
1991	861	148	1009

Figure 9.1 Mesothelioma deaths in Great Britain, by sex, 1968-91

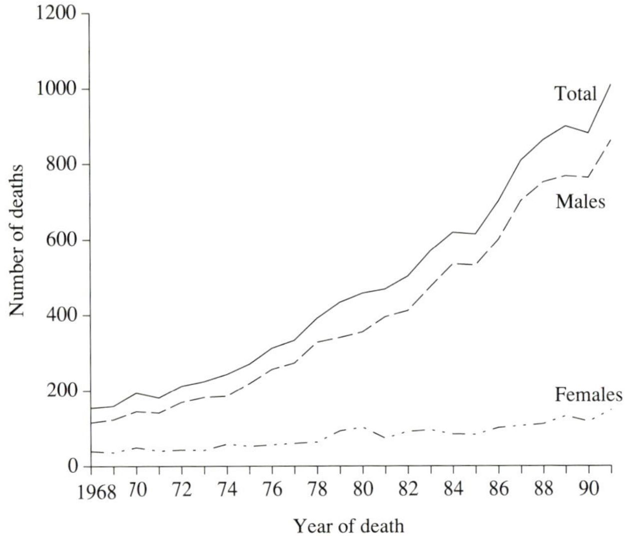

(Table 9.1 and Figure 9.1). This was interrupted in 1985 when the number of deaths fell from 618 to 614. In 1986 and 1987 there were steep increases of 14 and 15 per cent respectively. The rate of increase slowed down between 1987 and 1989, and finally fell by 2 per cent in 1990 from 899 to 880. This fall was followed by a steep increase in 1991 when the number of deaths rose by 15 per cent to 1,009.

During the study period, the number of mesothelioma deaths among males rose from 115 in 1968 to 861 in 1991, more than a seven-fold increase. For females there was a four-fold increase from 39 in 1968 to 148 in 1991. Because of the greater rate of increase in deaths among males, the proportion of total deaths which were male rose from 75 per cent in 1968 to 85 per cent in 1991. This is clearly shown in Figure 9.1.

9.2.3 Age at death

Age at death has been classified into six broad age-groups. Table 9.2 shows a break down of deaths from mesothelioma by sex and six four-year periods 1968-71 to 1988-91. Comparing the first and last four-year periods, the number of deaths increased in all age and sex groups except among females aged 15-44 years. The rate of increase was more pronounced in the older age-groups. In particular in 1968-71, 10 per cent of males and 13 per cent of females who died of mesothelioma were aged 75 years and over, whereas in 1988-91 the percentages were 20 and 26 respectively (see Figures 9.2 and 9.3).

Figures 9.4 and 9.5 show age-specific death rates (per million) for five broad age-groups, for males and females respectively. For males, with the exception of 15-44 year olds, the rates followed an upward trend in all age-groups throughout the study period. For the 45 years and over age-groups, rates increased steadily between the periods 1968-71 and 1972-75, whereas increases in rates between the remaining time periods were more pronounced among the 55 years and over age-groups, in particular among the 65 years and over age-groups.

Table 9.2 Mesothelioma deaths in Great Britain by age and sex , 1968-91

Age-group	Year of death					
	1968-71	1972-75	1976-79	1980-83	1984-87	1988-91
Males						
0-14		1				
15-44	33	35	58	70	80	87
45-54	83	146	212	223	292	379
55-64	215	260	362	531	770	914
65-74	139	239	401	541	750	1123
75+	54	73	164	270	474	639
Total	524	754	1197	1635	2366	3142
Females						
15-44	18	14	15	24	23	16
45-54	30	25	25	35	23	55
55-64	48	59	87	107	96	101
65-74	47	71	95	127	147	206
75+	21	26	50	68	85	130
Total	164	195	272	361	374	508

Figure 9.2 Male mesothelioma deaths in Great Britain by age, 1968-91

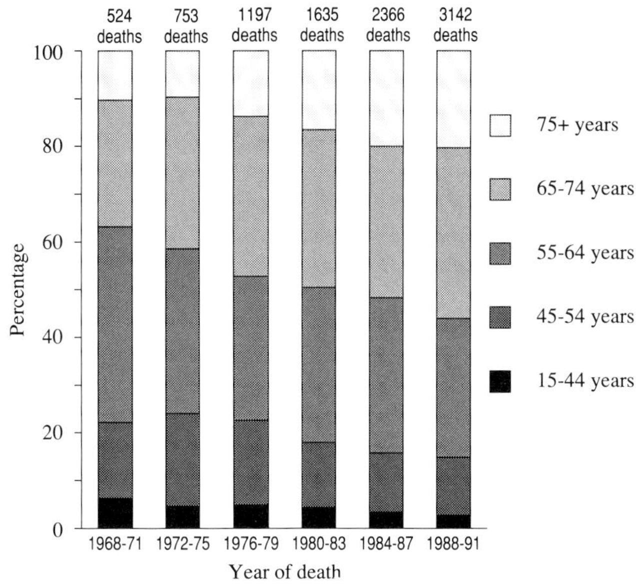

Figure 9.3 Female mesothelioma deaths in Great Britain by age, 1968-91

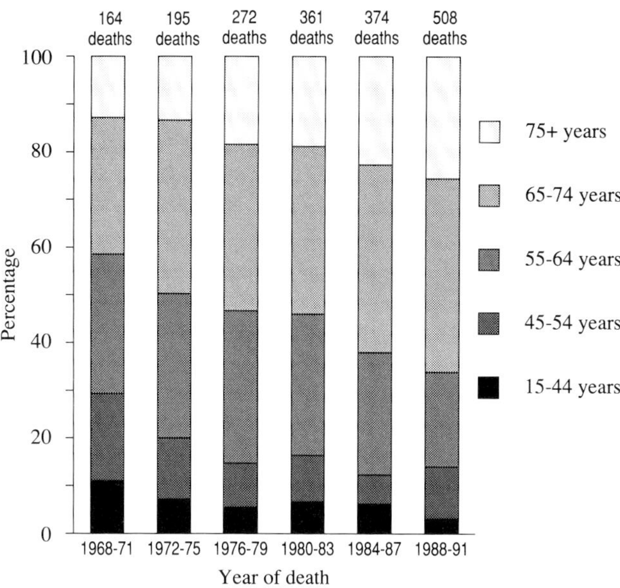

Figure 9.4 Mesothelioma age-specific death rates for men in Great Britain, 1968-91

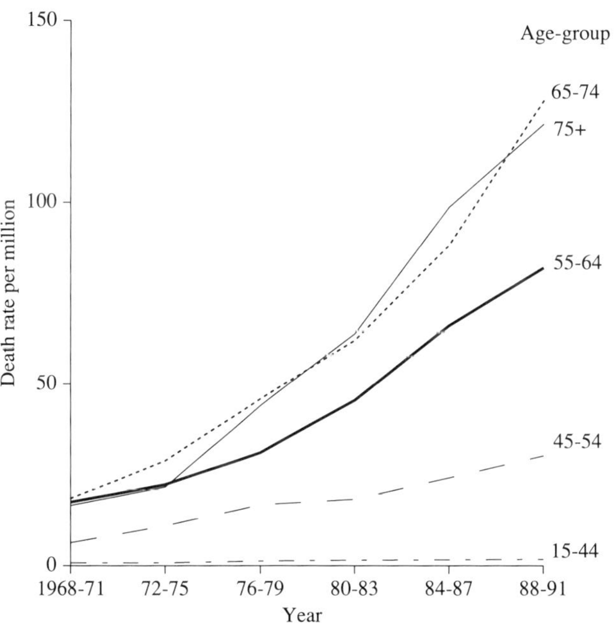

Figure 9.5 Mesothelioma age-specific death rates for women in Great Britain, 1968-91

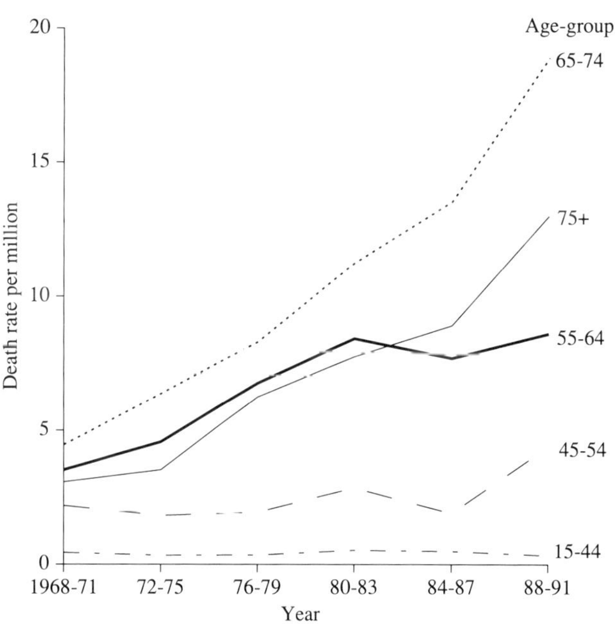

With smaller numbers, the pattern of female age-specific rates was more variable, showing flunctuations throughout the study period. Females in the 65 years and over age-group did however follow an upward trend.

9.2.4 Birth cohort, men

Only persons who were born between 1st July 1893 and 1st June 1958, who died of mesothelioma in Great Britain between 1968 and 1991, aged 15 to 89 years have been included in this analysis. Overall a total of 9,497 males and 1,817 females are included. Five-year age-specific death rates (15-19,20-24 ... 85-89) have been calculated for 13, five-year cohorts (1st July 1893-30 June 1898 to 1st July 1953-30th June 1958) using population estimates supplied by the OPCS and GRO(S).

Table 9.3 shows the number of mesothelioma deaths, and death rates per million population by five-year birth cohorts, for males. The data have been displayed in two different formats: Figure 9.6 shows age-specific mesothelioma death rates per million population, for males (alternate birth cohorts only), and Figure 9.7 shows for each age-group, rates for successive birth cohorts.

From the earliest cohort 1895 (1st July 1893-30th June 1898) through to cohort 1940 the rate of mesothelioma at each age group has been higher in the more recent cohort. Assuming the age-specific pattern is constant across cohorts, rates have increased by a factor of over 20 between the 1900 and the 1940 cohorts. The remaining three cohorts appear to break this pattern, displaying age-specific rates which are generally

Table 9.3 Number of male mesothelioma deaths and death rates per million in Great Britain by age, and birth cohort

Year of birth		Age at death														
		15-19	20-24	25-29	30-34	35-39	40-44	45-49	50-54	55-59	60-64	65-69	70-74	75-79	80-84	85-89
1893-98	Deaths												45	53	46	24
	Rate per million												22.31+	22.94	36.39	44.89
1898-03	Deaths											67	120	139	141	56
	Rate per million											20.45+	28.82	50.20	90.35	81.38
1903-08	Deaths										88	155	230	263	203	27
	Rate per million										20.55+	25.47	48.77	81.21	110.15	102.47+
1908-13	Deaths									69	189	288	392	395	81	
	Rate per million									14.37+	25.72	45.91	79.04	116.64	138.35+	
1913-18	Deaths								33	143	244	378	489	135		
	Rate per million								7.54+	19.73	37.11	66.62	111.01	142.29+		
1918-23	Deaths							27	96	226	439	638	204			
	Rate per million							5.04+	11.31	28.22	60.02	103.40	161.40+			
1923-28	Deaths						17	76	192	374	606	187				
	Rate per million						3.42+	9.27	24.24	49.82	91.02	120.71+				
1928-33	Deaths					7	22	91	189	413	150					
	Rate per million					1.46+	2.77	11.75	25.06	59.55	89.97+					
1933-38	Deaths				2	12	43	108	250	112						
	Rate per million				0.41+	1.51	5.52	14.15	35.04	65.79+						
1938-43	Deaths			4	5	18	62	160	76							
	Rate per million			0.79+	0.61	2.23	7.79	21.37	41.90+							
1943-48	Deaths		0	3	7	33	77	31								
	Rate per million		0+	0.30	0.70	3.34	8.19	14.50+								
1948-53	Deaths	0	0	4	4	19	12									
	Rate per million	0+	0	0.42	0.42	2.11	5.20+									
1953-58	Deaths	1	0	1	4	2										
	Rate per million	0.1	0	0.10	0.44	0.90+										

+ The youngest and oldest cohorts for each age group are truncated by the end of the observation period(1968-91), so the average age in the cell is shifted by about one year relative to the corresponding cells. This is shown in figure 9.6.

Figure 9.6 Male mesothelioma death rates in Great Britain by age, and birth cohort

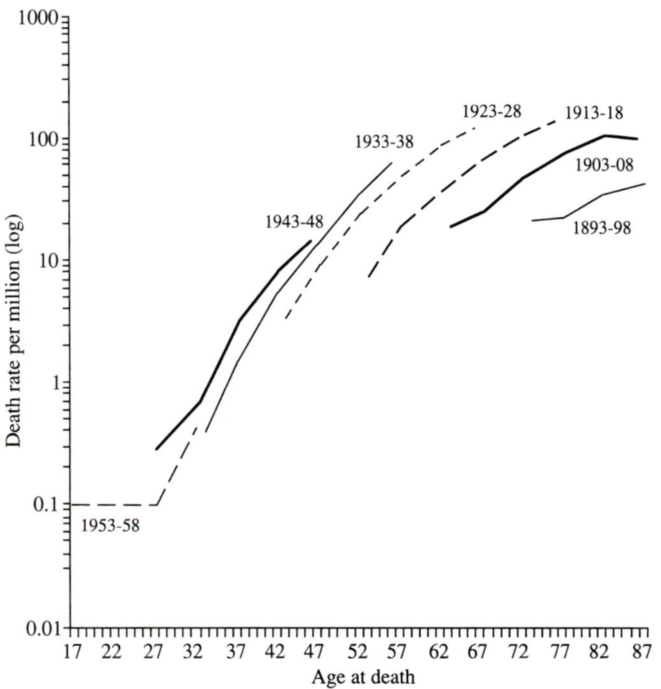

Note: For clarity observations with zero rates have been excluded, see Table 9.3

lower than those recorded in the immediately preceding cohorts.

The data in Table 9.4 show a comparison of the four latest male birth cohorts, 1940, 1945, 1950 and 1955. For each cohort and five-year age-group cell, the table shows the observed deaths, the total man years (in millions), the mean age of the observations contributing to the cell and the rate per million man years. For most cells the total man years represents 5 years worth of deaths, but for the youngest and oldest cells for each cohort the potential contributing man years are truncated by the ends of the observation period (1968 to 1991). This means that the total man years are less than for other cells, and that the average age in the cell is shifted about one year relative to corresponding complete cells. The comparison of these cohorts has been made by calculating expected deaths for 1940, 1950 and 1955 cohorts based on the observed rates for the 1945 cohort. The expected values for cells with truncated man years are adjusted for the difference in average age by linear interpolation of the observed 1945 cohort rates, on log scales for rate and age.

Figure 9.7 Male age-specific death rates in Great Britain, by birth cohort

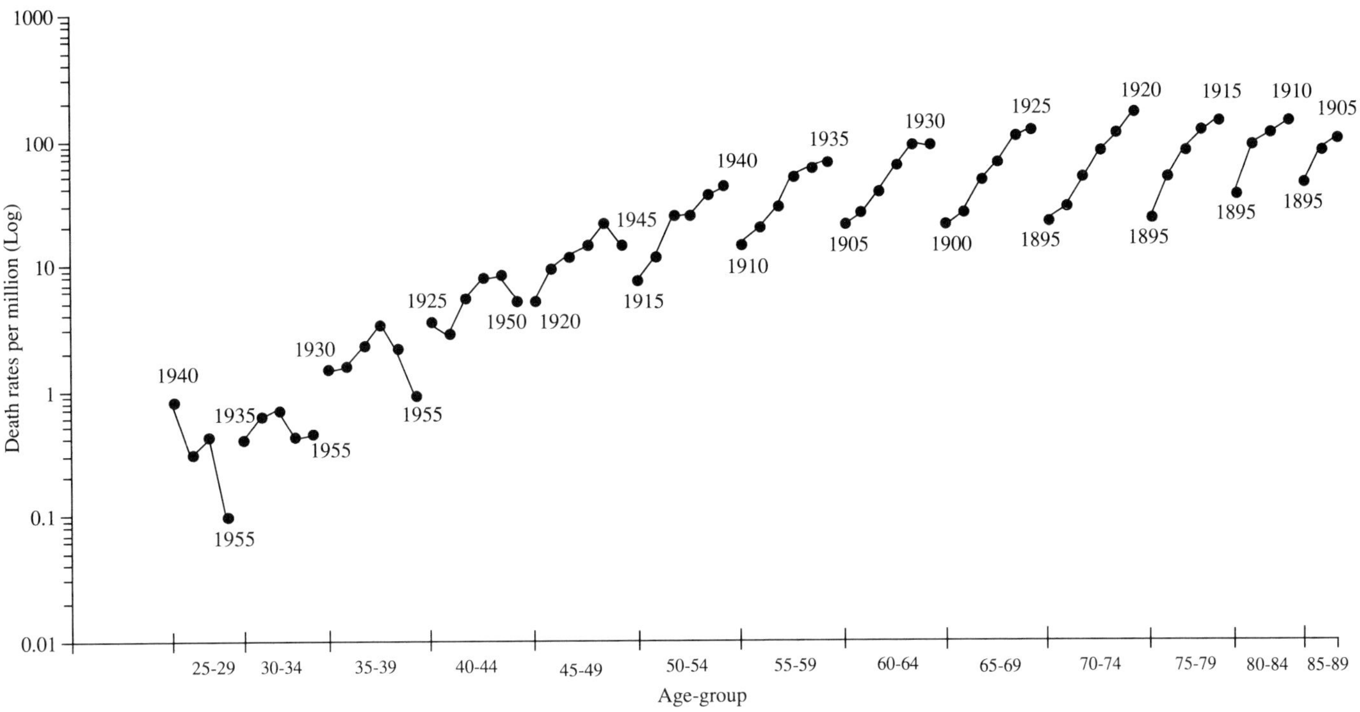

Notes; For each age-group, rates are shown for successive five-year birth cohorts. The first and last birth cohorts for each age-group are labelled.

Table 9.4 Detailed comparison of latest four male mesothelioma birth cohorts by age-group

Birth cohort		Age-groups 15-19	20-24	25-29	30-34	35-39	40-44	45-49
1940	Deaths (expected)*			4 (1.7)+	5 (5.7)	18 (27)	62 (65.2)	160 (132.0)+
	Man-years (millions)			5.042	8.218	8.075	7.958	7.486
	Mean age			27.67	32	37	42	47
	Death rate (per million)			0.79	0.61	2.23	7.79	21.37
1945	Deaths		0	3	7	33	77	31
	Man-years (millions)		6.322	10.161	10.038	9.88	9.399	2.139
	Mean age		22.67	27	32	37	42	45.67
	Death rate (per million)		0.00	0.30	0.70	3.34	8.19	14.50
1950	Deaths (expected)*	0 (0)	0 (0)	4 (2.8)	4 (6.6)	19 (30.1)	12 (15.1)+	
	Man-years (millions)	6	9.728	9.634	9.454	9.012	2.309	
	Mean age	17.67	22	27	32	37	40.67	
	Death rate (per million)	0.00	0.00	0.42	0.42	2.11	5.20	
1955	Deaths (expected)*	1 (0)	0 (0)	1 (2.8)	4 (6.4)	2 (5.0)+		
	Man-years (millions)	9.824	9.827	9.6	9.137	2.224		
	Mean age	17	22	27	32	35.67		
	Death rate (per million)	0.10	0.00	0.10	0.44	0.90		

* Expected deaths for the 1940, 1950 and 1955 cohorts calculated using the 1945 cohort observed rates (and assuming a zero rate for ages 15-19).
+ Adjustment for differences in mean age by linear interpolation using log(1945 rate) and log(age).

Total observed and expected deaths for all ages under 50 are shown in Table 9.5. On this basis, the 1940 cohort has the worst mortality experience, with 8 per cent more deaths than expected if 1945 cohort rates had applied. The 1950 cohort is 29 per cent better off than the 1945 cohort (39 observed/54.6 expected), and the 1955 cohort has rather less than 60 per cent of the deaths expected on the basis of the 1945 rates (8/14.2). On a formal test for linear trend the reduction across these four cohorts is statistically significant (p=0.006). These comparisons are effectively the same on all the relevant data, or when the small numbers in the under 30 age-groups are removed.

Table 9.5 Comparison of last four male birth cohorts (age at death up to 50)

Birth cohort	Total observed and (expected) deaths and O/E ratio All ages <50 Obs	Exp	O/E
1940	249	(231.6)	1.08
1945	151	(151.0)	1.00
1950	39	(54.6) [p=0.01]	0.71
1955	8	(14.2) [p=0.03]	0.57
		[p for trend = 0.006]	

Table 9.6 Number of female mesothelioma deaths and death rates per million in Great Britain by age, and birth cohort

Year of birth		Age at death														
		15-19	20-24	25-29	30-34	35-39	40-44	45-49	50-54	55-59	60-64	65-69	70-74	75-79	80-84	85-87
1893-98	Deaths												12	13	20	12
	Rate per million												3.58+	2.91	6.60	7.24
1898-03	Deaths											21	38	37	24	14
	Rate per million											4.89+	6.06	7.40	6.84	7.08
1903-08	Deaths										18	45	58	50	35	7
	Rate per million										3.64+	5.91	8.68	9.20	9.22	10.76+
1908-13	Deaths									12	42	73	80	80	11	
	Rate per million									2.30+	5.04	9.53	11.77	14.81	10.07+	
1913-18	Deaths								12	37	62	79	100	24		
	Rate per million								2.59+	4.70	8.27	11.42	16.85	16.85+		
1918-23	Deaths							7	26	44	73	125	29			
	Rate per million							1.26+	2.92	5.11	8.92	17.08	17.91+			
1923-28	Deaths						9	15	27	56	76	33				
	Rate per million						1.80+	1.81	3.33	7.14	10.56	18.84+				
1928-33	Deaths					1	8	7	17	49	18					
	Rate per million					0.21+	1.01	0.90	2.22	6.83	10.14+					
1933-38	Deaths				0	3	9	11	28	11						
	Rate per million				0+	0.39	1.18	1.45	3.91	6.38+						
1938-43	Deaths			0	3	7	16	22	9							
	Rate per million			0+	0.37	0.88	2.04	2.95	4.94+							
1943-48	Deaths		2	3	4	10	11	12								
	Rate per million		0.32+	0.30	0.41	1.02	1.18	5.64+								
1948-53	Deaths	1	0	1	5	5	2									
	Rate per million	0.18+	0	0.11	0.53	0.55	0.87+									
1953-58	Deaths	1	0	2	1	2										
	Rate per million	0.11	0	0.21	0.11	0.90+										

+ The youngest and oldest cohorts for each age group are truncated by the end of the observation period (1968-91), so the average age in the cell is shifted by about one year relative to the corresponding cells. This is shown in figure 9.8.

9.2.5 Birth cohort, women

Table 9.6 shows the number of mesothelioma deaths and death rates per million population by five-year birth cohorts, for females. Figure 9.8 displays age-specific mesothelioma death rates per million population by five-year birth cohorts (for clarity, alternate birth cohorts have been omitted), for females. The pattern shown here is very similar to that for males, although the trends are less strong, and the smaller numbers mean that the rates are more subject to statistical variability, especially at ages under 40. Here, exactly as for males, the rates increase steadily until the 1940 and 1945 cohorts, and then fall. The decrease for females in the 1950 and 1955 cohorts from the 1940/45 maximum is similar to that for males (30 per cent and 40 per cent respectively), though based on smaller numbers (14 deaths in the 1950 cohort and 6 in the 1955 cohort). The 1930 and 1935 cohorts depart slightly from the general pattern: the rates for the 1930 cohort fall between those for the 1920 and the 1925 cohorts, and those for the 1935 cohort are very close to the 1925 cohort rates.

Figure 9.8 Female mesothelioma death rates in Great Britain by age, and birth cohort

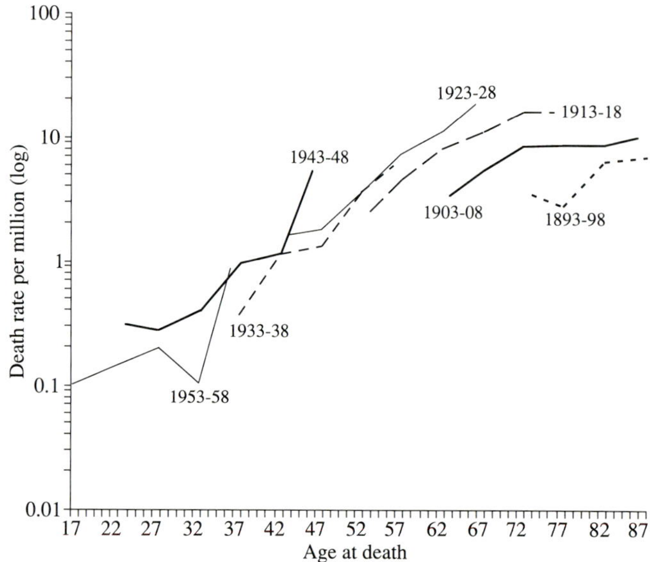

Note: For clarity observations with zero rates have been excluded, see Table 9.6

9.3 Site of cancer

Mesothelioma deaths by site of the neoplasm and by sex, for the period 1979,80-82-91 are shown in Table 9.7. Industrial action by registration officers in 1981 meant that some of the information normally requested to aid cause coding was unavailable, so 1981 has been excluded from the site analysis.

Table 9.7 Mesothelioma deaths in Great Britain by site of cancer and sex, 1979-80, 82-91

Year of death	Pleura	Peritoneum	Pleura and peritoneum	Site not specified	Total
Males					
1979	275	24	8	33	340
1980	294	25	5	31	355
1982	288	24	13	86	411
1983	393	33	6	42	474
1984	422	36	10	66	534
1985	398	38	14	81	531
1986	454	36	13	96	599
1987	525	33	15	129	702
1988	556	41	17	137	751
1989	549	42	14	162	767
1990	576	45	13	129	763
1991	642	39	14	166	861
Females					
1979	66	14	5	8	93
1980	75	18	3	6	102
1982	54	7	7	23	91
1983	75	8	3	9	95
1984	59	8	2	15	84
1985	55	10	2	16	83
1986	75	7	1	18	101
1987	81	11	4	10	106
1988	85	13	3	10	111
1989	91	12	1	28	132
1990	75	14	4	24	117
1991	104	10	3	31	148

Figure 9.9 Male mesothelioma deaths in Great Britain by site, 1979-80 and 1982-91

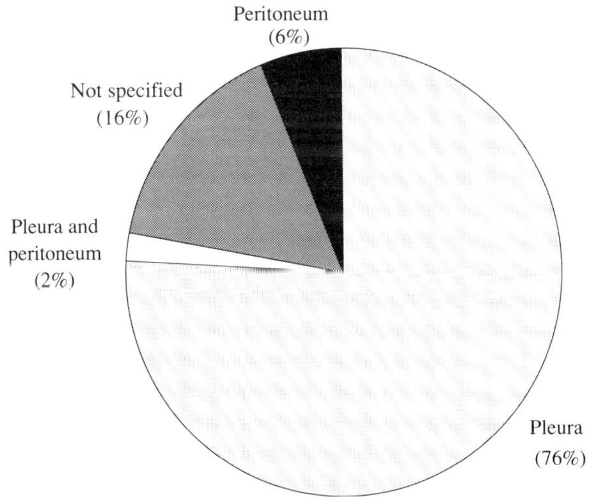

Figure 9.10 Female mesothelioma deaths in Great Britain by site, 1979-80 and 1982-91

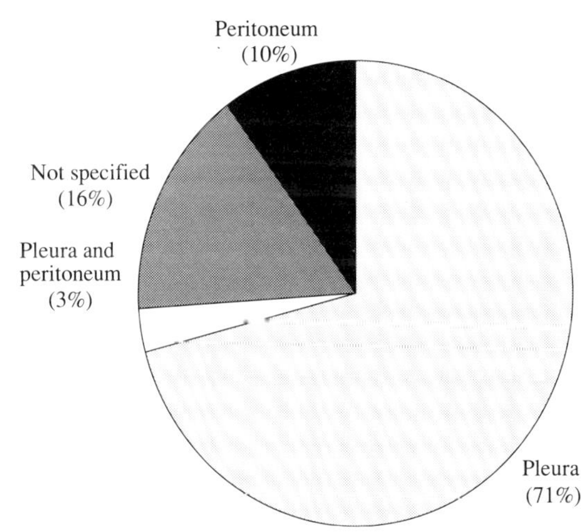

Mesothelioma of the pleura accounted for about three quarters of the deaths. A further 16 per cent of death certificates mentioning mesothelioma did not identify a neoplasm site, but it is likely that the majority of these will be pleural. The overall ratio of males to females with mesothelioma of the pleura, six to one was almost double that for mesothelioma of peritoneum, three to one (see Figures 9.9 and 9.10).

Table 9.8 gives the deaths rates (per million) for mesothelioma of the pleura by age for the period 1979-80, 82-91. The death rates for males were higher than for females in all age-groups, and the 65-74 year age-group recorded the highest rate for both sexes. The ratio of the pleural death rate of males to that of females increased markedly from three to one for the 15-44 year olds to eight to one for the 55-64 year olds. The highest ratio, ten to one, was among the 75 years and over age-group.

Table 9.8 Death rates (per million) for mesothelioma of the pleura in Great Britain, by age and sex, 1979-80, 82-91

Age group	Year of death 1979-80, 82-91	
	Males	Females
15-44	1.22	0.38
45-54	18.45	2.16
55-64	49.33	6.01
65-74	74.20	11.15
75+	63.84	6.04

Table 9.9 gives corresponding data for mesothelioma of the peritoneum. The death rates for males were again higher than those for females in each age-group, but the divergence with increasing age seen for mesothelioma of the pleura was absent.

Table 9.9 **Death rates (per million) for mesothelioma of the peritoneum in Great Britain, by age and sex, 1979-80, 82-91**

Age group	Year of death 1979-80, 82-91	
	Males	Females
15-44	0.17	0.02
45-54	2.66	0.35
55-64	4.12	1.03
65-74	4.11	1.45
75+	3.03	0.71

9.3.1 Underlying cause of death

Since 1986 the underlying cause of death as coded by OPCS using the ICD has been recorded on the register. Table 9.10 shows the underlying cause of death by sex for the period 1986-91. For males 55 per cent of mesothelioma deaths were coded as malignant neoplasm of the pleura, and a further 4.5 per cent as malignant neoplasm of the peritoneum. For females the corresponding figures were pleura 51 per cent and peritoneum 8 per cent. Apart from these sites, the most common codings are for disseminated cancer of unknown primary site (ICD code 199) and for cancer of the trachea, bronchus and lung (ICD code 162).

Table 9.10 **Mesothelioma deaths in Great Britain by underlying cause of death and sex, 1986-91**

Underlying cause of death	Males	Females
Malignant neoplasm of :		
- peritoneum (ICD 158)	202	57
- trachea, bronchus, lung (ICD 162)	670	120
- pleura (ICD 163)	2447	363
- other specified sites (ICD 140-157,160,161,164-198)	166	32
- unspecified sites (ICD 199)	778	120
Circulatory disease (ICD 390-459)	89	13
Respiratory disease (ICD 460-519)	41	3
Digestive disease (ICD 520-579)	18	3
Other (remaining ICD codes)	32	4
All causes	4443	715

In Chapters 4 to 7, any discussion of the asbestos-related diseases, cancer of the pleura and cancer of the peritoneum are based on underlying cause of death only. About 60 per cent of mesothelioma deaths recorded on the register will have been included in these analyses.

9.4 Occupation

9.4.1 Last full-time occupation

The mesothelioma register database provides for only one occupation code to be recorded for each death. For males and single women, this is always their own last full-time occupation (if recorded by registrar). For other women, if an occupation implying paid employment is recorded the corresponding code (operational codes between 1 and 348) is shown on the register, otherwise their husband's occupation code is taken. Any additional occupational information supplied by the certifier, e.g. other jobs with known asbestos exposure, are also recorded on the register. Most of the deaths on the register are due to asbestos exposure, but the long period of latency between first exposure and development of mesothelioma (seldom less than 15 years and can be as long as 60) means exposure may not have occurred in their last full-time job. Some deaths are not asbestos-related; mesothelioma has a natural background incidence generally assumed to produce about 100 cases per year.

9.4.2 Order groups

Each job description held on the register has been coded using OPCS operation codes. This section covers deaths in England and Wales, and Scotland during the period 1968-80 and 82-91, for all ages. Operational codes have been aggregated to form order groups. Supervisors and employers in the same job have been placed in the employee's order group.

Tables 9.11 and 9.12 show order groups for male and female mesothelioma deaths in Great Britain, and for England and Wales and Scotland separately. More detailed tables containing the number of deaths for each operation code are available on floppy disks.

Table 9.11 **Male mesothelioma deaths in England and Wales, and Scotland by occupational order group, 1968-80, 82-91**

Occupational order group	England & Wales	Scotland	Great Britain
1 Professional	203	14	217
2 Professional education and health	133	20	153
3 Literary, artistic and sports	28	2	30
4 Professional science	433	41	474
5 Managerial	609	62	671
6 Clerical	369	31	400
7 Selling	115	15	130
8 Security and protective service	169	19	188
9 Personal service	221	21	242
10 Agricultural workers	62	3	65
11 Materials processing	772	136	908
12 Metal and electrical	2601	377	2978
13 Assembly	327	45	372
14 Construction and mining	715	70	785
15 Transportation	686	66	752
16 Miscellaneous	529	82	611
17 Not stated	231	16	247
All occupations	8203	1020	9223

Table 9.12 **Female mesothelioma deaths (with own occupation recorded on death certificate), in England and Wales, and Scotland, by occupational order group, 1968-80, 82-91**

Occupational order group	England & Wales	Scotland	Great Britain
1 Professional	7	0	7
2 Professional education and health	48	9	57
3 Literary, artistic and sports	7	0	7
4 Professional science	6	0	6
5 Managerial	36	3	39
6 Clerical	102	6	108
7 Selling	26	2	28
8 Security and protective service	1	1	2
9 Personal service	120	9	129
10 Agricultural workers	1	0	1
11 Materials processing	120	5	125
12 Metal and electrical	10	3	13
13 Assembly	29	0	29
14 Construction and mining	3	0	3
15 Transportation	10	0	10
16 Miscellaneous	30	3	33
17 Not stated	108	13	121
All occupations	664	54	718

In Great Britain, a total of 9,941 mesothelioma deaths occurred in the specified period. Nearly 90 per cent of these took place in England and Wales. The pattern across occupational order groups is similar for England and Wales, and Scotland, so the analysis which follows has been completed for Great Britain as a whole.

9.4.3 Males

A total of 9,223 males died in Great Britain in the specified period. Deaths were recorded in every occupational order group, and in 307 (88 per cent) of the 350 occupational operation codes.

The order group metal and electrical recorded the highest number of mesothelioma deaths, almost one third of all male mesothelioma deaths in Great Britain. A further ten per cent of deaths fell into materials processing. Construction, and transportation consisted of 8.5 and 8.2 per cent of deaths respectively.

Within these order groups, certain operation codes stood out. In metal and electrical, metal working production fitters and fitter/machinists (598 deaths), plumbers, heating and ventilating fitters and gas fitters (393 deaths), machine tool operators (367 deaths) and metal plate workers, shipwrights and riveters (317 deaths) each formed over 10 per cent of deaths. Over one half of the deaths among materials processing were carpenters and joiners (519 deaths). In construction and mining over a third of deaths were construction workers nec (312 deaths). Brick layers and tile setters (111 deaths) and builders 'so described' (114 deaths) each formed a further ten per cent of deaths in this order group. About one quarter of deaths in transportation were store keepers and warehouse men (182 deaths) . Drivers of road goods vehicles (167 deaths) and stevedores and dockers (133 deaths) also recorded high numbers in this order group.

9.4.4 Females

A total of 1,801 females died from mesothelioma in the specified period, but only 718 deaths (40 per cent) had their own occupation recorded on the death certificate and have been included in the following analysis.

At least one death was recorded in each of the 17 order groups, and among 123 (35 per cent) of the 350 occupational operation codes.

Half of the occupation-coded female mesothelioma deaths were in the order groups, personal service, materials processing and clerical. Within personal service the operational codes cleaners, window cleaners, chimney sweeps and road sweepers (37 deaths) and other domestics and school helpers (33 deaths) had the highest number of deaths. Over a quarter of materials processing were sewers and embroiderers (32 deaths) and the clerical operational codes which stood out were other clerks and cashiers, not retail (61 deaths) and typists, shorthand writers and secretaries (25 deaths).

9.4.5 Job groups

This section analyses occupational mortality for England and Wales during the period 1979-80 and 1982-90. As in other chapters, the operational codes have been combined to form job groups (see Appendices 2 and 3 for details of job groups), and the main outcome measure is the proportional mortality ratio (PMR).

In the analysis of the earlier chapters, mesotheliomas were approximately identified by the ICD codes for cancers of the pleura (163) and peritoneum (158). Because the ICD is based on tumour site rather than origin, these codes will include some cancers which are not mesotheliomas, and will exclude some which are (Table 9.10 shows 790 mesotheliomas — about 15 per cent of the total — coded to cancer of the trachea, bronchus and lung). The mesothelioma register includes all death certificates which mention the word mesothelioma in part I or part II of the certificate, or in additional information supplied by the certifier. It therefore provides a more accurate — though not error-free — count of the mesothelioma total. A further difference is that the register holds deaths by year of death, and not year of registration as described in earlier chapters.

A total of 4,487 males and 804 females died from mesothelioma during the period 1979-80 and 1982-90, aged 16-74. A total of 364 (45 per cent) women had their own occupation recorded on their death draft, but only female death drafts which show an identifiable economically active occupation have been included in the following analysis (i.e. occupations with operation codes 1 to 348), a total of 305 deaths.

Table 9.13 lists job groups among males which have a significantly raised PMR based on more than 5 deaths. Job groups in Table 9.13 which consist of more than one operation code are expanded in Table 9.14.

Production fitters, carpenters, and plumbers and gas fitters each form four per cent or more of the total male mesothelioma deaths, and have rates which are significantly different from the average for all occupations. Plumbers and gas fitters, and carpenters have particularly high rates and are ranked third and fifth in the table.

The job group at highest risk of mesothelioma was metal plate workers, with a rate of over seven times the average. Over three quarters of these deaths had either boiler worker, plater or shipwright recorded as part of their job description on the death certificate.

Other job groups with particularly high risks include construction workers nec and vehicle body builders. Construction workers nec consists of three operational codes (see Table 9.14), only the operational code construction workers nec has a raised rate, and this is over six times the overall average. Over three quarters of the certificates for these deaths mentioned the term lagger or insulator. Vehicle body builders had a rate which was over six times the average. Nearly all these death certificates mentioned the term coach builder or coach finisher, and about one third of these referred to railways. PMRs for all occupational operational codes and job group codes will be available on floppy disks.

Table 9.13 Mortality of men aged 16-74 from mesothelioma in England and Wales by occupation, 1979-80, 82-90

Job group		Number of deaths	PMR (all men=100)	95% Confidence limits for PMR	
				Lower	Upper
146	Metal plate workers	110	700.4	575.5	844.2
153	Vehicle body builders	35	618.7	431.0	860.5
144	Plumbers and gas fitters	201	442.8	383.7	508.5
156	Coil winders	4	414.6	113.0	1061.5
104	Carpenters	258	365.7	322.4	413.1
137	Electricians	161	290.5	247.4	339.0
101	Upholsterers	19	283.3	170.6	442.4
174	Construction workers nec	187	255.6	220.3	295.0
194	Boiler operators	39	253.9	180.5	347.0
138	Electrical plant operators	18	253.5	150.3	400.7
027	Chemical engineers and scientists	18	248.4	147.2	392.6
145	Sheet metal workers	48	233.2	171.9	309.1
148	Scaffolders	11	225.6	112.6	403.7
132	Production fitters	304	216.3	192.7	242.1
030	Professional engineers nec	105	210.6	172.2	254.9
167	Plasterers	27	202.8	133.6	295.0
149	Welders	70	202.6	157.9	256.0
039	Managers in construction	40	196.8	140.6	268.0
191	Dockers and goods porters	69	195.1	151.8	246.9
143	Electrical engineers (so described)	39	187.0	133.0	255.7
037	Technicians nec	24	171.9	110.1	255.8
169	Builders etc	98	164.4	133.5	200.4
032	Laboratory technicians	27	164.2	108.2	238.9
031	Draughtspersons	28	160.6	106.7	232.1
124	Machine tool operators	179	133.0	114.2	154.0
160	Painters and decorators nec	100	131.0	106.5	159.3

Table 9.14 Mortality of men aged 16-74 from mesothelioma in England and Wales by occupation, 1979-80, 82-90

Job group	Number of deaths	PMR (all men =100)	95% confidence limits for PMR		Operation Code	Number of deaths	PMR (all men =100)	95% confidence limits for PMR	
			Lower	Upper				Lower	Upper
032 Laboratory technicians	27	164.2	108.2	238.9	080 Laboratory technicians	13	170.5	90.8	291.5
					081 Engineering technicians, technician engineers	9	155.7	71.2	295.5
					288 Laboratory assistants	5	164.5	53.4	383.9
174 Construction workers nec	187	255.6	220.3	295.0	312 Craftsmen's mates	0	0	0	81.1
					313 Building and civil engineering labourers	21	49.8	30.8	76.1
					316 Construction workers nec	166	628.3	539.8	735.6
027 Chemical engineers and scientists	18	248.4	147.2	392.6	066 Chemical scientists	13	229.1	122.0	391.7
					073 Chemical engineers	5	318.2	103.3	742.6
030 Professional engineers nec	105	210.6	172.2	254.9	068 Civil, structural, municipal mining and quarrying engineers	8	55.9	24.1	110.2
					069 Mechanical and aeronautical engineers	39	315.9	224.1	430.7
					070 Design and development engineers (mechanical)	13	299.9	143.0	483.6
					074 Production engineers	12	414.6	214.2	724.2
					075 Planning and quality control engineers	18	222.1	131.6	350.9
					076 Engineers nec	10	313.3	150.3	576.2
					078 Technologists nec	5	107.4	34.9	250.6
039 Managers in construction	40	196.8	140.6	268.0	092 Managers in building and contracting	20	131.4	85.4	210.9
					093 Clerks of works	20	391.8	239.3	605.1
191 Dockers and goods porters	69	195.1	151.8	246.9	334 Stevedores, dockers, inc foremen	57	293.3	226.6	385.8
					335 Goods porters, inc foremen	12	75.3	38.9	131.6
037 Technicians nec	24	171.9	110.1	255.8	088 Architectural and town planning technicians	0	0	0	408.4
					089 Buiding and civil engineering technicians	0	0	0	1685.7
					090 Technicians and related workers nec	24	186.9	119.8	278.1
169 Builders etc	98	164.4	133.5	200.4	304 Handymen, general building workers inc foremen	27	162.6	107.2	236.6
					305 Builders (so described)	71	165.1	129.0	208.3

For females, three job groups had significantly raised PMRs based on more than 3 deaths, chemical workers (including process workers nec) (PMR 883, 9 deaths), tailors and dress makers (PMR 336, 10 deaths) and sewers and embroiderers (including machinists nec) (PMR 198, 17 deaths). PMRs for all groups will be available on floppy disks.

Table 9.15 Mortality of women aged 16-74 from mesothelioma in England and Wales, by occupation (includes females with their own occupation recorded on the death draft), 1979-80, 82-90

Job group	Number of deaths	PMR (women's operation codes 1 to 348 = 100)	95% confidence intervals for PMR	
			Lower	Upper
075 Chemical workers	9	883	403.8	1676.2
098 Tailors and dressmakers	10	336.3	161.3	618.4
100 Sewers and embroiderers	17	197.5	115.0	316.2

9.5 Background to the Asbestos Survey

9.5.1 Study population

The asbestos survey population consists of a nationwide census of workers (covering England, Wales and Scotland), employed in premises or engaged in activities covered by the 1969 Asbestos Regulations, enrolled from 1971, and by the Asbestos (Licensing) Regulations(ALR) enrolled in 1984, workers entering employment subsequently and covered under these regulations and under the Control of Asbestos at Work (CAW) Regulations from 1988. For further details, see Annex 9.1.

9.5.2 Tracing of workers at NHSCR

Tracing criteria and numbers submitted for 'flagging' at NHSCR, under various rules applied during the currency of the changing regulations, are summarised in Table 9.16. A total of 57,402 of the 74,767 workers enrolled in the survey by June 1993 had been submitted for "flagging" at the time of this study, that is nearly all survey entrants coming in under the 1969 Regulations, and those entering under the Asbestos Licensing and CAW Regulations with recorded previous exposure or at their second medical examination, and CAW Regulation entrants recorded as 'strippers'.

A small number of survey members have not been submitted where available information is known to be insufficient for tracing. More substantial deficits as indicated by per cent of eligible members submitted appear where workers were entered onto the survey database, or had their second examination recorded, after January 1991 when the last batch covered by this analysis was submitted for 'flagging'. Examinations from 1989 are principally affected.

A trace rate of 98.7 per cent (representing 55,511 traced workers) was achieved for the study cohort as a whole. Trace rates broken down by the regulations under which members entered the survey are given in Table 9.17.

Of the traced survey members, 372 (0.7 per cent) were recorded as emigrants and were therefore assumed to be lost to follow-up and 5,327 (9.6 per cent) had died up to 1991.

Table 9.16 Tracing criteria

Regime	Flagging rule	Number submitted	% of eligible workers	Reason for deficit
1969 regs (1971-1983)	All new entrants flagged	40,620	99.9	Insufficient information for tracing
ALR (from 1984)	Flagged on entry if previous asbestos exposure recorded	5,832	99.6	Insufficient information for tracing
	Flagged at second medical examination otherwise	5,411	92.2	Records of 2nd exam. entered onto survey database after 1990
	Others (submitted under previous rules)	1,751	14.1	All ALR entrants were flagged at first. Rules restricting flagging to those with previous exposure of more than 1 exam. were introduced in 1986
CAW (1988)	All "strippers" flagged on entry	2,118	95.9	Late entry of records for new survey members onto database after 1990
	Other entrants flagged as under ALR rules	446	61.8	Records of 2nd exam. entered onto survey database after 1990
CAW (1989+)	All "strippers" flagged on entry	1,033	16.2	Records for new survey entrants entered onto database after 1990
	Other entrants flagged as under ALR rules	191	28.2	Records for new survey entrants, or 2nd exam., entered onto database after 1990

Table 9.17 Trace rates for asbestos workers

Form	AS2		ALR		MS75		Total	
	No.	%[1]	No.	%	No.	%	No.	%
Traced	42,549	98.7	9,926	99.0	3,036	97.1	55,511	98.7
Not traced	551		105		92		748	
Pending	59		676		408		1,143	
Not submitted	141[2]		9,208		8,016		17,365	
Total	43,300		19,915		11,552		74,767	

AS2 1969 Asbestos Regulations
ALR Asbestos (Licensing) Regulations 1983
MS75 Control of Asbestos at Work (CAW) Regulations 1987
1 Trace rate excludes cases pending or not submitted.
2 Not submitted due to insufficient information.

Table 9.18 ICD codes for cause of death groups

Cause group	8th revision 1970-1978	9th revision 1979-1991
All causes	1 -999.0	1.0-E999
All Malignant neoplasms	140.0-209	140.0-208.9
MN of lip, oral cavity and pharynx	140.0-149	140.0-149.9
MN of oesophagus	150	150.0-150.9
MN of stomach	151.0-151.9	151.0-151.9
MN of colon	153.0-153.9	153.0-153.9
MN of rectum	154.0-154.2	154.0-154.8
MN of liver (primary)	155.0-155.1	155.0-155.1
MN of larynx	161.0-161.9	161.0-161.9
MN of trachea, bronchus and lung	162.0-162.1	162.0-162.9
MN of ovary (female)	183.0-183.9	183.0-183.9
MN of bladder	188	188.0-188.9
MN of kidney	189.0-189.9	189.0-189.9
MN of lymphatic and haematopoietic tissue	200.0-209	200.0-208.9
Circulatory disease	390 -458.9	390 -459.9
Ischaemic heart disease	410 -414	410 -414.9
Respiratory disease	460 -519.9	460 -519.9

MN Malignant neoplasm

Table 9.19 Underlying causes of death for mesothelioma cases

8th revision	9th revision	Underlying cause	No. of cases
158.9	158.8, 158.9	MN of peritoneum	4
	162.9	MN of bronchus & lung, unspecified	5
163.0	163.9	MN of pleura	75
	164.1	MN of heart	1
171.1		MN of trunk	1
	195.2	MN of abdomen	4
199.1	199.0, 199.1	MN, site unspecified	42
	229.9	Benign neoplasm, site unspecified	1
	414.9	Ischaemic heart disease, unspecified	1
	496	Chronic airways obstruction	1
515.2	501	Asbestosis	3
		Total	183

9.5.3 Coding of death certificates

Death certificates were coded by OPCS to the underlying cause of death in accordance with the ICD revision applicable at the time of death. The eighth and ninth revisions cover the period of the study. A list of the ICD code ranges for the cause of death groups used in the mortality analyses is in Table 9.18.

All death certificates with a mention of mesothelioma have been given an additional code, and treated as having mesothelioma as the underlying cause of death. Death certificates coded to cancer of the peritoneum or pleura, neoplasm at unspecified sites, or with an ICD external cause of death code indicating exposure to asbestos, were also examined to identify any additional deaths which were very likely to have been mesotheliomas without this term being mentioned specifically. Four were found; their underlying causes of death were recorded as pleural adenocarcinoma (2), carcinomatosis peritonei, and metastatic abdominal carcinoma with unknown primary carcinoma (query mesothelioma) and previous asbestos exposure recorded. Original underlying cause of death codes for the mesotheliomas are given in Table 9.19.

Asbestosis in this study is counted where it has been mentioned anywhere on the death certificate, and in the absence of mesothelioma as defined above or lung cancer it is counted as underlying cause of death. Original underlying cause of death codes are given in Table 9.20.

9.5.4 Classification by job into industrial sector

Job codes recorded under the 1969 Asbestos Regulations were grouped into ten industrial sectors, one of which was 'insulation workers'. From 1 January 1986 the installation of asbestos insulation, along with asbestos spraying and the importation of crocidolite and amosite and their supply and use, was prohibited under the Asbestos (Prohibitions) Regulations 1985.[8] Under the CAW Regulations, an additional exposure categorisation was introduced, so that all exposed jobs were allocated to one of three 'exposure' sectors, namely 'manufacturing', comprising seven industrial sectors with

Table 9.20 Underlying causes of death for asbestosis cases

8th revision	9th revision	Underlying cause	No. of cases
	011.9	Pulmonary tuberculosis, unspecified	1
	151.9	MN of stomach, unspecified	1
	188.9	MN of bladder, unspecified	1
	195.1	MN of thorax	1
	195.2	MN of abdomen	1
	199.0 199.1	MN site unspecified	7
	239.1	Unspecified neoplasm of respiratory system	1
	255.4	Corticoadrenal overactivity	1
412.3	410 414.0 414.9	Ischaemic heart disease	11
	416.1 424.1 431 432.1	Circulatory disease	4
	466.0 485 491.2 496.0	Respiratory disease	11
515.2	501	Asbestosis	47
	505	Pneumoconiosis	1
	531.5	Gastric ulcer	2
		Total asbestoses (without mesothelioma or lung cancer)	**90**
		Mesothelioma, various sites	59
162.0 - 162.1	162.0 - 162.9	Lung cancer, excluding mesotheliomas	119
		Total asbestoses (with mesothelioma or lung cancer)	**178**

their job lists from the previous scheme, 'asbestos stripping/ removal', and 'other' activities, comprising the two remaining industrial sectors plus a group of 'miscellaneous process' jobs. A list of industrial sectors and their associated job codes is in Annex 9.2.

No job code was recorded for the ALR examinations. All subjects entering under these regulations have been allocated to the 'stripping' sector.

Study subjects were classified to the industrial sector for which their duration of employment was longest — estimated by the period over which they had the most recorded examinations plus any period of previous exposure — for the analyses by industrial sector presented here. 97.6 per cent of the study subjects were first examined in the sector in which they remained for the longest duration during follow-up.

9.5.5 Calculation of expected deaths

The date of start of person years at risk is determined by criteria used for 'flagging' the different groups in the survey at NHSCR. Person years at risk have been calculated from the date of first medical examination for all workers with one of the following:

- a date of first examination prior to 1984, or
- a history of exposure before entry into the study, or
- was a 'stripper' on entering under the CAW Regulations.

For the remainder, person years at risk is calculated from the date of the second examination, before which these survey members had not qualified for 'flagging' and therefore could not have been considered 'at risk' of having their death notified. The ALRs however who were 'flagged' early in the ALR regime, under rules that had been previously applied and who therefore had only a single examination, are considered 'at risk' from the date of this examination.

SMRs have been calculated in the usual way, for causes of death other than mesothelioma and asbestosis. Expected deaths were calculated by applying the appropriate age-time period-cause specific death rates for males and females and for England and Wales and for Scotland to the person years at risk in each cell. Mesothelioma and asbestosis are specific to asbestos exposure, and are not counted in national death registrations in a manner comparable to their definition in this study. The numbers of deaths from these causes have therefore been represented as a percentage of expected deaths from all causes (and are described as per cent excess deaths).

In the mortality tables, deaths re-attributed to mesothelioma as underlying cause have been excluded from the counts of observed deaths due to other specific causes. The numbers recoded in each category are given in Table 9.26. All but six of the mesotheliomas are included in the count of 'all malignant neoplasms'.

In cases of asbestosis where mesothelioma is also present or lung cancer has been recorded as the underlying cause of death, cases have been excluded from the numbers and percentages given for asbestosis in the tables, to avoid double counting of known asbestos related disease. The additional cases are noted in parentheses. Cases with asbestosis mentioned in conjunction with an underlying cause of death other than mesothelioma or lung cancer are counted in the SMR estimates relating to the other underlying cause of death, as listed in Table 9.19.

9.6 Results of the mortality analysis

9.6.1 Overall features

Tables 9.21 — 9.24 show numbers of workers included in the study along with their person years at risk, broken down by country of residence (England and Wales or Scotland) and sex, by industrial sector, by period of first exposure (with respect to the introduction of the various regulations as well as by decade) and by smoking status and age.

Table 9.21 Number in study and person years at risk (PYAR) by country and sex[1]

	Males		Females		Total	
	Number	PYAR	Number	PYAR	Number	PYAR
England and Wales	48,488	564,168	3,551	50,084	52,039	614,252
Scotland	3,426	34,184	41	588	3,467	34,773
Great Britain	51,914	598,352	3,592	50,672	55,506[2]	649,024

[1] Totals may not add up due to rounding.
[2] 5 workers were excluded from analysis due to insufficient data for calculating PYAR

Table 9.22 Number and PYAR by industrial sector[1]

	All workers	
	Number	PYAR
Manufacturing		
Textile manufacture	4,780	69,659
Asbestos, cement, mixture board & pipe manufacture	5,672	86,108
Manufacture of asbestos/rubber/ resin/bitumen mixture	12,030	155,384
Asbestos board & paper manufacture	2,103	28,723
Garment manufacture	340	4,310
Manufacture of dry mixes for insulation & 'plastering'	289	3,978
Maintenance workers all industries	5,211	69,787
Total for manufacturing sector	30,425	417,950
Asbestos stripping	12,448	65,508
Other		
Ship building repair and breaking	3,100	44,869
Building & construction	1,220	16,112
Miscellaneous processes	19	62
Total for other sector	4,339	61,042
Insulation workers ('69 Regs only)	7,800	102,944
Missing	494	1,580
Total	**55,506[2]**	**649,024**

[1] Totals may not add up due to rounding.
[2] 5 workers were excluded from analysis due to insufficient data for calculating PYAR.

Table 9.23 Number and PYAR by year of first exposure[1]

Year of first exposure	All workers	
	Number	PYAR
< 1930	120	1,406
1930 - 1939	1,142	15,289
1940 - 1949	2,757	36,559
1950 - 1959	5,720	74,278
1960 - 1969	11,682	156,956
Total < 1970	**21,421**	**284,488**
1970 - 1979	23,112	302,732
1980 - 1983	4,698	35,328
1984 - 1991	6,275	26,476
Total 1970 +	**34,085**	**364,536**
Total	**55,506[2]**	**649,024**

[1] Totals may not add up due to rounding.
[2] 5 workers were excluded from analysis due to insufficient data for calculating PYAR.

Table 9.24 Person years at risk by smoking status and age[1]

Age	Smokers:		Ex-smokers:		Non-smokers:		Total[2]
	Number in study:	%	Number in study:	%	Number in study:	%	Number in study:
	29,087	54.3%	12,273	22.9%	12,243	22.8%	5,506
	PYAR	%	PYAR	%	PYAR	%	PYAR
15-19	2,236	0.7	459	0.3	1,876	1.4	4,715
20-24	17,253	5.1	3,870	2.7	12,344	8.9	35,017
25-29	34,151	10.0	8,700	6.0	21,076	15.3	67,002
30-34	42,516	12.5	12,360	8.5	22,324	16.2	81,023
35-39	43,980	12.9	14,861	10.3	18,922	13.7	81,191
40-44	41,474	12.2	16,434	11.3	15,479	11.2	76,144
45-49	37,448	11.0	17,169	11.8	12,848	9.3	69,864
50-54	35,297	10.3	18,272	12.6	11,279	8.2	67,081
55-59	32,407	9.5	18,735	12.9	8,912	6.4	61,961
60-64	35,063	10.3	21,903	15.1	8,490	6.1	67,500
65-69	9,010	2.6	5,702	3.9	2,099	1.5	17,409
70-74	7,537	2.2	4,779	3.3	1,822	1.3	14,725
75-79	2,385	0.7	1,466	1.0	619	0.4	4,715
80-84	330	0.1	174	0.1	94	0.1	637
85+	18	0.0	16	0.0	7	0.0	47
All ages	341,107		144,898		138,192		648,977

[1] Totals may not add up due to rounding.
[2] Total column includes workers with missing smoking status

The SMRs for all causes and for all malignant neoplasms across the period of follow-up, for England and Wales and for Scotland, are given in Table 9.25. A clear healthy worker effect is operating only in the first decade of follow-up, in cancer as well as in non-cancer mortality, but has disappeared from 1981.

Known asbestos related diseases — mesothelioma and asbestosis — have been recorded on 7.4 per cent of death certificates in the cohort, with the 183 mesotheliomas accounting for 3.6 per cent of deaths, and asbestosis being recorded on an additional 209 (3.9 per cent) including 119 in association with lung cancer but not mesothelioma. Cancer deaths are significantly in excess overall (Table 9.26), with lung, stomach, bladder and liver cancers making significant contributions in different groups according to first exposure recorded as before or from the implementation of the 1969 Asbestos Regulations (Tables 9.27 and 9.28). In this cohort of workers as a whole, there has been an excess of about 380 cancers over the 21 years of the survey, with about 200 excess lung cancers along with the 183 mesotheliomas.

In contrast to the low SMRs for other causes amongst Scottish males, the SMR for all malignant neoplasms in this group was significantly and substantially raised (149.6). Lung cancer is the major contributor to this excess, represented by SMRs (based on Scottish rates) substantially but not significantly higher than for England and Wales for first exposures both before (169.9 versus 141.1 for England and Wales) and from (183.0 versus 119.5 respectively) 1970. SMRs for lung cancer for England and Wales and for Scotland for first exposure before and from 1970 are given in Table 9.29.

Data for England and Wales and for Scotland and that for males and females are considered together in the analyses which follow since no particular pattern of cancer or asbestosis incidence in these sub-groups of the cohort made a significant contribution to explaining the variation in the data.

9.6.2 Reliability of recorded date of first exposure

It is possible that exposure previous to a first examination, and therefore in many cases before 1970, may not have been reported accurately by all study subjects. As latencies of less than 10 years are unlikely for many cancers, including the major contributors being considered here, mesothelioma and lung cancer, and for asbestosis, deaths from these causes within ten years of first recorded exposure to asbestos are unlikely to be attributable to that exposure. Certainly in the cases of mesothelioma and asbestosis, an earlier exposure to asbestos must be suspected.

For the group of workers recorded as being first exposed after 1969, SMRs and estimates of excess deaths in what follow are based on mortality of workers with at least 10 years lapse from recorded first exposure.

9.6.3 Mortality from first exposure up to and after 1969

For workers first exposed to asbestos before the implementation of the 1969 Regulations (in 1970), the overall SMR for all malignant neoplasms is 130 (Table 9.27), and for workers first exposed only from 1970, the SMR for all malignant neoplasms in those with at least 10 years lapse from first exposure (latency) is 115 (Table 9.30). This reduction is statistically significant. While the majority of cancers are attributable to first exposures prior to 1970, an excess of about 40 cancers — including 8 mesotheliomas and 23 lung cancers — is associated with recorded first exposure after 1969 (Table 9.30). The introduction of the 1969 Asbestos Regulations has therefore reduced, but appears not to have eliminated, the excess risk for cancer associated with exposure to asbestos at work.

The lung cancer excesses make the largest contributions to the all cancer SMRs for first exposure before and from 1970, but there are also excesses of stomach cancer in particular, and of laryngeal and bladder cancers, in those recorded as

Text continues on page 144

Table 9.25 SMRs by time of death for all causes of death and for all malignant neoplasms

		Time of death				
		1971-75	1976-80	1981-85	1986-91	Total
England and Wales						
All causes	OBS	157	845	1,525	2,556	5,083
	SMR	79.7	95.1	108.2	107.3	104.2
	95%CI	(67.7-93.2)	(88.7-101.8)	(102.8-113.8)	(103.2-111.6)	(101.3-107.1)
All malignant neoplasms	OBS	44	304	547	917	1,812
	SMR	79.0	119.5	130.6	124.6	123.7
	95%CI	(57.4-106.0)	(106.2-133.9)	(119.8-142.1)	(116.6-133.0)	(118.0-129.6)
Scotland						
All causes	OBS	2	30	74	138	244
	SMR	40.0	66.1	99.4	91.6	88.5
	95%CI	(4.8-144.5)	(44.6-94.4)	(78.0-124.8)	(77.0-108.2)	(77.7-100.3)
All malignant neoplasms	OBS	0	15	32	62	109
	SMR	0.0	133.8	162.9	147.8	147.1
	95%CI	(0.0-230.4)	(74.9-220.7)	(111.4-230.0)	(113.3-189.5)	(120.0-177.4)

95% CI – 95% confidence interval for SMR.

Table 9.26 Mortality of all workers

(a) England and Wales, and Scotland - all workers

	Deaths	SMR	95% CI	MESO
All causes	5,327	103.4*	(101.2 -106.8)	
All malignant neoplasms	1,921	124.8**	(119.3 -130.5)	
MN of lip, oral cavity and pharynx	20	75.9	(46.4 -117.3)	
MN of oesophagus	46	76.9	(56.3 -102.6)	
MN of stomach	144	123.1*	(103.8 -144.9)	
MN of colon	88	88.8	(71.2 -109.4)	
MN of rectum	62	94.1	(72.1 -120.6)	
MN of liver (primary)	20	135.0	(82.5 -208.6)	
MN of larynx	14	96.1	(52.5 -161.2)	
MN of trachea, bronchus & lung	749	136.2**	(126.5 -146.4)	5
MN of ovary (female)	4	48.3	(13.1 -123.5)	
MN of bladder	65	129.6	(100.0 -165.1)	
MN of kidney	33	97.7	(67.2 -137.2)	
MN of lymphatic and haematopoietic tissue	101	94.0	(76.6 -114.2)	
Circulatory disease	2,344	96.7	(92.8 -100.7)	1
Ischaemic heart disease	1,721	98.5	(93.9 -103.3)	1
Respiratory disease	422	101.0	(91.5 -111.3)	4
	Deaths	%		
Mesothelioma	183	3.55	(3.07 -4.13)	
	Deaths	%		
Asbestosis	90 (178)	1.74	(1.41 -1.96)	

(b) England and Wales - Males

	Deaths	SMR	95% CI	MESO
All causes	4,806	103.9*	(101.0 -130.3)	
All malignant neoplasms	1,694	124.3**	(118.4 -130.3)	
MN of lip, oral cavity and pharynx	15	63.2	(35.4 -104.3)	
MN of oesophagus	45	83.2	(60.7 -111.3)	
MN of stomach	135	125.4*	(105.2 -148.5)	
MN of colon	76	86.9	(68.5 -108.8)	
MN of rectum	58	96.9	(73.6 -125.2)	
MN of liver (primary)	16	119.7	(68.4 -194.3)	
MN of larynx	13	96.2	(51.2 -164.5)	
MN of trachea, bronchus & lung	679	134.8**	(124.7 -145.4)	3
MN of ovary (female)	-	-	-	
MN of bladder	61	131.2*	(100.4 -168.5)	
MN of kidney	30	98.0	(66.1 -139.9)	
MN of lymphatic and haematopoietic tissue	86	88.8	(71.1 -109.7)	
Circulatory disease	2,155	97.7	(93.6 -102.0)	1
Ischaemic heart disease	1,602	100.1	(95.2 -105.2)	1
Respiratory disease	385	101.7	(91.6 -112.5)	4
	Deaths	%		
Mesothelioma	162	3.50	(2.98 -4.09)	
	Deaths	%		
Asbestosis	81 (166)	1.75	(1.39 -2.18)	

(c) England & Wales - Females

	OBS	SMR	95%CI	MESO
All causes	277	110.1	(97.3 -124.1)	
All malignant neoplasms	118	115.9	(96.0 -138.8)	
MN of lip, oral cavity and pharynx	1	101.5	(2.6 -562.8)	
MN of oesophagus	0	-	(0.0 -140.6)	
MN of stomach	3	75.2	(15.5 -219.7)	
MN of colon	9	124.8	(57.1 -237.0)	
MN of rectum	2	63.1	(7.6 -227.9)	
MN of liver (primary)	3	506.5*	(104.9 -1486.0)	
MN of larynx	0	0.0	(0.0 -998.6)	
MN of trachea, bronchus & lung	20	122.7	(72.6 -183.5)	1
MN of ovary (female)	4	48.8	(13.3 -124.9)	
MN of bladder	3	215.5	(44.5 -630.7)	
MN of kidney	1	73.1	(1.9 -406.7)	
MN of lymphatic and haematopoietic tissue	9	152.3	(69.6 -289.1)	
Circulatory disease	112	123.3	(101.5 -148.4)	
Ischaemic heart disease	58	115.1	(87.4 -148.8)	
Respiratory disease	18	99.1	(58.7 -156.7)	
	Deaths	%		
Mesothelioma	8	3.18	(1.37 -6.26)	
	Deaths	%		
Asbestosis	4 (0)	1.59	(0.43 -4.07)	

(d) Scotland - Males

	Deaths	SMR	95% CI	MESO
All causes	240	88.1	(77.4 -100.1)	
All malignant neoplasms	109	149.6**	(123.0 -180.7)	
MN of lip, oral cavity and pharynx	4	248.8	(67.7 -636.1)	
MN of oesophagus	1	27.9	(0.7 -156.1)	
MN of stomach	6	111.9	(41.2 -244.1)	
MN of colon	3	69.0	(14.2 -201.5)	
MN of rectum	2	70.7	(8.6 -255.3)	
MN of liver (primary)	1	118.5	(3.0 -663.3)	
MN of larynx	1	130.6	(3.3 -733.1)	
MN of trachea, bronchus & lung	50	172.5**	(128.1 -227.5)	1
MN of ovary (female)	-	-	-	
MN of bladder	1	44.0	(1.1 -245.4)	
MN of kidney	2	112.7	(13.7 -408.2)	
MN of lymphatic and haematopoietic tissue	6	129.5	(47.6 -282.1)	
Circulatory disease	74	57.9**	(45.5 -72.7)	
Ischaemic heart disease	58	60.9**	(46.3 -78.8)	
Respiratory disease	18	87.4	(51.8 -138.2)	
	Deaths	%		
Mesothelioma	13	4.78	(2.54 -8.17)	
	Deaths	%		
Asbestosis	5 (12)	1.84	(0.60 -4.29)	

(e) Scotland - Females

	OBS	SMR	95%CI	MESO
All causes	4	108.4	(29.5 -277.6)	
All malignant neoplasms	0	1.3	(0.0 -223.6)	
MN of lip, oral cavity and pharynx	0	0.0	(0.0 -29957.3)	
MN of oesophagus	0	0.0	(0.0 -7489.3)	
MN of stomach	0	0.1	(0.0 -4992.9)	
MN of colon	0	0.1	(0.0 -3328.6)	
MN of rectum	0	0.0	(0.0 -7489.3)	
MN of liver (primary)	0	0.0	(0.0 -29957.3)	
MN of larynx	0	0.0	-	
MN of trachea, bronchus & lung	0	0.3	(0.0 -1033.0)	
MN of ovary (female)	0	0.1	(0.0 -3328.6)	
MN of bladder	0	0.0	(0.0 -14978.7)	
MN of kidney	0	0.0	(0.0 -14978.7)	
MN of lymphatic and haematopoietic tissue	0	0.1	(0.0 -4279.6)	
Circulatory disease	3	202.7	(41.8 -592.4)	
Ischaemic heart disease	3	340.9	(70.3 -996.3)	
Respiratory disease	1	344.8	(8.7 -1921.2)	
	Deaths	%		
Mesothelioma	0	0.00	(0.00 -100.00)	
	Deaths	%		
Asbestosis	0	0.00	(0.00 -100.00)	

95% CI = conf. int. for SMR MESO = recodes to mesothelioma ** Significant at 1% level * Significant at 5% level

Table 9.27 All workers first exposed prior to 1970

	Deaths	SMR	95% CI	MESO
All causes	3,711	105.7**	(102.2-109.1)	
All malignant neoplasms	1,385	130.0**	(123.2-137.1)	
MN of lip, oral cavity and pharynx	16	91.6	(52.3-148.7)	
MN of oesophagus	31	75.2	(51.1-106.7)	
MN of stomach	96	115.2	(93.3-140.7)	
MN of colon	53	77.1	(57.8-100.9)	
MN of rectum	47	102.3	(75.1-136.0)	
MN of liver (primary)	17	171.7*	(100.0-274.9)	
MN of larynx	7	69.1	(27.8-142.2)	
MN of trachea, bronchus & lung	555	141.3**	(129.7-153.7)	5
MN of ovary (female)	2	37.5	(4.5-135.3)	
MN of bladder	44	121.3	(88.1-162.8)	
MN of kidney	24	105.5	(67.6-157.0)	
MN of lymphatic and haematopoietic tissue	62	91.6	(70.2-117.4)	
Circulatory disease	1,663	97.2	(92.6-102.1)	1
Ischaemic heart disease	1,208	98.4	(92.9-104.2)	1
Respiratory disease	318	105.5	(94.0-117.9)	4
	Deaths	%		
Mesothelioma	162	4.61	(3.93-5.38)	
	Deaths	%		
Asbestosis	85 (170)	2.42	(1.93-2.99)	

95% CI = 95% CI for SMR

** Significant at 1% level
* Significant at 5% level

Table 9.28 All workers first exposed 1970 or later

	Deaths	SMR	95% CI	MESO
All causes	1,615	98.5	(93.7-103.5)	
All malignant neoplasms	536	113.1**	(103.6-123.2)	
MN of lip, oral cavity and pharynx	4	45.1	(12.3-115.5)	
MN of oesophagus	15	80.8	(45.2-133.2)	
MN of stomach	48	142.6*	(105.1-189.0)	
MN of colon	35	115.3	(80.3-160.3)	
MN of rectum	15	75.2	(42.1-124.0)	
MN of liver (primary)	3	61.0	(12.6-178.2)	
MN of larynx	7	157.3	(63.2-324.1)	
MN of trachea, bronchus & lung	194	123.5**	(106.7-142.1)	
MN of ovary (female)	2	67.8	(8.2-244.9)	
MN of bladder	21	151.1	(93.5-230.9)	
MN of kidney	9	81.6	(37.3-154.8)	
MN of lymphatic and haematopoietic tissue	39	98.3	(69.9-134.3)	
Circulatory disease	681	95.3	(88.2-102.8)	
Ischaemic heart disease	513	98.8	(90.4-107.8)	
Respiratory disease	104	89.5	(73.1-108.5)	
Unknown cause of death	1			
	Deaths	%		
Mesothelioma	21	1.28	(0.79-1.96)	
	Deaths	%		
Asbestosis	5 (8)	0.30	(0.10-0.71)	

95% CI = 95% CI for SMR

** Significant at 1% level
* Significant at 5% level

Table 9.29 SMRs for lung cancer (ICD 162) for England and Wales, Scotland

First exposed	before 1970:		from 1970:	
	SMR	95% CI	SMR	95% CI
England and Wales	141.1	(129.4-153.7)	119.5	(102.5-138.5)
Scotland	169.9	(116.9-238.6)	183.0	(108.4-289.2)

Table 9.30 Mortality of all workers by latency and year of first exposure

Latency:	<10 years			10 years or more		
	Deaths	SMR	95% CI	Deaths	SMR	95% CI
Year of first exposure: < 1970						
All causes	58	82.5	(62.6-106.7)	3,653	106.1**	(102.7-109.6)
All malignant neoplasms	19	96.9	(58.4-151.4)	1,366	130.7**	(123.8-137.8)
MN of stomach	3	173.4	(35.8-506.8)	93	114.0	(92.0-139.6)
MN of liver	0	0.0	(0.0-2139.8)	17	174.2*	(101.5-278.9)
MN of larynx	0	0.0	(0.0-1997.2)	7	70.2	(28.2-144.7)
MN of trachea, bronchus & lung	4	53.1	(14.5-135.8)	551	143.0**	(132.4-157.0)
MN of bladder	1	185.2	(4.7-1031.8)	43	120.3	(131.2-155.6)
	Deaths	%		Deaths	%	
Mesothelioma	0	0.00	(0.00-5.25)	162	4.71	(4.01-5.49)
	Deaths	%		Deaths	%	
Asbestosis	0	0.00	(0.00-5.25)	85 (170)	2.47	(1.97-3.05)
Year of first exposure: 1970+						
All causes	692	93.0	(86.2-100.3)	923	103.1	(96.5-110.0)
All malignant neoplasms	225	110.9	(96.9-126.3)	311	114.7*	(102.1-128.4)
MN of stomach	15	100.1	(56.0-165.0)	33	176.4**	(121.4-247.7)
MN of liver	1	49.8	(1.3-277.2)	2	68.3	(8.3-246.6)
MN of larynx	6	324.3	(119.0-705.9)	1	38.5	(1.0-214.3)
MN of trachea, bronchus & lung	82	120.0	(95.4-149.0)	112	126.1*	(103.9-151.8)
MN of bladder	8	149.0	(64.3-293.5)	13	152.4	(81.1-260.6)
	Deaths	%		Deaths	%	
Mesothelioma	13	1.75	(0.93-2.99)	8	0.89	(0.39-1.76)
	Deaths	%		Deaths	%	
Asbestosis	2 (4)	0.27	(0.03-0.97)	3 (4)	0.33	(0.07-0.98)

** Significant at 1% level
* Significant at 5% level

first exposed from 1970. See Tables 9.27 and 9.28 for mortality from the full range of causes investigated for workers first exposed before and from 1970.

Cancer of the stomach is significantly in excess in workers first exposed after 1969 and with more than ten years recorded latency, and is appearing in 40-59 year olds in this group. Numbers are relatively low, with an overall excess of 14 recorded in those exposed after 1969. Interestingly, what may be considered a baseline rate for stomach cancer in this population, the SMR at latency of less than ten years is 100 for workers exposed after 1969, adding some weight to the interpretation that a genuine excess risk may have been observed here.

There was no excess of laryngeal cancer in those with reported first exposure prior to 1970. The excess in those recorded as first exposed from 1970 was confined to workers with less than ten years from reported first exposure (Table 9.30) and cannot therefore be considered as linked to asbestos exposure from 1970.

The excess of about four bladder cancers associated with first recorded exposure after 1969 and more than ten years latency (Table 9.30) was not significant itself. There was also a small excess for under ten years latency in this group. The presence of excess bladder cancers associated with first exposure in 1960-69, and therefore short latency, adds weight to the relevance of this observation, particularly as those affected from 1970 include younger age workers.

Other factors must be taken into account however in the interpretation of this observation and of the excess of stomach cancer noted above. There is little reason to believe that excesses not apparent in workers exposed before the 1969 Regulations came into force are likely to be associated with asbestos exposure experienced only after 1970, as exposure levels have in general fallen. As always in a situation of multiple testing some apparently 'significant' excesses or deficits can be expected to occur by chance, although the list of causes of death chosen for this study (Table 9.18) was based for the cancers on sites of *a priori* interest.

An excess, based on low numbers, of about seven primary liver cancers in the cohort is confined to workers first exposed prior to 1970, and is related to first exposures in 1950-69 and mainly to those aged over sixty at death.

9.7 Analysis by industrial sector

Annex 9.3 shows observed and expected deaths and SMRs for lung cancer and per cent excess deaths for mesothelioma and asbestosis for each industry sector covered in the survey, for workers with reported first exposure prior to 1970 and from 1970 only.

Insulation workers have suffered the highest raised risk for cancer (SMR 197.2, 394 deaths), and particularly for mesothelioma and lung cancer, although the excess risk for lung cancer and for asbestosis which was also apparent in this sector has been significantly reduced among workers exposed from 1970. The short latencies for the five mesotheliomas reported in those with first exposure post 1970, and the fact that all five were aged over 40 when recorded as first exposed, suggests the possibility of earlier unrecorded exposure.

For workers in textile manufacture (sector 1), SMRs for cancer, including lung cancer, have been lowered significantly for those exposed only after 1970. The same cannot be said however for those in the manufacture of asbestos, rubber, resin and bitumen mixtures (sector 3) who had significantly raised risk, taken across all periods of first exposure, for all malignant neoplasms and in particular for lung cancer.

Maintenance (sector 7) and construction workers (sector 10) exhibit higher ratios for lung cancer in workers reportedly exposed only after 1970 than those exposed before, but latencies are short so that misclassification of date of first exposures cannot be ruled out (see Annex 9.3). This is also likely to be the case for the six post 1969 mesotheliomas in maintenance workers. The five mesotheliomas in shipbuilding associated with first exposures after 1969 were in men aged 50-69. Again unreported earlier exposure before 1969 cannot be ruled out.

It is too early to draw any firm conclusions as regards the mortality of those brought into the survey through the Asbestos (Licensing) Regulations. With only eight years of follow-up a healthy worker effect is still apparent in the all cause SMR of 75.8 for the asbestos strippers, who entered the survey under these and the CAW Regulations, and the SMR for all malignant neoplasms does not differ significantly between those with reported first exposure either before 1970 (Annex 9.3 — the SMR is 114.5) or before or after the introduction of the 1983 Regulations (with SMRs of 101.6 based on 56 deaths and 102.5 based on 9 deaths respectively).

9.8 Analysis by date of first exposure, latency and age

Period of first exposure, latency and age at death are naturally highly correlated in this dataset, with one another and with duration of exposure also. The greatest excess mortality for mesothelioma in the cohort is at 40-49 years after first exposure, largely relating to first exposures in 1930-39, as shown in Figures 9.12 and 9.15. Figure 9.12 has deaths and person years at risk for the first ten years of latency excluded.

Lung cancer deaths in this cohort appear to be occurring across a broad range of latent intervals from ten years after first exposure upwards (Figure 9.14), and with excesses associated with first exposure from all periods covered (Figure 9.11).

For asbestosis rates rise steadily with increasing time from first exposure and, during this period of follow-up, relate particularly to exposures before 1959 (see Figures 9.13 and 9.16).

Figure 9.11 Lung cancer SMRs by date of first exposure

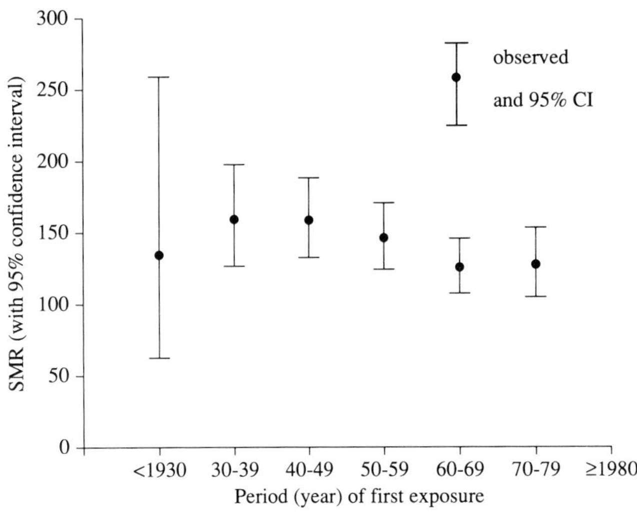

Data relating to latencies of <10 years are excluded.

Figure 9.12 Per cent excess deaths - mesothelioma by date of first exposure

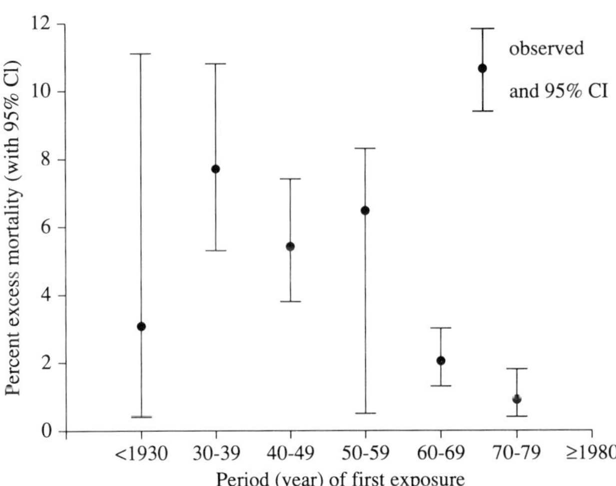

Deaths are expressed as a percentage of all cause expected mortality.
Data relating to latencies of <10 years are excluded.

Figure 9.13 Per cent excess deaths - asbestosis by date of first exposure

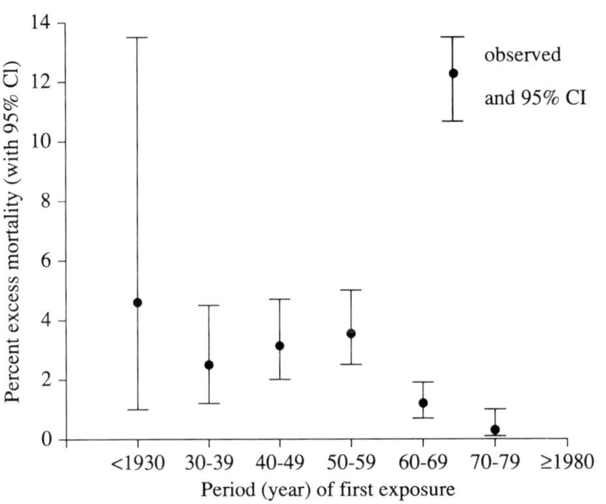

Deaths are expressed as a percentage of all cause expected mortality.
Asbestoses are counted without mesothelioma or lung cancer only.

Figure 9.14 Lung cancer SMRs by time from first exposure

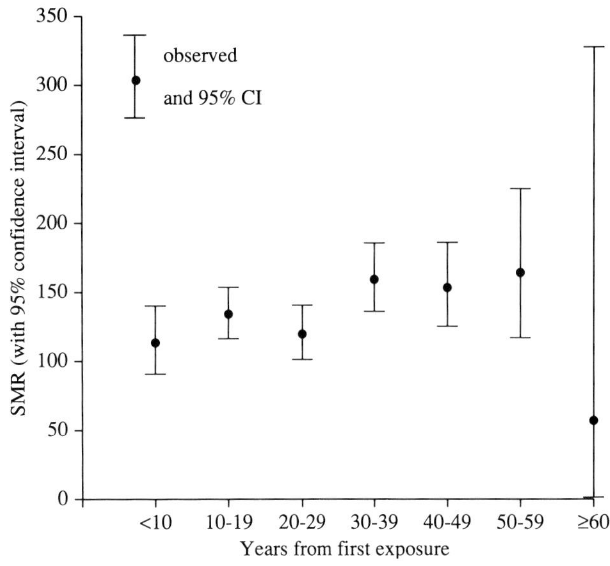

Figure 9.15 Per cent excess deaths - mesothelioma by time from first exposure

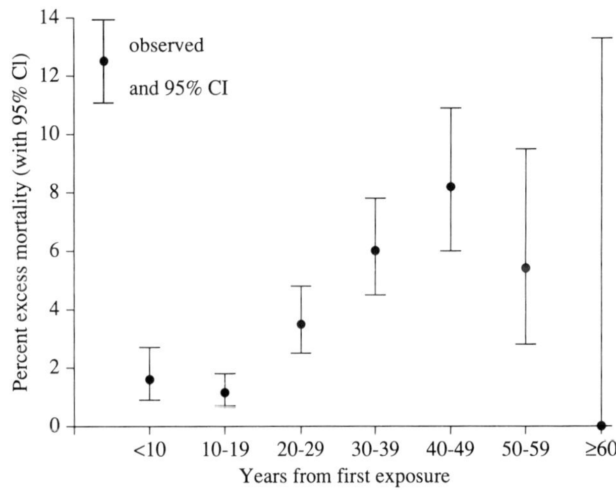

Deaths are expressed as a percentage of all cause expected mortality.

Figure 9.16 Per cent excess deaths - asbestosis by time from first exposure

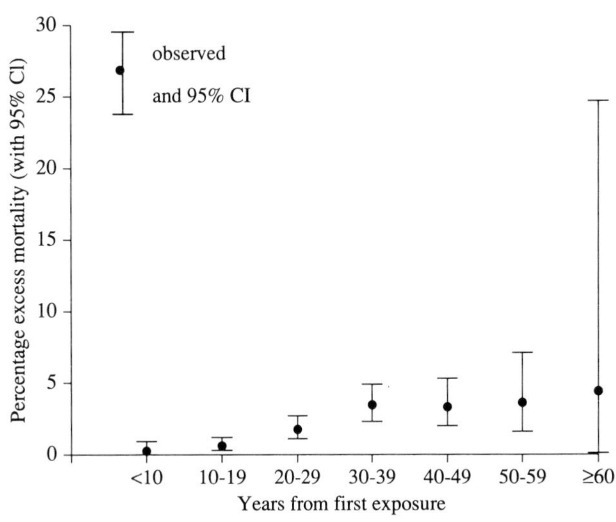

Deaths are expressed as percentage of all cause expected mortality.

9.9 Effect of smoking

Smoking status at last recorded examination for all workers in the study is shown in Table 9.31. The proportion of current smokers (54 per cent, see Table 9.24) in this cohort is higher than the national average (42 per cent, or 49 per cent for manual workers, averaged across the period of follow-up).[9]

As might be expected, those recorded at last examination in the survey as current smokers suffer the highest mortality, particularly from lung cancer, at three times the rate for ex-smokers and twenty times the rate for non-smokers. Rates of lung cancer in non-smokers are very low, although based on small numbers in this cohort, and are similar to rates which have been estimated across this period for non-smokers nationally in Great Britain and in the US.[10,11] The age profile of this cohort however is young. Although the SMR for smokers is not greatly elevated in comparison to overall SMRs for smokers estimated from national data, when broken down by age-group lung cancers are clearly in excess of national expectations for smokers in the age range covered by this cohort (see Figure 9.17).

Non-smokers have suffered, however, substantially higher rates of mortality from mesothelioma and asbestosis without lung cancer. This effect is confined to non-smokers first exposed before 1970 — 56 mesotheliomas and 26 asbestoses in this group accounted for an excess 11.75 per cent and 5.5 per cent of deaths (taken as a proportion of expected deaths from all causes in the general (smoking and non-smoking) population respectively.

References

1 Doll R. Mineral fibres in the non-occupational environment; concluding remarks. In: Bignon J, Peto J, Saracci R (eds). *Non-occupational exposures to mineral fibres.* Lyon: International Agency for Research on Cancer, 1989: 511-18 (IARC Scientific Publications No 90).

2 Newhouse ML, Thompson H. Mesothelioma of the pleura and peritoneum following exposure to asbestos in the London area. *Br Ind Med* 1965;22:158-164.

3 Wagner JC, Sleggs CA, Marchand P. Diffuse pleural mesothelioma and asbestos exposure in the North Western Cape Province. *Br J Ind Med* 1960;17:260-271.

4 Ministry of Labour, HM Factory Inspectorate. *Problems arising from the use of asbestos: Memorandum of the senior medical inspectors advisory panel.* London: Her Majesty's Stationery Office, 1967.

5 Asbestos Regulations 1969 (SI 1969/690). London: HMSO, 1969.

6 The Asbestos (Licensing) Regulations 1983 (SI 1983/1649). London: HMSO, 1983.

7 The Control of Asbestos at Work Regulations 1987 (SI 1987/2115). London: HMSO, 1987.

8 The Asbestos (Prohibitions) Regulations 1985 (SI 1985/910). London: HMSO, 1985.

9 OPCS Monitor, 26 November 1991. *Data from the General Household Survey.*

10 Stevens RG, Moolgavkar SH. A cohort analysis of lung cancer and smoking in British males. *Am J Epidem.* 1984; 119 No.4: 624-641.

11 Peto R, Lopez AD et al. Mortality from tobacco in developed countries: indirect estimation from national vital statistics. *Lancet* May 23 1992; 339: 1278. Data from the ACS CPS-II study.

Figure 9.17 Smokers - observed and expected deaths from lung cancer by age

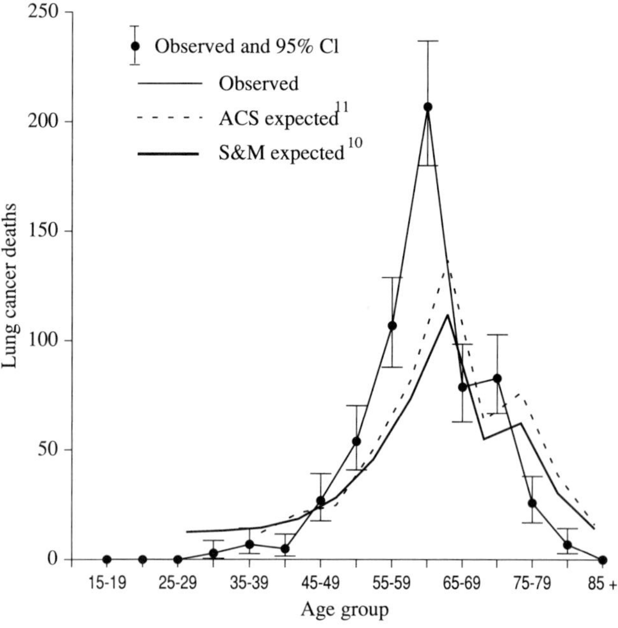

Table 9.31 Mortality of all workers by smoking status

Smoking status	Deaths	SMR	95%CI
Smokers			
All causes	3,484	129.1**	(124.9-133.5)
All malignant neoplasms	1,280	159.3**	(150.6-168.3)
MN of stomach	80	130.1*	(103.2-162.0)
MN of liver	9	117.0	(53.5-222.2)
MN of larynx	11	144.4	(72.1-258.3)
MN of trachea, bronchus & lung	601	288.6**	(191.8-225.8)
MN of bladder	42	160.4**	(115.6-216.8)
	Deaths	%	
Mesothelioma	95	3.52	(2.86-4.32)
	Deaths	%	
Asbestosis	61 (125)	2.26	(1.74-2.91)
Non - smokers			
All causes	450	63.1**	(57.3-69.3)
All malignant neoplasms	120	57.0**	(47.3-68.1)
MN of stomach	4	27.9**	(7.6-71.3)
MN of liver	3	145.6	(30.0-425.6)
MN of larynx	0	0.0	(0.0-160.2)
MN of trachea, bronchus & lung	7	10.3**	(4.1-21.1)
MN of bladder	4	65.3	(17.8-167.1)
	Deaths	%	
Mesothelioma	58	8.13	(6.18-10.51)
	Deaths	%	
Asbestosis	26 (42)	3.65	(2.38-5.34)
Ex - smokers			
All causes	1,216	78.0**	(73.7-82.6)
All malignant neoplasms	458	97.2	(88.4-106.6)
MN of stomach	49	133.6	(98.9-176.7)
MN of liver	8	175.8	(75.9-346.4)
MN of larynx	2	43.5	(5.3-157.1)
MN of trachea, bronchus & lung	118	67.7**	(56.0-81.0)
MN of bladder	17	105.8	(61.6-169.4)
	Deaths	%	
Mesothelioma	19	1.22	(0.73-1.90)
	Deaths	%	
Asbestosis	1 (7)	0.06	(0.00-0.36)

** Significant at 1% level
* Significant at 5% level

Annex 9.1 Background to the Asbestos Survey - Legislative framework

The 1969 Asbestos Regulations

A national survey of asbestos workers was set up at the time of the introduction of the 1969 Asbestos Regulations[5] based on workers employed in factories or workplaces which came within the scope of the Regulations. Recruitment of workplaces into the survey took place from 1971 through to the late 1970s.

All individuals employed at each workplace at its recruitment into the survey, plus workers subsequently employed there, were eligible for inclusion. Enrolled in the survey therefore was a population of workers, some of whom had worked with asbestos before the 1969 Regulations and some of whom had not, plus a prospective cohort population all of whom had worked only after the implementation of the legislation on 14 May 1970. Workers in the survey were seen at two yearly intervals for a voluntary medical examination, by a works medical officer in the larger workplaces or elsewhere by an Employment Medical Adviser — a doctor working for the HSE's Employment Medical Advisory Service. They also completed a questionnaire with details necessary for identification and details of smoking habit, occupational history and duration of exposure to asbestos. Examinations continued for as long as the worker remained in that particular employment.

The Asbestos (Licensing) Regulations (ALR) 1983 [6]

From 1 August 1984 any worker working with asbestos insulation or asbestos coating, in its application or removal, was required to be medically examined before starting this work, and at intervals of not more than two years thereafter, for as long as this work continued.

Everyone who was medically examined under the Asbestos (Licensing) Regulations became part of the survey. Similar data on personal details, date of examination, smoking habit and, if known, current and past exposure details were collected.

Data continued to be collected in its original form for workers in firms that had previously been participating in the survey and were still covered by the 1969 Asbestos Regulations, where these workers were not covered by the 1983 Asbestos (Licensing) Regulations.

The Control of Asbestos at Work (CAW) Regulations 1987[7]

Both the 1969 Asbestos Regulations and the Asbestos (Licensing) Regulations have been superceded by the Control of Asbestos at Work (CAW) Regulations. From 1 March 1988 the requirement for statutory medical examinations in the Asbestos (Licensing) Regulations was replaced by statutory examinations under these new regulations. The CAW regulations require examinations for all those occupationally exposed to asbestos above a certain action level, whatever the work activity. The ALR requirements covered workers engaged in asbestos stripping and coating only. The action level is defined as cumulative exposure to asbestos over a continuous 12-week period above a threshold of:

- 48 fibre-hours/millilitre of air for crocidolite or amosite, or
- 120 fibre-hours/millilitre of air for other asbestos, or
- a proportionate rate where both types of exposure were present.

Action levels are set at approximately one half of the 4-hour control limit level (0.2 fibres/ml air average over the period for crocidolite or amosite, 0.5 for other asbestos), applied cumulatively over the 12-week period, if a 40-hour working week is assumed.

A duty is placed on employers to have any of their employees covered by the CAW Regulations medically examined not more than two years prior to starting work with asbestos, and thereafter at intervals of not more than two years. Medical examinations, as for the ALR workers, are conducted by Employment Medical Advisers or appointed doctors. Data are collected at these examinations on personal details, smoking habit, past exposure to asbestos plus present exposure details and current job.

Since the introduction of the CAW Regulations, premises covered by the original 1969 Asbestos Regulations but where no workers reached the action level for statutory medicals could, if they so wished, continue with medical surveillance of their workers, but the requirement to forward data for inclusion in the Asbestos survey was discontinued.

Annex 9.2 List of jobs for each industrial sector

Textile manufacture

Raw material store
Raw material & finished product transport
Disintegrating/beating/opening/fibrising
Hopper feeding
Carding
Weaving
Spinning
Doubling/twisting
Braiding
Warping
Detritus handling
Inspecting
Supervising
Finished product store & despatch
** 'Fortex' process

Asbestos cement mixture board and pipe manufacture

Supervising
Raw material store
Raw material transport
Disintegrating
Mixing/beating
Wet board or pipe manufacturing
Wet board or pipe handling
Drying
Dry board or pipe handling
Machining or cutting
Sanding
Inspecting
Finished product/store/packing/despatch/transport
Detritus handling wet
Detritus handling dry
Other exposed workers

Manufacture of asbestos/rubber/resin/bitumen mixtures (for seals, friction materials and roofing felt)

Raw transport store
Transporting raw materials
Disintegrating
Handling raw fibre (bag tipping/weighing/mixing)
Pressing/moulding
Cutting/finishing/machining
Transporting finished product
Finished product storage/packing/despatching
Inspecting
Supervising
Other exposed workers

Asbestos board and paper manufacture

Raw material store
Raw material transport
Disintegrating/opening/fibrising
Mixing/beating
Handling wet mixture
Drying
Handling dry mixture
Cutting/machining
Store transport/packing/despatching products
Supervising
Detritus handling
Inspecting
Other exposed workers

Garment manufacture

Cloth store
Cutting out
Stitching
Transport of materials
Storing/packing/despatching
Inspecting
Supervising
Other exposed workers

Manufacture of dry mixes for insulation and plastering

Raw material stores
Raw material handling/bag tipping/weighing/mixing
Packaging
Stores and despatch
Other exposed workers

Maintenance workers all manufacturing sectors

Supervising
Fault finder/machine fitter/installation engineer/plant engineer
Labourer to plant engineers etc.
Carpenters/joiners
Electrician
Plumber
Other building trade craftsman, e.g. painter
Labourer to building trade craftsman
Ventilation plant servicing
Factory cleaning

* Insulation workers

Supervising
Lagging
Spraying
Mate
Mattress maker
Other exposed workers

** Asbestos stripping/removal

Supervising
Stripping, encapsulating
Other exposed workers (e.g. sampler, cleaner, scaffolder)

Ship building, repair and breaking

Asbestos storeman
Lagging
Boilermakers and installers
Carpenters/joiners

Plumbers
Engine fitter
Other exposed workers
Asbestos stripping
Cleaner
Shipbreaking

Building and construction

Heating engineers
Asbestos board cutting/fitting
Asbestos roofing construction and maintenance
Demolition
** Carpenter/joiners
** Plumber
** Other building trade craftsman, e.g. painter
** Labourer to building trade craftsman

Miscellaneous processes

Fitting clutch and brake pads
Machining/cutting asbestos board
Machining/cutting asbestos/resin board
Use of asbestos string/rope felt
** Other exposed workers (e.g. sampler cleaner scaffolder)

* Only applicable prior to the Control of Asbestos at Work Regulations.

** Only applicable after the Control of Asbestos at Work Regulations.

Annex 9.3 Mortality of all workers by latency and industrial sector

Year of first exposure:	First exposed prior to 1970						First exposed 1970 or later						Total					
Latency:	< 10 years			10 years or more			< 10 years			10 years or more			< 10 years			10 years or more		
	Deaths	SMR	95%CI	Deaths	SMR	95%CI	Deaths	SMR	95%CI	Deaths	SMR	95%CI	Deaths	SMR	95%CI	OBS	SMR	95%CI
Sector 1 - Textile manufacture																		
All causes	8	76.6	(33.1 -150.8)	402	118.4**	(107.0 -130.7)	63	90.0	(69.2 -115.2)	88	105.2	(84.4 -129.7)	71	88.3	(68.9 -111.3)	490	115.8**	(105.7 -126.7)
All malignant neoplasms	7	236.5	(95.1 -487.3)	131	125.1*	(104.6 -148.5)	14	72.6	(39.7 -121.8)	18	68.9	(40.8 -108.9)	21	22.2	(58.5 -144.3)	149	113.9	(96.4 -133.7)
MN of stomach	2	800.0	(96.9 -2889.9)	9	112.8	(51.6 -214.1)	1	74.6	(1.9 -415.8)	1	58.5	(1.5 -325.8)	3	1.6	(38.9 -551.4)	10	103.2	(49.5 -189.8)
MN of liver	0	0.0	(0.0 -14978.7)	3	340.9	(70.3 -996.3)	0	0.0	(0.0 -1664.3)	0	0.0	(0.0 -1069.9)	0	0.2	(0.0 -1497.9)	3	258.6	(53.3 -755.8)
MN of larynx	0	0.0	(0.0 -14978.7)	1	111.1	(2.8 -619.1)	1	625.0	(15.9 -3482.3)	0	0.0	(0.0 -1248.2)	1	0.2	(14.1 -3095.3)	1	87.7	(2.2 -488.7)
MN of trachea, bronchus & lung	1	92.6	(2.3 -515.9)	51	139.3*	(103.7 -183.1)	4	66.7	(18.2 -170.7)	7	87.2	(35.0 -179.6)	5	7.1	(22.9 -164.8)	58	129.9	(98.6 -167.9)
MN of bladder	1	1250.0	(31.7 -6964.5)	4	115.9	(31.6 -296.9)	1	217.4	(5.5 -1211.2)	1	133.3	(3.4 -742.9)	2	0.5	(44.9 -1337.9)	5	119.0	(38.7 -277.8)
	Deaths	%		Deaths	%		Deaths	%		Deaths	%		Deaths	%		Deaths	%	
Mesothelioma	0	0.00	(0.00 -35.31)	11	3.24	(1.62 -5.80)	0	0.00	(0.00 -5.27)	0	0.00	(0.00 -4.41)	0	0.00	(0.00 -4.48)	11	2.60	(1.30 -4.65)
	Deaths	%		Deaths	%		Deaths	%		Deaths	%		Deaths	%		Deaths	%	
Asbestosis	0	0.00	(0.00 -35.31)	5 (8)	1.47	(0.48 -3.44)	0 (1)	0.00	(0.00 -5.27)	0	0.00	(0.00 -4.41)	0 (1)	0.00	(0.00 -4.48)	5 (8)	1.18	(0.38 -2.76)
Sector 2 - Asbestos cement mixture board & pipe manufacture																		
All causes	10	72.0	(34.5 -132.4)	668	97.7	(90.4 -105.5)	101	117.8	(96.0 -143.2)	96	85.5	(69.3 -104.4)	111	111.4	(91.7 -134.2)	764	96.0	(89.2 -103.1)
All malignant neoplasms	4	103.1	(28.1 -264.0)	208	101.5	(88.2 -116.3)	31	134.3	(91.3 -190.6)	30	88.6	(59.8 -126.5)	35	129.8	(90.4 -180.6)	238	99.7	(87.4 -113.2)
MN of stomach	1	285.7	(7.2 -1591.9)	17	102.4	(59.7 -164.0)	3	170.5	(35.2 -498.1)	4	168.8	(46.0 -432.1)	4	189.6	(51.7 -485.4)	21	110.7	(68.5 -169.2)
MN of liver	0	0.0	(0.0 -9985.8)	2	114.3	(13.8 -412.8)	0	0.0	(0.0 -1361.7)	1	277.8	(7.0 -1547.7)	0	0.0	(0.0 -1198.3)	3	142.2	(29.3 -415.5)
MN of larynx	0	0.0	(0.0 -9985.8)	3	162.2	(33.4 -473.9)	1	476.2	(12.1 -2653.2)	0	0.0	(0.0 -936.2)	1	416.7	(10.6 -2321.5)	3	138.2	(28.5 -404.0)
MN of trachea, bronchus & lung	1	67.1	(1.7 -373.9)	85	111.0	(88.7 -137.3)	13	165.4	(88.1 -282.8)	10	89.9	(43.1 -165.4)	14	149.7	(81.9 -251.2)	95	108.3	(87.7 -132.5)
MN of bladder	0	0.0	(0.0 -2995.7)	4	54.0	(14.7 -138.2)	2	327.9	(39.7 -1184.4)	0	0.0	(0.0 -277.4)	2	281.7	(34.1 -1017.6)	4	47.1	(12.8 -120.6)
	Deaths	%		Deaths	%		Deaths	%		Deaths	%		Deaths	%		Deaths	%	
Mesothelioma	0	0.00	(0.00 -26.57)	18	2.63	(1.56 -4.16)	2	2.33	(0.28 -8.42)	1	0.89	(0.02 -4.96)	2	2.01	(0.24 -7.23)	19	2.39	(1.44 -3.73)
	Deaths	%		Deaths	%		Deaths	%		Deaths	%		Deaths	%		Deaths	%	
Asbestosis	0	0.00	(0.00 -26.57)	15(15)	2.19	(1.23 -3.62)	0 (1)	0.00	(0.00 -4.30)	0	0.00	(0.00 -3.29)	0 (1)	0.00	(0.00 -3.70)	15(15)	1.88	(2.84 -3.11)
Sector 3 - Manufacture of asbestos/rubber/resin/bitumen mixtures																		
All causes	22	107.0	(67.1 -162.0)	804	108.2*	(100.8 -116.0)	201	94.0	(81.4 -107.9)	323	124.6**	(111.2 -139.1)	223	95.1	(83.0 -108.4)	1,127	112.4**	(105.9 -119.2)
All malignant neoplasms	5	84.5	(27.4 -197.1)	289	125.6**	(111.3 -141.2)	66	109.3	(84.5 -139.0)	114	141.0**	(116.3 -169.3)	71	107.0	(83.6 -135.0)	403	129.6**	(117.1 -143.1)
MN of stomach	0	0.0	(0.0 -587.4)	19	107.6	(64.8 -168.0)	8	179.4	(77.4 -353.4)	11	198.6	(99.1 -355.3)	8	161.0	(69.5 -317.2)	30	129.3	(87.2 -184.6)
MN of liver	0	0.0	(0.0 -7489.3)	5	234.7	(76.2 -547.8)	1	172.4	(4.4 -960.6)	1	116.3	(2.9 -647.9)	1	161.3	(4.1 -898.6)	6	200.7	(73.6 -436.8)
MN of larynx	0	0.0	(0.0 -5991.5)	3	138.9	(28.6 -405.9)	2	363.6	(44.0 -1313.6)	1	129.9	(3.3 -723.6)	2	333.3	(40.4 -1204.1)	4	136.5	(37.2 -349.5)
MN of trachea, bronchus & lung	1	44.8	(1.1 -249.8)	110	131.5**	(108.1 - 158.5)	26	126.8	(82.8 -185.8)	44	166.4**	(120.9 -223.4)	27	118.8	(78.3 -172.8)	156	141.7**	(120.3 -165.8)
MN of bladder	0	0.0	(0.0 -1872.3)	16	206.5*	(118.0 -335.3)	2	124.2	(15.0 -448.7)	7	277.8*	(111.7 -572.3)	2	113.0	(13.7 -408.2)	23	224.0**	(142.0 -336.0)
	Deaths	%		Deaths	%		Deaths	%		Deaths	%		Deaths	%		Deaths	%	
Mesothelioma	0	0.00	(0.00 -17.95)	20	2.69	(1.64 -4.16)	1	0.47	(0.01 -2.60)	0	0.00	(0.00 -1.42)	1	0.43	(0.01 -2.38)	20	2.00	(1.22 -3.08)
	Deaths	%		Deaths	%		Deaths	%		Deaths	%		Deaths	%		Deaths	%	
Asbestosis	0	0.00	(0.00 -17.95)	5 (10)	0.67	(0.22 -1.57)	1 (0)	0.47	(0.01 -2.60)	0 (1)	0.00	(0.00 -1.42)	1 (0)	0.43	(0.01 -2.38)	5 (11)	0.50	(0.16 -1.16)
Sector 4 - Asbestos board and paper manufacture																		
All causes	2	46.7	(5.7 -168.8)	143	97.2	(81.9 -114.5)	40	75.5	(53.9 -102.8)	71	100.7	(78.6 -127.0)	42	73.4*	(52.9 -99.2)	214	98.3	(85.6 -112.4)
All malignant neoplasms	0	0.0	(0.0 -253.9)	55	124.9	(94.1 -162.6)	12	79.8	(41.2 -139.4)	24	114.0	(73.1 -169.6)	12	74.0	(38.2 -129.2)	79	121.4	(96.1 -151.3)
MN of stomach	0	0.0	(0.0 -2723.4)	4	116.3	(31.7 -297.7)	0	0.0	(0.0 -256.0)	2	137.0	(16.6 -494.8)	0	0.0	(0.0 -234.0)	6	122.4	(44.9 -266.5)
MN of liver	0	0.0	(0.0 -29957.3)	1	243.9	(6.2 -1358.9)	0	0.0	(0.0 -2304.4)	0	0.0	(0.0 -1426.5)	0	0.0	(0.0 -2139.8)	1	161.3	(4.1 -898.6)
MN of larynx	0	0.0	(0.0 -29957.3)	0	0.0	(0.0 -730.7)	0	0.0	(0.0 -2304.4)	0	0.0	(0.0 -1576.7)	0	0.0	(0.0 -2139.8)	0	0.0	(0.0 -499.3)
MN of trachea, bronchus & lung	0	0.0	(0.0 -624.1)	27	163.5*	(107.8 -237.9)	4	75.9	(20.7 -194.3)	8	116.1	(50.1 -228.8)	4	69.6	(19.0 -178.1)	35	149.6*	(104.2 -208.0)
MN of bladder	0	0.0	(0.0 -9985.8)	1	64.9	(1.6 -361.8)	0	0.0	(0.0 -696.7)	2	281.7	(34.1 -1017.6)	0	0.0	(0.0 -651.2)	3	133.3	(27.5 -389.7)
	Deaths	%		Deaths	%		Deaths	%		Deaths	%		Deaths	%		Deaths	%	
Mesothelioma	0	0.00	(0.00 -86.21)	4	2.72	(0.74 -6.96)	0	0.00	(0.00 -6.97)	0	0.00	(0.00 -5.23)	0	0.00	(0.00 -6.45)	4	1.84	(0.50 -4.70)
	Deaths	%		Deaths	%		Deaths	%		Deaths	%		Deaths	%		Deaths	%	
Asbestosis	0	0.00	(0.00 -86.21)	2 (5)	1.36	(0.16 -4.91)	0	0.00	(0.00 -6.97)	0	0.00	(0.00 -5.23)	0	0.00	(0.00 -6.45)	2 (5)	0.92	(0.11 -3.32)

Annex 9.3 - *continued*

Year of first exposure:	First exposed prior to 1970						First exposed 1970 or later						Total					
Latency:	< 10 years			10 years or more			< 10 years			10 years or more			< 10 years			10 years or more		
	Deaths	SMR	95%CI	Deaths	SMR	95%CI	Deaths	SMR	95%CI	Deaths	SMR	95%CI	Deaths	SMR	95%CI	Deaths	SMR	95%CI
Sector 5 - Garment manufacture																		
All causes	1	181.8	(4.6 -1013.0)	16	78.3	(44.8 -127.2)	1	27.5	(0.7 -153.5)	11	191.3	(95.5 -342.3)	2	47.8	(5.8 -172.8)	27	103.1	(68.0 -150.1)
All malignant neoplasms	0	0.0	(0.0 -1762.2)	4	59.9	(16.3 -153.3)	0	0.0	(0.0 -232.2)	6	272.7	(100.1 -593.6)	0	0.0	(0.0 -205.2)	10	112.6	(54.0 -207.1)
MN of stomach	0	0.0	(0.0 -29957.3)	0	0.0	(0.0 -696.7)	0	0.0	(0.0 -4279.6)	0	0.0	(0.0 -2723.4)	0	0.0	(0.0 -3744.7)	0	0.0	(0.0 -554.8)
MN of liver	0	0.0	(-)	1	2000.0	(50.7 -11143.2)	0	0.0	(0.0 -29957.3)	0	0.0	(0.0 -14978.7)	0	0.0	(0.0 -29957.3)	1	1428.6	(36.2 -7959.5)
MN of larynx	0	0.0	(-)	0	0.0	(0.0 -5991.5)	0	0.0	(0.0 -29957.3)	0	0.0	(0.0 -29957.3)	0	0.0	(0.0 -29957.3)	0	0.0	(0.0 -4992.9)
MN of trachea, bronchus & lung	0	0.0	(0.0 -4992.9)	1	50.8	(1.3 -282.8)	0	0.0	(0.0 -998.6)	2	370.4	(44.9 -1337.9)	0	0.0	(0.0 -832.1)	3	119.5	(24.6 -349.3)
MN of bladder	0	0.0	(-)	0	0.0	(0.0 -1576.7)	0	0.0	(0.0 -14978.7)	0	0.0	(0.0 -5991.5)	0	0.0	(0.0 -14978.7)	0	0.0	(0.0 -1248.2)
	Deaths	%		Deaths	%		Deaths	%		Deaths	%		Deaths	%		Deaths	%	
Mesothelioma	0	0.00	(0.00 -670.91)	0	0.00	(0.00 -18.06)	0	0.00	(0.00 -101.65)	0	0.00	(0.00 -64.17)	0	0.00	(0.00 -88.28)	0	0.00	(0.00 -14.09)
	Deaths	%		Deaths	%		Deaths	%		Deaths	%		Deaths	%		Deaths	%	
Asbestosis	0	0.00	(0.00 -670.91)	1 (0)	4.89	(1.18 -35.34)	0	0.00	(0.00 -101.65)	1 (0)	17.39	(4.21 -125.57)	0	0.00	(0.00 -88.28)	2 (0)	7.64	(0.92 -27.58)
Sector 6 - Manufacture of dry mixes for insulation & 'plastering'																		
All causes	0	0.0	(0.0 -434.2)	27	108.1	(71.3 -157.3)	8	147.1	(63.5 -289.8)	8	120.5	(52.0 -237.4)	8	130.5	(56.3 -257.1)	35	110.7	(77.1 -154.0)
All malignant neoplasms	0	0.0	(0.0 -1497.9)	10	134.2	(64.4 -246.9)	4	263.2	(71.7 -673.8)	2	98.5	(11.9 -355.9)	4	232.6	(63.4 -595.4)	12	126.6	(65.4 -221.1)
MN of stomach	0	0.0	(0.0 -14978.7)	0	0.0	(0.0 -516.5)	0	0.0	(0.0 -2496.4)	1	666.7	(16.9 -3714.4)	0	0.0	(0.0 -2139.8)	1	137.0	(3.5 -763.2)
MN of liver	0	0.0	(-)	0	0.0	(0.0 -4279.6)	0	0.0	(0.0 -29957.3)	0	0.0	(0.0 -14978.7)	0	0.0	(0.0 -29957.3)	0	0.0	(0.0 -3328.6)
MN of larynx	0	0.0	(-)	0	0.0	(0.0 -4279.6)	1	10000.0*	(253.7 -55716.2)	0	0.0	(0.0 -14978.7)	1	10000.0*	(253.7 -55716.2)	0	0.0	(0.0 -3328.6)
MN of trachea, bronchus & lung	0	0.0	(0.0 -3744.7)	7	247.3	(99.4 -509.6)	1	178.6	(4.5 -994.9)	0	0.0	(0.0 -421.9)	1	156.3	(4.0 -870.6)	7	197.7	(79.5 -407.4)
MN of bladder	0	0.0	(0.0 -29957.3)	1	384.6	(9.8 -2142.9)	0	0.0	(0.0 -7489.3)	0	0.0	(0.0 -4992.9)	0	0.0	(0.0 -5991.5)	1	312.5	(7.9 -1741.1)
	Deaths	%		Deaths	%		Deaths	%		Deaths	%		Deaths	%		Deaths	%	
Mesothelioma	0	0.00	(0.00 -534.78)	0	0.00	(0.00 -14.78)	0	0.00	(0.00 -67.83)	0	0.00	(0.00 -55.57)	0	0.00	(0.00 -60.20)	0	0.00	(0.00 -11.67)
	Deaths	%		Deaths	%		Deaths	%		Deaths	%		Deaths	%		Deaths	%	
Asbestosis	0	0.00	(0.00 -534.78)	1 (2)	4.00	(0.10 -22.31)	0	0.00	(0.00 -67.83)	0	0.00	(0.00 -55.57)	0	0.00	(0.00 -60.20)	1 (2)	3.16	(0.08 -17.62)
Sector 7 - Maintenance workers all manufacturing sectors																		
All causes	8	82.1	(35.5 -161.8)	455	97.1	(88.3 -106.6)	80	85.7	(67.9 -106.6)	122	102.0	(84.7 -121.8)	88	85.3	(68.4 -105.1)	577	98.1	(90.2 -106.5)
All malignant neoplasms	1	37.9	(1.0 -211.0)	144	102.3	(86.3 -120.5)	31	119.8	(81.4 -170.1)	35	97.3	(67.8 -135.3)	32	112.2	(76.8 -158.5)	179	101.3	(87.0 -117.3)
MN of stomach	0	0.0	(0.0 -1198.3)	9	79.6	(36.4 -151.2)	0	0.0	(0.0 -145.4)	4	153.8	(41.9 -393.9)	0	0.0	(0.0 -129.7)	13	93.5	(49.8 -159.9)
MN of liver	0	0.0	(0.0 -14978.7)	1	74.6	(1.9 -415.8)	0	0.0	(0.0 -1198.3)	0	0.0	(0.0 -768.1)	0	0.0	(0.0 -1109.5)	1	57.8	(1.5 -322.1)
MN of larynx	0	0.0	(0.0 -14978.7)	0	0.0	(0.0 -215.5)	0	0.0	(0.0 -1198.3)	0	0.0	(0.0 -832.1)	0	0.0	(0.0 -1109.5)	0	0.0	(0.0 -171.2)
MN of trachea, bronchus & lung	1	90.1	(2.3 -501.9)	57	106.2	(80.4 -137.6)	17	178.2	(103.8 -285.3)	11	88.4	(44.1 -158.1)	18	169.0*	(100.2 -267.1)	69	104.3	(81.2 -132.0)
MN of bladder	0	0.0	(0.0 -3744.7)	7	139.4	(56.1 -287.3)	0	0.0	(0.0 -399.4)	2	165.3	(20.0 -597.1)	0	0.0	(0.0 -360.9)	9	144.5	(66.1 -274.2)
	Deaths	%		Deaths	%		Deaths	%		Deaths	%		Deaths	%		Deaths	%	
Mesothelioma	0	0.00	(0.00 -37.89)	14	2.99	(1.63 -5.01)	2	2.14	(0.26 -7.73)	4	3.35	(0.91 -8.57)	2	1.94	(0.23 -7.00)	18	3.06	(1.81 -4.84)
	Deaths	%		Deaths	%		Deaths	%		Deaths	%		Deaths	%		Deaths	%	
Asbestosis	0	0.00	(0.00 -37.89)	4 (8)	0.85	(0.23 -2.19)	0 (1)	0.00	(1.17 -10.97)	0 (2)	0.00	(0.91 -8.57)	0 (1)	0.00	(1.06 -9.93)	4 (10)	0.68	(0.19 -1.74)
Sector 8 - Asbestos stripping/removal																		
All causes	0	0.0	(0.0 -7489.3)	92	76.0**	(61.2 -93.1)	53	80.1	(60.0 -104.8)	27	68.3*	(45.0 -99.3)	53	80.1	(60.0 -104.7)	119	74.1**	(61.4 -88.6)
All malignant neoplasms	0	0.0	(0.0 -29957.3)	43	114.5	(82.9 -154.2)	13	82.4	(43.9 -141.0)	9	85.2	(39.0 -161.8)	13	82.4	(43.9 -140.9)	52	108.1	(80.7 -141.7)
MN of stomach	0	0.0	(-)	2	77.5	(9.4 -280.0)	0	0.0	(0.0 -322.1)	1	149.3	(3.8 -831.6)	0	0.0	(0.0 -322.1)	3	92.3	(19.0 -269.8)
MN of liver	0	0.0	(-)	0	0.0	(0.0 -651.2)	0	0.0	(0.0 -1361.7)	0	0.0	(0.0 -2139.8)	0	0.0	(0.0 -1361.7)	0	0.0	(0.0 -499.3)
MN of larynx	0	0.0	(-)	0	0.0	(0.0 -680.8)	0	0.0	(0.0 -1997.2)	0	0.0	(0.0 -2723.4)	0	0.0	(0.0 -1997.2)	0	0.0	(0.0 -544.7)
MN of trachea, bronchus & lung	0	0.0	(-)	11	86.4	(43.1 -154.6)	1	24.0	(0.6 -133.9)	3	98.4	(20.3 -287.5)	1	24.0	(0.6 -133.9)	14	88.7	(48.5 -148.9)
MN of bladder	0	0.0	(-)	2	188.7	(22.9 -681.6)	0	0.0	(0.0 -881.1)	0	0.0	(0.0 -1152.2)	0	0.0	(0.0 -881.1)	2	151.5	(18.3 -547.3)
	Deaths	%		Deaths	%		Deaths	%		Deaths	%		Deaths	%		Deaths	%	
Mesothelioma	0	0.00	(0.00 -9225.00)	12	9.91	(5.12 -17.30)	1	1.51	(0.04 -8.42)	0	0.00	(0.00 -9.33)	1	0.00	(0.04 -8.41)	12	7.47	(3.86 -13.04)
	Deaths	%		Deaths	%		Deaths	%		Deaths	%		Deaths	%		Deaths	%	
Asbestosis	0	0.00	(0.00 -9225.00)	1 (9)	0.83	(0.02 -4.60)	0	0.00	(0.00 -5.58)	0	0.00	(0.00 -9.33)	0	0.00	(0.00 -9655.26)	1 (9)	0.62	(0.02 -3.47)

Annex 9.3 - *continued*

Year of first exposure:	First exposed prior to 1970						First exposed 1970 or later						Total					
Latency:	< 10 years			10 years or more			< 10 years			10 years or more			< 10 years			10 years or more		
	Deaths	SMR	95%CI	Deaths	SMR	95%CI	Deaths	SMR	95%CI	Deaths	SMR	95%CI	Deaths	SMR	95%CI	Deaths	SMR	95%CI
Sector 9 - Ship building, repair & breaking																		
All causes	3	59.9	(12.3 -175.0)	330	94.9	(84.8 -105.9)	36	74.5	(52.2 -103.1)	60	84.7	(64.7 -109.1)	39	73.1*	(52.0 -100.0)	390	93.2	(84.0 -103.0)
All malignant neoplasms	0	0.0	(0.0 -228.7)	132	126.7**	(106.0 -150.3)	12	91.0	(47.0 -158.9)	27	129.9	(85.6 -189.0)	12	82.8	(42.8 -144.6)	159	127.3**	(108.2 -148.6)
MN of stomach	0	0.0	(0.0 -2496.4)	12	143.5	(74.2 -250.7)	0	0.0	(0.0 -277.4)	4	263.2	(71.7 -673.8)	0	0.0	(0.0 -249.6)	16	161.9	(92.6 -263.0)
MN of liver	0	0.0	(0.0 -29957.3)	2	200.0	(24.2 -722.5)	0	0.0	(0.0 -2496.4)	0	0.0	(0.0 -1302.5)	0	0.0	(0.0 -2304.4)	2	162.6	(19.7 -587.4)
MN of larynx	0	0.0	(0.0 -29957.3)	0	0.0	(0.0 -290.8)	0	0.0	(0.0 -2304.4)	0	0.0	(0.0 -1426.5)	0	0.0	(0.0 -2139.8)	0	0.0	(0.0 -241.6)
MN of trachea, bronchus & lung	0	0.0	(0.0 -576.1)	42	105.2	(75.8 -142.2)	4	79.4	(21.6 -203.2)	8	109.3	(47.2 -215.3)	4	71.9	(19.6 -184.2)	50	105.8	(78.5 -139.5)
MN of bladder	0	0.0	(0.0 -7489.3)	4	107.8	(29.4 -276.1)	1	250.0	(6.3 -1392.9)	1	137.0	(3.5 -763.2)	1	227.3	(5.8 -1266.3)	5	112.6	(36.6 -262.8)
	Deaths	%		Deaths	%		Deaths	%		Deaths	%		Deaths	%		Deaths	%	
Mesothelioma	0	0.00	(0.00 -73.65)	14	4.03	(2.20 -6.75)	2	4.14	(0.50 -14.94)	3	4.24	(0.87 -12.38)	2	3.75	(0.45 -13.54)	17	4.06	(2.37 -6.50)
	Deaths	%		Deaths	%		Deaths	%		Deaths	%		Deaths	%		Deaths	%	
Asbestosis	0	0.00	(0.00 -73.65)	5 (13)	1.44	(0.63 -3.76)	1 (0)	2.07	(4.55 -27.02)	1 (0)	1.41	(3.11 -18.44)	1 (0)	1.87	(4.12 -24.48)	6 (13)	1.43	(0.53 -3.12)
Sector 10 - Building & construction																		
All causes	1	112.4	(2.9 -626.0)	59	105.3	(80.2 -135.9)	30	135.4	(91.4 -193.3)	28	97.0	(64.4 -140.2)	31	134.5	(91.4 -191.0)	87	102.5	(82.1 -126.4)
All malignant neoplasms	1	434.8	(11.0 -2422.4)	21	125.4	(77.7 -191.8)	10	169.2	(81.1 -311.2)	11	133.3	(66.6 -238.6)	11	179.2	(89.4 -320.6)	32	128.1	(87.6 -180.8)
MN of stomach	0	0.0	(0.0 -14978.7)	4	310.1	(84.5 -793.9)	1	208.3	(5.3 -1160.8)	1	169.5	(4.3 -944.3)	1	200.0	(5.1 -1114.3)	5	266.0	(86.4 -620.7)
MN of liver	0	0.0	(-)	0	0.0	(0.0 -1762.2)	0	0.0	(0.0 -4992.9)	0	0.0	(0.0 -3328.6)	0	0.0	(0.0 -4992.9)	0	0.0	(0.0 -1152.2)
MN of larynx	0	0.0	(-)	0	0.0	(0.0 -1762.2)	0	0.0	(0.0 -4992.9)	0	0.0	(0.0 -3744.7)	0	0.0	(0.0 -4992.9)	0	0.0	(0.0 -1198.3)
MN of trachea, bronchus & lung	0	0.0	(0.0 -3328.6)	8	130.5	(56.3 -257.1)	6	276.5*	(101.5 -601.8)	4	146.0	(39.8 -373.8)	6	265.5	(97.4 -577.8)	12	135.3	(69.9 -236.3)
MN of bladder	0	0.0	(0.0 -29957.3)	1	185.2	(4.7 -1031.8)	0	0.0	(0.0 -1762.2)	0	0.0	(0.0 -1033.0)	0	0.0	(0.0 -1664.3)	1	120.5	(3.1 -671.3)
	Deaths	%		Deaths	%		Deaths	%		Deaths	%		Deaths	%		Deaths	%	
Mesothelioma	0	0.00	(0.00 -414.61)	2	3.57	(0.43 -12.89)	0	0.00	(0.00 -16.66)	0	0.00	(0.00 -12.78)	0	0.00	(0.00 -16.02)	2	2.36	(0.29 -8.51)
	Deaths	%		Deaths	%		Deaths	%		Deaths	%		Deaths	%		Deaths	%	
Asbestosis	0	0.00	(0.00 -414.61)	0 (3)	0.00	(0.00 -6.59)	0	0.00	(0.00 -16.66)	0 (1)	0.00	(0.00 -12.78)	0	0.00	(0.00 -16.02)	0 (4)	0.00	(0.00 -4.35)
Sector 12 - Insulation workers																		
All causes	3	71.4	(14.7 -208.7)	655	133.8**	(123.6 -144.6)	77	96.0	(75.8 -120.0)	89	90.3	(72.5 -111.1)	80	94.8	(75.2 -118.0)	744	126.5**	(117.5 -136.0)
All malignant neoplasms	1	90.9	(2.3 -506.5)	328	221.4**	(197.8 -247.1)	30	141.6	(95.5 -202.1)	35	119.0	(82.9 -165.6)	31	139.1	(94.5 -197.4)	363	204.5**	(183.7 -226.9)
MN of stomach	0	0.0	(0.0 -3328.6)	17	149.6	(87.2 -239.6)	2	133.3	(16.1 -481.6)	4	201.0	(54.8 -514.7)	2	125.8	(15.2 -454.4)	21	157.3	(97.4 -240.5)
MN of liver	0	0.0	(0.0 -29957.3)	2	133.3	(16.1 -481.6)	0	0.0	(0.0 -1361.7)	0	0.0	(0.0 -907.8)	0	0.0	(0.0 -1302.5)	2	109.3	(13.2 -394.8)
MN of larynx	0	0.0	(0.0 -29957.3)	0	0.0	(0.0 -199.7)	1	526.3	(13.4 -2932.4)	0	0.0	(0.0 -1033.0)	1	500.0	(12.7 -2785.8)	0	0.0	(0.0 -167.4)
MN of trachea, bronchus & lung	0	0.0	(0.0 -748.9)	152	278.9**	(236.3 -326.9)	6	87.6	(32.1 -190.6)	15	157.9	(88.4 -260.4)	6	82.8	(30.4 -180.1)	169	264.1**	(225.8 -307.0)
MN of bladder	0	0.0	(0.0 -9985.8)	3	62.6	(12.9 -183.0)	2	377.4	(45.7 -1363.1)	0	0.0	(0.0 -344.3)	2	357.1	(43.3 -1290.1)	3	53.0	(10.9 -154.9)
	Deaths	%		Deaths	%		Deaths	%		Deaths	%		Deaths	%		Deaths	%	
Mesothelioma	0	0.00	(0.00 -87.86)	67	13.69	(10.61 -17.37)	5	6.24	(2.02 -14.56)	0	0.00	(0.00 -3.74)	5	5.93	(1.92 -13.83)	67	11.39	(8.83 -14.46)
	Deaths	%		Deaths	%		Deaths	%		Deaths	%		Deaths	%		Deaths	%	
Asbestosis	0	0.00	(0.00 -87.86)	46 (97)	9.40	(0.22 -2.09)	0 (1)	0.00	(1.36 -12.77)	1 (0)	1.01	(0.11 -3.32)	0 (1)	0.00	(1.29 -12.14)	47 (97)	7.99	(0.19 -1.74)

** Significant at 1% level * Significant at 5% level

Chapter 10 Monitoring occupational diseases

John Osman
John Hodgson
Sally Hutchings
Jacky Jones
Trevor Benn
Richard Elliott

Epidemiology and Medical Statistics Unit, HSE

10.1 Introduction

No single source of information is available in the UK on the nature and overall scale of occupational ill-health and it is now recognised that monitoring of occupational diseases has to be based on data from a range of sources.[1]

Other chapters in this supplement describe patterns of mortality and cancer incidence in different occupations and discuss possible work-related explanations for them. These data can be used to monitor a number of occupational diseases but many others are non-fatal and there are only a few well established occupational causes of cancer. The primary purpose of this chapter is to summarise other sources of information on the occurrence of occupational ill-health (though limited data on mortality and cancer incidence are included for completeness or if they are not covered elsewhere).

The first section summarises the methodology and main findings of a 1990 survey of self-reported work-related illness (a 'trailer' to the main Labour Force Survey — LFS). This is presented in some detail because although it gives only one perspective on the extent of occupational ill-health — that of individuals who believe they have an illness caused or made worse by work — it is unique in that it is derived from a representative sample of the England and Wales population and in contrast to other data is unrestricted in terms of the ranges of illness and occupation covered. The data and commentary presented in the second section are derived from, and largely replicate those published annually in updated form by the Health and Safety Executive (HSE), most recently as a supplement to the Health and Safety Commission's Annual Report for 1993/94.[2] Sources of information and the main tables of data are described first. These are followed by a detailed commentary on specific groups of hazards and diseases based on these sources and other relevant information including the LFS data.

10.2 Labour Force Survey

10.2.1 Methods

The LFS is a survey of households living at private addresses in Great Britain, carried out by the Office of Population Censuses and Surveys (OPCS) on behalf of the Employment Department. The main survey is used to collect data on trends in individuals' economic activity and at the time the data presented here were collected it was conducted on an annual basis. 'Trailer' questionnaires may be commissioned and added to the main LFS to obtain cross sectional data for other purposes. A full account of the basic survey methodology and sample selection can be found in the OPCS report on the 1988 and 1989 surveys.[3] The HSE trailer questionnaire was administered between March and May 1990 to a sample of about 40,000 England and Wales households and sought information about work place accidents and work-related illness either from survey subjects themselves or from 'proxies' (usually the spouse) if the survey subject was unavailable. The accident data are referred to briefly in Chapter 12 and have been reported in detail elsewhere[4] though data on the long term consequences of trauma (grouped with poisoning) are included here. For this survey a work-related illness was defined subjectively by each respondent (or proxy respondent) by their response to the question:

> 'In the past 12 months, have you suffered from any illness, disability or other physical problem that was caused or made worse by your work? Please include any work you have done in the past'.

A distinction was made in the survey between cases of illness directly caused by work, and those made worse. This distinction was based on responses to the question:

> 'Was your [complaint] caused by your work, or did your work simply make it worse?'

Reported occupations were coded by OPCS using the standard procedures for the LFS. The main occupational analyses shown in this report use a modified version of the occupation orders of the standard OPCS occupational classification (see Chapter 2). The modifications applied were chosen to bring together occupation units with similar work conditions, for example farm managers were classified to the 'Farming, fishing and forestry' occupation group rather than to managers; and labourers in identified sectors were grouped with the skilled manual workers in the same sector: thus labourers in engineering were included in occupation 'Processing (metal and electrical)'. Complete details of the changes which differ slightly from the revised occupational codings used elsewhere in this supplement can be supplied on request, and are given in a detailed HSE report on the survey.[5]

Respondents reporting a work-related illness were asked to record their doctor's diagnosis if they could recall this, or to describe the illness in their own words. Using this information each illness was coded by HSE staff familiar with the International Classification of Disease (ICD — see Chapter 2) using the ICD basic tabulation list with some additional categories to provide for specific occupational diagnoses (e.g. pneumoconiosis, vibration white finger). 'Stress' — described as such — was coded to the same category as depression and anxiety. 'Eyestrain' was given its own category, in which other non-specific visual symptoms ('sore eyes'; 'impaired vision', 'visual problems') were also placed. In the analyses reported here these cases have been grouped with headache.

Respondents who reported a work-related illness were asked to describe the job causing or exacerbating it. The main LFS asks subjects for their current job, their job one year ago, and, for those not currently employed, their most recent job within the last 3 years. The LFS thus gives an occupational breakdown of the recently working population (working in the last 3 years), but not for the remainder of the ever-worked. For the recently working, denominators for the calculation of occupational rates were calculated by assigning those who reported work-related illness to the occupation groups which gave rise to the illness, and distributing the remainder of the sample population across occupations according to their current or most recently held job within the last 3 years. This did not give 'true' rates, since the jobs to which illnesses were ascribed were not necessarily those currently or most recently held. However, such rates are a close approximation to the true rates for short term, non-persistent effects. More generally, if the relative sizes of the populations who have ever worked in the different occupation groups is close to the relative sizes of the populations currently or recently working in those groups, the rates will accurately reflect the relative risks between occupations. Calculating rates for disease reported from the population inactive for 3 or more years presented two problems. Firstly the LFS does not record an occupation for these people, 60 per cent of whom are retired. Secondly, the illnesses reported from this group were necessarily of a longer term nature, and the relative sizes of the occupational groups from which they had arisen were not necessarily the same as those observed in the currently employed population. In order to relate cases reported in this group to the approximate occupational distribution from which they arose, rate calculations were based on the 1981 Census. Thus cases of work related illness were assigned to the occupation groups which gave rise to the illness, and the remainder of the sample population was distributed across occupations in the proportions recorded in the 1981 Census (calculated separately for males and females). In this report respondents who were working at some time in the last 3 years' are referred to as 'recently working', and those who last worked more than 3 years ago as 'inactive 3+ years'. The LFS does not cover people resident in communal establishments. The main occupation group affected by this is the Armed Forces, for which less than a third of personnel are represented in the LFS sample. For this reason the Armed Forces have been excluded from the occupational rate calculations. The next most affected group is Nurses, of whom about 7 per cent are missed by the LFS sample. No adjustment has been made for the small effects this had on the estimation of overall prevalence (approximately 3 per cent) and prevalence rates and relative risks (1 per cent) for nurses, or for other groups in communal establishments.

The LFS accepts interview responses from 'proxy' respondents if a survey subject in a sampled household is not available and another household member is in a position to answer on their behalf. Proxy interviews are generally with the spouse of the survey subject. One in three (33 per cent) of 1990 LFS interviews were with proxies. Detailed comparisons between occupational groups were based on rates adjusted for the variation in the proportion of proxy responses and for any variation in age distribution between occupations. The adjustment of relative risks and occupational rates for differences in age and proxy response distributions was carried out by a form of indirect standardisation. For each of the 18 main disease groups, a log-linear model was fitted to the pattern of rates by age (16-44, 45-64/59, 65/60-74, 75+) and interview type (personal/proxy) in the whole sample. This model was then used to predict the 'expected' number of cases in each occupation. The occupation relative risk (RR) was calculated as the ratio of the number of observed cases to the number expected. The (adjusted) occupation rates were calculated by multiplying the average (all occupations) rate for the disease by the occupation RR. The effect of this adjustment was to estimate what each observed occupational rate would have been if the responses for that occupation had followed the average proportion of proxy responses and the average age distribution for the survey population as a whole. The 95 per cent confidence limits for prevalence estimates were calculated to show how they vary with size and are illustrated in Figure 10.1.

10.2.2 Results

The overall response rate to the main LFS was 82.3 per cent. One per cent of respondents refused to answer the 'trailer' questionnaire leaving 73,662 who did, of whom 33 per cent were 'proxy' respondents. Illness in the last year caused or made worse by work was reported by 4,377 respondents (5.9 per cent) resulting in an estimated national prevalence of 2.2 million affected individuals: 1.4 million males, 0.8 million females. Proxies reported less work related disease (4.5 per cent of interviews) than first person respondents (6.5 per cent). The effect of replacing proxy interviews by first person interviews would be to increase the number of reported cases by around 10 per cent.

Table 10.1 gives the estimated numbers of affected individuals by individual disease category. Separate estimates are given for the number of cases caused or made worse and the numbers caused only. Musculoskeletal disorders (including 'RSI') were the most common cause of ill health, accounting for 43 per cent (964,000) of the estimated total. Stress/ depression was the second most commonly reported work-related illness, accounting for an estimated 182,700 (8.1 per cent) cases. Other disease groups which recorded 5 per cent or more of the total cases were: 'Other' diseases (6.4 per cent); deafness, tinnitus and other ear conditions (5.4 per

Figure 10.1 Likely margin of error (95% CI) in total prevalence estimates as a function of size of estimate and sample numbers

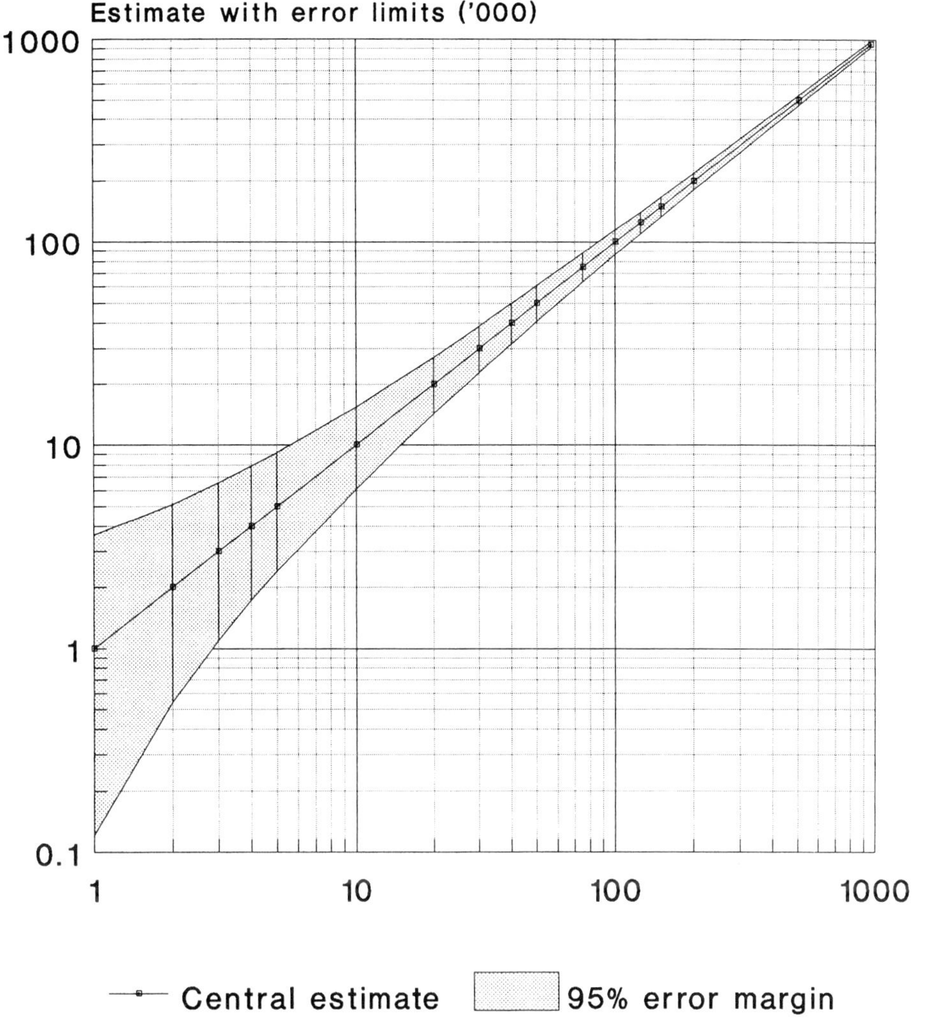

Table 10.1 Total self-reported work-related illness. Estimated England and Wales prevalence, by disease group, Spring 1990

Disease group	All cases (Thousands)	Cases caused (Thousands)
Stress/depression	182.7	104.9
Headache & 'eyestrain'	108.2	45.9
Deafness & ear conditions	121.4	103.1
Heart disease, hypertension & stroke	113.1	45.0
Vibration white finger	7.3	7.3
Varicose veins	15.6	3.0
Upper respiratory disease	60.8	20.3
Asthma	68.1	19.7
Lower respiratory disease	121.0	61.1
Pneumoconiosis	21.8	19.6
Skin disease	84.6	54.2
'RSI'	74.0	49.6
Musculoskeletal conditions	890.5	543.2
of which:		
musculoskeletal disorders of back	460.4	297.2
musculoskeletal disorders of upper limbs & neck	97.2	67.2
musculoskeletal disorders of lower limbs	90.1	55.1
Trauma & poisoning (long-term sequelae)	124.6	107.7
Eye conditions	34.2	22.4
Exhaustion, ME, symptoms	26.3	13.1
Infections	42.1	27.8
'Other' diseases	143.9	55.3
Total	2,240.2	1,303.2

Table 10.2 Prevalence rates of self-reported work-related illness (cases caused or made worse) by age and sex

Age group	Rates per 100 ever employed		Ratio M/F
	Males	Females	
16-44	6.1	4.6	1.3
45-59(f)/64(m)	10.0	5.8	1.7
60(f)/65(m)-74	9.3	3.5	2.7
75+	5.5	1.6	3.3

cent); lower respiratory disease (5.4 per cent); and heart disease, hypertension and stroke (5.0 per cent). Table 10.2 shows the overall prevalence rates for self-reported work-related illness broken down by age and sex. For both sexes the older working age group (age 45 to retirement age) shows the highest rate: 10 per cent for males and 5.8 per cent for females. The male rates are higher than those for females in all age groups, most markedly in the oldest age group (age 75+). Averaged over the whole population, males report 70 per cent more work-related illness than females. However, within occupations the sex differences are much smaller and the overall male to female rate ratio adjusted for occupation is 1.17.

Table 10.3 Total self-reported work-related illness. Estimated England and Wales prevalence, rates per 10,000 and relative risk, by occupation group

Occupational group		All causes				Cases caused	
		Prevalence ('000) (M+F)	Rates per 10,000		Relative risk (M+F)	Prevalence ('000) (M+F)	Relative risk (M+F)
			Males	Females			
18	Coal mining	67.8	4901	0	8.61++	57.4	11.92++
17	Construction	108.9	1067	0	1.87++	72.1	2.17++
14	Processing (metal & electrical)	302.6	978	558	1.68++	197.7	1.89++
19	Transport & materials moving	161.5	865	463	1.47++	104.9	1.65++
20	Other (mainly Labourers nec)	18.1	902	311	1.45+	12.7	1.74++
3	Nursing	86.6	757	753	1.34++	57.5	1.52++
13	Processing (other)	177.2	806	641	1.31++	116.3	1.47++
15	Painting	23.8	738	646	1.30	14.3	1.32
12	Farming, fishing & forestry	57.0	742	700	1.30++	38.0	1.50++
10	Security (excl. Armed Forces)	72.9	730	657	1.28	56.1	1.48++
2	Teaching	87.5	699	603	1.13	35.5	0.79
16	Repetitive assembly, inspection	69.4	754	530	1.08	45.7	1.24+
ALL OCCUPATIONS		**2240.0**	**713**	**423**	**1.00**	**1303.5**	**1.00**
5	Literary, artistic & sports	28.4	631	469	0.99	15.9	0.96
4	Other education & welfare	64.4	605	491	0.96	31.8	0.83
11	Catering, cleaning, hairdressing	227.8	610	427	0.81- -	127.8	0.76- -
6	Science & engineering	61.1	456	402	0.80-	34.2	0.77-
1	Professional	87.6	399	476	0.75- -	40.6	0.60- -
7	Managerial	131.3	412	362	0.70- -	62.9	0.59- -
8	Clerical	229.5	429	293	0.56- -	101.4	0.42- -
9	Selling	76.8	291	302	0.53- -	33.1	0.40- -

Rates per 10,000, adjusted for occupational differences in age and proportion of proxy responses

++/- - Rate significantly above/below average (P<.01)
+/- Rate significantly above/below average (P<.05)

Table 10.3 shows breakdowns of total reported work-related illness by occupational group. Five occupation groups account for 50 per cent of cases: processing (both sub-groups), clerical; catering, cleaning and hairdressing; and transport. For the clerical and catering, cleaning and hairdressing groups the large number of cases is due to the correspondingly large numbers in these occupations. The two groups have lower than average rates of reported illness. Table 10.3 also shows rates and prevalence estimates. The occupations are ordered by the relative risk of work-related disease based on all reported cases. The very high rates for coal miners are due in part to the distorting effect of the very rapid contraction of the coal mining work force over the last thirty years. The pattern of risk by occupation broadly follows the manual/non-manual split, with all but one of the non-manual occupations having below average rates, and all but one of the manual occupations having above average rates. The two exceptions are nursing (above average rates due to musculoskeletal conditions), and catering, cleaning and hairdressing (a group of manual occupations with work-related illness rates below average).

Table 10.4 lists the occupation groups with significantly ($p < 0.01$) raised relative risks of two or greater (based either on all reported cases or on cases 'caused' by work only) for each of the 18 disease categories. These associations correspond fairly closely to established occupational risk factors. Exceptions to this are the large excess risk of asthma for coal miners, perhaps involving mistaken description of other lower respiratory symptoms as 'asthma'; and the identification of heart disease as an occupational risk for professional occupations, in contradiction to epidemiological observation[6] that it is lower status occupations that, if any, entail higher risks for heart disease. Table 10.5 summarises the economic impact of the reported work-related illnesses. The total of 2.2 million cases fall into three roughly equal groups: approximately 726,000 of the affected individuals had had no job in the previous 12 months, but continued to suffer some illness related to a previously held job; 670,000 people suffered some illness during the previous year, but not to the extent of needing to take time off work. Just under 789,000 people were affected seriously enough to take time off work, or, in approximately 97,000 cases, were forced to leave a job because of its effect on their health.

10.2.3 Interpretation

As a record of individuals' opinions about the extent of work-related illness interpretation of the results of this survey is straight forward: it is clear that people perceive their working conditions as having an important direct influence on their health. This includes many non-severe cases: nearly half of the cases among those with a job in the last year took no sickness absence on account of their work-related illness. But most respondents (85 per cent) had consulted a doctor about their illness, implying that 1 in 20 of the average general practitioner's consultations involves—at least in the patient's view — a work-related condition.

However, because of their self-reported nature, these data cannot be taken directly and at face value as an indicator of the true overall extent of work-related illness. The reported levels for different diseases in different occupations will have been affected by common beliefs about associations between work and illness; the level of awareness within occupations of established and possible work-related effects; and the tendency for people to seek external explanations for illness. The inclusion of proxy responses will have

Table 10.4 Occupation groups with significantly (p<0.01) raised relative risks >2, for either all cases or cases caused, for each main disease category

Disease/Occupation	Rates per 10,000		Relative risk (all cases)	Relative risk (cases caused)
	Males	Females		
Stress/depression				
Teaching	197	162	3.75	4.46
Other education & welfare	126	50	*1.75*	2.37
Headache & 'eyestrain'				
Clerical	80	64	2.42	2.11
Professional	48	62	1.92	2.53
Deafness & ear conditions				
Coal mining	917	0	29.87	30.78
Processing (metal & electrical)	152	35	4.64	4.71
Processing (other)	68	58	2.09	2.05
Heart disease, hypertension & stroke				
Security (excl. Armed Forces)	107	0	3.24	*1.79*
Professional	72	19	2.02	2.63
Vibration white finger				
Processing (metal & electrical)	13	22	6.69	6.69
Varicose veins				
Selling	0	20	3.45	*2.34*
Upper respiratory disease				
Teaching	95	60	4.78	4.17
Asthma				
Coal mining	283	0	16.33	27.83
Processing (other)	46	42	2.60	*2.58*
Lower respiratory disease				
Coal mining	565	0	18.06	26.93
Other (mainly labourers nec)	191	0	5.05	8.25
Construction	105	0	3.34	3.61
Processing (metal & electrical)	101	86	3.21	3.12
Farming, fishing & forestry	99	0	2.37	2.65
Pneumoconiosis				
Coal mining	453	0	74.91	79.12
Construction	35	0	5.73	6.47
Processing (metal & electrical)	16	16	2.67	*2.11*
Skin disease				
Construction	58	0	2.74	*2.51*
Science & engineering	48	62	2.42	2.70
Catering, cleaning, hairdressing	14	43	1.84	2.02
'RSI'				
Repetitive assembly, inspection	48	71	3.27	4.52
Musculoskeletal disorders of back				
Coal mining	527	0	4.41	5.66
Nursing	445	397	3.40	3.84
Construction	289	0	2.42	2.95
Transport & materials moving	227	220	1.92	2.09
Musculoskeletal disorders of upper limbs & neck				
Transport & materials moving	65	0	2.38	2.89
Musculoskeletal disorders of lower limbs				
Coal mining	322	0	14.29	22.04
Farming, fishing & forestry	86	26	3.25	4.40
Transport & materials moving	53	28	2.29	2.86
Trauma & poisoning (long-term sequelae)				
Coal mining	212	0	7.22	8.16
Security (excl. Armed Forces)	124	65	3.94	4.54
Transport & materials moving	91	27	2.91	3.02
Construction	81	0	2.76	2.96
Processing (metal & electrical)	77	19	2.47	2.26
Processing (other)	88	15	1.98	2.09
Eye conditions				
Processing (other)	32	21	3.59	4.81
Exhaustion, ME, symptoms				
Teaching	43	42	6.51	*3.36*
Infections				
Nursing	0	38	3.39	4.29
Other education & welfare	33	34	3.25	4.15
'Other' diseases				
Coal mining	187	0	5.21	*7.34*

Note Relative risks shown in italics are not significant at a 1 per cent level but have been included for completeness.

Table 10.5 Estimated prevalence of work-related illness by economic impact

Economic impact category	Number of cases (Thousands)	Per cent
No job in last 12 months		
Retired	379.8	17.0
Other	346.5	15.5
With job in last 12 months		
Nil days lost	671.4	30.0
1-7 days lost	292.7	13.1
8-30 days lost	216.2	9.7
30+ days lost	183.0	8.2
Forced job change	96.7	4.3
Unknown/unspecified	53.7	2.4
Total	2,240.0	100

influenced the results though adjustments were made for the level of proxy response when making comparisons between occupational groups. It is difficult to understand how individuals could describe some diseases as being caused by work e.g. cardiovascular disease and most cancers (though they can be expected to provide more reliable information on the role of work in exacerbating some of these conditions). As well as the difficulty in attributing individual cases of some long latency diseases to a particular occupation, calculation of rates for these diseases had to take account of changing denominators for many occupations over the relevant time period and these could only be determined indirectly. This was a particular problem with the results for coal mining where there have been considerable reductions in the numbers employed over recent years.

Other factors suggest that, at least for some diseases, these data may provide a more reliable source of information on the occurrence of work-related cases in the general population. For certain complaints the association with work factors may be quite clear to the individual. In some cases this will be on the basis of the temporal relationship between the work and illness, e.g. headache and eye strain, dermatitis, stress. In others, the occupational link may be defined by the nature of the illness itself, e.g. vibration white finger, pneumoconiosis. Because of its population based nature, individuals in this survey would have expected no material gain from declaring their illness to be work-related, to some extent weighing against the possibility of over-reporting. The uneven distribution of the relative risks across diseases and occupational groups could also be taken as evidence against a general tendency for respondents to over-report work factors as causing or making worse their ill-health.

To look further at the validity of the data, other internal information from the survey was examined, including some derived from subsidiary questions. This information was used to rank particular diseases according to the reliability of individuals' perceptions of their work-relatedness and consisted of:

(a) the percentage of cases where other people at work were reported to be similarly affected;
(b) the percentage of the survey population lying in occupations with a rate greater than twice or less than half the overall average for each disease;
(c) the percentage of all reported cases, for each disease, which was said to be 'caused' by work;
(d) the percentage of cases where 'the people in charge at work' were said to have accepted that the illness was work-related.

Criteria (a), (c) and (d) are self-explanatory. Criterion (b) is a measure of the unevenness of spread of risk across the occupational groups. Specifically, occupational effects should be observed in some occupations and be absent in others, with consequently high scores on criterion (b).

Generally speaking, genuinely work-related cases will tend to return higher than average percentages on these four measures, and low percentages will indicate some weakness in the work/illness linkage. There are some exceptions to this of course: rare or isolated cases of work-related illness will score low on criterion (a), because sufferers will generally not know of other cases; and a genuinely work-related illness which affected one small occupational sector would score low on criterion (b), as would a genuinely occupational effect which was equally spread across occupation groups. Furthermore, a widespread belief that a particular kind of illness was a specific risk in some occupation will produce high scores on criteria (a), (b) and (c) — and possibly (d) as well — regardless of whether that belief is well-founded or not. Nevertheless, taken together these indicators do provide a suggestive picture. The rankings of the 18 disease groupings used in this survey for each of these four criteria are shown in Table 10.6.

The possible patterns of ranking for diseases on these four criteria can be classified in relation to four extreme possibilities:

(1) the disease ranks consistently high:
(2) the ranks all lie around the average;
(3) the ranks are inconsistent at one extreme or the other; and
(4) the ranks are consistently low.

Figure 10.2 illustrates how the 18 disease groups fall in relation to these four extremes. The position of each disease on the diagram is determined by the number of times it appears in either the top third or bottom third of the four rankings shown in Table 10.6. Each kite-shaped cell of the diagram corresponds to one possible combination of these values. Thus skin disease, which never appears in either the

Table 10.6 Comparative rankings of disease groups by four indicators of the nature of the work/illness link

Disease groups ranked by:							
(a)		(b)		(c)		(d)	
% of cases (caused) where others at work are similarly affected		% of population with cases caused rate >2x or < 0.5x overall average for each disease		cases caused as % of all cases		% of cases (caused) where the people in charge at work accept that the illness is work-related	
100	Vibration white finger	95	Pneumoconiosis	100	Vibration white finger	100	Vibration white finger
100	Pneumoconiosis	89	Vibration white finger	89	Pneumoconiosis	100	Pneumoconiosis
90	Deafness	84	Deafness	87	Trauma and poisoning	91	Headache and 'eyestrain'
82	Upper respiratory	81	Lower respiratory	85	Deafness	90	Deafness
78	Lower respiratory	81	Trauma and poisoning	67	'RSI'	88	Trauma and poisoning
76	Headache and 'eyestrain'	76	Asthma	67	Infections	82	'RSI'
75	Stress	75	Headache and 'eyestrain'	65	Skin	81	Musculoskeletal
67	Varicose veins	68	Varicose veins	64	Eye conditions	80	Asthma
67	Asthma	61	Musculoskeletal	61	Musculoskeletal	79	Stress
65	Eye conditions	56	Eye conditions	57	Stress	79	Lower respiratory
59	Infections	54	Skin	51	Lower respiratory	78	Upper respiratory
59	Skin	49	Exhaustion, ME, symptoms	49	Exhaustion, ME, symptoms	75	Skin
58	Musculoskeletal	42	Other	43	Headache and 'eyestrain'	73	Eye conditions
54	Heart	37	'RSI'	39	Heart	71	Infections
54	'RSI'	35	Infections	38	Other	71	Other
43	Other	30	Upper respiratory	34	Upper respiratory	69	Heart
42	Exhaustion, ME, symptoms	21	Stress	30	Asthma	67	Exhaustion, ME, symptoms
30	Trauma and poisoning	18	Heart	20	Varicose veins	0	Varicose veins

top third or the bottom third of the four rankings is placed in the bottom left hand (0,0) cell of the diagram. Where the position of a disease on the diagram would be changed by a slight shift in the rankings, its name has been moved towards the cell boundary in the appropriate direction. Thus, skin disease, which comes close to having one top third ranking (on criterion (c)) and also to having a bottom third ranking (criteria (a) and (d)), is placed in the top right hand corner of the (0,0) cell.

Most of the disease groups fall fairly clearly into one of the four categories defined above. For vibration white finger, pneumoconiosis and deafness, all criteria are consistent with a genuine work-related effect. Long term consequences of trauma and poisoning should probably be included with this group also since its low position on criterion (a) is to be expected, because accident victims will in general not know of others 'similarly affected'. At the other extreme heart disease, hypertension and stroke and 'other' diseases show no features on these criteria to contradict the *a priori* arguments advanced earlier that these cases are unlikely to constitute evidence of genuine work-related illness.

Five disease groups have average rankings across all — or most — criteria; asthma, musculoskeletal conditions, skin diseases, stress and eye conditions. These five groups represent the average evidential strength of the survey: the baseline from which the 'stronger' and 'weaker' evidence diseases are assessed.

One disease — (RSI) — has conflicting criteria, ranking low on criteria (a) and (b), and high on criteria (c) and (d). An interpretation that resolves this conflict is that RSI is an uncommon individual response to working conditions that can be found in a wide range of occupations. Under this view, the low rankings on criteria (a) and (b) are expected, and RSI would move towards the 'stronger evidence' group.

Headache and 'eyestrain' and lower respiratory disease lie in the area between the consistently high and the average groups. The acute nature of headache and 'eyestrain' makes it inherently more likely that the affected individuals will be able reliably to identify their illness as work-related. This is not the case for lower respiratory disease, and additional data on the pattern of rates by time in/since work in the full report of the Survey casts some doubt on the reliability of cases reported in this disease group.

Infections, varicose veins and exhaustion etc lie between the average group and the consistently low group, and upper respiratory disease lies between the consistently low and the conflicting groups.

Tables showing sample numbers, rates per 10,000 and relative risks by occupational group for all cases (broken down by recently working and inactive > 3 years), and for cases caused by work will be available later for each of the 18 disease groups shown in Table 10.1.

10.3 Other sources of information and description of main tables

10.3.1 The Industrial Injuries Scheme

The Industrial Injuries Scheme (IIS) administered by the Department of Social Security (DSS) compensates workers (or their dependants) injured or killed by an accident at work or suffering from a prescribed disease. The self-employed are not covered by this scheme. Diseases are only 'prescribed' in connection with defined occupations or occupational conditions. For example, tuberculosis is a prescribed disease, but only in respect of individuals whose occupation involves contact with a source of tuberculous infection.

Figure 10.2 Classification of disease groups by the relative strength and consistency of internal survey evidence of work-relatedness

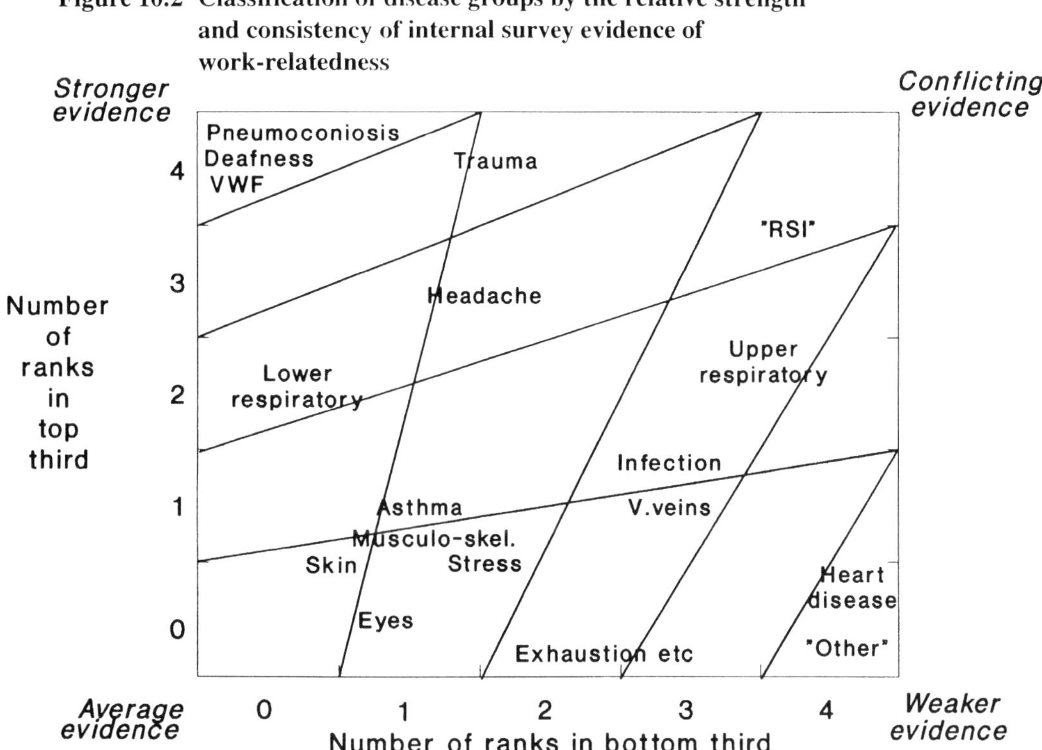

Diseases are only prescribed if some occupational cause is well established, and if terms of prescription can be framed in such a way that most cases falling within the terms will be of genuine occupational origin. Tables 10.8 and 10.9 show the list of currently prescribed diseases. Full details of the terms of prescription are available in DSS publications.[7]

Where there is a long delay between the cause of a disease and its appearance, it is difficult both to identify and prove occupational causation, and to frame satisfactory terms of prescription. Even when this is done, the numbers of awards will probably understate the disease's incidence, because individuals may be unaware of the possible occupational origin of their disease or the availability of compensation, the latter applying to shorter latency diseases as well.

Three principal benefits have been payable under the IIS:

Injury Benefit. Until March 1983, a special, higher rate of sickness benefit was payable to people absent from work because of prescribed disease. The abolition of this special rate of benefit from April 1983 means that information from this source is unavailable for the major part of the period covered by this Decennial Supplement but in 1980/81 and 1981/82 there were respectively, 7,146 and 6,365 total spells of sickness absence qualifying for this benefit. Dermatitis was the cause of 3,927 and 3,415 spells in these successive years, the other major diseases giving rise to this benefit being inflammation of tendons of the hand, forearm, or associated tendon sheaths (2,407 and 2,275 spells); and beat conditions of the hand, knee and elbow (618 and 524 spells). There had been a steady decline in the total of spells qualifying for benefit since earlier years (spells in 1972/3 totalling 16,239 with a distribution across specific diseases broadly similar to that described for the later period).

Until April 1988, Industrial Death Benefit was paid to a worker's dependants where death was caused or materially accelerated by a prescribed disease. Data on Death Benefit awards for pneumoconioses and asbestos related diseases are included in the text and associated figures on specific disease groups (see below). Data for 1978 to 1987 for other prescribed diseases are summarised in Table 10.7. Awards are counted in the actual year of death from 1983 onwards. However for years prior to 1983, notifications delayed beyond the year following the year of death were allocated to the year prior to the year of notification, rather than to the year of death, with the proportion of these late notifications varying from year to year.

Disablement Benefit. This is the only continuing source of data on the incidence of occupational disease available from the IIS. It is paid in cases where a prescribed disease has led to some long-term disability. For claims lodged after 1 October 1986, and for all diseases except pneumoconiosis, byssinosis and mesothelioma, benefit is only paid if disablement is assessed at 14 per cent or more.

The figures for awards of disablement benefit derive from different sources for two groups of the prescribed diseases. Figures for diseases where compensation is assessed by a 'Special Medical Board' (SMB) are compiled on a calendar year basis and have been available for some years (Table 10.8). Statistics for the other prescribed diseases (Table 10.9), with the exception of occupational deafness, are currently compiled for years starting on 1 October and have been available on this basis since October 1983, when statistical record keeping was reorganised within the DSS following the abolition of Industrial Injury Benefit. Before this date, figures for total awards of disablement benefit for these prescribed diseases were not counted separately from awards of injury benefit.

Figures for occupational deafness in Table 10.9 are quoted for calendar years, that is January to December, not October (of the previous year) to September as is the case for the other prescribed diseases in this table. The data prior to 1986-87 are unreliable and have not been presented. Table 10.10 shows (for diseases other than those assessed by SMBs) the proportion assessed as having 1-13 per cent disability (no benefit) and 14 per cent or greater disability (benefit paid). Table 10.11 shows a break down of pneumoconiosis and byssinosis awards by industry to which the disease was attributed and Table 10.12 an age and percentage disablement break down of the pneumoconiosis data for coal mining, asbestos workers and other industries.

10.3.2 Statutory reporting of occupational diseases

Prior to April 1986, certain industrial diseases were notifiable under the Factories Act. Only small numbers of cases were notified — 64 in 1980 including 39 of chrome ulceration and 11 of lead poisoning. The Reporting of Injuries, Diseases and Dangerous Occurrences Regulations (RIDDOR) were introduced in April 1986 largely in response to the loss of data on both accidents and diseases following abolition of Injury Benefit. The regulations require employers to report all cases of a defined list of diseases occurring among their employees where:

(a) they receive a doctor's written diagnosis; and

(b) the affected employee's current job involves the work activity specifically associated with the disease.

The list of reportable diseases mirrors to a large extent that of the list of prescribed diseases current at that time but with some important exceptions (dermatitis, deafness and musculoskeletal conditions). Data for the years 1986/7 to 1993/4 are shown in Table 10.13. Comparison of these figures with those for Disablement Benefit in Tables 10.8 and 10.9 suggests substantial under-reporting under RIDDOR particularly for diseases with long induction periods (for example, the pneumoconioses and occupational cancers). The criteria which need to be fulfilled in order for a case to be reported together with the reporting mechanism itself are thought to be largely responsible for this under-reporting.*

* Proposals for improving RIDDOR, including proposals for improving the reporting of occupational diseases are under review at the time of writing this supplement.

Text continues on page 168

Table 10.7 Deaths resulting in award of Industrial Death Benefit etc, by scheme and main disease, 1978-87*

	1978	1979	1980	1981	1982	1983	1984	1985	1986	1987
Industrial Injuries Scheme										
Asbestosis	41	74	78	65	70	92	79	87	101	104
Other pneumoconiosis	548	535	510	474	482	482	367	436	375	338
Byssinosis	14	9	12	10	9	10	5	12	11	6
Farmer's lung	-	3	2	2	3	8	3	5	4	1
Papilloma of the bladder	8	10	15	7	12	13	9	9	9	15
Mesothelioma	109	131	133	175	190	202	250	289	292	339
Other prescribed diseases	16	16	13	23	16	33	30	45	40	47
Total II Scheme	736	778	763	756	782	840	743	883	832	850
Pneumoconiosis, Byssinosis and Miscellaneous Diseases Benefit Scheme										
Asbestosis	2	-	2	1						
Other pneumoconiosis	63	61	67	40	44	38	48	31	24	38
Byssinosis	1	1	-	-						
Other diseases	15	15	12	9	10	11	19	9	21	20
Total PBMDB scheme	81	77	81	50	54	49	67	40	45	58
Certification that death was due to the disease (Workers' Compensation scheme)										
Other pneumoconiosis	54	60	66	68	48	60	50	40	40	22
Total WC scheme	54	60	66	68	48	60	50	40	40	22
Total all schemes	**871**	**915**	**910**	**874**	**884**	**949**	**860**	**963**	**917**	**930**
of which pneumoconiosis (including asbestosis and byssinosis)	723	740	735	658	653	682	549	606	551	508

* Death benefit is not payable after April 10, 1988: 1987 is the last full year of data.

Source: DSS

Table 10.8 Prescribed industrial diseases assessed by Special Medical Boards: new cases of assessed disablement by disease, 1980-93

Disease No.	Disease	1980	1981	1982	1983	1984	1985	1986	1987	1988	1989	1990	1991	1992	1993
B6	Farmers' lung	14	12	11	8	4	6	11	8	15	13	7	5 (1)	5	3
C15	Poisoning by nitrous fumes	-	-	4	1	-	-	-	3	-	-	-	-	1	-
C17	Beryllium poisoning	-	1	2	1	-	-	2	4	3	-	2	1	-	1
C18	Cadmium poisoning	-	2	3	4	1	2	3	3	2	-	2	5	4	1
C22b	Primary carcinoma of bronchus or lung in nickel workers	-	-	-	1	5	2	3	-	-	-	1	2	1	-
D1	Pneumoconiosis (a)	728	734	733	670	577	702	747	652	562	661	709 (7)	751 (8)	765 (10)	853 (17)
D2	Byssinosis (a)	156	108	133	72	56	37	26	23	13	15	18 (11)	7 (2)	4 (1)	5 (3)
D3	Diffuse mesothelioma	-	93	123	148	201	245	305	399	479	441	462 (14)	519 (21)	551 (13)	608 (19)
D7	Occupational asthma (b,c)	..	..	95	183	137	166	166	220	222	220	216 (49)	293 (57)	553 (115)	510(120)
D8	Primary carcinoma of the lung with accompanying evidence of one or both of (1) asbestosis (2) bilateral diffuse pleural thickening (c,d)	..	..	..	..	..	8	34	55	59	54	58	55 (1)	54	72 (2)
D9	Bilateral pleural thickening (c)	..	..	..	..	..	61	111	115	114	125	146 (1)	149	160 (3)	172 (2)
D10	Primary carcinoma of the lung (c,e)	..	..	..	..	..	..	..	-	-	4	5	4	5 (1)	2
D11	Primary carcinoma of the lung with accompanying silicosis (c)	..	..	..	..	..	..	..	..	..	..	..	..	..	1
D12	Chronic bronchitis and/or emphysema (c)	..	..	..	..	..	..	..	..	..	..	..	..	..	1560
Total		898	950	1104	1088	981	1229	1408	1482	1469	1533	1626	1791	2103	3788

(a) See also tables 10.11 and 10.12

Source: DSS

(b) See also table 10.17

(c) The following diseases were prescribed after January 1 1982.

Disease No.	Date Prescribed
D7	March 29 1982
D8	April 1 1985
D9	April 1 1985
D10	April 1 1987
D11	April 19 1993
D12	Sept 13 1993

(d) Previously classified as 'Lung cancer in asbestos workers'. New classification used from April 1993

(e) Previously classified as 'Lung cancer'.

.. Not applicable.

Figures in brackets show the number of females (data available from 1990 only). Where no figure is given, all cases were male.

Table 10.9 Prescribed industrial diseases other than those assessed by Special Medical Boards(a): new cases of assessed disablement by disease, 1983/84 to 1992/93(b)

Disease No.	Disease	1983/84	1984/85	1985/86	1986/87	1987/88	1988/89	1989/90	1990/91	1991/92	1992/93(p)
Conditions due to physical agents (physical cause)											
A1	Radiation effects	1	-	1	-	2	-	-	8	3	3 (1)
A2	Heat cataract	1	-	1	2	2	3	7	10	3	5 (1)
A3	Decompression sickness	1	-	4	-	1	1	2	2	1	3
A4	Cramp of hand or forearm	3	3	3	13	11	14	18	46 (29)	52 (36)	116 (95)
A5	Beat hand	64	73	79	14	22	11	5	16 (3)	17 (4)	16 (4)
A6	Beat knee	29	47	82	37	138	97	74	151 (2)	269 (5)	202 (4)
A7	Beat elbow	38	60	59	6	11	4	16	20 (1)	31 (5)	38 (2)
A8	Inflammation of tendons of the hand, forearm or associated tendon sheaths (Tenosynovitis)	337	390	619	376	322	294	423	556 (421)	649 (518)	911 (733)
A9	Miner's nystagmus	-	1	-	-	1	-	-	-	1	1
A10	Occupational deafness (d)	n/a	n/a	n/a	1202	1261	1170	1128	1041 (n/a)	972 (n/a)	901 (n/a)
A11	Vibration white finger (c)	..	3	641	1366	1673	1056	2601	5403 (41)	2369 (22)	1447 (21)
A12	Carpal tunnel syndrome (c)	..	..	..	..	..	..	..	..	..	20 (5)
Conditions due to biological agents (caused by animal, plant or other living agent)											
B1	Anthrax	-	-	1	-	-	-	-	1	-	-
B3	Infection by leptospira	2	-	-	1	-	-	2	-	1	1
B5	Tuberculosis	6	7	3	13	3	5	-	3 (3)	3 (1)	6 (4)
B7	Brucellosis	3	-	-	2	-	1	2	1 (1)	1 (1)	1
B8	Viral hepatitis	3	5	9	5	3	1	1	2 (1)	4 (3)	1 (1)
B9	Infection by streptococcus suis (c)	1	-	-	3	-	-	-	1	-	-
B10a	Avian chlamydiosis (c)	..	..	..	..	..	-	-	1	-	1 (1)
B10b	Ovine chlamydiosis (c)	..	..	..	..	..	-	-	-	1	-
B11	Q fever (c)	..	..	..	..	..	-	-	1	-	1
B12	ORF (c)	..	..	..	..	..	..	..	..	2	-
Conditions due to chemical agents											
C1	Poisoning by lead or compounds of lead	2	2	2	3	1	-	-	2	2 (2)	1 (1)
C2	Poisoning by manganese	-	-	-	-	-	-	-	1	-	2
C3	Poisoning by phosphose or an inorganic compound of phosphorus	-	-	-	-	-	-	-	-	-	2
C4	Poisoning by arsenic	-	-	-	-	-	-	1	-	-	-
C5	Poisoning by mercury or compound of mercury	-	-	-	3	-	-	-	-	-	-
C6	Poisoning by carbon disulphide	-	-	-	-	1	-	-	-	-	1
C7	Poisoning by benzene or a homologue of benzene	-	2	1	3	-	3	1	-	5 (2)	1
C8	Poisoning by nitro-, amino-, or chloro-benzene or homologues	-	-	-	-	-	13	3	-	-	1
C9	Poisoning by dinitrophenol	-	-	-	-	-	-	-	1	-	-
C10	Poisoning by tetrachlorethane	-	-	-	-	-	-	-	-	-	-
C13	Poisoning by chlorinated napthalene	-	-	-	-	-	-	1	-	-	-
C14	Poisoning by nickel carbonyl	-	-	-	-	-	-	-	-	-	1
C19	Acrylamide monomer	-	-	-	-	-	-	-	-	1 (1)	-
C20	Dystrophy of the cornea (including ulceration of the corneal surface) of the eye	-	-	-	1	-	-	1	-	1	-
C21a	Localised new growth of skin	1	2	2	4	3	2	4	5	6	4
C21b	Squamous celled carcinoma of skin	1	-	2	4	3	2	5	-	2	1
C22a	Carcinoma of the mucous membrane of the nose or associated air passages	-	-	-	-	-	-	-	-	-	-
C23	Papilloma of the bladder	5	5	5	21	21	7	8	16	21 (1)	26 (4)
C24b	Osteolysis of the terminal phalanges of the fingers	-	-	-	-	-	-	-	-	-	-
C25	Occupational vitiligo	3	-	-	2	-	1	-	-	-	-
C26	Liver/kidney damage due to carbon tetrachloride (c)	..	..	..	-	-	-	-	1	-	-
C28	Central nervous system dysfunction & associated gastro-intestinal disorders due to exposure to chloromethane	..	..	..	-	-	-	-	-	-	-
C29	Peripheral neuropathy due to exposure to n-hexane or methyl n-butyl keytone (c)	..	..	..	..	-	-	1	-	1	-
Miscellaneous conditions											
D4	Inflammation/ulceration of mucous membrane of upper respiratory tract or mouth	12	9	17	36	19	15	22	13 (5)	75 (8)	494 (33)
D5	Dermatitis	611	619	785	464	368	285	301	434 (149)	411 (140)	419 (152)
D6	Adeno-carcinoma of nasal cavity/nasal carcinoma	2	1	5	2	5	2	5	1	-	3 (1)
Total		1126	1229	2321	3583	3871	2987	4632	7737	4904	4630

(P) These figures are provisional

Source:DSS

Figures in brackets show the number of females (data available from 1990-91 only). Where no figure is given, all cases were male.

(a) See Table 10.8

(b) Years starting October 1

(c) The following diseases were prescribed after October 1 1983:

Disease no	Date prescribed
B9	October 3 1983
A11	April 1 1985
C26, C29	January 4 1988
B10a, B10b, B11	July 19 1989
B12	September 26 1991
A12	April 19 1993

(d) Figures for occupational deafness are based on calendar years to 1993, that is Jan-Dec 1987 to Jan-Dec 1993. Figures prior to 1987 are not available.

Awards for occupational deafness are made only for disablement at 20 per cent or more.

.. Not applicable

n/a Not available

Table 10.10 Prescribed industrial diseases other than those assessed by Special Medical Boards (a) new cases of assessed disablement by percentage assessment and disease 1986-87 to 1992-93 (b)

Disease no.	Disease	Claims assessed in 1986/87			Claims assessed in 1987/88			Claims assessed in 1988/89			Claims assessed in 1989/90		Claims assessed in 1990/91		Claims assessed in 1991/92		Claims assessed in 1992/93 (p)	
		Old rules payment	New rules assessment		Old rules payment	New rules assessment		Old rules payment	New rules assessment									
			1-13% (no benefit)	14%+ (Benefit paid)		1-13% (no benefit)	14%+ (Benefit paid)		1-13% (no benefit)	14%+ (Benefit paid)	1-13% (no benefit)	14%+ (Benefit paid)	1-13% (no benefit)	14%+ (Benefit paid)	1-13% (no benefit)	14%+ (Benefit paid)	1-13% (no benefit)	14%+ (Benefit paid)
Conditions due to physical agents (physical cause)																		
A1	Radiation effects	-	-	-	-	2	-	-	-	-	-	-	7	1	1	2	1	2 (1)
A2	Heat cataract	2	-	-	-	1	1	-	2	1	4	3	4	6	3	-	4 (1)	1
A3	Decompression sickness	-	-	-	1	-	-	-	1	-	-	2	-	2	1	-	2	1
A4	Cramp of hand or forearm	11	2	-	3	7	1	1	13	-	17	1	41 (26)	5 (3)	45 (31)	7 (5)	97 (77)	19 (18)
A5	Beat hand	14	-	-	1	21	-	1	10	-	5	-	16 (3)	-	16 (4)	1	15 (3)	1 (1)
A6	Beat knee	24	13	-	10	122	6	3	86	8	72	2	148 (2)	3	265 (5)	4	197 (3)	5 (1)
A7	Beat elbow	5	1	-	4	6	1	-	4	-	15	1	20 (1)	-	30 (4)	1 (1)	36 (2)	2
A8	Inflammation of tendons of the hand, forearm or associated tendon sheaths (Tenosynovitis)	285	87	4	49	255	18	7	281	6	388	35	492 (371)	64 (50)	544 (435)	105 (83)	686 (551)	225 (182)
A9	Miner's nystagmus	-	-	-	-	1	-	-	-	-	-	-	-	-	1	-	1	-
A10	Occupational deafness (d)	1202	..	..	1261	..	..	1170	..	..	..	1128	..	1041	..	972	..	901
A11	Vibration white finger (c)	1055	300	11	140	1396	137	17	926	113	2566	35	5372 (39)	31 (2)	2343 (22)	26	1390 (18)	57 (3)
A12	Carpal tunnel syndrome (c)	..	..	..	..	..	..	..	..	..	..	..	..	..	..	..	15 (2)	5 (3)
Conditions due to biological agents (caused by animal, plant or other living agent)																		
B1	Anthrax	-	-	-	-	-	-	-	-	-	-	-	1	-	-	-	-	-
B3	Infection by leptospira	-	1	-	-	-	-	-	-	-	1	1	-	-	-	1	1	-
B5	Tuberculosis	8	1	4	1	1	1	-	4	1	-	-	2 (2)	1 (1)	2 (1)	1	1 (1)	5 (3)
B7	Brucellosis	2	-	-	-	-	-	-	1	-	-	2	1 (1)	-	-	1 (1)	-	1
B8	Viral hepatitis	3	2	-	-	1	2	-	1	-	-	1	-	2 (1)	3 (2)	1 (1)	-	1 (1)
B9	Infection by streptococcus suis (c)	3	-	-	-	-	-	-	-	-	-	-	-	1	-	-	-	-
B10a	Avian chlamydiosis (c)	..	..	..	..	..	..	-	-	-	-	-	-	1	-	-	1 (1)	-
B10b	Ovine chlamydiosis (c)	..	..	..	..	..	..	-	-	-	-	-	-	-	-	1	-	-
B11	Q fever (c)	..	..	..	..	..	..	-	-	-	-	-	-	1	-	-	1	-
B12	ORF (c)	..	..	..	..	..	..	..	..	..	..	..	..	..	2	-	-	-
Conditions due to chemical agents																		
C1	Poisoning by lead or compounds of lead	3	-	-	-	1	-	-	-	-	-	-	2	-	2	-	-	1 (1)
C2	Poisoning by manganese	..	-	-	-	-	-	-	-	-	-	-	-	1	-	-	1	1
C3	Poisoning by phosphose or an inorganic compound of phosphorus														-	-	1	1
C4	Poisoning by arsenic	-	-	-	-	-	-	-	-	-	-	1	-	-	-	-	-	-
C5	Poisoning by mercury or compound of mercury	3	-	-	-	-	-	-	-	-	-	-	-	-	-	-	-	-
C6	Poisoning by carbon disulphide	-	-	-	-	1	-	-	-	-	-	-	-	-	-	-	-	1
C7	Poisoning by benzene or a homologue of benzene	3	-	-	-	-	-	1	2	-	1	-	-	-	3 (2)	2	1	-
C8	Poisoning by nitro-, amino-, or chloro-benzene or homologues	-	-	-	-	-	-	1	12	-	2	1	-	-	-	-	1	-
C9	Poisoning by dinitrophenol	-	..	-	-	-	-	-	-	-	-	-	1	-	-	-	-	-
C13	Poisoning by chlorinated napthalene	-	-	-	-	-	-	-	-	-	-	1	-	-	-	-	-	-
C14	Poisoning by nickel carbonyl	-	-	-	-	-	-	-	-	-	-	-	-	-	-	-	1	-
C19	Poisoning by acrylamide monomer										..	..	..	..	1	-	-	-
C20	Dystrophy of the cornea (including ulceration of the corneal surface) of the eye	1	-	-	-	-	-	-	-	-	-	1	-	-	1	-	-	-
C21a	Localised new growth of skin	3	1	-	1	1	1	-	2	-	3	1	5	-	5	1	2	2
C21b	Squamous celled carcinoma of skin	3	1	-	2	1	-	-	2	-	4	1	-	-	1	1	1	-
C23	Papilloma of the bladder	17	-	4	9	4	8	1	4	2	4	4	3	13	9	12 (1)	5 (1)	21 (3)
C25	Occupational vitiligo	1	-	1	-	-	-	-	1	-	-	-	-	-	-	-	-	-
C26	Kidney/Liver damage due to carbon tetrachloride (c)	..	-	-	-	-	-	-	-	-	-	-	-	1	-	-	-	-
C29	Poisoning by peripheral neuropathy due to exposure to n-hexane or methyl n-butyl keytone (c)	..	..	..	-	-	-	-	-	-	-	1	-	-	-	1	-	-
Miscellaneous conditions																		
D4	Inflammation/ulceration of mucous membrane of upper respiratory tract or mouth	25	11	-	-	18	1	-	14	1	21	1	13 (5)	-	74 (8)	1	486 (33)	8
D5	Dermatitis	354	105	5	53	305	10	15	267	3	294	7	408 (140)	26 (9)	379 (128)	32 (12)	374 (136)	45 (16)
D6	Adeno-carcinoma of nasal cavity/nasal carcinoma	2	-	-	-	1	4	-	1	1	3	2	1	-	-	-	1	2 (1)
Total		3029	525	29	1535	2145	191	1217	1634	136	3400	1232	6537	1200	3731	1173	3322	1308

See footnotes to table 10.9

Source : DSS

Table 10.11 Pneumoconiosis and Byssinosis: new cases diagnosed by Special Medical Boards (Respiratory Diseases)(a) by industry to which the disease was attributed (b), 1980-93

	1980	1981	1982	1983	1984	1985	1986	1987	1988	1989	1990	1991	1992	1993
Industrial Injuries Scheme Cases														
Pneumoconiosis														
Coal mining	461	493	467	402	330	364	357	325	299	339	344	379	383	395
Other mining and quarrying:														
Slate	47	27	24	12	8	7	11	6	3	8	5	3	2	7
Other-except refractories	8	15	13	5	7	1	12	12	9	3	9	2	2	5
Asbestos (c)	144	140	172	199	186	273	312	247	202	268	306 (7)	330 (7)	354 (9)	418 (16)
Foundry workers														
Iron foundry workers	17	9	10	10	13	17	17	13	12	10	9	11	5	6
Steel foundry workers	4	2	3	7	-	1	1	5	6	6	4	1	2	2
Non-ferrous foundry workers	-	2	4	-	1	-	1	1	2	-	3	-	-	-
Steel dressers	3	3	2	5	3	6	2	2	3	2	-	2	2	2
Pottery manufacture	18	10	17	14	9	14	10	18	11	9	6	8 (1)	4 (1)	6 (1)
Refractories (d)	6	5	3	5	5	3	6	3	6	4	7	-	2	-
Other attributable industries	20	28	18	11	15	16	18	20	9	12	16	15	9	12
Total	728	734	733	670	577	702	747	652	562	661	709	751	765	853
Byssinosis														
Cotton	148	108	124	67	53	36	25	23	13	15	*- breakdown not available*			
Flax	-	-	9	5	3	1	1	-	-	-	*- breakdown not available*			
Total	148	108	133	72	56	37	26	23	13	15	18	7	4	5
Cases diagnosed by Medical Appeal Tribunals														
Pneumoconiosis (excluding asbestosis)	25	26	25	25	30	21	28	36	32	26	*- breakdown not available*			
Asbestosis	6	13	13	13	14	28	17	35	23	12	*- breakdown not available*			
Byssinosis	2	2	2	2	7	-	1	2	2	-	*- breakdown not available*			
Total	33	41	40	40	51	49	46	73	57	38	*- breakdown not available*			
PBMDB scheme cases (e)														
Pneumoconiosis and Byssinosis	62	40	25	44	30	18	17	28	20	18	14	26	17	21
Overall total: Pneumoconiosis and Byssinosis	971	923	931	826	714	806	836	776	652	732	741	784	786	879

Source: DSS

(a) Formerly known as Pneumoconiosis Medical Panels.
(b) The industry to which the disease is attributable is in some cases defined occupationally.
(c) Cases where mesothelioma was also diagnosed are excluded, and shown in Table 10.8
(d) Including the mining, quarrying and processing of refractory material.
(e) The figures of Pneumoconiosis, Byssinosis and Miscellaneous Diseases Benefits scheme cases refer to years ending September 30.

Figures in brackets show the number of females (data available from 1990 except for PBMDB cases). Where no figure is given, all cases were male.

Table 10.12 Pneumoconiosis: new Industrial Injuries Scheme cases diagnosed by Special Medical Boards (Respiratory Diseases) (a) in coal mining, asbestos and other industries, by age and percentage disablement (b), 1990-93

Age	Percentage disablement assessed 10 or less	20 30 40	50 60 70	80 90 100	Total	Age	Percentage disablement assessed 10 or less	20 30 40	50 60 70	80 90 100	Total
Coal mining						**Asbestos workers**(c)					
1990						1990					
Under 45	1	0	0	0	1	Under 45	1	0	0	1	2
45 - 64	40	22	3	1	66	45 - 64	29	98	7	12	146
65 +	117	135	21	4	277	65 +	29	104	14	11	158
Total	158	157	24	5	344	Total	59	202	21	24	306
1991						1991					
Under 45	1	0	0	0	1	Under 45	1	0	1	1	3
45 - 64	41	22	1	0	64	45 - 64	35	97	4	8	144
65 +	120	165	22	7	314	65 +	33	122	10	18	183
Total	162	187	23	7	379	Total	69	219	15	27	330
1992						1992					
Under 45	1	0	0	0	1	Under 45	2	3	0	0	5
45 - 64	53	22	2	0	77	45 - 64	40	106	2	11	159
65 +	114	160	26	5	305	65 +	42	109	13	26	190
Total	168	182	28	5	383	Total	84	218	15	37	354
1993						1993					
Under 45	3	0	0	0	3	Under 45	0	0	0	1	1
45 - 64	25	23	0	0	48	45 - 64	49	113	6	5	173
65 +	118	189	26	11	344	65 +	44	170	12	18	244
Total	146	212	26	11	395	Total	93	283	18	24	418
Other						**Total**					
1990						1990					
Under 45	3	1	0	0	4	Under 45	5	1	0	1	7
45 - 64	8	7	1	0	16	45 - 64	77	127	11	13	228
65 +	14	21	3	1	39	65 +	160	260	38	16	474
Total	25	29	4	1	59	Total	242	388	49	30	709
1991						1991					
Under 45	0	1	1	0	2	Under 45	2	1	2	1	6
45 - 64	8	7	0	1	16	45 - 64	84	126	5	9	224
65 +	8	14	2	0	24	65 +	161	301	34	25	521
Total	16	22	3	1	42	Total	247	428	41	35	751
1992						1992					
Under 45	0	0	0	0	0	Under 45	3	3	0	0	6
45 - 64	5	2	0	0	7	45 - 64	98	130	4	11	243
65 +	7	12	1	1	21	65 +	163	281	40	32	516
Total	12	14	1	1	28	Total	264	414	44	43	765
1993						1993					
Under 45	0	1	0	0	1	Under 45	3	1	0	1	5
45 - 64	9	5	0	1	15	45 - 64	83	141	6	6	236
65 +	2	15	5	2	24	65 +	164	374	43	31	612
Total	11	21	5	3	40	Total	250	516	49	38	853

(a) Formerly known as Pneumoconiosis Medical Panels.

Source: DSS

(b) Under a special provision a person found to be suffering from pneumoconiosis qualifies for a pension at the 10% rate even if he or she has no discernible respiratory disablement arising from the disease.

(c) Cases where mesothelioma was also diagnosed are excluded and shown in Table 10.8

Table 10.13 Cases of occupational disease reported under RIDDOR, 1986/87 to 1993/94 (a)

Disease		1986/87	1987/88	1988/89	1989/90	1990/91	1991/92	1992/93	1993/94 (p)	Corresponding DSS PD Number
Poisoning by										
1a	Acrylamide	-	-	-	-	-	-	-	-	C19
1b	Arsenic	-	1	2	-	-	-	-	1	C4
1c	Benzene	-	1	-	1	1	1	1	-	C7
1d	Beryllium	-	-	-	3	-	-	-	-	C17
1e	Cadmium	1	1	2	1	3	1	-	-	C18
1f	Carbon disulphide	-	-	-	-	-	-	-	-	C6
1g	Diethylene dioxide	-	-	-	-	-	-	-	-	C11
1h	Ethylene oxide	-	-	-	-	-	-	-		
1i	Lead	3	5	6	4	1	4	-	7	C1
1j	Manganese	-	-	-	-	-	1	-	-	C2
1k	Mercury	2	1	-	-	-	-	1	1	C5
1l	Methyl bromide	-	2	-	1	1	-	-	1	C12
1m	Nitrochlorobenzene	3	-	2	1	2	1	1	-	C8
1n	Oxides of nitrogen	-	1	-	1	-	-	1	-	C15
1o	Phosphorus	4	2	3	2	1	1	1	8	C3
2	Chrome ulcer	11	19	14	6	13	11	2	17	*
3	Folliculitis	5	1	1	-	1	1	-	2	*
4	Acne	-	1	-	1	-	-	-	3	*
5	Skin cancer	3	-	1	4	4	2	-	1	C21
6	Radiation skin injury	-	6	2	1	4	1	7	5	part A1
7	Occupational asthma	70	45	61	57	70	68	70	90	D7
8	Extrinsic alveolitis	4	13	7	5	7	2	8	3	B6
9	Pneumoconiosis	13	5	4	6	4	3	3	1	part D1
10	Byssinosis	-	-	1	2	-	2	2	-	D2
11	Mesothelioma	7	13	9	4	13	9	8	7	D3
12	Lung cancer (asbestos)	1	1	-	-	-	-	-	-	D8
13	Asbestosis	11	14	-	10	4	5	8	3	part D1
14	Lung cancer (nickel)	-	-	-	-	-	-	-	-	C22b
15	Leptospirosis	5	12	7	9	2	14	10	10	B3
16	Hepatitis	28	25	23	20	23	41	17	17	B8
17	Tuberculosis	14	11	7	7	9	9	12	14	B5
18	Pathogenic infection	20	6	16	15	20	28	18	19	
19	Anthrax	-	-	-	-	-	1	-	-	B1
20	Bone cancer	-	-	-	-	-	-	-	-	part A1
21	Blood dyscrasia	1	-	-	-	-	3	3	-	
22	Cataract	3	7	-	-	2	-	-	-	A2
23	Decompression sickness	-	25	71	34	2	42	18	13	A3
24	Barotrauma	-	-	1	-	-	-	1	-	
25	Nasal/sinus cancer	-	1	-	2	2	-	2	-	C22a\D6
26	Angiosarcoma	-	-	-	-	-	-	2	-	C24a
27	Urinary tract cancer	-	6	1	4	-	3	-	1	C23
28	Vibration white finger	70	111	68	120	120	131	136	108	A11
Total		279	336	309	321	309	385	332	332	

(P) Provisional
(a) Years starting April
* These three RIDDOR categories form part of DSS PD D5 (dermatitis), not separately identifiable in DSS figures. Dermatitis in general is not reportable under RIDDOR.

The data in this table record the extent of the employer reporting of diseases scheduled under RIDDOR, not the incidence of the diseases themselves. Comparison with other information sources - themselves incomplete - show that the number of cases reported under RIDDOR clearly understates the real incidence of work-related disease.

Table 10.14 Death certificates mentioning specified asbestos-related disease 1968-91

Disease	Year of death																							
	1968	1969	1970	1971	1972	1973	1974	1975	1976	1977	1978	1979	1980	1981	1982	1983	1984	1985	1986	1987	1988	1989	1990	1991p
Asbestosis																								
A Together with lung cancer	25	24	26	32	44	43	33	49	53	59	60	46	56	77	75	60	60	66	84	59	78	75	76	57
B Together with mesothelioma	32	27	40	29	40	30	65	50	74	53	85	76	69	65	79	89	86	87	65	109	89	97	120	86
C Alone or together with other diseases	23	27	21	33	24	34	41	48	63	73	49	56	46	60	53	61	69	74	82	85	74	82	88	106
Total A + C	48	51	47	65	68	77	74	97	116	132	109	102	102	137	128	121	129	140	166	144	152	157	164	163
Total asbestosis deaths (A+B+C)	80	78	87	94	108	107	139	147	190	185	194	178	171	202	207	210	215	227	231	253	241	254	284	249
Mesothelioma																								
Of pleura	112	115	138	131	154	162	181	207	230	260	281	341	369	326	342	468	481	453	529	606	641	640	651	746
Of peritoneum	17	19	18	15	22	25	24	19	32	25	33	38	43	27	31	41	44	48	43	44	54	54	59	49
Of pleura and peritoneum	4	6	4	7	3	10	5	6	7	11	23	13	8	7	20	9	12	16	14	19	20	15	17	17
Site not specified	21	19	34	28	33	27	33	38	43	37	54	41	37	108	109	51	81	97	114	139	147	190	153	197
D Total mesothelioma deaths (includes B above)	154	159	194	181	212	224	243	270	312	333	391	433	457	468	502	569	618	614	700	808	862	899	880	1009
Males	115	123	145	141	169	182	185	218	256	273	328	340	355	395	411	474	534	531	599	702	751	767	763	861
Females	39	36	49	40	43	42	58	52	56	60	63	93	102	73	91	95	84	83	101	106	111	132	117	148
Total number of deaths (A+C+D)	202	210	241	246	280	301	317	367	428	465	500	535	559	605	630	690	747	754	866	952	1014	1056	1044	1172

p provisional

10.3.3 Asbestosis and mesothelioma recorded on death certificates

The figures in Table 10.14 are derived from information recorded on death certificates. They show the numbers of death certificates issued each year on which either asbestosis or mesothelioma (or both) are mentioned. Some death certificates mentioning both conditions do so in ways which suggest that the word 'asbestosis' or mesothelioma, rather than the existence of an asbestos induced lung fibrosis, which is what the word should strictly mean.

Consequently, the trends in deaths from asbestosis *per se* are probably better reflected by the figures for asbestosis without mention of mesothelioma, rather than the total of certificates mentioning asbestosis. Chapter 9 contains more detailed information on asbestos related diseases.

10.3.4 Other deaths from occupational lung disease

Deaths from pneumoconiosis, byssinosis and farmers' lung, recorded on death certificates as the underlying cause of death and obtained from mortality data supplied by OPCS and the General Register Office for Scotland, are displayed in Table 10.15.

Other sources of data are described in the detailed commentary that now follows.

10.4 Detailed commentary on specific diseases

10.4.1 Pneumoconiosis (other than asbestosis)

The rules governing the award of Disablement Benefit for pneumoconiosis have not been affected either by the abolition of Injury Benefit or by the restriction of benefit to cases with higher levels of disability. Knowledge of the disease and of the arrangements for compensation are widespread within the main affected industries: mining, quarrying, foundries and potteries. The figures for compensated cases can therefore be expected to give a reasonably accurate reflection of the incidence of disease. This is borne out by the similar trends shown by the three available series: Disablement Benefit, Death Benefit and deaths with pneumoconiosis as their registered underlying cause (Figure 10.3).

Table 10.15 Deaths due to occupationally related lung disease, other than asbestos (a), 1981-92

	Year at death											
	1981	1982	1983	1984	1985	1986	1987	1988	1989	1990	1991	1992
Pneumoconiosis (other than asbestos)	341	314 (2)	317 (2)	314 (3)	324 (3)	337 (3)	279 (3)	281 (1)	318 (2)	328 (2)	287	274 (1)
Byssinosis	26	22 (11)	33 (16)	24 (13)	25 (19)	29 (15)	25 (17)	22 (13)	25 (17)	19 (12)	16 (13)	21 (14)
Farmer's lung and other occupational allergic alveolitis	13	15 (1)	15 (1)	10	7 (1)	15 (2)	16 (2)	9	8 (1)	6	8 (1)	4 (1)
Total	380	351 (14)	365 (19)	348 (16)	356 (23)	381 (20)	320 (22)	312 (14)	351 (20)	353 (14)	311 (14)	299 (16)

(a) The data in this table are derived from death certificates.
The figure is the number of deaths coded to the disease as underlying cause.
Figures in brackets show the number of females. Where no figure is given, all cases were male.

Source OPCS

Figure 10.3 Pneumoconiosis (other than asbestosis), Great Britain, 1971-93

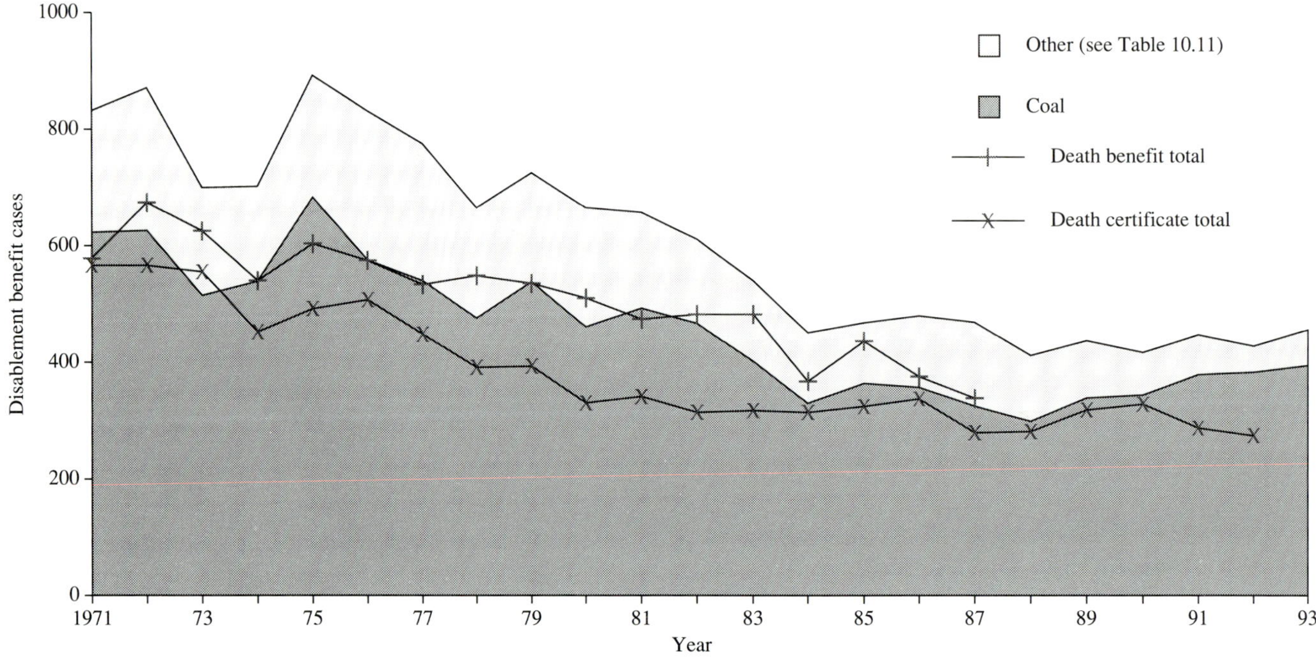

From 1990 disablement figures do not include cases awarded by medical appeals tribunals (see Table 10.11)

The pneumoconioses have a long latency between exposure and onset of disease. Only in exceptional cases will the disease be produced in less than 10 years, and most cases appear between 15 and 30 years from first exposure. This means that the cases now coming forward largely reflect the working conditions of 10 and more years ago.

As might be expected, coal mining still accounts for the majority of compensated cases (91 per cent) outside the asbestos industry, most of these occurring amongst the retired. The proportions of workers diagnosed before retirement age is higher in the other industries affected, although absolute numbers are much lower.

Due to the long and variable delay from first exposure to the onset of detectable disability, the broad trends of the figures is more informative than any detailed fluctuations from year to year in drawing conclusions about changes in the incidence of these diseases; and, by implication, changes in the conditions that produced them. In these broad terms, there had been a long term decline in the incidence of pneumoconiosis other than asbestosis until 1988 since when there has been a slight but continuing increase in the numbers of cases counted, principally due to a small but constant increase in the numbers of compensated coal miners. The much smaller numbers of compensated workers in other industries have fluctuated over the same time period.

10.4.2 Asbestos-related disease

Four diseases have been unequivocally linked to asbestos exposure: asbestosis, mesothelioma, lung cancer and pleural thickening. By definition, every case of asbestosis is due to asbestos; the association with mesothelioma is also very strong, though there is thought to be a low 'natural' background incidence. For lung cancer the situation is different, since the predominant cause of this cancer is smoking, and asbestos exposure increases the risk of disease both in smokers and non-smokers (though, in absolute terms, much more so for smokers than non-smokers). Lung cancer is a prescribed disease in connection with asbestos provided the individual shows some other clinical sign of asbestos exposure (asbestosis, or pleural thickening), as well as evidence of occupational asbestos exposure. All 3 of these diseases display long delays from first exposure to diagnosis.

For asbestosis (Figure 10.4), Disablement Benefit awards show a continuing, but erratic upward trend. They fluctuated between 100 and 200 per year while rising slowly through the 1970s and early 1980s, but then rose to a peak of 329 in 1986, falling back to 225 in 1988, and rising again to a maximum of 418 in 1993. Throughout this period awards of death benefit also grew from around 70 in the early 1970s to just over 100 in 1987, the last full year for which claims could be made. Death certificates mentioning asbestosis (excluding those also mentioning mesothelioma), show a somewhat stronger increase from similar levels in the mid-1970s to around 163 in 1991.

The Disablement Benefit data show that relatively higher numbers of younger (pre-retirement age) workers are affected by asbestosis than is the case for pneumoconiosis in the coal mining industry (Table 10.12).

Analysis of claims for disablement benefit in the last three years (1991 to 1993) shows that 82 per cent of current diagnoses still relate to dates of first exposure up to the end of the 1950s, with first exposures peaking between 40 and 50 years ago, although latencies of up to 60 years and over are also apparent. However, 13 per cent relate to more recent

Figure 10.4 Asbestosis, Great Britain, 1971-93

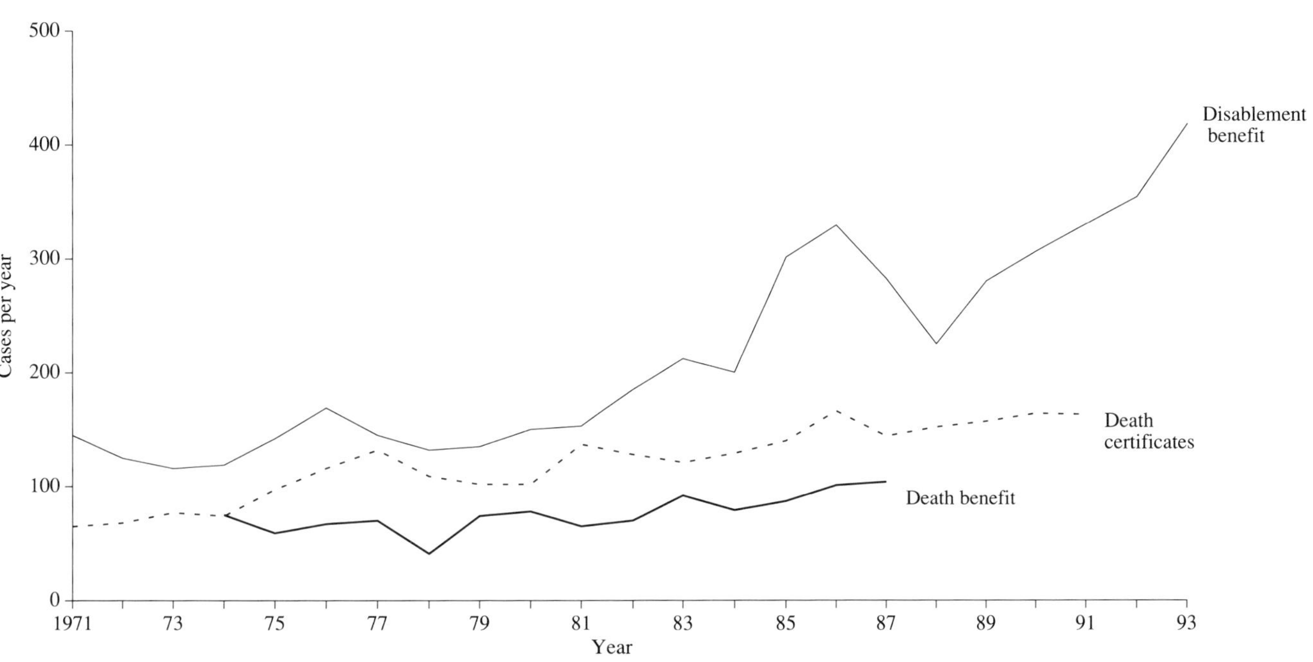

From 1990 disablement benefit figures do not include cases awarded by medical appeals tribunals (see Table 10.11)

exposure in the 1960s, reflecting shorter latencies of between 20 and 30 years, and a small number (5 per cent) — which reflects however a relatively short latency period for the people involved — relate to first exposures after the implementation of the 1969 Asbestos Regulations in 1970.

Deaths from mesothelioma (Figure 10.5) show a steady upward trend between 1971 and 1984. This was interrupted in 1985 when the number of deaths fell from 626 to 615. In 1986 and 1987 there were steep increases of 14 per cent and 15 per cent respectively. The rate of increase slowed down between 1987 and 1989 and finally fell by 2 per cent in 1990 from 899 to 881. This fall was followed by a steep increase in 1991 when the number of deaths rose by just under 15 per cent to 1,009. The numbers of IIS awards of compensation for mesothelioma have continued to rise steeply in recent years, though they fall well short of the numbers recorded on death certificates (Figure 10.5). Although both sources are imperfect, the death certificate series (Table 10.14) probably gives a more reliable picture of trends in the incidence of this disease than the numbers of Disablement Benefit awards (Table 10. 8), since death certification will not be affected by changes in compensation rules or their application, nor by changes in individuals' propensity to claim compensation. A detailed analysis of mesothelioma mortality data and the initial results of a survey of the mortality experience specific groups of asbestos workers in Great Britain are presented in Chapter 9.

Asbestos-related lung cancer as a prescribed disease gave rise to an average of 56 awards/year over the period from 1983 to 1992 with an increase to 72 cases in 1993. Studies of particular groups of asbestos exposed workers suggest that the numbers of excess lung cancers produced is greater than the number of mesotheliomas (though there is considerable variation in the size of the excess from study to study). This suggests that the actual number of lung cancer cases attributable to asbestos exposure is many higher than those receiving benefit. Many of these cases may not be recognised as such by the sufferers or by their doctors. There is no clinical feature by which lung cancers caused by asbestos can be definitively distinguished from cases in which asbestos has not been involved.

Bilateral diffuse pleural thickening, prescribed from the same date as lung cancer in asbestos workers, is another disease commonly associated with asbestos exposure, which can lead to impairment of lung function. Awards of Disablement Benefit for this condition are currently running at about two to three times the rate recorded for asbestos related lung cancers.

10.4.3 Other occupational cancers

For a small number of cancers in addition to asbestos-related lung cancer and mesothelioma, a clear occupational link has been established. This has been possible where workplace exposure to a known carcinogen is involved, or where a clear association has been established for example between occurrence of a rare cancer and work in a specific industry or job. In many cases the occupational link has been given recognition in the designation of the cancer as a prescribed disease, for example bladder cancer associated with work exposure to aromatic amines, nasal and lung cancers (nickel) and nasal cancer (in hardwood and leather workers), angiosarcoma of the liver (vinyl chloride), and skin cancer (mineral oil and some other substances). The numbers of awards for these cancers are summarised in Figure 10.6. As for asbestos-related lung cancers, most cases attributable to these causes are likely to remain undetected and uncompensated.

Figure 10.5 Mesothelioma, Great Britain, 1971-93

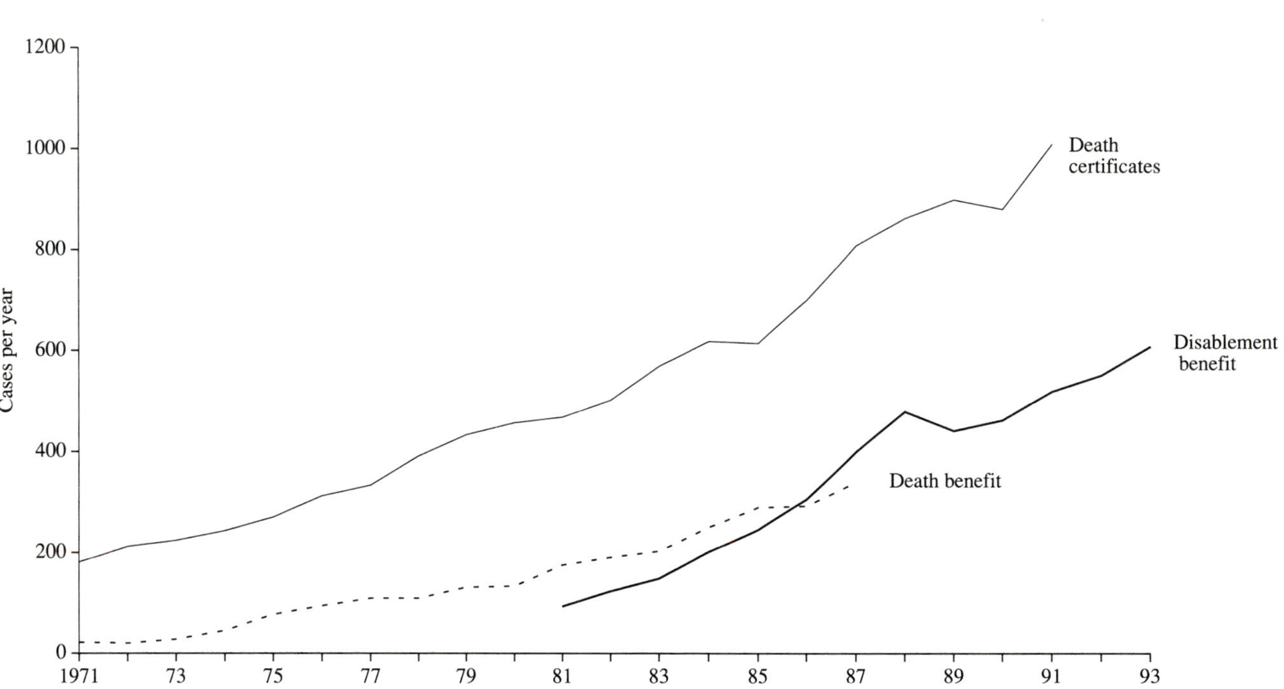

Note: 1987 is last full year for Death benefit.
Disablement Benefit figures not available for years to 1980.

Figure 10.6 Occupational cancer other than mesothelioma, Great Britain, 1971/72 to 1992/93

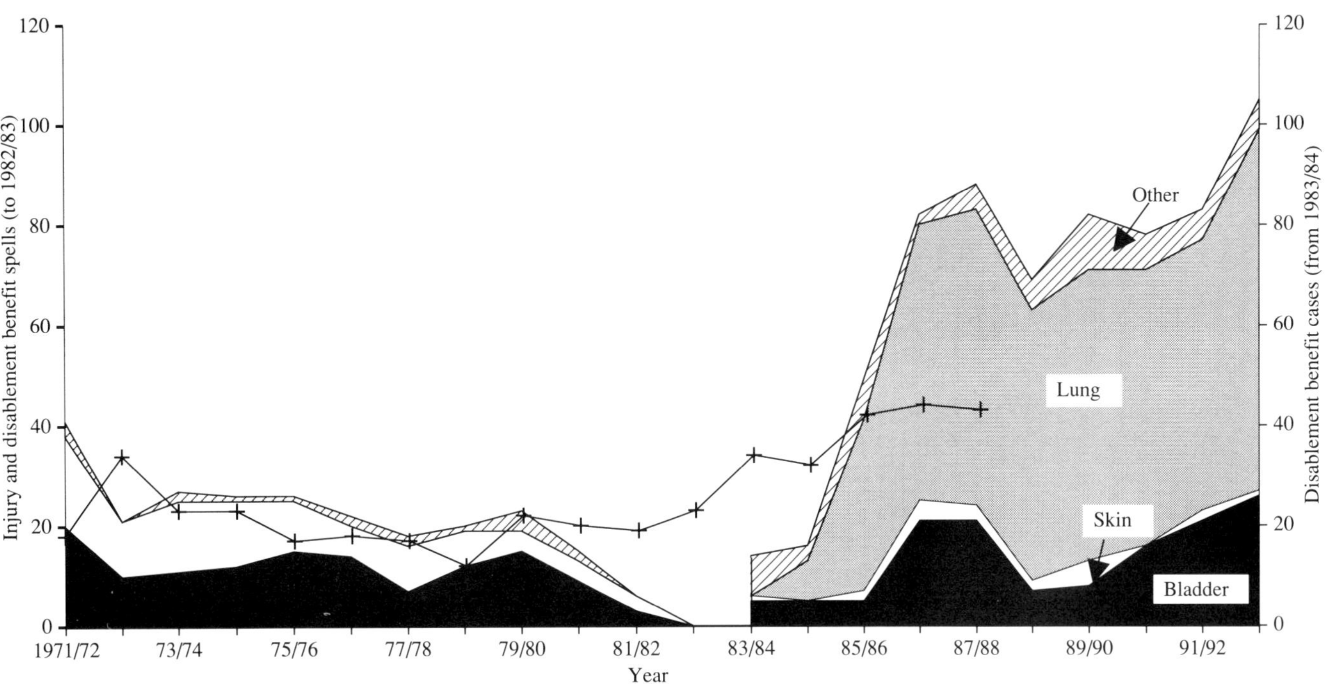

Note: Figures included for the lung cancers (PDs C22b, D8, D10 and D11) are for calendar years, 1972 to 1993

—+— Death benefit: all cancers

10.4.4 Surveillance of respiratory diseases

Since the beginning of 1989, the Epidemiological Research Unit at the National Heart and Lung Institute, in collaboration with the British Thoracic Society and the Society of Occupational Medicine, and funded by HSE, has operated a reporting scheme for cases of occupationally-related respiratory disease seen for the first time by occupational and chest physicians throughout the United Kingdom.

The scheme, which was introduced to supplement other sources of data on occupational lung disease, is known as SWORD — 'Surveillance of Work-related and Occupational Respiratory Disease'. The SWORD team has published detailed analyses of the 1989-93 data for the full range of respiratory diseases covered,[8,9] as well as a special study of the asthma cases for 1989 and 1990.[10]

A summary of the figures for 1989-93 is shown in Table 10.16. Interpretation of the figures for this period in terms of time trends is not straightforward because of changes in the diagnostic categories and in the methods of data collection. From the beginning of 1992, some chest physicians now only send in reports for one month each year, the resulting figures being grossed up by 12 in the annual estimated totals. The remainder of the chest physicians and all the participating occupational physicians form a core group who continue to record all cases. The doubling of the estimated numbers for 1992 and 1993 compared with previous years is probably a consequence of improved notification from the sampled physicians, for whom the work of reporting is now less onerous.

Table 10.16 Surveillance of work-related and occupational respiratory disease (SWORD)

	Total cases 1989-93			Cases seen during 1993 by SWORD participants				
	1989-91 Annual avg.	1992 Estimated	1993 Estimated	% Male	Age distribution (%)			
					16-29	30-44	45-59	60+
Allergic alveolitis	37	97	115	67	7	33	40	20
Asthma	509	1047	912	70	25	38	31	6
Bronchitis	43	133	60	92	3	3	31	63
Building-related illness	15	11	23	25	12	38	50	0
Byssinosis	11	4	6	17	0	0	50	50
Infectious diseases	39	53	51	81	7	47	40	6
Inhalation accidents	193	251	295	82	19	39	38	4
Lung cancer	45	146	63	100	0	0	17	83
Malignant mesothelioma	303	723	700	93	1	7	32	60
Benign pleural disease	313	681	766	97	0	3	40	57
Pneumoconiosis	298	418	385	94	1	3	26	70
Other	41	71	80	90	26	41	15	18
Total	1847	3635	3456	85	10	19	33	38

During 1993, however, greater care was taken to ensure that only cases first seen during the month in question were used in statistical estimates and it is possible that the drop in the total estimates from the previous year can be, at least partially, attributed to this. Notification may be more complete in some areas than in others — see the section on asthma below for further discussion. It is likely that only the more serious cases will be seen by a chest physician. Less serious cases may be seen by occupational physicians but only where physician based occupational health services are provided.

Analyses of the latest year's figures show that diseases of long latency (lung cancer, mesothelioma, non-malignant pleural disease and pneumoconiosis) accounted for 55 per cent of all cases, most of which were attributable to asbestos exposure; asthma accounted for 26 per cent and inhalation accidents for 9 per cent. The proportion of cases in each disease group was similar to previous years. Over 80 per cent of the total number of cases were male; this pattern was repeated across most diseases with the exception of byssinosis and building-related illness. The four diseases of long latency were predominantly in the 45 and over age group while the more acute types of illness were spread more evenly across age groups.

10.4.5 Occupational asthma

Summary data for Disablement Benefit for occupational asthma are detailed in Table 10.8. Benefit became payable for this condition when linked with a specified range of substances (agents 1 to 7 in the table) from March 1982. From September 1986, seven new categories of sensitising agents (agents 8 to 14) were added to the list, and a further ten categories were added in September 1991 (15 to 24); these prescription changes can be seen in the data summary shown in Figure 10.7. The list now includes an 'open category', under which benefit can be paid for occupational asthma caused by an agent not specifically listed, provided a causal link can be shown in each case.

Occupational asthma has a much more rapid onset than the pneumoconioses, and awards can be expected to reflect working conditions within a much shorter time-scale. However, the numbers of compensated cases in the early years of prescription may be affected by the spread of knowledge of the possibility of compensation and by the fact that awards can be made retrospectively within ten years of exposure to prescribed conditions.

After a fall in the second full year of prescription, the number of cases due to agents in the original list grew gradually until 1990 before rising sharply for the next two years. However, 1993 saw a drop in the number of cases in this group from 296 to 275. Agents added in 1986 gave rise to an additional 49 cases in the following year; this number fell gradually until also experiencing a sharp rise to a peak of 72 in 1992. Cases in this group fell to 50 in 1993. In contrast to these falls, the most recently prescribed group of agents accounted for 185 cases in both 1992 and 1993.

Within this group the open category was by far the most numerous, accounting for 129 cases in 1992 and 140 in 1993. However, these large numbers may be a transient effect of claims being made for existing cases that did not previously qualify. After the open category, five main agent categories account for most of the cases — isocyanates, hardening

Figure 10.7 Occupational asthma by sensitising agent group, Great Britain, 1982-93

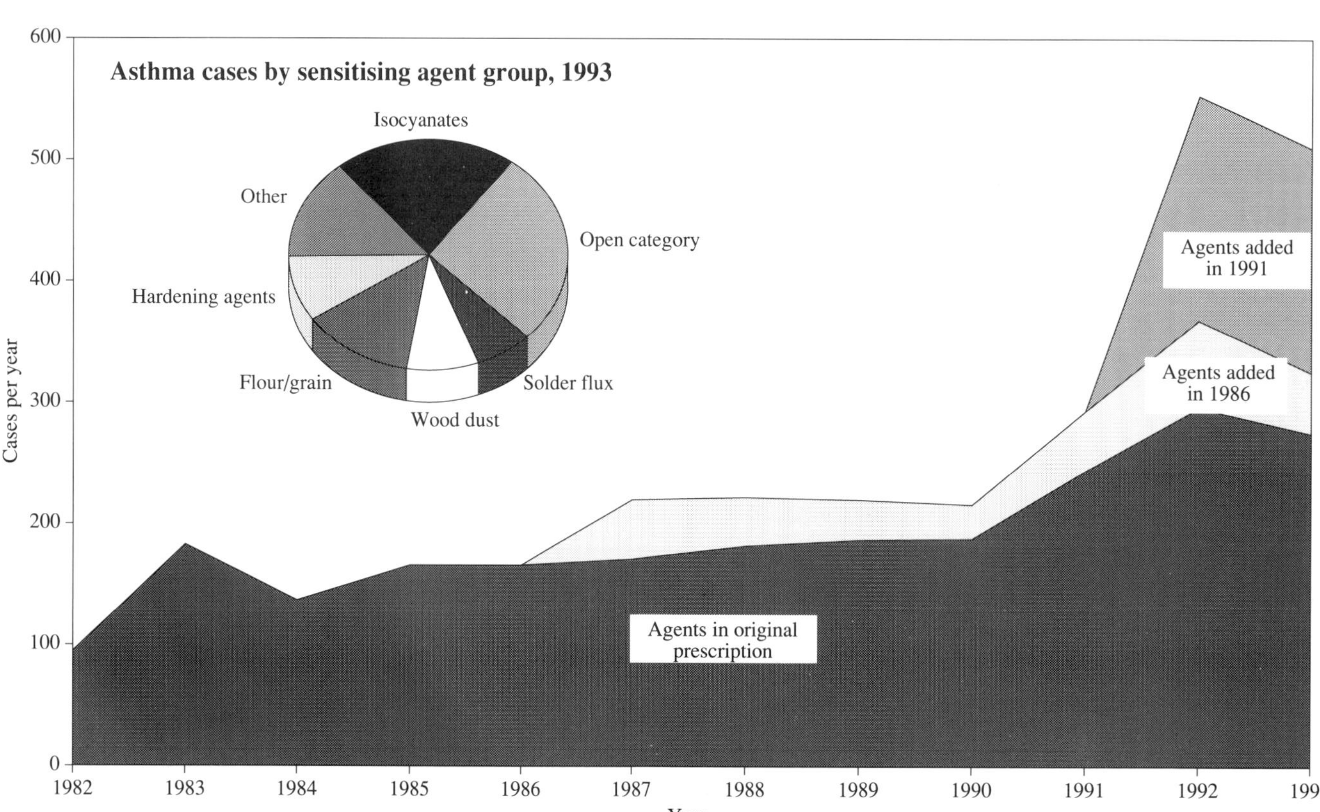

agents, soldering flux, flour/grain and wood dust. (A breakdown of the 1993 cases by agent group can be seen in Figure 10.7 (a)). Cases attributable to each of these agents fell in 1993 with the exception of flour/grain cases which rose from 60 to 66. The total number of cases in 1993 was 510 which represents an 8 per cent drop on the previous year's figure.

Table 10.17 shows a breakdown of occupational asthma by percentage disability. The majority of cases up to 1986 were assessed at 13 per cent disability or less. Since the rules governing disablement benefit were changed in October 1986 such that all cases falling in this group no longer qualify for benefit, there has been a marked shift in assessment distribution; three quarters of all cases from 1987 onwards have been assessed at 14 per cent or more. Few cases were assessed at 50 per cent or more in either period. The drop in the total number of cases during 1993 reflected a fall in the total number of cases assessed at all levels of disability, although the proportion of cases assessed at 14 per cent or greater rose to 82 per cent; this was mainly as a result of a 13 point rise in the proportion of open category cases being assessed at the higher level to just over 83 per cent. There was little change in the disablement assessment distribution in any of the other categories.

Under the SWORD scheme an average of some 500 new cases of asthma were reported each year from 1989 to 1991. In 1992 the estimated annual number was roughly doubled to 1,047, but fell slightly during 1993 to 912. The increase in SWORD cases during 1992 is probably attributable to improved reporting arrangements and it is also possible that the decrease in 1993 may be a result of further slight methodological changes, although Disablement Benefit figures do suggest some lowering of incidence rates. Analysis of SWORD asthma rates for 1989-90 showed strong regional variations which were only partially explained by the geographical distribution of industry and were thought to be the result of variations in the completeness of reporting.[10] It is possible that there is still some degree of incompleteness in the national estimates despite the new improved reporting arrangements.

Male workers accounted for 70 per cent of the asthma cases seen by physicians involved in the SWORD scheme during 1993. Nearly all cases were aged below 60. The excess in the number of SWORD cases compared to those assessed as disabled during the period 1989-91 could be linked to the low number of SWORD cases attributable to agents then prescribed — about half of the total cases. Despite the introduction of the open category, however, the total number of assessed cases during 1992-93 was still only 54 per cent of the SWORD estimate. Examination of the detailed 1989-91 SWORD figures shows that the pattern of sensitising agents was broadly similar to that of the assessed cases, although isocyanate cases were more prevalent in SWORD cases (44 per cent) than amongst assessed cases (33 per cent). Occupations with greatly increased risk (more than 15 times average incidence) included spray painters and workers involved in chemical or plastics processing.[10]

The LFS gave an estimated prevalence of 68,000 cases of self-reported work-related asthma, of which only 20,000 were thought to be caused rather than made worse by work.

10.4.6 Byssinosis

Byssinosis is an illness associated with exposure to cotton dust with both acute and, in some cases, long-term effects. The numbers of cases have decreased steadily, although changes in the compensation rules, most recently in 1979, have periodically produced sharp increases in the numbers of compensated cases. The numbers of death certificates with byssinosis recorded as the underlying cause of death (these are only separately identifiable from 1979), have remained constant at around 25 deaths per year (Figure 10.8 and Table 10.15). The numbers of byssinosis cases recorded by SWORD fell from 23 in 1989 to no more than 6 cases per year subsequently.

10.4.7 Extrinsic allergic alveolitis

Farmers' lung is an allergic reaction to fungal spores, particularly those which grow in mouldy hay. Similar conditions are suffered by other groups of workers — e.g. mushroom pickers — with exposures to organic material. The general term for these diseases is extrinsic allergic alveolitis. Few cases — around 10 per year — are recorded through the compensation system (Figure 10.9). One explanation for this will be that many farmers are self-employed and therefore cannot claim benefit under the IIS. The numbers of deaths ascribed to farmers' lung (and related conditions), is of the same order of magnitude, which suggests, since the disease rarely progresses to a life-threatening level, that there are substantially more cases than those receiving compensation. The SWORD figures show an average of 37 cases of this type of disease (allergic alveolitis) for the 3 years 1989 to 1991, rising to 97 in 1992 (including a few cases related to non-occupational causes e.g. pigeon fanciers lung) and 115 in 1993 with the improved reporting scheme.

10.4.8 Upper respiratory and mouth disorders

Numbers of assessed cases of inflammation or ulceration of the upper respiratory tract or mouth have increased, from 10 to 40 cases annually in the period 1986-91 to 75 in 1991-92 and 494 in 1992-93. Almost all (98 per cent of those in 1992-93) have less than 14 per cent disability from this disease alone, though it is likely that some will qualify for benefit by aggregation with other diseases or injuries. Any occupation exposed to harmful dust, liquid or vapour could qualify under the terms of prescription. The increase in numbers almost certainly reflects an increase in claims following a take up campaign by trade unions, mainly in one region, rather than a true increase in the incidence of the disease.

10.4.9 Chronic bronchitis and emphysema

From September 1993 these conditions became prescribed for coal-miners with a specified level of lung function impairment, definite evidence of coal dust retention on a chest radiograph and a minimum of 20 years underground exposure to coal dust. Data are available for a short period only and do

Tect continues on page 176

Table 10.17 Occupational asthma: new cases of assessed disablement, by causative agent and percentage disability, 1982-93

Agent	1982	1983	1984	1985	1986	1987	1988	1989	1990	1991	1992	1993
1 Isocyanates	39	74	51	46	48	60	64	72	73	95	121	108
of which: 13% or less	35	58	43	33	31	29	20	20	16	24	25	16
50% or more	1	-	2	-	1	-	2	3	1	3	2	5
2 Platinum salts	3	9	4	9	12	9	12	6	5	3	3	1
of which: 13% or less	3	9	4	8	10	4	7	-	2	2	2	-
50% or more	-	-	-	-	-	-	-	-	-	-	-	-
3 Hardening agents	5	12	14	19	28	18	31	24	22	34	64	47
of which: 13% or less	4	10	9	13	13	9	9	9	2	7	11	9
50% or more	-	1	-	1	1	-	2	1	-	1	3	3
4 Soldering flux	21	24	27	25	20	21	24	30	23	35	37	34
of which: 13% or less	17	15	19	18	9	8	1	2	2	5	3	6
50% or more	-	2	1	-	1	1	2	3	1	-	3	2
5 Proteolytic enzymes	4	3	1	6	-	6	2	3	3	3	1	6
of which: 13% or less	4	2	-	2	-	1	-	-	1	-	1	3
50% or more	-	-	-	-	-	1	-	1	-	-	-	-
6 Animals/insects	4	7	8	7	12	7	9	9	7	10	10	13
of which: 13% or less	4	6	6	3	11	2	2	3	4	-	3	6
50% or more	-	-	-	1	-	-	-	-	-	-	1	-
7 Flour/grain	19	54	32	54	46	50	40	43	55	64	60	66
of which: 13% or less	19	34	24	34	28	23	8	15	8	15	6	13
50% or more	-	1	1	2	1	3	1	2	3	1	6	1
8 Antibiotics						30	6	4	2	5	8	1
of which: 13% or less						13	-	1	-	1	3	-
50% or more						-	-	-	-	-	1	-
10 Wood dusts						15	28	25	23	40	52	43
of which: 13% or less						6	8	4	5	9	9	8
50% or more						-	-	1	2	2	5	1
11 Ispaghula						-	-	1	-	-	-	-
of which: 13% or less						-	-	-	-	-	-	-
50% or more						-	-	-	-	-	-	-
12 Castor bean dust						-	-	-	-	1	-	1
of which: 13% or less						-	-	-	-	-	-	-
50% or more						-	-	-	-	-	-	-
13 Ipecacuanha						-	1	-	-	-	-	1
of which: 13% or less						-	-	-	-	-	-	-
50% or more						-	-	-	-	-	-	-
14 Azodicarbonamide						4	5	3	3	3	12	4
of which: 13% or less						1	1	1	1	3	3	1
50% or more						-	-	1	-	-	-	-
15 Animals and insects (larval forms)											2	1
of which: 13% or less											1	-
50% or more											-	-

Table 10.17 - *continued*

Agent	1982	1983	1984	1985	1986	1987	1988	1989	1990	1991	1992	1993
16 Glutaraldehyde											13	13
of which: 13% or less											5	2
50% or more											-	-
17 Persulphate salts and henna											6	1
of which: 13% or less											1	-
50% or more											-	-
18 Crustaceans											7	6
of which: 13% or less											4	-
50% or more											-	-
19 Reactive dyes											5	4
of which: 13% or less											2	-
50% or more											-	-
20 Soya bean dust											2	-
of which: 13% or less											-	-
50% or more											-	-
21 Tea dust											5	1
of which: 13% or less											-	-
50% or more											-	-
23 Fumes from stainless steel welding											16	19
of which: 13% or less											2	4
50% or more											1	2
24 Open category											129	140
of which: 13% or less											40	25
50% or more											8	7
Totals: Agents 15 to 24 (c)											185	185
Agents 1 to 7 (a)	95	183	137	166	166	171	182	187	188	244	296	275
Agents 8 to 14 (b)	-	-	-	-	-	49	40	33	28	49	72	50
All agents	95	183	137	166	166	220	222	220	216	293	553	510
of which: 13% or less	86	134	105	111	102	96	56	55	41	66	121	93
50% or more	1	4	4	4	4	5	7	12	7	7	30	21

Note: There have been no awards for the following agent;
(9) Cimetidine (prescribed 1.9.86), (22) Green coffee bean dust (prescribed 20.9.91)
From 1.10.86 cases with 13% or less disability do not qualify for benefit.
(a) Agents prescribed from the start of the prescription.
(b) Agents added to prescribed list with effect from 1 September 1986.
(c) Agents added to prescribed list with effect from 26 September 1991.

Source: DSS

Figure 10.8 Byssinosis, Great Britain, 1972-93

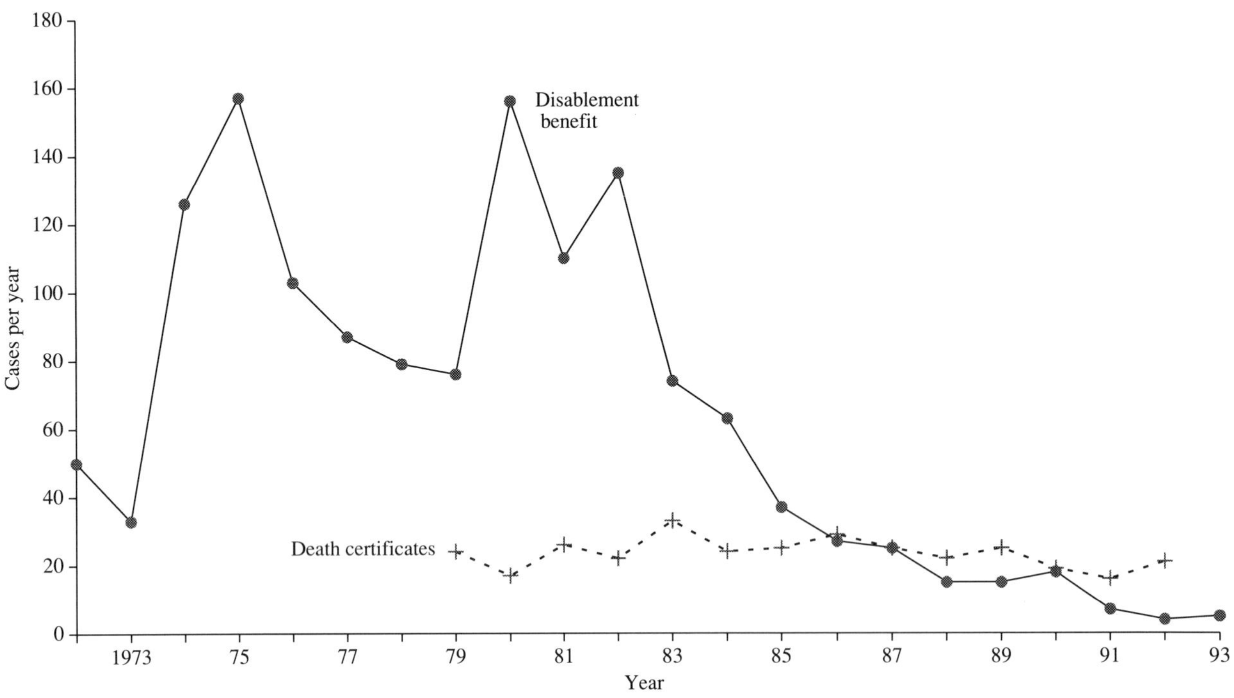

Figure 10.9 Extrinsic allergic alveolitis, Great Britain, 1971-93

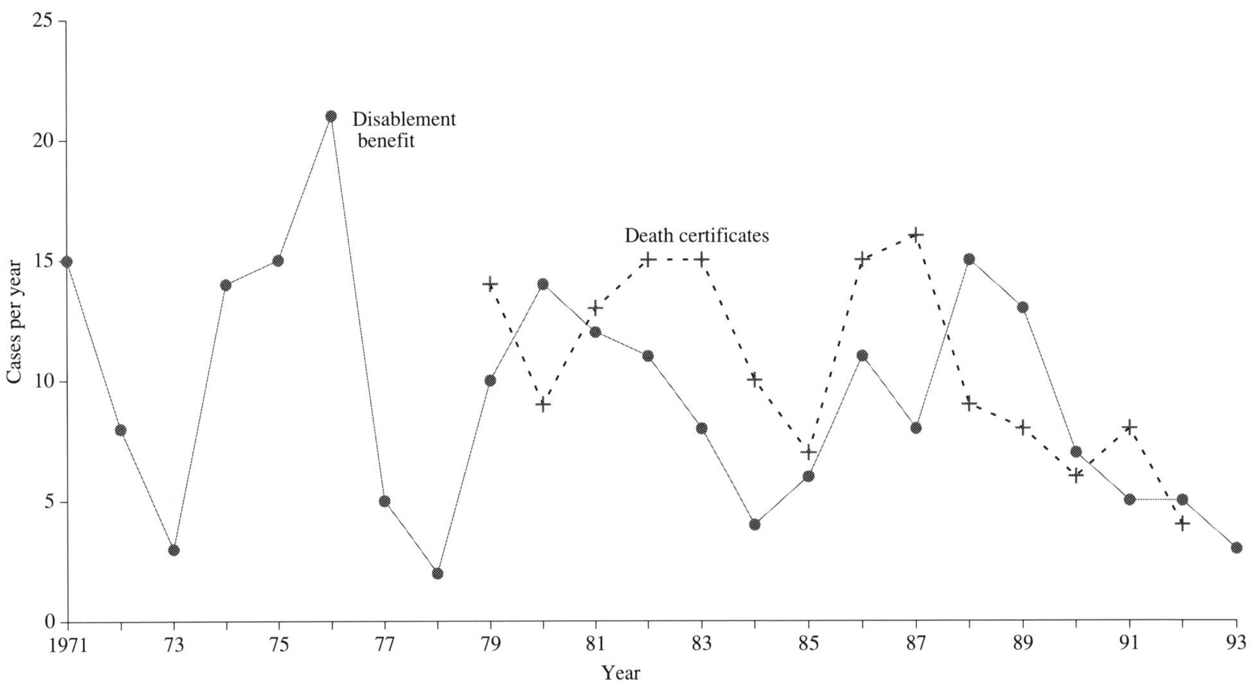

not merit comment at this stage but are given for completeness in Table 10.8. The number of cases identified in the early history of a newly prescribed disease often represents a backlog of cases who have become eligible for benefit and it may be some years before a significant trend emerges.

10.4.10 Sick building syndrome

The term 'sick building syndrome' refers to a range of relatively minor symptoms which can be caused by the indoor environment. No one specific causative factor has been identified but factors which could be implicated include air-conditioning and humidification systems, poor lighting and air pollution.[11] Typical reported symptoms included nasal problems, eye irritation, dry skin and tiredness. Because these symptoms are common in the general population, and for the most part not serious, they are rarely reported to a doctor and it is difficult to assess the extent of the problem with any real certainty.

The SWORD register estimated 11 cases of building related sickness in 1992 and 23 in 1993. These numbers were similar to those obtained under the previous reporting arrangements in 1989-1991, although a special investigation into outbreaks in three buildings during 1989 yielded a further 190 cases. Because of the nature of the disease the majority of cases are unlikely to be seen by either an occupational or chest physician, and so these figures will certainly underestimate the true incidence. The LFS yields an estimate of 2,500 people affected in a twelve month period, but this is still likely to be much lower than the true figure, as people may well suffer building related symptoms without associating them with their working environment.

10.4.11 Infections

There are few useful sources of statistics on occupational infections. The Communicable Disease Surveillance Centre (CDSC) can provide the best estimates of the overall incidence of infections but at present the proportion attributable to occupation is rarely known.

Some infections, such as hepatitis, tuberculosis, brucellosis and leptospirosis, are prescribed diseases. However, not only will mild cases of brucellosis and leptospirosis go unrecognised but all four infections will rarely lead to disability in the long term and the incidence will not be reflected in payments of disability benefit. Figure 10.10 and the data in Table 10.9, therefore, should be interpreted very cautiously.

10.4.12 Dermatitis

The risk of dermatitis caused by an allergic or irritant reaction to substances used or handled at work is present in a wide range of jobs. However, for the workforce as a whole, the prevalence has fallen as conditions have improved and as the number of 'dirty' jobs has contracted. Figure 10.11 shows that the annual number of cases of compensated dermatitis (strictly, the number of spells of sickness absence due to dermatitis for which Industrial Injury Benefit was paid) fell from over 10,000 in 1971/72 to about 2,000 in 1982/83 (the 10 months to March 1983, the final period for which Injury Benefit was normally payable). Since then, the numbers of disablement benefit cases have fluctuated as the compensation rates have been redefined; in particular the introduction of the 14 per cent rule is likely to have affected uptake. In the 7 years since the introduction of this rule, only 127 of the 2,258 cases had sufficient disablement from dermatitis to qualify for benefit.

There are 4 other sources for data on occupational dermatitis, two of which are based on records of general practitioner consultations.

The 'Morbidity Statistics from General Practice' surveys (MSGP) in 1955/56, 1970/71 and 1981/82[12-14] each give data on consultations for occupational dermatitis, though the definitions used were not exactly the same in all 3 surveys. The surveys estimated 11,000 cases of occupational dermatitis in England and Wales in 1955-56, 32,000 in 1970/71 and 100,000 in 1981/82.

In the first 6 months of 1989, the HSE commissioned a survey based on 73 GPs throughout the UK who recorded the number of cases of occupational dermatitis that they saw in this period. Using this we can estimate that in 1989 there were approximately 38,000 cases of dermatitis caused by work and a further 25,000 cases whose condition was made worse by their work.

The 1990 LFS yielded estimates of the number of people with dermatitis or other skin diseases which they believed had been caused or made worse by work, these estimates being 54,000 and 30,000 respectively. Over 90 per cent of the cases reported arose as a result of a recent job (in last 3 years), suggesting that this is not a persistent condition. Among respondents reporting a skin complaint which was caused or made worse by a recent job, three occupational groups had

Figure 10.10 Occupational infections, Great Britain, 1971/72 to 1992/93

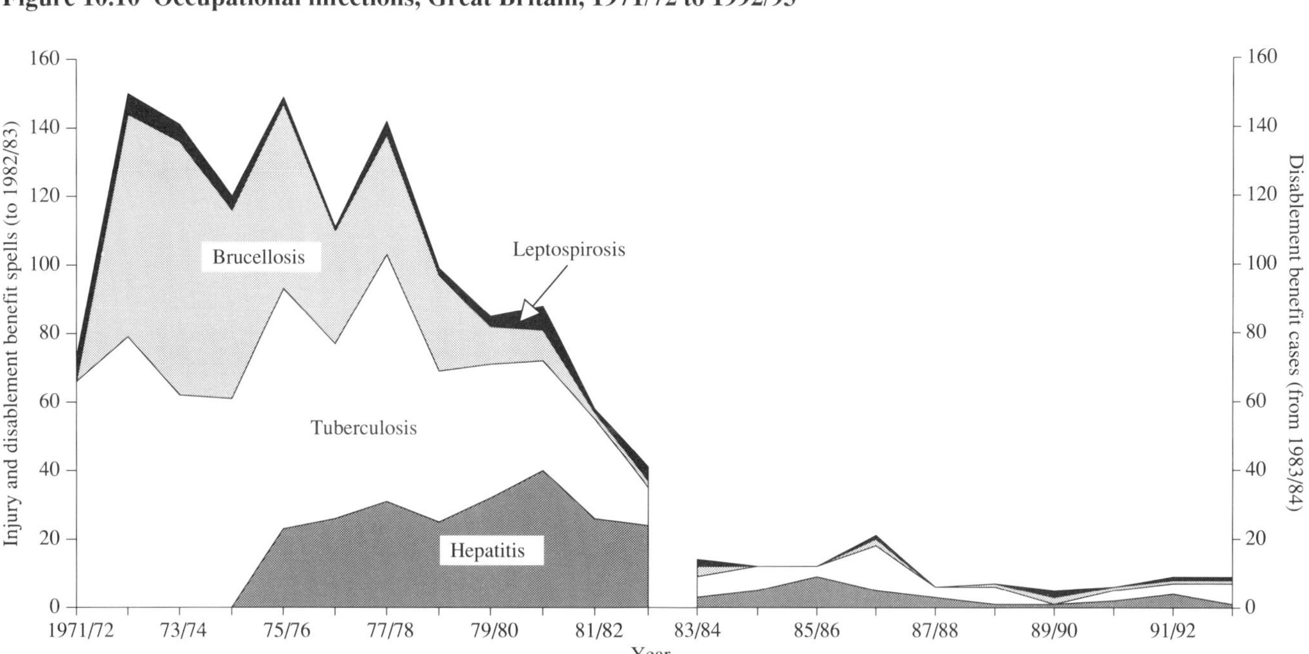

Figure 10.11 Occupational dermatitis, Great Britain, 1971/72 to 1992/93

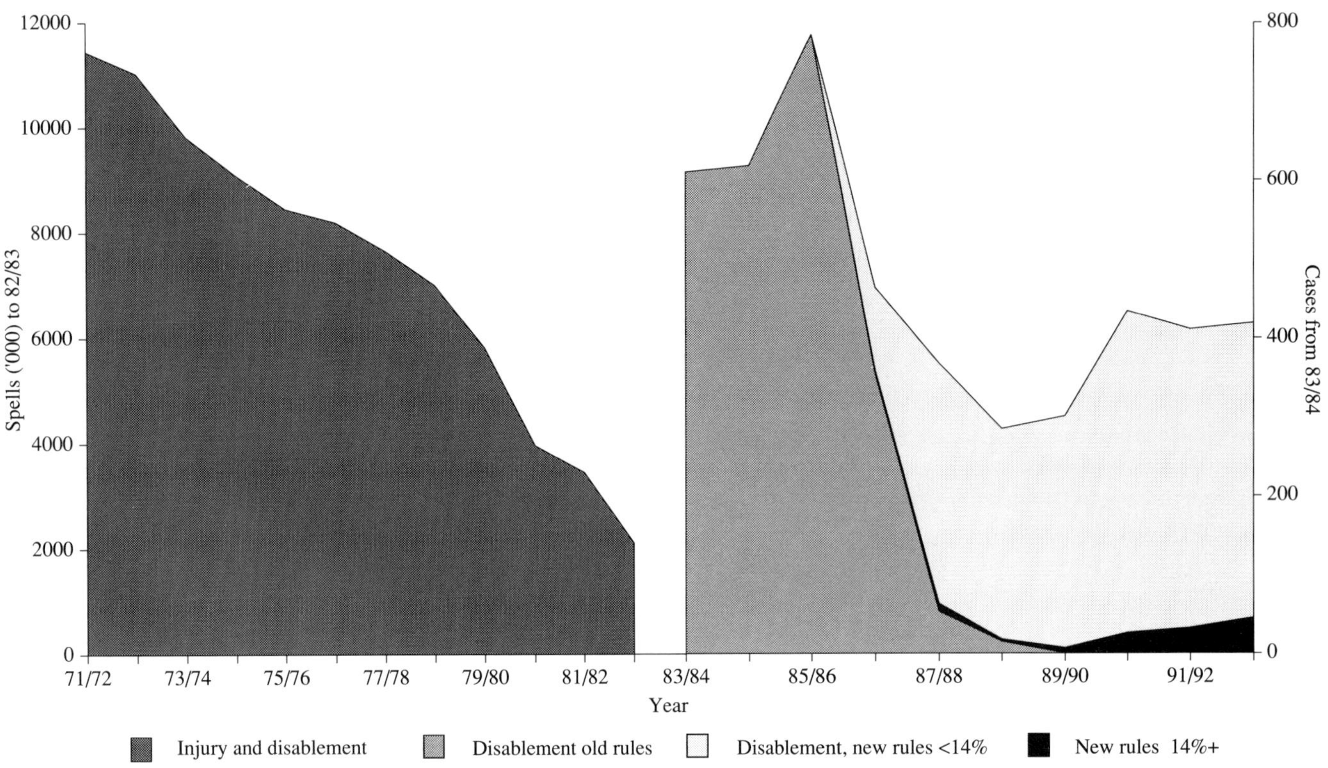

rates significantly above the average for this complaint: construction, science and engineering, and catering, cleaning and hairdressing. One occupation showed a significantly raised rate in an economically inactive group (inactive in the last 3 years): metal and electrical work. This rate was about 6 times the average rate for the economically inactive.

A surveillance scheme for occupational skin disease (EPI-DERM) has been running since February 1993 and information for the first five quarters (to April 1994) is now available. All consultant dermatologists in the UK have been asked to report cases of skin disease thought to have been caused, or made worse, by work. Of 312 consultants 210 (67 per cent) have sent information on cases received. In all 2,811 cases have been reported.

Cases are routinely examined by diagnosis (type of skin disease) and the substance(s) believed to be responsible. The large majority of cases were contact dermatitis (79 per cent) with the next most frequent category being neoplasia (13 per cent) of which the largest group (177/368) was basal cell carcinoma. Sixty squamous cell carcinomas have been reported and two melanomas. Other neoplasias were either keratoses or of unspecified type. There were also 73 cases of contact urticaria, 55 of traumatic skin lesions, 54 infections, mainly tinea and warts, 19 cases of nail dystrophy or paronychia and 14 of folliculitis or acne. In addition there were 62 cases of 'other' dermatoses, including conditions such as viliigo, concrete burns and psoriasis. When all cases were considered together the agent most frequently reported was ultra-violet radiation (350 cases) associated with neoplasia. Only six of the neoplasia were associated with oils. Amongst those with occupationally related contact dermatitis the most frequent substances were those shown in

Table 10.18 Suspected agents most frequently reported: cases of contact dermatitis in UK, February 1993 - April 1994

Males		Females	
Petroleum oils	141 (11.8%)	Nickel	240 (23.3%)
Rubber	132 (11.0%)	Unsp. irritants	148 (14.4%)
Unsp. irritants	117 (9.8%)	Wet work	146 (14.2%)
Chromes	101 (8.4%)	Hairdressing chemicals	128 (12.4%)
Cutting oils	61 (5.1%)	Rubber	122 (11.9%)
All agents	1196	All agents	1029

Table 10.18. Rubber and unspecified irritants appear in the list for both men and women, however nickel was by far the substance most frequently reported for women (240/1,029). Amongst the men, petroleum oils and cutting oils were both in the top five substances and chromes were also reported for many cases. The five most frequent agents were found in more than three quarters of the female cases, but less than half the male cases, reflecting the much wider range of occupational exposure to chemicals in employment undertaken by men.

Personal characteristics (age and sex) and the occupation of cases reported to EPI-DERM can be compared with those reported to the LFS, identifying groups at higher risk. This comparison is not valid for neoplasias for which, in EPI-DERM, the job recorded is that in which exposure occurred, rather than the present occupation. Indeed neoplasias are reported for men and women who are no longer working; the comparative figures from the LFS would not include people who have retired.

Table 10.19 presents rates for contact dermatitis by age and sex. It is evident that young women are at particularly high risk. For men, the rates increase gradually with age.

Table 10.19 Rates of contact dermatitis by age and sex in the UK, February 1993 - April 1994

Age group	Rates (per 10,000)*	
	Males	Females
Under 30 years	0.69	1.38
30-44 years	0.79	0.66
45-59 years	0.91	0.63
60 years and over	1.03	0.74
All ages	0.82	0.90

* Cases in EPI-DERM compared with numbers in the Labour Force Survey

Table 10.20 shows rates for the 10 occupations most at risk, considering only occupations in which at least 10 cases have been reported to EPI-DERM. The rate/10,000 for labourers in manufacturing may in part be artefactual, with this rather imprecise code being used more frequently for the EPI-DERM cases than for respondents to the LFS.

Table 10.20 Rates of contact dermatitis by occupation in the UK, February 1993 - April 1994

Recorded occupation (Standard Occupational Classification)		Rate (per 10,000)*
919	Other labourers in making and processing industries nec	32.6
660	Hairdressers	16.7
661	Beauticians	8.6
840	Machine tool operatives	8.3
561	Printers	8.2
820	Chemical plant operatives	5.7
912	Labourers in engineering	5.5
620	Chefs and cooks	5.1
821	Paper, wood, plant operatives	5.1
643	Dental nurses	5.0
	All occupations	0.9

* Cases reported to EPI-DERM compared with numbers in the same occupation in the Labour Force Survey

However, the rates for other occupations shown in Table 10.20 appear likely to reflect a real increase in risk. For each of these occupations the numbers of cases reported was more than 5 times that expected from the rate of reporting for all occupations.

From May 1994, the EPI-DERM scheme has been extended to occupational physicians and 686 have undertaken to report cases at the end of July 1994. This broadening of the reporting base will provide better future estimates of the incidence of occupationally related skin disease, although cases treated by occupational health nurses and general practitioners will still not be counted.

10.4.13 Musculoskeletal disorders

Musculoskeletal conditions affect a very large number of people, both in work and out of it. For example, nearly half of all working age adults will experience some low back pain in any 6-week period.[15] A limited number of specifically work-related musculoskeletal disorders are prescribed under the IIS. In the last full year in which injury benefit could be awarded (1981/82) there were 2,828 injury and/or disablement benefit cases, and in the last year before restriction of disablement benefit to cases with 14 per cent disability (1985/86) there were 842 disablement benefit cases (Figure 10.12 and Table 10.9).

There has been a rise in the number of cases assessed as having disability due to cramp of the hand or forearm, and inflammation of the tendons of the hand, forearm or associated tendon sheaths (tenosynovitis) in recent years. The number of cases of the latter has more than doubled between 1989/90 and 1992/93. Much of the increase may be attributed to increased awareness of the possible work related nature of these conditions, but the proportion assessed as having 14 per cent or greater disability for these two prescribed diseases combined rose from 8 per cent in 1989/90 to 16 per cent in 1990/91 and to 24 per cent in 1992/93 (Table 10.10).

The GP based survey commissioned by HSE in 1989, referred to in the preceding section, also recorded cases of carpal tunnel syndrome (CTS) — symptoms caused by the entrapment or compression of nerves in the wrist — which can be caused by repetitive twisting and gripping. The participating GPs judged that about half of the cases of CTS which they saw were either caused or exacerbated by the patients' work. On this basis the observed rates of work-related CTS were 0.8 per 1,000 in women and 0.4 per 1,000 in men. This would imply a national annual incidence of 20,000 work-related cases for which medical advice was sought.

The LFS trailer results show that musculo-skeletal disorders were the most commonly recorded self-reported work-related illness. The estimated number of prevalent cases 'caused' by work was 593,000 of which half were back disorders, 19 per cent upper limb disorders, 10 per cent lower limb disorders and 21 per cent other or unspecified.

High risk occupations reported for lower limb and back disorders in the LFS included coal mining, construction and transport and materials moving, with a particularly high risk for lower limb disorders among coal miners. The risk for back disorders among nurses was over 3 times the average risk, and the security (excluding armed forces) and farming, fishing and forestry groups recorded 3 fold risks for lower limb conditions and moderately raised risks for back disorders. Repetitive assembly and inspection was a high risk occupation for the disease group 'Repetitive Strain Injuries' (tenosynovitis, bursitis, RSI, tennis elbow, carpal tunnel syndrome and frozen shoulder), with about three times the average risk. For the remaining upper limb disorders, transport and materials moving (men only) and processing (metal and electrical) were recorded as high risk occupations.

10.4.14 Occupational deafness

Although in recent years the numbers of assessed claims for Disablement Benefit for occupational deafness have been

Figure 10.12 Musculoskeletal disorders ('000), Great Britain, 1971/72 to 1992/93

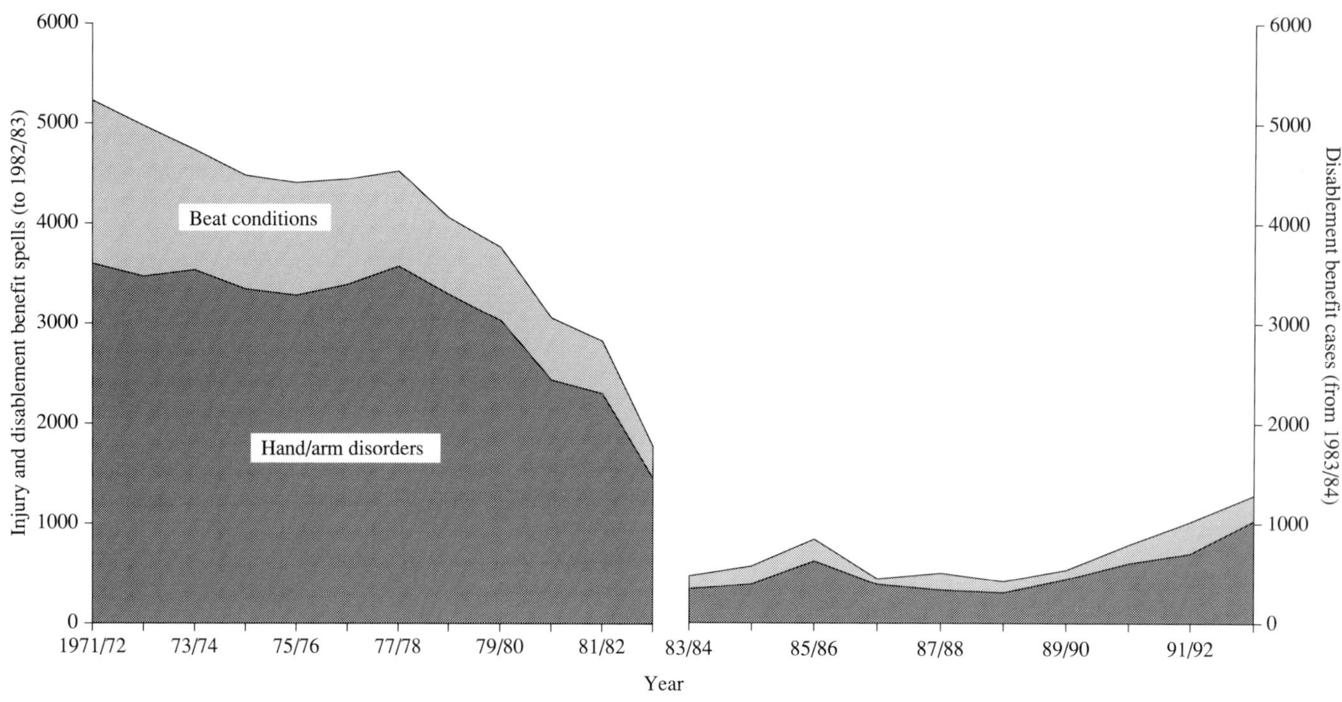

overtaken by those for vibration white finger and more recently tenosynovitis, it remains the third most frequent category in the Prescribed Disease statistics. A change in qualifying conditions was introduced on 1 October 1983; from that date, claimants need only have worked for 10 years in prescribed noisy conditions — previously it was 20 years. An earlier widening of the terms of prescription took place in 1979, and the additional claims due to this reached a peak in 1981. Figures for recent years show a gradual decline from 1,261 assessed claims in 1988 to 901 in 1992/93.

Figure 10.13 Occupational deafness - disablement benefit, Great Britain, 1975 to 1993

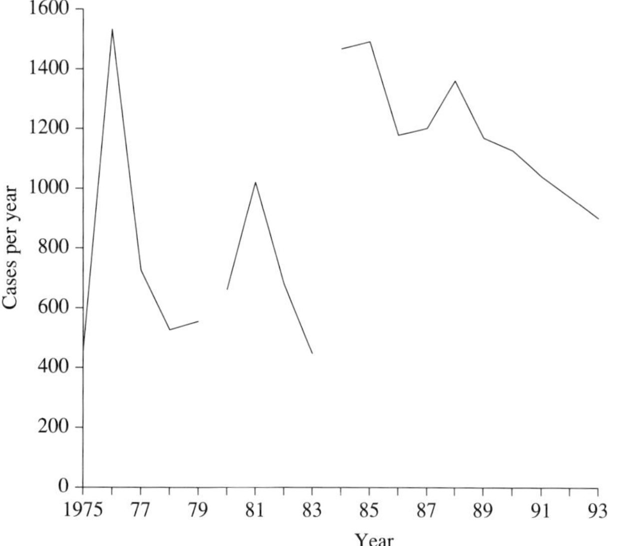

Breaks in the graph indicate changes to prescription rules

The figures for 1984 to 1986 include some unsuccessful appeals and references to Medical Appeals Tribunals and are therefore unreliable. In 1987 to 1989 these were being recorded at a rate of about 21% of number of assessed cases.

Figure 10.13 and Table 10. 9 show the annual numbers of new claimants assessed as having a level of disability of 20 per cent or greater. DSS estimate a total of 13,000 people were receiving disablement benefit for occupational deafness in April 1992.

Results from population surveys, however, suggest that the total number of people with occupational deafness may be much larger than the number who receive DSS awards. In the OPCS Disability Survey between 1985 and 1988,[16] deafness was the largest category of 'industrial disease' (described as such by respondents), affecting an estimated 52,500 people in England and Wales.

Estimates from the LFS are larger still, suggesting that they may include complaints below the threshold of severity set by the OPCS Disability Survey. An estimated 103,100 people in England and Wales had deafness, tinnitus or other ear conditions which they thought had been caused by their work. A further 18,300 thought they had deafness or ear conditions which had been made worse by work.

The occupation at highest risk for deafness and ear conditions in the LFS was coal mining, with a risk estimated to be between 7 and 20 times the average. The uncertainty attaching to this figure is caused by the difficulty in estimating the population at risk for a condition caused by long term exposure in an industry where the numbers employed have shrunk very markedly. Other high risk occupations included those in the processing (metal and electrical) group, with over 4 times the average risk, and the processing (other) group with twice the average risk. Within these groups, jobs such as sheet workers and riveters, and workers in foundries, forging, textile and chemical processing were particularly at risk, reflecting the generally high noise exposures in such jobs.

10.4.15 Vibration white finger (VWF)

Vibration white finger (Table 10.9) is a disorder of the blood supply to the fingers and hand which can be caused by regular use of vibrating hand-held tools. The damage caused by vibration is chronic rather than immediate, and recently diagnosed cases will be the product of at least 5 years', and in some cases more than 20 years' exposure.

Following the prescription of this disease in 1985, numbers of assessed cases rose rapidly, overtaking deafness to reach a peak of 5,403 in 1990/91. Numbers fell to 1,447 in 1992/93, suggesting the clearance of a backlog of existing cases, but VWF remains the commonest prescribed disease.

Very few VWF cases have been assessed at 14 per cent or more disability. Numbers of such cases did rise to 57 in 1992/93 but this represents only 4 per cent of total cases. It is possible, however, that some cases may have reached the benefit threshold by aggregation of their VWF disability with that due to another prescribed disease — usually occupational deafness. It is also believed that compensation agreements between some employers, their insurers and trade unions, together with the practice of some solicitors of advertising their services in handling compensation claims, may have contributed to the increase in DSS claims up to 1990/91.

The LFS yields a prevalence estimate of 7,300 VWF sufferers. This is relatively low compared to the DSS assessed claim figures, which represent a total of 16,500 cases when summed since 1984/85. The low LFS estimate may be partly due to sampling error, and partly to the fact that respondents only specified their most serious work disorder. Higher estimates can be obtained by multiplying estimated numbers of workers exposed to hand transmitted vibration[17,18] by typical prevalence rates obtained from industrial studies.

Such a technique is liable however to overestimate the true prevalence since industrial studies tend to concentrate on workplaces where exposures or prevalence are especially high. Combining estimates from various sources suggests a best estimate of around 20,000 prevalent cases.[19]

10.4.16 Carpal tunnel syndrome

Carpal tunnel syndrome, if caused by use of hand held vibrating tools, was made a prescribed disease in April 1993. Twenty cases were assessed as having some level of disability (14 per cent or greater in 5 of them) in this short period of observation which is included only for completeness, interpretation being impossible at such an early stage.

10.4.17 Acute poisoning

Acute poisoning by chemicals at work is reportable under RIDDOR as an industrial accident (and, for some substances, also as a reportable illness). Around 2,000 cases are reported annually, with 10 to 25 fatalities. In about half of these 2,000 cases it is possible to identify an agent responsible for the poisoning (for example as a gassing or chemical burn). Gases, including carbon monoxide, are responsible for over a quarter of recorded such poisonings, followed by acids and caustic alkalis (recorded in another 25 per cent of cases), petroleum products and solvents (about 10 per cent between them), and alcohols, pesticides and metals which have accounted for 2-3 per cent of poisonings each since 1986.

A study commissioned by HSE of cases of poisoning by industrial chemicals in 1985 based on a 10 per cent sample of attendances at NHS Accident and Emergency departments showed that 6 per cent of attendances for toxic substance related ill-health arose from work place exposures. This implies an annual national total of about 14,000 cases. The commonest categories of substance were acids, alkalis, irritant vapours and solvents. There were no deaths in the sample, and the discharge rate was higher than for other types of poisoning, suggesting a higher proportion of precautionary attendances

A surveillance study of acute poisonings from pesticides commissioned by the HSE and conducted by the West Midlands Poison Unit (WMPU) has produced some data on general incidence of pesticide related ill-health. Over a period of 28 months between June 1991 and September 1993 about 70 cases a year of 'confirmed' or 'likely' poisonings, as defined according to a system based on that used by HSE's Pesticide Incidents Appraisal Panel (PIAP), were reported under the WMPU Green Card scheme in the West Midlands and Trent regions. Most had mild or moderate symptoms only (64 per cent and 31 per cent respectively, 'moderate' being defined as evidence of organ damage without signs of organ failure).

Suspected pesticide poisoning incidents reviewed by PIAP can be differentiated as occupational or non-occupational. In the two years from April 1992 to March 1994 there have been 84 and 88 (provisional figures) pesticide poisoning incidents respectively involving 119/132 people investigated by HSE and assessed by PIAP. Of these, 12/17 incidents respectively (involving 15/23 workers) occurred in an occupational setting. Excluding cases for whom a decision is still outstanding, over 70 per cent of these had symptoms that 'confirmed' exposure or were thought 'likely' to relate to pesticide exposure (based on 1992/93 figures for whom decisions have been made in 89 per cent of all cases), represented by 8 workers affected in that year. In all 27 per cent of all 'confirmed' (occupational and other) cases were working with the pesticide or in close proximity to the operator. However, the remainder (73 per cent) were members of the public affected on adjacent property.

10.4.18 Exposure to lead

The Control of Lead at Work Regulations 1980 require regular medical examination of all workers with significant exposure to lead, by an Appointed Doctor or an Employment Medical Adviser (EMA). The examinations include measurement of workers' blood-lead levels. Annual returns from Appointed Doctors and EMAs give summary statistics for each work place based on the maximum blood lead level recorded for each worker under surveillance.

Figure 10.14 Blood lead levels: males, Great Britain, 1982 to 1992/93

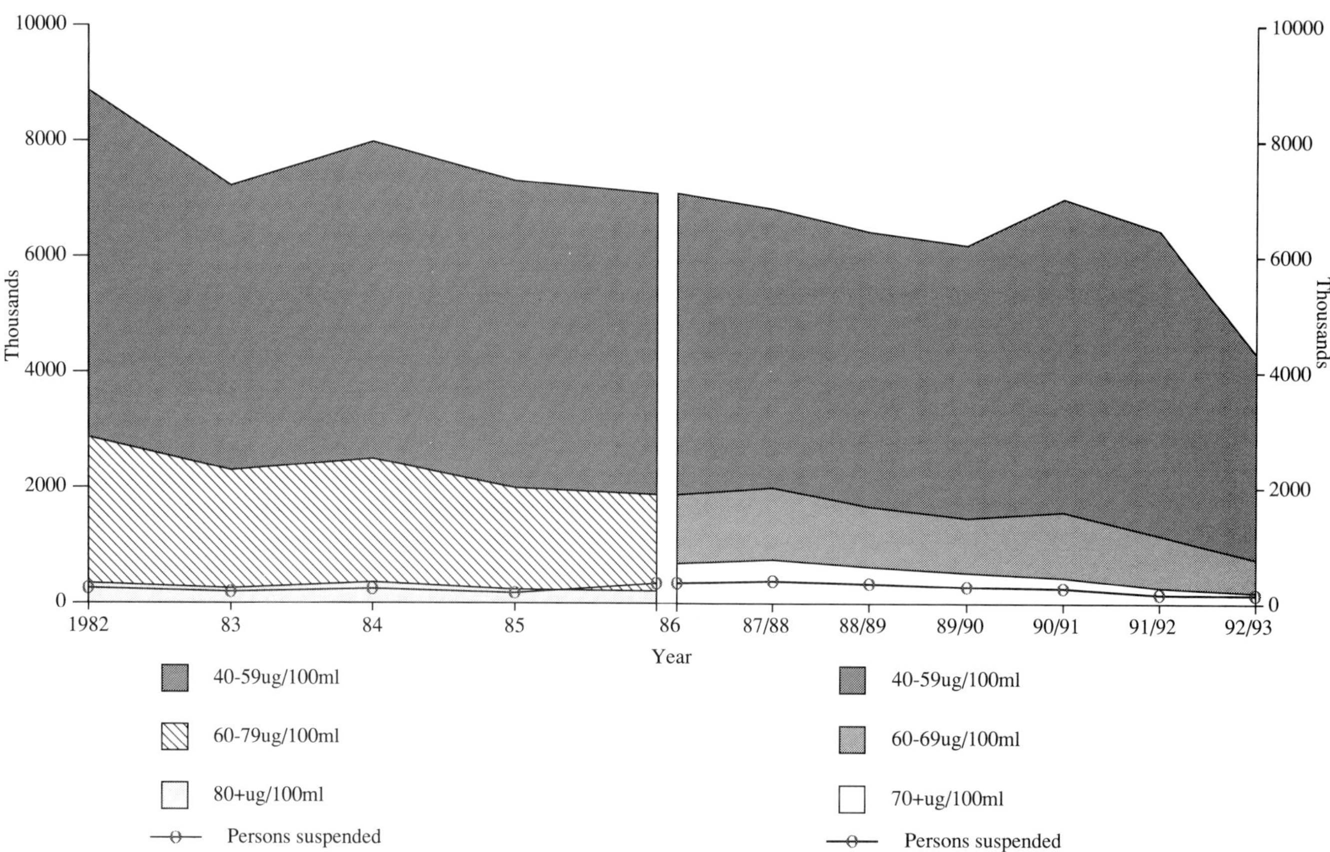

Figures 10.14 and 10.15 are based on the returns since 1982 and Table 10.21 refers to the period 1987/88 to 1992/93. The figures refer to calendar years for 1982 to 1986 and to financial years (April 1 to March 31) from 1987/88 onwards. The Approved Code of Practice made under the regulations prescribes that when a male worker's blood-lead level exceeds a certain limit (79µg/100ml from 1982 to 1985 but lowered to 69µg/100ml in 1986) a repeat test will be made, and if this is still over the limit he will be suspended from working with lead. For females of reproductive capacity a lower limit of 39µg/100ml is prescribed, above which they should be suspended from work. The number of males under surveillance has fallen for the second successive year, from 25,400 in 1990/91 to 24,000 in 1991/92 and 20,000 in 1992/93. The number of men with a blood-lead level over 69µg/100ml has fallen in five successive years from 762 (3 per cent of males) in 1987/88 to 196 (1 per cent) in 1992/93. Numbers in the 60-69µug/100ml range have similarly fallen, whilst there have been compensatory increases in the proportion below 40µg/100ml.

Figure 10.15 Blood lead levels: females, Great Britain, 1982 to 1992/93

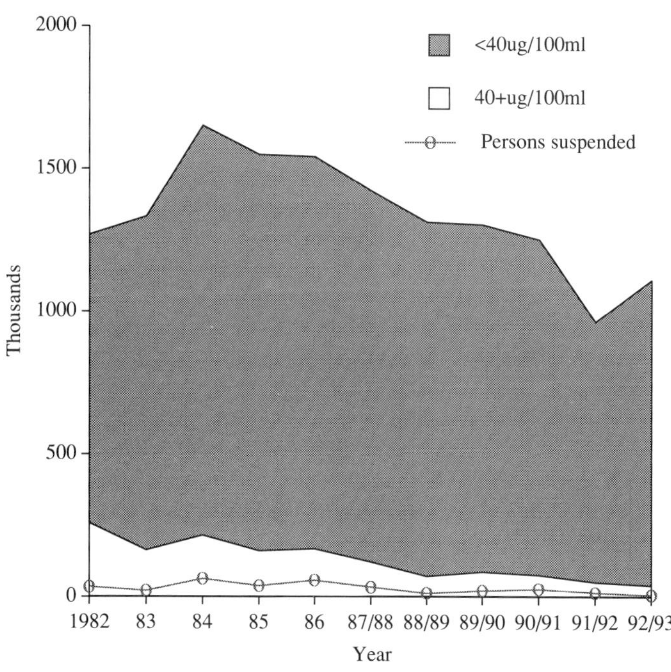

Numbers of men suspended from work have fallen, though the proportion of workers with high levels who were suspended showed a tendency to increase; 75 per cent of those over 69 µg/100ml were suspended in 1992/93 compared with 55 per cent in the previous year and 51 per cent in 1987/88. A worker with a high reading need not be suspended if a repeat measurement is below the limit though he would still be recorded in the statistics as being above the limit.

The number of females under surveillance rose to 1,112 in 1992/93 from 964 in the previous year. However this followed several years of falling figures, from 1,423 in 1987/88. The numbers with blood-lead levels above 39µg/100ml fell from 123 (9 per cent of females) in 1987/88 to 41 (4 per cent of females) in 1992/93. The numbers of females suspended from work in recent years were small and subject to fluctuations; there were 5 such in 1992/93 compared with 13 the previous year.

Table 10.21 Lead workers under medical surveillance, 1987/88 to 1992/93

Maximum measured blood-lead µg/100ml	1987/88		1988/89		1989/90		1990/91		1991/92		1992/93	
	Males	Females	Males	Females	Males	Females	Males	Females	Males	Females	Males	Females
<40	15310	1300	16820	1240	16195	1216	18447	1175	17531	912	15677	1071
40-59	4819	97	4751	66	4713	82	5413	68	5266	45	3550	37
60-69	1241	21	1038	6	947	6	1132	7	905	1	578	3
70-79	523	4	441	1	366	0	274	1	192	0	141	1
80 and over	239	1	196	0	168	0	178	1	84	6	55	0
Total under surveillance	22132	1423	23246	1313	22389	1304	25444	1252	23978	964	20001	1112
of which: 70 and over	762	5	637	1	534	0	452	2	276	6	196	1
Individuals suspended	388	33	340	12	286	21	260	26	153	13	147	5

Table 10.22 shows the distribution of blood-lead levels by industrial sector in 1992/93. Smelting, refining and casting, the lead battery industry and the manufacture of lead compounds accounted for most of the males under surveillance. The main areas of employment for women were the manufacture of lead compounds, potteries glazes and transfers, glass making, and smelting refining and casting.

The sectors having the highest proportions of men over 69µg/100ml were the demolition and scrap industries. However the numbers of men under surveillance in these sectors were relatively small and the numbers of these that had high levels are prone to the fluctuations that affect small counts. In absolute terms the lead battery industry accounted for the greatest number of men over the limit, but this partly reflected the high number who were employed in this sector. Following improvements in hygiene and control, the proportion of men in the lead batteries sector who were over 69 µg/100ml has fallen from 9 per cent in 1988/89 to 2 per cent in 1992/93.

For women the data are given as numbers rather than percentages since the numbers who exceed the 39µg/100ml level were very small. Just over a half of the 41 women who were over this level were employed in the lead batteries sector.

10.4.19 Occupational exposure to ionising radiation

There are few cases of compensated or reported radiation injury in any year, reflecting the careful control of occupational radiation exposure in the UK. There is widespread monitoring of radiation exposure for potentially exposed groups of workers and these monitoring data can be used as a guide to trends in exposure and therefore risk. In its most recent report[20] of radiation exposure of the UK population, the National Radiological Protection Board (NRPB) estimated that about 280,000 workers are exposed to ionising radiations while at work, of which some 120,000 are exposed mainly to natural sources of radiation such as cosmic radiation and the radioactive decay products of radon gas.

Employers must make arrangements with approved dosimetry services (ADS) to carry out systematic dose assessments for each of their employees who they have designated as a classified person under the Ionising Radiations Regulations 1985. In practice, many other workers are also routinely monitored for other reasons.

ADS submit annual summaries of doses recorded for classified workers to HSE's Central Index of Dose Information (CIDI), which is operated under contract by NRPB. Statistical summaries of this data have been published for each year

Table 10.22 Lead workers under medical surveillance, by sex, blood-lead level and industry sector, 1992/93

Sector	Males						Females		
	Percentage in blood-lead category (ug/100ml)						Number in category		
	<40	40-49	50-59	60-69	70+	Total under surveillance	<40	40+	Total under surveillance
Smelting, refining, alloying, casting	74.8	13.7	7.9	2.8	0.9	5398	136	4	140
Lead battery industry	55.7	21.2	14.3	6.7	2.1	3759	81	22	103
Badge and jewellery enamelling and other vitreous enamelling	92.9	2.0	3.1	2.0	0.0	98	46	5	51
Glass making	70.5	17.2	8.4	3.3	0.6	952	154	4	158
Manufacture of pigments and colours	91.7	5.1	1.6	1.2	0.4	565	71	0	71
Potteries, glazes and transfers	87.1	7.7	3.9	1.4	0.0	286	174	5	179
Manufacture of inorganic and organic lead compounds	95.7	2.6	1.1	0.6	0.0	2438	182	0	182
Shipbuilding, repairing and breaking	95.0	4.0	0.0	1.0	0.0	99	0	0	0
Demolition industry	75.9	11.2	6.9	3.2	2.8	650	1	0	1
Painting buildings and vehicles	91.3	4.0	1.3	1.9	1.5	893	2	0	2
Work with metallic lead and lead containing alloys	84.0	9.1	4.6	2.0	0.3	1767	120	1	121
Other processes	91.3	5.3	1.9	1.0	0.5	2951	103	0	103
Scrap industry	69.0	11.7	3.5	9.0	6.9	145	1	0	1
All sectors	78.4	11.3	6.5	2.9	1.0	20001	1071	41	1112

from 1986 to 1992. HSE has also published a review trends in dose information reported to CIDI over the period 1986-91.[21]

The statistical summary for 1992 shows that doses were reported for 58,173 persons in that year with a mean whole-body dose, after standard corrections, of 1 mSv. The corresponding figures for 1986 were 56,044 persons and a mean dose of 2.3 mSv. Over the period 1986-91 there was a more than 10-fold reduction in the number of classified persons with an annual whole-body dose above the principal investigation level of 15 mSv; including a pro-rata substitution for 'notional doses' only 143 persons (less than 0.3 per cent of the total) had a reported dose greater than this in 1992.

The occupational categories of nuclear reactor operations (6,688 persons), nuclear reactor maintenance (9,632 persons), nuclear fuel fabrication (3,937 persons) and nuclear fuel reprocessing (5,439 persons) accounted for nearly 45 per cent of classified persons in 1992. Mean doses for these categories were 1.1 mSv, 1.2 mSv, 1.8 mSv and 2.0 mSv respectively, after correcting for 'notional doses'.

The occupational group with the highest average annual dose in 1992 was non-coal mining underground, with an average annual dose of 14.7 mSv, excluding 'notional doses'. Industrial radiography can be hazardous, particularly when carried out on site. 5,295 classified persons were employed in this occupation during 1992 with a mean dose of 0.9 mSv, after correcting for 'notional doses'; just over 0.4 per cent had annual doses greater than 15 mSv.

10.5 Conclusions

This chapter has reviewed data on the scale of occupational ill health in the UK from a range of sources. These sources cover a wide spectrum of information, from self reports of individuals' perceptions of the relationship between their work and any type of ill-health, to assessments made by doctors with special expertise in the diagnosis of specific occupational diseases. The information described has already contributed to a fundamental review by HSE of occupation health risks in the UK.[22] Nevertheless, HSE wishes to continue broadening its base of statistical data on occupational disease and welcomes any suggestions for how this might be done, either through the use of any existing occupational ill-health data of which it is currently unaware; by analysis of relevant information collected for other purposes; or the development of new sources of information such as reporting schemes and ad-hoc surveys.

References

1. Carter JT. There's a lot of it about? *Br J Ind Med* 1991; 48: 289-91.
2. *Health and Safety Commission. Annual Report 1993/94: Statistical Supplement.* Sudbury: HSE Books, 1994.
3. Office of Population Censuses and Surveys. *Labour Force Survey 1988 and 1989.* LFS No 8. London: HMSO, 1991.
4. Stevens G. Workplace injury: a view from HSE's trailer to the 1990 Labour Force Survey. *Employment Gazette* 1992; 100: 621-38.
5. Hodgson JT, Jones JR, Elliott RC, Osman J. *Self-reported work-related illness: results from a trailer questionnaire on the 1990 Labour Force Survey in England and Wales.* Sudbury: HSE Books, 1993.
6. Marmot MG, McDowall M. Mortality decline and widening social inequalities. *Lancet* 1986; ii: 174-76.
7. Industrial Injuries Advisory Council. *Periodic report 1993.* London: HMSO, 1993.
8. Meredith SK, McDonald JC. Work related respiratory disease in the UK 1989-1992: a report on the SWORD project. *Occupational Medicine* 1994; 44: 183-189.
9. Sallie BA, Ross DJ, Meredith SK, McDonald JC. SWORD 1993. Surveillance of work related and occupational respiratory disease in the UK. *Occupational Medicine* 1944; 44: 177-182
10. Meredith SK. Reported incidence of occupational asthma in the UK 1989-90. *Journal of Epidemiology and Community Health* 1993; 47: 459-463.
11. Raw GJ. Sick building syndrome: a review of the evidence on causes and solutions. HSE Contract Research Report 42/1992. Sudbury: HSE Books, 1992.
12. Logan WP, Cushion AA. *Morbidity statistics from general practice, Vol 1, (General), 1955-56.* London: HMSO, 1958.
13. Royal College of General Practitioners, Office of Population Censuses and Surveys and Department of Health and Social Security. *Morbidity statistics from general practice 1970-71: socio-economic analyses.* London: HMSO, 1982.
14. Royal College of General Practitioners, Office of Population Censuses and Surveys and Department of Health and Social Security. *Morbidity statistics from general practice 1981-82: third national study.* London: HMSO, 1986.
15. Verbrugge LM, Ascione FJ. Exploring the iceberg. Common symptoms and how people care for them. *Med Care* 1987; 25: 481-486.
16. Martin J, Meltzer H et al. *The prevalence of disability among adults. OPCS surveys of disability in Britain, Report 1.* London: HMSO, 1988.
17. Kyriakides K. *Survey of exposure to hand-arm vibration in Great Britain* HSE Research Paper 26. Sudbury: HSE Books, 1988.
18. Bednall AW. *Survey of exposure to hand-arm vibration in Great Britain: mines and quarries.* HSE Research Paper 29. Sudbury: HSE Books, 1991.
19. Benn T. *Estimation of the prevalence of hand-arm vibration syndrome in Great Britain. Proceedings of the Institute of Acoustics* 1993; 15: 463-470.
20. Hughes JS, O'Riordan MC. *Radiation exposure of the UK population—1993 review.* NRPB R263. London: HMSO, 1993.
21. *Central index of dose information. Summary of statistics for 1992.* Sudbury: HSE Books, 1994.
22. Anon. Actions on health — an HSE policy statement. *Occupational Health Review.* 1994; 48: 24-27.

Chapter 11 Occupational injuries at work

Jacqui Cooper

Statistical Services Unit, HSE

11.1 Sources and definitions

The statistics presented in this chapter relate to injuries reported under the Reporting of Injuries, Diseases and Dangerous Occurrences Regulations 1985 (RIDDOR). These Regulations came into effect on 1st April 1986 and replaced the preceding Notification of Accidents and Dangerous Occurrences Regulations 1980 (NADOR). Changes in definitions, particularly the definition of a major injury, mean that many of the statistics derived from reports made under RIDDOR are not comparable with those previously reported under NADOR. With the exception of the overall numbers of fatal injuries, this chapter therefore covers the period 1986/87 to 1991/92. Figures for 1992/93 which were not finalised at the time of publication can be found in the statistical supplement to the 1993/94 Health and Safety Commission's Annual Report.[1] Statistics after 1985 are given on a financial year basis. For definitions, please see Annex 11.1.

Under RIDDOR, as well as fatal and major injuries the 'responsible person' has a duty to report injuries resulting in an absence from normal work of over 3 days. In the case of a reportable injury to an employee, the responsible person would be the employer. For a self-employed sub-contractor it would be the main employer, or contractor. Depending on the type of premises, the report has to be made to one of the Inspectorates in the Health and Safety Executive (HSE) or to local authorities.

Injuries to employees (including trainees), the self-employed and members of the public are reportable if they are judged to have arisen from work activity, although injuries to members of the public are not covered here. The statistics also exclude:

- accidents giving rise to three or fewer days off work;
- assaults on staff;
- road traffic accidents involving people travelling in the course of their work, which are covered by road traffic legislation;
- accidents reportable under separate merchant shipping, civil aviation and air navigation legislation;
- fatal injuries to the self-employed except when they are working at premises under the control of someone else at work;
- those accidents notified under the Poisonous Substances in Agriculture Regulations 1984 until their repeal;
- accidents to members of the armed forces.

The figures for fatal injuries are based on RIDDOR statistics and other sources of information to ensure that the statistics on fatalities are virtually complete. It is possible, however, that a small number of fatalities may be missed either due to the particular circumstances of the accident or a lengthy gap between the accident and the ensuing fatality. This should not affect the consistency of the data over time.

In contrast to the comprehensive information on fatalities, there is considerable under-reporting under RIDDOR which, unlike systems based on compensation and insurance, relies on direct reporting by employers, and their familiarity with the reporting regulations. The HSE commissioned a trailer to the 1990 Labour Force Survey (LFS) which asked questions on work related accidents and ill-health to complement the data regularly collected from employers so as to provide a more comprehensive picture of work related injuries and ill-health. It appears from the LFS that less than a third of reportable non-fatal injuries at work are being reported under RIDDOR. Earlier crude estimates had suggested a reporting rate of no more than 50 per cent.

The LFS trailer depends, of course, on respondents' correct recall of events which took place during the previous year. However, analysis of levels of injury from other sources of data, such as the General Household Survey (GHS) and comparisons with other European countries, broadly supported the LFS findings. The LFS showed there is little variation in the injury incidence rates for employees and self-employed people. These results, when compared with the incidence rates based on reported injuries, show that while the reporting level of injuries to employees is about 30 per cent, it is less than 5 per cent for self-employed people. The commentary in this chapter therefore excludes non-fatal injuries to the self-employed. The numbers of all reported incidents are included in the reference tables which will be available on disk.

In addition to these variations in reporting levels by employment status, the LFS results also showed variations in reporting levels between the main employment sectors. The reporting level is highest in the energy sector (about 80 per cent), just over 40 per cent in construction and just below 40 per cent in manufacturing. The level is much lower in services (23 per cent) and agriculture (17 per cent). Despite these levels of under-reporting, reported accidents are a good indicator of the relative order of risk by industry. Furthermore, the physical types of injury in reported accidents are representative of those indicated by the survey. This means that reported injuries, though incomplete, still reflect the overall picture of all accidents in terms of relative risks and physical injuries.

A further set of questions on work-related injuries was asked in the 1993/94 LFS and will now be included each year. When the analysis of these questions is available, it will provide a global figure for injuries including those not reportable under RIDDOR. The repeated use of the LFS will also provide an opportunity to examine any changes in the level of reporting, although it is thought unlikely that reporting patterns have influenced the trends in non-fatal rates.

As the information recorded on the injury databases of the various Inspectorates and enforcing authorities within HSE differs, it is only possible to provide statistics for kind and nature of injury for those injuries reported to HSE's Field Operations Divisions and local authorities. Injuries reported to local authorities are forwarded to HSE on a voluntary basis so that a combined database can be compiled. Injuries reported under RIDDOR to the Railways, Explosives and Mines Inspectorates, and the Offshore Safety and Nuclear Safety Divisions are therefore included only in the overall figures, Tables 11.1 to 11.6, and not in the subsequent more detailed text and tables.

11.2 The overall picture

11.2.1 Fatal injuries

In each year since 1986/87, over 350 people have been killed at work, with the number fluctuating from a high of 609 in 1988/89, which included the 167 fatalities in the Piper Alpha disaster, to a low of 368 in 1991/92. This compares with an

Table 11.1 Fatal injuries and fatal injury rates to employees and the self-employed, 1961 to 1991/92, as reported to all enforcing authorities

	Numbers			Rates (per 100,000 workers)		
	Employees	Self-employed	Total	Employees	Self-employed	Total
1961	1,228	n/r	..	5.6	n/r	..
1971	780	n/r	..	3.6	n/r	..
1981	441	54	495	2.1	2.6	2.1
1982	472	48	520	2.3	2.3	2.3
1983	448	65	513	2.2	3.0	2.3
1984	438	60	498	2.1	2.5	2.1
1985	400	71	471	1.9	2.8	2.0
1986/87	355	52	407	1.7	2.0	1.7
1987/88	361	84	445	1.7	3.0	1.8
1988/89	529*	80	609	2.4*	2.7	2.5*
1989/90	370	105	475	1.7	3.3	1.9
1990/91	346	87	433	1.6	2.7	1.7
1991/92	297	71	368	1.4	2.3	1.5

n/r Not reportable
* Includes the 167 fatalities in the Piper Alpha disaster.

Table 11.2 Fatal injuries to employees and the self-employed by kind of accident, 1986/87 to 1991/92, as reported to HSE's Field Operations Division (a) and local authorities

	Number of injuries					
	1986/87	1987/88	1988/89	1989/90	1990/91	1991/92
Contact with moving machinery or material being machined	37	26	33	24	34	20
Struck by moving including flying/falling object	36	60	44	69	53	41
Struck by moving vehicle	51	61	81	53	73	62
Fall from a height	107	111	122	158	108	104
Trapped by something collapsing or overturning	42	44	33	54	29	29
Drowning or asphyxiation	23	15	18	17	12	13
Contact with electricity or electrical discharge	28	24	22	27	28	21
Other kinds of accident	25	42	34	39	48	35
Injuries not classified by kind	10	9	1	-	-	-
Total	359	392	388	441	385	325

(a) Excludes deaths reported to HSE's Quarries Inspectorate for years prior to 1990/91.

Table 11.3 Fatal injuries to employees and the self-employed by nature of injury, 1987/88 to 1991/92, as reported to HSE's Field Operation Division (a) and local authorities

Number of injuries	1987/88	1988/89	1989/90	1990/91	1991/92
Fractures	103	61	73	63	47
Concussion and internal injuries	23	25	30	18	19
Contusions	63	46	43	52	63
Injuries of more than one nature	45	81	66	47	52
Poisoning, gassing and asphyxiation	23	32	23	33	32
Other natures of injury	77	58	79	64	40
Injury not known	58	85	127	108	72
Total	392	388	441	385	325

(a) Excludes deaths reported to HSE's Quarries Inspectorate for years prior to 1990/91.

Table 11.4 Major injuries to employees by kind of accident, and nature of injury, 1986/87 - 1991/92 combined

Kind of accident	
Percentage caused by:	
Slip, trip or fall on same level	30
Falls from a height	22
Struck by moving including flying/falling object	12
Contact with moving machinery or material being machined	10
Handling, lifting or carrying	7
Other causes (each accounting for less than 5 per cent of accidents)	19
Nature of injury	
Percentage resulting in:	
Fracture	71
Amputation	9
Other causes (each accounting for less than 5 per cent of accidents)	20

Table 11.5 Over 3-day injuries to employees by kind of accident and nature of injury, 1986/87 - 1991/92 combined

Kind of accident	
Percentage caused by:	
Handling, lifting or carrying	35
Slip, trip or fall on same level	19
Being struck by moving including flying/falling object	15
Falls from a height	8
Striking against something fixed or stationary	7
Contact with moving machinery or material being machined	5
Other causes (each accounting for less than 5 per cent of accidents)	11
Nature of injury	
Percentage resulting in:	
Sprains and strains	38
Contusions	18
Superficial injuries	11
Lacerations and open wounds	8
Fractures	8
Other causes (each accounting for less than 5 per cent of accidents)	17

Table 11.6 Fatal injury rates in the agricultural sector for employees (selected SIC 80 classes) and the self-employed, 1986/87 to 1991/92

SIC 80	1986/87	1987/88	1988/89	1989/90	1990/91	1991/92
Employees						
01 Agriculture and horticulture	8.1	6.9	6.4	8.3	8.1	5.9
02 Forestry	26.3	8.8	17.5	0.0	33.3	38.5
0 Agriculture, forestry and fishing (a)	8.6	6.8	7.0	8.1	9.0	6.7
Self-employed	6.9	12.7	10.3	12.3	10.9	13.0

(a) Excludes sea fishing.

average of 500 fatalities each year in the first half of the decade. As the numbers of people in employment have fluctuated over this period, the numbers are better considered in the form of the fatal injury rate which has fallen from over 2 per hundred thousand at the beginning of the decade and is now less than half the rate in the early 1970s and less than a quarter of that in the early 1960s. However, this reduction is largely accounted for by changes in the pattern of employment as there has been a fall in the numbers of people working in the more hazardous sectors of industry such as mining and manufacturing and a growth in employment in the service sector.

The majority of fatal injuries are to employees, around 80 per cent in each of the last three years. However, the overall fatal injury rate for the self-employed is higher than that for employees, as there are proportionately more self-employed people in those sectors of industry which are more dangerous to work in, namely agriculture and construction (see Figure 11.1).

The most common cause of fatal injury, which accounted for nearly a third of all fatalities in each of the last six years was a fall from a height, over 80 per cent of these being a fall of over 2 metres. One in six fatal injuries were caused by being struck by a moving vehicle and one in eight by being struck by a moving object (see Table 11.2).

Reflecting the kinds of accident most likely to lead to fatal injury, the majority of deaths were the result of fractures,

Figure 11.1 Fatal injury incidence rate per 100,000 employees - all industry, 1961 to 1991/92

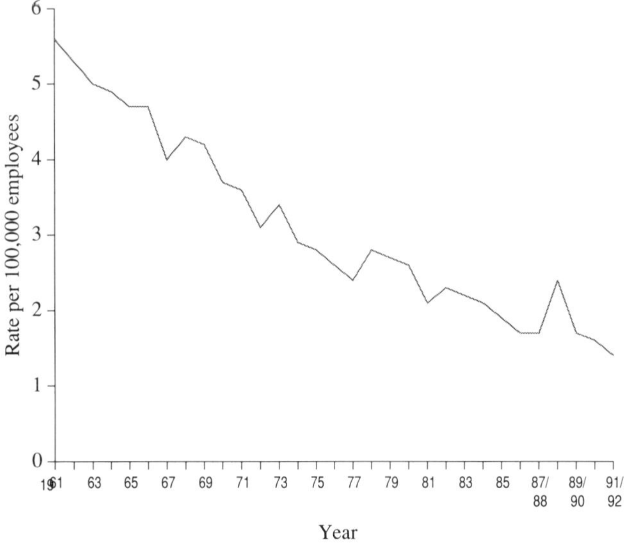

All enforcing authorities

contusions and multiple injuries. These three types of injury accounted for 61 per cent of deaths over the five-year period, for those fatalities where the nature of the injury was recorded. A further 10 per cent were caused by poisoning, gassing or asphyxiation and 8 per cent by concussion and/or internal injuries. (See Table 11.3)

11.2.2 Major injuries

There were around 20,000 major injuries to employees in each of the five years 1986/87 to 1990/91 and a lower number (17,600) in 1991/92. The major injury rate dropped over the period, from 99 per hundred thousand employees in 1986/87 to 82 per hundred thousand in 1991/92 (see Figure 11.2). This was totally attributable to the change in employment patterns, rather than to improvements in safety standards.

Figure 11.2 Major injury incidence rate per 100,000 employees, 1986/87 to 1991/92

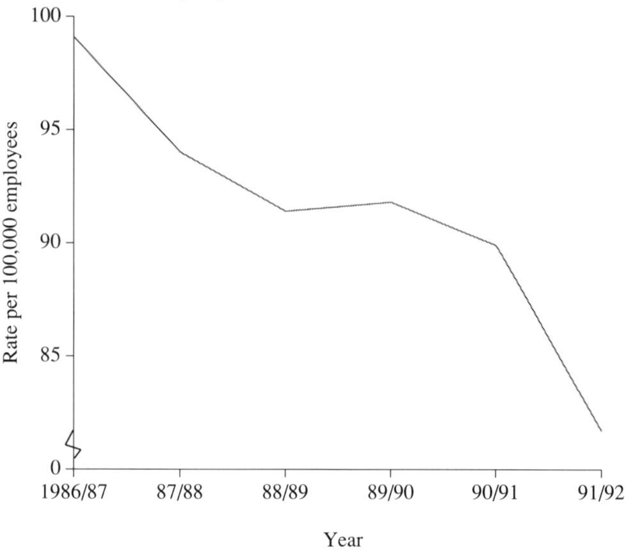

All enforcing authorities

Three in ten major accidents were the results of slips, trips or falls on the same level, with falls from a height accounting for a further two out of ten. Coming into contact with moving machinery or material being machined and being struck by a moving including flying or falling object both accounted for around 10 per cent of accidents each.

Fractures were by far the most common type of major injury, accounting for over 70 per cent of major injuries over the six year period. Amputations, mainly of the fingers, were the next most common outcome of an accident (see Table 11.4).

11.2.3 Over 3-day injuries

On average, there were over 160,000 over 3-day injuries to employees reported each year between 1986/87 and 1991/92. As with major injuries, the over 3-day injury rate has been falling over the six-year period from 760 per hundred thousand employees to 710 in 1991/92, but the improvement was again attributable to changes in the pattern of employment (see Figure 11.3).

Injuries caused by handling, lifting or carrying accounted for 35 per cent of all over 3-day injuries over the six-year period, with a similar proportion (38 per cent) of injuries resulting in sprains and strains. Slips, trips or falls on the same level were also a common cause of over 3-day injuries (19 per cent) with a further 15 per cent caused by being struck by a moving, including a flying or falling, object. Contusions and superficial injuries together accounted for 30 per cent of over 3-day injuries (see Table 11.5).

Figure 11.3 Over 3-day injury incidence rate per 100,000 employees, 1986/87 to 1991/92

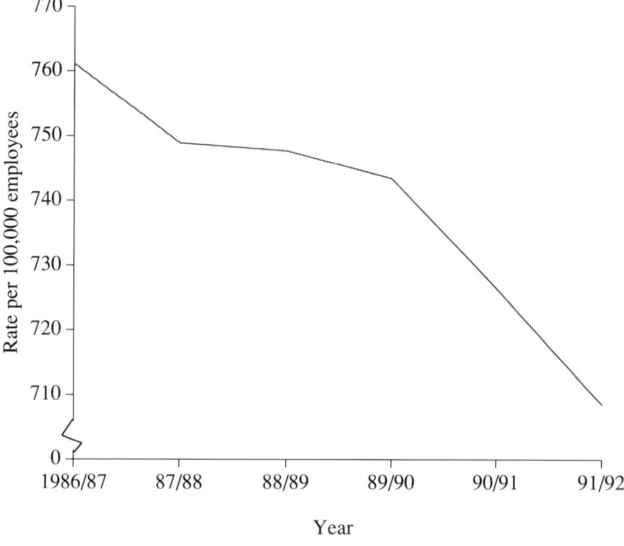

All enforcing authorities

The overall distribution of severity, kind and nature of accident masks the significantly different patterns found between the industrial sectors, resulting from the differing nature of work in each area. The next sections look at injuries in each of the five main industrial sectors: agriculture, energy, manufacturing, construction and services.

11.3 Agriculture

The agricultural sector, SIC division O, includes three main industries: agriculture and horticulture, forestry and fish farming. The majority of employees are employed in agriculture and horticulture. Approximately half of the total workforce in the agricultural sector is self-employed.

11.3.1 Fatal Injuries

Over 20 employees were fatally injured each year in the agricultural sector from 1986/87 to 1990/91 and 18 in 1991/92. Nearly all of these were people working in agriculture and horticulture. The fatal injury rate has fluctuated around 7 per hundred thousand to 8 per hundred thousand employees, nearly five times as high as the rate for all industries. As Table 11.6 shows, the fatal injury rate in forestry is several times higher than the rate for the agricultural sector as a whole, but this reflects a small number of deaths in a workforce of less than 8 thousand in 1991/92 (see Table 11.6).

Compared with all industries, the proportions of deaths of employees caused by coming into contact with electricity or electrical discharge and moving machinery or material being machined were twice as high in agriculture. There were also proportionately more deaths caused by drowning or asphyxiation, but fewer caused by falling from a height.

A similar, although not identical, pattern holds for the self-employed: proportionately more deaths were caused by

Figure 11.4 Fatal injuries to employees: percentage of total injuries for selected natures of injury by sector, 1987/88 to 1991/92

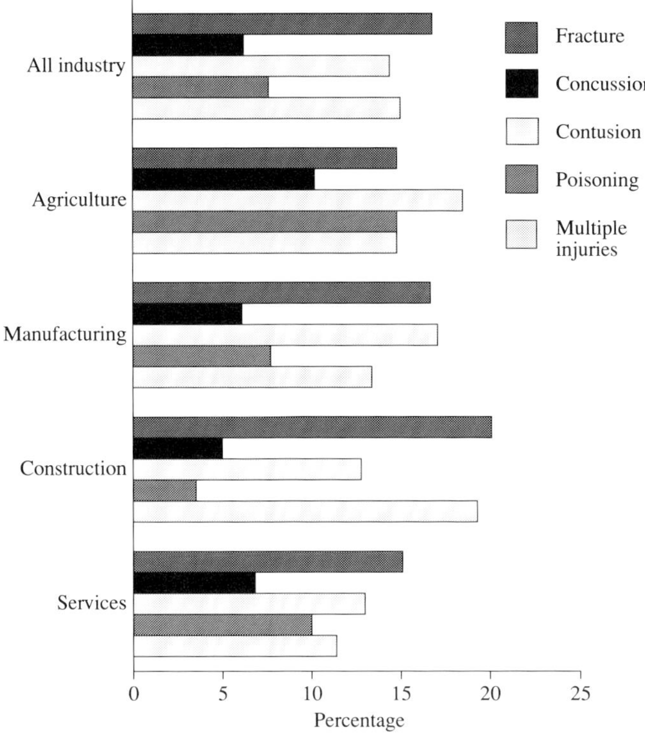

Based on reports to FOD Inspectorates and local authorities

contact with moving machinery or material being machined but also a higher proportion of self-employed workers in agriculture were fatally injured by being struck by a moving vehicle.

As a result of the kinds of accident in which agricultural workers are more likely to be involved, proportionately more deaths were caused by contact with electricity and poisoning, gassing or asphyxiation than in other types of industry (see Figure 11.4).

11.3.2 Major Injuries

The reported rate of major injuries in agriculture was two to three times as great as the figure for all industries between 1986/87 to 1991/92, and this is a sector with low levels of reporting. Again, the rate in forestry was higher than for the rest of agriculture, with this industry having one of the highest rates of injury overall outside the energy sector.

Sixteen per cent of major injuries to employees were caused by both contact with moving machinery or material being machined and being struck by a moving object compared with 10 per cent and 12 per cent for all industries respectively. Not surprisingly, over half of all injuries to employees involving an animal were to workers in the agricultural sector. Proportionately fewer reported major injuries in agriculture were caused by slips, trips or falls on the same level.

A higher proportion of major injuries in agriculture were amputations, 13 per cent compared with 9 per cent for industry as a whole. Fractures accounted for a smaller proportion, 65 per cent compared with 71 per cent.

11.3.3 Over 3-day injuries

As in industry as a whole, the most common kind of over 3-day injury was a result of handling, lifting or carrying but this only accounted for one in five over 3-day injuries in agriculture compared with one in three in industry as a whole. In contrast to the distribution of fatal and major injuries where falls from a height caused a smaller proportion of injuries in agriculture, for over 3-day injuries 11 per cent were caused by a fall, compared with 8 per cent for industry as a whole. The proportions caused by contact with moving machinery or being struck by a moving object remained above the overall level, together accounting for 27 per cent of injuries in agriculture as compared with 20 per cent in all industry. There were proportionately fewer reported minor over 3-day injuries such as sprains and strains and superficial injuries in agriculture, but higher proportions of fractures, lacerations and open wounds and contusions. This may be a reflection of the types of injury which are actually reported.

11.4 Energy

SIC division 1, energy and water supply industries, includes coal mining, extraction and processing of minerals and natural gas, production and distribution of gas, and electricity and the water supply industry. The number of people employed in this sector fell steadily over the period 1986/87 to 1991/92, mainly due to a reduction in the number of people employed in mining and to a lesser extent energy production, although there was an offset to this in that the numbers employed in oil and gas extraction increased.

11.4.1 Fatal injuries

The fatality rate in this sector has fluctuated around 6 per hundred thousand employees, with the exception of 1988/89 when 167 people were killed in the Piper Alpha disaster. Of the thirty odd fatalities each year, most deaths occurred in the coal industry and in the extraction of minerals and natural gas (see Figure 11.5).

Figure 11.5 Average fatal injury incidence rate for employees, 1986/87 to 1991/92

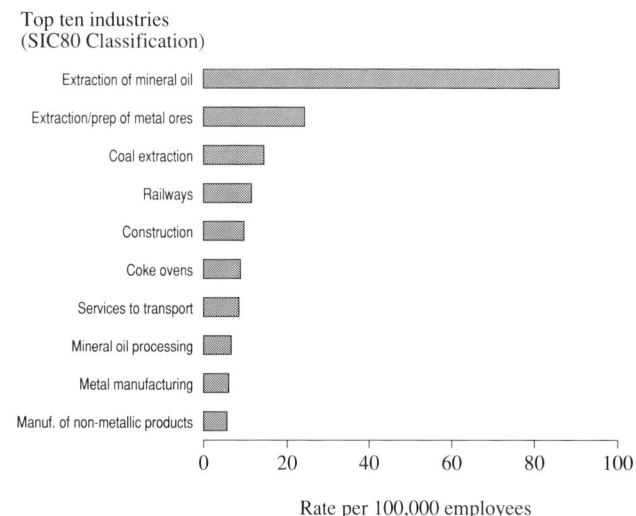

All enforcing authorities

With the number of fatalities in this sector being low, individual incidents causing a few deaths in any one year can shift the overall distribution considerably. Furthermore, most fatal injuries are not reported to HSE's Field Operations Division and therefore comparable detailed information is not available. It is therefore not particularly useful to compare the distributions of deaths by kind of accident and nature of injury in the energy sector with other industries.

11.4.2 Major injuries

The major injury rate in the energy sector fell each year from 243 per hundred thousand employees in 1988/89 to 176 in 1991/92 although this figure is still more than twice that for all industries. However, the reduction in the rate largely reflects the fall in the number of people employed in the mining industry, where the injury rate has generally been over 600 per hundred thousand employees.

The most common cause of major injury in the energy sector, as in all industries, was a slip, trip or fall. This kind of accident accounted for 27 per cent in the energy sector compared with 30 per cent overall. Injuries caused by contact with electricity or electrical discharge were more common in the energy sector (9 per cent compared with 2 per cent), whilst there were fewer injuries involving machinery in this sector. As in all industries, fractures were the most common type of major injury in the energy sector, but they accounted for 68 per cent of injuries in energy, compared with 71 per cent overall. Reflecting the type of accidents which caused major injuries, there were proportionately three times as many burns in the energy sector and half the level of amputations.

11.4.3 Over 3-day injuries

Over 3-day injury rates in the energy sector have also fallen over the 6 year period. Whilst the reduction is in part due to the changing mix of work done in the energy sector with a reduction in the number of people employed in coal mining, there have also been reductions in the overall 3-day injury rates in industries such as extraction of minerals and natural gas, injuries reported under RIDDOR in the nuclear fuel production industry, electricity and gas production and distribution and the water supply industry as well as in mining itself.

The distribution of over 3-day injuries in the energy sector was generally similar to that for all industries, again the exception being fewer machinery accidents. The proportion of handling injuries was also higher in the energy sector (39 per cent) than in industry as a whole (35 per cent). Consequently, more accidents in the energy sector resulted in sprains and strains (47 per cent compared with 38 per cent overall) and fewer in lacerations, fractures and contusions.

11.5 Manufacturing

The manufacturing sector is a diverse one, covering three SIC divisions. Metal goods, engineering and vehicle industries (SIC division 3) accounted for 45 per cent of the total number of employees in manufacturing in 1991/92. A further 41 per cent worked in 'other manufacturing industries' (SIC division 4) the largest of which were food, drink and tobacco and the manufacture of paper and paper products, printing and publishing. The third SIC division includes metal manufacturing, the chemical industry and the extraction of minerals and ores other than fuels. Over the six-year period from 1986/87 to 1991/92, the number of people employed in manufacturing dropped by over 10 per cent. By 1991/92, the proportion of employees working in manufacturing had dropped to 21 per cent of all employees, a reduction of 3 per cent from 1986/87.

11.5.1 Fatal injuries

In the six-year period 1986/87 to 1991/92, 566 employees in manufacturing suffered a fatal accident at work. Over a third of these occurred in SIC division 2 where the fatality rate was on average two to three times higher than the overall rate for manufacturing. The rate was particularly high in those industries dealing with the extraction and preparation of metalliferous ores and the manufacturing of non-metallic mineral products (see Table 11.7).

Not surprisingly, accidents caused by coming into contact with moving machinery or material being machined were much more common in the manufacturing sector than in most other sectors although on a par with the proportion in the agricultural sector. The proportion of deaths caused by machinery was over twice as high in manufacturing (17 per cent) compared with industry as a whole. More employees were also killed by being struck by a moving object (20 per cent compared with 14 per cent), again reflecting the type of work carried out in this sector. Accidents which are more likely to occur outside such as being struck by a moving vehicle or falling from a height were less common in manufacturing.

Only about 6 per cent of the self employed work in manufacturing, although there were on average 6 deaths a year between 1986/87 and 1991/92. The most common cause of fatalities to the self-employed was a fall from a height.

Although the causes of fatalities are different in the manufacturing sector, the nature of fatal injuries suffered show a fairly similar pattern to industry as a whole: predominantly contusions, fractures and multiple injuries.

11.5.2 Major injuries and over 3-day injuries

Rates of major and over 3-day injuries were approximately fifty per cent higher in manufacturing compared with all industries and twice as high in SIC division 2 (see Figure 11.6).

There was a similar picture for the kinds of non-fatal injuries occurring in the manufacturing sector as for fatal injuries. The proportions of both major and over 3-day injuries caused by machinery were twice as high in the manufacturing sector compared with all industry, accounting for one in five major injuries and one in ten over 3-day injuries. As with fatal injuries, falls from a height were responsible for a smaller proportion of injuries. For non-fatal injuries the proportion caused by moving objects was in line with the overall industry average. Slips, trips and falls caused relatively fewer non-fatal injuries, particularly major ones (see Figure 11.7).

Table 11.7 Fatal injury rates for employees in the manufacturing industry, 1986/87 to 1991/92

SIC 80	Description	1986/87	1987/88	1988/89	1989/90	1990/91	1991/92
2	Extraction of minerals and ores other than fuels, manufacture of metals, mineral products and chemicals	5.9	6.1	5.0	4.7	3.7	3.8
	within which						
21/23	Extraction of and preparation of metalliferous ores	40.8	38.1	25.0	8.3	17.5	18.5
22	Metal manufacturing	6.5	11.2	6.0	3.3	3.2	6.7
24	Manufacturing of non-metallic mineral products	5.6	4.2	5.3	8.9	5.1	4.8
3	Metal goods, engineering and vehicle industries	1.6	1.5	1.2	2.0	1.1	1.3
	within which						
32	Mechanical engineering	2.4	2.0	1.0	2.4	1.9	2.2
36	Manufacture of transport equipment other than motor vehicles	2.3	1.7	1.8	2.5	1.2	1.9
4	Other manufacturing industries	1.4	1.1	1.5	1.4	1.8	0.9
	within which						
41/42	Food, drink and tobacco	2.2	1.3	1.3	1.5	2.3	1.2
43	Textile industry	3.4	0.4	1.8	1.0	3.2	0.6
2-4	Total manufacturing industries	2.1	1.9	1.8	2.1	1.8	1.5

Figure 11.6 Average major injury incidence rate for employees, 1986/87 to 1991/92

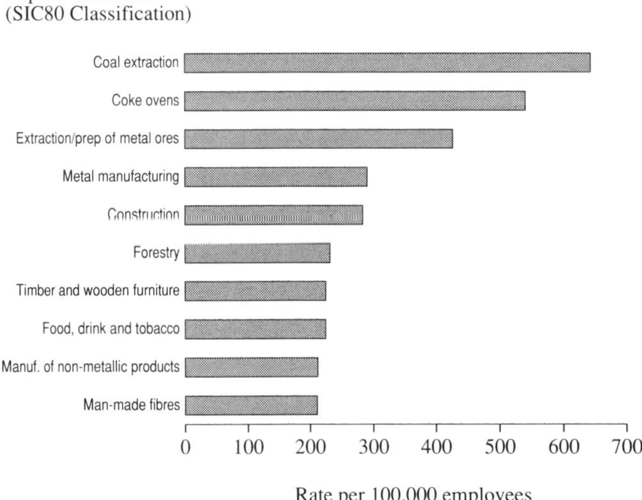

All enforcing authorities

Over two thirds of accidents causing amputations occurred in manufacturing, accounting for 16 per cent of major injuries in manufacturing compared with 9 per cent for industry as a whole.

Figure 11.7 Major injuries to employees: percentage of total injuries for selected kinds of accident by sector, 1986/87 to 1991/92

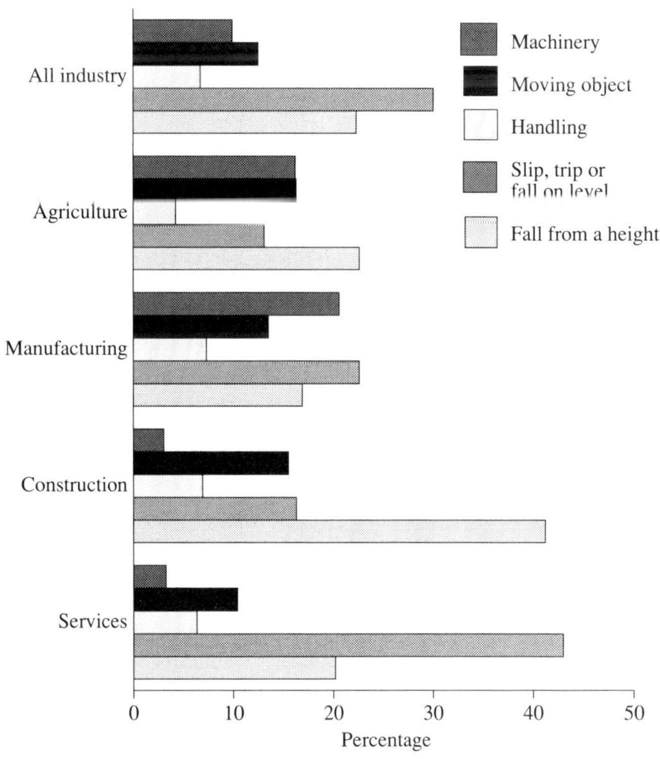

Based on reports to FOD Inspectorates and local authorities

11.6 Construction

11.6.1 Fatal injuries

Construction is a hazardous industry in which to work. One quarter of all fatal injuries to employees over the period 1986/87 to 1991/92 were to workers in the construction industry. Although the construction fatal injury rate dropped each year from 10.3 per hundred thousand workers in 1987/88 to 8.8 in 1991/92, it remained the highest of all sectors. On average, five workers are killed in the construction industry every fortnight (see Figure 11.8).

Figure 11.8 Fatal injury incidence rate per 100,000 employees by industrial sector, 1981 to 1991/92

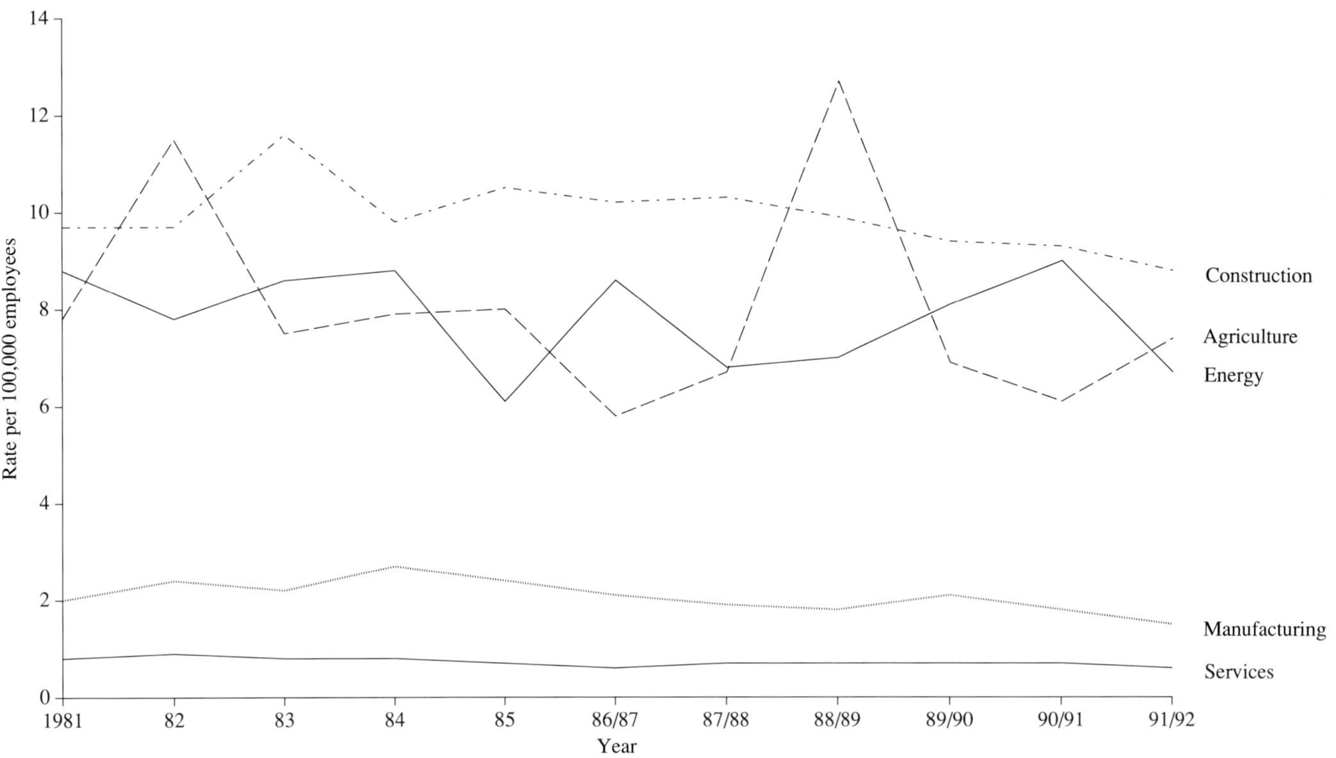

All enforcing authorities

Nearly half of the fatal injuries in construction were caused by a fall from a height, most of these being from heights of over 2 metres. Other common causes of fatality included being trapped by something collapsing or overturning (responsible for 14 per cent of deaths) and being struck by a moving vehicle, although the contribution this latter factor made was in line with the proportion for industry as a whole (see Figure 11.9).

Because of the kinds of accident occurring in construction, where the nature of injury was recorded, there were proportionately more deaths from multiple injuries and fractures and fewer due to contusions.

The self-employed fatality rate has fluctuated between 2.5 per hundred thousand and 7.5 per hundred thousand over the six-year period. However, the definition of a self-employed person in the construction industry is far from clear cut and this may have contributed to the variation in the rate. Nearly two thirds of fatal injuries to the self-employed were caused by falls from a height, and again the majority of these were falls from a height of over 2 metres.

11.6.2 Major injuries

The rate of reported major injuries in construction was also the highest of the main employment sectors, although there were reductions in 1990/91 and 1991/92 compared with the previous year.

Figure 11.9 Fatal injuries to employees: percentage of total injuries for selected kinds of accident by sector, 1986/87 to 1991/92

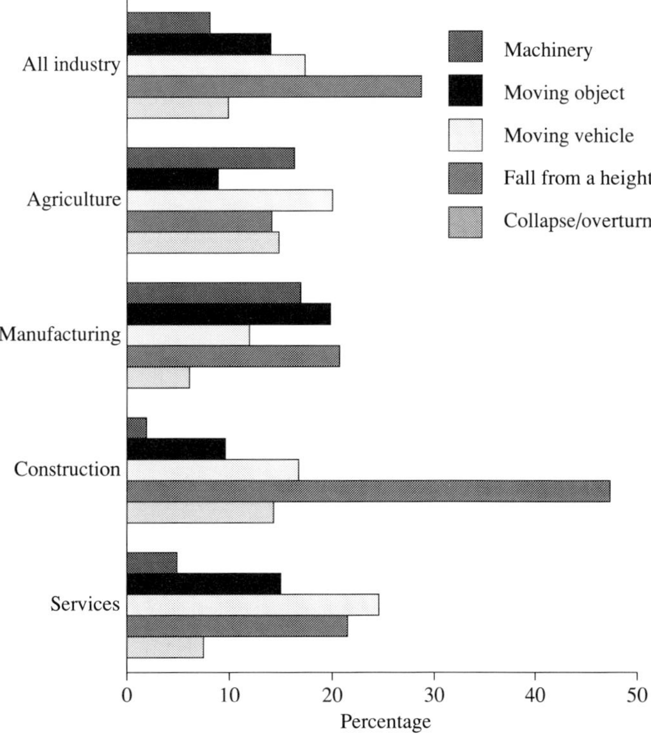

Based on reports to FOD Inspectorates and local authorities

Falls from a height accounted for four in ten major injuries in construction, double the proportion for industry as a whole and a further 15 per cent were caused by being struck by a moving object, some of which would have been falling objects. Relatively fewer major injuries were reported in construction as the result of a slip, trip or fall on the same level, about half the proportion in all sectors of industry. As with fatal injuries, fractures were a more commonly reported type of major injury in construction.

11.6.3 Over 3-day injuries

For over 3-day injuries, as with industry as a whole, over a third were caused by handling, lifting or carrying. Again, falls from a height caused proportionately more over 3-day injuries in construction (14 per cent compared with 8 per cent for all sectors), whilst slips, trips and falls accounted for 16 per cent of injuries in the construction industry compared with 19 per cent overall.

11.7 Service industries

The service sector covers a wide range of industries varying from transport and communication, including railways, which had the fourth highest fatality rate of all industrial divisions over the period 1986/87 to 1991/92, to office based activities which have virtually no fatal injuries and low rates of non-fatal injuries. Specifically, the service sector includes: distribution, hotels and catering; repairs (SIC division 6); transport and communication (SIC division 7); banking, finance, insurance, business services and leasing (SIC division 8); and other services, primarily public administration, education and health services (SIC division 9).

Local authorities receive only around one in seven reports on injuries to employees. As a result, the real risk of reportable injuries in the service sector is substantially higher than reported injuries suggest. Rates are further understated because of the extent of part-time working in the services sector where over 25 per cent of employees work part-time compared with less than 10 per cent in manufacturing.[2]

11.7.1 Fatal injuries

On average, 100 employees suffered a fatal injury in the service sector each year from 1986/87 to 1991/92. The fatal incidence rate has remained fairly constant at 0.6 to 0.7 per hundred thousand employees. Of the deaths in the service sector over 40 per cent occurred in the transport and communications services, primarily on the railways (92) and other inland transport (94), with supporting services to transport accounting for a further 45. One quarter of fatalities were to employees in SIC division 9 (other services), with two thirds of these being in SIC classes 91/92:- public administration, national defence, compulsory social security and sanitary services.

Being struck by a moving vehicle accounted for a quarter of fatal injuries to employees in the service sector (compared with 17 per cent overall), whilst fewer injuries were caused by falls from a height (21 per cent compared with 29 per cent). The proportions of fatalities caused by contact with machinery or being trapped by something collapsing or overturning were lower in the service sector than in industry as a whole, whilst 20 deaths occurred as a result of exposure to fire between 1986/87 and 1991/92 compared with the figure of 10 which would have been expected given the distribution of kinds of accident for all employees.

A comparison of the nature of fatalities in the service sector with the rest of industry shows that the nature of injury was less frequently recorded in the service sector, but when these injuries were excluded, the types of physical injury sustained were similar in the service sector to the overall distribution.

The fatality rate for the self-employed in the service sector overall has been comparable to that for employees, but most of the deaths occurred in other services, giving a higher fatality rate for that division for the self-employed than for employees.

Of the fatalities to the self-employed in the service sector between 1986/87 and 1991/92, a third were caused by a fall from a height compared with forty per cent overall and there were also relatively fewer deaths caused by machinery accidents or moving objects. Although numerically small, fatal injuries caused by exposure to fire or explosion and contact with harmful substances or electricity were proportionately higher in the service sector compared with the rest of industry.

11.7.2 Major injuries

The major injury rate in the service sector averaged 54 injuries per hundred thousand employees between 1986/87 and 1991/92, compared with 91 for industry as a whole. However, the average rate in the transport and communication sector was on a par with the rate for all industries and was substantially higher for employees on the railways and in supporting services to transport.

Slips, trips and falls accounted for 43 per cent of all major injuries in the service sector compared with 30 per cent overall, whilst machinery accidents and falls from a height were less common causes of injury. Consequently, a higher proportion of accidents in the service sector resulted in fractures (78 per cent compared with 71 per cent for all industries) and amputations were less common in the services sector.

11.7.3 Over 3-day injuries

The reported rate of over 3-day injuries in the service sector between 1986/87 to 1991/92 was about two thirds of that for industry as a whole. However, as the over 3-day injury rate for all industries has fallen over the period, due to changes in employment patterns, and the rate in the service sector has tended to rise, the gap between the two rates, although still substantial, has been reduced. Rates of reported over 3-day injuries within the service sector were also highest in the SIC division covering transport and communication industries, again noticeably on the railways and in supporting services to transport and also in postal services and telecommunications. Within the other sectors, rates were generally below the overall rate, although this is the sector with the lowest level of reported accidents. With the exception of the repair of consumer goods and vehicles industry, rates in the divisions other than transport and communications were below those for all industries.

Proportionately more over 3-day injuries in the service sector resulted from handling, lifting or carrying (37 per cent compared with 35 per cent) and slips, trips and falls (23 per cent compared with 19 per cent). Smaller proportions were causing by machinery and moving objects. Consequently, a higher proportion of accidents in the service sector resulted in sprains and strains; they accounted for 45 per cent of over 3-day injuries in the service sector and 38 per cent overall (see Figure 11.10).

Figure 11.10 Over 3-day injuries to employees: percentage of total injuries for selected kinds of accident by sector,1986/87 to 1991/92

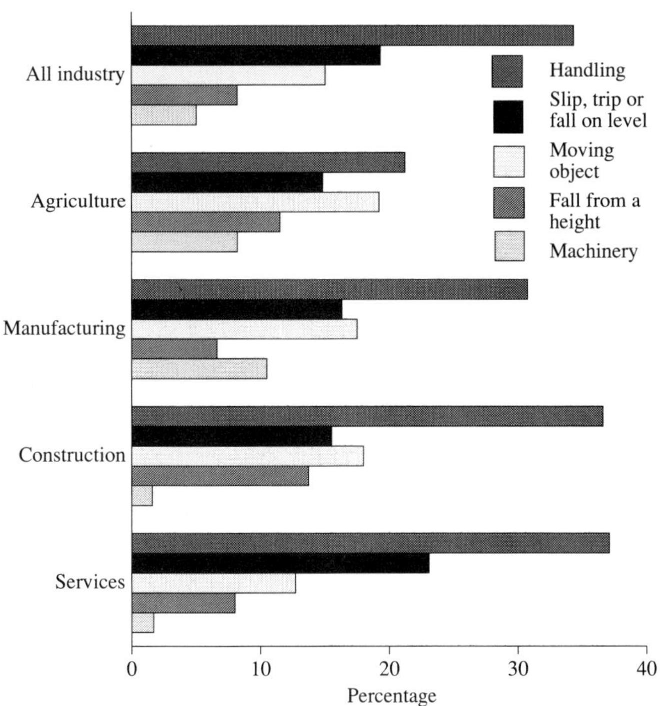

Based on reports to FOD Inspectorates and local authorities

References

1 *HSC Annual Report 1993/94 and Statistical Supplement.*
2 *Local Authorities Report on Health and Safety in Service Industries.*

Annex 1 Definitions of terms used

Fatal injuries include all deaths occurring up to a year after the accident.

Major injuries, based on Regulation 3(2) of RIDDOR, include the following:

(a) fracture of the skull, spine or pelvis;
(b) fracture of any bone:
(i) in the arm or wrist but not in the hand; or
(ii) in the leg or ankle, but not in the foot;
(c) amputation of:
(i) a hand or foot; or
(ii) a finger, thumb or toe, or any part thereof if the joint or bone is completely severed;
(d) the loss of sight of an eye, penetrating injury or chemical or hot metal burn to an eye;
(e) injury including burns requiring immediate medical treatment, or loss of consciousness, resulting from an electric shock from any electrical circuit or equipment, whether or not due to direct contact;
(f) loss of consciousness resulting from lack of oxygen;
(g) decompression sickness requiring immediate medical treatment (unless 1981 Diving Operations at Work Regulations apply);
(h) acute illness or loss of consciousness resulting from absorption of substance by inhalation, ingestion or through the skin;
(i) acute illness requiring medical treatment where there is reason to believe that this resulted from exposure to a pathogen or infected material;
(j) any other injury which results in the person injured being admitted immediately into hospital for more than 24 hours.

Chapter 12 Smoking, drinking and occupation

Richard Elliott

Epidemiology and Medical Statistics Unit, HSE

12.1 Introduction

Smoking and drinking are two very common personal habits which are likely to have an influence on morbidity and mortality. One or both are commonly invoked as possible explanations for differences in the incidence or prevalence of disease between occupational groups although supporting data are often lacking. The main purpose of this section is to provide data from a common source on smoking and drinking by occupation which may be used in a general sense in the interpretation of statistics on occupational ill health. The fully detailed data are contained in the tables on disk but the majority of the tables are reproduced in simplified form in the volume. The text describes the source and analysis of the data, comments on the main variations in smoking and drinking habit by occupation and offers some guidance on the use of the data contained in the tables.

The possible confounding effects of smoking and alcohol consumption in the interpretation of occupational mortality data are described where relevant in Chapters 4 to 6. Reference should be made to this chapter for the data on the smoking and alcohol consumption rates of the occupations concerned.

Smoking and drinking are influenced strongly by social factors other than occupation. Marital status and having children, for example, have been shown to exert a major effect.[1,2] In turn, differences in the rate of smoking and drinking by age have been interpreted as being due in part to these factors. Thus, when differences in the pattern of smoking or drinking by occupation are observed, it is necessary to ask to what extent these effects are directly generated by the occupation itself and in what measure they are indirectly generated by the effect of the occupation on the whole of the lifestyle of the workers concerned. For instance, if members of an occupational group are likely to marry later in their careers, this may indirectly raise the observed alcohol consumption figures in the younger age groups. Attention is drawn, where appropriate in the text, to other sources of data on smoking and alcohol consumption which may assist in this interpretation.

There are three main differences between the presentation and interpretation of data on smoking and drinking in the present context. The first is that information on smoking habits has been presented and discussed in previous volumes of the decennial supplement on Occupational Mortality[3,4] and to some extent a content and style of presentation has already been established. This volume attempts to build upon that framework and expand upon it. The pattern is less well established for data on alcohol consumption. Although some data, derived from the 1982 General Household Survey (GHS) were presented in the last volume,[4] the classification of alcohol consumption used then has not come into general use and there is a need to present the data in a form which facilitates comparisons with other contemporary literature. The presentation of the alcohol consumption data does not therefore attempt to follow a pre-existing format.

Secondly, the effect of age and sex on alcohol consumption is more pronounced than is the case for smoking. Not only is this the case for the population in general but there is also marked variation in the pattern of drinking by age and sex between occupations. There is thus a need to present a more detailed analysis of alcohol consumption than smoking habit to illustrate the occupational differences.

Finally, there are obvious differences in the interpretation of the data for practical purposes. Smoking is well known to be associated with the development of several diseases which are of interest in the occupational health context, such as lung cancer and chronic bronchitis, and the treatment of smoking habit as a confounding factor in the interpretation of statistics on such conditions is well understood. In contrast, there are few occupational diseases where alcohol consumption is a simple confounding variable. The inter-relationship between alcohol consumption and occupational illness is probably much more complex and at present poorly understood.

The effect of all these considerations is that the presentation of the smoking data is largely an attempt to continue an established and understood pattern of reporting while that of alcohol consumption is new to this series of reports and can be seen as a baseline upon which future reports can build.

12.2 Sources and general methods

The source of data on smoking and drinking habits used previously in the decennial supplement series was the GHS, which at present asks about these subjects every two years. The decision to continue to use this source for smoking data seemed obvious. In the case of alcohol consumption ad hoc surveys have been carried out by OPCS, the most recent of which was conducted in 1989 and reported as *Drinking in England and Wales in the Late 1980s*.[2] These surveys probably obtain more accurate estimates of individual alcohol consumption than the GHS and are certainly more comprehensive in the details collected. For example, they include questions on patterns of drinking and on reducing or ceasing drinking. Unfortunately the sample size is too small for anything more than a superficial analysis by occupational order. Furthermore, it was considered desirable to use a source which would be repeated and would enable continuity to be preserved in future analyses. The GHS was therefore also chosen as the source of data on alcohol consumption.

The GHS sample is almost 20,000 adult subjects per year. This would be adequate for an analysis of smoking and drinking by occupational order but insufficient to allow robust estimates at unit level. Two years' data, from 1988 and 1990, were combined to give a sample of approximately 16,500 men and 18,500 women over 16 for whom current or last occupation, smoking status and alcohol consumption were known. More years could be added to obtain a larger sample for detailed analyses but this would detract from the value of the results as a cross sectional picture at the end of the decade for comparison with future repetitions.

Fifty was chosen as the minimum sample size which would give reliable estimates of the prevalence of smoking or alcohol consumption. Since it appeared at the outset that sex differences were such that estimates based on mixed samples would rarely be meaningful, the decision was taken to restrict analysis by occupational units to those where the sample contained at least 50 men or 50 women. This condition was satisfied by 83 occupations for men and 46 for women, or 106 for one or both sexes.

The occupation codes used in the GHS are the OPCS 1980 Classification of Occupations Operational Codes and this classification was used in the analysis. There are fewer codes than in the full list of OPCS occupational units mainly because the operational codes do not distinguish foremen and other types of supervisor from the corresponding occupation which they supervise. If they did so, we would have combined the units for the analysis in any case.

All subjects in the GHS-derived sample, whether or not they would be included in one of the occupational unit samples, were grouped into occupational orders as in Table 12.1.

Since these orders are based on the operational codes, they are not identical to the standard OPCS classification orders, the main difference being that senior grades in certain occupations, such as the armed services and the police, will be classed in the order corresponding to that occupation rather than in order 5 (managerial). Where this may significantly affect the interpretation of the data, attention will be drawn to the relevant points in the discussion.

For occupational orders, a breakdown by age, in 4 bands, and sex was supplied by OPCS. For the selected occupational units the age breakdown was limited to a division into ages 16-64 and 65+ as the majority of these sample sizes were too small to withstand further stratification. The remaining analyses were conducted by the Epidemiology and Medical Statistics Unit of HSE.

12.3 Smoking status - general points

The analysis is restricted to cigarette smoking only as other forms of tobacco smoking make such a small contribution to the overall use of tobacco. Smoking habit has been treated as a simple dichotomous variable, i.e. an individual is or is not a current smoker. Ex-smokers are thus grouped together with those who have never smoked. Examination of the distribution of non-current smokers into the two components would give some idea of trends in smoking habits by occupation and this approach was followed in the 1970-72 decennial supplement.[3] The value of this information in determining past smoking rates by occupation is very limited and this analysis was not pursued in the present volume.

The proportion of workers who currently smoke cigarettes has been chosen as the single most useful indicator of smoking status for an occupational group. This follows from the analyses presented in the 1970-72 decennial supplement[3] which showed a high degree of correlation between this index and the SMR for lung cancer across occupational orders, indicating that it was a useful indicator of the biological effect of smoking. It was reported in that volume that there was little correlation between the mean cigarette consumption of the smokers in each order and the lung cancer mortality, mainly because the cigarette consumption by order was fairly narrowly distributed around a weekday mean of 17 and a weekend mean of 20 cigarettes per day. In

Table 12.1 Percentages smoking by age - male

No.	Occupational order	16-24	25-44	45-64	65+	All ages
1	Professional	29.1	23.5	19.8	15	21.5
2	Prof. ed. & health	45.8	21.4	17.6	14.4	19.5
3	Arts & sport	28	36.3	28.3	8.3	29.9
4	Professional science	20.6	19.6	15.7	12.6	17.5
5	Managerial	28.7	30.9	26.7	17.2	26.5
6	Clerical	25.3	33.2	32.6	23.6	28.9
7	Selling	26.1	32.5	25.9	17.2	26.9
8	Security	33.3	31.5	33.9	30.9	32.2
9	Personal service	34.4	52.6	42.5	32	40.6
10	Agric. workers	35.7	38.3	40.8	23	34.6
11	Materials proc.	40.1	41.4	35.1	15.9	35.2
12	Metal & electrical	36	36.8	36.3	26.1	34.5
13	Assembly	43.1	47.5	35	26.2	38.8
14	Construction	48.7	52.7	42.2	28.3	44.6
15	Transportation	40.7	52	41.1	30.4	42.9
16	Miscellaneous	45.9	67.9	44	29.6	46.0
17	N/A	60	28.6	50	75	50.0
M	Missing occ					
	All occupations	34.3	36.3	31.7	22.7	32.1

other words, the proportion of men who smoke varies much more by occupation than the cigarette consumption of the smokers.

The argument that it is valid to characterise the relative smoking habits of occupational groups on the basis of proportion of smokers alone, must be distinguished from the idea that smoking could generally be treated as a dichotomous variable for epidemiogical purposes. Although the mean cigarette consumption of smokers is similar for all occupations, it is probable that the distribution of cigarette consumption among individuals within each occupation is wide. Adjustments for smoking in specific studies should, if possible, treat smoking as a continuous variable in order to maximise the use of the available data.

The index of smoking status developed in the previous decennial supplements is the proportional current smoking ratio (PCSR). This is 100 times observed smokers/expected smokers where the expected number is calculated from the age specific smoking rates for the whole population sample and the numbers in each age band in the occupational sample. It is thus indirectly age standardised. A value of 100 indicates that the proportion of smokers is the same as that expected for an age matched sample of the whole population. Tables 12.1 and 12.2 show the percentages of male and female smokers respectively by occupational order in four age bands. The full versions of these tables giving the numbers of smokers and an overall table combining the data for men and women will be available on disk. The two most striking features of these results are the variation in smoking rates by age between occupations and the general decrease in smoking across all occupations in those of post retirement age.

Previous analyses of smoking by age in the GHS series have suggested that there is a general trend towards a decline in the proportion of the population who smoke with increasing age. This is more evident in married people and even more pronounced for those with dependent children. This trend is seen in the totals for all occupations at all ages in women and

Figure 12.1 Percentage of current smokers by age

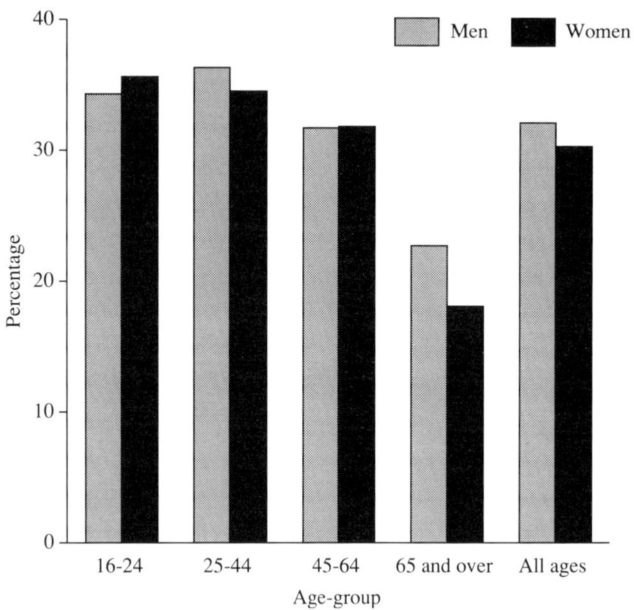

at all but the youngest age group in men. The trend is not evident between age groups 16-24 and 25-44 in men and this may be because the youngest age group contains a particularly high proportion of single subjects with no children (see Figure 12.1). The GHS reports for 1988 and 1990 show that single men show an increasing trend in smoking prevalence with age in these age bands. Part of the decrease in the proportion of people smoking with age could be due to selective survival of smokers to older ages, but this can not explain the variations seen with marital status and other social factors.

Since previous analyses have not been carried out by occupation, the marked occupational variations of the trend with age seen here have not been noted. This variation is discussed in more detail in the notes on specific occupational orders and units. As has been suggested above, the differences between

Table 12.2. Percentages smoking by age - female

No.	Occupational order	16-24	25-44	45-64	65+	All ages
1	Professional	31.5	19	13.9	13.2	18.5
2	Prof. ed. & health	34.7	22.9	22.2	15.3	22.2
3	Arts & sport	31.7	26.8	23.8	17.9	25.9
4	Professional science	15	14.8	4.5	40	14.2
5	Managerial	34.4	35.4	31.5	16.5	29.7
6	Clerical	30.3	27.2	26.2	15.8	25.6
7	Selling	34.4	38.7	32.5	17.7	31.7
8	Security	33.3	41.2	45.8	19.2	35.6
9	Personal service	40.9	47.1	40.3	20.9	37.0
10	Agric. workers	25	31.9	28.3	11.5	23.2
11	Materials proc.	53.9	49	38.8	17.3	36.8
12	Metal & electrical	60	46	30.6	23.8	34.0
13	Assembly	44.8	51.1	39.5	17.2	38.7
14	Construction	66.7	66.7	-	0	57.1
15	Transportation	36.4	36.5	36.5	14.8	30.7
16	Miscellaneous	0	63.6	50	34.5	42.4
17	N/A	0	18.8	13.3	13.3	14.3
M	Missing occ					
	All occupations	35.6	34.5	31.8	18.1	30.3

occupations may be partly due to differences in typical age at marriage rather than a direct influence of occupation on smoking behaviour.

The marked decline in the prevalence of smoking in those over 65 has been hinted at by previous GHS presentations which show a less striking decline in the over 60s.[1] Since the prevalence of smoking at all ages has been falling over the last two decades, the decline cannot be a reflection of past smoking habits but must be explained by cessation of smoking in the older age groups and perhaps selective survival of the non-smokers. It might be thought that, at least in relative terms, the distribution of smoking prevalence by occupation in today's retired workers would reflect the distribution in the forty to fifty year olds of two decades ago, which might be relevant to the interpretation of current statistics on lung cancer and other smoking-related diseases. However, the decline at age 65 and over is so large that any residual effects of past smoking habits would be masked so no inferences about the past distribution of smoking habit should be drawn from these data. Part of this decline may again be due to selective survival of non-smokers into old age, but the data suggest that the decline becomes much more marked close to retirement age.

The reasons for ceasing to smoke at or near retirement age could include reduced income on retirement and health reasons. The timing of the decline would suggest that the first is of major importance. Against this, it could be argued that there is no indication of an earlier decline in women, who retire at age 60 or over, but this would possibly be obscured by the fact that in most households the main income provider is male. A stronger counter-argument is that previous GHS analyses have shown that the unemployed and the economically inactive (at lower ages) have generally higher smoking rates than those in work, suggesting that reduced income is unlikely to be the only determinant of the effect seen in those over 65.

Tables 12.3 and 12.4 show, for men and women respectively, the total percentages smoking and PCSRs, defined above, for ages 16-64 and 65+ by occupational order. Figure 12.2 also shows this information. Note that as PCSRs are age-specific they may only be compared within an age band. Thus, for example, the higher PCSR for women aged 65+ in order 1 does not indicate that the absolute smoking rate in this age group is higher than that of the 16-64 year olds. It is actually considerably less. It should also be noted that the PCSRs for women over 65 may be distorted by the low sample size in several of the occupational orders.

Figure 12.2 shows that there is a general inverse relationship between smoking rates and social class as defined by occupation but the most prominent feature of the tables is the division between manual and non-manual occupational orders with the former having higher rates of smoking. This is seen most easily in Table 12.3 where a definite transition takes place between orders 8 and 9. The transition is less obvious in women because orders such as 5 (managerial) and 7 (selling), which in terms of socio-economic status are intermediate between the professional and the manual occupations, have smoking ratios near to those of the population in general. It is possible that the sex difference between the PCSRs in these occupations reflects a difference in the types of industry in which each sex is likely to be employed in these occupations. This is discussed in more depth in the notes below on specific occupations.

12.4 Alcohol consumption - general points

Whereas smoking can be treated, for the reasons described, as a dichotomous variable alcohol consumption within any population has a much more continuous distribution. The level of consumption must therefore be taken into consideration but it is rather simplistic to expect that mean alcohol consumption of an occupational group would be a very useful

Text continues on page 202

Table 12.3 Male proportional current smoking ratios

No.	Occupational order	Age 16-64					Age 65+				
		Sample	Observed	Expected	% smoking	PCSR	Sample	Observed	Expected	% smoking	PCSR
1	Professional	1034	235	356.2	22.7	66.0	200	30	45.4	15	66.1
2	Prof. ed. & health	708	145	241.4	20.5	60.1	139	20	31.6	14.4	63.4
3	Arts & sport	180	59	62.4	32.8	94.6	24	2	5.4	8.3	36.7
4	Professional science	950	175	327.4	18.4	53.4	167	21	37.9	12.6	55.4
5	Managerial	1613	465	548.8	28.8	84.7	413	71	93.8	17.2	75.7
6	Clerical	883	268	302.4	30.4	88.6	246	58	55.8	23.6	103.9
7	Selling	636	182	218.6	28.6	83.3	116	20	26.3	17.2	75.9
8	Security	432	140	150.0	32.4	93.4	68	21	15.4	30.9	136.0
9	Personal service	661	284	224.9	43	126.3	178	57	40.4	32	141.1
10	Agric. workers	302	116	103.4	38.4	112.2	100	23	22.7	23	101.3
11	Materials proc.	1068	416	366.1	39	113.6	207	33	47.0	15.9	70.2
12	Metal & electrical	1947	710	666.5	36.5	106.5	445	116	101.0	26.1	114.8
13	Assembly	532	223	181.6	41.9	122.8	130	34	29.5	26.2	115.2
14	Construction	930	449	318.9	48.3	140.8	212	60	48.1	28.3	124.7
15	Transportation	1364	626	465.2	45.9	134.6	332	101	75.4	30.4	134.0
16	Miscellaneous	193	105	65.8	54.4	159.5	98	29	22.2	29.6	130.3
17	N/A	16	7	5.5	43.8	126.8	4	3	0.9	75	330.4
M	Missing occ	241	56				0	0			
	All occupations	13449	4605	4605.0	34.2	100.0	3079	699	699.0	22.7	100.0

Table 12.4 Female proportional current smoking ratios

No.	Occupational order	Age 16-64 Sample	Observed	Expected	% smoking	PCSR	Age 65+ Sample	Observed	Expected	% smoking	PCSR
1	Professional	460	88	155.9	19.1	56.4	53	7	9.6	13.2	73.1
2	Prof. ed. & health	1766	417	593.8	23.6	70.2	365	56	66.0	15.3	84.9
3	Arts & sport	165	45	56.2	27.3	80.0	28	5	5.1	17.9	98.8
4	Professional science	143	19	49.2	13.3	38.6	5	2	0.9	40	221.3
5	Managerial	734	248	246.0	33.8	100.8	224	37	40.5	16.5	91.4
6	Clerical	4301	1185	1456.3	27.6	81.4	835	132	150.9	15.8	87.5
7	Selling	1590	566	538.7	35.6	105.1	440	78	79.5	17.7	98.1
8	Security	61	26	20.4	42.6	127.3	26	5	4.7	19.2	106.4
9	Personal service	3290	1424	1104.2	43.3	129.0	1283	268	231.9	20.9	115.6
10	Agric. workers	124	36	41.6	29	86.5	61	7	11.0	11.5	63.5
11	Materials proc.	857	393	288.2	45.9	136.4	398	69	71.9	17.3	95.9
12	Metal & electrical	145	60	48.6	41.4	123.6	105	25	19.0	23.8	131.7
13	Assembly	749	339	251.6	45.3	134.8	227	39	41.0	17.2	95.0
14	Construction	6	4	2.1	66.7	190.2	1	0	0.2	0	0.0
15	Transportation	148	54	49.9	36.5	108.3	54	8	9.8	14.8	82.0
16	Miscellaneous	30	15	9.9	50	150.8	29	10	5.2	34.5	190.8
17	N/A	34	5	11.4	14.7	44.0	15	2	2.7	13.3	73.8
M	Missing occ	241	56				284	31			
	All occupations	14603	4924	4924.0	33.7	100.0	4149	750	750.0	18.1	100.0

Figure 12.2 PCSRs by occupational order

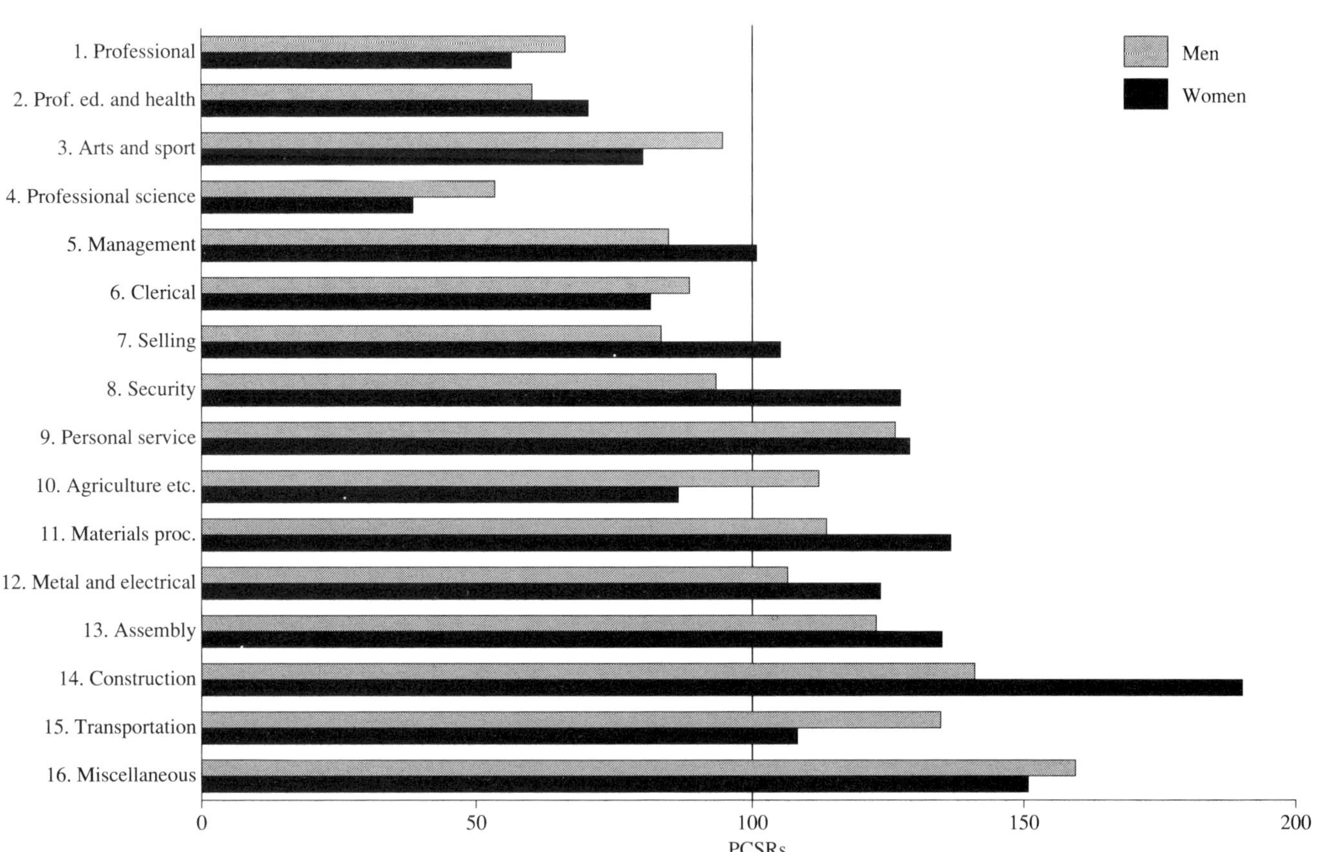

Table 12.5 Male smoking by occupational unit

No.	Occupational unit	All ages		Age 16-64		Age 65+	
		% smoking	PCSR	% smoking	PCSR	% smoking	PCSR
001	Judges, barristers, advocates, solicitors	14.3	44.5	13.8	40.3	16.7	73.4
002	Chartered and certified accountants	15.1	47.0	15.4	44.9	13.6	60.1
006	Financial managers	17.9	55.6	20.0	58.4	0.0	0.0
007	Underwriters, brokers, investment analysts	27.0	84.2	28.1	82.1	20.0	88.1
012	Systems analysts, computer programmers	24.8	77.4	25.2	73.5	0.0	0.0
013	Marketing & sales managers & executives	25.4	79.3	26.8	78.2	16.7	73.4
016	Buyers & purchasing officers (not retail)	25.0	77.9	29.5	86.3	8.3	36.7
021	General admin - national government (HEO to principal)	17.5	54.7	17.1	49.9	18.8	82.6
033	Teachers (other than higher education)	16.0	49.8	17.5	51.1	8.7	38.3
039	Welfare occupations (not elsewhere classified)	28.0	87.4	28.5	82.9	25.0	110.1
041	Medical practitioners	13.3	41.5	12.0	35.0	20.0	88.1
043	Nurse administrators, nurses	37.3	116.1	36.4	106.2	42.9	188.8
069	Mechanical & aeronautical engineers	15.7	48.9	18.5	54.1	4.8	21.0
071	Electrical engineers	7.8	24.3	6.4	18.6	11.8	51.8
072	Electronic engineers	19.2	59.8	20.6	60.1	0.0	0.0
075	Planning & quality control engineers	21.6	67.2	17.1	50.1	31.3	137.7
079	Draughtsmen	19.8	61.7	22.0	64.1	7.1	31.5
080	Laboratory technicians	19.0	59.2	18.8	55.0	20.0	88.1
081	Engineering technicians, technician engineers	19.7	61.3	20.4	59.6	16.7	73.4
091	Production, works & maintenance managers/works foremen	26.4	82.3	27.9	81.5	19.4	85.3
092	Managers in building & contracting	26.5	82.7	28.4	83.0	16.7	73.4
095	Transport managers	16.9	52.7	16.1	46.9	22.2	97.9
097	Managers in warehousing & materials handling nec	27.0	84.1	33.3	97.4	0.0	0.0
099	Office managers nec	24.4	76.0	23.9	69.8	26.8	118.0
101	Other proprietors & managers (sales)	28.0	87.2	31.5	91.9	15.4	67.8
103	Publicans	33.3	103.9	35.7	104.3	26.7	117.5
104	Restauranteurs	33.3	103.9	35.4	103.3	14.3	62.9
107	Farmers, horticulturists, farm managers	19.9	62.0	25.7	75.1	8.5	37.2
111	Managers nec	21.6	67.2	25.0	73.0	9.1	40.0
115	Other clerks & cashiers (not retail)	24.8	77.4	24.9	72.7	24.6	108.3
119	Office machine operators	30.8	95.9	30.6	89.4	33.3	146.8
123	Postmen, mail sorters	41.7	130.0	43.3	126.4	29.4	129.6
124	Messengers	21.7	67.7	23.7	69.3	10.0	44.0
125	Shop salesmen & assistants	23.2	72.4	23.9	69.7	18.2	80.1
128	Roundsmen, van salesmen	32.8	102.3	31.0	90.6	44.4	195.8
133	Sales representatives	25.2	78.5	26.4	77.1	17.6	77.7
134	Sales reps (property & services), other agents	24.8	77.3	26.5	77.5	18.5	81.6
135	Armed forces personnel (all ranks)	31.6	98.4	33.7	98.4	0.0	0.0
137	Police officers (all ranks)	24.0	74.9	24.5	71.5	14.3	62.9
138	Fire service personnel (all ranks)	21.2	65.9	21.3	62.1	20.0	88.1
140	Security guards & officers, patrolmen, watchmen	44.1	137.4	46.2	135.0	38.1	167.8
143	Chefs, cooks	43.0	134.1	44.6	130.2	20.0	88.1
145	Barmen, barmaids	54.2	169.0	53.6	156.5	66.7	293.7
157	Caretakers	40.2	125.4	50.0	146.0	24.2	106.8
158	Cleaners, window cleaners, chimney sweeps & road sweepers	45.4	141.4	48.5	141.7	37.9	167.1
165	Service workers nec	37.5	116.9	40.0	116.8	30.0	132.1
166	Farm workers	33.0	102.9	35.7	104.3	24.0	105.7
168	Gardeners and groundsmen	32.5	101.3	37.3	109.0	20.7	91.1
184	Chemical, gas and petroleum process plant operators	34.8	108.5	41.4	121.0	10.5	46.4
186	Butchers, meat cutters	40.5	126.2	42.5	124.0	16.7	73.4
202	Food and drink nec	38.5	119.9	40.7	119.0	27.3	120.1
207	Printing machine minders & assistants	34.5	107.5	39.1	114.3	16.7	73.4
214	Carpenters, joiners	32.7	101.9	37.5	109.5	7.7	33.9
241	Machine tool operators	36.2	112.9	38.5	112.3	28.1	123.9
248	Metal working production fitters and fitters/machinists	34.6	107.8	37.9	110.8	22.1	97.4
249	Motor mechanics, auto engineers	37.8	117.9	37.8	110.4	37.9	167.1
253	Electricians, electrical maintenance fitters	31.7	98.9	33.2	97.0	23.3	102.4
256	Telephone fitters	27.7	86.4	27.8	81.1	27.3	120.1
259	Other electronic maintenance engineers	25.8	80.4	26.8	78.2	16.7	73.4
260	Plumbers, heating & ventilation fitters, gas fitters	30.3	94.5	31.4	91.8	21.1	92.7
261	Sheet metal workers	40.4	125.7	43.8	127.8	22.2	97.9
265	Welders	40.7	127.0	39.8	116.3	45.5	200.2
276	Other metal, jewellery, electrical production workers	32.9	102.5	35.5	112.3	12.5	55.1
282	Painters & decorators nec, french polishers	44.6	138.9	46.6	136.2	34.2	150.7
286	Metal, electrical goods	30.6	95.3	34.6	101.0	22.5	99.1
287	Packers, bottlers, canners, fillers	42.5	132.4	44.3	129.3	36.8	162.3
300	Bricklayers, tile setters	41.1	127.9	40.9	119.6	41.7	183.5
302	Plasterers	50.0	155.8	50.0	146.0	50.0	220.2
303	Roofers, glaziers	50.6	157.8	51.3	149.9	33.3	146.8
304	Handymen, general building workers	36.0	112.0	37.7	110.0	30.0	132.1

Table 12.5 - *continued*

No.	Occupational unit	All ages % smoking	All ages PCSR	Age 16-64 % smoking	Age 16-64 PCSR	Age 65+ % smoking	Age 65+ PCSR
305	Builders (so described)	41.7	130.0	46.6	136.1	5.6	24.5
313	Building & civil engineering labourers	55.6	173.3	60.9	177.8	32.3	142.1
314	Face trained coalmining workers	35.7	111.3	42.5	124.2	24.5	108.0
316	Construction workers	39.4	122.7	43.3	126.4	21.7	95.8
325	Bus & coach drivers	47.4	147.6	53.3	155.8	25.0	110.1
326	Drivers of road goods vehicles	44.1	137.3	47.4	138.3	26.0	114.6
327	Other motor drivers	41.8	130.3	44.3	129.3	27.3	120.1
330	Mechanical plant drivers, operators (earth moving & civil engineering)	42.4	132.2	46.7	136.3	0.0	0.0
332	Fork lift, mechanical truck drivers	45.9	142.9	48.0	140.1	27.3	120.1
333	Storekeepers, warehousemen	37.1	115.7	39.9	116.4	28.7	126.5
335	Goods porters	31.8	99.1	37.0	108.0	6.7	29.4
338	Workers in transport operating, materials moving & storing & related nec	58.5	182.2	58.8	171.8	57.1	251.7
346	General labourers, other	49.0	152.6	55.7	162.7	32.1	141.6

Table 12.6 Female smoking by occupational unit

No.	Occupational unit	All ages % smoking	All ages PCSR	Age 16-64 % smoking	Age 16-64 PCSR	Age 65+ % smoking	Age 65+ PCSR
013	Marketing & sales managers & executives	10.2	33.6	10.3	30.7	0.0	0.0
033	Teachers (other than higher education)	12.6	41.6	13.1	38.8	10.3	56.7
039	Welfare occupations (not elsewhere classified)	32	105.9	34	100.8	22.6	124.9
043	Nurse administrators, nurses	30	99.2	32.6	96.7	17.6	97.4
053	Professional & related in education, welfare & health nec	11	36.2	14.5	43.1	0.0	0.0
099	Office managers nec	30.8	101.7	31.9	94.6	20.0	110.6
101	Other proprietors & managers (sales)	29.6	97.7	34.3	101.7	17.0	93.9
102	Hotel & residential club managers	20	66.1	27.6	81.8	11.5	63.8
103	Publicans	51.8	171.1	60.5	179.5	33.3	184.4
104	Restauranteurs	31.3	103.3	36	106.9	15.4	85.1
111	Managers nec	25.9	85.5	30	89.0	0.0	0.0
112	Civil service executive officers	23.6	78.0	22.7	67.4	33.3	184.4
115	Other clerks & cashiers (not retail)	25.4	83.8	27.5	81.5	15.2	84.2
116	Retail shop cashiers, check-out, cash & wrap operators	37.5	123.9	39	115.7	25.9	143.4
117	Receptionists	26.9	88.9	26.6	78.8	31.3	172.9
118	Typists, shorthand writers, secretaries	22.4	74.1	24.2	71.6	13.9	76.8
119	Office machine operators	28.5	94.2	30.1	89.4	10.5	58.2
121	Telephone operators	32.3	106.8	38.7	114.7	12.5	69.2
125	Shop salesmen & assistants	30.8	101.9	34.7	102.9	18.0	99.6
126	Shelf fillers	39.2	129.6	40.4	119.9	25.0	138.3
133	Sales representatives	33.3	110.2	36.2	107.3	0.0	0.0
134	Sales reps (property & services), other agents	35.4	117.1	37.1	110.2	22.2	122.9
143	Chefs, cooks	34.2	112.9	43.2	128.2	17.2	95.0
144	Waiters, waitresses	35.7	118.1	36.1	107.0	34.1	188.9
145	Barmen, barmaids	62.7	207.1	64.1	190.1	50.0	276.6
146	Counter hands, assistants	38.8	128.2	44.6	132.4	26.2	144.7
147	Kitchen porters	40.6	134.1	47.7	141.5	24.0	132.8
150	Nursery nurses	16	52.9	16.4	48.8	0.0	0.0
151	Other domestic & school helpers	28.3	93.7	34.2	101.6	16.1	89.0
156	Hospital, ward orderlies	45.5	150.4	47.3	140.3	30.0	166.0
158	Cleaners, window cleaners, chimney sweeps & road sweepers	38.3	126.5	48.2	143.1	20.7	114.5
159	Hairdressers, barbers	38.1	126.0	40.6	120.3	20.7	114.5
162	Launderers, dry cleaners, pressers	38.8	128.2	46.7	138.4	24.4	134.9
165	Service workers nec	32.2	106.3	35	103.8	13.3	73.8
166	Farm workers	17.5	57.8	23.2	68.8	10.6	58.9
175	Leather cutters & sewers, footwear lasters, makers, finishers	31.4	103.7	38.7	114.8	20.0	110.6
202	Food and drink nec	49	161.9	53	157.2	26.7	147.5
210	Tailors, tailoresses, dressmakers	17.4	57.5	25	74.1	13.3	73.8
212	Sewers, embroiderers	38.2	126.2	44.9	133.3	15.8	87.3
227	Paper goods and printing	36.8	121.8	43.6	129.3	22.2	122.9
241	Machine tool operators	37	122.4	58.3	173.0	20.0	110.6
283	Assemblers (electrical, electronic)	38.1	125.8	41.7	123.8	21.4	118.5
285	Assemblers (vehicles & other metal goods)	45.3	149.7	50	148.3	14.3	79.0
286	Metal, electrical goods	31.7	104.8	43.5	128.9	16.7	92.2
287	Packers, bottlers, canners, fillers	38.5	127.2	44.1	130.8	19.2	106.2
299	Painting, assembling & related occupations nec	53.8	178.0	62.7	186.1	21.4	118.5
333	Storekeepers, warehousemen	30.7	101.4	33.8	100.2	22.2	122.9

Figure 12.3 Categories of alcohol consumption for men and women. Bands 3 and 4 combined constitute 'more than sensible' levels of drinking

	1. None	2. Sensible	3. Intermediate	4. Unsafe
Men	0 units/wk	0-21 units/wk	22-50 units/wk	>50 units/wk
Women	0 units/wk	0-14 units/wk	15-35 units/wk	>35 units/wk

indicator of drinking status. A moderate mean consumption level could indicate either that the majority of group members drank moderately or that the majority drank only little but a substantial minority were very heavy drinkers.

The method used in the last decennial supplement on Occupational Mortality[2] involved classifying drinking status using a combination of the quantity of alcohol consumed on a typical occasion and the frequency of such occasions. Current accepted practice is to classify subjects by estimating the average number of units of alcohol consumed per week. A unit of alcohol is equal to a half pint of beer, lager or cider, one glass of wine, one single measure of spirits or any equivalent quantity of alcohol. Categories of drinking do not rely on arbitrary choices as to what constitutes a low, medium or high level of consumption but on boundaries which are now accepted by the majority of medical authorities as constituting sensible and/or safe limits of drinking. These were developed by the World Health Organisation in consultation with many medical organisations an attempt to unify the various differing recommendations existing in the 1970s.

For men, it is considered that the consumption of up to 21 units of alcohol per week is unlikely to lead to physical illness or psychological dependency. This therefore defines sensible drinking. From 22 to 50 units per week the risk of ill health is likely to increase and more than 50 units is considered to be definitely associated with a high risk of alcohol related diseases and is considered unsafe. Because of physiological differences the comparable levels for women are lower and the boundary levels are 14 and 35 units. Therefore in the following analyses alcohol consumption is based on classification into four bands as shown in Figure 12.3. These have the merit that although the absolute levels are different for men and women they are considered to represent similar levels of effect, thus functionally meaningful overall summary tables for the two sexes can be prepared by summing the male and female figures.

The analyses by occupational order and age group are divided into the 4 bands indicated in Figure 12.3. Simplified versions of the tables for men and women, showing the sample size in each age group and the percentage breakdown by drinking category, are Tables 12.7 and 12.8.

Age standardised proportional current drinking ratios (PCDRs) have been calculated using the methodology described above for PCSRs for subjects in the age groups 16-64 and 65 and over. For men these have been calculated for all subjects drinking more than 21 units per week (bands 3 and 4 combined) and for those drinking more than 50 units per week. The corresponding levels for women are 14 and 35 units. These represent the proportions of subjects drinking more than is considered sensible and more than is safe. These results for the occupational orders are presented as Tables 12.9 and 12.10.

The mean weekly alcohol consumption of the occupational orders is also presented by age groups in Tables 12.11 and 12.12 respectively. In these tables, a proportional alcohol consumption ratio (PACR) is presented for ages 16-64 and 65 plus. This is simply the mean consumption for the occupational group of interest expressed as a percentage of the population mean alcohol consumption for the whole of the population of the same age band. It is not therefore fully age standardised. Although I have commented above on the crudity of mean alcohol consumption as an indicator of the drinking status of occupational groups, it can be seen by comparing Tables 12.8 and 12.10 that for men at least there is a fairly close correlation between the PCDRs for more than 21 units per week (more than sensible) and the PACRs in the age band 16-64. This is rather less apparent in the female tables.

Tables 12.13 and 12.14 show the PCDRs by occupational units for which a sample size of more than 50 men or 50 women was available. These omit the data for ages 65 and over as many of the sample sizes in that age group are too small for meaningful discussion. The mean weekly alcohol consumption and PACRs by occupational unit for men and women are presented as Tables 12.15 and 12.16.

Figure 12.4 shows that for the whole population two thirds of men drink within the sensible limit with a further 20 per cent drinking between 22 and 50 units per week. 6.4 per cent are teetotal and 7 per cent drink more than 50 units per week. For women there is a downward displacement of this pattern with 11 per cent being teetotal, 78 per cent drinking sensibly, 8.9 per cent drinking between 14 and 35 units per week and less than 2 per cent drinking at definitely unsafe levels. Since the band definitions are different for the two sexes this means that women not only tend to drink less in absolute terms but also relative to the levels which are considered safe and sensible.

In both sexes the number of non-drinkers roughly doubles between age bands 45-64 and 65+. There is thus an apparent tendency to give up drinking at or near retirement age similar to that noted above for smoking. There is also a fall in the percentages drinking in the upper two bands suggesting a reduction in alcohol consumption amongst those who continue to drink. The majority of the population, however, still drink within the sensible limits for their sex. These bands are, of course, quite wide and the diagram shows that there is a

Figure 12.4 Percentage distribution of alcohol consumption

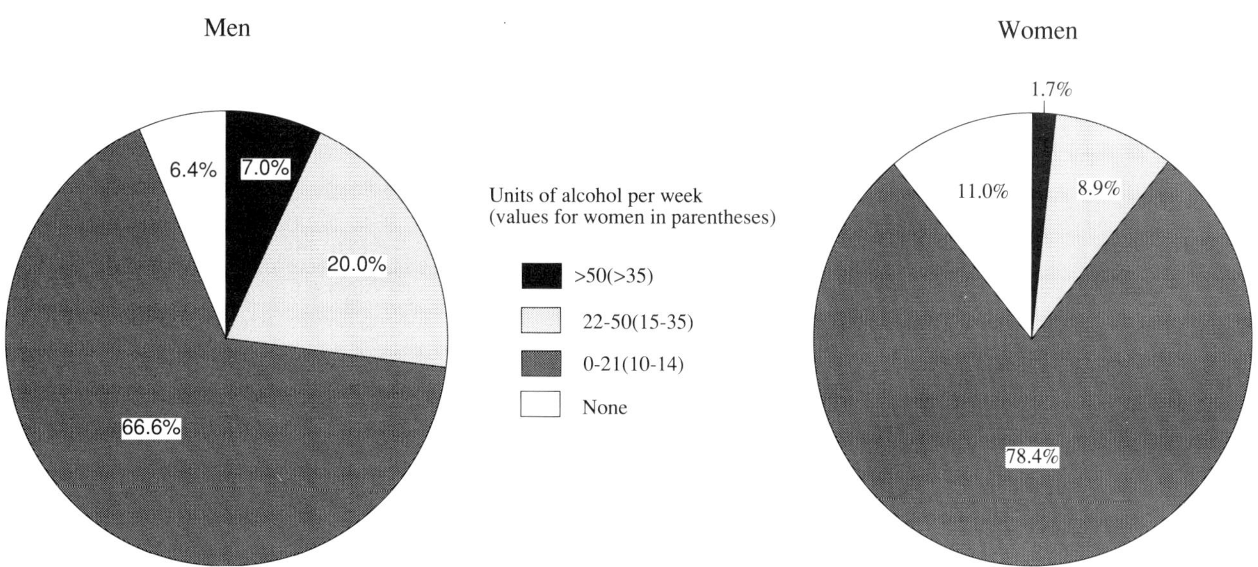

Figure 12.5 Mean alcohol consumption (units/week)

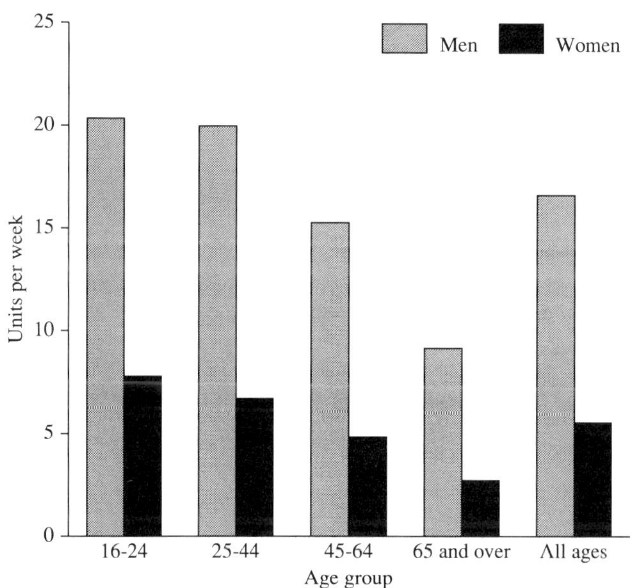

considerable reduction in the absolute level of alcohol consumption in both sexes between ages 45-64 and 65+. Therefore, the majority of the population, even if still drinking within the sensible limits, must reduce their alcohol consumption at around retirement age. The points made in the discussion of smoking in the over 65s are probably equally relevant here.

As can be seen from Figures 12.5, the mean alcohol consumption for the whole population is highest in the youngest age group and declines with age. The same pattern is seen for the majority of occupational orders. This general trend is well known and is discussed in both previous GHS reports and in *Drinking in England and Wales in the late 1980s*.[2] The latter publication discusses the relationship of this decline to factors such as marriage and having children in some detail. It is obviously related to socio-economic factors and is not purely an age related effect. This must be borne in mind when considering the differences in drinking pattern by age between different occupational groups although the data presented here do not permit any formal discussion of the matter.

The downward trend in mean alcohol consumption with age raises the question whether it is amongst the young that the heaviest drinkers will be found. To a large extent the grouped analyses by age (Tables 12.7 and 12.8) support this conclusion. Almost 11 per cent of all men aged 16-24 drink more than the safe limit of 50 units of alcohol per week and several of the highest percentages for occupational orders are also to be found in this age band. This percentage declines with age suggesting that most of these heavy drinkers will reduce their consumption as they grow older, although the data presented here do not preclude the possibility that the alcohol consumption of a small minority of very heavy drinkers may increase with age. It is also not possible to determine the relative extent to which subjects leave the heavier drinking categories by reducing their drinking levels or by ceasing to drink entirely. A similar age pattern is seen in women although the proportion drinking at very high levels is much lower at all ages.

Neither in the tables of PCDRs (Tables 12.9 and 12.10) or of mean alcohol consumption (Tables 12.11 and 12.12) do we see so pronounced a division between manual and non-manual workers as was observed for smoking status, although there is a fairly clear split in Table 12.10 of PCDRs for men aged 16-64 drinking more than 50 units per week. This division is also seen in Figure 12.6 which also illustrates that there is less evidence of a general trend with social class than is the case for smoking. The extremely high PCDR for women in security occupations is based on very small numbers and should be treated with caution. For those drinking more than 21 units per week (Figure 12.7) the position is less clear and is complicated by two factors. One is that the number drinking more than 21 units includes the number drinking more than 50 units therefore the PCDRs are not

Text continues on page 212

Table 12.7 Percentages drinking by age - male

No.	Occupational order	Age 16-24					Age 25-44					Age 45-64					Age 65+					All ages				
		Sample no.	None (%)	0-21 units (%)	22-50 units (%)	>50 units (%)	Sample no.	None (%)	0-21 units (%)	22-50 units (%)	>50 units (%)	Sample no.	None (%)	0-21 units (%)	22-50 units (%)	>50 units (%)	Sample no.	None (%)	0-21 units (%)	22-50 units (%)	>50 units (%)	Sample no.	None (%)	0-21 units (%)	22-50 units (%)	>50 units (%)
1	Professional	103	2.9	55.3	25.2	16.5	566	2.1	64.3	26	7.6	362	4.1	67.7	21	7.2	198	5.6	74.7	18.2	1.5	1229	3.3	66.2	23.2	7.2
2	Prof. ed. & health	24	8.3	45.8	25	20.8	359	5.3	67.4	22	5.3	322	5.9	75.2	15.5	3.4	139	11.5	74.8	12.2	1.4	844	6.6	71.0	18.0	4.4
3	Arts & sport	25	12	76	12	0	102	3.9	53.9	34.3	7.8	53	5.7	73.6	17	3.8	24	8.3	66.7	20.8	4.2	204	5.9	63.2	25.5	5.4
4	Professional science	125	4	63.2	21.6	11.2	507	3.2	67.7	23.9	5.3	317	4.7	75.7	16.1	3.5	166	5.4	81.9	10.8	1.8	1115	4.0	71.6	19.5	4.9
5	Managerial	95	3.2	56.8	26.3	13.7	768	2.9	62.5	26.4	8.2	744	5.5	68.3	20.2	6	411	12.9	71.3	12.7	3.2	2018	5.9	66.2	21.3	6.6
6	Clerical	292	7.9	67.8	18.2	6.2	323	6.8	57.9	27.6	7.7	259	6.9	74.5	17	1.5	244	13.1	76.2	10.2	0.4	1118	8.5	68.3	18.9	4.3
7	Selling	204	9.8	69.1	18.1	2.9	253	4	63.2	26.1	6.7	172	5.8	71.5	18.6	4.1	116	8.6	79.3	9.5	2.6	745	6.7	69.3	19.6	4.4
8	Security	44	0	65.9	29.5	4.5	257	1.6	71.6	20.2	6.6	127	3.9	70.1	20.5	5.5	68	19.1	72.1	4.4	4.4	496	4.4	70.8	19.0	5.8
9	Personal service	223	5.8	64.1	17	13	210	8.6	59	21.9	10.5	225	8.9	61.3	20.4	9.3	177	15.3	70.1	13.6	1.1	835	9.3	63.4	18.4	8.9
10	Agric. workers	81	9.9	67.9	18.5	3.7	119	5	65.5	22.7	6.7	97	10.3	70.1	14.4	5.2	100	13	72	13	2	397	9.3	68.8	17.4	4.5
11	Materials proc.	204	5.4	55.9	24	14.7	489	3.9	62.6	24.7	8.8	371	8.1	72.5	15.6	3.8	206	17.5	72.8	7.8	1.9	1270	7.6	66.1	19.2	7.2
12	Metal & electrical	353	4	61.5	24.1	10.5	877	2.3	61.7	26	10	705	7.2	67	19.6	6.2	445	12.8	76.4	8.8	2	2380	6.0	66.0	20.6	7.5
13	Assembly	113	8	53.1	25.7	13.3	217	2.8	64.5	22.1	10.6	197	9.6	60.9	22.3	7.1	130	8.5	74.6	13.1	3.8	657	6.8	63.5	21.0	8.7
14	Construction	157	4.5	54.1	24.8	16.6	439	2.7	53.3	27.8	16.2	331	4.8	61.6	23.3	10.3	212	14.6	69.8	11.8	3.8	1139	5.8	58.9	23.1	12.2
15	Transportation	195	4.6	57.4	21.5	16.4	603	4.5	65	20.9	9.6	556	5.2	71.8	16.2	6.8	333	14.4	73.9	8.4	3.3	1687	6.7	68.1	17.0	8.2
16	Miscellaneous	35	5.7	71.4	22.9	0	80	3.8	62.5	25	8.8	74	12.2	67.6	10.8	9.5	98	15.3	67.3	16.3	1	287	10.1	66.6	18.1	5.2
17	N/A	5	20	60	0	20	6	16.7	33.3	50	0	4	50	50	0	0	4	75	25			19	36.8	42.1	15.8	5.3
M	Missing occ	214					25					2										241				
	All occupations	2278	5.8	61.5	21.7	10.9	6175	3.6	62.9	24.8	8.7	4916	6.3	69.2	18.6	5.9	3071	12.6	73.9	11.2	2.3	16440	6.4	66.6	20.0	7.0

Table 12.8 Percentages drinking by age - female

No.	Occupational order	Age 16-24					Age 25-44					Age 45-64					Age 65+					All ages				
		Sample no.	None (%)	0-14 units (%)	15-34 units (%)	>35 units (%)	Sample no.	None (%)	0-14 units (%)	15-34 units (%)	>35 units (%)	Sample no.	None (%)	0-14 units (%)	15-34 units (%)	>35 units (%)	Sample no.	None (%)	0-14 units (%)	15-34 units (%)	>35 units (%)	Sample no.	None (%)	0-14 units (%)	15-34 units (%)	>35 units (%)
1	Professional	54	5.6	68.5	18.5	7.4	282	3.9	77	15.2	3.9	121	4.1	8.26	11.6	1.7	51	13.7	80.4	5.9	0	508	5.1	77.8	13.8	3.3
2	Prof. ed. & health	142	7.7	76.8	12	3.5	996	5.1	84.2	9.2	1.4	623	10.8	79.5	9.1	0.6	365	17.3	76.4	4.7	1.6	2126	9.0	81.0	8.6	1.4
3	Arts & sport	41	2.4	73.2	22	2.4	82	3.7	74.4	17.1	4.9	42	11.9	69	19	0	28	17.9	75	7.1	0	193	7.3	73.1	17.1	2.6
4	Professional science	39	12.8	69.2	15.4	2.6	79	6.3	77.2	15.2	1.3	21	4.8	85.7	9.5	0	5	40	60	0	0	144	9.0	75.7	13.9	1.4
5	Managerial	60	6.7	80	11.7	1.7	384	5.7	73.4	18	2.9	285	8.8	78.2	10.5	2.5	224	17	74.6	8	0.4	953	9.3	75.6	13.0	2.1
6	Clerical	931	5.5	76.8	14.4	3.7	1955	5.7	80.7	11.4	2.2	1391	6.9	82.7	8.8	1.6	834	15	78.1	6.4	0.6	5111	7.5	80.1	10.4	2.0
7	Selling	383	7	81.7	9.7	1.6	675	6.5	83.6	8.4	1.5	521	9.2	80.8	9	1	438	19.9	77.2	2.5	0.5	2017	10.2	81.1	7.5	1.1
8	Security	3	0	100	0	0	34	0	79.4	14.7	5.9	24	0	75	8.3	16.7	26	15.4	65.4	19.2	0	87	4.6	74.7	13.8	6.9
9	Personal service	533	7.5	76.5	12.4	3.6	1392	6.5	81.3	9.8	2.5	1348	13.5	78.6	6.5	1.3	1276	27	70.4	2.4	0.2	4549	14.4	76.9	7.0	1.6
10	Agric. workers	23	4.3	87	8.7	0	47	6.4	85.1	8.5	0	53	17	79.2	3.8	0	61	27.9	67.2	3.3	1.6	184	16.3	77.7	5.4	0.5
11	Materials proc.	151	9.3	69.5	15.9	5.3	365	7.9	78.4	10.7	3	338	18.6	73.7	7.1	0.6	396	26	72.5	1	0.5	1250	16.7	74.2	7.3	1.8
12	Metal & electrical	19	5.3	78.9	10.5	5.3	63	6.3	77.8	14.3	1.6	61	21.3	73.8	4.9	0	105	25.7	71.4	2.9	0	248	18.1	74.2	6.9	0.8
13	Assembly	133	11.3	72.2	12.8	3.8	309	9.1	79.3	10.7	1	306	16.7	77.5	5.2	0.7	226	20.4	77	2.2	0.4	974	14.4	77.2	7.3	1.1
14	Construction	3	0	100	0	0	3	0	66.7	33.3	0	0	0	0	0	0	1	0	100	0	0	7	0.0	85.7	14.3	0.0
15	Transportation	22	22.7	45.5	31.8	0	74	8.1	79.7	8.1	4.1	52	15.4	76.9	7.7	0	54	22.2	75.9	1.9	0	202	15.3	74.3	8.9	1.5
16	Miscellaneous	3	33.3	33.3	33.3	0	11	27.3	54.5	9.1	9.1	16	12.5	81.3	6.3	0	29	17.2	79.3	3.4	0	59	18.6	72.9	6.8	1.7
17	N/A	3	0	66.7	33.3	0	16	6.3	87.5	0	6.3	15	20	73.3	6.7	0	15	33.3	66.7	0	0	49	18.4	75.5	4.1	2.0
M	Missing occ	304					113					107					284					808				
	All occupations	2543	7.0	76.4	13.3	3.3	6767	6.1	80.7	11.0	2.2	5217	11.1	79.6	8.1	1.3	4134	21.6	74.2	3.7	0.5	18661	11.0	78.4	8.9	1.7

Figure 12.6 PCDRs (more than safe level) for ages 16-64 by occupational order

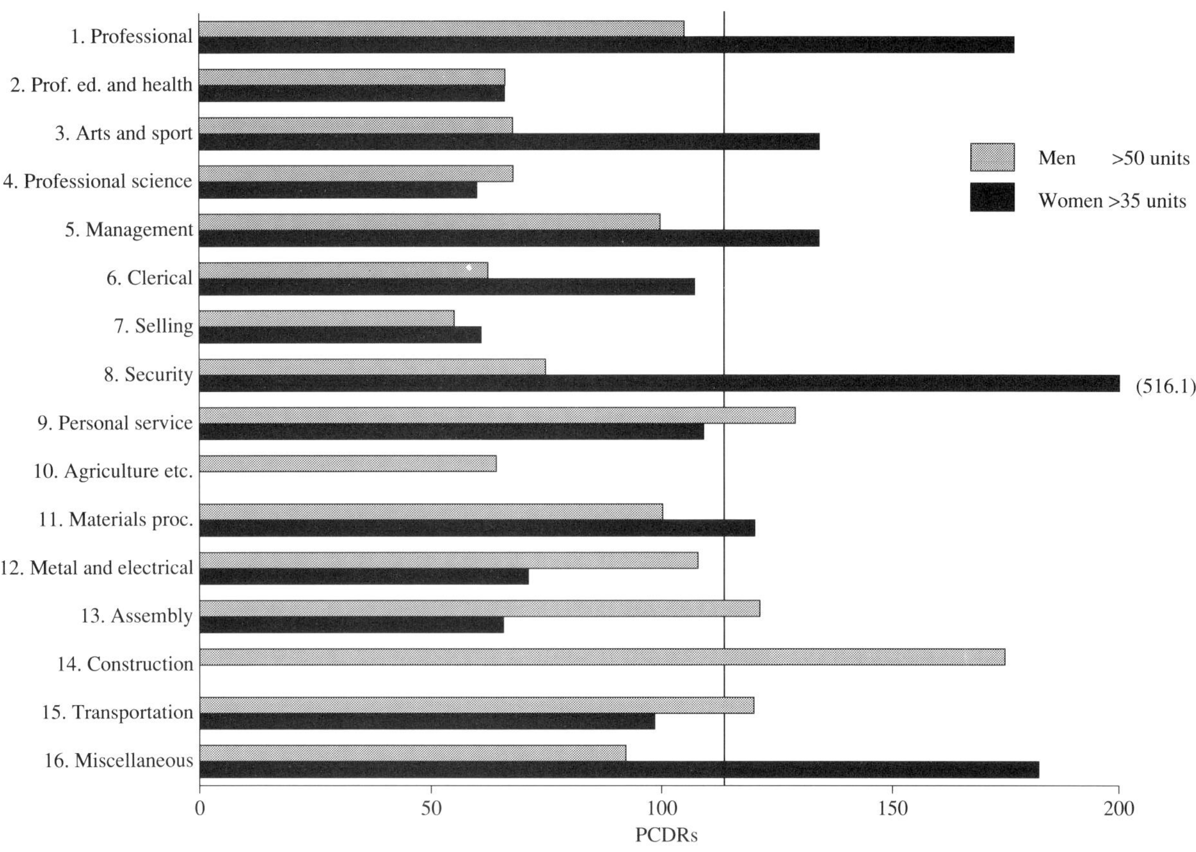

Figure 12.7 PCDRs (more than sensible level) for ages 16-64 by occupational order

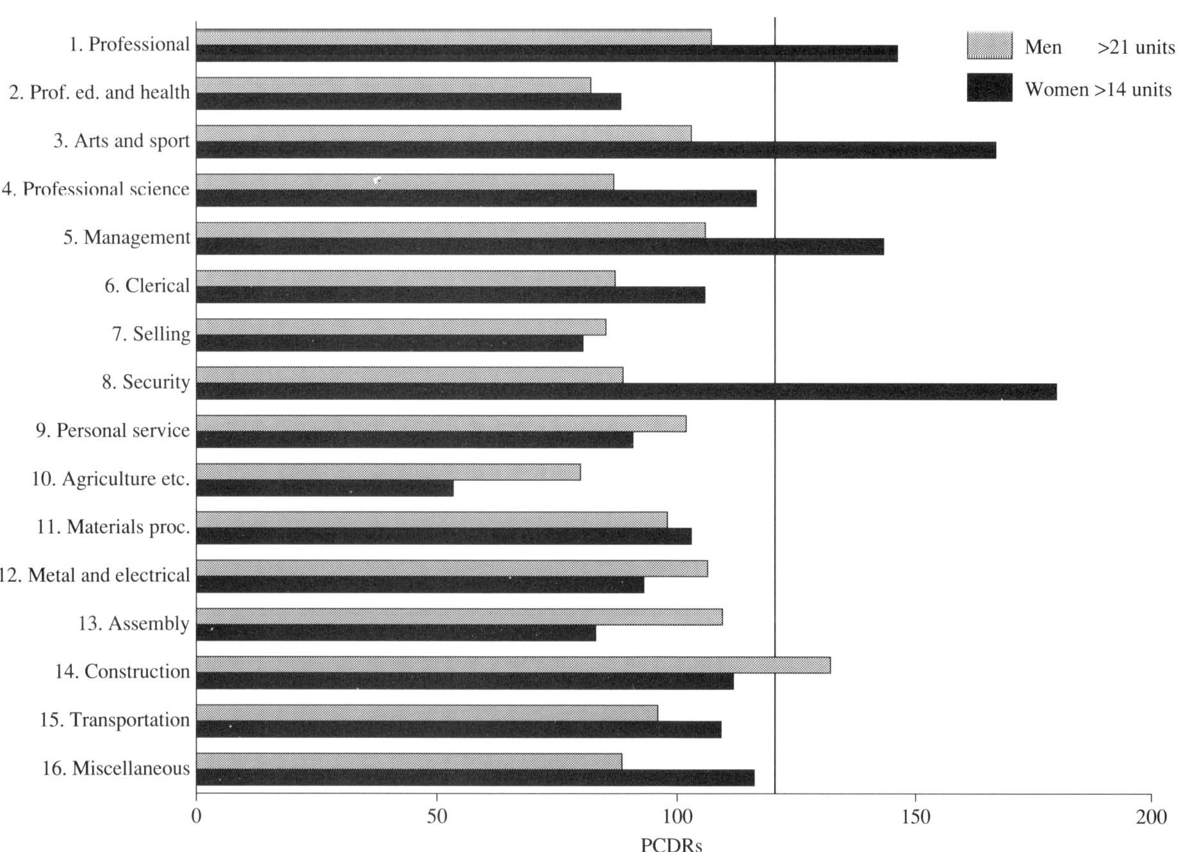

Table 12.9 Male proportional current drinking ratios

No.	Occupational order	Age 16-64 Sample no.	> 21 units Obs.	Exp.	PCDR	> 50 units Obs.	Exp.	PCDR	Age 65+ Sample no.	> 21 units Obs.	Exp.	PCDR	> 50 units Obs.	Exp.	PCDR
1	Professional	1031	335	312.1	107.3	86	82.0	104.9	198	39	26.8	145.4	3	4.6	65.5
2	Prof. ed. & health	705	170	207.1	82.1	35	52.9	66.1	139	19	18.8	100.9	2	3.2	62.2
3	Arts & sport	180	57	55.3	103.0	10	14.8	67.8	24	6	3.3	184.6	1	0.6	180.2
4	Professional science	949	251	288.5	87.0	52	76.6	67.9	166	21	22.5	93.4	3	3.8	78.2
5	Managerial	1607	499	470.8	106.0	121	121.3	99.8	411	65	55.7	116.8	13	9.5	136.8
6	Clerical	874	233	267.0	87.3	47	75.3	62.4	244	26	33.1	78.7	1	5.6	17.7
7	Selling	629	165	193.5	85.3	30	54.4	55.1	116	14	15.7	89.1	3	2.7	111.9
8	Security	428	117	131.7	88.9	26	34.7	74.9	68	6	9.2	65.1	3	1.6	190.8
9	Personal service	658	202	198.3	101.9	72	55.9	128.8	177	26	24.0	108.4	2	4.1	48.9
10	Agric. workers	297	72	90.1	79.9	16	24.9	64.2	100	15	13.5	110.7	2	2.3	86.5
11	Materials proc.	1064	315	321.4	98.0	87	86.8	100.3	206	20	27.9	71.7	4	4.8	84.0
12	Metal & electrical	1935	620	581.9	106.5	169	156.6	107.9	445	48	60.3	79.6	9	10.3	87.5
13	Assembly	527	173	157.9	109.6	52	42.9	121.3	130	22	17.6	124.9	5	3.0	166.4
14	Construction	927	369	279.5	132.0	131	74.9	174.8	212	33	28.7	114.9	8	4.9	163.2
15	Transportation	1354	386	402.0	96.0	128	106.7	120.0	333	39	45.1	86.5	11	7.7	142.9
16	Miscellaneous	189	50	56.4	88.7	14	15.2	92.4	98	17	13.3	128.1	1	2.3	44.1
17	N/A	15	4	4.6	86.5	1	1.3	76.7	4	0	0.5	0.0	0	0.1	0.0
M	Missing occ	241	25			7									
	All occupations	13369	4018	4018.0	100.0	1077	1077.0	100.0	3071	416	416.0	100.0	71	71.0	100.0

Table 12.10 Female proportional current drinking ratios

No.	Occupational order	Age 16-64 Sample no.	> 14 units Obs.	Exp.	PCDR	> 35 units Obs.	Exp.	PCDR	Age 65+ Sample no.	> 14 units Obs.	Exp.	PCDR	> 35 units Obs.	Exp.	PCDR
1	Professional	457	84	57.6	146.0	17	9.6	176.6	51	3	2.2	138.2	0	0.3	0.0
2	Prof. ed. & health	1761	189	213.5	88.5	23	34.9	66.0	365	23	15.5	148.0	6	1.9	323.6
3	Arts & sport	165	36	21.6	166.9	5	3.7	134.0	28	2	1.2	167.8	0	0.1	0.0
4	Professional science	139	22	18.9	116.5	2	3.3	60.0	5	0	0.2	0.0	0	0.0	0.0
5	Managerial	729	125	87.3	143.1	19	14.2	134.0	224	19	9.5	199.2	1	1.1	87.9
6	Clerical	4277	575	542.9	105.9	99	92.3	107.2	834	58	35.5	163.4	5	4.2	118.0
7	Selling	1579	162	201.5	80.4	21	34.5	60.9	438	13	18.6	69.7	2	2.2	89.9
8	Security	61	13	7.2	179.7	6	1.2	516.1	26	5	1.1	451.7	0	0.1	0.0
9	Personal service	3273	362	398.4	90.9	72	65.9	109.2	1276	33	54.3	60.7	3	6.5	46.3
10	Agric. workers	123	8	15.0	53.4	0	2.5	0.0	61	3	2.6	115.5	1	0.3	322.7
11	Materials proc.	854	108	104.9	103.0	21	17.5	120.2	396	6	16.9	35.6	2	2.0	99.4
12	Metal & electrical	143	16	17.2	93.1	2	2.8	71.1	105	3	4.5	67.1	0	0.5	0.0
13	Assembly	748	76	91.5	83.1	10	15.2	65.7	226	6	9.6	62.4	1	1.1	87.1
14	Construction	6	1	0.9	111.8	0	0.2	0.0	1	0	0.0	0.0	0	0.0	0.0
15	Transportation	148	20	18.3	109.3	3	3.0	98.5	54	1	2.3	43.5	0	0.3	0.0
16	Miscellaneous	30	4	3.4	116.1	1	0.5	182.4	29	1	1.2	81.0	0	0.1	0.0
17	N/A	34	3	4.0	74.7	1	0.6	154.5	15	0	0.6	0.0	0	0.1	0.0
M	Missing occ	524	27			4			284				1		
	All occupations	14527	1804	1804.0	100.0	302	302.0	100.0	4134	176	176.0	100.0	21	21.0	100.0

Table 12.11 Mean alcohol consumption (units/week) - male

No.	Occupational order	16 - 24	25 - 44	45 - 64	65+	16 - 64	Total	PACR	
								16-64	65+
1	Professional	25.52	19.74	17.07	11.02	19.38	18.03	105.9	120.6
2	Prof. ed. & health	28.84	16.6	12.6	9.57	15.19	14.26	83.0	104.7
3	Arts & sport	10.58	21.93	14.76	10.47	18.24	17.33	99.7	114.6
4	Professional science	20.75	16.83	12.86	10.54	16.02	15.21	87.5	115.3
5	Managerial	24.46	20.16	16.16	10.27	18.56	16.87	101.4	112.4
6	Clerical	15.93	20.44	11.64	6.82	16.32	14.25	89.2	74.6
7	Selling	12.93	19.96	14.43	9.32	16.17	15.1	88.4	102.0
8	Security	19.12	16.33	14.08	8.11	15.95	14.87	87.2	88.7
9	Personal service	20.87	19.55	18.64	8.82	19.69	17.38	107.6	96.5
10	Agric. workers	13.2	16.56	13.81	7.94	14.75	13.03	80.6	86.9
11	Materials proc.	23.23	19.58	12.97	7.15	17.98	16.22	98.3	78.2
12	Metal & electrical	21.43	21.45	15.6	8.53	19.32	17.3	105.6	93.3
13	Assembly	22.37	20.18	16.95	11.72	19.45	17.92	106.3	128.2
14	Construction	24.96	26.83	20.31	9.97	24.18	21.54	132.1	109.1
15	Transportation	23.84	19.22	14.67	8.75	18.02	16.19	98.5	95.7
16	Miscellaneous	12.81	21.45	13.56	8.81	16.76	14.04	91.6	96.4
17	N/A	25.51	19.71	2.26	0.15	17.12	13.55	93.6	1.6
M	Missing occ								
	All orders	20.34	19.96	15.25	9.14	18.3	16.58	100	100

Table 12.12 Mean alcohol consumption (units/week) - female

No.	Occupational order	16 - 24	25 - 44	45 - 64	65+	16 - 64	Total	PACR	
								16-64	65+
1	Professional	10.9	8.74	6.74	4.54	8.46	8.07	136.2	168.1
2	Prof. ed. & health	8.19	6.09	4.98	3.63	5.87	5.48	94.5	134.4
3	Arts & sport	11.19	9.96	6.52	2.63	9.39	8.41	151.2	97.4
4	Professional science	8.15	7.46	4.98	3.37	7.28	7.14	117.2	124.8
5	Managerial	7.75	8.59	6.42	4.31	7.67	6.88	123.5	159.6
6	Clerical	8.18	6.89	5.42	3.61	6.69	6.19	107.7	133.7
7	Selling	6.01	5.53	4.94	2.53	5.45	4.82	87.8	93.7
8	Security	6.93	7.81	11.19	6.5	9.1	8.32	146.5	240.7
9	Personal service	7.28	6.47	4.23	1.87	5.68	4.61	91.5	69.3
10	Agric. workers	4.4	4.59	2.2	2.47	3.53	3.18	56.8	91.5
11	Materials proc.	8.71	6.49	4.11	1.82	5.94	4.63	95.7	67.4
12	Metal & electrical	8.98	6.15	3.12	2.08	5.23	3.9	84.2	77.0
13	Assembly	8.45	5.65	3.29	2.35	5.18	4.52	83.4	87.0
14	Construction	4.53	6.67	-	13.5	5.6	6.73	90.2	500.0
15	Transportation	8.64	7.9	3.87	2.5	6.59	5.5	106.1	92.6
16	Miscellaneous	9.53	8.22	3.68	3.08	5.93	4.53	95.5	114.1
17	N/A	8.63	7.34	2.84	1.39	5.47	4.22	88.1	51.5
M	Missing occ								
	All orders	7.78	6.69	4.83	2.7	6.21	5.53	100	100

Table 12.13 PCDRs by occupational unit - male

No.	Occupational unit	All ages Sample	> 21 units No.	> 21 units PCDR	> 50 units No.	> 50 units PCDR	Age 16-64 Sample	> 21 units No.	> 21 units PCDR	> 50 units No.	> 50 units PCDR
001	Judges, barristers, advocates, solicitors	68	26	141.8	4	84.2	56	22	130.7	4	88.7
002	Chartered and certified accountants	126	41	120.6	9	102.3	104	36	115.2	9	107.4
006	Financial managers	56	21	139.0	3	76.7	50	20	133.1	3	74.5
007	Underwriters, brokers, investment analysts	74	22	110.2	11	212.9	64	21	109.2	11	213.4
012	Systems analysts, computer programmers	149	47	117.0	14	134.6	147	47	106.4	14	118.2
013	Marketing & sales managers & executives	227	80	130.7	19	119.9	198	74	124.4	19	119.1
016	Buyers & purchasing officers (not retail)	55	11	74.2	3	78.1	43	10	77.4	3	86.6
021	General admin - national government (HEO to principal)	57	12	78.1	2	50.2	41	11	89.3	2	60.6
033	Teachers (other than higher education)	269	59	81.3	10	53.2	223	54	80.6	9	50.1
039	Welfare occupations (not elsewhere classified)	82	23	104.0	4	69.9	74	23	103.4	4	67.1
041	Medical practitioners	58	12	76.7	1	24.7	48	10	69.3	1	25.9
043	Nurse administrators, nurses	51	9	65.4	0	0.0	44	8	60.5	0	0.0
069	Mechanical & aeronautical engineers	102	23	83.6	6	84.2	81	22	90.4	6	91.9
071	Electrical engineers	64	23	133.2	5	111.9	47	18	127.4	3	79.2
072	Electronic engineers	73	15	76.2	3	58.9	68	14	68.5	3	54.8
075	Planning & quality control engineers	50	13	96.4	2	57.3	35	9	85.6	2	70.9
079	Draughtsmen	96	24	92.7	5	74.6	82	22	89.3	5	75.7
080	Laboratory technicians	80	16	74.2	2	35.8	70	16	76.1	2	35.5
081	Engineering technicians, technician engineers	61	10	60.8	2	47.0	49	10	67.9	2	50.7
091	Production, works & maintenance managers/ works foremen	354	99	103.7	28	113.3	292	88	100.3	25	106.3
092	Managers in building & contracting	112	41	135.7	8	102.3	94	37	131.0	7	92.4
095	Transport managers	65	19	108.4	1	22.0	56	18	106.9	1	22.2
097	Managers in warehousing & materials handling nec	63	20	117.7	8	181.8	51	17	110.9	8	194.7
099	Office managers nec	331	98	109.8	15	64.9	275	90	108.9	15	67.7
101	Other proprietors & managers (sales)	478	133	103.2	29	86.9	374	123	109.4	27	89.6
103	Publicans	57	33	214.7	14	351.7	42	30	237.7	13	384.2
104	Restauranteurs	72	15	77.2	6	119.3	65	15	76.8	6	114.6
107	Farmers, horticulturists, farm managers	208	26	46.3	5	34.4	139	20	47.9	3	26.8
111	Managers nec	102	29	105.4	8	112.3	80	20	83.2	6	93.1
115	Other clerks & cashiers (not retail)	729	152	77.3	26	51.1	547	134	81.5	25	56.7
119	Office machine operators	51	18	130.9	4	112.3	48	18	124.8	4	103.4
123	Postmen, mail sorters	150	46	113.7	11	105.0	133	45	112.6	11	102.7
124	Messengers	65	10	57.0	2	44.1	55	8	48.4	2	45.1
125	Shop salesmen & assistants	196	34	64.3	3	21.9	174	32	61.2	2	14.3
128	Roundsmen, van salesmen	67	18	99.6	8	171.0	58	16	91.8	8	171.2
133	Sales representatives	247	67	100.6	12	69.6	213	64	100.0	10	58.3
134	Sales reps (property & services), other agents	125	33	97.9	5	57.3	98	29	98.5	5	63.3
135	Armed forces personnel (all ranks)	94	31	122.3	9	137.1	88	30	113.4	8	112.8
137	Police officers (all ranks)	153	40	96.9	10	93.6	146	40	91.2	10	85.0
138	Fire service personnel (all ranks)	51	15	109.1	1	28.1	46	15	108.5	1	27.0
140	Security guards & officers, patrolmen, watchmen	160	31	71.8	8	71.6	118	27	76.1	7	73.6
143	Chefs, cooks	78	26	123.6	8	146.9	73	26	118.5	8	136.0
145	Barmen, barmaids	59	27	169.7	14	339.8	56	26	154.5	14	310.3
157	Caretakers	87	20	85.2	6	98.8	54	16	98.6	6	137.9
158	Cleaners, window cleaners, chimney sweeps & road sweepers	194	50	95.6	14	103.3	136	44	107.6	14	127.8
165	Service workers nec	79	20	93.9	8	145.0	60	18	99.8	8	165.5
166	Farm workers	106	22	77.0	4	54.0	81	18	73.9	4	61.3
168	Gardeners and groundsmen	200	41	76.0	8	57.3	142	34	79.7	7	61.2
184	Chemical, gas and petroleum process plant operators	89	22	91.7	6	96.5	70	21	99.8	6	106.4
186	Butchers, meat cutters	78	27	128.3	11	202.0	72	25	115.5	9	155.2
202	Food and drink nec	66	14	78.6	2	43.4	55	13	78.6	2	45.1
207	Printing machine minders & assistants	58	21	134.2	5	123.5	46	20	144.7	5	134.9
214	Carpenters, joiners	322	95	109.4	27	120.1	271	91	111.7	26	119.1
241	Machine tool operators	149	36	89.6	9	86.5	117	34	96.7	8	84.9
248	Metal working production fitters and fitters/machinists	447	119	98.7	24	76.9	353	108	101.8	22	77.4
249	Motor mechanics, auto engineers	230	52	83.8	12	74.7	201	48	79.5	12	74.1
253	Electricians, electrical maintenance fitters	289	99	127.0	27	133.8	246	95	128.5	27	136.2
256	Telephone fitters	83	30	134.0	7	120.8	72	30	138.6	7	120.7
259	Other electronic maintenance engineers	61	14	85.1	3	70.4	55	14	84.7	3	67.7
260	Plumbers, heating & ventilation fitters, gas fitters	179	56	116.0	11	88.0	159	53	110.9	11	85.9
261	Sheet metal workers	56	18	119.2	8	204.6	47	16	113.3	7	184.9
265	Welders	133	53	147.8	20	215.3	111	49	146.9	19	212.5
276	Other metal, jewellery, electrical production workers	147	35	88.3	11	107.2	115	34	98.4	11	118.7
282	Painters & decorators nec, french polishers	230	82	132.2	23	143.2	192	76	131.7	20	129.3
286	Metal, electrical goods	121	31	95.0	8	94.7	81	26	106.8	8	122.6
287	Packers, bottlers, canners, fillers	80	19	88.1	5	89.5	61	15	81.8	4	81.4
300	Bricklayers, tile setters	151	60	147.3	26	246.6	127	56	146.7	25	244.4
302	Plasterers	56	23	152.3	9	230.2	54	23	141.7	9	206.9
303	Roofers, glaziers	79	37	173.7	15	271.9	76	35	153.2	15	245.0
304	Handymen, general building workers	89	23	95.8	7	112.6	69	19	91.6	5	90.0

Table 12.13 - *continued*

No.	Occupational unit	All ages					Age 16-64				
		Sample	> 21 units		> 50 units		Sample	> 21 units		> 50 units	
			No.	PCDR	No.	PCDR		No.	PCDR	No.	PCDR
305	Builders (so described)	150	55	135.9	15	143.2	132	52	131.1	15	141.1
313	Building & civil engineering labourers	166	57	127.3	16	138.0	136	55	134.6	16	146.0
314	Face trained coalmining workers	141	53	139.4	13	132.0	87	41	156.8	10	142.7
316	Construction workers	127	36	105.1	14	157.9	104	33	105.6	13	155.2
325	Bus & coach drivers	96	14	54.1	9	134.3	76	14	61.3	9	147.0
326	Drivers of road goods vehicles	470	112	88.4	34	103.6	397	103	86.3	34	106.3
327	Other motor drivers	152	42	102.4	6	56.5	129	39	100.6	5	48.1
330	Mechanical plant drivers, operators (earth moving & civil engineering)	65	23	131.2	4	88.1	59	23	129.7	4	84.2
332	Fork lift, mechanical truck drivers	108	46	157.9	18	238.7	97	45	154.4	18	230.3
333	Storekeepers, warehousemen	411	98	88.4	32	111.5	310	85	91.2	26	104.1
335	Goods porters	86	23	99.2	11	183.2	71	22	103.1	11	192.3
338	Workers in transport operating, materials moving & storing & related nec	64	17	98.5	8	179.0	50	15	99.8	7	173.8
346	General labourers, other	193	48	92.2	11	81.6	137	37	89.9	10	90.6

Table 12.14 PCDRS by occupational unit - female

No.	Occupational unit	All ages					Age 16-64				
		Sample	> 14 units		> 35 units		Sample	> 14 units		> 35 units	
			No.	PCDR	No.	PCDR		No.	PCDR	No.	PCDR
013	Marketing & sales managers & executives	59	19	306.0	2	195.8	58	19	263.8	2	165.9
033	Teachers (other than higher education)	677	63	88.4	8	68.3	560	59	84.8	7	60.1
039	Welfare occupations (not elsewhere classified)	179	28	148.6	7	225.9	149	25	135.1	5	161.4
043	Nurse administrators, nurses	820	74	85.7	10	70.5	678	67	79.6	8	56.8
053	Professional & related in education, welfare & health nec	74	7	89.9	0	0.0	55	4	58.6	0	0.0
099	Office managers nec	156	27	164.4	2	74.1	141	26	148.5	2	68.2
101	Other proprietors & managers (sales)	388	48	117.5	7	104.2	282	40	114.2	7	119.4
102	Hotel & residential club managers	55	10	172.7	2	210.1	29	7	194.4	2	331.7
103	Publicans	55	11	190.0	2	210.1	37	9	195.9	1	130.0
104	Restauranteurs	110	17	146.8	4	210.1	84	15	143.8	4	229.1
111	Managers nec	58	10	163.8	0	0.0	50	9	144.9	0	0.0
112	Civil service executive officers	72	13	171.5	1	80.2	66	13	158.6	1	72.9
115	Other clerks & cashiers (not retail)	2848	368	122.8	59	119.7	2355	327	111.8	55	112.3
116	Retail shop cashiers, check-out, cash & wrap operators	228	15	62.5	4	101.4	201	15	60.1	4	95.7
117	Receptionists	224	29	123.0	4	103.2	208	28	108.4	4	92.5
118	Typists, shorthand writers, secretaries	1241	159	121.7	27	125.7	1033	146	113.8	26	121.1
119	Office machine operators	227	28	117.2	5	127.3	208	27	104.5	5	115.6
121	Telephone operators	97	7	68.6	4	238.2	73	6	66.2	4	263.6
125	Shop salesmen & assistants	1762	140	75.5	17	55.7	1352	127	75.6	15	53.4
126	Shelf fillers	51	3	55.9	1	113.3	47	3	51.4	1	102.3
133	Sales representatives	52	9	164.4	3	333.3	48	9	151.0	3	300.6
134	Sales reps (property & services), other agents	78	13	158.3	1	74.1	70	13	149.5	1	68.7
143	Chefs, cooks	284	26	87.0	6	122.1	185	23	100.1	5	130.0
144	Waiters, waitresses	232	24	98.3	6	149.4	191	22	92.8	6	151.1
145	Barmen, barmaids	232	63	258.0	17	423.3	208	62	240.0	17	393.1
146	Counter hands, assistants	408	29	67.5	5	70.8	279	28	80.8	5	86.2
147	Kitchen porters	247	16	61.5	3	70.2	173	14	65.2	3	83.4
150	Nursery nurses	74	4	51.4	1	78.1	72	4	44.7	1	66.8
151	Other domestic & school helpers	971	60	58.7	11	65.4	656	50	61.4	10	73.3
156	Hospital, ward orderlies	288	30	99.0	5	100.3	258	29	90.5	5	93.2
158	Cleaners, window cleaners, chimney sweeps & road sweepers	1194	75	59.7	10	48.4	761	68	72.0	9	56.9
159	Hairdressers, barbers	235	32	129.4	7	172.1	206	32	125.1	7	163.5
162	Launderers, dry cleaners, pressers	116	7	57.3	1	49.8	75	5	53.7	1	64.1
165	Service workers nec	115	10	82.6	1	50.2	100	10	80.5	1	48.1
166	Farm workers	103	5	46.1	1	56.1	56	2	28.8	0	0.0
175	Leather cutters & sewers, footwear lasters, makers, finishers	51	5	93.1	0	0.0	31	4	103.9	0	0.0
202	Food and drink nec	97	12	117.5	3	178.7	82	11	108.0	2	117.3
210	Tailors, tailoresses, dressmakers	69	3	41.3	0	0.0	24	2	67.1	0	0.0
212	Sewers, embroiderers	410	38	88.1	6	84.5	316	38	96.8	6	91.3
227	Paper goods and printing	56	8	135.7	1	103.2	38	8	169.5	1	126.6
241	Machine tool operators	54	3	52.8	0	0.0	24	2	67.1	0	0.0
283	Assemblers (electrical, electronic)	155	17	104.2	1	37.3	127	16	101.5	1	37.9
285	Assemblers (vehicles & other metal goods)	53	7	125.5	1	109.0	46	7	122.5	1	104.6
286	Metal, electrical goods	82	5	57.9	1	70.5	46	5	87.5	1	104.6
287	Packers, bottlers, canners, fillers	437	34	73.9	5	66.1	339	30	71.3	4	56.8
299	Painting, assembling & related occupations nec	65	3	43.8	0	0.0	51	3	47.4	0	0.0
333	Storekeepers, warehousemen	101	11	103.5	1	57.2	74	10	108.8	1	65.0

Table 12.15 Mean alcohol consumption (units/week) - male

No.	Occupational unit	All ages			Age 16-64		
		Sample	Cons.	PACR	Sample	Cons.	PACR
001	Judges, barristers, advocates, solicitors	68	20.2	121.9	56	21.6	117.9
002	Chartered and certified accountants	126	17.9	108.0	104	19.6	107.0
006	Financial managers	56	18.4	110.9	50	19.6	107.0
007	Underwriters, brokers, investment analysts	74	23.2	140.2	64	25.5	139.1
012	Systems analysts, computer programmers	149	20.0	120.9	147	20.2	110.5
013	Marketing & sales managers & executives	227	20.4	122.7	198	21.5	117.3
016	Buyers & purchasing officers (not retail)	55	14.6	87.8	43	16.3	89.2
021	General admin - national government (HEO to principal)	57	13.4	80.8	41	15.6	85.1
033	Teachers (other than higher education)	269	13.8	83.2	223	14.9	81.1
039	Welfare occupations (not elsewhere classified)	82	16.0	96.4	74	17.5	95.5
041	Medical practitioners	58	14.7	88.4	48	13.7	74.8
043	Nurse administrators, nurses	51	11.3	67.9	44	11.4	62.5
069	Mechanical & aeronautical engineers	102	17.1	103.1	81	19.1	104.3
071	Electrical engineers	64	18.5	111.8	47	18.7	102.1
072	Electronic engineers	73	14.1	84.8	68	13.9	75.7
075	Planning & quality control engineers	50	14.5	87.4	35	14.9	81.5
079	Draughtsmen	96	13.8	83.5	82	14.5	79.0
080	Laboratory technicians	80	12.9	77.7	70	13.8	75.2
081	Engineering technicians, technician engineers	61	12.3	74.1	49	13.8	75.5
091	Production, works & maintenance managers/works foremen	354	17.8	107.6	292	18.9	103.4
092	Managers in building & contracting	112	19.0	114.8	94	20.3	111.1
095	Transport managers	65	14.0	84.3	56	15.3	83.4
097	Managers in warehousing & materials handling nec	63	20.0	120.7	51	21.5	117.3
099	Office managers nec	331	16.4	99.0	275	17.7	96.8
101	Other proprietors & managers (sales)	478	15.7	94.6	374	17.9	98.0
103	Publicans	57	37.7	227.3	42	45.5	248.7
104	Restauranteurs	72	16.2	97.4	65	17.4	95.3
107	Farmers, horticulturists, farm managers	208	9.2	55.2	139	10.0	54.6
111	Managers nec	102	18.7	112.8	80	19.7	107.5
115	Other clerks & cashiers (not retail)	729	13.2	79.7	547	15.3	83.7
119	Office machine operators	51	20.0	120.6	48	21.2	115.6
123	Postmen, mail sorters	150	17.3	104.3	133	18.9	103.2
124	Messengers	65	12.4	74.5	55	13.5	73.6
125	Shop salesmen & assistants	196	10.9	65.4	174	11.3	61.6
128	Roundsmen, van salesmen	67	21.6	130.5	58	23.7	129.2
133	Sales representatives	247	16.6	100.1	213	17.5	95.8
134	Sales reps (property & services), other agents	125	15.4	92.7	98	16.6	90.9
135	Armed forces personnel (all ranks)	94	19.4	117.1	88	20.1	109.7
137	Police officers (all ranks)	153	15.9	96.1	146	16.5	90.0
138	Fire service personnel (all ranks)	51	15.1	91.3	46	16.7	91.1
140	Security guards & officers, patrolmen, watchmen	160	12.3	74.1	118	13.6	74.4
143	Chefs, cooks	78	19.0	114.8	73	20.2	110.5
145	Barmen, barmaids	59	31.9	192.5	56	32.8	179.0
157	Caretakers	87	15.7	94.5	54	21.0	114.5
158	Cleaners, window cleaners, chimney sweeps & road sweepers	194	15.8	95.1	136	19.2	104.8
165	Service workers nec	79	16.7	100.7	60	19.8	108.1
166	Farm workers	106	11.6	69.7	81	13.4	73.3
168	Gardeners and groundsmen	200	12.6	75.8	142	14.5	79.4
184	Chemical, gas and petroleum process plant operators	89	15.5	93.4	70	18.1	98.9
186	Butchers, meat cutters	78	23.7	143.2	72	23.6	128.9
202	Food and drink nec	66	12.3	74.0	55	14.0	76.7
207	Printing machine minders & assistants	58	18.8	113.1	46	22.0	119.9
214	Carpenters, joiners	322	17.3	104.5	271	19.6	107.0
241	Machine tool operators	149	15.3	92.5	117	16.8	91.9
248	Metal working production fitters and fitters/machinists	447	16.5	99.5	353	18.4	100.4
249	Motor mechanics, auto engineers	230	14.3	86.4	201	15.2	83.1
253	Electricians, electrical maintenance fitters	289	19.0	114.5	246	21.3	116.1
256	Telephone fitters	83	19.3	116.6	72	21.8	119.3
259	Other electronic maintenance engineers	61	14.1	85.0	55	15.3	83.3
260	Plumbers, heating & ventilation fitters, gas fitters	179	17.4	104.8	159	18.5	101.1
261	Sheet metal workers	56	22.1	133.5	47	23.6	129.2
265	Welders	133	23.6	142.0	111	26.1	142.6
276	Other metal, jewellery, electrical production workers	147	17.5	105.6	115	20.7	113.0
282	Painters & decorators nec, french polishers	230	20.7	124.5	192	21.7	118.5
286	Metal, electrical goods	121	14.9	90.1	81	18.0	98.4
287	Packers, bottlers, canners, fillers	80	15.7	94.9	61	17.4	94.9
300	Bricklayers, tile setters	151	23.2	139.7	127	25.4	138.8
302	Plasterers	56	26.9	162.2	54	27.7	151.5
303	Roofers, glaziers	79	29.8	179.9	76	30.1	164.6
304	Handymen, general building workers	89	16.1	97.3	69	16.9	92.5

Table 12.15 - *continued*

No.	Occupational unit	All ages			Age 16-64		
		Sample	Cons.	PACR	Sample	Cons.	PACR
305	Builders (so described)	150	20.3	122.2	132	22.1	120.6
313	Building & civil engineering labourers	166	21.6	130.4	136	25.0	136.3
314	Face trained coalmining workers	141	20.2	121.8	87	24.8	135.7
316	Construction workers	127	19.1	115.3	104	21.4	116.8
325	Bus & coach drivers	96	11.3	68.2	76	13.2	72.3
326	Drivers of road goods vehicles	470	16.9	101.8	397	18.5	100.9
327	Other motor drivers	152	13.6	82.3	129	14.2	77.3
330	Mechanical plant drivers, operators (earth moving & civil engineering)	65	17.6	106.3	59	18.6	101.5
332	Fork lift, mechanical truck drivers	108	24.8	149.6	97	26.7	145.8
333	Storekeepers, warehousemen	411	15.0	90.7	310	16.9	92.6
335	Goods porters	86	17.2	103.6	71	19.7	107.5
338	Workers in transport operating, materials moving & storing & related nec	64	20.5	123.9	50	24.1	131.5
346	General labourers, other	193	15.0	90.5	137	17.1	93.4

Table 12.16 Mean alcohol consumption (units/week) - female

No.	Occupational unit	All ages			Age 16-64		
		Sample	Cons.	PACR	Sample	Cons.	PACR
013	Marketing & sales managers & executives	59	11.6	210.5	58	11.7	188.7
033	Teachers (other than higher education)	677	5.6	102.0	560	6.1	98.9
039	Welfare occupations (not elsewhere classified)	179	7.5	136.0	149	7.7	124.3
043	Nurse administrators, nurses	820	5.0	91.0	678	5.5	88.1
053	Professional & related in education, welfare & health nec	74	4.4	79.0	55	4.7	75.4
099	Office managers nec	156	7.4	133.8	141	7.7	123.8
101	Other proprietors & managers (sales)	388	5.8	104.7	282	6.7	107.1
102	Hotel & residential club managers	55	8.0	144.8	29	10.5	169.7
103	Publicans	55	8.5	153.7	37	10.1	162.2
104	Restauranteurs	110	6.9	125.0	84	7.3	117.2
111	Managers nec	58	6.9	124.8	50	7.2	115.8
112	Civil service executive officers	72	7.7	138.5	66	8.3	134.1
115	Other clerks & cashiers (not retail)	2848	6.3	113.4	2355	6.8	109.7
116	Retail shop cashiers, check-out, cash & wrap operators	228	4.7	85.7	201	5.0	81.2
117	Receptionists	224	6.2	112.5	208	6.6	105.8
118	Typists, shorthand writers, secretaries	1241	6.4	116.3	1033	7.0	112.1
119	Office machine operators	227	6.4	115.0	208	6.7	108.2
121	Telephone operators	97	5.4	97.3	73	5.3	85.8
125	Shop salesmen & assistants	1762	4.5	81.4	1352	5.1	82.4
126	Shelf fillers	51	4.6	82.8	47	4.9	79.4
133	Sales representatives	52	8.6	154.8	48	8.8	141.1
134	Sales reps (property & services), other agents	78	8.2	148.6	70	8.5	136.9
143	Chefs, cooks	284	4.5	81.0	185	5.9	94.8
144	Waiters, waitresses	232	5.3	96.6	191	6.0	97.3
145	Barmen, barmaids	232	11.5	208.0	208	12.3	197.7
146	Counter hands, assistants	408	4.2	75.4	279	5.2	84.1
147	Kitchen porters	247	4.4	79.0	173	5.5	88.1
150	Nursery nurses	74	4.4	79.6	72	4.5	72.6
151	Other domestic & school helpers	971	3.8	69.4	656	4.8	77.0
156	Hospital, ward orderlies	288	5.5	99.8	258	5.9	95.2
158	Cleaners, window cleaners, chimney sweeps & road sweepers	1194	3.5	62.6	761	4.5	72.8
159	Hairdressers, barbers	235	5.7	102.5	206	6.2	100.2
162	Launderers, dry cleaners, pressers	116	4.8	86.4	75	6.4	103.7
165	Service workers nec	115	4.9	88.1	100	5.5	88.4
166	Farm workers	103	2.6	47.2	56	2.4	38.0
175	Leather cutters & sewers, footwear lasters, makers, finishers	51	3.8	68.2	31	4.9	79.2
202	Food and drink nec	97	5.8	105.1	82	6.0	97.3
210	Tailors, tailoresses, dressmakers	69	2.3	42.3	24	3.7	59.1
212	Sewers, embroiderers	410	4.8	86.4	316	5.8	93.1
227	Paper goods and printing	56	6.0	108.3	38	8.3	134.1
241	Machine tool operators	54	2.8	50.8	24	3.9	62.2
283	Assemblers (electrical, electronic)	155	5.3	94.9	127	6.0	96.9
285	Assemblers (vehicles & other metal goods)	53	5.9	106.1	46	6.8	108.7
286	Metal, electrical goods	82	3.9	70.3	46	5.7	91.0
287	Packers, bottlers, canners, fillers	437	4.3	78.1	339	4.8	77.1
299	Painting, assembling & related occupations nec	65	3.3	60.4	51	3.7	60.1
333	Storekeepers, warehousemen	101	4.9	88.8	74	6.1	97.4

independent. There are several occupations for which the percentage of subjects drinking more than 21 but less than 50 units is quite high but the overall PCDR for more than 21 units is reduced because the percentage drinking more than 50 units is low. The PCDRs for 21-50 units per week, not presented because they have a less definable functional significance, would show even less evidence of a manual/ non-manual divide.

Secondly, there is wide variation in the drinking status of the non-manual occupational orders. Even within those with low PCDRs there are often particular age groups or specific occupations with much higher rates of alcohol consumption than the order as a whole. For example men aged 16-64 in order 2 have low PCDRs at both the 'greater than sensible' and 'greater than safe' levels but it can be seen from Table 12.5 that in the youngest age group this order has the highest heavy drinking rates in the table. In women aged 16-64, the PCDRs for more than 14 units per week show some evidence of a reversed trend, with higher drinking rates in the non-manual workers. It must be remembered that this relates to a lower limit for 'sensible' drinking in women. It is possible that many women in occupations of high socio-economic status drink similar or even slightly lower quantities than their male colleagues but are nevertheless classified in a higher drinking band, while in the lower social orders the difference between male and female mean alcohol consumption is greater and the women are thus included in the same, or a lower drinking band. This point will be made again in the discussion on some individual occupations. The concept is supported by Table 12.12 which shows that the PACRs are higher for women in the non-manual than the manual occupational orders.

12.5 Notes on specific occupations

Tables 12.5 and 12.6 show the percentages smoking and PCSRs for those occupations for which a sample of 50 men or 50 women respectively was available. The PCDRs are shown in Tables 12.13 and 12.14. Overall tables for men and women combined will be available on disk. In them, the observed and expected numbers smoking in the two sexes have been added whether or not both sexes satisfied the selection criteria described above. This limits the confidence with which the data may be interpreted and caution should be exercised in using the overall figures for occupations which do not appear in both of the single sex tables. The PACRs by occupational unit are shown in Tables 12.15 and 12.16.

The following discussion is organised by occupational order, since the detailed age breakdown is not available at unit level, with comments on the occupational units of interest within the order where appropriate.

12.5.1 Order 1 - Professional

This order includes the legal profession, the financial professions and senior managers and administrators. It is male dominated. The overall proportion of smokers is low and the age trends beyond age 24 are similar to the general population but the percentage of smokers in the 16-24 year old group is relatively high, especially in women. In men there is a marked variation in smoking habit by occupational unit within the order. The legal profession and accountants have two of the lowest PCSRs amongst men while the executive occupations contain about twice as many smokers but remain below the general population average. The only female occupational unit large enough for presentation is marketing and sales managers and executives, who have a very low PCSR. In the unpublished data on the female occupations which failed to qualify there appears to be even wider variation than in men.

The overall PCDRs for both 'more than sensible' and 'more than safe' drinking are raised for both sexes but markedly so for women. The unit tables for men show that the modest elevation in both PCDRs conceals important differences between specific occupations. For example the legal profession has a PCDR of 142 for more than 21 units per week but this falls to 84 for more than 50 units. On the other hand the respective PCDRs for underwriters are 110 and 213. It is unfortunate that only order 013 was large enough to be represented in the female tables. Several of the (unpublished) PCDRs for more than the safe level of 35 units in women are very high indeed although based on very small samples. For the legal profession, for example, the figure is 693 based on 3 out of a sample of 25 drinking at this level against 0.4 expected.

Another feature of this order is that the tables of drinking by age show that heavy drinking in the young is well above the population average with 16.5 per cent of men and 7.4 per cent of women drinking more than the safe limit for their sex. The PACR of women in this order is much higher than that of men. One possible explanation of the very high PCDRs seen for some occupational units compared with men is that some women in the high drinking occupations may drink similar quantities to their male colleagues but, since the 'safe' limit for women is lower, they are classified in a higher drinking category.

12.5.2 Order 2 - Professional education and health

In addition to the major contributions made by teachers, doctors and nurses the sample for this order contained a significant number of 'not elsewhere classified' education and health workers. None of the several professions related to medicine qualified for inclusion in the analyses by unit. The overall proportion of smokers in this order is again low but the excess of smokers in the youngest age band is even more pronounced than in order 1. This is more so in men, whose percentage smoking at age 16-24 is well above the population mean and is the third highest for all the occupational orders. Since this age band would contain almost no doctors, who in any case have a very low smoking rate, and as the overall PCSRs for teachers are also low the nursing profession suggests itself as a possible source of the excess of young smokers. The unit tables are not analysed in detail by age but it can be seen that in those of working age the PCSR for female nurses is 96.7 and that of male nurses is 116.1. These are amongst the highest for all professional occupations.

The low PCDRs for both sexes conceal some disturbing underlying facts seen in Tables 12.7 and 12.8 showing the breakdown by age. The percentage of men aged 16-24 drinking more than 50 units per week in this order is almost 21, which is the highest of any occupational order by a considerable margin. In terms of absolute numbers, however, this is not large as the proportion of very young men in the order is low. For women the proportion of heavy drinkers is close to the population average of 3.3 per cent. The profession with the highest rate of drinking at the 'greater than sensible' level in men and at both levels in women is unit 39, welfare occupations not elsewhere classified. Teachers and nurses are much nearer to the rates for the order as a whole and doctors rather lower. In absolute terms, however, the mean alcohol consumption of doctors is higher than that of teachers or nurses.

The points made above illustrate the need to consider all aspects of the data available to get a true impression of the drinking patterns of an occupational order or unit. Even when this is done, as has already been stressed, the reasons for the differences seen between occupations may not be directly explained by the occupation. For example one would expect differences in age structure of the workforce and in age at marriage between doctors, nurses and teachers.

12.5.3 Order 3 - Arts and sport

None of the occupations were of sufficient size to be included in the analysis by unit. Sports professionals are likely to make only a small contribution to the data for the order. The main contributions come from the performing arts and journalism. The age distribution of smoking does not differ greatly from the general population and such apparent differences as exist should be treated with great caution as they are based on very small sample sizes. The PCSRs for both sexes are higher than for any of the other professional orders.

The high alcohol consumption of this order is perhaps not surprising when one considers that it includes groups, such as journalists, traditionally believed to have a high alcohol intake. None of the constituent occupational units was large enough to be represented in the analysis for either sex.

12.5.4 Order 4 - Professional science

For women this is a small occupational order and none of the constituent units satisfied the criteria for separate analysis. For men the units represented are mainly from various branches of engineering. For both sexes the percentages smoking and PCSRs for those under age 65 are the lowest of all orders. The elevated PCSR for women over 65 is based on a very small sample. The only one of the available occupational units to differ markedly from the overall picture was electrical engineers whose PCSR at ages 16-64 and at all ages are around one third of those of other constituent units, including electronic engineers, based on similar sample sizes.

The PCDRs for men are low. Again the only anomaly is unit 071, electrical engineers, with all ages PCDRs of 133 and 112 for the two levels, compared with values of 76 and 59 for electronic engineers. This is in the opposite direction to the findings on smoking status. No female occupational unit fulfilled the inclusion criteria. There is unpublished evidence that the sample for the order includes 45 laboratory technicians, who had high PCDRs, which would probably explain the PCDR for women drinking more than 14 units. Male laboratory technicians do not have high PCDRs.

12.5.5 Order 5 - Managerial

This order, as defined by OPCS. is based to a significant extent on social class distinctions and as a result contains a number of anomalies from a strictly occupational point of view. For example publicans are included in this order rather than in order 9 with bar staff. Farmers are also included whereas farm workers are in order 10. In both cases the distinction is based largely on self-employed status. These occupations will be discussed further under orders 9 and 10 but it can be seen from the occupational unit tables for men that those classified as managerial have lower rates of smoking than the corresponding non-managerial occupation. The situation for women is less clear.

The PCSRs for women in order 5 are close to 100 while those for men are around 15 per cent lower than the corresponding female figures. The male data are more nearly related to those for the 4 professional orders and quite different from the manual occupations while the data for women fall into an intermediate category. This difference is probably partly explained by differences in the socio-economic status of the sexes in order 5. The available data for occupational units suggests that for women the order is very much dominated by managers of offices and shops while for men there is a relatively much higher proportion of works and site managers. The suggestion thus arises that men in order 5 tend to be managers of larger establishments than women and have a relatively higher socio-economic status whereas that of women may be more closely related to the workers they manage. However, it should be noted that even for the specific units of office and sales managers (99 and 101) the proportion of women smoking is higher than amongst their male counterparts. It is impossible to ascertain whether this may also reflect a tendency for women managers to be lower down the socio-economic scale.

The male PCDRs are close to 100 but those for women are notably elevated. The explanation may be the same as that advanced above. The anomalously classified occupations of publicans, hotel managers, restaurateurs and farmers will be discussed in the relevant orders below. Otherwise the only notable exception to the overall picture in the unit data is male unit 97, managers in warehousing etc., nec, who have very high PCDRs at the upper drinking level.

12.5.6 Order 6 - Clerical

This order has a large preponderance of female workers and accounts for almost 30 per cent of all working women. The unit making the largest contribution to the data for both sexes is number 115, Other clerks and cashiers (not retail). The PCSRs for both sexes are intermediate between the ranges for professional and manual occupations but that for women of working age is distinctly lower than for men. It is interesting that the female smoking rate is much lower than for the orders on either side in the socio-economic scale while for men it is a little higher. The available data for occupational units provides less material on which to base speculation about the reasons for the sex difference than is the case for order 5. There is also a distinct difference in the age distribution of smoking by sex. In women the trend is one of decreasing proportion smoking with increasing age similar to, but less marked than, that seen in the professional orders. In men the proportion smoking is higher in mid-life than at the extremes, which is more like the distribution in the manual occupations.

In the limited range of occupational units available for analysis two stand out as markedly different from the order as a whole. These are female retail shop cashiers (116) with a PCSR of 124 (all ages) and male postmen (123) whose PCSR is 130. Neither of these contain large enough samples to significantly influence the overall findings for the order. Unit 116 includes check-out operators and it might be felt that it would be better included in order 7. It is certainly true that the smoking rate for women would be more in keeping with that order.

The male PCDRs are also fairly low but those for women are close to 100. The largest female constituent unit, 115 (other cashiers - not retail), has somewhat higher PCDRs than the order as a whole. Male postmen, who have a high rate of smoking, have only slightly elevated PCDRs although they are high compared with the order as a whole.

12.5.7 Order 7 - Selling

This is again a female dominated order but the distribution of specific occupational units differs by sex. The majority of women in the order are shop sales assistants while for men about half are sales representatives. The distribution of smoking by age and sex is not unlike that in order 5 and a similar argument may be applied to suggest that the higher rate of smoking in women may be largely explained by socio-economic differences. A similar caveat also applies in that the proportion smoking in unit 125 (shop sales assistants) is markedly higher in women than men. The unit which differs most from the overall pattern is female shelf fillers (126) with an all ages PCSR of 130.

The PCDRs for the order are low for both sexes and the only notable exception amongst the individual units is female sales representatives (133) with all ages PCDRs of 164 and 333 based on a fairly small sample. Male sales representatives have PCDRs close to 100 for more than 21 units and much lower for more than 50 units thus these data do not support the popular image of the travelling salesman as a heavy drinker.

12.5.8 Order 8 - Security

This is a relatively small and male dominated order in which none of the female occupational units were large enough to meet the criteria for separate presentation. A few words are appropriate, however, about the composition of the order between sexes in view of the overall difference between the smoking rates by sex. The major contributions by unit in men are police officers (137) and security guards etc. (140) which together account for about 75 per cent of the order. Amongst women the highest contribution , almost 50 per cent, is from armed forces personnel. The interesting point is that this appears not to be the explanation for the much higher percentage smoking amongst women. The limited data available for the female occupational units (not presented) suggest that smoking rates in female military personnel are similar to their male counterparts but the smoking rate amongst female police officers is much higher. In men it can be seen that the order is rather heterogeneous with regard to smoking habit. The PCSR for policemen is 75 while that for unit 140 (security guards, watchmen, etc.) is 137.

The difference in drinking habits between the sexes is even more marked. The PCDRs for men are low and similar to those for the professional occupational orders while those for women are the highest of all the orders. The male occupational units for which data are available are armed forces personnel (135), police officers (137), firemen (138) and security guards, watchmen, etc., (140) The PCDRs for security guards etc., who make up about half of the sample, are lower than for the order as a whole and those of police and fire service personnel are rather higher but close to 100. Those for the armed services are notably high. The female units were too small to meet the criteria for presentation but the unpublished data available for services personnel, policewomen and security guards all show the same pattern as for the whole order.

12.5.9 Order 9 - Personal service

This is a female dominated order and the fact that the composition by occupational units appears more diverse in women is probably a consequence of the larger sample size in all of the female constituent units meaning that more units will meet the eligibility criteria for analysis. It is at this point in the occupational order tables that the non-manual/manual division in smoking habit can be seen to really begin for both sexes. The PCSRs for the sexes are similar as is the age distribution, with a marked increase in smoking rate from age 16-24 to 25-44 and a decline thereafter. This pattern is seen for most of the manual occupational orders which also tend to start off from higher smoking levels at age 16-24 than the non-manual orders (with the notable exception of education and health professionals).

This order includes occupational unit 145, barmen and barmaids who have some of the highest PCSRs of all occupations. These would be reduced somewhat, particularly for men, if 'publicans' (who are classified in order 5) were included with other bar staff. In both occupational units the proportion of smoking amongst women is higher but the effect is more marked in unit 103 (publicans) as male publicans actually have a PCSR near to 100. Perhaps this once again reflects a social status difference by sex as described under order 5.

This order has the second highest PCDR for more than 50 units per week for men, although the PCDR at the lower level of drinking is close to 100. For women the PCDR at the upper level is modestly elevated and that at the lower level is only 91. Much of the elevation in men is contributed by the units of chefs and cooks (143) and barmen (145), both of which have very high PCDRs at both levels, although the other male occupational units represented in the tables also have high PCDRs at the higher drinking level. Publicans (103), included in order 5, have even higher PCDRs than barmen. This is in contrast to the findings for smoking.

Among women, barmaids have even higher PCDRs than their male counterparts, although it must be remembered that this is in respect of lower absolute levels of alcohol consumption. Hairdressers (159) also high drinking ratios at both levels and those of chefs/cooks (143) and waitresses (144) are raised at the higher level. Publicans (103), hotel/club manageresses (102) and restaurateurs (104), from order 5, also have high PCDRs but rather less than barmaids. This finding is similar to that for female smokers in these occupations. Although individually based on small observed numbers these PCDRs are consistent between the three units mentioned. The other female service occupations in Table 12.14 all have PCDRs well below 100.

12.5.10 Order 10 - Agricultural workers

This order includes all farm workers and foremen, horticultural workers, foresters, etc. but as described above self-employed persons describing themselves as 'farmer' (unit 107) will come to be included in order 5. Including unit 107 in order 10 would reduce the PCSRs to some extent for both sexes and on this basis it is probably true to say that smoking habit amongst male farmers differs little from the general population. Female agricultural workers appear to smoke rather less at all ages although this is based on a relatively small sample size.

The PCDRs and PACRs for both sexes are well below 100. This is also the case for the constituent units of farm workers (166) of both sexes and male gardeners and groundsmen (168). The same pattern is seen in the unpublished data available for smaller units. Male farmers (107), currently included in order 5, also conform to the general pattern of low alcohol consumption.

12.5.11 Order 11 - Materials processing (other than metal and electrical). Order 12 - Materials processing (metal and electrical). Order 13 - Repetitive assembly, painting, etc.

These three orders together include the majority of manual workers from the whole of manufacturing industry. It would not be expected that there would be major differences in smoking status between them, as whole orders, and this is indeed the case. The main interest in these orders is in the occupational unit tables which show a wide range of variation by specific occupation. It is hoped that these tables may prove of value in assessing data on the comparative rates of smoking related disease by occupation but because of the selection criteria imposed it is likely that in many cases the occupation of interest will not be included. The unit with the highest all ages PCSR for men in orders 11-13 is 282, painters and decorators etc., confirming other sources which attribute high rates of smoking to this occupation. Unit 299, which includes painting as a manufacturing finishing process, has the highest PCSR for women.

There is rather more diversity between the orders with regard to drinking habits. The PCDR at the 'more than safe' level for men in order 11 is close to 100 and that of order 12 only a little higher but the corresponding figure for order 13 is notably higher. Among women the PCDR for more than 35 units a week is high in order 11 but well below 100 in orders 12 and 13. These disparities are much less apparent in the PCDRs for the 'more than sensible' drinking level or in the mean alcohol consumption data.

The explanation for the differences between the orders is not obvious in the unit tables (12.13 and 12.14). This is not unexpected as so many occupations are missing because of small sample size. It is worth observing that male painters and decorators (282), who make up the largest unit contributing to order 13, have high PCDRs well above those of the order as a whole. This unit also contains a high percentage of smokers. The other male units with high drinking ratios are butchers (186), printing machine minders (207) and carpenters and joiners (214) in order 11 and electricians (253), telephone fitters (256), sheet metal workers (261) and welders (265) in order 12. The other units represented in all three orders have PCDRs well below 100.

The representation of specific units for women is even less complete and no explanation for the high PCDR in unit 11 presents itself. The high values for units 202 (food and drink, not elsewhere classified) and 227 (paper goods and printing) are based on very small observed numbers and could not contribute significantly to the value for the whole order. No other female units present any striking findings.

12.5.12 Order 14 - Construction

This can be considered to be an exclusively male order as the sample size for women is meaninglessly small. The overall PCSR for ages 16-64 is the highest for all the occupational orders except 16, 'miscellaneous', and a large proportion of the latter order consists of general labourers who may well work in construction. There is some variation between specific building trades but all have high proportions of men smoking. Note that this order also includes coal miners but unit 314 (face-trained coal miners) has the lowest all ages PCSR in the order and thus its inclusion does not alter the general findings.

The male PCDRs at both drinking levels are by far the highest of all the orders as is the PACR for men aged 16-64. These findings are consistent across all the units represented with the exception of handymen and general building workers (304) for whom the figures are rather lower than 100. Table 12.7 shows that drinking at high levels starts at a young age in this order and although it falls from age 45 onwards the percentage remains higher than the general population.

12.5.13 Order 15 - Transportation, storage, etc.

This order is again male dominated. Only one female occupation is represented in the unit tables (333, storekeepers).

The overall PCSR for men of 135 is not far below that of construction workers. The main difference is that there is a rather lower percentage smoking in the lowest age band. This could mean either that transport workers continue to take up smoking over a longer period of their lives or that fewer of the young are now starting to smoke. The unit tables for men suggest that drivers, and particularly bus drivers, are more likely to smoke than non-driving occupations, such as porters and storekeepers, who are included in this order. This may reflect to some extent the ease with which it is possible to smoke while driving. The highest PCSR is that for unit 338, which, at 182, is the highest for any male occupational unit. Unfortunately this is one of the 'not elsewhere classified' units and its composition is uncertain.

Men have a high PCDR for drinking more than 50 units a week although the rest of the data for both sexes in the order are unremarkable. Table 12.13 shows that male bus and coach drivers (325) have a high proportion of drinkers at the >50 units a week level. Their PCDRs at the >21 units a week level are low and, since the data giving rise to these figures include the numbers drinking more than 50 units, the proportion drinking 22-50 units a week (not presented) is markedly low. Bus drivers thus show some evidence of a bimodal distribution of drinking habit with peaks in both the sensible and unsafe bands. The mean weekly alcohol consumption and PACR of this unit are much lower than 100, demonstrating how this measure on its own can be misleading as an indicator of the drinking status of a population.

The findings for bus drivers raise important questions concerning possible implications for road safety. Presumably their drinking takes place out of working hours but it could still lead to appreciable pre-shift alcohol levels in the heaviest drinkers. Data in *Drinking in England and Wales in the late 1980s* shows that the percentage of workers drinking during working hours in order 15 is low, but a detailed analysis by occupational unit would be precluded by the smaller sample size compared with the GHS.

The other occupational units involved in driving vehicles on the roads tend to have low PCDRs at the upper level although unit 330 (mechanical plant drivers) has a fairly high PCDR for more than 21 units a week. Fork lift truck drivers (332) have markedly high PCDRs at both levels. Of the non-vehicle operating occupations storekeepers, etc. (333), which is the largest such unit, do not have raised PCDRs but goods porters (335) and those 'not elsewhere classified' have high PCDRs at greater than 50 units a week.

The only female unit represented in Table 12.14 is storekeepers etc. (333) who comprise more than two thirds of the order and whose drinking habits are similar to the order as a whole.

12.5.14 Order 16 - Miscellaneous

Little can be said about this order other than that it is composed of several units of unskilled workers and general labourers. Some of these are classifiable to the industry in which they work but the largest group, unit 346, are not classified at all. As an order these workers have the highest PCSR of all, although men have PCDRs below 100. The high PCDR for women drinking more than 35 units a week is based on only one observed subject.

12.6 Conclusion

These data form the most comprehensive statistics on smoking and alcohol consumption by occupation yet published in the decennial supplement series. The detailed breakdown by age and occupational order reveals many occupationally related phenomena which are not apparent in the previous analyses of the same data in the GHS reports, such as the very high rates of smoking and drinking amongst the younger members of some of the professional occupations. It is hoped that the data at occupational unit level will be of assistance in interpreting the findings of morbidity and mortality studies, at least for many of the commoner occupations. The data are derived from an ongoing series of surveys, therefore, if they prove valuable, it will be possible to repeat the analyses and build upon this foundation in future volumes of the decennial supplement.

References

1 Office of Population Censuses and Surveys, *General Household Survey 1988*. London: HMSO, 1990.

2 Goddard E, *Drinking in England and Wales in the late 1980s*, London: HMSO, 1991.

3 Office of Population Censuses and Surveys, *Occupational mortality 1970-72. Decennial supplement*. London: HMSO, 1978.

4 Office of Population Censuses and Surveys, *Occupational mortality 1979-80, 82-83. Decennial supplement*. London: HMSO, 1986.

Chapter 13 Occupation and sickness absence

Simon Clarke
Richard Elliott
John Osman

Epidemiology and Medical Statistics Unit, HSE

13.1 Introduction

Sickness absence has been defined as absence from work which employees attribute to sickness or injury and which the employer accepts as such.[1] It is generally accepted that sickness absence in this context may include absence for reasons other than employee sickness, such as illness among other family members or childcare difficulties. Indeed sickness absence is thought to have a multifactorial aetiology where sickness or injury may be only one of many causes.[2] There is no comprehensive source of data matching up to the definition but the data presented here, which are derived from self-reporting of incapacity to work through sickness or injury, may be considered to broadly reflect sickness absence. This is the first occasion on which data on sickness absence have been published in the Decennial Supplement series, although previous volumes have contained data on self-reported long-standing illness derived from the General Household Survey[3] (GHS).

The UK Labour Force Surveys (LFS) 1987-1991 provide the data source for this section. Summary data on sickness absence from the annual LFS have previously been published in the *Employment Gazette*. The data presented in this section are a re-analysis and more detailed presentation. Although the LFS data can not be regarded as an accurate measure of ill-health, whether occupational in origin or not, they do provide a useful and comprehensive national information source on sickness absence by age, sex, occupation and time period. The breakdown of sickness absence data in this manner could prove valuable not only in the context of occupational health but also could be of interest in estimating the economic cost of sickness absence and in planning welfare provision for different industrial and occupational groups.

The subsequent commentary describes the data and suggests possible explanations for some of the more notable features. The commentary first considers general points regarding sex and age and then comments on effects by occupational group.

13.2 Data sources and methods

A description of the general methodology of the LFS can be found in Chapter 2. The data presented here are obtained from a series of questions designed to ascertain from the interviewees:

- the number of days they were unable to work last week (the reference week) because of sickness or injury, including those days they would not normally work;
- the length of their last period of absence from work through sickness or injury; and
- whether they have any health problems that limit their work activity and if so which was the most problematic.

The details of the questions have varied slightly from year to year, most importantly with regard to whether the reference week includes Sunday, which limits the extent to which temporal trends can be examined.

The tables present estimates, scaled up from the LFS sample, to represent the whole working population. Scaled up values below 10,000 in any stratum are not considered reliable because they are based on small numbers of respondents. This would be true of the majority of estimates if analysed fully by year, age, sex and occupational group. To prevent this the time dimension was removed from most of the analyses by adding the figures across the years 1987-1991. Figure 13.1 shows the temporal trends in those reporting at least 1 day sick in the reference week. These plots show little variation across the years 1987-1991 among males, with a possible increasing trend in sickness absence in females. However, only between the highest and lowest reported

Figure 13.1 Percentage sickness absence of at least one day in reference week for all occupations and ages by sex and year

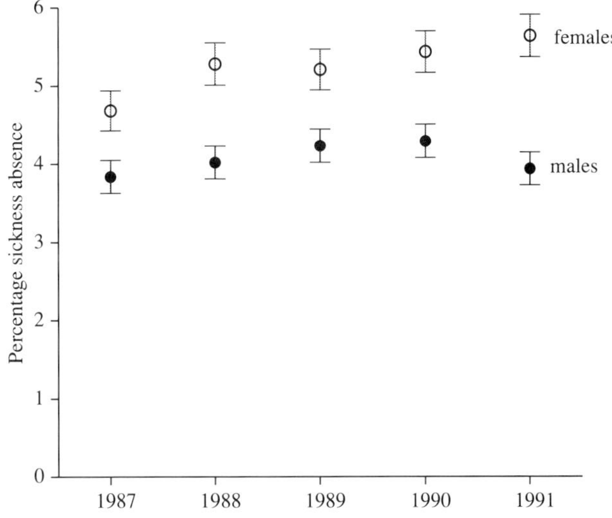

figures (1987 and 1991) do the 95% confidence limits fail to overlap and the difference between these estimates is less than 1 per cent. On this basis, it was considered that summation across years was acceptable when examining the data by occupation, sex and age group. Although caution should be exercised in interpreting small differences between strata which might be explained by a difference in the temporal trends within the strata concerned.

The data in the section are presented in four tables. More detailed tables are available on the floppy disk that complement this publication. General points on the content, and where appropriate, comments on construction of the tables are presented below:

(i) Table 13.1 shows the percentage of those reporting at least 1 day sick in the reference week for the combined years 1987-1991 by sex, age and occupational group.

Table 13.1 Percentage with 1 or more days sickness absence in the reference week: 1987-1991 combined by occupational group, age and sex

Ages	Males		Females		Males		Females		Males		Females	
	% absent	No. absent	% absent	No. absent	% absent	No. absent	% absent	No. absent	% absent	No. absent	% absent	No. absent
	All occupational groups				**I. Professional & related - managers & administrators**				**II. Professional education & health**			
16-19	4.1	162,000	5.3	217,000	4.2	2,200	4.2	2,000	4.1	700	4.4	4500
20-24	4.4	336,200	5.8	400,000	2.7	13,100	5.2	16,900	4.0	4,800	5.5	35,800
25-29	3.8	310,700	5.4	337,500	2.7	21,600	4.7	20,300	3.5	10,700	5.0	49,600
30-34	3.8	265,400	5.0	271,800	2.2	16,500	4.6	14,900	4.2	17,400	4.7	46,300
35-39	3.5	232,900	4.5	256,200	2.5	17,800	4.0	9,800	2.8	14,900	4.9	53,600
40-44	3.6	246,300	4.9	314,800	3.0	21,200	3.3	8,300	3.5	19,500	5.0	59,700
45-49	3.7	204,100	5.3	275,600	2.1	9,700	4.8	8,800	2.2	9,300	5.7	51,600
50-54	4.1	198,900	5.5	241,000	2.3	8,700	2.4	2,700	2.5	8,700	6.2	45,600
55-59	5.2	214,000	5.9	201,300	2.7	8,800	4.3	4,200	5.5	14,100	6.3	28,400
60-64	6.6	178,400	NA	NA	4.4	7,300	NA	NA	5.0	7,200	NA	NA
Total	4.1	2,348,900	5.3	2,515,200	2.6	126,900	4.3	87,900	3.4	107,300	5.3	375,200
	III. Literary artistic & sports				**IV. Professional & related - science etc**				**V. Managerial**			
16-19	3.6	1,400	6.2	2,000	3.0	4,100	3.9	1,000	3.7	2,900	5.6	3,500
20-24	3.2	2,700	3.3	2,600	3.5	19,100	4.1	5,400	2.3	10,600	5.8	19,900
25-29	4.2	3,600	4.0	3,500	2.6	17,400	5.8	7,900	2.4	19,200	3.9	13,500
30-34	2.2	1,800	3.3	1,800	3.3	20,600	6.2	5,600	2.9	24,400	3.3	8,700
35-39	1.2	700	4.8	2,200	2.7	15,800	1.4	1,000	2.3	22,500	4.5	12,000
40-44	2.4	1,800	4.5	2,100	2.7	15,500	5.7	4,500	2.0	21,500	4.0	12,000
45-49	3.7	1,900	8.4	2,300	3.3	14,400	6.1	2,900	2.3	20,400	4.1	10,300
50-54	1.0	400	4.8	1,500	3.3	11,500	7.1	2,200	2.2	15,500	3.6	6,400
55-59	5.0	1,500	0.0	0	5.1	14,300	11.4	2,300	3.2	16,500	2.0	2,100
60-64	4.9	1,000	NA	NA	6.8	10,000	NA	NA	4.4	12,200	NA	NA
Total	3.0	16,700	4.3	18,000	3.3	142,700	5.2	32,900	2.5	165,700	4.2	88,500
	VI. Clerical & related				**VII. Selling**				**VIII. Security etc.**			
16-19	2.9	18,000	5.4	81,300	2.8	18,100	3.5	32,300	9.5	2,400	0.0	0
20-24	4.9	45,000	5.7	157,300	3.2	15,500	4.7	29,300	4.1	8,600	6.7	3,300
25-29	4.0	26,400	5.0	109,600	2.8	11,500	5.0	26,400	3.4	12,500	5.9	2,300
30-34	4.1	19,400	4.8	77,700	2.7	7,700	5.2	23,800	4.3	13,400	8.8	2,500
35-39	3.4	13,200	4.0	67,500	3.0	6,700	2.6	12,900	5.3	12,700	4.7	1,300
40-44	4.6	17,400	4.4	84,00	2.5	6,600	3.2	18,500	4.7	10,100	4.4	1,100
45-49	4.5	13,200	3.9	58,900	2.0	3,600	4.0	18,500	4.3	7,400	0.8	100
50-54	4.7	15,300	4.6	56,000	3.1	4,800	4.6	18,300	2.7	3,400	7.8	1,400
55-59	3.9	12,800	4.9	48,000	3.1	4,300	3.7	11,200	4.7	4,900	5.6	800
60-64	5.4	12,300	NA	NA	3.8	3,600	NA	NA	4.4	4,000	NA	NA
Total	4.2	193,000	4.8	740,500	2.9	82,400	4.0	191,100	4.3	79,500	5.7	12,800
	IX. Personal service				**X. Farming, fishing & related**				**XI. Processing etc. (excluding metals and electrical)**			
16-19	3.7	15,800	5.2	44,300	3.8	4,400	2.8	1,100	4.7	19,600	8.5	19,000
20-24	4.8	19,400	4.8	48,800	3.8	7,900	5.7	2,700	5.2	41,700	9.2	34,000
25-29	4.4	13,000	4.3	39,100	4.4	7,800	4.1	1,500	5.3	35,200	11.8	30,000
30-34	4.2	9,100	4.7	50,400	4.2	5,600	5.8	1,500	4.8	23,300	7.8	16,200
35-39	5.5	9,600	4.6	57,200	4.7	4,300	5.1	1,400	4.8	21,300	7.5	16,100
40-44	3.7	6,600	5.6	81,500	3.2	3,000	2.1	700	4.7	20,300	8.3	20,000
45-49	5.3	9,800	6.0	79,200	2.8	2,200	4.6	1,100	6.1	23,000	8.6	17,200
50-54	5.4	10,400	5.8	69,000	4.9	4,100	6.9	1,500	4.5	15,500	7.3	13,100
55-59	5.3	11,000	6.7	69,700	7.8	7,200	1.9	400	5.2	14,900	10.0	13,700
60-64	7.6	15500	NA	NA	6.1	5,500	NA	NA	10.3	19,400	NA	NA
Total	4.8	120,100	5.4	539,200	4.5	52,000	4.3	11,800	5.3	234,100	8.9	179,200

Table 13.1 - *continued*

Ages	Males % absent	Males No. absent	Females % absent	Females No. absent	Males % absent	Males No. absent	Females % absent	Females No. absent	Males % absent	Males No. absent	Females % absent	Females No. absent
	XII. Processing etc. (metals & electrical)				**XIII. Printing, assembling etc.**				**XIV. Construction, mining etc. n/e**			
16-19	4.5	31,200	9.5	5,100	3.7	7,000	9.4	17,200	7.4	13,100	31.6	700
20-24	4.4	60,000	6.9	6,600	5.5	19,100	10.2	31,200	5.6	20,100	9.3	500
25-29	3.9	54,600	6.8	4,400	4.9	14,100	10.7	22,900	6.8	21,900	8.1	400
30-34	3.7	44,800	11.3	6,300	5.0	10,900	8.6	13,700	7.1	17,800	0.0	0
35-39	4.3	49,600	7.1	3,600	4.3	8,500	7.6	13,800	5.5	11,000	0.0	0
40-44	4.3	47,100	7.1	4,400	4.3	8,600	6.2	13,200	6.0	14,500	0.0	0
45-49	4.1	37,500	9.0	5,300	4.8	9,100	8.1	15,000	4.4	9,300	17.4	400
50-54	6.0	46,700	7.3	4,000	5.5	10,300	9.0	14,600	4.9	9,100	29.4	500
55-59	6.0	37,700	9.6	3,500	5.5	9,600	9.4	13,400	7.1	11,900	0.0	0
60-64	7.9	31,000	NA	NA	6.7	7,400	NA	NA	9.8	10,800	NA	NA
Total	4.6	440,100	8.1	43,100	5.0	104,400	8.9	155,000	6.3	139,400	9.0	2,500
	XV. Transport, operating etc.				**XVI. Miscellaneous**							
16-19	6.3	18,200	7.4	2,200	4.2	3,100	6.1	700				
20-24	5.4	40,800	7.8	4,800	5.4	7,000	5.7	800				
25-29	4.2	34,200	9.5	5,300	5.9	7,100	11.2	700				
30-34	3.8	25,400	4.2	2,100	9.3	7300	0.0	0				
35-39	3.5	21,100	4.6	2,000	4.5	2,800	17.5	1,000				
40-44	3.7	26,800	7.2	4,000	6.6	4,700	7.4	700				
45-49	4.5	28,300	4.8	2,000	6.4	4,700	13.9	2,100				
50-54	5.0	29,300	11.1	3,900	6.9	4,900	3.3	300				
55-59	7.9	40,700	5.2	1,700	4.4	2,900	19.7	1,600				
60-64	7.1	24,800	NA	NA	9.6	6,100	NA	NA				
Total	4.9	289,600	6.9	28,000	6.2	50,500	9.4	7,900				

Scaled up values are presented to the nearest 100 and values below 10,000 are considered unreliable. All calculations were performed on unrounded data and therefore some values may not add up as expected.

(ii) Table 13.2 presents estimates of the percentage of working days lost due to sickness absence in 1987-1991 by sex, age and occupational group. This is complicated by the fact that the LFS questions ask subjects to report all days they were unable to work in the reference week, irrespective of whether they were normal working days. The questions do not ascertain the normal length of the subject's working week. It was therefore assumed that the average length of the working week for all subjects was 5 days and the denominator for the estimates was obtained by multiplication of the total numbers in each category by 5. It follows that those reporting 6 or 7 days absence must be assumed to have lost only 5 working days in the week. Furthermore, the estimate assumes that when 5 or fewer days unable to work were reported these days were all working days. Consequently the percentages in this table represent a very crude estimate of working days lost.

(iii) Table 13.3 shows, for those reporting a last period of sickness (question 2), the percentage breakdown by duration of that sickness period 1987-1991 by sex, age and occupational group.

(iv) Table 13.4 presents the percentage reporting work limiting health conditions (question 3) and the relative proportion of different health conditions reported by this population. The subjects were asked to select from a list of 11 broad categories of illness, defined by the LFS designers, as listed in the table. Where a respondent reported more than one work limiting health condition, only the condition that was most limiting is included in the analysis. The data are for the years 1987-1991 and broken down by sex and occupational group. Low numbers prevented any age breakdown in this table. After 1988 the further category of 'difficulty in seeing' was added to the questionnaire. Consequently calculation of the denominator for this condition used totals reporting a limiting health condition in the years 1989-1991 and not 1987-1991 as for the other conditions.

13.3 General comments

The generally accepted multifactorial nature of sickness absence is borne out by many of the trends by age and sex in the data presented. Women show higher rates of sickness absence then men and a relatively greater proportion of short-term sickness absence than men (see Figures 13.2 and 13.3). However, a lower percentage of women than men report health problems that limit their work (see Figure 13.7). This could, in part, be explained by general differences in the health experience of men and women whether working or not, for example due to minor gynaecological problems. It is also likely that even within occupational orders the kind of work undertaken by men and women may differ, not only in terms of physical demands but also in responsibilities and commitments. These differences may lead to sickness absence, as defined above, even if they do not actually cause ill

Text continues on page 227

Table 13.2 Estimates of the percentage working days lost through sickness absence: 1987-1991 combined by occupational group, age and sex

Ages	Males		Females		Males		Females		Males		Females	
	% absent	No. absent	% absent	No. absent	% absent	No. absent	% absent	No. absent	% absent	No. absent	% absent	No. absent
	All occupational groups				**I. Professional & related**				**II. Professional education & health**			
16-19	2.1	420,500	2.6	535,300	2.2	5,600	1.6	3,900	2.9	2,600	2.4	12,500
20-24	2.4	916,000	3.2	1,110,400	1.1	26,600	2.6	41,900	2.6	15,700	3.2	104,300
25-29	2.2	896,700	3.2	992,000	1.2	47,700	2.4	52,700	1.3	19,500	3.1	150,800
30-34	2.3	820,200	3.2	863,800	1.1	41,400	2.8	46,600	2.3	47,400	3.2	157,300
35-39	2.2	738,400	3.0	848,400	1.4	49,400	2.6	32,000	1.8	48,000	3.4	186,600
40-44	2.5	869,400	3.4	1,097,800	1.7	60,600	1.6	19,500	1.9	53,000	3.4	203,500
45-49	2.7	761,500	3.9	1,019,900	1.4	32,800	3.2	29,000	1.6	33,900	4.4	201,500
50-54	3.1	766,500	4.2	929,400	1.7	32,100	1.8	10,200	1.9	33,400	4.9	177,500
55-59	4.2	858,000	4.8	811,800	1.9	31,100	2.5	12,400	4.7	60,000	4.8	108,100
60-64	5.5	745,400	NA	NA	2.9	24,500	NA	NA	3.9	28,200	NA	NA
Total	2.7	7,792,800	3.4	8,208,800	1.5	351,800	2.4	247,900	2.2	341,700	3.7	1,302,300
	III. Literary artistic & sports				**IV. Professional & related - science etc.**				**V. Managerial**			
16-19	0.9	1,700	2.4	3,800	1.6	11,000	2.3	3,100	2.2	8,500	3.1	9,700
20-24	1.6	7,000	2.9	11,300	1.8	49,100	1.1	7,200	0.9	20,100	3.2	55,200
25-29	1.8	7,500	1.7	7,600	1.2	41,200	2.6	17,900	1.3	52,300	2.4	41,800
30-34	1.0	4,000	1.9	5,100	1.6	51,400	3.5	15,900	1.6	66,900	1.9	26,000
35-39	0.7	2,100	2.6	6,100	1.5	44,400	0.9	3,100	1.4	68,800	3.1	41,400
40-44	1.1	4,000	2.4	5,700	1.5	43,300	3.7	14,900	1.2	68,700	2.7	41,300
45-49	2.4	6,000	5.3	7,100	2.2	48,700	4.0	9,500	1.5	64,000	3.3	42,000
50-54	0.4	800	3.8	5,700	1.8	31,400	3.4	5,300	1.5	53,400	2.8	24,800
55-59	2.2	3,200	0.0	0	3.6	50,900	9.8	10,000	2.6	65,400	1.8	9,200
60-64	4.2	4,100	NA	NA	4.7	34,200	NA	NA	3.7	52,100	NA	NA
Total	1.4	40,500	2.5	523,00	1.9	405,700	2.7	86,800	1.6	520,200	2.8	291,400
	VI. Clerical & related				**VII. Selling**				**VIII. Security etc.**			
16-19	1.3	41,800	2.5	189,800	1.3	41,800	1.8	83,000	6.2	7,900	0.0	0
20-24	2.7	122,800	3.0	410,900	1.4	34,800	2.6	81,500	1.8	19,000	4.4	10,900
25-29	2.5	82,100	2.8	308,000	1.6	31,900	3.0	80,200	2.1	38,500	4.2	8,300
30-34	2.9	66,800	2.9	236,700	1.5	21,100	3.1	69,900	3.4	53,400	6.3	9,200
35-39	1.9	36,800	2.5	209,800	1.7	19,400	1.7	40,400	3.5	41,500	4.2	5,800
40-44	3.5	65,800	3.0	281,400	1.9	24,600	2.5	71,700	4.0	42,600	2.4	3,000
45-49	3.0	44,600	2.7	205,300	1.6	14,500	2.7	62,100	3.7	31,300	0.3	300
50-54	3.9	63,100	3.3	204,200	2.3	17,400	3.3	65,500	2.5	15,300	6.5	5700
55-59	2.9	47,700	3.9	189,600	2.8	19,300	3.0	44,000	4.1	21,100	3.8	2,600
60-64	4.7	53,400	NA	NA	3.1	14,900	NA	NA	3.4	15,600	NA	NA
Total	2.7	624,900	2.9	2,235,500	1.7	239,700	2.5	598,300	3.1	286,200	4.0	45,700
	IX. Catering, cleaning etc.				**X. Farming, fishing & related**				**XI. Processing etc. (excluding metals & electrical)**			
16-19	1.9	41,000	2.5	104,900	1.7	9,700	1.6	3,100	2.5	51,500	4.9	54,400
20-24	2.3	45,300	3.0	150,600	1.8	18,100	3.6	8,500	2.9	118,000	5.6	102,800
25-29	2.7	39,600	2.6	117,200	2.7	23,700	2.0	3,800	3.5	115,800	7.6	96,900
30-34	2.8	29,800	2.9	156,800	3.4	23,100	1.8	2,300	3.2	77,900	5.9	61,200
35-39	3.6	31,700	3.1	189,400	3.2	14,600	2.6	3,500	3.2	72,100	4.7	49,900
40-44	3.0	26,300	4.1	299,500	2.4	11,100	2.1	3,400	3.9	83,900	5.7	68,800
45-49	4.0	37,100	4.7	310,600	2.1	8,300	3.6	4,300	5.0	93,300	5.7	57,200
50-54	4.4	42,700	4.7	277,900	4.2	17,300	6.9	7,600	3.8	65,000	6.3	56,400
55-59	4.3	44,500	5.6	292,500	6.0	27,400	1.9	1,800	4.5	64,000	9.0	61,500
60-64	7.1	71,800	NA	NA	5.1	22,800	NA	NA	9.5	88,800	NA	NA
Total	3.3	409,800	3.8	1,899,500	3.0	176,100	2.8	38,400	3.8	830,200	6.0	609,100
	XII. Processing etc. (metals & electrical)				**XIII. Printing, assembling etc.**				**XIV. Construction, mining etc. n/e**			
16-19	2.7	91,100	3.4	9,200	1.9	17,600	5.2	47,400	3.2	28,600	16.0	1,800
20-24	2.6	178,800	4.7	22,200	3.0	52,400	5.5	84,800	3.4	61,400	3.7	1100
25-29	2.4	171,100	2.8	9,000	2.9	40,900	7.1	75,700	4.2	68,200	6.4	1,800
30-34	2.3	136,700	7.6	21,400	3.2	34,800	6.1	48,600	4.2	52,600	0.0	0
35-39	2.8	160,200	5.4	13,800	2.9	28,800	5.8	52,500	3.9	38,400	0.0	0
40-44	3.0	166,300	4.7	14,400	3.7	37,700	4.9	52,500	4.5	53,500	0.0	0
45-49	3.2	143,800	6.9	20,500	3.7	35,400	6.0	55,900	3.4	35,600	10.4	1,100
50-54	4.7	182,300	5.8	16,000	4.6	43,000	6.7	54,700	4.2	39,100	29.4	2,400
55-59	4.7	146,200	5.5	9,800	5.0	43,500	7.7	54,900	6.1	50,900	0.0	0
60-64	6.5	127,500	NA	NA	5.6	30,800	NA	NA	8.2	45,100	NA	NA
Total	3.1	1,504,000	5.1	136,300	3.5	364,900	6.0	527,000	4.3	473,500	5.8	8,100

Table 13.2 - *continued*

Ages	Males % absent	Males No. absent	Females % absent	Females No. absent	Males % absent	Males No. absent	Females % absent	Females No. absent	Males % absent	Males No. absent	Females % absent	Females No. absent
	XV. Transport, operating etc.				**XVI. Miscellaneous**							
16-19	3.3	48,300	4.3	6,600	3.2	11,700	3.7	2,100				
20-24	3.2	120400	4.3	13500	3.5	22,900	5.7	4,000				
25-29	2.4	98,100	6.5	18,100	3.1	18,600	6.7	2,200				
30-34	2.6	89,400	2.1	5,300	6.0	23,500	0.0	0				
35-39	2.4	71,400	3.4	7,600	3.2	9,800	13.7	3,800				
40-44	3.0	107,200	5.3	14,500	4.4	15,700	7.4	3,600				
45-49	3.6	112,300	3.5	7,100	5.2	19,200	8.8	6,600				
50-54	3.7	106,100	8.6	15,200	6.6	23,000	0.7	300				
55-59	6.4	165,200	4.6	7,600	3.6	12,000	18.7	7,500				
60-64	6.2	108,100	NA	NA	6.8	21,600	NA	NA				
Total	3.5	1,026,500	4.7	95,300	4.4	178,200	7.2	30,100				

Scaled up values are presented to the nearest 100 and values below 10,000 are considered unreliable. All calculations were performed on unrounded data and therefore some values may not add up as expected.

Table 13.3 Duration of last spell of sickness absence: 1987-1991 by occupational group, age and sex

Ages	Males: Total reporting sickness period	Percentage absent: 1-3 days	4-6 days	1-2 weeks	2-3 weeks	3-4 weeks	4-8 weeks	> 8 weeks	Females: Total reporting sickness period	Percentage absent: 1-3 days	4-6 days	1-2 weeks	2-3 weeks	3-4 weeks	4-8 weeks	> 8 weeks
	All occupational groups															
16-19	160,100	63.3	16.7	5.8	4.5	1.7	5.0	3.0	216,100	65.4	15.2	8.9	3.4	1.3	2.5	3.2
20-24	334,800	59.1	13.9	12.0	4.3	1.6	3.6	5.5	397,700	58.4	14.9	9.8	3.5	2.6	5.2	5.7
25-29	308,700	55.7	15.9	8.4	4.9	2.0	4.8	8.3	334,600	54.8	13.9	8.6	5.5	4.4	3.9	8.8
30-34	264,600	50.8	15.3	9.9	6.8	3.1	5.2	8.8	270,500	45.5	17.8	9.7	4.9	3.3	6.9	11.8
35-39	230,300	48.1	16.2	9.9	4.2	4.0	7.0	10.7	255,100	43.3	14.3	11.0	6.2	3.6	8.3	13.2
40-44	245,500	40.0	13.8	11.0	6.1	4.5	10.1	14.5	313,400	38.6	15.9	9.1	4.8	4.8	10.7	16.1
45-49	203,400	32.2	15.8	7.9	5.6	4.9	9.2	24.4	273,700	34.2	12.9	8.6	5.9	4.0	13.0	21.5
50-54	199,100	29.8	11.4	11.8	6.2	5.3	11.6	23.9	241,000	30.5	11.9	10.8	6.6	5.2	10.2	24.7
55-59	212,900	25.9	9.8	8.1	6.9	3.7	10.2	35.3	200,500	24.1	13.3	10.2	7.6	6.2	10.0	28.5
60-64	177,400	19.5	9.7	9.4	8.0	4.9	9.9	38.5	NA	NA	NA	NA	NA	NA	NA	NA
Total	2,336,800	44.0	14.0	9.6	5.6	3.4	7.3	16.0	2,502,500	45.0	14.5	9.6	5.2	3.9	7.7	14.0
	I. Professional & related - managers & administrators															
16-19	2,200	70.6	0.0	0.0	14.8	0.0	0.0	14.5	2,000	81.7	18.4	0.0	0.0	0.0	0.0	0.0
20-24	13,100	79.1	10.2	0.0	10.8	0.0	0.0	0.0	16,900	62.8	8.6	20.2	6.4	0.0	2.0	0.0
25-29	21,200	75.9	16.2	0.0	3.6	4.2	0.0	0.0	20,300	60.9	9.1	5.8	5.7	0.0	4.3	14.2
30-34	16,500	68.7	9.0	15.7	4.3	0.0	2.4	0.0	14,900	48.1	15.0	6.7	3.4	5.8	5.4	15.6
35-39	17,800	62.0	20.4	7.2	0.0	1.9	4.0	4.6	9,800	41.4	15.6	15.7	6.3	3.8	10.3	7.0
40-44	20,900	55.9	19.3	1.8	1.8	7.0	7.0	7.3	8,300	75.4	8.0	10.0	0.0	0.0	6.5	0.0
45-49	9,700	31.1	36.1	7.5	7.6	0.0	3.7	13.9	8,800	40.0	21.3	4.2	13.1	0.0	7.8	13.6
50-54	8,700	35.2	29.5	11.0	3.3	0.0	8.8	12.2	2,700	33.9	21.0	45.2	0.0	0.0	0.0	0.0
55-59	8,800	42.4	4.2	12.5	15.8	7.9	3.3	13.9	4,200	54.3	8.3	7.3	10.7	0.0	0.0	19.3
60-64	7,100	33.8	8.7	5.1	4.9	6.7	12.9	28.1	NA	NA	NA	NA	NA	NA	NA	NA
Total	126,000	58.9	16.7	5.9	5.0	3.1	3.9	6.6	87,900	55.5	12.4	11.2	5.7	1.4	4.8	9.0
	II. Professional education & health															
16-19	700	49.3	0.0	0.0	50.7	0.0	0.0	0.0	4,800	58.2	7.3	17.9	8.2	0.0	0.0	8.4
20-24	4,400	36.9	10.7	22.5	11.3	0.0	0.0	18.6	35,500	49.0	13.8	12.2	6.8	3.4	3.6	11.1
25-29	10,700	72.8	10.4	16.8	0.0	0.0	0.0	0.0	49,200	53.3	12.4	9.9	6.0	2.8	5.6	10.0
30-34	17,000	59.0	13.9	7.2	6.3	4.5	3.8	5.3	45,500	36.0	20.4	13.4	4.1	5.4	7.1	13.7
35-39	14,600	49.1	13.7	7.3	2.6	2.1	15.0	10.2	53,600	38.1	17.0	10.6	7.9	3.9	8.3	14.3
40-44	19,800	55.6	18.9	10.3	3.5	2.0	5.4	4.3	59,400	39.7	15.1	9.9	2.3	3.5	9.2	20.3
45-49	9,300	38.8	12.9	11.8	3.6	3.4	7.5	21.9	51,200	28.8	12.1	6.8	4.6	5.4	14.2	28.1
50-54	8,700	28.8	9.1	4.5	8.6	8.1	8.9	32.0	46,100	27.6	12.6	9.9	6.9	6.0	9.3	27.8
55-59	14,100	23.3	9.3	2.4	5.8	6.3	13.4	39.5	28,400	27.6	13.0	6.2	4.9	5.8	11.8	30.8
60-64	7,200	24.5	10.6	9.5	10.0	0.0	10.2	35.2	NA	NA	NA	NA	NA	NA	NA	NA
Total	106,500	46.1	12.9	9.1	5.3	3.2	7.5	16.0	373,600	38.0	14.5	10.0	5.4	4.4	8.6	19.0

Table 13.3 - *continued*

Ages	Males								Females							
	Total reporting sickness period	Percentage absent:							Total reporting sickness period	Percentage absent:						
		1-3 days	4-6 days	1-2 weeks	2-3 weeks	3-4 weeks	4-8 weeks	> 8 weeks		1-3 days	4-6 days	1-2 weeks	2-3 weeks	3-4 weeks	4-8 weeks	> 8 weeks
	III. Literary artistic & sports															
16-19	1,400	75.0	25.0	0.0	0.0	0.0	0.0	0.0	2,000	100.1	0.0	0.0	0.0	0.0	0.0	0.0
20-24	2,700	60.3	0.0	0.0	0.0	0.0	39.6	0.0	2,600	40.3	16.3	0.0	12.6	0.0	0.0	30.7
25-29	3,600	80.1	0.0	7.3	0.0	0.0	0.0	12.5	3,500	81.7	0.0	8.3	0.0	0.0	0.0	10.1
30-34	1,800	64.8	18.9	16.4	0.0	0.0	0.0	0.0	1,800	58.8	19.6	0.0	0.0	21.5	0.0	0.0
35-39	700	54.0	0.0	0.0	0.0	46.0	0.0	0.0	2,200	62.0	0.0	0.0	23.0	0.0	15.0	0.0
40-44	1,800	69.3	0.0	0.0	21.7	0.0	9.1	0.0	2,100	50.9	22.6	0.0	0.0	0.0	0.0	26.4
45-49	1,900	53.7	0.0	0.0	0.0	33.1	13.2	0.0	2,300	52.5	15.2	0.0	0.0	0.0	32.3	0.0
50-54	400	100.0	0.0	0.0	0.0	0.0	0.0	0.0	1,500	30.0	22.4	0.0	0.0	29.3	0.0	18.2
55-59	1,500	76.7	23.3	0.0	0.0	0.0	0.0	0.0	0	0.0	0.0	0.0	0.0	0.0	0.0	0.0
60-64	1,000	38.2	0.0	34.9	26.9	0.0	0.0	0.0	NA	NA	NA	NA	NA	NA	NA	NA
Total	16,700	67.3	6.2	5.4	3.9	5.7	8.9	2.7	18,000	61.4	10.7	1.6	4.7	4.5	6.0	11.0
	IV. Professional & related - science etc.															
16-19	4,100	59.2	20.6	0.0	10.7	0.0	9.5	0.0	1,000	100.0	0.0	0.0	0.0	0.0	0.0	0.0
20-24	19,100	70.7	9.1	13.1	2.5	0.0	2.3	2.4	5,400	91.3	0.0	0.0	8.6	0.0	0.0	0.0
25-29	17,400	69.1	12.6	7.7	5.2	0.0	1.0	4.3	7,900	52.6	16.8	15.2	11.1	4.3	0.0	0.0
30-34	20,300	67.2	15.8	7.6	3.8	0.0	1.7	3.9	5,600	66.1	14.3	6.9	0.0	0.0	6.4	6.3
35-39	15,800	57.9	19.3	16.2	0.0	0.0	6.5	0.0	1,000	32.2	0.0	32.7	0.0	0.0	0.0	35.1
40-44	15,500	55.7	18.7	5.1	4.0	0.0	10.3	6.2	4,500	60.1	5.7	7.1	0.0	8.8	10.1	8.2
45-49	14,400	41.1	9.2	7.5	6.9	5.4	3.8	26.0	2,900	63.7	13.5	0.0	0.0	0.0	12.2	10.6
50-54	11,500	58.8	9.9	3.0	8.0	3.2	11.4	5.9	2,200	81.6	0.0	0.0	0.0	0.0	0.0	18.4
55-59	14,300	38.8	7.6	6.6	3.0	3.1	12.0	28.9	2,300	17.5	14.4	0.0	0.0	33.0	0.0	35.2
60-64	10,000	43.4	11.0	10.4	0.0	3.8	10.6	20.8	NA	NA	NA	NA	NA	NA	NA	NA
Total	142,400	57.5	13.0	8.5	3.9	1.4	6.0	9.5	32,900	63.5	9.5	6.8	4.1	4.6	3.6	8.0
	V. Managerial															
16-19	2,900	35.5	36.7	13.7	0.0	0.0	14.1	0.0	3,500	58.8	8.9	19.5	12.8	0.0	0.0	0.0
20-24	10,600	80.9	15.3	0.0	0.0	0.0	0.0	3.8	19,900	59.3	8.0	7.3	5.9	4.3	8.9	6.3
25-29	18,800	58.4	27.8	4.5	2.1	0.0	7.1	0.0	13,500	46.6	18.1	7.2	2.7	9.9	8.7	6.8
30-34	24,400	64.6	9.5	9.7	1.5	2.5	4.6	7.6	8,700	56.0	13.2	4.1	11.0	0.0	2.2	13.5
35-39	21,800	52.0	12.3	4.3	6.6	12.5	8.3	4.0	12,000	44.8	8.3	8.1	2.7	9.6	11.7	15.0
40-44	21,500	50.4	11.5	7.3	6.3	8.8	6.6	9.1	11,700	40.1	24.9	2.6	3.1	0.0	8.3	21.1
45-49	20,400	47.6	24.6	10.3	3.4	5.3	6.6	2.2	10,300	24.3	10.1	19.2	6.9	3.3	10.8	25.4
50-54	15,200	41.6	13.9	6.9	7.1	2.4	9.2	18.9	6,400	35.1	11.5	12.4	4.6	11.6	18.1	6.7
55-59	16,500	28.9	13.4	9.6	7.4	2.0	6.6	32.1	2,100	15.7	0.0	0.0	55.8	0.0	0.0	28.4
60-64	12,200	17.4	5.3	9.8	5.7	11.6	18.7	31.3	NA	NA	NA	NA	NA	NA	NA	NA
Total	164,200	49.6	15.4	7.3	4.4	5.1	7.5	10.7	88,100	45.6	12.7	8.5	6.6	5.0	8.8	12.8
	VI. Clerical & related															
16-19	18,000	70.9	15.6	9.0	0.0	2.3	2.2	0.0	81,000	70.0	17.3	5.9	1.6	0.0	2.3	2.9
20-24	44,500	59.0	14.9	9.4	1.9	2.1	6.1	6.5	156,300	63.7	15.8	8.1	1.8	2.6	3.5	4.5
25-29	26,400	56.4	11.4	8.4	10.7	1.1	6.1	5.9	107,900	56.9	14.5	8.6	6.0	3.5	3.5	7.0
30-34	19,800	42.2	30.4	8.2	7.0	2.1	2.4	7.7	77,500	47.0	19.0	9.1	5.7	3.2	5.8	10.1
35-39	13,200	61.4	8.9	9.4	5.4	0.0	5.8	9.2	67,000	49.3	16.2	10.7	2.8	4.5	7.6	8.8
40-44	17,400	33.9	9.0	13.9	7.0	10.2	15.1	11.0	84,100	42.8	13.9	9.8	4.3	3.7	11.2	14.4
45-49	13,200	47.1	5.3	0.0	2.7	2.7	10.9	31.2	58,100	40.2	10.8	8.5	8.3	1.8	13.1	17.3
50-54	15,400	22.5	5.0	16.6	7.8	11.3	6.8	30.0	56,800	38.2	11.4	9.6	7.4	4.5	12.6	16.4
55-59	12,800	28.5	4.6	3.3	2.6	9.5	9.8	41.7	48,000	25.6	17.3	12.0	9.6	6.2	11.3	18.1
60-64	12,300	12.6	10.2	8.7	8.8	10.3	5.5	43.9	NA	NA	NA	NA	NA	NA	NA	NA
Total	193,000	47.2	12.7	9.0	5.2	4.4	6.7	14.8	736,700	51.6	15.3	8.9	4.6	3.1	6.8	9.6
	VII. Selling															
16-19	17,300	70.7	14.0	4.5	0.0	0.0	7.6	3.2	31,900	64.1	14.7	9.4	1.2	4.5	1.3	4.8
20-24	15,500	56.8	26.0	8.4	3.0	2.6	0.0	3.2	29,300	53.7	17.3	19.1	0.0	1.1	6.1	2.7
25-29	11,500	65.1	15.7	11.3	0.0	0.0	4.3	3.5	26,400	58.1	11.3	5.5	3.1	9.1	1.0	11.9
30-34	7,700	57.0	13.2	8.1	5.9	5.3	5.5	5.0	23,800	52.9	13.5	10.9	4.9	3.6	8.0	6.2
35-39	6,700	59.9	21.9	12.7	0.0	5.5	0.0	0.0	12,500	45.4	17.9	12.1	2.9	2.4	7.6	11.6
40-44	6,600	34.5	10.9	11.1	10.8	6.0	14.8	12.1	18,200	27.1	24.3	5.8	11.6	5.6	7.5	18.1
45-49	3,600	14.2	20.8	9.5	8.3	0.0	27.6	19.6	18,500	46.3	12.2	5.2	5.9	4.0	8.4	18.0
50-54	4,800	39.1	18.3	7.2	0.0	7.3	14.6	13.6	18,300	39.5	12.2	13.8	0.0	7.5	11.1	15.9
55-59	4,300	8.4	22.4	17.2	0.0	0.0	9.2	42.8	11,200	28.8	22.1	13.7	0.0	9.4	10.1	16.0
60-64	3,600	10.9	9.4	9.8	0.0	9.2	0.0	60.6	NA	NA	NA	NA	NA	NA	NA	NA
Total	81,600	51.9	17.6	9.0	2.4	2.8	6.5	9.8	190,100	49.3	15.6	10.6	3.1	5.0	6.0	10.4

Table 13.3 - *continued*

Ages	Males								Females							
	Total reporting sickness period	Percentage absent:							Total reporting sickness period	Percentage absent:						
		1-3 days	4-6 days	1-2 weeks	2-3 weeks	3-4 weeks	4-8 weeks	> 8 weeks		1-3 days	4-6 days	1-2 weeks	2-3 weeks	3-4 weeks	4-8 weeks	> 8 weeks
	VIII. Security etc.															
16-19	2,400	46.6	0.0	0.0	0.0	0.0	0.0	53.4	0	0.0	0.0	0.0	0.0	0.0	0.0	0.0
20-24	8,600	57.3	10.8	23.5	3.8	4.7	0.0	0.0	3,300	70.3	0.0	0.0	9.0	0.0	0.0	20.7
25-29	12,200	55.6	11.5	3.5	0.0	2.8	10.3	16.2	2,300	46.7	0.0	0.0	18.4	0.0	0.0	34.8
30-34	13400	25.4	18.7	14.3	18.4	8.7	5.3	9.2	2,500	36.1	0.0	16.5	0.0	0.0	16.4	31.1
35-39	12,300	42.7	22.9	0.0	3.5	0.0	3.6	27.2	1,300	0.0	0.0	71.3	0.0	0.0	0.0	28.7
40-44	10,100	24.7	0.0	24.4	10.9	7.0	7.4	25.7	1,100	36.1	40.5	0.0	23.5	0.0	0.0	0.0
45-49	7,400	29.3	6.8	0.0	4.3	0.0	10.2	49.5	100	0.0	100.0	0.0	0.0	0.0	0.0	0.0
50-54	3,400	12.1	8.8	27.5	0.0	9.6	0.0	41.9	1,400	52.0	0.0	23.8	0.0	0.0	0.0	24.3
55-59	4,900	18.7	0.0	16.3	15.6	0.0	8.5	40.8	800	39.7	0.0	0.0	0.0	0.0	0.0	60.2
60-64	4,000	29.2	0.0	15.7	0.0	7.7	10.0	37.3	NA	NA	NA	NA	NA	NA	NA	NA
Total	78,800	36.4	10.8	11.7	6.9	4.1	6.0	24.2	12,800	44.7	4.5	13.0	7.7	0.0	3.3	26.8
	IX. Personal service															
16-19	15,000	60.6	17.8	2.6	8.9	5.0	2.4	2.9	43,900	65.0	12.7	9.3	6.1	0.8	4.1	2.1
20-24	19,400	71.1	9.6	4.7	2.5	0.0	4.6	7.5	48,200	51.1	16.7	11.7	4.7	2.6	7.4	5.8
25-29	12,600	46.2	25.2	6.3	3.1	0.0	3.4	15.7	38,300	54.5	11.3	11.5	6.9	5.0	4.1	6.6
30-34	9,100	39.7	14.5	14.1	20.0	3.4	4.2	4.1	50,000	45.9	17.0	11.6	3.0	1.9	6.8	13.8
35-39	9,600	50.5	12.4	13.2	0.0	9.0	3.8	11.2	56,800	43.8	10.4	9.8	9.5	1.8	8.7	16.0
40-44	6,400	40.1	6.1	21.6	0.0	0.0	3.0	29.2	81,500	33.6	16.8	9.3	5.1	5.8	12.0	17.3
45-49	9,800	26.0	14.6	11.1	7.7	0.0	15.0	25.5	78,500	28.5	12.7	11.4	5.8	3.2	12.8	25.6
50-54	9,600	23.8	3.6	19.0	4.1	8.6	14.9	26.1	67,600	23.7	12.0	13.2	6.8	6.0	10.2	28.1
55-59	11,000	24.4	20.9	11.9	6.5	0.0	11.0	25.3	69,400	21.2	10.0	12.7	5.9	7.5	8.9	33.8
60-64	15,200	6.4	19.5	17.1	12.2	0.0	7.8	37.0	NA	NA	NA	NA	NA	NA	NA	NA
Total	117,500	41.0	15.0	10.9	6.6	2.3	6.7	17.5	534,300	37.9	13.3	11.2	6.0	4.1	9.0	18.5
	X. Farming, fishing & related															
16-19	4,400	74.0	26.0	0.0	0.0	0.0	0.0	0.0	1,100	68.7	0.0	0.0	0.0	0.0	0.0	31.2
20-24	7,900	76.0	6.4	0.0	12.1	0.0	5.5	0.0	2,700	54.5	15.9	0.0	0.0	0.0	0.0	29.7
25-29	7,800	62.0	15.4	0.0	5.6	0.0	10.6	6.4	1,500	68.5	0.0	0.0	0.0	0.0	31.3	0.0
30-34	5,600	28.1	10.9	19.6	10.9	0.0	11.2	19.3	1,500	100.1	0.0	0.0	0.0	0.0	0.0	0.0
35-39	4,300	38.3	9.8	15.6	9.4	0.0	8.7	18.3	1,400	73.1	0.0	0.0	0.0	0.0	26.9	0.0
40-44	3,000	42.4	12.9	22.1	0.0	0.0	11.8	10.8	700	0.0	0.0	0.0	0.0	0.0	0.0	100.0
45-49	2,200	34.8	14.9	0.0	16.5	17.4	0.0	16.3	1,100	38.3	0.0	0.0	0.0	61.7	0.0	0.0
50-54	4,100	29.2	0.0	8.6	9.0	0.0	8.9	44.3	1,500	0.0	0.0	0.0	0.0	0.0	0.0	100.0
55-59	7,200	34.1	0.0	10.4	0.0	0.0	26.9	28.5	400	0.0	0.0	0.0	0.0	0.0	0.0	100.0
60-64	5,500	26.3	6.4	7.8	12.6	6.7	6.2	34.1	NA	NA	NA	NA	NA	NA	NA	NA
Total	52,000	47.0	9.5	7.6	7.4	1.4	10.1	16.9	11,800	52.1	3.6	0.0	0.0	5.8	7.2	31.3
	XI. Processing etc. (excluding metals & electrical)															
16-19	19,600	66.1	12.8	8.0	6.4	0.0	4.8	1.7	19,000	57.2	17.8	8.1	7.3	0.0	3.8	5.8
20-24	41,700	54.5	9.8	21.7	3.1	3.2	1.9	5.8	33,600	52.7	17.5	10.5	3.3	2.5	7.0	6.4
25-29	35,200	44.0	18.8	17.8	6.2	2.4	1.1	9.7	30,000	50.8	17.6	9.1	5.7	5.5	4.5	6.9
30-34	23,100	46.9	12.8	9.9	3.2	1.7	11.1	14.5	16,200	35.5	17.1	4.8	7.0	2.2	14.4	18.9
35-39	21,300	45.5	18.2	10.0	5.1	1.8	7.7	11.7	16,100	46.2	10.3	11.5	5.7	3.8	8.5	14.1
40-44	20,300	25.6	8.6	12.7	6.2	5.0	20.4	21.6	20,000	39.3	16.4	3.3	6.9	3.3	18.5	12.3
45-49	23,000	23.8	17.2	8.4	2.1	4.6	9.5	34.3	17,200	40.7	17.2	5.4	0.0	8.2	18.0	10.4
50-54	15,500	23.1	9.7	15.3	10.5	2.1	15.7	23.6	13,100	19.8	8.0	11.9	12.7	4.2	11.1	32.4
55-59	14,900	17.3	6.9	19.3	9.2	0.0	9.7	37.6	13,700	10.8	21.9	12.0	8.2	2.8	9.4	35.0
60-64	19,100	10.9	3.8	9.5	10.1	10.2	9.3	46.2	NA	NA	NA	NA	NA	NA	NA	NA
Total	233,600	38.8	12.4	14.1	5.6	3.1	7.8	18.1	178,900	42.5	16.3	8.5	5.8	3.6	9.9	13.4
	XII. Processing etc. (metals & electrical)															
16-19	30,800	58.0	18.2	3.5	4.9	2.6	7.9	5.0	5,100	63.1	20.6	9.9	0.0	0.0	6.4	0.0
20-24	59,600	53.2	17.6	12.7	5.9	1.6	4.6	4.4	6,600	43.4	18.9	17.1	0.0	6.6	14.0	0.0
25-29	54,300	48.1	12.7	10.7	6.3	3.1	8.8	10.3	4,400	70.0	8.7	11.9	0.0	0.0	0.0	9.3
30-34	44,600	51.4	15.3	6.3	6.7	2.6	6.8	10.8	6,300	48.2	16.6	5.7	4.5	0.0	12.7	12.3
35-39	48,900	43.3	16.1	12.0	2.9	4.9	9.1	11.7	3,600	30.0	44.0	0.0	16.2	0.0	0.0	9.8
40-44	47,100	40.4	14.0	10.3	10.6	2.2	6.6	16.0	4,400	41.8	7.1	9.1	15.8	0.0	6.6	19.6
45-49	37,500	29.0	15.2	9.1	6.3	6.8	8.3	25.3	5,300	38.7	13.2	18.2	8.2	0.0	7.8	13.9
50-54	47,500	27.7	9.7	15.3	4.6	8.4	13.1	21.3	4,000	39.3	7.4	0.0	11.4	0.0	8.8	33.0
55-59	37,400	28.8	11.3	5.7	7.9	2.5	13.8	29.8	3,000	31.1	13.3	13.7	12.0	0.0	0.0	29.8
60-64	31,000	23.2	12.4	2.2	11.8	1.2	6.1	43.2	NA	NA	NA	NA	NA	NA	NA	NA
Total	438,500	41.2	14.3	9.5	6.6	3.6	8.4	16.4	42,700	46.0	16.5	10.0	6.6	1.0	7.3	12.6

Table 13.3 - *continued*

Ages	Males								Females							
	Total reporting sickness period	Percentage absent:							Total reporting sickness period	Percentage absent:						
		1-3 days	4-6 days	1-2 weeks	2-3 weeks	3-4 weeks	4-8 weeks	> 8 weeks		1-3 days	4-6 days	1-2 weeks	2-3 weeks	3-4 weeks	4-8 weeks	> 8 weeks
	XIII. Printing, assembling etc.															
16-19	7,000	65.6	19.3	5.3	9.8	0.0	0.0	0.0	17,200	52.8	16.6	17.8	4.3	6.6	0.0	2.0
20-24	19,100	62.6	15.0	9.8	2.6	4.2	0.0	5.7	30,800	59.6	16.4	2.7	4.5	1.6	9.5	5.8
25-29	14,500	53.5	18.4	7.6	9.1	0.0	2.6	8.7	22,900	46.1	20.9	4.6	4.8	6.7	4.1	12.7
30-34	10,900	53.9	18.2	3.6	3.1	8.1	3.9	9.1	13,700	42.0	24.4	8.9	7.3	4.3	6.1	6.9
35-39	8,100	37.0	14.5	12.9	4.4	9.2	4.7	17.2	13,800	31.2	16.1	16.0	5.3	3.6	7.0	20.8
40-44	8,200	10.3	7.8	32.5	8.8	8.4	18.3	13.9	13,200	24.6	16.2	21.1	3.8	18.1	10.9	5.4
45-49	8,700	23.9	25.3	9.3	7.2	0.0	7.9	26.4	15,000	26.7	18.4	2.1	3.4	9.6	16.0	23.7
50-54	10,300	20.1	12.2	10.9	9.7	0.0	13.7	33.4	14,700	33.9	15.8	4.9	8.2	0.0	5.4	31.8
55-59	9,600	11.5	22.5	3.9	3.6	8.0	11.6	38.9	13,400	28.0	6.8	2.7	8.5	3.2	15.1	35.6
60-64	7,400	19.3	10.4	16.9	11.0	9.7	5.4	27.2	NA	NA	NA	NA	NA	NA	NA	NA
Total	103,800	39.2	16.5	10.6	6.5	4.4	6.1	16.7	154,700	41.4	17.1	8.1	5.4	5.5	8.0	14.6
	XIV. Construction, mining etc. n/e															
16-19	13,100	73.7	16.3	5.4	2.2	2.5	0.0	0.0	700	100.1	0.0	0.0	0.0	0.0	0.0	0.0
20-24	20,100	55.9	12.9	6.6	10.4	0.0	7.0	7.2	500	100.2	0.0	0.0	0.0	0.0	0.0	0.0
25-29	21,900	47.0	4.0	8.5	5.9	7.6	9.6	17.4	400	100.0	0.0	0.0	0.0	0.0	0.0	0.0
30-34	17,800	52.2	9.4	6.9	13.4	3.1	2.3	12.8	0	0.0	0.0	0.0	0.0	0.0	0.0	0.0
35-39	11,000	33.2	24.4	14.0	9.6	0.0	3.4	15.5	0	0.0	0.0	0.0	0.0	0.0	0.0	0.0
40-44	14,500	41.8	10.8	7.8	2.2	2.7	13.4	21.3	0	0.0	0.0	0.0	0.0	0.0	0.0	0.0
45-49	9,300	34.9	12.7	11.3	4.9	4.7	7.5	24.0	400	100.3	0.0	0.0	0.0	0.0	0.0	0.0
50-54	9,100	22.3	4.1	5.0	8.0	4.0	25.7	30.9	500	0.0	0.0	0.0	0.0	0.0	0.0	100.0
55-59	11,100	9.6	7.8	13.4	10.3	6.5	9.5	43.0	0	0.0	0.0	0.0	0.0	0.0	0.0	0.0
60-64	10,800	18.9	5.6	14.3	6.5	0.0	17.1	37.5	NA	NA	NA	NA	NA	NA	NA	NA
Total	138,600	42.2	10.5	8.9	7.6	3.2	8.8	18.9	2500	80.9	0.0	0.0	0.0	0.0	0.0	19.2
	XV. Transport, operating etc.															
16-19	18,200	56.9	14.4	12.9	5.2	0.0	8.5	2.1	2,200	51.8	15.2	33.0	0.0	0.0	0.0	0.0
20-24	40,800	52.2	14.8	18.0	3.4	1.1	1.9	8.6	5,200	59.4	9.4	0.0	7.6	6.3	7.7	9.6
25-29	34,200	55.8	21.4	5.9	3.5	1.4	2.3	9.8	5,300	41.2	27.6	15.0	0.0	0.0	0.0	16.1
30-34	25,400	38.2	15.8	13.1	6.0	5.8	8.7	12.5	2,100	51.9	19.4	13.4	15.2	0.0	0.0	0.0
35-39	21,100	44.8	10.1	9.6	9.4	3.1	7.1	15.9	2,000	35.5	0.0	0.0	19.8	0.0	0.0	44.6
40-44	26,800	26.1	23.8	8.2	5.0	1.1	13.4	22.5	3,600	27.5	14.2	18.0	12.3	10.8	6.8	10.4
45-49	27,900	24.2	11.4	7.5	9.2	4.7	13.6	29.5	2,000	39.1	18.1	32.2	0.0	0.0	0.0	10.7
50-54	29,700	32.0	18.3	10.6	5.0	4.2	7.1	22.9	3,900	10.1	19.8	0.0	9.6	0.0	14.8	45.7
55-59	40,700	26.5	6.3	5.6	6.2	4.5	6.3	44.6	1,700	28.5	18.8	0.0	0.0	0.0	19.0	33.7
60-64	24,800	15.5	9.1	9.3	5.7	4.7	9.3	46.5	NA	NA	NA	NA	NA	NA	NA	NA
Total	289,600	37.2	14.5	10.0	5.7	3.1	7.3	22.3	28,000	38.7	16.6	11.0	6.9	2.6	5.5	18.6
	XVI. Miscellaneous															
16-19	3,100	36.1	42.5	0.0	0.0	13.0	8.4	0.0	700	50.1	0.0	0.0	0.0	0.0	49.9	0.0
20-24	7,000	51.0	13.3	17.6	0.0	0.0	11.1	6.9	800	0.0	0.0	46.0	0.0	53.8	0.0	0.0
25-29	6,700	54.7	32.8	0.0	0.0	0.0	5.1	7.4	700	50.3	0.0	0.0	0.0	49.9	0.0	0.0
30-34	7,300	37.6	25.2	21.7	6.6	0.0	0.0	8.9	0	0.0	0.0	0.0	0.0	0.0	0.0	0.0
35-39	2,800	33.0	27.0	15.0	12.2	0.0	0.0	12.8	1,000	36.2	0.0	36.4	0.0	0.0	27.2	0.0
40-44	4,700	46.2	8.3	14.9	0.0	16.4	0.0	14.2	700	0.0	0.0	0.0	0.0	50.5	0.0	49.5
45-49	4,700	24.0	23.0	8.5	0.0	23.8	6.6	14.0	2,100	45.3	0.0	0.0	17.0	0.0	17.9	19.8
50-54	4,900	6.9	12.1	8.1	6.3	0.0	16.4	50.1	300	100.0	0.0	0.0	0.0	0.0	0.0	0.0
55-59	2,900	12.2	33.3	0.0	11.6	4.3	9.9	28.7	1,600	0.0	0.0	0.0	49.8	0.0	25.3	24.9
60-64	5,700	24.8	18.6	6.3	0.0	0.0	30.7	19.6	NA	NA	NA	NA	NA	NA	NA	NA
Total	49,700	35.0	22.3	10.3	2.9	4.9	9.1	15.5	7,900	29.2	0.0	9.2	14.4	14.7	17.7	14.8

Scaled up values are presented to the nearest 100 and values below 10,000 are considered unreliable. All calculations were performed on unrounded data and therefore some values may not add up as expected.

Table 13.4 Health problems which limit work: Percentage with limiting health problem and relative percentage by condition, 1987-1991 combined by occupational group and sex

	All occupational groups		I. Professional & related		II. Professional education & health		III. Literary artistic & sports		IV. Professional & related - science		V. Managerial	
	Males	Females	Males	Females	Males	Females	Males	Females	Males	Females	Males	Females
All persons of working age	57,737,100	47,834,400	4,843,100	2,027,400	3,122,900	7,102,900	561,600	414,400	4,362,100	632,000	6,604,400	2,113,800
All with health problems which limit work	4,615,600	3,448,800	276,200	90,000	215,700	382,300	37,200	22,300	255,500	31,700	409,600	109,700
Percent with health problems which limit work	8.0	7.2	5.7	4.4	6.9	5.4	6.6	5.4	5.9	5.0	6.2	5.2
Distribution of different health problems (as a percentage of all those reported)												
Problems or disability connected with arms, legs, hands, feet, back or neck	34.4	37.7	31.7	42.6	33.7	43.0	33.4	29.9	30.2	37.6	37.7	42.6
Difficulty in seeing **	5.9	2.1	7.3	1.2	6.9	2.1	5.8	8.4	6.9	3.9	6.4	3.1
Difficulty in hearing	3.6	3.3	4.3	2.7	3.9	2.8	1.0	3.3	3.3	6.5	2.3	2.3
Skin conditions, allergies	4.5	6.3	4.4	7.0	4.8	5.2	4.9	15.2	6.0	7.2	3.7	3.8
Chest or breathing problems,asthma or bronchitis	11.1	10.5	10.4	11.1	8.1	8.6	9.4	12.3	11.0	13.9	9.3	14.4
Heart, blood pressure or blood circulation problems	10.3	6.9	12.1	4.9	10.6	7.3	13.3	4.2	10.9	5.7	13.0	4.8
Stomach, liver, kidney or digestive problems	3.5	3.4	3.3	4.2	3.9	2.2	5.0	3.3	2.9	0.8	3.3	4.1
Diabetes	4.0	2.2	5.2	2.7	4.1	1.5	4.2	0.0	5.8	3.9	4.5	4.3
Depression, bad nerves	1.8	3.3	1.2	0.8	2.5	3.2	3.5	1.4	1.6	2.1	1.1	2.0
Epilepsy	2.4	2.2	2.7	2.3	2.3	1.3	1.8	2.1	2.1	0.0	0.9	0.6
Other health problems or disabilities	5.6	7.2	4.0	9.1	5.4	8.5	7.7	15.4	5.7	6.5	5.8	8.8

	VI. Clerical & related		VII. Selling		VIII. Security etc.		IX. Personal service		X. Farming, fishing & related		XI. Processing etc. (excl. metals & elec.)	
	Males	Females	Males	Females	Males	Females	Males	Females	Males	Females	Males	Females
All persons of working age	4,624,500	15,367,800	2,871,400	4,754,100	1,858,900	226,900	2,477,600	10,072,200	1,161,700	272700	4,427,000	2,021,100
All with health problems which limit work	442,800	928,900	186,800	362,500	130,600	17,800	298,100	1,016,900	109,000	16,600	400,000	194,500
Percent with health problems which limit work	9.6	6.0	6.5	7.6	7.0	7.9	12.0	10.1	9.4	6.1	9.0	9.6
Distribution of different health problems (as a percentage of all those reported)												
Problems or disability connected with arms, legs, hands, feet, back or neck	32.8	38.3	25.9	35.2	42.0	41.6	30.0	36.1	34.7	35.0	32.1	36.5
Difficulty in seeing **	7.5	2.3	9.7	1.5	6.2	11.5	5.8	1.8	3.4	2.3	5.8	1.0
Difficulty in hearing	2.4	3.9	2.9	2.3	2.9	0.0	4.4	3.0	4.8	4.5	4.5	5.4
Skin conditions, allergies	5.6	6.4	6.5	7.6	2.3	0.0	5.0	6.1	3.0	4.3	4.2	6.6
Chest or breathing problems,asthma or bronchitis	11.5	10.1	17.4	13.6	4.9	10.6	12.8	9.3	11.3	21.5	12.0	9.6
Heart, blood pressure or blood circulation problems	8.3	5.6	6.9	6.8	15.4	3.2	9.9	8.6	7.5	8.6	9.7	6.2
Stomach, liver, kidney or digestive problems	3.0	3.7	2.7	2.8	2.6	12.3	3.8	3.6	2.5	0.0	4.7	4.1
Diabetes	4.4	2.3	4.7	1.6	3.4	0.0	3.8	2.5	2.8	0.0	4.6	2.9
Depression, bad nerves	2.0	2.5	2.2	3.4	1.7	3.8	2.5	4.4	2.3	0.0	1.1	4.0
Epilepsy	2.8	2.3	2.9	2.5	2.1	0.0	3.8	2.5	5.5	2.3	3.3	2.6
Other health problems or disabilities	6.8	7.1	7.0	7.4	4.0	3.5	6.9	6.9	6.5	10.2	4.8	5.8

	XII. Processing etc. (metals & electrical)		XIII. Printing, assembling etc.		XIV. Construction, mining etc. n/e		XV. Transport, operating etc.		XVI. Miscellaneous	
	Males	Females	Males	Females	Males	Females	Males	Females	Males	Females
All persons of working age	9,599,800	530,900	2,097,200	1,748,200	2,222,500	27,900	5,923,000	406,300	808,100	83,500
All with health problems which limit work	695,900	48,300	233,400	173,200	205,200	2,800	601,400	38,300	106,200	11,600
Percent with health problems which limit work	7.2	9.1	11.1	9.9	9.2	9.9	10.2	9.4	13.1	13.9
Distribution of different health problems (as a percentage of all those reported)										
Problems or disability connected with arms, legs, hands, feet, back or neck	38.6	41.2	32.7	32.9	37.5	22.3	36.7	37.8	31.4	44.4
Difficulty in seeing **	4.6	1.6	5.7	2.0	4.0	13.7	4.6	5.4	4.6	0.0
Difficulty in hearing	4.5	2.9	3.9	3.9	3.2	0.0	3.5	0.6	5.6	0.0
Skin conditions, allergies	4.9	7.1	4.8	7.6	3.8	19.2	3.8	7.0	3.9	0.0
Chest or breathing problems,asthma or bronchitis	11.0	15.4	12.9	11.5	12.6	0.0	11.0	15.9	8.7	8.8
Heart, blood pressure or blood circulation problems	9.2	5.1	11.1	8.2	8.1	17.5	11.4	3.5	9.4	9.2
Stomach, liver, kidney or digestive problems	2.9	4.2	4.4	2.8	4.7	17.4	4.1	3.2	3.2	5.8
Diabetes	3.6	1.2	1.8	1.5	2.9	12.2	3.5	2.6	2.8	0.0
Depression, bad nerves	1.7	3.1	1.8	2.6	2.2	0.0	1.6	3.2	3.4	7.3
Epilepsy	1.3	0.5	1.9	2.8	2.8	0.0	2.0	1.0	5.3	2.3
Other health problems or disabilities	5.2	2.6	5.8	7.1	4.3	0.0	4.9	7.6	9.9	10.6

Scaled up values are presented to the nearest 100 and values below 10,000 are considered unreliable. All calculations were performed on unrounded data and therefore some values may not add up as expected.
** Added after 1988. Consequently the denominator for this percentage is calculated from the totals reporting a health limiting condition, 1989-1991.

Figure 13.2 Percentage reporting at least one day absence in the previous week by occupational order and sex

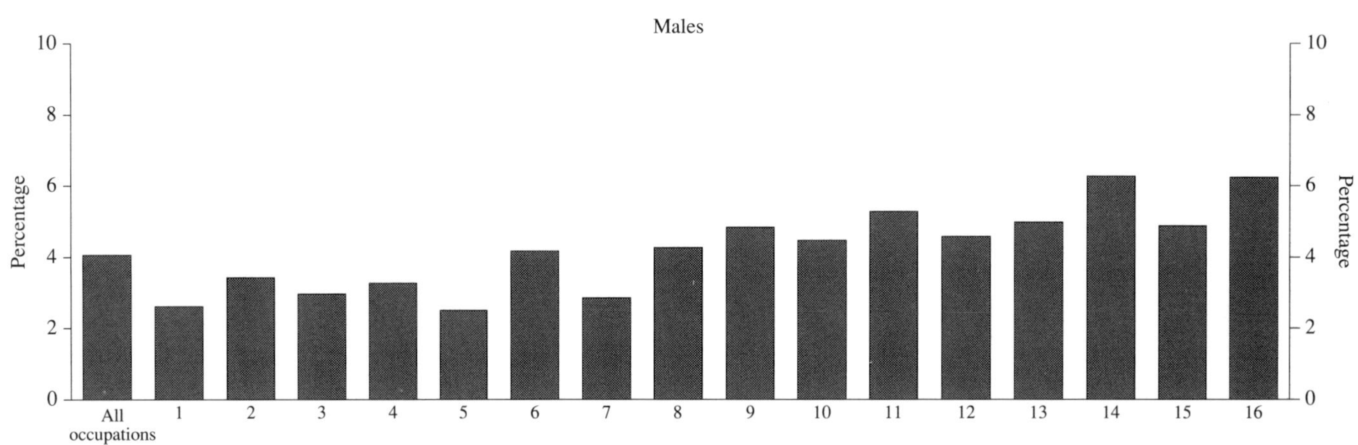

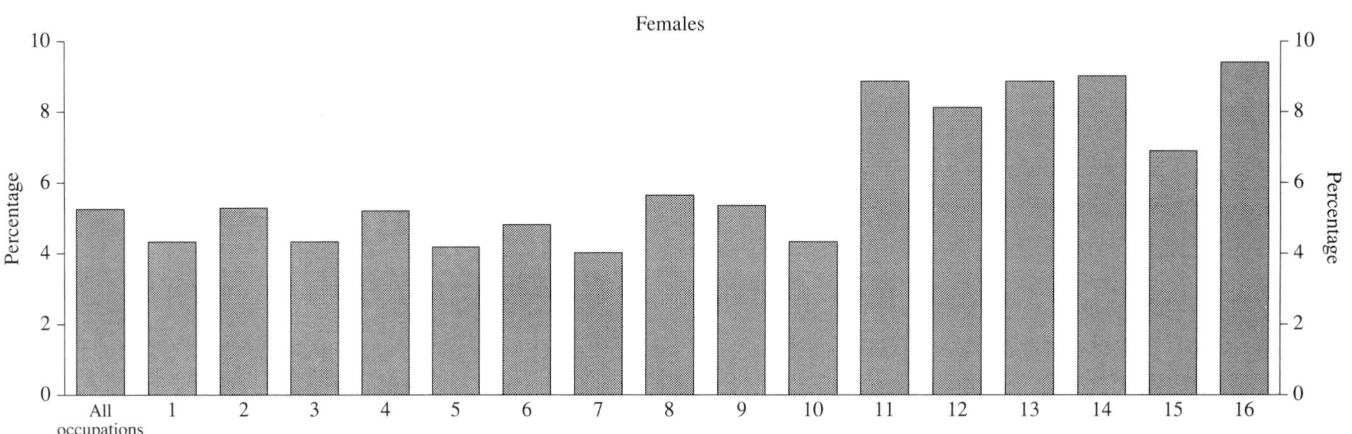

1. Professional
2. Prof. ed. and health
3. Arts and sport
4. Professional science
5. Management
6. Clerical
7. Selling
8. Security
9. Personal service
10. Agriclture etc.
11. Materials proc
12. Metal and electrical
13. Assembly
14. Construction
15. Transportation
16. Miscellaneous

Figure 13.3 Length of last absence period for all those reporting their last period of absence

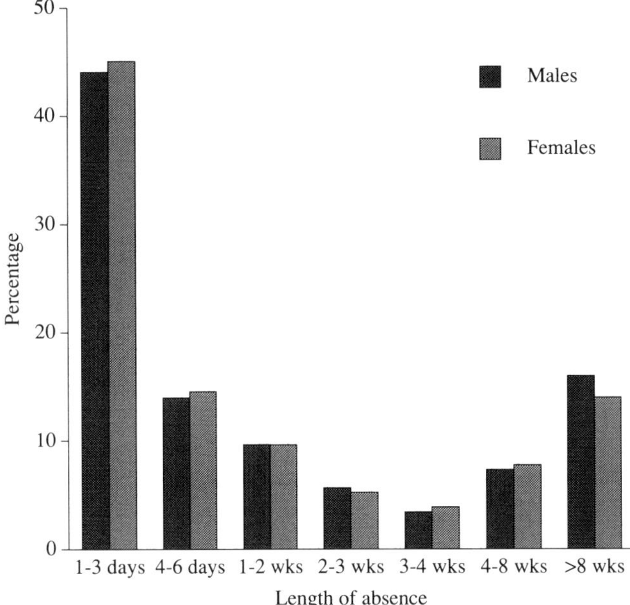

health. Finally the generally greater share of childcare and family responsibilities undertaken by women[4] may also contribute to the higher rates of sickness absence. Some of these reasons for absence might be expected to result in spells of shorter duration than those directly attributable to illness or injury and there is some evidence that the ratio of short-term to long-term sickness absence is higher in women than men (see Figure 13.4 and Table 13.3). Examples of all of these possible differences will be discussed under the headings for specific orders.

Socio-economic factors are known to influence sickness absence.[1,2] Socio-economic status, which is related to occupation, has been shown to be inversely related to health. In general, lower socio-economic status correlates with poor health whereas high socio-economic status tends to be associated with better health.[5] Greater sickness absence would therefore be expected in occupations with low socio-economic status. Women generally tend to work in occupations with lower socio-economic status than men, although for married women this effect is probably offset by the fact that their socio-economic status is defined by their husband's occupation.

Examination of the data by age shows that sickness absence is high in the younger age bands and highest in the older age bands (see Figure 13.5). However, these age bands differ markedly in the relative length of sickness absence periods. Young age bands having a very high percentage of short-term absence relative to long-term absence, whereas in the older age bands the trend is toward more long-term absence (see Figure 13.6). This indicates that the aetiology of sickness absence varies across age strata. Possibly chronic health conditions, more prevalent with age, are the reason behind this trend in the older age bands. Among the young age bands sickness absence may be dependent on a number of other factors not directly related to health, such as fewer commitments and responsibilities both personal and occupational, lower social status and perhaps a higher incidence of minor accidents.

Examination of the length of sickness absence indicates that a very high proportion of sickness absence spells are of less than 6 days and the majority are in the 1-3 day category (see Figure 13.3). The ability to self-certify sickness periods of

Figure 13.4 Ratio of absence lengths, 1-6 days: 8 or more weeks by sex and occupational order

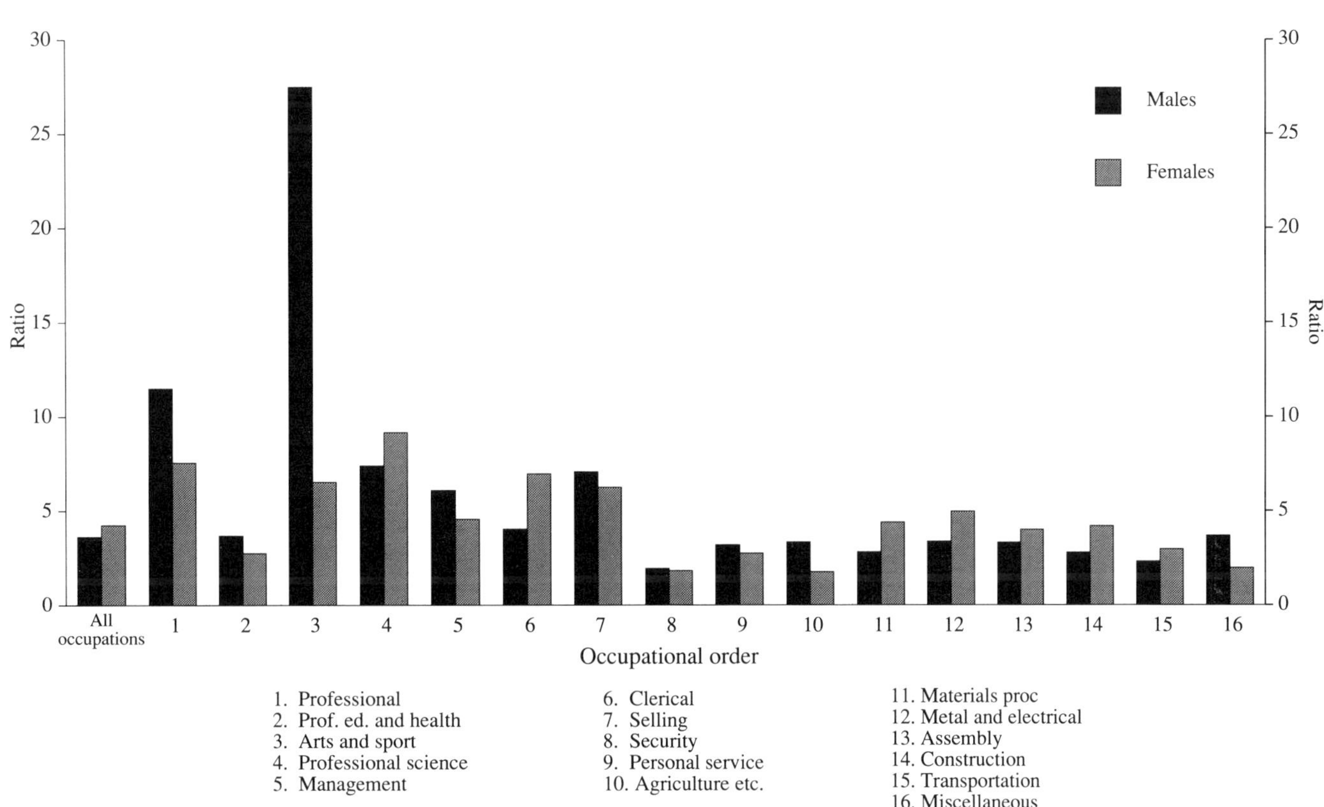

Figure 13.5 Percentage reporting at least one day absence by age

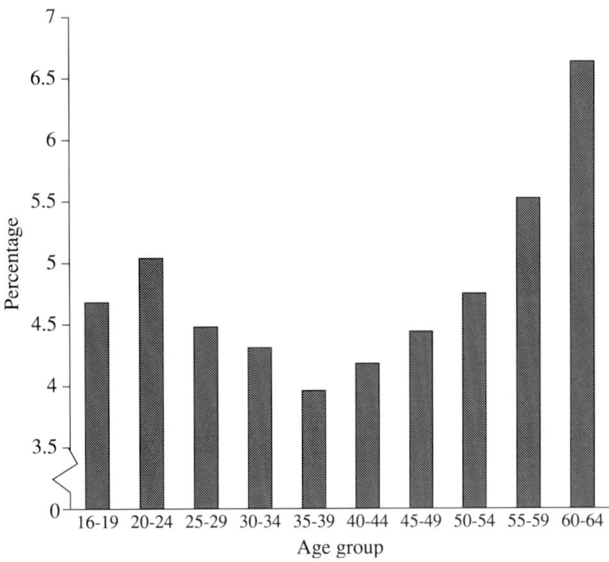

Figure 13.6 Ratio of absence lengths, 1-6 days: 8 or more weeks by age

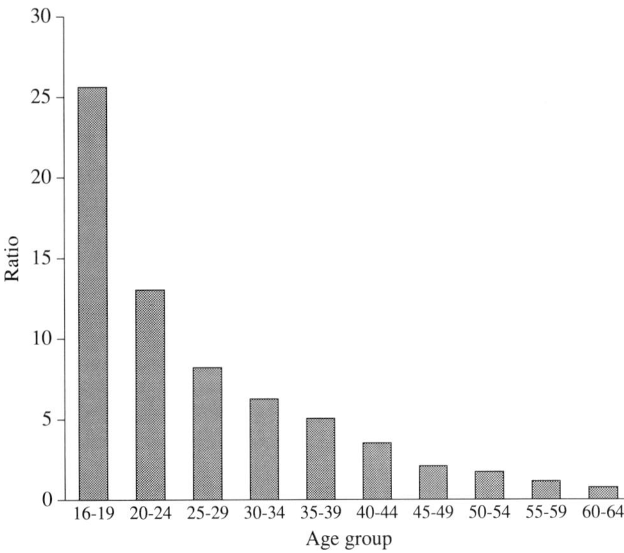

less than 7 days probably has a large influence on this skewed distribution. As well as responsibilities outside the working environment previously mentioned other factors such as job dissatisfaction, stress or lack of general well-being may increase the relative incidence of short-term sickness absence.

Table 13.4 shows the data on work-limiting health conditions. Musculoskeletal conditions, respiratory disease and cardiovascular disorders are the most common limiting conditions. Comments on the differences seen in specific occupations, where appropriate, are made in the paragraphs on the relevant occupations. It is interesting to compare the overall pattern with the results from a recent HSE survey of self-reported work-related health in which subjects were asked to report conditions caused or made worse by work.[6] The relative order of complaints and conditions, although not identical, is crudely similar except for conditions related to stress and depression. This suggests that the overall distribution of illness perceived as caused by work is similar to the distribution of illness, arising from whatever cause, which limits ability to work. In the HSE survey 'stress/depression' rated second after musculoskeletal conditions as a problem which was directly caused by or made worse by work. However, the figures presented here suggest that depression and 'bad nerves' are not major limitations on work relative to other conditions. It may well be that depression and 'bad nerves' are not perceived as a major limitation on work ability, although it is considered that the problem is caused or made worse by work. Possibly it is more acceptable to people to admit that work causes them stress or depression than to report that their mental state affects work performance.

13.4 Comments by occupational order

13.4.1 Order 1 – Professional and related

With respect to self-reported sickness absence this is one of the healthiest occupational orders. Relative to the other occupational orders sickness absence is low for both sexes and among all age groups (see Figure 13.2). The relative length of sickness absence appears distributed more towards shorter than longer spells of absence than for most other occupational orders (see Figure 13.4) and the percentages reporting health conditions that limit work are the lowest for any occupational order (see Figure 13.7).

This order includes the legal profession, the financial professions and senior managers and administrators. The low sickness absence figures may relate to favourable working conditions and generally good health status among these workers. The comparatively high socio-economic status of these workers would be expected to be correlated with better than average health. The relative low levels of long-term sickness, suggesting low levels of chronic debilitating diseases, and the low level of reported work limiting health conditions support this expectation.

13.4.2 Order 2 – Professional education and health

This occupational order comprises teachers, doctors, nurses and other health and welfare professionals. Approximately twice the number of females to males are employed in these occupations in the UK. The percentage sickness absence of at least 1 day in the reference week for both males and females is close to the average for all the working population (see Figure 13.2). Compared to all occupational orders the distribution of length of sickness absence appears more distributed towards longer spells of sickness absence for the females only (see Figure 13.4). This may represent possible chronic health problems among female dominated occupations within this order, the most prominent of which is nursing. However, the percentage reporting health limiting problems especially among females is low compared with the average for all occupational orders (see Figure 13.7). The relative distribution of 'most problematic health limiting condition' shows a high percentage of problems or disability connected with arms, legs, hands, feet or neck among females compared with other orders. It is possible that a significant proportion of the longer periods of sickness absence found in women are related to these musculoskeletal

Figure 13.7 Percentage reporting work limiting health conditions by sex and occupational order

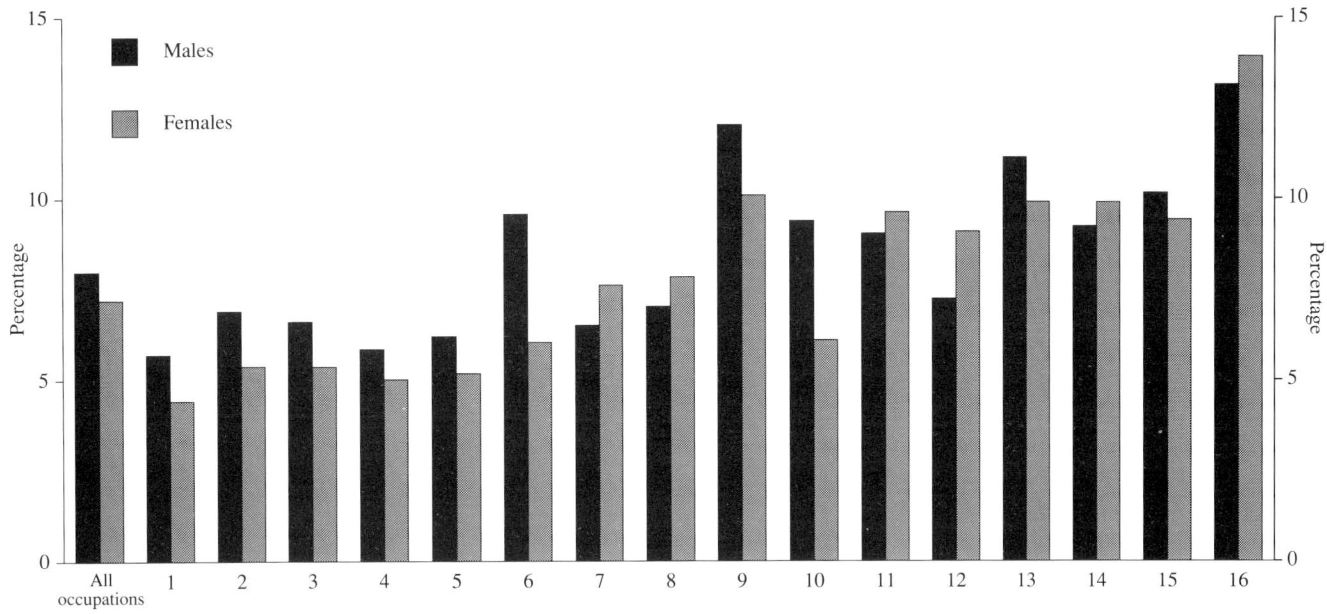

See Figure 13.2 for key to occupational orders.

problems. There is other evidence to suggest that these problems are more common among nurses who are one of the largest female dominated occupational units in this order.[6]

13.4.3 Order 3 – Arts and Sport

Occupations in this order include sports professionals, photographers, journalists and performing artists, with the latter two groups comprising the majority of workers in this order. The occupational order is small preventing firm conclusions on many aspects of the data because of statistical uncertainty. Generally sickness absence appears to be low among both males and females (see Figure 13.2) and the percentage reporting limiting health conditions is also relatively low (see Figure 13.7).

13.4.4 Order 4 – Professional science

This male dominated order of engineers and scientists has relatively low levels of sickness absence (see Figure 13.2). A high relative proportion of short-term sickness compared to long-term sickness suggests a low prevalence of chronic diseases and disabling conditions in this order (see Figure 13.4). The comparatively low percentage reporting work limiting health conditions supports this hypothesis (see Figure 13.7). This effect is possibly related to the relatively high socio-economic standing of the occupations within this order.

13.4.5 Order 5 – Managerial

The main criteria for inclusion in this occupational order is managerial status. As such the order represents workers from a diverse range of occupations who are distinguished from other workers in similar occupations by higher socio-economic status and greater responsibility and control over their work.

The very low sickness absence in this group (see Figure 13.2) may be a reflection of greater responsibility at work and higher socio-economic status. The distribution of length of sickness absence among men appears to be roughly similar to the average for all occupations. However, among women there is relatively more long-term sickness (see Figure 13.4). The reasons behind this pattern are unclear. No indication is provided from examining the data on work limiting health problems. Among both sexes the prevalence of these conditions is low and no unusual distribution across specific conditions is evident.

13.4.6 Order 6 – Clerical

This is a large female dominated occupational order and contributes nearly a third of all the data for females and a significant proportion of the male data. Consequently the pattern in the data would not be expected to vary significantly from the average for all occupations.

Despite the overall similarity of the data to the general pattern seen for all occupations some interesting points are evident. There is little difference between male and female sickness absence in this group compared to other occupational orders (see Figure 13.2). This may be because the occupations concerned are female dominated resulting in increased willingness on the part of employers to implement flexible working arrangements that accommodate the greater family responsibilities of female employees. There may also be greater parity in socio-economic status between the sexes in this order resulting in generally similar working conditions and health profiles. The percentage reporting work limiting health problems appears quite high among men relative to all occupational groups and quite low among women (see Figure 13.7). However, the relative proportions of work-limiting health problems appear to be similar to those affecting all occupational groups in both sexes. This may simply be a reflection of the parity in socio-economic status between sexes in this order.

13.4.7 Order 7 – Selling

This is again a female dominated order but the distribution of specific occupations by sex within the order differs markedly. The GHS data quoted in Chapter 12, smoking and alcohol, show that the majority of women in the order are shop assistants while for men about half are sales representatives. This, unlike the previous order, suggests a significant difference in socio-economic status between the sexes.

Interestingly, the percentage sickness absence is very low both for males and females in this order (see Figure 13.2). The causes of this effect may vary between the sexes. The highly competitive nature of employment as a sales representative, which is a predominantly male occupation, might reduce sickness absence in men. Among women, a relatively high percentage of part-time employment, for example in sales assistants, may alleviate much of the pressure felt from the increased burdens of family responsibility. Among both sexes the tendency is towards greater short-term rather than long-term sickness when compared to all occupational groups (see Figure 13.4). This suggests relatively low levels of chronic health problems. Further support for this possibility comes from the low level of reported work limiting health conditions among men. However, among women the rates are slightly above the average for all occupational groupings (see Figure 13.7). The higher rates in women than men are possibly related to socio-economic differences, although the relative size of this effect in women conflicts with the sickness absence pattern discussed above. No clues as to the causes of this effect are evident in the relative proportions of 'most problematic work limiting health condition' among women, which appear similar to the proportions presented for women in all occupational groups.

13.4.8 Order 8 – Security

This is a small and very male dominated occupational order, consequently little statistical confidence can be placed on the female data. The level of sickness absence among these workers is similar to the average for all occupational groups (see Figure 13.2). The data on length of sickness period in men appear to suggest a relatively high proportion of long-term sickness absence which may indicate a high prevalence of chronic health conditions (see Figure 13.4). However, the small numbers suggest that this may be a sampling effect. The percentage reporting work limiting health conditions in men is below the average for all occupations, which supports this view (see Figure 13.7). Among health conditions that were considered most limiting to work the relative proportion of heart conditions among men was high. A recent HSE survey of self-reported work related illness[6] found that among all occupational orders security produced, for heart disease, the strongest relationship for either being caused by or made worse by work. Both these sets of data suggest a high proportion of heart disease in this occupational order. This cannot be explained by smoking habit as the data (see Chapter 12) show that men in this order smoke proportionally less than the general population. Furthermore, Table 13.4 shows that the percentage who report respiratory problems as limiting their ability to work is low.

13.4.9 Order 9 – Personal service

This order comprises catering workers, including bar staff, cleaners, caretakers and other professions providing personal services. The order is female dominated and the occupations within the order suggest little difference between the socio-economic status of males and females unlike many of the previous occupational orders.

Sickness absence appears similar to the average level for all occupations for females and slightly above the average for males, resulting in similar levels for both sexes (see Figure 13.2). This could be because for males, relative to other occupational orders, this order contains a high proportion of workers of low socio-economic status. The data on length of sickness period suggest a relatively high proportion of long-term sickness absence which may indicate a high prevalence of chronic health conditions (see Figure 13.4). This is further supported by the high percentage of both sexes reporting work limiting health conditions (see Figure 13.7). This trend may be related to low socio-economic status and lifestyle factors such as smoking and drinking that are relatively highly prevalent in this occupational order as shown in Chapter 12.

13.4.10 Order 10 – Agricultural workers

This relatively small occupational order includes horticulturists, foresters, fishermen, farm workers and foremen, although self-employed workers in these categories would be included in order 5. Sickness absence appears to be similar to the average for all occupations in men and slightly less than the average in women (see Figure 13.2). Indeed the sexes in this order have similar rates of sickness absence which again may be related to the similarity in socio-economic status between sexes. The data on length of sickness absence suggest that there is a relatively high proportion of long-term sickness absence (see Figure 13.4). Also a relatively high percentage of work limiting conditions are reported in men (see Figure 13.7). This suggests a high prevalence of chronic health conditions among these workers. However, the greater physical requirements of these occupations could result in work incapacity occurring with more minor deterioration in health than would be expected in other occupational orders. The smoking and drinking data suggest that these habits are an unlikely explanation for chronic ill-health in this order.

13.4.11 Order 11 – Materials processing (other than metal and electrical)

This large order of manual workers encompasses a wide range of occupations including wood, paper, glass, plastics and textile workers and those employed in the food processing industry. This order has one of the highest percentages of sickness absence, particularly among female employees (see Figure 13.2). This could, in part, be due to female dominated occupational units within this order, e.g. textile workers, having lower socio-economic status relative to their male counterparts. A relatively high proportion of long-term sickness among both sexes suggests a high prevalence of chronic health conditions in this order. This is supported by the high proportion of workers reporting health conditions that limit their work (see Figure 13.7) and may be partly accounted for by the low socio-economic status of workers in this order.

13.4.12 Order 12 – Materials processing (metal and electrical)

This is a large male dominated order of manual workers in manufacturing industries. The level of sickness absence is high but lower than the two large orders 11 and 13 which contain the majority of the remaining industrial manual workers (see Figure 13.2). Among males, the percentage reporting work limiting health conditions is less than the average for all occupations and considerably less than orders 11 and 13 (see Figure 13.7). There is no obvious explanation for this in the data but orders 11-13 are all rather heterogeneous and it is difficult to comment in more detail in the absence of information on occupational units.

13.4.13 Order 13 – Repetitive assembly, painting, etc.

This is a large order containing workers involved in repetitive work such as painting, assembling and packaging. This order has a high level of sickness absence for both sexes (see Figure 13.2). The proportion of long-term sickness absence is also high relative to short-term sickness indicating a high prevalence of chronic health problems (see Figure 13.4). This is supported by the very high percentages of workers reporting work limiting health conditions (see Figure 13.7). This type of repetitive work has been associated with upper limb disorders and high levels of occupational stress, yet neither of the work limiting health categories representing these conditions appears significantly high relative to other occupational orders. Indeed, the relative proportions of different health limiting conditions appears not to differ substantially from that observed for all occupational groups.

13.4.14 Order 14 – Construction

A very male dominated occupational order, where the number of females in most categories is too small to allow meaningful interpretation of the data. Levels of sickness absence are very high (see Figure 13.2) and there is a relatively high proportion of long-term sickness (see Figure 13.4) and a high percentage of reported work limiting conditions (see Figure 13.7). Workers in this order are known to have relatively unhealthy lifestyles. For example, the prevalences of smoking and drinking to excess are very high compared to other occupations, as shown in Chapter 12. This, along with their low socio-economic status and the physically demanding nature of much of the work in this order, may be partly responsible for the very high figures for long-term sickness.

13.4.15 Order 15 – Transportation, storage etc.

This is a male dominated order with high rates of sickness absence. The high proportion of long-term sickness indicates a high prevalence of chronic health problems (see Figure 13.4), as does the relatively high percentage reporting work limiting conditions. The low socio-economic status of the these workers may provide some explanation of these figures. A certain level of health may also be required by many of the driving occupations included in this order. Consequently, compared to other occupations, less severe health conditions may limit work activities and result in greater sickness absence.

13.4.16 Order 16 – Miscellaneous

This small order comprises an ad hoc assembly of mostly unskilled workers and general labourers not classifiable to the other orders. Sickness absence rates are very high (see Figure 13.2) as is the percentage reporting health limiting conditions (see Figure 13.7). The order is of very low socio-economic status but it is uncertain whether the low socio-economic status of the workers is responsible for their ill-health or whether in some cases chronic illness may have limited their ability to secure a job with higher status. Speculation on the causes of the high sickness rates is of limited value because of the ill-defined nature of many jobs in this order.

13.5 Conclusions

The data presented represent one of the most comprehensive and detailed published sources of UK sickness absence data to date. However, some caution must be exercised in interpretation of the data because of the possible unreliability of the smaller scaled up values. Despite this caveat several general trends are clearly evident. A clear difference between sexes is evident with higher sickness absence rates and a greater proportion of short-term sickness among female workers. Increasing age is related to an increase in the duration of the absence spell, whereas the rate of sickness absence appears high in both the younger and older members of the work-force. General trends by occupational order appear to reflect the socio-economic standing of the orders. With higher rates of sickness absence, a greater proportion of long-term sickness and a greater percentage of work limiting health conditions all appearing more common in orders with on the whole lower socio-economic status. Other underlying fluctuations by occupational order may be related to specific facets of the occupations or the populations involved.

References

1 Taylor PJ. Absenteeism definitions and statistics. In *Encyclopaedia of Occupational Health and Safety*. London: ILO, 1983: 16.

2 Searle ST, Ch.1. — *Sickness absence in occupational health practice, 3rd edition,* Ed. Waldron HA, London: Butterworths, 1989.

3 Office of Population Censuses and Surveys. *Occupational mortality. Decennial Supplement 1979-80, 1981-82. London:* HMSO,1984.

4 Steers RM, Rhodes SR. Major influences on employee attendance: a process model. *Journal of Applied Psychology* 1978;63:391-407.

5 Davey Smith G, Shipley MJ, Rose G. Magnitude and causes of socio-economic differentials in mortality: further evidence from the Whitehall Study. *Journal of Epidemiology and Community Health 1990;*44:265-270.

6 Hodgson JT, Jones JR, Elliott RC, Osman J. *Self-reported work-related illness.* HSE Research Paper No. 33. London: HMSO, 1993.

Chapter 14 Occupation and fertility

Penny Babb

Fertility Statistics Unit, OPCS

14.1 Introduction

The investigation of the deleterious effects of work on the reproductive system has focused on a range of occupational hazards. Conflicting results have been obtained from studies on the relation between work exposures during pregnancy and pregnancy outcome.[1] Epidemiological study has addressed a number of adverse outcomes: infertility, spontaneous abortions, congenital malformations, low birthweight, chromosomal abnormalities and childhood cancer. Only a relatively few occupational hazards have been demonstrated to have an effect on human reproduction. Numerous other exposures, however, have been proposed as potentially hazardous. Hunter's Diseases of Occupations[2] includes possible reproductive hazards under four major categories: chemical, physical, microbiological and psychosocial. Examples are shown in Table 14.1.

Table 14.1 Potential hazards leading to adverse reproductive outcomes

Hazard category	Examples
Chemical	
Metals	Lead, mercury
Pesticides	Chlordecone, DBCP, dieldrin
Therapeutic agents	Oestrogens, thalidomide
Sterilizing agents	Ethylene oxide
Other chemicals	Styrene, carbon disulphide, anaesthetic gases, e.g. nitrous oxide
Physical	Radiation, noise, vibration,heavy physical work
Microbiological	Viruses, bacteria
Psychosocial	Tobacco, alcohol, marijuana

A wide variety of occupations and industries contain the possible risk of adverse reproductive outcomes: operating theatre staff; laboratory work; industries manufacturing oral contraceptives and cytotoxic drugs; nurses and pharmacists handling such chemicals; medical radiographers; workers handling fumigants and fungicides, such as ethylene oxide, in the food, textile and agricultural industries; and metal workers, for example, smelter workers exposed to lead, copper, and cadmium.[3]

Testing cannot keep pace with the development of new chemicals. Many chemicals in use today pose a potential threat to health. Epidemiological studies provide further vital evidence on toxic hazards. They are hampered by a lack of reliable baseline information on the frequencies of adverse outcomes such as spontaneous abortions and certain childhood cancers.[4] Deciphering the occupational effects is further complicated by multiple exposures, particularly where synergistic effects occur, and non-occupational exposures, such as from psychosocial hazards identified above. The analysis of national vital statistics may aid these investigations by providing some information on possible hazardous occupations and an indication of background levels of the outcomes which should be expected.[5]

14.2 Vital registration data

McDowall[5] has identified three main advantages of vital statistics to the occupational epidemiology of reproductive outcomes. First, the national coverage of events enables rare adverse reproductive outcomes to be examined by parents' occupation. Second, it also provides comparability with other occupational data through the use of standard classifications. Lastly, the data are also relatively cheap and readily available. These vital statistics contain important disadvantages, however. While risk factors such as mother's age, number of children and social class (as defined by occupation) are available from such data, other factors, for example, the psychosocial hazards noted above, are not. The occupation of both parents is only coded for a 10 per cent sample of birth registrations but for all stillbirths and infant deaths. Since the father's details are not collected for births outside marriage registered by the mother only, the father's occupation is not available for these births — 8 per cent of all live births in 1992. Information on the mother's occupation was first requested in 1986. In that year, only a third of all live birth and a quarter of stillbirth and infant death registrations contained mother's occupation. The coverage improved in subsequent years, possibly helped by the computerisation of register offices. By 1990, however, mother's occupational details were still not available for a half of live births and two thirds of stillbirths and infant deaths.[6]

Quality cannot be inferred from coverage. Whilst father's occupation is available for the majority of births, it reflects only a job title and gives a limited indication of exposure to hazards.[5] Information on mother's occupation is especially unreliable. The occupational data may be biased by quoting 'preferred' occupations, while some women may not give a job title that they do not want recorded on their child's birth registration. Further inaccuracies may occur where the informant at birth registration is not the job holder.[6]

14.3 Methods

The sample of births in the following study were live births registered between 1980 and 1990 and are analysed by father's occupation and by mother's occupation in the years 1986-90, (father's occupational details are only available on marital and joint registrations i.e. sole registrations are excluded in analyses by father's occupation). Data for 1981 have been excluded since they are incomplete following industrial action taken by registrars in that year. The 10 per cent sampling of the registrations has led to small numbers in many occupations. To overcome this problem, the occupations have been grouped according to similar exposures using the 'Southampton Classification' described in Appendices 2 and 3.

The data on births in England and Wales by occupation, have been indirectly standardised for father's age, to give the standardised fertility ratios (SFR) for each occupational group. The SFR is a measure of fertility reflecting actual numbers of births rather than ability of women to conceive. It reflects the number of births which occurred in each occupation as a percentage of the number of births that would have occurred had the occupation experienced the age-specific rates of England and Wales. Since the SFR uses the male population from the 1981 Census, a mismatch exists between the 'population at risk' and those fathers in each occupation, and is a particular problem where differences occur in the reporting of occupations. This problem may be overcome to some degree in the future by using the OPCS Longitudinal Study which links vital registration data with census records.

The analysis of occupational fertility is complicated by timing effects of both marriage and childbearing. Consequently, the analysis includes an examination of mean age of mother at childbirth. The proportion of births outside marriage that were jointly registered by both parents is also presented by father's occupation at registration. An analysis of sex ratios by occupation and social class will be included in a later OPCS volume on child health.

14.4 Occupational fertility levels

The number of births and standardised fertility ratios by father's occupation, in 1980/82, in England and Wales for occupational groups with significantly low levels of fertility are summarised in Table 14.2. Occupations such as national government administrators and managers in building and contracting are unlikely to contain hazardous exposures. The low fertility may reflect other characteristics of the men employed in these jobs.

The table highlights several occupations with low levels of fertility that may be due to exposure to hazardous substances or conditions at work, for example, laboratory technicians (032), four groups of workers in the printing industry (094, 095, 096, 111), and oilers and greasers (154). McDowall[5] reports on an investigation of birth registrations by father's occupation in the period 1980-82. He also examines other adverse reproductive outcomes at the occupational unit level. Differences will exist between the findings of the current study and the McDowall study because of the exposure groupings used here. The low fertility of laboratory technicians was also identified in the McDowall study, however, for whom evidence has been found of a raised risk from congenital malformations (to female laboratory workers and male workers exposed to solvents).[5]

Table 14.2 Father's occupations showing significantly* low standardised fertility ratios, 1980/82, England and Wales

Job group		SFR
	England and Wales	100
008	General administrators	65
012	Vocational trainers, social scientists, etc.	87
032	Laboratory technicians	74
037	Technicians nec.	78
039	Managers in construction	66
053	Office workers and cashiers	89
054	Postal workers	90
055	Petrol pump attendants	34
058	Security workers	34
060	Other service personnel	73
067	Tannery workers	62
081	Paper cutters	55
087	Man-made fibre makers	49
094	Compositors	76
095	Printing plate preparers	55
096	Printing machine minders	68
109	Other woodworkers	76
111	Other makers of paper goods	50
118	Annealers, hardners, temperers (metal)	41
122	Centre lathe turners	80
134	Aircraft engine fitters	51
140	Electric cable and line workers	75
154	Oilers and greasers	44
162	Instrument assemblers	25
164	Packers and sorters	62
186	Mechanical plant drivers	77
191	Dockers and goods porters	66

* Statistically significantly different from 100 at 95% confidence level.

14.5 Mean age of mother at childbirth and proportion of jointly registered births outside marriage

The mean ages of mothers at childbirth were calculated for births by father's and mother's occupation. The professional occupations tended to have higher mean ages than in manual occupations. This is in agreement with previous analyses of mean age and social class (as defined by occupation).[7] The overall trend in both groups was an increase over the 1980s. Occupations with particularly high mean ages include: teachers in higher education (010), lawyers (001), and teachers (011) (see Table 14.3). Those occupations with the lowest mean ages were packers and sorters (164), painters and decorators (160), sewers and embroiderers (100), and leather and shoe workers (068).

Table 14.3 Mean age of mother at childbirth by occupation, England and Wales

Job group		Father's occupation 1988-90	Mother's occupation 1989-90
001	Lawyers	31.5	31.5
008	General administrators	31.0	31.1
010	Teachers in higher education	32.0	32.6
011	Teachers nec	31.3	31.5
012	Vocational trainers, social scientists etc.	29.8	31.2
015	Doctors	30.4	31.2
068	Leather and shoe workers	26.1	24.2
100	Sewers and embroiderers	25.4	24.2
160	Painters and decorators nec.	25.5	24.1
164	Packers and sorters	24.1	23.5

The proportion of births outside marriage almost doubled during the 1980s. Three quarters of births outside marriage in 1992 were jointly registered after rising from around half of births outside marriage in 1982.[8] The percentage of jointly registered births also rose in most occupational groups. However, large differences can be seen in the proportions of jointly registered births between occupations. The lowest proportion occurred to clergy (014) which was stable at less than 1 per cent. Other low proportions occurred to doctors (015) and dentists (016) (3.6 and 3.5 per cent in 1988-90, respectively). The highest proportions in 1988-90 were to case and box makers (106, 43.8 per cent), scaffolders (148, 40.2 per cent), painters and decorators nec (160, 38.3 per cent), construction workers nec (174, 37.8 per cent), roofers and glaziers (168, 36.8 per cent) and plasterers (167, 33.4 per cent).

14.6 Limitations of the analysis

The results of this analysis need to be viewed with caution. They can only be used to highlight possible areas of interest because of the limitations of both the method and the occupational vital registration data. Many non-occupational factors will have had an impact on the levels of fertility observed and so low fertility can only be deemed to have a potential occupational cause when associated with strong epidemiological evidence.

References

1 Baxter PJ, Waldron HA. The introduction and monitoring of occupational diseases. In: HA Waldron (ed). *Occupational Health Practice*. Third edition. London: Butterworths, 1989.

2 McCloy EC. Reproduction and work. In: PAB Raffle, WR Lee, RI McCallum, and R Murray (eds). *Hunter's Diseases of Occupations, Section IX*. London: Hodder and Stoughton, 1987.

3 Hemminki K, Vainio H. Occupational epidemiology and reproduction. In: JM Harrington (ed). *Recent advances in occupational health*, No.2. Chapter 9. Edinburgh: Churchill Livingstone, 1984.

4 Slovak AJM. Reproductive health hazards in the workplace. In: JK Howard and FH Tyrer (eds). *Textbook of occupational medicine*. Edinburgh: Churchill Livingstone, 1987.

5 McDowall ME. *Occupational reproductive epidemiology: the use of routinely collected statistics in England and Wales 1980-82*. SMPS No.50. London: HMSO, 1985.

6 Cooper JS, Botting B. Analysing fertility and infant mortality by mother's social class as defined by occupation. *Population Trends:* HMSO 1992;70: 15-21.

7 Department of Health. *On the state of the public health*. Chapter 1: Vital statistics. London: HMSO, 1993.

8 Babb P. Birth statistics. *Population Trends:* HMSO 1994;74: 8-11.

Chapter 15 Ad hoc occupational mortality studies

Morris Greenberg

Health Statistics, OPCS

15.1 Introduction

The period after the Second World War saw a surge in interest in occupational health and safety in the UK, not experienced since the Earl Grey Reform Bill of 1832. The latter promptly led to the inception of the Factories Acts, which were to continue unaltered in principle for the next 140 years, though undergoing progressive accretion and becoming in the process unwieldy. The Health and Safety at Work Act of 1974, no less promptly, consolidated the previous Acts and introduced radical changes in terms of responsibility, unifying the requirements for health and safety at work and extending cover to all employed and self-employed persons. This was accompanied by an expansion in epidemiological research into the health of specific groups of workers, individual factory populations and whole industries.

15.2 The development of an OPCS service to external research workers

To its traditional activity of routinely analysing occupational mortality data, OPCS started to provide services to external research workers. These included tracing members of study populations, creating study populations and extracting death registrations. Members of the office found themselves providing guidance to external researchers on the conduct of studies on an informal basis, and from time to time participated in them. During the period 1960-93, the 130 occupational health studies initiated, constituted a third of all the medical studies assisted by OPCS. Sizes of studies, in terms of numbers of subjects flagged by OPCS, range from under 500 (22 studies) to over 50,000 (4 studies) with 54 studies in the 1,000-4,999 range. By 1993, the cumulative number of subjects flagged exceeded two million. The steepest gradient of growth corresponded with the period 1971-1985. The titles and descriptions of the studies listed in Annex 15.1 and précised in Annex 15.2 indicate the nature and urgency of the problems under investigation and the results known to 1993. The increase in epidemiological activity and the increasingly sophisticated questions asked, have prompted the development of facilities within OPCS to link registers and record systems.

15.3 Reasons for the studies

Studies of the health experiences of occupation groups with which OPCS has been involved have been prompted in several ways.

15.3.1 Anecdotal reports

A number of studies were initiated by suspicions expressed by a number of sources. The study of antimony smelters (MR3)* was prompted by allegations in the course of a Coroner's enquiry, that there had been an excess of cases of lung cancer at a factory that had ceased production several years previously. The number of cases was small, and a simple review at a similar factory still in production, also reported a small inconclusive cluster. A study of newspaper printers (MR128) owed its origin to a previous study prompted by a report from a urologist that he had recently seen three cases of bladder cancer in printers. (In the event there was no excess of bladder cancer, but there were significant excesses of other cancers.)

Confirmatory studies were mounted in response to suspicions, based on few cases, that exposure to bis-Chloromethyl ether caused lung cancer (MR7) and to vinyl chloride monomer caused angiosarcoma of the liver (MR24). Media reports of clusters of cases of childhood malignancy in the vicinity of nuclear establishments led to a series of studies of children and of their parents (MR208, 210, 211, 212, 305).

15.3.2 Health surveillance in industries with known hazards

Industries with acknowledged health risks have monitored the mortality pattern of their workers to evaluate the residual problems after process and material changes had been made, in order to develop a strategy for further intervention. Where feasible, these studies were designed to produce dose response relationships between the critical endpoints and direct or surrogate measures of exposure to the critical dust, chemical or physical agent.

The coal, steel, nuclear, rubber, and asbestos industries, by using OPCS facilities were able to monitor the mortality experience of their workforces and to quantify the effects of their hazardous processes and agents.

The coal industry for example, having first studied the physical effects of dust exposure using radiographic, physiological and symptomatic criteria of clinical significance, then investigated the long-term health consequences of major effects and of those minor changes believed to be of no immediate clinical significance. While poor survival was associated with progressive massive fibrosis, the lesser effects were found to be associated with excess mortality and

* Numbers in parentheses in the text refer to medical research studies listed in Annex 15.1.

cancer morbidity (MR12A). Similarly, the rubber and cable making industry, which had observed an excess risk of bladder cancer in operatives exposed to certain additives, reviewed the situation after their withdrawal from use. Using OPCS death register and cancer register facilities, it was possible to confirm that persons employed after the date of withdrawal no longer experienced an excess bladder cancer risk. However, other cancer excesses were found to be operating. The responsible agent could not be identified, but the associated processes that required urgent increased environmental control were indicated.

15.3.3 Hypotheses generated by the occupational mortality decennial supplement

The decennial supplement analyses have regularly thrown up suspicions of abnormal health experience in particular occupational groups. These hypotheses have instigated special studies designed to test them, and where positive further investigations to identify the causal agent. Butchers having repeatedly been found to experience an excess of lung cancer, a small population of butchers was selected to be the subject of a prospective study (MR247). This found several other neoplasms also to be in excess and a national study of papilloma virus in lung cancer patients was designed to determine whether it played a role in the causation of the tumour.

15.3.4 Reassessment of the human health hazard in the light of new information

Data on various radiobiological effects and received doses by subjects in critical radiation exposure studies have been re-estimated and have periodically undergone review over the past few decades. As the dose response relationships have been subject to controversy, the risk estimates have been considered provisional and subject to confirmation. The concern for the occupational and public health effects of the low doses that occur widely, urgently required the testing of the risk estimates calculated by standard setting bodies. Nuclear industry populations exposed to a range of internal and external radiation doses, were studied by individual installations and aggregated to provide greater statistical power (MR47) to test the risk estimates proposed by international standards setting bodies. This was supplemented by studies conducted on Cornish tin miners (MR142) and on Coal miners (MR319), whose exposures were deemed to approximate to the low levels operating domestically.

15.4 Facilities offered by OPCS to external researchers

15.4.1 Data sources

OPCS routinely collects large volumes of named data of potential use for research into occupational and general environmental health that are described in greater detail in Chapter 2. These include draft entries for births, marriages and deaths, and cancer registrations, for which indexes are constructed. In addition, it collects and stores census forms and data from ad hoc and regular health and social surveys such as the General Household Survey. From its Longitudinal Study analyses, it can offer better estimates of mortality and cancer morbidity for working populations than are available from routine vital statistics, and though based on a small sample, by aggregation can provide adequate occupational populations for study.

Where it has been important to study the health experience of persons exposed to specific occupational or environmental agents, for whom nominal rolls are not readily obtainable, populations have been derived from census records and the 1939 register.

This was done in the following studies:

Health experience of persons living in the vicinity of power lines and sub-stations (MR182);
Persons living in the vicinity of nuclear establishments;[1]
National studies of butchers (MR302) and of cooks (MR303);
The health of Cornish tin miners exposed to radon (MR142).

Because of the confidential nature of the census data, for which there is a hundred-year embargo on any data which might identify individuals, their use is strictly controlled and only anonymised data derived from them may be provided to external researchers, or published. This is achieved by providing censored data, or by the preliminary analyses being conducted in-house and the anonymised results being reviewed in collaboration with the outside researchers. In studies of children born in a particular locality during a specific period, local birth registers have been searched to identify the population of interest, as well as to provide control groups for comparison (MR210).

15.4.2 Tracing facilities

At the time of their analyses, researchers need to know whether a subject is still alive or is dead or is otherwise lost from further study. OPCS, by means of the Central Health Register Inquiry System (CHRIS), previously the National Health Service Register (NHSCR), which it maintains, can provide the vital status of individuals entitled to National Health Service care in England and Wales, and who are registered with a doctor.

Older studies in the UK, which involved local field worker initiatives, were expensive of time and resources, and their efficiency compared unfavourably with those assisted by NHSCR. An early example of this is provided by the mortality study of antimony workers (MR3). With the best of local resources, over some 10 years only an unacceptably low trace of some 70 per cent was achieved. With NHSCR assistance, despite the imperfect nature of the records, a trace rate of 99% was achieved overall, and never less than 96% in any subset.

Occasionally a researcher will need to contact a subject's doctor or the surviving subjects, to conduct interviews or examinations. The tracing facility has made it possible for the researcher to make contact with subjects via the appropriate Family Health Service Authority and thence the general practitioner, without breaching confidentiality.

For studies that include populations resident in Scotland (MR124, 12 B-D, 119) and Northern Ireland, and where members of study populations may move between parts of the United Kingdom, OPCS can coordinate tracing and flagging with the Scottish General Register Office and NHSCR. In the case of Northern Ireland, the researcher operates through the Central Services Agency and the General Register Office, Belfast.

When the endpoint of a study is death, copies of the draft death registration can be provided to researchers. When cancer registration is the endpoint, subject to certain conditions of confidentiality, similarly notification to the researcher can be arranged.

The International Classification of Diseases provides comprehensive lists of causes of death, diseases and injuries, in standard nomenclature. No less importantly, it provides rules for the coding the primary cause of death from the multiple entries commonly found on the death certificate. Recently, computer programming has permitted the majority of death registrations to be coded automatically. Research workers acting as their own nosologists may operate to a different standard from that operated by OPCS, and are at risk of introducing bias, rendering invalid comparisons with published national and regional rates and with studies from other countries conforming to the WHO standard. OPCS will routinely provide coding to the latest revision of the ICD for the cause of death on the death registration copies it provides researchers. To assist in the analysis of studies that extend over a number of years, by arrangement deaths may be coded to earlier recent revisions and for multiple causes.

Statutory registers, personnel records, pay records, factory records, and pension and death benefit records have been used singly or in combination to reconstruct historic rolls of employees in factories or in processes. Such data will have been adequate for the purposes for which they were originally collected, but were infrequently compiled and stored for ease of retrieval or with an eye to research. As a consequence, OPCS staff have acquired considerable skill in the matching of inadequately identified persons with their entries in its registers.

15.4.3 Characteristics of research workers and institutions assisted by OPCS

The Chief Medical Statistician has always confined assistance to bona fide research workers. These have included the single handed research worker, variously a general practitioner (MR863), a hospital consultant (MR75) or a public health specialist (MR190). The researchers may also constitute a substantial team covering a range of fields of expertise (eg pathology, mineralogy, statistics, medicine), and based in several academic units (MR44). Where in a research institute an individual constitutes the prime mover, it is common for the project to transfer to another institution with the career moves of that individual, sometimes continuing into retirement. (When responsibility for a project transfers to a new researcher, fresh agreements to cover confidentiality are required to be completed.)

15.5 Varieties of study designs and analyses

Study designs and analyses are largely dictated by the population and events data available and the questions being asked of them. Over the years, occupational and environmental epidemiologists have applied more sophisticated statistical techniques the better to exploit the data and to permit the answering of more complex questions. Where multiple agents or factors have been suspected of being associated perhaps causally with the development of disease, as in the study of rubber workers (MR88), regression models and life tables were employed to evaluate the power of association between specific malignancies and individual agents.

15.5.1 Prospective studies

Prospective studies start by identifying groups of persons and proceed to follow them to the endpoints of death or cancer registration. For an occupational group it is ideal to have the nominal roll of all persons employed at the start up of the factory or the process of interest. Commonly, the record is only adequate on or after a particular date. From these records, the researcher may limit analysis to persons employed for an arbitrary minimum period of six months or a year or even 20 years, sometimes in a conscious attempt to avoid the perceived ‘diluting’ effects of short periods of exposure.

Alternatively, a person years adjustment is applied when calculating expected values from the rates being used for comparison. The analysis may be lagged for an appropriate period to allow for latency in development of the disease of interest.

In the study of rubber workers (MR864), the population consisted of men aged 35-65 in employment on 1st February 1967 and the analyses were made in 1972, 1974 and 1976. The age at which subjects are censored from analysis of mortality of cancer registration, has varied over the years. Early on, only persons of working age were reported on. More recently the post-retirement period has been considered worthy of study. Because of the lack of confidence about the precise cause of death in the elderly, a cut-off point is still used but it has advanced beyond the age of 65. In this study, subjects who were about to retire in 1967, would be rising 70 at the time of the first analysis, and still susceptible to developing the tumours of interest. The population consisted of men whose periods of employment involved exposure to the hazardous agents being studied, and men employed only subsequent to the withdrawal of these agents from compounding.

The term ‘historic prospective’ has been applied by some to studies such as that of women gas mask assemblers (MR44), where the population in employment in 1939, was assembled and analysed in 1980.

15.5.2 Retrospective studies

These studies start with the endpoint and look back to study the characteristics of the population from which they were derived. The cause specific registers that have been created by OPCS for individual customers have prompted such study. Certain registers contain mortality data only, some include cancer registration data.

15.5.3 Mortality

Researchers have requested to be provided with copies of death registrations in England and Wales, mentioning specific causes, effectively creating a disease register. The first specific cause register was initiated in 1921, when death registrations featuring bladder cancer, papilloma of the bladder, and any tumours of the urinary organs, were copied to the Chester Beatty Institute for Cancer Research. By 1981, when the register was closed, a unique body of data had been assembled and a wealth of reports had been published. The data have remained of interest for epidemiological research with publication appearing some 10 years later.[2] The original association of bladder cancer with occupational exposure in dye manufacture reported by Rehn in 1895, was extended by studies based on the register.[3,4,5] Among the agents or occupations found to be associated with an increased risk of bladder cancer were, boot and shoe manufacture and repair, rubber and cable making, certain chemical work, textile work, glasswork, pest control (exposure to a specific rodenticide), deck and engine room crew, a railway workers, electrical and electronic workers, foodworkers engaged in bread and flour confectionery and in the extraction of oils and fats, workers exposed to paint and pigments, benzene, cutting oils and tobacco work.

15.5.4 Cause specific cancer registers

Registers have been assembled that contain both mortality notifications and cancer registrations to investigate certain hazards.

The mesothelioma register was set up by HM Factory Inspectorate in 1967 with the assistance of OPCS, following the earlier report from South Africa of the appearance of this tumour in asbestos miners and persons living in the neighbourhood of mines, to investigate the hypothesis that occupational and environmental exposures to asbestos contributed importantly to the development of malignant mesothelial tumours in the UK. A number of analyses confirmed the pre-eminence of asbestos in the causation of the disease, identified the most hazardous occupational exposures, and have monitored the evolution of the epidemic. In addition, para-occupational and environmental exposures not previously considered to be significantly hazardous were identified as in need of intervention.[6,7]

When a few cases of angiosarcoma of the liver were reported in workers who had been heavily exposed to vinyl chloride monomer in the polymerisation process, an urgent investigation was set in motion to confirm the observation and to evaluate the effectiveness of the new process controls. This was used to supplement the ad hoc national study of VCM workers (MR24).

Nasal cancer. The finding of an excess of nasal cancer in groups of furniture (MR137), boot and shoe workers (MR146) and nickel smelters (MR56), prompted a search for other occupational associations. For the years 1963-67, all cases registered as having malignant neoplasms of the nose, nasal cavities, middle ear and accessory sinuses were extracted to form a register. The researchers subsequently confirmed the diagnoses and investigated smoking habits and occupations.[8] They found a number of occupational orders to be at special risk.

15.6 The importance of OPCS assisted studies

The facilities provided by OPCS for studying specific cause mortality and cancer morbidity, and for creating populations for study and for tracing their members, compare favourably with the best available elsewhere. They constitute a powerful and versatile resource for the investigation and monitoring of occupational and environmental health. As a consequence, the conduct and publication of OPCS assisted studies present a valued contribution in this field. While the United States has a population some five times greater than the UK, with the complexities of the US administration system and population mobility, studies that require a high efficiency of tracing and national coverage, present considerable difficulties. The Scandinavian countries have developed centralised well linked record systems, lending themselves to the efficient study of their populations. However the individual population bases vary from 8-19 per cent of that for England and Wales (overall 45 per cent), limiting their ability to detect with confidence the lesser orders of hazard of concern today.

In most European countries, the administrative complexities of death registration, and the confidential nature of vital statistics considered as public record in the UK, militate against the ready conduct of population studies.

As there are few studies specifically addressed to subsets of members of the general public exposed to well or ill-defined environmental agents, the derivation of most environmental quality standards depends on interpolation from occupational studies.

The differences in social, genetic, and environmental background that obtain in developed and developing countries notwithstanding, findings by epidemiologists in the UK have general application. The observations of UK scientists in the occupational health field, have been taken into consideration by national and international bodies involved in establishing occupational and environmental exposure standards. Outstanding examples include studies of: cancers in rubber workers (MR864); mortality in coal workers (MR12A, 12B, 12C); nasal cancer in woodworkers (MR137) and in boot and shoe operatives (MR146); cancer in asbestos workers (MR44).

That UK findings are of more than parochial interest is confirmed not only by their frequent citation abroad, but also by their sponsoring from abroad. The European Coal and Steel Community, has funded outstanding studies of workers in these industries (MR12A, 12B, 12C) and the European Commission is involved with the Royal Society of Chemistry in the investigation of the health experience of professional chemists (MR136).

Over the years, excesses of occupational mortality and cancer incidence have become increasingly less tolerated. Where once an acceptable excess mortality of less than two per hundred was considered enlightened, for occupational and general population exposures, excess risks of less than one per hundred thousand may be required to be achieved. Populations with the statistical power to detect such orders with confidence can exceed any one nation's resources. Faut de mieux, attempts have been made to remedy this deficiency by applying meta-analytical techniques to collections of studies. (The petroleum industry has conducted such an analysis including the results of a UK study in investigating the leukaemogenesis of benzene (MR321).

An alternative preferable strategy that has been adopted has been to conduct coordinated multinational studies. World Health Organisation through its International Agency for Research on Cancer (IARC), has led several such initiatives, to which the UK has made important contributions.

With the decision to substitute man made mineral fibre for carcinogenic natural fibres such as asbestos, experimental studies having shown that most fibres could elicit similar effects at cellular level, it was considered urgent that a European industry study be conducted. The UK has made an important contribution to this collaborative study designed to detect carcinogenic effects (MR109).

The term 'dioxin' is loosely used for a range of chemical compounds sharing similar high toxicity in particular species and highly versatile in their effects. They are products of combustion and have been produced as the active agent or as a trace by product in a range of materials used variously as herbicides and as dialectrics. Traces are to be found in adult and infant tissues and in breast milk, worldwide. While some of the effects of some of these compounds have been studied in man, the effects at the low levels present in the general public remain to be evaluated. IARC is coordinating an international study to which the UK is contributing a subset (MR186). A further international study being coordinated by IARC, to which there is a UK contribution, is of cancer in welders (MR274), where the results of previous studies have been inconsistent.

Transnational companies studying the mortality of workers exposed in particular processes requiring to boost the power of their inquiry, have combined their UK data in studies of carbon black workers (MR154) and of workers exposed to cutting fluids (MR361).

15.7 The future

It is foreseeable that the requirement for epidemiological studies of populations in the investigation of acute and chronic exposures to chemical and physical agents will continue and increase in volume.

The Health and Safety Executive, conscious of the value of the analysis of routine environmental and health data acquired from statutory examinations, set up such programmes in 1970 for asbestos workers (MR5) and in 1974 for lead workers (MR33). (Inter alia, this led to the innovation of the standardisation of examination criteria.)

Latterly, European daughter directives have required data derived from the surveillance of workers in particular hazardous occupations to be preserved for at least 30 years. When the decision is made to utilise these valuable data to provide indicators of residual risks, OPCS will be well placed to assist.

In the non-occupational field, a number of environmental agents deemed potentially to have important adverse health effects, are required to be monitored regularly. The significance of the various qualitative and quantitative measures of exposure could be tested for their acute and long-term health significance by linking them with the various health indicators archived by OPCS.

The development of systems to automate the input and analysis of data, has contributed to make the epidemiological study of suspect occupational and environmental hazards more economic. The new computerised central index will improve tracing rates, facilitate flagging and reduce duplication of records. Further economies could be achieved through developments in data handling and record linkage that would be justified by more extensive use of OPCS facilities. These economies would make its services more attractive to its UK users. OPCS' unique facilities constitute a valuable international resource.

References

1 Cook-Mozaffari P, Ashwood PJ, Vincent FL, Foreman D and Alderson M. *Cancer incidence and mortality in the vicinity of nuclear installations England and Wales 1959-80.* OPCS Studies on Medical and Population subjects No. 51.

2 Dolin PJ and Cook-Mozaffari P. Occupation and bladder cancer: a death certificate study. *Br J Cancer* 1992; 66: 568-578.

3 Case RAM, Hosker ME, McDonald DB and Pearson JT. Tumours of the urinary bladder in workman engaged in the manufacture and use of certain dyestuff intermediates in the British chemical industry. Part I. *Br J Ind Med* 1945; 11: 75-104.

4 Case RAM and Pearson JT. Tumours of the urinary bladder etc. Part II. *Br J Ind Med* 1954; 11: 213-216.

5 Davies JM, Thomas HF and Manson D. Bladder tumours among rodent operatives handling ANTU. *Br Med J* 1982; 927-31.

6 Greenberg M and Lloyd Davies TA. Mesothelioma register 1967-68. *Br J Ind Med* 1974; 31: 91-104.

7 Jones RD and Thomas PG. Incidence of mesothelioma in Britain. *Lancet* 1986; 1: 1275.

8 Acheson ED, Cowdell RH and Rang EH. Nasa; cancer in England and Wales: an occupational survey *Br J Ind Med* 1981; 38: 218-224.

Annex 15.1 A classification of ad hoc studies assisted by OPCS

1. Electromagnetic radiation

Ionising: occupational

- 47 NRPB: National Registry of Radiation Workers (NRRW)
- 110 UK Atomic Energy Authority (UKAEA) mortality study
- 138 Central Electricity Generating Board (CEGB): Classified radiation worker mortality study
- 142 A study of mortality in Cornish tin miners
- 155 A prospective mortality study of iron ore miners in Cumberland
- 176 Mortality study of workers at British Nuclear Fuels
- 183 Mortality of employees at the Atomic Weapons Research Establishment 1951-82
- 185 The long-term health experience of the UK participants in the UK atmospheric nuclear weapons tests
- 248 Mortality and cancer morbidity in Cornish tin miners
- 314 Mortality and cancer morbidity of Royal Naval submariners

319 Mortality of coal miners exposed to radon and thoron daughters
859 UK radium luminiser worker study

Ionising: environmental
208 Leukaemia and lymphoma among young persons near Sellafield Nuclear plant
210 Study of children born to mothers resident in Seascale 1950-1983
211 A cohort study of girls attending Calder Girls' School in Seascale, West Cumbria
212 A cohort study of children at Local Authority schools in Seascale 1953-1984
242 Study of children attending Singing Surf School in Seascale since 1950
305 Childhood cancer in West Berkshire and North Hampshire

Non-ionising
182 Mortality of persons resident in the vicinity of Electric transmission facilities
354 Brain tumours in radar research workers

2. Mineral fibre

Asbestos
5 National survey of asbestos workers
20 Royal Naval Dockyard mortality studies
42 Mortality of workers at an asbestos factory in Leeds
44 Mortality studies of female gas mask assemblers exposed to asbestos
48 Mortality of workers at acetylene production plants (crocidolite)
63 Mortality of workers manufacturing friction material using asbestos
75 The mortality of shipyard workers with pleural plaques
83 Mortality of persons compensated for asbestosis
108 Cohort study of workers exposed to amosite asbestos
139 Mortality of gas mask assemblers
177 A study of asbestos workers
189 Mortality of chrysotile asbestos cement workers
251 Long-term health effects of working in a building containing friable structural asbestos
419 An analysis of asbestos-induced lung disease
852 Mortality of workers in a London asbestos factory
Mesothelioma Register
Pattern of mortality of parents of patients dying of malignant pleural or peritoneal mesothelioma

Other mineral fibre
109 Mortality of man-made mineral fibre workers
175 Study of glass-reinforced plastic laminators exposed to styrene

3. Rubber and cable making industry studies

88 British Rubber Manufacturers Association research project
205 A follow-up study of workers in a chemical company manufacturing methylene-bis-ortho-chloraniline
221 Mortality study of rubber cable workers
326 Mortality of workers exposed to mercaptobenzothiazole
363 Cancer mortality in the British rubber industry
864 Survey of occupational cancer in the rubber and cable making industries

4. Studies of workers exposed to benzene, polyaromatic hydrocarbons and petrochemicals

12C Mortality of coke oven workers
124 Mortality of employees at various petrochemical plants
146 Mortality of boot and shoe workers [Benzene based adhesives]
154 Longitudinal mortality and morbidity study in the carbon black industry
172 Study of artificial leather cloth workers [Benzene exposure]
226 A study of leukaemia and other cancers among benzene workers
321 Institute of Petroleum epidemiology study
361 Mortality in men exposed to cutting fluids
426 A cohort study of rubber adhesive workers

5. Dust diseases of the lung

12A The effect of coal workers' pneumoconiosis on mortality
12B Mortality of men employed in the British steel industry
30 Mortality of talc workers
84 Mortality of slate workers in Gwynedd
97 Mortality study of foundry workers
193 Mortality of pottery workers
233 Mortality study of cement workers
256 Morbidity and mortality in silica sand quarry workers

6. Nasal cancer studies

137 Study of Buckinghamshire furniture workers
146 Mortality of boot and shoe workers
147 Mortality of tanners

7. Cotton workers

14 Mortality of workers in the British cotton industry
61 A mortality study of workers in Lancashire cotton mills

8. 'Dioxins'

152 Pentachlorophenol workers
186 Cancer registration and mortality studies of workers exposed to chloracnegenic agents
187 Mortality of herbicide manufacturers and sprayers
222 Mortality study of capacitor workers
250 Mortality and cancer incidence of workers making phenoxy herbicides

9. Lead workers

33 Mortality study of lead workers undergoing statutory examination
344 Mortality of lead oxide workers

10. Chromium and its salts

41 Chrome pigment worker health studies
54 Mortality of bichromate workers
863 Mortality of chrome platers in the West Riding

11. Nickel and its salts

43 Mortality study of nickel platers
56 Mortality of nickel workers
69 Mortality of nickel workers
95 Mortality of Nickel-Cadmium battery workers

12. Cadmium and its salts

95 Mortality of Nickel-Cadmium workers
118 Mortality of cadmium workers

13. Chemical carcinogenic agents or processes

3 A mortality study of workers at a plant processing antimony
24 A study of workers exposed to vinyl chloride monomer
55 Mortality of workers manufacturing chlorinated toluenes
59 Mortality study of British pathologists
65 A follow-up study of fertiliser workers [Nitrates]
70 Cancer mortality in male hairdressers
74 Mortality of millers
79 Mortality of acrylonitrile workers
96 A study of exposure to hydrazine
111 Mortality of pharmaceutical workers
119 Mortality of anthraquinone workers
125 Cancer incidence in relation to immune stimulation
129 Formaldehyde in the British chemical industry
132 Cohort study of the effects of exposure to styrene in industrial workers
165 Mortality and cancer morbidity among semiconductor workers
168 A cohort study of workers exposed to formaldehyde in the British chemical industry
171 Cancer of the larynx in mustard gas workers
175 Study of glass-reinforced plastic laminators exposed to styrene
178 A study of workers exposed to bis-chloromethyl ether in a methylating process
188 Employees at a factory manufacturing 4,4'-bipyridyl
190 Mortality study of pest control officers
205 A follow-up study of workers in a chemical company manufacturing methylene-bis-ortho-chloraniline (MOCA)
237 Mortality of workers manufacturing 1,2 Dibromoethane (Ethylene dibromide-EDB)
245 Cohort study of workers exposed to spinning finishes [Formaldehyde]
246 A study of workers exposed to ethylene oxide
247 Mortality study of butchers and slaughtermen
260 A study of fertiliser workers [Nitrate]
273 Mortality in Liverpool University
299 Mortality of cooks
274 Mortality of welders
286 A 50 year mortality follow-up study of brewery workers
288 Retrospective study of employees who had worked in a pharmaceutical development experimental plant
302 1961 Census-based study of butchers
315 Follow-up of persons who became ill after eating contaminated bread [Epping jaundice]
316 Mortality and cancer morbidity of production workers in the UK flexible polyurethane foam industry
352 Indigo worker mortality study
370 Colorectal cancer among polypropylene workers
373 A mortality study of aniline workers
384 A cohort study of employees in perspex plants
396 A study of workers exposed to mineral acids
406 A cohort study of employees exposed to methylene chloride in the manufacture of triacetate film base
409 Follow-up study of photographic chemical workers
416 International study of cancer risks in biology research workers
869 Survey of workers manufacturing di-amino-di-phenyl methane (DADPM)

14. Miscellaneous

52 Mortality of persons statutorily notified as suffering from chemical poisoning
72 A cohort study of mortality in serving and ex-firemen in England from 1965-1979
113 Mortality of North sea divers
136 Mortality of professional chemists in England and Wales 1965-1989
262 Stress and health study of civil servants
310 Mortality study of workers exposed to cyanide
324 Mortality in a group of workers exposed to CS_2
Survey of rayon workers exposed to carbon disulphide

A study of British Airways pilots

Annex 15.2 Brief summaries on OPCS studies of occupational health

A mortality study of workers at a plant processing antimony (MR3)

Reason for the study
In 1961, at an inquest on the death from lung cancer of a man formerly employed at a Lancashire antimony smelting plant it was alleged that further deaths had occurred, following which 3 more deaths were identified over the previous 8 years. Investigation at a similar plant in the Newcastle area also found 4 deaths from lung cancer during this period.

Study population
All personnel at the Newcastle factory who were employed at or entered employment subsequent to 1st January 1961.

Nature of the study
A pilot exercise employing local facilities failed to trace an adequate proportion of the population.

In 1971, using NHSCR an overall trace rate of 99 per cent was achieved and never less than 96 per cent for individual subsets. Coded copies of death certificates were provided. Data were analysed by occupation within the curtilage of the factory and by occupation within the smelting plant, and latterly by years of employment and years employed, and by age group.

Results
The study population were exposed to a number of substances in addition to antimony, including arsenic, arsenic oxides and polycyclic aromatic hydrocarbons. With the first analyses based on 1971, all cause mortality for the total population and individual groups, of course was below that predicted from National and from local figures (healthy worker effect). However, overall 10 cases of lung cancer were found in men where 8 would have been expected from local figures. Nine of these cases were found in antimony workers where 5.7 were to be expected, but 8 of these occurred in men aged 45-64 in whom only 4.5 were expected. The data were suggestive though not conclusive for an excess of lung cancer in workers in the antimony process, most pronounced in those employed prior to 1961.

Publications
1. A report on mortality from 1962 to 92 has been submitted for publication. Earlier references to the survey and abstracts of results have appeared in abstract in annual reports of HM Chief Inspector of Factories. As a consequence its findings were studied by standard setting bodies.

2. Jones R,D. Survey of Antimony Workers: Mortality 1961-1992. *Occ Environ Med* (In press 1994)

National survey of asbestos workers (MR5)

Reasons for the study
To evaluate the benefits of the 1969 Asbestos Regulations and subsequent related legislation, a National study of asbestos workers was mounted to include, clinical, physiological, radiological, smoking, occupation and environmental hygiene data. To complete the survey, the mortality pattern of subjects was studied.

Study population
Men and women working with asbestos in England and Wales and Scotland and who were subject to the regulations qualified for examination.

Nature of the study
Subjects were traced and flagged at NHSCR and coded death certificates and cancer registrations provided by OPCS. Mortality was to be compared with national rates and related to a number of variables including, age, sex, industry sector, job, measures of asbestos exposure, smoking history, clinical and radiographic changes and progression of signs.

Results
The first analysis carried out on data accumulated to 1981 was restricted to men employed in England and Wales. 1,123 deaths overall were reported (E=1285.51), with respiratory cancer in 186 (E=147.9, SMR=126, $p<0.01$). 35 deaths were due to malignant mesothelioma (19 per cent of all respiratory cancer deaths). Lung cancer mortality increased with years of exposure and was most noticeable in insulation workers for whom 9.8 per cent of death certificates mentioned mesothelioma. Results of a more recent analysis are given in Chapter 9.

Publications
1. Hodgson, J.T & Jones, R.D (1986) 'Mortality of asbestos workers in England and Wales: 1971-81'.See also chapter 9 of this supplement. *Br J Ind Med*, 43:158-164.

Mortality study of workers exposed to chloromethyl ether (MR7)

Reason for the study
The unexpected observation of a cluster of cases of lung cancer in a group of men exposed to chloromethyl ether containing traces of bis-chloromethyl ether, led to suspicions of a carcinogenic hazard in the process. Animals exposed to chloromethyl ether developed cancers of the skin, lung and olfactory epithelium.

Study population
1. Jarrow: 1196 men were identified by 1972 who had been at risk of exposure since 1960.

2. Pontyclun: 571 men exposed between 1948 and 1971.

Nature of the study
Subjects were traced and flagged at NHSCR and coded death certificates provided. Mortality patterns were studied in relation to Regional rates and to level and duration of exposure. The data for Jarrow are also being included by Rohm and Haas, Bristol, Pennsylvania, with findings at their plants in France and Italy.

Results
An excess of lung cancer was observed at Pontyclun which was related to level of exposure rather than duration. No further cases have been reported in workers starting after 1972, when the process was changed. For Jarrow, a deficit of cases of lung cancer was reported: this was still noted for high and low levels of exposure in the most recent analysis (September 1987).

Publications
1. McCallum, R.I, Wooley, V. & Petrie, A. (1983) 'Lung cancer associated with Chloromethyl Methyl Ether: an investigation at two factories in the United Kingdom'. *Br J Ind Med*, 40: 384-389.

2. McCallum, I.R & Wooley, V. (1985) *Lung cancer associated with Chloromethyl Methyl ether manufacture. Deaths in employees at factories A & B up to 31st December 1984* Report: Department of Occupational Health and Hygiene, University of Newcastle.

3. Gowers, D.S & DeFonso, R. (1987) *Jarrow update*. Internal report: Rohm & Haas.

The effect of coal workers' pneumoconiosis on mortality (MR12A)

Reasons for the study
Minor degrees of radiological abnormalities, increased respiratory symptoms and small ventilatory function decrement not considered clinically significant had been demonstrated to be interrelated and related to dust exposure. Their effects on life expectation had not been evaluated.

Study population
94 per cent of 26,363 miners employed at 20 collieries in England and Wales who had been the subject of a series of surveys between 1953 and 1958.

Nature of the study
Subjects were traced and flagged at NHSCR and coded death certificates provided. Mortality from all causes was compared with Regional rates, and life tables were constructed to study survival times in selected subsets.

Results
In general, SMRs for overall mortality were less than 100, though there were significant variations between regions. Poorer survival prospects were associated with the development of Progressive Massive Fibrosis and of profusions of small radiological opacities category 1 in 'Simple Pneumoconiosis'. Exposure to mine dust was associated with increased mortality from certain causes such as gastrointestinal cancer.

Publications
1. Jacobsen, M (1976) *Dust exposure, lung diseases and coalminers' mortality*. PhD Thesis, Edinburgh: The University.

2. Jacobsen, M. (1979) 'Lung cancer and coalworkers' pneumoconiosis'. *Br Med J*, ii: 208

3. Miller, B.G, Jacobsen, M. & Steele, R.C (1981) *Coalminers' mortality in relation to radiological category, lung function and exposure to airborne dust*. Final report on the CEC contract 7426-16/8/001 and Institute of Occupational Medicine, Edinburgh report TM/81/10.

4. Miller, B. G. & Jacobsen, M. (1985) 'Dust exposure, pneumoconiosis and coalminers mortality'. *Br J Ind Med*, 42:723-733.

Mortality of men employed in the British Steel Industry (MR12B)

Reasons for the study
The industry was nationalised in 1967 bringing together a variety of organisations and trades. As little was known about the health of workers in the industry overall or in its sub-units, it was decided to establish a statistical base for such study.

Study population
All males who had been continuously employed from 1st January 1963 to 1st August 1967, for whom adequate records could be obtained.

Nature of the study
Those employees lost to observation were traced at NHSCR and all employees were flagged. Coded copies of causes of death were provided.

Occupational histories were sought for all subjects from company records, and for those still in employment a variety of additional information was sought including, a detailed occupational history, smoking habits, respiratory symptoms, general health, and ventilatory function. The mortality experience of steel workers was compared with all men in the appropriate regions, using man years at risk analysis and standard mortality ratios as summaries. Additionally, comparisons were made between subsets and overall experience.

Results of the study
Of an apparent qualifying total of some 87,000 men, it was decided to analyse the deaths of 9,298 out of a total of 81,253 at risk. The overall SMR was favourable (78). The fraction of deaths due to lung cancer was statistically significantly high (relative standardised mortality ratio = 114, $p<0.001$). Other malignancies were also raised but to a much lesser extent. Analysis of mortality of some 33,000 men with adequate occupational histories, showed a relatively high mortality for blast-furnacemen who also had a marked lung cancer excess. Other trades also showed significant variations.

Publications
1. Jacobsen, M., Collings, P. L., Darby, A., Hurley, J. F., Jack, A. K. & Steele, R. C. (1979) *Mortality of men employed in the British Steel Industry: a prospective study.* (Final report of CEC contract 6244-008/8/108) Institute of Occupational Medicine Report No TM/79/14. Edinburgh.

2. Jacobsen, M., Collings, P. L., Fanning, D. M., Hurley, J. F. & Steele, R. C. (1982) *Mortality of workers in the British Steel and Coke industries.* (Final report on CEC contracts 7246-24/8/001 & 7246.24.009) Institute of Occupational Medicine Report No TM/82/06. Edinburgh.

3. Hurley, J. F., Miller, B. G. & Jacobsen, M. (1990) *Mortality 1967-1977 of industrial workers and ex-workers from the British Steel Industry: further analyses.* Institute of Occupational Medicine report No TM/90/07. Edinburgh.

Mortality of coke oven workers (MR12C)

Reasons for the study
The processes of carbonizing coal and the distillation of the liquid product, generate volatile and non-volatile carcinogenic agents. Analyses of death certificates for England and Wales for the periods 1949-1953 and 1959-1963, indicated excesses of lung cancer in the relevant category of workers. The nationalized Coal and Steel industries decided to analyse the mortality patterns of their workforce engaged in carbonizing coal.

Study population
All males for whom there were adequate records, employed on 1st January 1967 at factories involved in coal tar distillation, patent fuel manufacture and coke oven work.

Nature of the study
Subjects were traced and flagged at the NHSCRs and death certificates provided. Mortality patterns of workers were analysed within specific jobs and specific plants and by exposure histories. Comparisons were made as appropriate with overall rates for Scotland and for England and Wales, and for social class.

The latest review used more powerful analytical methods and employed improved job classification, and estimated individual exposures to Benzene Soluble Fraction.

Results
1. The initial NHSCR trace rate was 98.3 per cent. For the patent fuel worker group, the proportion of deaths from cancer at various sites was unusually high. For coke workers there was overall an excess of lung cancer. Deaths among tar distillers were too few for conclusions to be drawn.

2. An analysis of some 2,800 coke workers employed by BSC who had worked for at least 18 months prior to 1967, showed a lung cancer death rate 25 per cent higher than predicted from National rates overall, but comparable with rates for industrial workers. 'All cancer' deaths compared with those for all British men. For oven workers, the age adjusted lung cancer death rate increased with time worked at ovens.

3. A 20-year follow up of the total population to 31st July 1987 had a trace rate of 98.2 per cent for the subset analysed in (1) and 97.1 per cent for that in (2). Using general population regional rates for comparison, lung cancer SMRs for these subsets were 125 and 127 respectively (statistically significant). There was a statistically significant excess of lung cancer deaths increasing with the length of time employed. Review of leukaemia mortality did not find excesses in subsets considered to have been exposed to benzene: at 4 tar distilleries and a benzene refinery, no deaths from leukaemia were observed.

Publications
1. Institute of Occupational Medicine, Edinburgh, Reports.

1977 TM/77/13	1986 TM/86/01
1978 TM/78/01	1987 TM/87/20
1982 TM/82/06	1991 TM/91/02

2. Hurley, J. F., Archibald, R. McL., Collings, P. L., Fanning, D. M., Jacobsen, M. & Steel, R. C. (1983) 'Mortality of coke workers in Britain'. *Am J Ind Med* 4:691-704.

3. McLaren, W.M and Hurley, J. F. (1987) 'Mortality of tar distillation workers'. *Scand J Work Environ and Health* 13: 404-411.

4. Hurley, J. F., Cherrie, J. W. and MacLaren, W. (1991). 'Exposure to Benzene and mortality from Leukaemia, results from coke oven and other coal products workers'. *Br J Ind Med*, 48:502-503.

Mortality of workers in the British cotton industry (MR14)

Reasons for the study
There has been controversy as to whether the Respiratory signs and symptoms experienced by cotton workers lead to chronic disabling diseases and subsequently to an adverse mortality experience. Consequently, the then Department of Employment's Industrial Health Advisory Committee and its Advisory Panel on respiratory Diseases in Cotton Workers, recommended that a mortality study be carried out.

Study population
3,884 workers who had been examined between 1966 and 1970 in a longitudinal morbidity study of UK cotton workers.

Nature of the study
Subjects were traced at NHSCR and flagged and coded death certificates were provided. Overall and individual cause mortality were compared with values predicted from National mortality rates. Mortality experience was related to years of employment, byssinosis symptoms and smoking habit.

Results
For subjects for whom there were some examination data there were statistically significant deficits found for mortality overall, and for diseases of the respiratory system and for all malignancies. There was a tendency for mortality to improve with increase in the period of follow up. For 340 workers for whom there were no examination data, the picture was different with statistically significant excesses for all cause mortality and for circulatory disease and for cardiovascular disease. Subjects who had reported byssinosis symptoms had slightly raised mortality from respiratory diseases overall. Mortality from lung cancer was lower than expected and decreased with length of service.

Publication
1. Hodgson, J. T. and Jones, R. D. (1990) 'Mortality of workers in the British cotton industry in 1968-1984'. *Scand J Environ Health*, 16: 113-20.

A study of nickel/chromium platers (MR19)

Reasons for the study
There has been concern for the cancer risk of workers exposed to nickel, chromium and various of their compounds. The process of plating generates a fine mist of electrolyte. As the existing data for evaluating the health of platers have generated suspicious findings rather than firm conclusions, it was decided to study a further group.

Study population
Workers who had started employment in a large plating works between 1946 and 1975 and who had been employed for a minimum of 6 months and for whom adequate records existed.

Nature of the study
Subjects were traced by NHSCR at 31st December 1983 and copies of coded death certificates provided by OPCS. The mortality experience was compared with that predicted from England and Wales rates according to age, sex, calendar year and

man years. Regression models and life tables were employed to test the hypothesis of chromium having no effect on mortality. Chrome exposure was expressed both as cumulative duration of exposure in any plating job, and as cumulative duration of plating bath exposure.

Results
Statistically significant excesses of mortality were found among males for: carcinoma of the stomach (0=21, E=11.3 p<0.05); primary liver cancer (0=4, E=0.6 p<0.01); carcinoma of the nose and nasal sinuses (0=2, E=0.2 p< 0.05); carcinoma of the lung (0=63, E=40.0 p<0.001).

Regression modelling indicated that chrome bath workers most heavily exposed to mist were at greater risk from lung cancer. Duration of work at the bath provided a significant positive association. Nickel exposure was not found to be an important confounding factor.

Publication

1. Sorahan, T., Burges, DC. L. and Waterhouse, J. A. H. (1987) 'A mortality study of nickel/chromium platers'. *Br J Ind Med*, 44:250-258.

Royal Naval Dockyard Mortality Studies (MR20)

Reasons for the studies
Earlier studies having shown that not only workers manipulating asbestos products were affected but also other tradesmen working near them were suffering, the significance of minor effects on the chest radiograph and minor decrements of lung function on life expectation required to be evaluated.

Study populations
The workforce at Devonport, Chatham, Portsmouth and Rosyth Royal Naval Dockyards was studied over various periods of time and in various combinations.

Nature of the studies
Tracing and flagging were carried out by NHSCRs and death certificates provided.

1. A mortality study of 6,292 men employed at Devonport on 1st January 1947, followed to the end of 1978.

2. A mortality study of all 32,462 people employed at the three English RN dockyards for at least 6 months and followed up for 17 years from 1972/3. A subset of males from Devonport who had been examined by symptom questionnaire and chest radiograph formed an interim analysis.

Results
In the first study a trace rate of 99.7 per cent was achieved. Of 1,043 deaths, 31 were due to malignant mesothelioma and 14 to asbestosis or pulmonary fibrosis. There was a deficit of lung cancer.

In the second study, over 97 per cent were traced. Some 700 men had died of lung cancer and 130 of mesothelioma. Smoking was a highly significant variable in predicting death from lung cancer, stomach cancer and ischaemic heart disease. Duration of high exposure to asbestos was the best predictor of mesothelioma. Respiratory symptom complaint was associated with excess overall mortality. Although a considerable number of malignant mesotheliomas was found, there was no excess of lung cancer.

Publications

1. Rossiter, C. E. and Coles, R. M. (1980) 'HM Dockyard Devonport: 1947 mortality study'. In *Biological effects of mineral fibres*. Ed. Wagner, J. C. IARC Scientific Publications No 3. Inserm Symposia Series Volume 92. IARC, Lyon. Vol 2. 713-721.

2. Rossiter, C. E., Coles, R. M. and Jackaman, I. (1983) *HM Naval Base, Devonport: lung cancer and mesothelioma case control studies*. Paper presented to the ILO Pneumoconiosis Conference, September 1983.

3. Toll, S. (1984) *Royal Naval Dockyard Mortality Study*. Report: MRC Clinical Research Centre and Leicester Polytechnic.

A study of workers exposed to vinyl chloride monomer (VCM) (MR24)

Reason for the study
The reports in 1974 of positive animal carcinogenicity tests at low level exposure coinciding with the report of three cases of angiosarcoma of the liver in three heavily exposed VCM workers in one American company prompted a national industry study to investigate the extent of the effect of VCM exposure.

Study population
1. In the first round, a total of 7,409 workers were identified as being potentially exposed to VCM.

2. 5,560 persons first potentially exposed between 1940 and 1974 for a minimum of one year and for at least 25 per cent of the time in specific jobs.

Nature of the study
In both phases of the study subjects were traced and flagged at NHSCR and coded copies of death certificates provided. Predicted values for specific cause mortality were calculated from National rates, adjusted for person years. The population in this study were included in the International Agency for Research on Cancer multi-centre study.

Results
A trace rate of 98.9 per cent was achieved and 780 deaths identified. There was an overall deficit in mortality (0=381, E=429 p<0.001), mainly contributed to by circulatory and respiratory diseases which were significantly in deficit. There was a significant excess of non-metastatic liver cancer (0=11, E=1.94 p<0.001). The most heavily exposed persons had a significant excess of liver tumours (0=7, E=0.38 p<0.001).

Publications

1. Fox, A. J. and Collier, P. F. (1977) 'Mortality experience of workers exposed to vinyl chloride monomer in the manufacture of polyvinyl chloride in Great Britain'. *Br J Ind Med*, 34: 1-10.

2. Simonato, L. L'Abbe, K.A. Andersen, A Belli, S. et al 1991 "A collaborative study of cancer incidence and mortality among vinyl chloride workers. *Scand J Work Environ Health*, 17:159-69.

3. Jones, R. D., Smith, D. M. and Thomas, P. G. (1988) 'A mortality study of vinyl chloride monomer workers employed in the United Kingdom 1940-74'. *Scand J Work Environ Health*, 14: 153-60.

Mortality study of carbon black workers (MR25)

Reasons for the study
Industrial carbon black contains traces of absorbed hydrocarbons. These are extractable with difficulty but have been found in experimental animals to include powerful carcinogens. Substantial amounts of carbon black have been used in the printing and rubber industries in which lung cancer excesses have been reported in the past indicating a need to study the possible carcinogenicity of this agent.

Study population
1,422 male manual workers employed for at least 1 year between 1947 and 1974 at five carbon black factories for whom adequate personnel records were available.

Nature of the study
Subjects were traced and flagged at the NHSCRs and coded causes of death were provided. Expected numbers of deaths were calculated for person years at risk using Regional specific rates and National rates as appropriate. Duration of employment was compared for lung cancer cases and matched controls.

Results
An overall trace of 99.8 per cent was achieved. For the factories for which data were probably complete, using National rates for comparison, overall mortality was less than predicted as was death from all cancers. There were however 25 deaths from respiratory cancer (E=16.2, SMR=154) and 3 from bladder cancer (E=1.3, SMR=233).

Adjusting for local rates these excesses persisted but were still not statistically significant. (The extent of incompleteness of the nominal roll might have been sufficient to lead to the understatement of adverse effects.)

Publication

1. Hodgson, J. T. and Jones, R. D. (1985) 'A mortality study of carbon black workers employed at five United Kingdom factories between 1947 and 1980'. *Arch Environ Health*, 40: 261-268.

Mortality of talc workers (MR30)

Reasons for the study
Talc, a platey form of hydrated magnesium silicate, has been found to include fibrous mineral in certain specimens. Animal studies and studies of humans exposed to non-asbestos fibrous minerals have shown some of them to be carcinogenic. This prompted a human epidemiological study of workers exposed to talc of a grade typically containing no mineral fibre.

Study population
Workers employed between 1953 and 1973 in a factory compounding and packing cosmetic talc and using talc in tablet manufacture.

Nature of the study
Subjects were traced and flagged at NHSCR and coded copies of death registrations provided to the researchers. Observed values for individual cause mortality, with special attention to malignancies, will be compared with expected values derived from National and Regional rates.

Mortality study of lead workers undergoing statutory examination (MR33)

Reasons for the study
A range of effects of lead have been identified in exposed workers that may be detected in the absence of overt disease. Statutory surveillance of lead workers provides a number of standardised measures of exposure and effects in well characterised populations, offering the possibility of studying long-term health consequences of exposure.

Study population
All workers in England, Wales and Scotland required to undergo statutory surveillance from 1973 to 1979.

Nature of the study
Subjects have been traced and flagged at NHSCR and coded death certificates obtained.

Chrome pigment worker health studies (MR41)

Reasons for the studies
Experimental studies with chromium and certain of its compounds have shown them to be carcinogenic. Workers exposed to chrome ores and zinc and lead chromates have been shown to experience lung cancer excesses. A study was proposed to determine whether it was the zinc chromate that was responsible rather than the lead salt.

Study population
All male blue collar workers employed at three factories for at least a year by 30th June 1975.

Nature of the study
All workers for whom there were adequate records were traced at NHSCR and flagged, and coded causes of death were provided. Men were divided into sets by year of entry and analyses were related to duration of service and by notional exposure grades. Observed deaths were compared with expected values derived from National rates.

Results
The first analysis showed that by mid-1977, at factory A there was a significant increased mortality from lung cancer for workers with high and medium exposures who had started employment 1932-54. No excess was observed for men with low exposures and with less than 1 year's service. No excess lung cancer was found among men first employed after 1954. At factory B a lung cancer excess was found associated with high and medium exposures. At factory C where zinc chromate was not produced, no lung cancer excess was found.

By 1981, a 99.7 per cent trace was achieved on 1152 men in the study. It was concluded that there was no support for lead chromate being considered carcinogenic but it was noted that lung cancer was being observed after an unusually short latent period.

In a separate analysis of 57 chromate workers who had been diagnosed as having had lead poisoning, it was noted that there was a significant excess of deaths from nephritis and cerebrovascular diseases.

Publications

1. Davies, J. M. (1978) 'Lung cancer mortality of workers making chrome pigment'. *Lancet*, (i): 384.
2. Davies, J. M. (1984) 'Lung cancer mortality among workers making lead chromate and zinc chromate at three English factories'. *Br J Ind Med*, 41: 158-169.
3. Davies, J. M. (1984) 'Long term mortality study of chromate pigment workers who suffered lead poisoning'. *Br J Ind Med*, 41: 170-178.

Mortality of workers at an asbestos factory in Leeds (MR42)

Reasons for the study
The data on this factory had originally been collected with a Rochdale population (MR177). As they had had a relatively short-term exposure to asbestos, it was decided to follow them separately.

Study population
Some 500 women employed during and shortly after the Second World War at a Leeds asbestos factory.

Nature of the study
Subjects have been traced through NHSCR and flagged, and coded death certificates have been provided to the researchers. The pattern of mortality is to be compared with National and Regional rates.

Results
A first test analysis showed an abnormally low overall mortality rate. Before any further analyses are attempted, a careful review of the data base is to be conducted.

Mortality study of nickel platers (MR43)

Reasons for the study
Nickel had been suspected as being a carcinogen, and its compounds have been shown to be experimental carcinogens. A human cancer risk has been reported in certain smelting processes where a number of confounding factors exist. Studies of platers previously had been confounded by exposure to chromium and its salts.

Study population
929 workers at a factory operating since 1945, who had worked solely in nickel plating.

Nature of the study
Subjects were traced at NHSCR and flagged and coded causes of death were provided. SMRs were calculated using National rates and adjusting for man years. Local rates were also applied. Mortality experience was related to exposure in terms of job held and period of employment.

Results
The first analysis of deaths showed that in the group with longer exposure, there was a significant excess of stomach cancer and non-malignant respiratory disease. Of the 8 deaths from lung cancer, 4 were in nickel bath workers who died unusually young.

Publication

1. Burges, D. C. L. (1980) 'Mortality study of nickel platers'. In: *Nickel Toxicology*. Eds Brown, S. S. and Sunderman, F. W. Academic Press, New York.

Mortality studies of female gas mask assemblers exposed to asbestos (MR44)

A Study of Nottingham gas mask workers exposed to crocidolite asbestos 1939-1944

Reason for the study
The report in 1965 of a case of malignant mesothelioma in a woman who had assembled gas masks containing crocidolite at a factory in Nottingham during the Second World War, prompted a study at this factory, to compare mortality from this species of asbestos with other species.

Study population
All woman weighing or assembling filters, and others working in the same room, to a total of 535, were identified from company records.

Nature of the study
Subjects were traced and flagged at NHSCR and death certificates were provided. Exposure category was determined for each subject from a variety of records, as, 'definitely exposed', 'definitely not exposed' and 'not known', as was duration of exposure. Deaths were analysed for the period 1951-77.

Results
As a result of inadequate identification details, a trace of only 93 per cent was achieved. Overall mortality was slightly below predicted values, but 'all cancers' were significantly raised (0=64, E=36.2), as was 'lung cancer' (0=10, E=3.7). Among women definitely exposed to asbestos, there were 5 deaths from ovarian carcinoma (E=3.7), but none among those not exposed (E=0.63). There were 12 deaths from malignant mesothelioma.

Publication

1. Wignall, B. K. and Fox, A. J. (1982) 'Mortality of female gas mask assemblers'. *Br J Ind Med*, 39: 34-38

B Mortality of two groups of women exposed to chrysotile and to crocidolite in the assembly of gas masks

Reason for study
To evaluate the relative hazards of a serpentine and an amphibole asbestos. The opportunity was taken of studying separate groups of women assembling civilian gasmasks (chrysotile) and military (crocidolite).

Study populations
The 1939 National Register was used to identify women in towns where gas masks were manufactured (570 in Blackburn, chrysotile; and 757 in Leyland and Preston, crocidolite) who were so employed.

Nature of the study
Subjects employed in September 1939, were identified, then traced and flagged at NHSCR. Their death certificates were extracted to 30 June 1980 and provided to the researchers in anonymised form for analysis.

Expected values for specific cause deaths were calculated adjusting for person years from England and Wales rates, and by area for the years 1968-78.

Results
There were significant excesses of lung cancer and of ovarian cancer (P=0.01) for the Leyland workers,

together with 5 malignant mesotheliomas. The Blackburn group had no specific cause cancer excess other than a case of malignant mesothelioma.

Publication

1. Acheson, E. D., Gardner, M. J., Pippard, E. C. and Grime, L. P. (1982). 'Mortality of two groups of women who manufactured gas masks from chrysotile and crocidolite asbestos: a 40-year follow up'. *Br J Indust Med*, 39: 344-348.

C A study of gas mask workers

Study population

This has combined all the gas mask workers including the Nottingham population, the Birmingham population MR139, and the Blackburn, Leyland/Bolton populations.

Nature of the study

All subjects were traced and flagged, and coded death certificates provided to the researchers. In this study, in addition to the simple mortality analyses, where possible, lung tissue was analysed to identify the type and quantity of asbestos fibre present in lung parenchyma.

Results

By 1990, 55 deaths from malignant mesothelioma had been identified since 1965 among workers predominantly exposed to crocidolite, whereas only 6 were to be found among workers exposed to chrysotile. In general, malignant mesotheliomas were associated with heavier exposures to asbestos, though some cases occurred in women outside the department and with only marginal contact.

Publications

1. Jones, J. S. P., Pooley, F. D. and Smith, P. G. (1976) 'Factory populations exposed to crocidolite asbestos — a continuing survey'.In: *Environmental Pollution and Carcinogenic risks*. Eds. Rosenfeld, C. and Davis, W. Inserm Symposia Series Vol 52. IARC Scientific Publications No 13.

2. Jones, J. S. P., Smith, P. G., Pooley, F. D. et al (1981) *The consequences of exposure to asbestos dust in a wartime gas-mask factory*. Proceedings of the 1979 Lyon Conference on the Biological effects of mineral fibres. Inserm Symposia vol 91. IARC Scientific Publications No 30.

3. Gibbs, A. R., Jones, J. S. P., Bolton, V. G., Pooley, F. D. and Griffiths, D. M. (1991) *Mesothelioma in wartime gas mask workers*. Presentation at the Fourth International Conference on Environmental lung disease. Montreal.

NRPB: National Registry of Radiation Workers (NRRW) (MR47)

Reason for the study

To study the association between fatal neoplasms and measured low level radiation exposure.

Study population

Initially four groups of consenting workers employed by UKAEA, BNFL, CEGB, MoD and SSEB:

Radiation workers employed on or after 1st January 1976;
Ex-radiation still in employment 1st January 1976;
Ex-radiation workers who had left or died prior to 1st January 1976;
Those who became radiation workers after 1st January 1976.

Subsequently further groups were included of industrial radiographers, Science Research Council workers and workers at the Radiochemical Centre.

Nature of the study

Subjects were traced at the NHSCRs for England and Wales and for Scotland, with assistance from National Insurance records. They were flagged and coded copies of death registrations were provided. The first analysis of mortality of 95,217 radiation workers was made at 31st December 1988, comparing observed deaths with numbers predicted from National rates, and calculating SMRs, adjusting for person years. In addition an internal analysis was made, taking into account a number of personal and occupational characteristics, to study trends in mortality with recorded doses of external radiation. To allow for latency, doses were lagged for 2 years for leukaemia and 10 years for other selected neoplasms as the cause of death. At a later date, internal doses will be taken into consideration.

Results

Overall, a 99.8 per cent trace was achieved. Although mortality was lower than for the general population, for cancer and in particular leukaemia and multiple myeloma there was evidence for an association with external radiation exposure. The SMR for thyroid cancer was significantly raised at 303 (0=9, E=2.97), but there was no trend with external dose.

Analyses internal to the NRRW cohort provided evidence of an increasing trend in risk with dose for leukaemia (p=0.03, one-sided test) and, to a lesser extent, for all malignant neoplasms (p=0.10).

Lifetime risk for all neoplasms was 10.0 per cent per Sv (90 per cent C. I.=<0 - 24 per cent) and 76 per cent per Sv for leukaemia (0.07 - 2.4 per cent). These are higher than ICRP estimates, but the confidence intervals are large and the ICRP values fall well within them.

Publication

1. Kendall, G. M., Muirhead, C. R., MacGibbon, B. H., et al (1992) 'Mortality and occupational exposure to radiation: first analysis of the National Registry for Radiation Workers'. *Br Med J*, 304: 220-225.

Mortality of workers at acetylene production plants (MR48)

[Crocidolite asbestos exposure]

Reason for study

Acetylene cylinders incorporated a porous plug including crocidolite asbestos from 1930 or earlier until 1972, that required disturbance in servicing. When the potential hazard was appreciated it was decided to evaluate the risk of the practice.

Study population

All workers for whom there were records employed for at least one shift who started between 1935 and 1976 at plants scattered throughout the United Kingdom.

Nature of the study

Subjects were traced and flagged at NHSCR and coded death certificates provided. All cause and specific cause mortality were studied for all the plants. Predicted values were based on Scotland or England and Wales figures as appropriate: for one plant in the West Midlands regional figures were also used.

Results

Overall there was a 98.2 per cent successful trace, with 99.5 per cent for the largest and oldest plant with the most complete set of records. No cases of malignant mesothelioma were found but there was an excess overall of death from cancers of the lung, stomach, and pancreas. This varied between plants and the significance of the findings is uncertain.

Publication

1. Newhouse, M. L., Matthews, G., Sheikh, K., Knight, K. L., Oakes, D. and Sullivan, K. R. (1988) 'Mortality of workers at acetylene plants'. *Br J Ind Med*, 45: 63-69

Mortality of persons statutorily notified as suffering from chemical poisoning (MR52)

Reason for the study

The criteria of effect in statutory notification of chemical poisoning vary from ulceration of the skin or nasal septum in chrome workers to acute or chronic systemic toxicity in lead poisoning. The question arose as to whether after the ostensible recovery from the condition there might be long-term health consequences from what were commonly heavy exposures.

Study populations

Workers notified as poisoned by lead, chrome, arsenic, benzene, trichloroethylene, mercury and nickel carbonyl.

Nature of the study

Subjects were traced and flagged at NHSCR and coded death registrations provided. The population was divided by agent into subsets according to year of notification. Observed values are to be compared with those values derived from National rates.

Mortality of bichromate workers (MR54)

Reasons for the study

Studies in the UK and abroad have suggested that there was a risk of a lung cancer excess in workers exposed to chromate dust. A study was requested to measure the risk in a UK population and to evaluate the benefits of process modification.

Study population

Past and present workers employed for at least 1 year between 1948 and 1977 at three factories, who had had a chest X-ray and for whom there was adequate information.

Nature of the study

Subjects were traced and flagged at NHSCR and copies of coded draft death entries were provided. Observed deaths were compared with predicted values calculated from National rates according to man years, age, sex and calendar period. Multivariate analysis was also carried out to disentangle the effects of duration of employment and the effects of plant modification.

Results

Despite the incompleteness of records, a trace rate of 97.1 per cent was achieved overall, with no less that 95.3 per cent in any sub-group. By 1977, a lung cancer excess was found (0=116, E=48.0 p<0.001).

In the one factory still operating, the relative risk for lung cancer had been reduced from over 3.0 before plant modification to 1.8. Duration of employment was the major independent factor associated with the risk of lung cancer. The analysis at 1988 confirmed the reduction in lung cancer risk and indicated a nasal cancer risk.

Publications

1. Alderson, M. R., Rattan, N. S. and Bidstrup, P. L. (1981) 'Health of workmen in the bichromate producing industry'. *Br J Ind Med*, 38: 117-124.

2. Davies, J. M. (1984) 'Lung cancer mortality among workers making lead chromate at three English factories'. *Br J Ind Med*, 41: 158-169.

3. Davies J. M. (1984) 'Long-term mortality of chomate pigment workers who suffered lead poisoning'. *Br J Ind Med*, 41: 170-178.

4. Davies, J. M., Easton, D. F. and Bidstrup, P. L. (1991) 'Mortality from respiratory cancer and other causes in the United Kingdom chromate workers'. *Br J Ind Med*, 48: 299-313.

Mortality of workers manufacturing chlorinated toluenes (MR55)

Reasons for the study

In the process of synthesizing chlorinated toluenes, a range of other compounds are produced including benzotrichloride. A report of lung cancer excess in a group of exposed workers, animal experimental evidence of a carcinogenic effect of benzotrichloride, and suggestive findings from a proportional mortality study for an excess of lung tumours and lymphomas in a group of exposed UK workers, prompted a prospective mortality study.

Study population

Workers employed for a minimum of 6 months since 1st January 1961 and 31st December 1975: this included a survivor group with a potential for a maximum exposure of 24 years prior to 1st January 1961.

Nature of the study

Subjects were traced and flagged at NHSCR and coded death certificates provided. Mortality was analysed initially for the period 1961-1976 and subsequently for 1961-1984. SMRs for 1961-1976 were calculated using National rates and those for Merseyside, and an analysis conducted based on regression models in life tables. For the period 1961-1984, SMRs were based on National rates, and a 'nested case control study' was conducted.

Results

1961-1976: Using National rates, there were statistically significant excesses of malignancies of the digestive and respiratory systems and of cancer overall. When Merseyside rates were used, there was no longer a lung cancer excess. Years of employment and level of exposure to chlorinated toluene affected the cancer mortality pattern.

1961-1984: Using National rates, SMRs of 138 for 'all causes', 163 for 'all cancers' and 129 for non-malignant disease were calculated. Statistically significant excesses were found for lung cancer (0=26, E=14.5) and Hodgkin's disease (0=3, E=0.4). Workers exposed to chlorinated toluenes had significant excesses of lung cancer in the period. 1977-1984 ($p<0.05$) and in the period 1961-84 ($p<0.001$).

Publications

1. Sorahan, T., Waterhouse, J. A. H., Cooke, M. A., Smith, E. M. B., Jackson, J. R. and Tempkin, L. (1983) 'A mortality study of workers in a factory manufacturing chlorinated toluenes'. *Annals Occup Hyg* 27:173-182.

2. Sorahan, T. and Cathcart, M. (1989) 'Lung cancer mortality among workers in a factory manufacturing chlorinated toluenes: 1961-1984'. *Br J Ind Med* 46: 425-427.

Mortality of nickel workers (MR56)

Reasons for the study

In 1932, ten cases of nasal cancer (9 fatal), had been diagnosed in workers at a nickel refining plant in Wales. Subsequent reviews of this population were made to evaluate the effects of reducing exposure to dust and to identify the causal agent.

Study population

Initial mortality studies at this plant attempted to review all men employed or likely to have been employed for 5 or more years since 1902. Successive studies have discovered more subjects in the 'older' group, as well as accumulating larger numbers overall. Latterly a subset has been identified where exposure was to soluble nickel compounds.

Nature of the study

Subjects were flagged and traced at NHSCR and coded death certificates provided. Observed deaths were compared with expected values derived from National rates and related to period of exposure, years of exposure, year of first employment, department of employment and the changing environmental conditions.

Results

For the subset in chemical manufacture, the population size was too small to detect other than gross hazard

Using National rates for comparison, the lung cancer rate for refinery workers was significantly raised, possibly to 2½ times the normal risk for men. When local rates were used this excess was no longer statistically significant.

Initially it was thought that the cancer risk was eliminated by 1925, but it was found to have continued beyond that date, though reducing both for nasal cancer and for lung cancer, more markedly for the former. By 31st December 1984, it was possible to view 1930 as the watershed. Workers employed prior to that date experienced 172 deaths from lung cancer (Expected=43.77 England and Wales figures), and 74 deaths from nasal cancer with 0.35 expected. For those employed after 1930, there were 44 deaths from lung cancer with 35 expected 0:E 95 per cent CI=91-168), and 1 death from nasal cancer with 0.18 expected (0:E 95 per cent CI=14-3028).

Publications

1. Doll, R., Morgan, L. G. and Speizer, F. (1970) 'Cancers of the lung and nasal sinuses in nickel workers'. *Br J Cancer*, 24: 623-632.

2. Doll, R., Matthews, J. D. and Morgan, L. G. (1977) 'Cancers of the lung and nasal sinuses in nickel workers: a reassessment of the period of risk'. *Br J Ind Med*, 34: 102-105.

3. Cuckle, H., Doll, R. and Morgan, L. G. (1980). 'Mortality study of men working with soluble nickel compounds'. In: *Nickel toxicology*, Ed. Brown, S. S. and Sunderman, F. M. Academic Press, p11.

4. Easton, D. F., Peto, J., Morgan, L. G., Metcalfe, L. F., Usher, V. and Doll, R. (1990) 'Respiratory cancer mortality in Welsh nickel refiners — which nickel compounds are responsible?' In: *Nickel and human health-current perspectives. Advance in environmental science and technology*. Ed. Niebor, E. and Altio, A. John Wiley, New York.

5. Doll, R. [Chairman] (1990) 'Report of the International Committee on nickel carcinogenicity in man'. *Scand J Work Environ Health*, 16: 1-82.

Mortality study of British pathologists (MR59)

Reasons for the study

A number of agents met with in the pathology laboratory are known to be carcinogenic. Previous studies of pathologists having suggested certain neoplasms to be occurring to excess, a confirmatory study was indicated.

Study population

2720 male and female Members of the Royal College of Pathologists alive on 31st December 1973 and those joining subsequently up to 31st December 1980.

Nature of the study

Subjects were traced and flagged at NHSCR and coded death certificates provided. Predicted values for deaths for all cause mortality and for specific causes were calculated from National and Regional rates adjusted for person years.

Results

99.5 per cent were traced at NHSCR and 96 per cent of death certificates were obtained. Overall mortality for men was extremely favourable, that for women less so. Previous observation of excess mortality from suicide, accident, poisoning and violence were confirmed for men and women. A non-Hodgkin's lymphoma excess was not confirmed. An excess of brain tumours were found among the men that remains to be explained. No malignancy associated with formalin exposure was reported.

Publication

1. Harrington, J. M and Oakes, D. (1984) 'Mortality study of British pathologists 1974-80'. *Br J Ind Med*, 41: 188-191

A mortality study of workers in Lancashire cotton mills (MR61)

Reasons for the study

The Registrar General has repeatedly noted an excess of deaths from cardiovascular disease and nephritis in cotton workers, with higher death rates for those in dustier jobs. It was thought that this excess was due in part to confounding by respiratory disease. Latterly, excesses of oral and pharyngeal cancers have been confirmed for woman cotton workers.

Study population
1,586 Lancashire cotton workers who had participated in a longitudinal study of respiratory morbidity were considered for analysis.

Nature of the study
Subjects were traced by the NHSCR and coded copies of death certificates provided. Observed numbers of cases were analysed by cause, comparing them with values that would be predicted from National and by Southeast Lancashire conurbation and Greater Manchester rates.

Results
Despite inadequate identification data an overall trace of 94 per cent was achieved. Where full details were available this rose to 99 per cent. For males and females overall mortality was less than expected and no specific cause was in excess. Length of exposure did not lead to increased mortality. The subsets currently in dustier jobs or with higher prevalences of byssinosis did not have higher mortality. There was no evidence that mortality was associated with byssinosis but men with bronchitis had higher mortality than those without. It was not possible to confirm excesses reported in previous publications.

Publication

1. Berry, G. and Molyneux, M. K. P. (1981) 'A mortality study of workers in Lancashire cotton mills'. *Chest.* 79S: 11S-15S.

Case-control study of lung cancer and of prostatic cancer and exposure to cadmium compounds and other chemicals (MR62)

Reasons for the study
Cadmium and its compounds have been found to be carcinogenic in the laboratory. Epidemiological studies have provided some reasonable evidence for a human lung cancer hazard, but have been less persuasive for there being a prostatic cancer hazard. Interpretation has been affected by mixed occupation exposures to other carcinogens.

Study population
The persons included in 3 UK studies of workers exposed to cadmium (MR 62, 95, 118).

Nature of the study
Tracing and flagging have been carried out at NHSCR and coded copies of death registrations provided to the researchers. An assessment will be made of historical occupational exposures for each job, department and for each calendar year. Estimated exposures of employees dying of lung cancer and of prostatic cancer will be compared with those of four survivors matched as closely as possible for year of starting employment, sex, plant and year of birth (nested case control study).

This UK study will contribute to an international collaborative study.

Mortality of workers manufacturing friction materials using asbestos (MR63)

Reason for study
To determine the health effects of exposure to chrysotile asbestos as used at a particular factory and to compare them with the effects of crocidolite employed at two periods of time.

Study population
13,460 workers employed on or after 1941.

Nature of the study
Subjects were traced and flagged at NHSCR and coded death certificates were provided. For men and for women, observed all cause and specific cause mortality was compared with values predicted from National rates. These were related to length of employment and year of commencement of employment, and after 20 years of employment. A secondary study was conducted to relate malignant pleural mesothelioma to exposure to crocidolite using matched controls.

Results
It was concluded that an excess of mortality was not detectable in this study. An association between crocidolite exposure and mesothelioma was significant at the 0.06 level.

Publications

1. Newhouse, M. L., Berry, G. and Skidmore, J. W. (1982) 'A mortality study of workers manufacturing friction materials with chrysotile asbestos'. In: Walton, W. H. ed. *Inhaled particles V*. Pergamon, Oxford.

2. Newhouse, M. L., Berry, G. and Skidmore, J. W. (1982) 'A mortality study of workers manufacturing friction materials with chrysotile asbestos'. *Ann Occup Hyg*, 26: 899-909

3. Berry, G. and Newhouse, M. L. (1983) 'Mortality of workers manufacturing friction materials using asbestos'. *Br J Ind Med*, 40: 1-7

4. Newhouse, M. L. and Sullivan, K. (1989) 'A mortality study of workers manufacturing friction materials: 1941-1986'. *Br J Ind Med*, 46: 176-179

A follow-up study of fertiliser workers (MR65)

Reason for the study
As a result of experimental studies, certain nitrosamines were found to be powerful carcinogens and as a consequence were suspected as human hazards. There was concern that endogenous synthesis might be enhanced following an increased intake of inorganic nitrate. It was decided that study of cancer mortality in fertiliser workers would be a good test of this hypothesis.

Study population
All male fertiliser workers identified from census schedules by OPCS from the 1961 and 1971 censuses of England and Wales.

Nature of the study
Subjects were flagged at NHSCR and death certificates were abstracted. (Confidentiality of the census data was preserved by their handling being restricted to OPCS staff and information provided to outside researchers not permitting the identification of individuals.) Observed mortality for malignancies of oesophagus, stomach, intestine, rectum, liver, lung and bladder (posited as markers for nitrosamine carcinogenicity), were compared with expected values derived from England and Wales rates. Analyses were made at 1978 and 1985. The 1971 group were also compared with expected values derived from working men in the OPCS Longitudinal Survey.

Results

1. 1978 analysis. For the 1961 group, there was low mortality for all causes, all cancers and circulatory diseases. There was no excess cancer at any site, nor was a dose response observed. The 1971 group had an excess of deaths from 'all cancers', but no individual cause differed significantly from expected values derived from National rates or those for working men.

2. 1985 analysis. The 1961 group findings were replicated. Evidence for an association between cancer mortality and nitrate dust exposure was strengthened in the follow-up of the 1971 group, but there was a negative association with nitrate exposure and digestive organ cancers.

Publications

1. Fraser, P., Chilvers, C. and Goldblatt, P. (1982) 'Census based mortality study of fertiliser manufacturers'. *Br J Ind Med*, 39: 323-29

2. Fraser, P., Chilvers, C., Day. D and Goldblatt, P. (1989) 'Further results from a census based mortality study of fertiliser manufacturers'. *Br J Ind Med*, 46: 38-42.

Mortality of nickel workers (MR69)

Reasons for the study
The nasal cancer excess originally observed in workers in the Mond process was initially attributed to nickel carbonyl. Studies of workers in other types of refinery showed excesses of respiratory cancers, and several agents were suspected at various stages of refining, including nickel and its compounds, arsenic and fibrous minerals including asbestos and mordenite. A factory was identified where nickel alloys were produced from metallic nickel and other metals including Iron, copper, cobalt, chrome and molybdenum.

Study population
1,925 men were identified from company records who had worked in the operating area at the plant for at least 5 years before 1st April 1978.

Nature of the study
Subjects were traced and flagged by NHSCR and coded death certificates provided. The rates of mortality of males resident in Herefordshire towns as well as National rates were used to calculate expected values.

For those workers not qualifying for inclusion in the population, medical records and local registries were searched for nasal sinus cancer.

Results
A 98.9 per cent trace was achieved and 117 deaths identified. When the group was subdivided according to level and duration of exposure to atmospheric nickel, lung cancer excess was greatest for those with the highest exposure and for those observed for 10 or more years after first employment, but statistical significance was not observed. Because of the small numbers involved and the short history of the factory, the findings neither confirm or refute the observations in other studies.

Publications

1. Cox, J. E., Doll, R., Scott, W. A. and Smith, S. (1981) 'Mortality of nickel workers: experience of men working with metallic nickel'. *Br J Ind Med*, 38: 235-239.

Cancer mortality in male hairdressers (MR70)

Reasons for the study
When certain hair dyes proved positive on mutagenicity testing, this led to concern that hairdressers might have been at risk of cancer excess. A number of studies had produced contradictory results so a study of national data was set up to test the hypotheses generated.

Study population
A 10 per cent sample of persons identified as 'hairdressers' was drawn from the 1961 census, and their status determined from the NHSCR records. Identification details were removed before the data on causes of death left OPCS to preserve confidentiality.

Results
91.6 per cent of the 1,999 hairdressers in the census sample were traced, for whom there were 504 deaths. There was no support for there being an occupational cancer risk for male hairdressers. As hair dye use was not ascertainable for individuals there was a limit to the confidence that might be placed in the conclusions to be derived from this study.

Publications
1. Alderson, M. (1980) 'Cancer mortality in male hairdressers'. *J Epidem Community Med*; 34: 182-185

A cohort study of mortality in serving and ex-Firemen in England from 1965-1979 (MR72)

Reasons for the study
Suspecting that the greater involvement of synthetic polymers in fires might be adding to hazard, a further study of fire fighters was mounted.

Study population
Samples of active or retired members of 5 brigades, selected as representative of city and of rural brigades, with a minimum of 1 year's service, and who were in employment on or after 1st April and until 31st December 1979.

Nature of the study
Subjects were flagged at NHSCR and copies of coded death certificates provided.

Standardised mortality ratios were calculated from Regional rates, and for the period 1971-75 were based on the rates derived from the OPCS Longitudinal Study.

Results
Overall mortality was favourable. A small increase in the risk of death from ischaemic heart disease was seen in some brigades. In general there was no evidence of an increased risk of death from respiratory disease. One region only showed slight evidence for an excess of lung cancer and other respiratory diseases. A follow-up analysis by 5 year intervals of ischaemic heart disease in firemen 1965-1986, showed a clear downward trend for certain regions. Mortality from ischaemic heart disease in firemen overall was less than for the general population. No deaths from ischaemic heart disease were certified in 1980-1986 for 'never smokers', whereas for 'ex-smokers' they made up one third of all deaths, and for 'current smokers' one fifth.

Publications
1. Donnan, S. P. B. (1982) *Study of the causes of death in firemen*. Research report No 20. Central Fire Brigades Advisory Councils for England and Wales and for Scotland. Joint Committee on Fire Research.

2. Donnan, S. P. D. (1987) *A cohort study of mortality in serving and ex-firemen in England — 1965-1986: a follow-up study*. J. C. F. R. Research report No 29. Home Office Scientific Research and Development Branch.

Mortality of millers (MR74)

Reasons for the study
Millers are exposed to a range of biologically active agents derived from the flora and fauna accompanying grain, and the chemical agents used to treat grain and flour. A study of nasal carcinoma and occupation suggested that millers might experience an excess. Raised risks of reticulosarcoma, lymphoma and alimentary carcinoma were reported in persons with gluten enteropathies.

Study population
327 out of 348 persons identified in the 1961 census as millers or bakers.

Nature of the study
Subjects identified in the census as millers were traced and flagged at NHSCR and coded copies of anonymised death entries were provided to the researcher. Observed numbers of all cause and specific cause mortality were compared with predictions based on National rates, adjusted for person years at risk.

Results
A 94 per cent trace was achieved. 156 deaths were observed with 165.0 expected. No deaths were due to nasal cancer. Twelve cases of lung cancer were reported (E=15.0, 95 CI=0.4-1.4). All malignancies totalled 33 with 37.5 predicted.

Publications
1. Alderson, M. R. (1987) 'Mortality of millers and bakers', *Br J Cancer*, 55: 695-696.

The mortality of shipyard workers with pleural plaques (MR75)

Reasons for the study
The appearance of pleural plaques in the chest radiograph when changes consistent with interstitial pulmonary fibrosis are not discernible, has been treated as no more than a stigma of exposure to asbestos of no adverse prognostic significance. This asseveration was questioned after a preliminary study suggested that shipyard workers with pleural plaques might have an excess of respiratory malignancies.

Study population
(i) 429 shipyard workers whose chest radiographs showed pleural plaque formation and various control populations.

(ii) Separately, 156 shipyard workers from Barrow with pleural plaques, matched with 156 men from Carlisle without plaques or known asbestos exposures.

Nature of the study
Subjects and controls lost to the researchers were traced and flagged at NHSCR and coded copies of death certificates provided. Deaths from malignant mesothelioma, bronchial carcinoma and other causes were compared between 'cases' and 'controls' and numbers expected from National rates.

Results
(i) Subjects with plaques experienced 19 deaths from bronchial carcinoma (controls=4) and 23 deaths from malignant mesothelioma (controls=0), differences that were highly significant. All cause deaths were 127 for 'cases' and 74 for the 'controls'.

(ii) In a preliminary analysis of data accumulated subsequently, the evidence of bronchial carcinoma was similar in cases with pleural plaques and their controls.

Publications
1. Edge, J. R. (1976) 'Asbestos related disease in Barrow-in-Furness'. *Environmental Research*, 11: 244-247.

2. Edge, J.R. (1979) *Incidence of bronchial carcinoma in shipyard workers with pleural plaques*. Annuals of the NY Acad Sci, 330: 289-294.

Mortality of acrylonitrile (AN) workers (MR79)

Reasons for the study
Animal studies have shown AN to be carcinogenic, but for a variety of reasons, human epidemiological studies have been inconclusive. A further study was mounted to help resolve the problem.

Study population
1,111 men employed at six factories in England, Wales, Scotland and Northern Ireland, between 1950 and 1968.

Nature of the study
Subjects were traced and flagged by the three appropriate offices in Scotland, Northern Ireland and England and Wales, and coded death certificates provided. Those who had worked for a year or more were separated from those working shorter periods. Expected numbers of deaths were calculated from the appropriate rates adjusting for person years at risk. An internal case control analysis was carried out for cases of lung cancer.

Results
A trace rate of 99.8 per cent was achieved. For those exposed for 1 year or more, there was a significant deficit of deaths from non-malignant disease overall. There was a significant excess of cancer of the stomach (0=5, E=1.9 p=0.05) and of accidental death (0=3, E=0.7 p=0.05). The age group 55-64 experienced non-significant excesses of respiratory disease and of accidents and a statistically significant excess of stomach cancer (0=3, E=0.7 p=0.05). The age group 15-44 had a statistically significant excess of lung cancer (0=3, E=0.7 p=0.05).

Publications
1. Werner, J. B. and Carter, J. T. (1981) 'Mortality of United Kingdom acrylonitrile polymerisation workers'. *Br J Ind Med*, 38: 247-253.

Cancer mortality of workers in a UK steel foundry: 1946-85 (MR81)

Reasons for the study
Fumes and dusts arising from fuels and moulding agents in foundries contain confirmed and suspected carcinogens. Studies of foundry workers, predominantly in iron foundries, have found excesses of cancer mortality though not invariably. It was decided to study workers in a group of steel foundries to determine if there were cancer risks and if so to investigate the significant exposures.

Study population
All male operatives who had started work 1946-1965 in 9 English steel foundries and who had been employed for at least one year. Later a Scottish foundry was added.

Nature of the study
Subjects were traced by NHSCRs and the National Social Security index, and coded death certificates were provided by OPCS. Mortality was related to job classification. Standardised mortality ratios were calculated using National rates, and regression models in life tables were constructed to investigate associations between adverse effects and duration of employment in the industry and in particular jobs.

Results
Over the period of study, there were statistically significant excesses of lung and of stomach cancers. There was an increased risk of lung cancer from working in the foundry and fettling areas: there was weaker association for stomach cancer and work in the foundry area.

Publications

1. Fletcher, A. C. (1982) 'A study of mortality in British steel foundries'. *Journal of Research SCRATA*. Number 58/59 Sept/Dec. 2-31.

2. Fletcher, A.C and Ades, A. (1984) 'Lung cancer mortality in a cohort of English Foundry Workers'. *Scand J Work Environ Health*, 10: 7-16.

3. Sorahan, T. and Cooke, M. A. (1989) 'Cancer mortality in a cohort of United Kingdom steel foundry workers'. *Br J Ind Med*, 46: 74-81.

Mortality of persons compensated for asbestosis (MR83)

Reasons for the study
To investigate the natural history and prognostic significance of signs and symptoms in patients diagnosed as having compensatable asbestosis.

Study population
155 patients diagnosed with asbestosis during 1968 and 1974 and followed up during 1978 and 1979 by a clinical research unit.

Nature of the study
The study population was followed clinically where possible on a regular basis. Members were traced where necessary and flagged at NHSCR and coded death certificates were provided. Comparison was made with numbers predicted from national rates adjusting for person years at risk.

Discriminant analysis was used to investigate the predictive value of various parameters. Researchers were also provided with the identity of the Family Practitioner Committee with which patients were registered for those lost to follow up to facilitate further contact.

Results
By the end of August 1979, 59 deaths had occurred. There was an excess mortality overall after 5 and 10 years of observation, due to excesses of deaths from lung cancer, respiratory disease and mesothelioma. There was no rise in mortality with increased profusion of small opacities in the chest radiograph. Three cases of lung cancer were recorded where 0.74 were expected ($p<0.03$) where the profusion of small opacities were category O. Death appeared to diminish with time. The present of finger clubbing and a reduced FEV_1 seemed of value of predicting premature death, unrelated to level or duration of exposure.

Publications

1. Coutts, I. I. (1982) 'Observations on the natural history of asbestosis'. MD thesis, University of Liverpool.

2. Coutts, I. I., Gilson, J. C., Kerr, I. M., Parkes, W. R. and Turner-Warwick M. (1987) 'Mortality in cases of asbestosis diagnosed by a pneumoconiosis medical panel'. *Thorax*, 41: 111-116.

3. Coutts, I. I., Parkes, W. R. and Turner-Warwick, M. (1987) 'The significance of finger clubbing in asbestosis'. *Thorax*, 42: 117-119.

Mortality of slate workers in Gwynedd (MR84)

Reason for the study
Local authorities suspected an excess of mortality in a group of slate quarriers in whom a respiratory survey had found evidence of radiographic, physiological and symptomatic abnormalities. An initial proportional mortality study suggested an abnormal mortality pattern meriting further investigation.

Study population
725 men whose only exposure to hazardous dust had been in the slate industry in Gwynedd. As controls, 530 men were chosen with no occupational dust exposure.

Nature of the study
To relate mortality experience to measures of slate dust exposure, and to its effects on ventilatory capacity and on the radiograph.

Results
A small excess of overall mortality was found in men exposed to slate dust compared with controls. This was largely attributed to cigarette smoking. There was no relationship between mortality and lung function decrement except for ex-smokers who had purely worked with slate. Of the 129 deaths in slate workers, 36 were due to lung cancer (31.9 per cent) compared with 14 cases (18.1 per cent) in the 73 deaths in controls. The authors did not deem there to be significant differences in mortality between the two groups.

Publications

1. Oldham, P. D., Bevan, C., Elwood, P. C. and Hodges, N. G. (1986) 'Mortality of slate workers in North Wales'. *Br J Ind Med*, 43: 550-555.

British Rubber Manufacturers Association (BRMA) Research Project (MR88)

Reasons for the study
Studies in the UK and abroad had reported excesses of cancers of the bronchus, stomach and at other sites among workers in the rubber and cable making industry, despite known carcinogenic antioxidants having been withdrawn. The industry decided to investigate the incidence of a wide range of diseases in its various sectors.

Study population
Some 37,000 males who had been employed in 13 factories who had first entered the industry between 1946 and 1960 with a minimum of 12 months employment.

Nature of the study
Individuals no longer employed on 31st December 1970 were traced using the NHSCR and National Insurance files and flagged. Coded copies of death certificates were provided. Observed numbers of deaths were compared with those predicted from National and Regional rates adjusting for man years at risk. The population was divided into three groups according to year of entry into the industry.

Results
In the analysis of the study to 1970, a 98.6 per cent trace was achieved. By 1985, there was statistically significantly increased mortality from carcinomas of the pharynx, oesophagus, stomach, and bronchus. Regression models and life tables indicated positive associations between (i) the risk of stomach cancer and dust exposures, and (ii) the risk of lung cancer and fume or solvent exposure.

Publications

1. Sorahan, T., Parkes, H. G., Veys, C. A. and Waterhouse, J. A. H (1986). 'Cancer mortality in the British rubber industry: 1946-8'. *Br J Ind Med* 43: 363-373

2. Sorahan, T., Parkes, H. G., Veys, C. A. et al (1988) 'Mortality in the rubber industry 1946-85'. *Br J Ind Med*, 46: 1-11

Mortality of Nickel-Cadmium battery workers (MR95)

Reasons for the study
A number of diseases have been associated with exposure to nickel and cadmium. This study was set up to test the hypotheses of associations with carcinoma of the prostate, carcinomas of the respiratory system, hypertension, nephritis and nephrosis, other genitourinary disorders and non-malignant respiratory disease.

Study population
Some 3,000 workers exposed to cadmium oxide dust for at least one month who had started employment between 1923 and 1975.

Nature of the study
Subjects were traced and flagged at NHSCR and coded death certificates were provided. Jobs were coded according to level of cadmium oxide exposure. Mortality experience was compared with predictions based on National rates adjusted for age, sex and calendar year.

A further analysis involving regression models in life tables was employed to test the null hypothesis of no effect of cadmium oxide on outcome.

Results
No convincing evidence was found of occupational risks for any of the diseases previously reported.

Publications

1. Sorahan, T. (1982) *A mortality study of nickel-cadmium battery workers*. Proceedings of the Third International Cadmium Conference, Miami, 1982. London, Cadmium Association. 138-141.

2. Sorahan, T. and Waterhouse, J. A. H. (1983) 'Mortality study of nickel-cadmium battery workers by the method of regression models in life tables'. *Br J Ind Med*, 40: 293-300.

3. Sorahan, T., Adams, R. G. and Waterhouse, J. A. H. (1983). 'An analysis of mortality from nephritis and nephrosis among nickel-cadmium battery workers'. *J Occup Med*, 25: 609-612.

4. Sorahan, T. (1987). 'Mortality from lung cancer among a cohort of nickel cadmium battery workers: 1946-84'. *Br J Ind Med*, 44: 803-809.

A study of exposure to hydrazine (MR96)

Reasons for study
Following experimental exposure to hydrazine, mice had an increased incidence of lung and liver tumours and malignant lymphomas, and rats developed lung tumours. This prompted a study to see if it was also carcinogenic to man.

Study population
Records were available from 427 men employed for at least 6 months at a factory between 1945 and its closing in 1970. Only 68 had substantial exposure, 170 were not directly exposed and 189 were only exposed to hydrazine for a proportion of their time.

Nature of the study
Workers were flagged at the NHSCR and coded copies of death certificates and of cancer registrations were provided. All cause and specific cause mortality were analysed by category of exposure, duration of exposure and years since first exposed.

Results
A 95 per cent trace was achieved. No obvious hazard was seen. There were 2 deaths from lung cancer where 1.61 were expected from England and Wales rates in those with the highest level of exposure who had been employed for more than 10 years.

The study continues and a further analysis will be conducted when there is a substantial increase in data.

Publications
1. Wald, N., Boreham, J., Doll, R. and Bonsall, J. (1984) 'Occupational exposure to hydrazine and subsequent risk of cancer'. *Br J Ind Med*, 41: 31-34.

Mortality study of foundry workers (MR97)

Reasons for the study
Workers in foundries are exposed to a range of agents, some of which are known carcinogens. It was decided to study the mortality experience of workers at iron, steel and non-ferrous foundries. In particular to evaluate the long term significance of minor changes, of symptomatology, in the chest radiograph and in ventilatory function, generally considered not be of clinical importance.

Study population
Participants in a study of a sample of foundrymen in England and Wales and Scotland and their controls in 1964/65, in which details were available of respiratory symptoms, illness, chest radiographic abnormalities and ventilatory function.

Nature of the study
Foundrymen and their controls have been traced and flagged at NHSCR, and coded copies of death certificates provided to the researchers. The mortality pattern of foundrymen will be compared with workers at control factories and with National rates. Comparisons will be made within the study population by process and job. The consequences of respiratory symptoms, radiographic changes and minor ventilatory changes will be investigated.

Cohort study of workers exposed to amosite asbestos (MR108)

Reasons for the study
Experimentally the three commonly used forms of asbestos differ little in their biological effects. Limited human exposure data suggested that the carcinogenic powers of amosite might fall midway between chrysotile and crocidolite. The availability of a population of factory workers exposed largely to amosite (minimal exposure to crocidolite and chrysotile but some exposure to crystalline silica) prompted their study.

Study population
The total workforce at a plant manufacturing board and panels between 1945 and 31st December 1978.

Nature of the study
(i) Subjects were traced and flagged at NHSCR and copies of coded death certificates provided. The mortality pattern observed at 1980 was compared with that expected from National rates and analysed according to the dustiness of the job.

(ii) The population was followed to 1986 and re-analysed.

Results
(i) At the end of 1980, only 422 deaths had occurred out of a total of 5,969 persons. There was a doubling of risk of death from lung cancer (0=57, E=29). For the years 1981-1983, excess mortality was related to the degree of dustiness of work and not to years of exposure. There were 5 cases of malignant mesothelioma, but no significant excess was seen for other tumours.

(ii) Lung cancer death increased to 107 (E=59.2), the doubling having continued.

Publications
1. Acheson, E. D., Bennett, C., Gardner, M. J. and Winter, P. D. (1981) 'Mesothelioma in a factory using amosite and chrysotile asbestos'. *Lancet*, 2: 1404 1405.

2. Acheson, E. D,. Gardner, M. J., Winter, P. D. and Bennett, C. (1984) 'Cancer in a factory using amosite asbestos'. *Int J Epidemiol*, 13:3-10.

[As a supplementary exercise, the NHSCR notified the researchers of the Family Practitioner Committee Area for the men. Through the FPCs, general practitioners were contacted with material for onward dispatch as appropriate to their ex-Amosite worker patients, providing advice on quitting smoking, and requesting information on current smoking habit.]

Mortality of UK man-made mineral fibre workers (MR109)

Reasons for the study
Experimental studies with a range of man-made mineral fibres showed them to share many properties of carcinogenic natural mineral fibres. To evaluate the hazard to a man, studies were carried out in a number of European countries and co-ordinated by the WHO, International Agency for Research in Cancer.

Study populations
A Workers employed for at least 1 year at a factory in England between April 1946 and 31st December 1978. Glass wool was the predominant product, though other material was manufactured.

B Workers employed from 1956 to 31st December 1978 at a glass filament factory.

Nature of the study
Population A was traced predominantly by NHSCR and flagged and coded death certificates provided.

Population B was traced at the Central Services Agency, Belfast or by local initiatives with death certificates provided by the General Register Office, Belfast.

Expected numbers of deaths were calculated from National and local rates adjusting for person years.

Results
Factory A: A trace of 98.3 per cent was achieved for men and 95.8 per cent for women. No deaths from mesothelioma were observed. For males there was a lung cancer excess (0=69, E=55, SMR=126 95% CI=0.98-1.59). There were few lung cancer cases for women (0=7, E=5 - SMR=1.52 95% CI=0.60-3.09). There was no relation between lung cancer and duration and level of exposure.

Factory B: A trace of 98.7 per cent was achieved for men and 98.6 per cent for women. The mortality pattern conformed to National and local experience.

Publications
1. Gardner, M. J., Winter, P. D., Pannett, B., Simpson, M. J. C., Hamilton, C. and Acheson, E. D. (1986) 'Mortality study of workers in the man-made mineral fibres production industry in the United Kingdom'. *Scand J Work Environ Health*, 12, Suppl 1, 85-93.

UK Atomic Energy Authority (UKAEA) mortality study (MR110)

Reason for the study
To monitor the health of employees.

Study population
A total of 49,456 records were obtained from UKAEA and other files of persons who had been employed at their 7 establishments at any time before 31st December 1979.

Nature of the study
Subjects were traced and flagged at NHSCR and coded copies of death registrations were provided to the researchers. Observed deaths were compared with numbers predicted from National rates.

Results
An overall 99.5 per cent trace rate was achieved and 99.9 per cent for radiation monitored workers. Overall mortality was much as expected for an occupational group in England and Wales. Mortality from all malignant neoplasms and diseases of the nervous, circulatory, respiratory, digestive and genitourinary systems, was below expected: numbers of cancers of stomach, lung, bladder and brain, were significantly low.

When mortality and recorded radiation exposure were studied, cancer of the prostate related significantly to cumulative exposure whether latency was or was not assumed: it was more marked where cumulative exposure exceeded 50mSv and there was exposure to tritium and other radionuclides.

A follow-up analysis to 31 December 1986 of 1,506 deaths confirmed the prostatic cancer excess in men monitored for internal contamination with radionuclides. For women, an excess of uterine cancer was found statistically significantly associated with external radiation exposure. Overall cancer morbidity was less marked than mortality; site-specific analyses were similar to mortality.

Publications

1. Beral, V., Inskip, K., Fraser, P., Booth, M., Coleman D. and Rose, G. (1985) 'Mortality of employees of the United Kingdom Atomic Energy Authority, 1946-1979'. *Br Med J*, 291: 440-447.

2. Fraser, P., Booth, M., Beral, V. M. Inskip, H., Firsht, S. and Speak, S. (1985) 'Collection and validation of data in the United Kingdom Atomic Energy Authority mortality study'. *Br Med J*, 291: 435-439.

3. Inskip, H., Beral, V., Fraser, P., Booth, M., Coleman, D and Brown, A. (1987) 'Further assessment of the effects of occupational radiation exposure in the United Kingdom Atomic Energy Authority mortality study'. *Br J Ind Med*, 44: 149-160.

4. Carpenter, L., Beral, V., Fraser, P. and Booth, M. (1990) 'Health related selection and death rates in the United Kingdom Atomic Energy Authority workforce'. *Br J Ind Med*, 47: 248-258.

5. Fraser, P., Carpenter, L., Maconochie, N., Higgins, C., Booth, M. E. Beral, V. (1993). 'Cancer mortality and morbidity in employees of the United Kingdom Atomic Energy Authority'. *Br J Cancer*, 67: 615-624

Mortality of pharmaceutical workers (MR111)

Reason for study

Many raw materials and processes employed in this industry potentially expose workers to a range of toxic and carcinogenic materials, yet their health experience had only been patchily studied.

Study population

Men and women aged 35 and over identified as employed in the industry in the 1961 (1,472) and 1971 (2,102) censuses.

Nature of the study

Subjects were traced and flagged at NHSCR after identification by OPCS, and coded copies of death registrations were provided and the data anonymised. Cause of death was studied by sector of industry and job within sector at the time of census. Numbers of observed deaths by cause, were compared with numbers predicted from National rates adjusted for person years at risk.

Results

The trace rate for the 1961 group was 90.0 per cent and for the 1971 group 96.1 per cent. There was a consistent deficit for most causes of death analysed. Where excesses occurred, they were not restricted to specific jobs or industry grouping, or were too few to be analysed for statistical significance.

Publications

1. Harrington, J. M. and Goldblatt, P. (1986). 'Census-based mortality study of pharmaceutical industry workers'. *Br J Ind Med*, 43: 206-211.

Mortality of North Sea Divers (MR113)

Reason for the study

A decompression sickness control register has been investigating long term morbidity in divers. It was decided to extend the investigation to study their mortality pattern.

Study population

North Sea Divers routinely examined under the diving regulations together with a number of others.

Nature of the study

Subjects have been traced and flagged at NHSCR and copies of death certificates are provided.

Results

By the first quarter of 1991 there had been a total of 48 deaths, but no specific mortality pattern had emerged, other than a number of accidents.

Publications

Progress reports have been given to the MRC Decompression Sickness Panel.

Mortality of cadmium workers (MR118)

Reasons for the study

While a number of acute and chronic effects of cadmium and its compounds have been determined, certain effects have been reported but not confirmed. A study was mounted to investigate whether workers engaged in a cadmium process were at greater risk of chronic renal and respiratory diseases, cancers of the prostate, stomach and lung, and hypertension and related diseases.

Study population

6,995 men born before 1940 employed for at least one year, on or in the vicinity of a cadmium process between 1942 and 1970 at 17 major UK plants.

Nature of the study

Subjects were traced and flagged at NHSCR and coded causes of death provided. Deaths occurring between 1943 and the end of 1979 were analysed initially and updated to 1984 and 1989. Expected values were derived from National and Regional rates, adjusted for man years at risk. Exposures were classified by level and duration of exposure. In updates, a case reference study was conducted on one subset employed in a foundry and matched pair regression analysis was conducted.

Results

In the initial study there were no excess deaths due to prostatic carcinoma, cerebrovascular disease or renal disease. There was a statistically significant excess of deaths from bronchitis, strongly related to duration and intensity of exposure to cadmium. 199 deaths due to lung cancer were observed where 185.6 were expected: the excess was not statistically significant, nor was there a relationship with levels of exposure.

In 1984 the results confirmed the initial findings except in the case of lung cancer where was now a stronger indication of an excess risk (SMR = 115, 95% CI = 101-129). Logistic regression analysis of data derived from a large subset employed at a foundry did not show the lung cancer excess to be related to duration of employment with cadmium.

The following 5 year update showed a further fall in prostatic cancer mortality (SMR = 75, 95% CI = 53 - 103). Overall, the lung cancer mortality excess was of borderline significance (SMR 112, 95% CI = 100-124), with a non-significant trend for exposure categories. Stomach cancer mortality remained significantly increased, but this was still not related to duration or intensity of exposure. The bronchitis excess related to intensity of exposure was still seen.

Publications

1. Armstrong, B. G. and Kazantzis, G. (1983) 'The mortality of cadmium workers'. *Lancet*, i: 1424-1427.

2. Ades, A.E. and Kazantzis, G. (1988) 'Lung cancer in a non-ferrous smelter: the role of cadmium'. *Br J Ind Med*, 45: 434-442.

3. Kazantzis, G., Lam, T. H. and Sullivan, K. R. (1988). 'Mortality of cadmium exposed workers: a five year update.' *Scand J Work Environ Health*, 14: 220-223.

4. Kazantzis, G. and Blanks, R. G. (1992). *A mortality study of cadmium exposed workers.* Cadmium 92: Edited proceedings 7th International Cadmium Conference, New Orleans, USA. Eds. Cooke, M.E., Hiscock, S. A., Morrow, Hiand Volpe, R. A. Cadmium Association, London, pp 150-157.

Mortality of anthraquinone workers (MR119)

Reasons for the study

Substituted anthraqinones have a long history of use for dyestuffs. A number of these agents have proved positive on mutagenicity testing and are carcinogenic in rats, meriting a study of production workers.

Study population

Workers at two plants, one in Scotland, the other in Lancashire, engaged in the production of anthraquinone dyes who had been employed for a minimum of 6 months at any time between 1st January 1956 and 31st December 1965.

Nature of the study

Subjects were traced and flagged at the NHSCRs and coded death certificates provided. Members were followed to the 30th June 1980 and numbers of deaths compared with Scottish or England and Wales rates as appropriate.

Results

At the Scottish plant there was a trace rate of 98.4 per cent and there was no statistically significant excess of cancer overall. Three works departments showed statistically significant excesses of 'all cancers' though numbers were small. Of six men with oesophageal cancer all worked in the engineering department.

Publications

1. Gardiner, J. S., Walker, S. A. and MacLean, A. J. (1982) 'A retrospective mortality study of substituted anthraquinone dyestuffs workers'. *Br J Ind Med* 39: 355-360.

[This study was extended and the preliminary results presented at a conference in 1988. The results of these studies have been presented to the management and to the workforce].

Mortality study of employees at various petrochemical plants (MR124)

Reasons for the study
Prior to embarking on developing an innovative data base for the health surveillance of all its employees, a major oil company conducted a pilot survey of a group of workers exposed to known hazards to assess their health experience and to investigate the design of the data base.

Study population
Persons employed for a minimum of one year at several plants where there was exposure to Benzene or Butadiene and where personnel records went back to 1920.

Nature of the study
Subjects were traced and flagged at NHSCR and coded copies of draft death registrations were provided. Other data available for individuals included, chemical exposures, process, plant, category of exposure, period of exposure and duration of exposure. Patterns of mortality observed are to be compared with values expected from National and Regional rates, and related to specific chemical exposure patterns. Analyses will also consider years of employment and the time since last employed.

Results
A preliminary analysis based on one refinery of men ever exposed to benzene or butadiene, showed a deficit of 'all cause mortality' (0=1734, E=1903.53). For 'all cancers', 463 were observed where 457.43 were expected. No excess of 'leukaemia' was observed (0=7, E=11.12). There were excesses of laryngeal carcinoma (0=10, E=5.03) and carcinoma of the oesophagus (0=18, E=11.71), but there were no trends seen with years of exposure or time since first employment.

A study of cancer incidence in relation to immune stimulation (MR125)

Reason for the study
An increased risk of certain tumours has been observed in persons who are immunosuppressed, and has been suspected in persons who have undergone desensitisation. It was decided to study a group of persons required to undergo immunisation.

Study population
All persons ever employed at a microbiological research establishment, where a variety of virulent organisms were handled, who had been routinely immunised.

Nature of the study
Subjects were traced and flagged at NHSCR and coded copies of draft death registrations and cancer registrations were provided to the researcher. Outcomes will be related to the immunisation histories of individuals. A parallel study is being conducted at the National Cancer Institute (USA).

Historical prospective study of the Printing Industry (MR128)

Reasons for the study
Excesses of bladder cancer and other tumours have been reported anecdotally, in mortality data analyses and in individual factory studies. Printing ink formulae have commonly included traces of animal carcinogens, some with effects on the urothelium. A study was set up to detect abnormalities in health experience in the industry and its sections and to identify if possible the causal agent.

Study population
Male members of the Manchester branches of two unions employed for at least 6 months in the period 1949-1963 for whom there were adequate records.

Nature of the study
Subjects were traced and flagged at NHSCR and coded death certificates provided to the researchers. Initially, mortality in the two unions were compared with rates for England and Wales, with and without adjustment for the generally higher mortality in the Manchester area. Subsequently a case control study was carried out to investigate the association of abnormal health pattern with time period and work site.

Results
Trace rates of 98.4 and 97.2 per cent was achieved for the two unions. The results of the study do not support the hypothesis of an occupational risk of bladder cancer in the printing industry. However, men involved in newspaper printing have raised lung cancer mortality (SMR=179, 95 CI: 144-218). A nested case control study found evidence of a 'dose-response' effect, with lung cancer rates increasing with duration of work as a newspaper printer. This, together with evidence of biological feasibility, suggests a real occupational risk, though smoking cannot be excluded as a factor. Consistent with the findings of the cohort study, a review of the literature strongly suggests an increased risk of cancer of the buccal cavity and pharynx in printers.

Publications

1. Leon, D. (1989) *Mortality of male members of two printing trades unions*. Report: Department of Epidemiology and Population Science, London School of Hygiene.

2. Leon, D., Thomas, P. and Hutchings (1989) *Lung cancer in newspaper machine men*. Report: Department of Epidemiology and Population Science, London School of Hygiene, and the Epidemiology and Medical Statistics Unit, HSE.

3. Leon, D. A. (1991). 'Mortality of men in the printing industry'. PhD thesis. University of London.

4. Leon, D. A. (1994), 'Mortality in the British printing industry: a historical cohort of trade union members in Manchester'; *Occ and Environ Med*, 51: 79-86.

5. Leon, D. A., Thomas, P. & Hutchings, S (1994), 'Lung cancer among newspaper printers exposed to ink mist: a study of trade union members in Manchester, England'; *Occ and Environ Med*, 51: 87-94.

Formaldehyde in the British Chemical Industry (MR129)

Reason for the study
A report of nasal cancer occurring in rats heavily exposed to formaldehyde led to concern that a carcinogenic effect might have been operating in heavily exposed workers.

Study population
A total of 7,680 men from 6 factory populations who had started work before 1965 and for whom adequate data were available from personnel records. (Persons employed 1966-70 were included in study MR168.)

Method of study
The vital status of individuals on 31st December 1981 was determined by NHSCR supplemented by the National Insurance Index, and coded death certificates were provided by OPCS. An index of formaldehyde exposure was derived for individuals. Observed mortality was compared with expected numbers derived from National and local rates.

Results
An overall trace rate of 98 per cent was achieved. No deaths from nasal cancer were recorded among the 1,619 death certificates. Overall, no significant excess of lung cancer death was observed when local rates were used for comparison. One factory had a high mortality for lung cancer when compared with National rates (SMR=124, 95% confidence limits 104-148). In this group there was a trend with increasing exposure at the borderline of statistical significance: on subsequent review, there was no longer a trend with duration of exposure.

There were no excesses of deaths from leukaemia or tumours reported in other studies. However the number of men exposed to high levels for more than 5 years and followed for at least 20 years was too small to provide reassurance.

Publications

1. Acheson, E. D., Gardner, M. J., Pannett, B., Osmond, C. and Taylor, C. P. (1984). 'Formaldehyde in the British Chemical Industry'. *Lancet*, i: 611-616.

2. Acheson, E. D., Barnes, H. R., Gardner, M. J., Osmond, C., Pannett, B. and Taylor, C. P. (1984). 'Formaldehyde process workers and lung cancer'. *Lancet*, i: 1066-1067.

3. Acheson, E. D., Barnes, H. R., Gardner, M. J., Osmond, C., Pannett, B. and Taylor, C. P. (1984). 'Cohort study of formaldehyde workers'. *Lancet*, ii: 403.

4. Gardner, M. J., Osmond, C. and Pannett, B. (1985). 'Formaldehyde and lung cancer'. *Lancet*, ii: 1366-1367.

5. Gardner, M. J., Osmond C., Pannett, B. and Acheson, E. D. (1986). 'Formaldehyde, lung cancer and bronchitis'. *Lancet*, ii: 437-438.

Cohort study of the effects of exposure to styrene in industrial workers (MR132)

Reasons for the study
The report of angiosarcomas developing in a group of workers heavily exposed to vinyl chloride and the demonstration of its carcinogenic potency in laboratory animals exposed at low concentration, drew suspicion to its related compound styrene (vinyl benzene). On testing, it and its metabolites were found to be mutagenic and carcinogenic and earlier epidemiological studies suggested that it might cause leukaemia or lymphoma. The annual production is substantial and workers have been heavily exposed to its vapours in certain applications.

Study population
Workers exposed to styrene at eight factories where it was used to manufacture resin moulded articles reinforced with glass fibre during 1947-82.

Nature of the study
Subjects were traced and flagged at NHSCR and coded copies of death certificates and cancer registrations provided by OPCS. Indices of levels of exposure to styrene were estimated. Mortality and cancer morbidity patterns were related to predicted values based on national rates, and to styrene exposure levels.

Results
Of a total of 7,949 workers (95.2 per cent of the nominal roll), 97.4 per cent were traced. There was overall a deficit of cancers and particularly of lymphoid and haemopoietic cancer (6 deaths observed, 14.9 expected). Excesses were found for some tumours but these did not reach conventional significance.

Publications
1. Coggon, D., Osmond, C., Pannett, B., Simmonds, S., Winter, P. D. and Acheson, E. D. (1987). 'Mortality of workers exposed to styrene in the manufacture of glass-reinforced plastics'. *Scand J Work Environ Health*, 13: 94-99.

Mortality of professional chemists in England and Wales 1965-1989 (MR136)

Reasons for the study
There had been a number of reports that chemists, despite overall health advantages, had experienced significant excesses of certain malignancies, the sites of which were not consistent between studies. As the problem was recognised to transcend national barriers, the Commission of the European Communities assisted in the study.

Study population
14,884 males who were members of the Royal Society of Chemistry on 1st January 1965.

Nature of the study
Subjects were traced and flagged at NHSCR and coded copies of death registrations were provided for men who had died between 1965 and 1989. Numbers of deaths observed were compared with values predicted from National rates for Social Classes 1 and 2, obtained from the OPCS Occupational Mortality Supplement for 1970-72.

Results
A 99.9 per cent trace was achieved at NHSCR and death certificates obtained for 99.15 per cent of the 4,012 decedents. There was an overall deficit of cancers (0=918, E=1195.77); for specific cancers, significant excesses and deficits were found. Insofar as the study was testing hypotheses, it confirmed excesses reported in some studies, including carcinoma of the pancreas, malignant melanoma, lymphatic and haemopoietic cancer, large intestine cancer, and carcinoma of the kidney. 11 cases of primary pleural neoplasm were reported, 8 of which were diagnosed as malignant mesothelioma. New findings, were excess carcinoma of the duodenum (0=3, E=0.58), mental disorder especially presenile dementia (0=15, E=5.93), disorders of the nervous system (0=72, E=45.58) and paralysis agitans (0=38, E=18.46).

Publications
1. Hunter, W. J., Henman, B. A., Bartlett, D. M. and Le Geyt, I. P. (In press). 'Mortality of professional chemists in England and Wales 1965-89'. *Am J Ind Med*, 22:

Study of Buckinghamshire furniture workers (MR137)

Reasons for the study
A raised incidence of adenocarcinoma of the nasal cavity and ethmoid sinuses in furniture workers had been observed in the UK and confirmed repeatedly elsewhere. However as some of these studies reported cancer excesses at other anatomical sites, the population data were reviewed.

Study population
5,108 men with adequate records, born before 1 January 1940, who had worked prior to 31st December 1968 for any period of time at one of 9 High Wycombe furniture factories.

Nature of the study
The population was traced by NHSCR and the National Insurance files, and coded death certificates provided. These were analysed according to, the nature of the employment, duration of employment and interval since first employed and compared with National rates.

Results
Apart from nasal cancer, no significant increase in cancer mortality or any other specific cause mortality was found. Nor was there a trend of increasing mortality with increasing dustiness of exposure, for cancer at any site.

Publications
1. Rang, E. H. and Acheson, E. D. (1981). 'Cancer in furniture workers'. *Int J Epidemiol*, 10: 253-261.

2. Acheson, E. D., Winter, P. D., Hadfield, E. and Macbeth, R. G. (1982). 'Is nasal cancer in the Buckinghamshire furniture industry declining?' *Nature*, 299: 263-265.

3. Acheson, E. D., Pippard, E. C. and Winter, P. D. (1984). 'The mortality of English furniture makers'. *Scand J Work Env Health*, 10: 211-217.

Central Electricity Generating Board (CEGB): Classified radiation worker mortality study (MR138)

Reason for the study
To contribute to knowledge of the health effects of low doses of ionizing radiation and establish the appropriateness of ICRP risk estimates upon which radiological protection standards are based.

Study population
1. Some 4,000 persons who ceased to be classified workers before 1st January 1976 when the National Registry for Radiation Workers (NRRW, Study MR47) was established.

2. The study was extended to cover everyone who has ever been a classified worker within the CEGB (now Nuclear Electric). The total study population numbers over 20,000. Information on mortality within the population is obtained from the NRPB (MR47) for those persons included on the NRRW. OPCS is extending MR138 to cover those post-1976 employees who have not agreed to inclusion on the NRRW.

Nature of the study
Subjects were traced and flagged at NHSCR and coded copies of death registrations were provided to researchers. Numbers of deaths observed for specific causes will be compared with expected values derived from England and Wales rates. The findings will be related to employment histories and to radiation exposure records.

Publication
1. Bonnell, J. A., Harte, G. A. and Rubin, C. A. (1983). 'Progress to date in the Central Electricity Generating Board study of low-level radiation effects in classified workers'. In: Proceedings of the IAEA conference on 'Biological effects of low-level radiation'. Venice, April 1983.

Mortality of gas mask assemblers (MR139)

Reasons for the study
It was decided to determine the effects of working with crocidolite asbestos alone. A small group of women working in wartime at a factory assembling gas mask filters where crocidolite was exclusively used was identified.

Study population
A list of 46 names of women employed at this factory was obtained.

Nature of the study
Subjects were traced at NHSCR and flagged. The mortality of this small group will be compared with that of other groups of women assembling gas mask filters using various types and mixtures of asbestos. More detailed studies of individual cases will be carried out as opportunity presents.

It will be kept under surveillance and reported with the other studies (MR44).

A study of mortality in Cornish tin miners (MR142)

Reasons for the study
From 1878 onwards, studies of miners exposed to substantial levels of radon and its daughter products have shown there to be an excess risk of death from bronchial carcinoma. High levels of radon have been found consistently in the atmosphere of Cornish tin mines prompting a study of the mortality pattern in miners.

Study population
1,333 tin miners (a total of 27,631 man years at risk) were identified from the NHS Central Register, by searching records around the Cornish tin mines. Job descriptions were obtained from the registration books. These were subsequently compared with records obtained in an earlier study.

Nature of the study
Subjects were flagged at NHSCR and death certificates retrieved. Observed numbers of deaths from various causes were compared with predicted values derived from National rates adjusted for age, man years, and for the calendar period October 1939 to 31st December 1976.

Results
Overall mortality was significantly raised (0=936, E=768.8 SMR=122). For non-malignant respiratory disease the SMR was 165: of the 185 deaths observed, 93 were from silicosis and 56 from silicotuberculosis. There was excess mortality from bronchial carcinoma (SMR=117) and gastric carcinoma (SMR=141). When analysed by occupation, underground workers fared worse than surface workers.

Publications
1. Fox, A. J., Goldblatt, P. and Kinlen, L. (1981). 'A study of mortality of Cornish tin miners'. *Br J Ind Med*, 38: 378-380.

Mortality of boot and shoe workers (MR146)

Reasons for the study
Nasal cancer had been recognised as a hazard in the manufacture and repair of boots and shoes, but there were suspicions that there might have been in addition, an excess bladder cancer risk and risks of other malignancies, perhaps associated with the use of benzene based adhesives.

Study population
People identified in the 1939 register as being currently employed as boot and shoe operatives in Rushden, Stafford and Street, areas traditionally associated with the industry.

Nature of the study
Subjects were identified, traced and flagged at NHSCR and anonymised data provided to the researchers, including job titles. These titles were divided into 5 categories to correspond with the dustiness of work. Observed deaths at 31st December 1982 were compared with predicted values derived from National rates. For certain causes, local rates were used.

Results
At the end of 1982, a 97.5 per cent trace was achieved. For 'all causes', 'all malignancies' and 'non-malignant diseases', all five groups of workers had favourable mortality experience. The anticipated excess of nasal cancer was observed (0=10, E=1.37), with a significant excess for workers in the finishing room. There was a deficit of other respiratory cancers. A leukaemia excess was observed in one town (0=7, E=3). In another town, death from aplastic anaemia occurred in two men (ages 24 and 26), in a department with similar exposures to benzene. In two towns, there was an overall excess of rectal cancer (0=61, E=47.6), which achieved significance in particular processes.

Publications
1. Pippard, E. C. and Acheson E. D., (1985). 'The mortality of boot and shoe makers, with special reference to cancer'. *Scand J Work Environ Health*, 11: 249-255.

Mortality of tanners (MR147)

Reasons for the study
The observation of a nasal cancer excess in boot and shoe makers and repairers exposed to mixed dusts and fumes led to the study of workers manufacturing leather to test the hypothesis that leather dust was the causal agent. Previous studies had not been consistent.

Study population
833 men were identified from the 1939 Register as employed in two districts as leather tanners in September of that year.

Method
The identified tanners were traced at NHSCR at 31st December 1982 and coded copies of death certificates provided. The numbers of cases of specific causes of death were compared with predicted values based on England and Wales rates.

Results
A 98.6 per cent trace rate was achieved. 'Vegetable' tanners showed a statistically significant deficit of all cause mortality and no statistical excesses of mortality from cancer at any site were found. For 'chrome' tanners, there was an excess of 'other cause' mortality, largely due to diseases of the circulatory and respiratory systems, but this did not achieve statistical significance.

Publications
1. Pippard, E. C., Acheson, E. D. and Winter, P. D. (1985). 'The mortality of tanners'. *Br J Ind Med*; 42: 285-287.

Study of welders (MR149)

Reasons for the study
Analysis of welding fumes would suggest the potential for welders to be at risk of excess respiratory hazards (malignant and non-malignant). The results of epidemiological studies have however been inconsistent. The report of a raised SMR for lung cancer in welders in the OPCS Occupational Mortality Decennial Supplement for 1970-72 prompted a further study.

Study population
Welders at a shipyard employed between 1940 and 1968 for whom adequate personnel records were available.

Nature of the study
Subjects were traced by NHSCR with some supplementary tracing through the National Insurance records. Coded death certificates were provided to the researchers.

Observed deaths were compared with expected values derived from national and local rates adjusting for man years. Further comparisons were made between welders, who are most exposed to fume, platers who are least exposed to welding fume, and burners exposed to other fumes.

Results
By December 1982, of 3,480 persons on the roll, a trace rate of 99.5 per cent was achieved. Of the 543 deaths, 13 were due to mesothelioma confirming a general exposure to asbestos. For the welders, there was suggestive evidence for an increased risk of lung cancer and death from pneumonia.

Publications
1. Newhouse, M. L., Oakes, D. and Woolley, A. J. (1985). 'Mortality of welders and other craftsmen at a shipyard in N. E. England'. *Br J Ind Med*, 42: 406-410.

Pentachlorophenol (PCP) Workers (MR152)

Reasons for the study
The synthesis of pentachlorophenol may lead to contamination with 'dioxins' which are known to be animal carcinogens. At a factory manufacturing this chemical between 1950 and 1978, two cases of non-Hodgkin's lymphoma, a tumour which has been reported as being associated with 'dioxin' exposure, were reported.

Study population
All 158 workers employed during the period of PCP production.

Nature of the study
All workers were flagged at NHSCR and coded copies of death certificates and cancer registrations were provided. Expected values were calculated from regional and national rates, and registration rates, to 1981, adjusted for man years at risk. When it was discovered that carcinogens were involved in other processes and that there was some interchange of employees, the study was extended to involve all other workers employed for over 3 months in other processes on the site. Data on the PCP exposed workforce have been included in the International Agency for Research on Cancer (IARC) phenoxy herbicide and chlorophenols register. Further, this study will be linked with MR 186 and will contribute to an international collaborative study (IARC, LYONS). All cases of lymphoma will be further investigated.

Results
Results of the local study have not been published separately from those on the IARC register.

Longitudinal mortality and morbidity study in the Carbon Black industry (MR154)

Reasons for the study
Industrial carbon black contains traces of absorbed hydrocarbons. These are extractable with difficulty but have been found in experimental studies to include powerful carcinogens. As human population studies had been inconclusive and as a threshold had not been established for these agents, the European Trade Association sponsored a collaborative international study to combine data from production workers at 19 plants in 8 countries.

Study population
Employees at a UK factory in production since 1950.

Nature of the study
For the mortality aspect of the investigation, workers were traced and flagged at NHSCR and coded copies of death registrations were provided. (A separate portion of the study measured respiratory health and exposures to carbon black, SO_2, and the urinary metabolites of polyaromatic hydrocarbons.) Observed numbers of deaths will be compared with expected figures calculated from National rates adjusted for person years.

Results
An interim analysis of the UK data was considered inconclusive. The next analysis will be combined with the other European data.

A prospective mortality study of iron ore miners in Cumberland (MR155)

Reasons for the study
An earlier study of these ore miners using limited data indicated there to be an excess risk of death from lung cancer. A follow up study was mounted to test the original observation and to see if other adverse effects might result from exposure to the mine dusts, radon and its daughter products.

Study population
Men resident September 1939 in Ennerdale and Whitehaven who were identified on OPCS census records as active or retired iron ore miners.

Nature of the study
Cases were traced and flagged by OPCS and anonymised details of their mortality experience provided to the researchers. Comparisons were made with National rates and with urban and rural area specific rates for the period up to 1982. Analyses were also made according to whether employment at September 1939 was 'above ground', 'below ground' or 'unstated'.

Results
Lung cancer excess was found overall, confirming the conclusions reached in the previous study. It was found to be greatest in underground workers. There had been a decline in lung cancer in latter years to which improvement in ventilation might have contributed.

Survey of Rayon Workers exposed to carbon disulphide (CS_2) (MR157)

Reasons for the study
To investigate suspicions that long term exposure to CS_2 might cause cardiovascular disease, neuropsychiatric disorders and other diseases.

Populations
Men employed at a rayon factory for at least a year any time between 1st January 1945 and 31st December 1949.

Nature of the study
1. Analyses of mortality by occupation, of men aged 45-64 dying between 1st January 1950 and 31st December 1964 who had been employed for more than 10 years. Subjects were traced and flagged at NHSCR and death certificates provided. Expected values for deaths were derived from national rates, and comparisons were made with other occupations within the factory.

2. The population was reconstructed and traced and flagged as before. Jobs were coded to maximal CS_2 exposure, and personal sampling for CS_2 was carried out to obtain indices of exposure. Analysis of mortality was extended beyond age 64.

Results
1. A 97 per cent trace was achieved. Exposure to CS_2 was associated with 2½ times the coronary artery death rate of other rayon workers. The strongest evidence for the effect of long term low level exposure was seen during the 1940s: it was only slight for the period 1958-1962.

2. A 97 per cent trace was again achieved. The pattern of mortality of men aged 45-64 for 1950-1982 was similar to 1950-1964. Over the age of 65 mortality tended to decline with increasing exposure, suggesting a reversible cardiotoxic effect.

Publications

1. Tiller, J. R., Schilling, R. S. F. and Morris, J. N. (1968). 'Occupational toxic factor in mortality from coronary heart disease'. *Br Med J*, iv: 407-11.

2. Sweetnam, P. M., Taylor, S. W. C. and Elwood, P. C. (1987). 'Exposure to carbon disulphide and ischaemic heart disease in a viscose rayon factory'. *Br J Ind Med*, 44: 220-227.

Mortality and cancer morbidity among semiconductor workers (MR165)

Reasons for the study
Safety representatives from a factory which had been producing semiconductors since 1965, were concerned about an apparent cluster of cancers in the 1970s. A preliminary scrutiny at the local cancer registry found 3 cases of malignant melanoma. Workers were potentially exposed to a range of chemicals and physical agents including ultraviolet radiation which has been causally associated with malignant melanoma.

Study population
A group of workers in employment on 1st January 1970 together with a group starting employment between 1st January 1970 and 31st December 1979, all with a minimum period of employment of 1 month.

Nature of the study
Most of the study population was traced by NHSCR: National Insurance and local records accounted for virtually all the remainder. Coded death certificates were provided. SMRs were calculated for deaths for cancer at various sites and for other causes using National rates for expected values. For cancer morbidity, standardized registration ratios were calculated for various tumours using rates for the West Midlands.

A case control study was carried out comparing each case of interest with 4 controls selected at random from the study population matched for a number of variables.

A follow-up study of deaths and cancer registrations to 31st December 1989 was analysed as before.

Results
A 100 per cent trace was achieved for those adequately identified and the remainder were accounted for. Overall no excesses of deaths or cancer registrations were observed. An excess of malignant melanomas of borderline statistical significance was found (based on 3 cases), but it was not possible to attribute it unequivocally to the effect of occupational exposure. In the follow-up analysis, there were largely deficits of cause specific mortality and for cancer registrations. The previous finding of a statistical excess of malignant melanomas was not sustained (3 cases observed, 1.50 expected).

Publications

1. Sorahan, T., Waterhouse, J. A. H., McKiernan, M. J. and Aston, R. H. R. (1985). 'Cancer incidence and cancer mortality in a cohort of semiconductor workers'. *Br J Ind Med*, 42: 546-550.

2. Sorahan, T., Pope, D. J. and McKiernan, M. J. (1992). 'Cancer incidence and cancer mortality in a cohort of semiconductor workers: an update'. *Br J Ind Med*, 49: 215-216.

A cohort study of workers exposed to formaldehyde in the British Chemical Industry: an update (MR168)

Study population
To the 7,660 men first employed before 1965 (included in MR129), were added 6,357 men employed after 1964.

Nature of the study
The follow up of mortality and Cancer registration from 1st January 1941 to 31st December 1989 was through NHSCR. Exposure to formaldehyde was estimated as previously and the status of cohort members was checked.

Results
Only one death from nasal cancer occurred out of 3,201 reported deaths (E=1.7). This was in a man with low exposure and 37 years after first exposure. There were no cases of nasopharyngeal cancer (E=1.3). A slight excess of lung cancer (0=402, E=359) was not consistently related to exposure level estimates: the high exposure group had the highest SMR (0=124, 95% C.I=107-144) but was largely based on one factory.

For all cancers combined, there was an excess of 14 before 1965 but a deficit of 3 per cent for those joining subsequently. Rectal cancer showed a statistically significant excess in pre-1965 men (0=45, E=320, SMR 141, 95% C.I=103-188. There was an excess of mortality from respiratory disease (0=360, E=314, SMR 115, 95% CI= 103-127): This was contributed to by three of the factories, one of which contributed heavily and had statistically significant excesses for, all cause mortality, all malignant tumours and lung cancer in high exposure men.

For men employed after 1964 the only statistically significant excess found was for lung cancer in men with high exposures (0=3, E=0.5, SMR= 552, 95%, CI= 114-1610). It was concluded by the authors that the results do not yet justify firm conclusions about the carcinogenicity of formaldehyde.

Cancer of the larynx in mustard gas workers (MR171)

Reasons for the study
A patient with laryngeal cancer told his surgeon that some of his fellow workers producing mustard gas had a similar condition.

Study population
Persons identified from a batch of factory records as having been employed in the mustard gas process. Between the first and second analysis it was possible to assemble a more comprehensive roll.

Nature of the study
The first round was essentially a mortality study using England and Wales death rates for determining expected values. An attempt was also made to identify non-fatal cases of neoplasm and to compare observed figures with expected values derived from the regional cancer register.

The second round was more extensive. Subjects were traced through the NHSCR and flagged, with coded death certificates and cancer registrations being provided. SMRs were calculated for a variety of diseases and malignancies, using National, Merseyside and urban Cheshire rates. In addition, a case control study was carried out and a cohort analysis to study the association with duration of employment and time since first employment.

Results
Because of limited identification details, a trace of only 84 per cent was achieved by the end of 1974. 45 deaths were due to malignant neoplasm where 39 were expected. Excesses were observed for: Larynx and trachea (0=3, E=0.40 p<0.02); Lung and pleura (0=21, E=13 NS); Pancreas (0=4, E=1.5 NS); Kidney and adrenal (0=2, E=0.6 NS); Brain and other parts of the nervous system (0=3, E=1.1 NS). One case certified as dying from Bronchopneumonia had a recurrent laryngeal carcinoma.

Seven workers developed carcinoma of the larynx where 0.75 were predicted. Two men sustained pharyngeal malignancies which were successfully treated to be followed later by a further malignancy of different site and type.

Of non-malignant causes of disease, pneumonia (0=14, E=7) and Accidents, poisonings and violence (0=12), were statistically significantly in excess (p<0.05).

By the end of 1984, despite missing identification details from the historic records, a 95 per cent trace was achieved. Using National rates for comparison, large and highly significant excesses were seen for carcinoma of the larynx (0=11, E=4.04 p=0.003), pharynx (0=15, E=2.73 p=0.001), other upper nasopharyngeal sites (0=12, E=4.29 p=0.002). For lung cancer, 200 deaths were observed where 138.9 would have been expected (p=0.001). There were significant excesses of deaths from non-malignant respiratory diseases. The risk for cancers of pharynx and lung were related to duration of employment. Using Cheshire rates did not significantly affect the findings.

Publications

1. Manning, K. P., Skegg, D. C. G., Stell, P. M. and Doll, R. (1981). 'Cancer of the larynx and other occupational hazards of mustard gas workers'. *Clin Otolaryngol*, 6: 165-170.

2. Easton, D. F., Peto, J and Doll, R. (1988). 'Cancers of the respiratory tract in mustard gas workers'. *Br J Ind Med*, 45: 652-659.

Study of artificial leather cloth workers (MR172)

Reasons for the study
Artificial leather cloth workers may have been exposed to heavy concentrations of benzene. In a study of two groups of such workers identified from the 1939 census, a few cases of leukaemia were observed as expected, but in addition there was an excess of deaths from stomach cancer. This latter finding being unexpected, it was decided to study another group of leather cloth workers for confirmation.

Study populations
1. Workers employed in artificial cloth manufacture in 1939 in the districts of Hyde and Todmorden. (An attempt to expand this population by including another area produced too few additions.)

2. Workers employed in leather cloth and oilcloth manufacture in two other areas in 1939.

Nature of the study
The populations derived from the 1939 National Register, anonymised to preserve confidentiality, were traced and flagged at NHSCR and coded death certificates were supplied to the researchers. Mortality observed was compared with that predicted from National rates adjusting for person years at risk.

Results
In population 1 there were excesses of leukaemia (0=2, E=1) and of stomach cancer (0=16, E=7.2).

When last reviewed, population 2 showed a lesser excess of stomach cancer (0=10, E=7.6).

Although the results gave some support for benzene being a multi system carcinogen, with the non-availability of a large enough population to provide adequate power to the study to measure lesser excesses with confidence, it was not deemed worthwhile to pursue the study further.

Study of glass-reinforced plastic laminators exposed to styrene (MR175)

Reasons for the study
In a UK study of workers exposed to styrene (vinyl benzene) there were suggestions of an excess of lymphomas.

Study population
Workers employed at a number of small companies including boat builders, selected for availability of adequate work and personnel records, and situated in a convenient area. They will have entered industry on or after 1st January 1950.

Nature of the study
Subjects will be traced and flagged by NHSCR, and coded death certificates and cancer registrations (appropriately secured) provided for analysis. Comparisons will be made with national patterns, and outcomes related to exposure history. (The data will be combined with that from other European countries in a collaborative study.)

Publication

1. Kogevinas, M., Ferro, G., Sarraci, R., Anderson, A et al (1993) Cancer mortality in an international cohort of workers exposed to styrene. In: *Butadiene and styrene: assessment of health hazards*. IARC Scientific publications No 127. IARC, Lyon

Mortality studies of workers at British Nuclear Fuels Ltd (MR176)

Reasons for the study
To monitor the health of employees and to evaluate the risk assessment for low level exposure to ionising radiation.

Study population
1. Initially this included all persons employed from the opening of Windscale and Calder works in 1949 to 31st December 1975 totalling 16,000, including Classified and non-Classified radiation workers, current employees and retired persons (MR49).

2. A subset of the Windscale population involved in the 1957 accidental discharge of radioactive material were analysed for mortality and cancer morbidity.

3. Studies on BNFL workers in Springfields (MR92) and Capenhurst (MR99) were initiated in 1979.

4. In 1980, the populations were merged and expanded to include people employed on or after 1 January 1976, to which are added new employees.

Nature of the study
Subjects are traced at NHSCR and flagged, with coded death entries being provided to the researchers. The mortality pattern of employees has been studied separately at each plant and related to years at risk and exposure levels.

After the initial analysis, cancer registration data will also be analysed. Further studies based on Sellafield workers will consider their internal exposure to plutonium, and will focus on those exposed to greater than 500Sv in their working lifetime. Some of the data will contribute to the National Register (NRRW), some to the Nuclear Industry Combined Epidemiological Analysis (NICAE), and some to a collaborative study co-ordinated by IARC, Lyon.

Results
In the initial analysis of Windscale data, of the nominal roll of 14,327, 96 per cent were traced. The cancer mortality rate was in deficit for the 2,277 overall deaths. Only for 'ill defined and secondary sites', was there a statistically significant excess (0=30, E=19.7).

Subsequent analyses showed there to be statistically significant excesses of 'liver and gall bladder cancers', 'lung cancer' and 'Hodgkin's disease'. There were non-statistically significant excesses of 'myeloma' (0=7, E=4.2), 'prostatic cancer' (0=19, E=15.8) and 'leukaemia' (0=10, E=2.2).

There was a positive association between accumulated dose and death rate for, 'bladder cancer', 'multiple myeloma', 'lymphatic and haemopoietic neoplasms', but not for 'leukaemia'.

The findings were thought to be compatible with the ICRP cancer risk predictions, but a nil effect, or one ten times greater than observed could not be excluded.

Publications

1. Clough, E. (1982). 'Further report on the BNFL Radiation-Mortality Study'. *J Soc Radiol Protection*, 3: 18-20.

2. Smith, P. G. and Douglas, A. J. (1986). 'Mortality of workers at the Sellafield plant of British Nuclear Fuels'. *Br Med J*, 293: 845-854.

3. Binks, K. and McElvenny, D. (1990). 'Mortality and cancer registration experience of the Sellafield workers known to be involved in the 1957 Windscale incident'. European Nuclear Conference 23-28 September 1990, Lyon, France.

4. Tagg, B., Lawson, A. W. and Binks, K. (1991). 'Occupational exposure and health at Sellafield UK since 1948'. Occupational Radiation Protection, British Nuclear Energy Society, 29th April-3rd May 1991, Guernsey.

A study of asbestos workers (MR177)

Reason for the study
This constitutes the continuation of a series of studies into mortality and morbidity at an asbestos factory begun in 1928.

Study population
This has been modified at each analysis, with the identification of further eligible subjects employed in the dust risk areas of the factory for whom some indices of dust exposure were available.

Nature of the study
Subjects were traced at NHSCR and flagged. Coded copies of death certificates were provided. Observed all cause and specific cause mortality were related to values predicted from National rates and analysed according to years of exposure, years exposed, department employed in and dust levels to which exposed.

Results
In Peto et al (1977) a 98 per cent trace rate was achieved at NHSCR. Those persons first employed before 1933 had a statistically significant increase in mortality from lung tumours and non-malignant respiratory diseases, more marked for those with more than 10 years of exposure.

When the population was divided into groups of those exposed between 1933 and 1950 and those from 1951 onwards, there was still excess mortality, particularly fifteen and more years after first exposure, though less than prior to 1933.

Publications

1. Peto, J., Doll, R., Howard, S. V., Kinlen, L. J. and Lewinsohn, H. C. (1977). 'A mortality study among workers in an English asbestos factory'. *Br J Ind Med*, 34: 169-173.

2. Peto, J., Doll, R., Hermon, C., Binns, W., Clayton, R. and Goffe, T. (1985). 'Relationships of mortality to measures of environmental asbestos pollution in an asbestos textiles factory'. *Ann Occup Hyg*, 29: 305-355.

A study of workers exposed to Bis-chloromethyl ether in a methylation process (MR178)

Reasons for the study
A case of lung cancer was reported in 1978 in a man of 39 who had been employed in a methylation process, though the process had been terminated. The following year a second case was diagnosed in a man of 35 who had previously smoked heavily. That same year saw a report of 20 cases of lung cancer in a German factory where BCME exposure was involved. When a third case was diagnosed in a man of 39, a study of workers who had been previously exposed was set up. Exposures were potentially to a mixture of known and suspected carcinogens.

Study population
398 past and present male workers who had been employed for more than 6 months since the start of the factory in 1950.

Nature of the study
Subjects were traced and flagged by NHSCR and coded death certificates provided. Patterns of mortality were studied in relation to involvement with the chloromethylation process and other agents. Predicted values were derived from National rates. Qualitative indices of exposure were derived for various compounds and there was some information on smoking.

Results
A 96 per cent trace was achieved. By June 1985, there was an excess of lung cancer (0=11, E=4.2 $p<0.001$) and of leukaemia (0=2, E=0.29 $p=0.05$). Regular employment in the chloromethylation process was associated with a significant risk of lung cancer: other cancers were present numerically to excess, but not significantly so. No other agent was associated with added risk.

By 1989, there were 14 deaths from lung cancer with 5.6 expected ($p<0.001$).

Publications

1. Roe, F. J. C. (1985). 'Chloromethylation: three lung cancer deaths in young men' (letter). *Lancet*, 2: 268.

2. Roe, F. J. C., Fry, J. S. and Lee, P. N. (1985). *Chloromethylation as an occupational hazard.* Report on an epidemiological study of workers in a chemical factory: Appendix i.

3. Fry, J. S., Lee, P. N. and Roe, F. J. C. (1989). *Second report on an epidemiological study of workers in a chemical factory.*

Mortality of persons resident in the vicinity of Electric transmission facilities (MR182)

Reasons for the study
Some studies had reported an association between residence near power transmission facilities with the development of acute myeloid leukaemia, other lymphatic cancers and suicide. Though reports have not been consistent, the extent to which there is a general population exposure to this form of man-made electromagnetic radiation prompted a further study.

Study population
7,920 persons resident at the time of the 1971 census in the vicinity of electricity substations and overhead power cables, selected from a sample in East Anglia.

Nature of the study
Subjects were traced and flagged at NHSCR and death certificates were analysed. The mortality pattern was compared where possible with that predicted from East Anglian rates. The work was carried out at OPCS, entirely 'in house' and anonymity preserved.

Results of the study
A 96 per cent trace was achieved.

For the 814 deaths (fewer than expected), there were overall no obvious major hazards. The only statistically significant excess mortality was for lung cancer in women overall and in particular in those living closest to the installations. The significance of these observations is not certain. There were no excess deaths observed for myeloid leukaemia and other lymphoid cancers.

Publication

1. McDowall, M. L. (1986). 'Mortality of persons resident in the vicinity of electric transmission facilities'. *Brit J Cancer*, 53: 271-279.

Mortality of employees at the Atomic Weapons Research Establishment 1951-82 (MR183)

Study population
A total of 22,552 personnel records were available for persons employed by AWRE at any one of the establishment's 5 sites between 1st January 1951 and 31st December 1982.

Nature of the study
Subjects were traced and flagged by NHSCR and coded death certificates provided. A mortality study of employees, compared those with radiation records with those without, and further classifying exposed workers according to whether there was exposure to unsealed sources. Cancer registrations were also studied to check the completeness of cancer identification on death certificates.

Results
99.7 per cent of the population were traced, and a total of 3,115 deaths identified. For 9,389 workers monitored for their exposure to radiation, the mortality experience overall was similar to that of the non-monitored workers other than for excesses of 'prostatic cancer' and 'ill-defined and secondary sites'.

Of 3,742 with a radiation record and possible internal exposure to radionuclides, there was overall no cancer excess, but after a 10 year lag, death rates from prostatic and renal cancers were twice the national average. For the sub-set with the relatively high exposure to external radiation and monitored for possible exposure to multiple radionuclides, when exposures were lagged by 10 years, there was a significant trend of excess mortality from malignant neoplasm overall with increasing cumulative exposure to external radiation for this group: lung, and to a lesser extent prostatic cancers, were the main contributors.

Publications

1. Beral, V., Fraser, P., Carpenter, L., Booth, M., Brown, A. and Rose, G. 'Mortality of employees of the Atomic Weapons Establishment, 1951-82'. *Brit Med J*, 1988; 297: 757-770.

The long term health experience of the UK participants in the UK atmospheric nuclear weapons tests (MR185)

Reasons for the study
During the 1950s and early 60s, British servicemen and civilians attended bomb tests in Australia and the Pacific. In 1982, investigative journalists collected data whose analysis gave rise to suspicion that this population might have experienced excesses of leukaemia and reticuloendothelial system tumours.

Study population
As large a sample of attendants at the tests as could be identified from official records, together with a similar sized control group.

Nature of the study
Subjects were traced and flagged at NHSCR and coded copies of draft death registrations and cancer registrations were provided. For a proportion of the population, there were measures of radiation exposure. Mortality and cancer morbidity were studied and comparisons made with national rates and with experience of control populations.

For the subset that had been monitored, radiation levels were studied in relation to outcome.

Results
The analysis to 1st January 1984 reported that participation in the tests was not found to have detectable effects on life expectancy, nor on total risk of developing cancer. There were suggestions that there might be risks of developing leukaemia and multiple myeloma, but this was not related to unusual radiation exposure. Controls experienced

significant excesses of carcinomas of prostate and of the kidney: prostatic cancer excesses had been reported in studies of UK and US radiation workers.

Publications

1. Darby, S. C., Kendall, G. M., Fell, T. P. et al (1988). *Mortality and Cancer Incidence in the UK Participants in UK Atmospheric Nuclear Weapon Tests and Experimental Programmes.* NRPB R214. London, HMSO.

2. Darby, S. C., Kendall, G. M., Fell, T. P. et al (1988). 'A summary of mortality and incidence of cancer in men from the United Kingdom's atmospheric nuclear weapons tests and experimental programme'. *Br Med J*, 296: 332-38.

3. Darby, S. C., Kendall, G. M., Fell, T. P. et al (1990). 'Mortality among United Kingdom Servicemen who served abroad in the 1950s and 1960s'. *Br J Ind Med*, 47: 793-804.

4. Darby, S. C., O'Haggan, J. A., Kendall, G. M., Doll, R., Fell, T. P. and Muirhead, C. R. (1991). 'Completeness of follow-up in a cohort study of mortality using the United Kingdom National Health Service Registers and records held by the Department of Social Security'. *J Epidem Community Health*, 45: 65-70.

Cancer registration and mortality studies of workers exposed to chloracnegenic agents (MR186)

Reason for the study

To increase the power of studies investigating the 'dioxine' hazard in humans.

Study population

This study covers two separate populations. One consists of workers involved in the production of the herbicide 2,4,5-T (with potential exposure to 'dioxin'), the other consists of two groups, of workers manufacturing dichloro-and trichloroaniline.

Nature of the study

All subjects have been traced and flagged at NHSCR and coded death cancer registrations provided to the researchers. Observed values for specific cause mortality and cancer registration will be compared with values derived from National and Regional rates. Those persons affected by chloracne will be treated as a special subset and mortality and morbidity related to this event and to duration of employment.

The data derived from this study and from MR 152 will be contributing to an international collaborative IARC study.

Mortality of herbicide manufacturers and sprayers (MR187)

Reasons for the study

There had been concern about the possible carcinigenicity of phenoxy herbicides, some of which are contaminated by 'dioxins'. 2, 4, 5-T, and its precursor in manufacture, 2, 4, 5-trichlorophenol, had been linked with soft tissue sarcoma and lymphoma. It was decided to study workers predominantly exposed to another phenoxy herbicide (MCPA) to see if they were at risk.

Study population

All men employed at the factory or at spray depots between 1st January 1947 and 31st December 1975.

Nature of the study

Individual exposures were classified as 'high', 'low' or 'background'. Members of the study cohort were traced through NHSCR, and where necessary the National Insurance index, and their vital status was determined at 31st December 1983. Coded causes of death and cancer registrations were provided. Expected values were based on National rates and on aggregated Rural rates for 1968-78.

Results

A trace rate of 98.3 per cent was achieved. Overall mortality from cancer, heart disease and respiratory diseases was less than predicted from National rates. Applying rural rates, the cancer deficit changed to a non-statistically significant slight excess. Only 1 death was reported as due to a soft tissue sarcoma (E=0.6). Three workers had nasal cancer (E=0.7), an association not previously reported. It was concluded that the study did not exclude a cancer risk, but for MCPA the soft tissue sarcoma risk was less than that indicated by earlier studies for 2, 4, 5-T and was small. The study will be extended (MR250).

Publication

1. Coggon, D., Pannett, B., Winter, P. D., Acheson, E. D. and Bonsall, J. (1986). 'Mortality of workers exposed to 2-methyl-4-chlorophenoxyacetic acid'. *Scand J Work Environ Health*, 12: 448-54.

Employees at plants manufacturing 4, 4'-bipyridyl (MR188)

Reason for the study

A number of workers at plants manufacturing 4,4'-bipyridyl were found to have skin lesions ranging from metaplasia to squamous carcinoma. A toxicological investigation had suggested that the source of the risk was a tarry by product of an old plant. Although the method of production had changed, and a review of mortality and morbidity data available gave no cause for concern, it was considered prudent to conduct a detailed survey of workers at these plants.

Study population

A total of 761 employees who had worked at the bipyridyl plants were identified from personnel records for whom adequate identification and occupational details were available. The population has been expanded by the addition of new starters.

Nature of the study

Subjects for whom information on the company files was incomplete were traced at NHSCR. Records were flagged and coded death certificates provided. Standard mortality ratios were calculated for various causes using rates for males in England and Wales and local rates adjusting for man years. Mortality and cancer incidence were analysed by occupational group, plant worked at, and period of service. The effects of ten and fifteen year latency periods were studied. A case control study of lung cancer was conducted in an attempt to identify the source of hazard.

Results

A trace of 96.6 per cent was achieved.

All cause mortality adjusted for local rates gave an SMR of 78. The SMR for 'all cancers' was 107 and for lung cancer 124 (not significant). Calculated for a latency period of 15 years the SMR for lung cancer of 211 was significant at the 5 per cent level. The lung cancer cases were found to be concentrated in specific processes but a causal relationship with an agent was not determined.

Publication

1. Paddle, G. M., Osborn, A. J. and Parker, G. D. J. (1991). 'Mortality of employees in plants manufacturing 4, 4'-bipyridyl'. *Scand J Work Environ Health*, 17: 175-178.

Mortality of Chrysotile asbestos cement workers (MR189)

Reasons for the study

As a result of the majority of groups of workers studied formally having had mixed exposures, there has been controversy about the relative hazards of the various varieties of asbestos. Because of their design, the findings from earlier studies of chrysotile workers had been inconclusive.

Study population

All workers employed at an asbestos cement factory 1941-1983 using chrysotile predominantly.

Nature of the study

Subjects were traced at NHSCR and at the National Insurance register, and flagged. Coded copies of death certificates were sent to the researchers. A standard cohort mortality study was carried out adjusting for man-years to produce predicted numbers of specific cause deaths from National and from Regional rates. Results were related to job held and asbestos dust exposure levels.

Results

By the end of 1984 no excess of lung cancer was observed. One case of malignant pleural mesothelioma was found 7 years after the start of exposure in a man who had only worked for 4 years. There was one death from lung cancer with asbestosis in a man 33 years after first exposure and 3 short episodes of work of durations, 8, 4 and 1 month. The authors discounted the significance of the asbestos exposures in these cases.

As the duration of exposure in this population is relatively short, it is planned to repeat the analyses when more substantial data have accumulated.

Publications

1. Gardner, M. J., Winter, P. D., Pannett, B. and Powell, C. A. (1986). 'Follow up study of workers manufacturing chrysotile asbestos cement products'. *Br J Ind Med*, 43: 726-732.

2. Gardner, M. J. and Powell, C. A. (1986). 'Mortality of asbestos cement workers using almost exclusively chrysotile fibre'. *J Soc Occ Med*, 36: 124-126.

3. Gardner, M. J. and Powell, C. A. (1987). 'A review of mortality findings in studies of asbestos cement workers using predominantly chrysotile fibre'. In: *The biological effects of chrysotile*. Lippincott.

Mortality study of pest control officers (MR190)

Reasons for the study

Several anecdotal reports of bladder cancer in rodent operatives attracted attention to the potential hazard of alpha-naphthyl thiourea (ANTU), which may contain up to 0.2 per cent beta-naphthylamine, a known urothelial carcinogen, or may yield it on metabolism.

Nature of the study

Four separate approaches were adopted.

1. All death certificates on the bladder cancer mortality register between 1951 and 1980 (serviced by OPCS) were reviewed.

2. 275 operatives known to MAFF were reviewed for bladder cancer and occupational exposures.

3. 360 operatives listed by local authorities were reviewed for their bladder cancer experience.

4. Some 1,200 pest control officers employed by local authorities for at least 6 months between 1974 and 1981 in England and Wales were flagged by OPCS for study of their mortality pattern and its association with the variety of pesticides with which they come in contact. A 10 per cent random sample was sent self-administered questionnaires to obtain supplementary information on other relevant factors.

Results

1. 28 rodent operatives were identified with bladder tumour of whom 4 had handled ANTU.

2. No bladder tumour deaths were found.

3. 4 cases of bladder tumour were identified, though they were previously known of.

4. The data remain to be analysed.

Publications

1. Thomas, H. F. and Donaldson, L. J. (1986). 'Profile of local authority pest officers'. *J Roy Soc Health*, 6: 204-206.

Mortality in pottery workers (MR193)

Reasons for the study
An industry study of morbidity in pottery workers, noted increases in respiratory symptoms, slight radiological changes and minor decrements of lung function in the absence of overt signs of disease. The opportunity was taken to evaluate the long term health consequences of pottery dust exposure, including the putative carcinogenic effects of exposure to mixed dusts containing crystalline silica.

Study population
6,192 well characterised participants in a cross sectional morbidity study.

Nature of the study
National Insurance records and NHSCR were used to trace subjects and coded death certificates were provided. Numbers of observed deaths overall and by disease were compared with predicted values derived from National rates and from Regional rates adjusted for person years exposure, age and sex. For lung cancer, adjustments were made for cigarette smoking habit. Mortality was also related to the effects found in the morbidity study and to measures of dust exposure.

Results
A trace rate of 99.9 per cent was achieved in the preliminary analysis of mortality related to dust exposure. For men aged over 60 at the time of the survey dying between 1970 and July 1985, there was an excess of lung cancer whether National or local rates were used for comparison, and when adjusted for smoking habit. This seemed to be associated with cumulative exposure to respirable quartz. An excess of stomach cancer when National rates were employed was not seen when local rates were employed.

Publications

1. Winter, P. D., Gardner, M. J., Fletcher, A. C. and Jones, R. D. (1990). 'A mortality follow-up study of pottery workers; preliminary findings on lung cancer'. In: *Occupational exposure to silica and cancer risk*. I.A.R.C. Scientific Publications No. 97, 83-94. I.A.R.C. Lyons.

A follow up study of workers in a chemical company manufacturing methylene-bis-ortho-chloraniline (MBOCA) (MR205)

Reasons for the study
MBOCA has been widely used to prepare hard polyurethane products. It is chemically similar to benzidine, a known cause of human bladder cancer. Experimental studies had confirmed its carcinogenic properties, and some cases of bladder cancer had been reported at a manufacturing plant. It was therefore considered essential to review the health experience of workers exposed to this agent.

Study population
Workers employed for at least 6 months since the start of the process at a company in 1965.

Nature of the study
Qualifying members have been flagged at the NHSCR and coded causes of death and cancer registrations will be analysed according to exposure and work history. The mortality pattern will be compared with predictions based on National rates.

Results
An initial pilot study indicated a substantial collection of data available, sufficient to consider the study as likely to be of value.

Leukaemia and lymphoma among young persons near Sellafield Nuclear Plant (MR208)

Reasons for the study
Following the report of an excess of leukaemia in children living near Sellafield, the Black committee recommended a series of local studies including that of childhood leukaemia and its associations.

Study population
52 cases of leukaemia, 22 cases of non-Hodgkin's lymphoma and 23 cases of Hodgkin's disease in young persons under the age of 25 identified in West Cumbria Health District between 1950 and 1985.

Nature of the study
Cases were matched by sex and date of birth with area and local controls from the same birth register. Comparisons were made from birth and medical records, from questionnaires to parents of cases and controls, and from employment and radiation dose records held by employers.

Results
There is a suggestion that the recorded whole body radiation to fathers during employment at Sellafield is associated with the development of leukaemia among their children, with the highest risk for the higher cumulative doses prior to conception.

Publications

1. Gardner, M. J., Snee, M. P., Hall, A. J., Powell, C. A., Downes, S. and Terrell, J. D. (1990). 'Results of case-control study of leukaemia and lymphoma among young people near Sellafield nuclear plant in West Cumbria'. *Br Med J*, 300: 423-429.

2. Gardner, M. J., Hall, A. J., Snee, M. P., Downes, S., Powell, C. A., and Terrell, J. D. (1990). 'Methods and basic data of case-control study of leukaemia and lymphoma among young people near Sellafield nuclear plant in West Cumbria'. *Br Med J*, 300: 429-434.

Study of children born to mothers resident in Seascale 1950-1983 (MR210)

Reasons for the study
The report of the Black Committee into the cluster of childhood cancers in the vicinity of Seascale recommended a study of cancer incidence and mortality of all children born since 1950 to mothers resident in Seascale at the time of birth.

Study population
1,068 children born to mothers resident in Seascale 1950-83.

Nature of the study
Local birth registers were searched to identify children born to mothers living in Seascale at the appropriate period. They were traced and flagged at NHSCR and death certificates extracted. Expected numbers of specific cause deaths were calculated from national rates.

Results
As NHS numbers were always available a 100 per cent trace was achieved. There was a deficit of deaths due to non-malignant diseases, with 12 reported against 21.55 predicted. There was a significant excess of malignant disease, with 9 observed against 1.60 predicted, with the 95 per cent confidence interval for the observed to expected ratio being 2.58-10.69. Five of these deaths were from leukaemia where 0.53 were predicted (95 CI of observed to predicted ratio 3.04-21.84).

Publications

1. Gardner, M. J., Hall, A. J., Downes, S. and Terrell, J. D. (1987). 'Follow up study of children born to mothers resident in Seascale, West Cumbria (birth cohort)'. *Br J Ind Med*, 295: 822-827.

A cohort study of girls attending Calder Girls' School in Seascale, West Cumbria (MR211)

Reason for the study
Recommendation of the Black Committee for the study of children living in Seascale but not having been born there.

Study population
111 children on the school register 1951-1967, not already included in the Seascale birth cohort study.

Nature of the study
Subjects were traced and flagged at NHSCR and death certificates and cancer registrations provided for the researchers. Predicted values for mortality and cancer registrations were derived from national rates.

Results
A trace of 86.5 per cent was obtained. No deaths (0.82 would have been expected) or cancer registrations were reported. Further attempts at tracing will be made before further analyses are attempted.

Publications

1. Gardner, M. J., Hall, A. J., Downes, S. and Terrell, J. D. (1987). 'Follow up study of chil-

dren born elsewhere but attending school in Seascale, West Cumbria (schools cohort)'. *Br J Ind Med*, 295: 819-822.

A cohort study of children at Local Authority schools in Seascale 1953-1984 (MR212)

Reasons for the study
Recommended by the Black Committee of Enquiry into childhood cancer in West Cumbria.

Study population
1,311 children on the admissions registers of Seascale Local Authority schools 1950-1984, not included in the Seascale birth cohort study (MR210).

Nature of the study
Subjects were traced and flagged at NHSCR and coded death certificates and cancer registrations provided to the researchers. Numbers of cases at 30th June 1986 were compared with values predicted from National rates.

Results
94 per cent were traced. Ten deaths were reported from all causes, with 10.42 predicted. A case of malignant melanoma was reported in a man of 33 who had moved out of the district. The remaining nine deaths were classified under 'injury and poisoning'.

3 cases of non-fatal cancer were registered (predicted = 2.04) in the period 1971-84. (Attempts will be made to reduce the numbers of records untraced before further analyses are attempted.)

Publications

1. Gardner, M. J., Hall, A. J., Downes, S. and Terrell, J. D. (1987). 'Follow up study of children born elsewhere but attending schools in Seascale, West Cumbria (schools cohort)'. *Br J Ind Med*, 295: 819-822.

Mortality study of rubber cable workers (MR221)

Reasons for the study
Studies of rubber workers showed that despite the withdrawal of carcinogenic antioxidants with correction of the bladder cancer excess, an excess risk of cancer at a number of other anatomical sites existed. Though the processes of cable making tend to be enclosed, it was deemed desirable to ascertain whether this was sufficient to protect workers.

Study population
Workers employed at a cable factory with personnel records going back to the time of its founding at the turn of the century.

Nature of the study
The study members were traced and flagged at NHSCR and coded copies of death certificates provided. Comparisons will be made between groups within the study and with predicted values derived from National rates adjusted for man years.

Mortality study of capacitor workers (MR222)

Reasons for the study
Polychlorinated biphenyls (PCBs) had been extensively used as a dialectric in capacitors. Initially exposure to them was only associated with the development of a form of acne. Latterly, the observation that PCBs commonly were contaminated with 'dioxins' led to concern. These agents were found in experimental studies on several species, to be powerful toxic agents with properties that included mutagenicity, carcinogenicity and fetotoxicity.

Study population
1,379 women and 905 men who had worked for at least 12 months at a factory manufacturing capacitors incorporating PCBs between 1955 and 1980.

Nature of the study
Subjects were traced and flagged at NHSCR and coded copies of draft entries in the death register were provided to the researchers. Values for expected individual causes of death will be derived from National rates adjusting for person years. Mortality was related to sex, order of exposure, years of employment and duration of follow up.

Results
In a preliminary analysis in 1990, there were 111 deaths for women (SMR=97, 95% CI=78-115) of which "all neoplasms" totalled 42 (SMR=130, 95% CI=91-169).

For men, there was a significant reduction in all cause mortality with 179 deaths (SMR=84, 95% CI=72-97 p=<0.05). 21 deaths from malignancy gave an SMR of 165 (95% CI=102-252 p=<0.005). Men had 29 deaths from non-malignant respiratory disease (SMR=188, 95% CI=126-270 p=<0.005). The statistically significant excesses of mortality were found in those exposed to very high and to medium levels of PCB, and were related to years of follow up. At this stage and in the absence of smoking histories no firm conclusions could be drawn.

A study of leukaemia and other cancers among benzene workers (MR226)

Reasons for the study
Several studies of heavily exposed benzene workers had shown them to be at increased risk of leukaemia. It was decided to attempt to determine the dose response relationship, as the results of studies at lower level exposure had been inconsistent.

Study population
Some 3,000 persons were identified for whom benzene exposure levels were available at the inception of the study.

Nature of the study
Subjects were traced at the NHSCR and at the National Insurance register and flagged. Coded copies of death certificates and cancer registrations were provided to the researchers. Mortality experience was compared with national rates and between benzene exposure groups.

Results
An initial (unpublished) analysis did not provide evidence of abnormal mortality or cancer morbidity. There was uncertainty about the validity of these initial 'negative' findings: further analyses will be carried out when adequate data have accumulated.

Mortality study of cement workers (MR233)

Reason for the study
When a study of a UK population of cement workers showed an excess of gastric carcinoma (McDowall, 1984), the trade association and the relevant unions requested a study of a further group of cement workers to evaluate the hypotheses of association between cement dust exposure and the development of lung cancer and gastric cancer.

Study population
Workers employed at 6 works in 1939 and alive on 1st January 1948, not previously studied.

Nature of the study
Subjects were traced and flagged at NHSCR, and coded causes of death provided to the researchers. Predicted values for individual cause mortality were derived from National and local rates and standard mortality ratios were calculated for certain conditions and classified by occupation for various periods of the study between 1948 and 1981.

Results
Of 967 men still alive on 1st January 1948, 676 had died by 31st December 1984. Overall, mortality was below the value predicted from England and Wales rates. There was a significant deficit of deaths from 'all respiratory diseases'. Those with high potential exposures to dust had raised, but not statistically significant ratios for 'all respiratory diseases', deficits of lung cancer mortality, and no significant excess of gastric carcinoma.

Publications

1. McDowall, M. E., Snashall, D. and Farebrother, M. J. B. (1986). 'Lung and stomach cancer mortality in cement workers'. Report submitted to the trade association and four unions. Referred to in an editorial in the *British Journal of Industrial Medicine*, 43, 1986; 43: 505-506.

Mortality of workers manufacturing 1, 2 Dibromoethane (Ethylene dibromide — EDB) (MR237)

Reasons for the study
EDB is useful as a pesticide and as an antiknock compound. Studies have shown EDB and a related compound to be experimental carcinogens and to depress fertility. A mortality study was mounted at 2 factories where the compound was produced.

Study population
115 men for whom adequate records were available from factory A (53 of whom had regular exposure to EDB), which had operated between 1940 and 1973 and where EDB production ceased in 1970. Some 320 men with a minimum of 4 years exposure from factory B (a maximum of 55 of whom had regular EDB exposure), which had more complete records and had started up in 1954 and was still in production.

Nature of the study
Subjects were traced and flagged at NHSCR, and coded death certificates provided. Observed deaths were compared with those predicted from local rates. Standardised mortality rates were calculated and presented according to orders of exposure.

Results
The results of initial analyses have been inconclusive largely as a result of the small numbers involved, with no statistically significant differences between observed and predicted values overall or between workers exposed to different extents.

Publications

1. Turner, D. and Barry, P. S. I. (1978). 'An epidemiological study of workers in plants manufacturing ethylene dibromide'. Presentation at the XIXth International Congress on Occupational Health, Dubrovnik, Yugoslavia, 25-30 Sept.

2. Oxley, G. R. (1985). *A proportional mortality study on two groups of former employees having had exposure to 1, 2 dibromoethane*. Associated Octel Company Report.

Study of children attending Singing Surf School in Seascale since 1950 (MR242)

Reasons for the study
Recommendations of the Black Committee for a study of Seascale children, of which a small local school comprised a subset.

Study population
124 children born since 1950 admitted to a preparatory school 1949-1968, not already included in the Seascale birth cohort study (MR210).

Nature of the study
Children were traced and flagged at NHSCR and coded copies of death certificates and cancer registrations extracted. Mortality to 20th June 1986 was studied to provide measurements of incidence and mortality from cancer among children who had lived in Seascale for some part of their childhood although not born there. Comparisons were with National rates.

Results
A trace of 90 per cent was obtained. No deaths were reported whereas 1.45 were predicted. One case of in situ carcinoma of the cervix was reported in a woman of 29 living outside Cumbria.

Publications
1. Gardner, M. J., Hall, A. J. Downes, S. and Terrell, J. D. (1987). 'Follow up study of children born elsewhere but attending schools in Seascale, West Cumbria (schools cohort)'. *Br J Ind Med*, 295: 819-822.

Cohort study of workers exposed to spinning finishes (MR245)

Reasons for the study
Chloracetamide-N-methylol (CAM), a formaldehyde releasing biocide had been included in spinning finishes. Experimental evidence of the carcinogenic potential of formaldehyde led to CAM being tested and found to be clastogenic. The product was withdrawn in 1984 and because a preliminary look at company health records was insufficiently informative, a follow up study of all exposed workers was planned.

Study population
All male staff employed on 1st January 1976 at a particular factory where CAM had been used 1974-1984.

Nature of the study
Using the company record systems subjects were identified and notified to NHSCR for tracing. Coded death certificates were provided and PMRs were calculated for certain causes and SMRs, allowing for man years of exposure.

Results
The small numbers involved and the limited period of follow up militated against firm conclusions being reached. In the event, excesses of tumours of the large intestine (0=4, E=1.8) and kidney (0=4), E=0.6) were noted.

A study of workers exposed to ethylene oxide (EtO) (MR246)

Reasons for the study
EtO, widely used as a sterilisant, has been recognised as an experimental mutagen and carcinogen. A tenfold excess of leukaemia and an inconsistent excess of other neoplasms, were reported in a study of Swedish hospital steriliser workers using EtO, and workers involved in manufacture prompting a further study.

Study population
Workers for whom adequate records were available at three companies manufacturing EtO and derivatives, and workers in 8 selected hospital sterilising units.

Nature of the study
The population was largely traced at NHSCR and flagged and coded death certificates were provided. Observed cases were compared with numbers predicted from National rates, allowing for person years, age, sex, cause and calendar year.

Results
A 98.7 per cent trace rate was achieved, with losses mainly due to inadequate identification data in the basic records. Total carcinoma mortality was only slightly higher than expected. For non-Hodgkin's sarcoma, 4 deaths were observed when 1.62 were expected. No important increase in mortality was seen from other causes.

Publications
1. Coggon, D. (1986). 'Ethylene oxide-a new human carcinogen?' *Lancet*, i.

2. Gardner, M. J., Coggon, D., Pannett, B. and Harris, E. C. (1989). 'Workers exposed to ethylene oxide: a follow-up study'. *Br J Ind Med*, 46: 860-865.

3. Shore, R. E., Gardner, M. J. & Pannett, B. (1993). 'Ethylene oxide: an assessment of the epidemiological evidence on carcinogenicity'. *Br J Ind Med*, 50: 971-997.

Mortality of butchers and slaughtermen (MR247)

Reasons for the study
Historically there was a myth that work as a butcher was protective against tuberculosis, though the early Registrar General reports found them to be very unhealthy. More recent reviews of routine vital statistics have shown butchers to suffer an excess of lung cancer. This study was mounted to investigate this finding, and should the excess be confirmed, to identify the associated factors in a group of workers.

Study population
610 workers satisfying certain entry criteria employed at two factories slaughtering and processing pig meat, and at abattoirs and meat distribution centres handling beef, pork and lambs.

Nature of the study
Subjects were traced and flagged at NHSCR and coded copies of death certificates provided. The numbers of deaths from various causes observed were compared with values predicted from National and from local rates.

Two parallel studies are being conducted. One is a mortality study of butchers identified in the 1961 census. The other involves identifying all lung cancer deaths over a period of 2 years in England and Wales and testing pathological material for papilloma virus DNA.

Results
A 99 per cent trace was achieved. A lung cancer excess was found overall (0=42, E=31.7) using National rates, the excess was still present when local rates were used. When individual plants were studied, at one, no lung cancer excess was observed employing National rates; when local rates were used an excess was seen. Exposure to warm meat was associated with a greater lung cancer excess (0=22, E=11.9: SMR=184 with 95 confidence limits 115-279). Excess was found for stomach cancer and smaller excesses for liver cancer, lymphoma, myeloma and leukaemia.

Publications
1. Coggan, D., Pannett, B., Pippard, E. C. and Winter, P. D. (1989). 'Lung cancer in the meat industry'. Br J Ind Med, 46: 188-191.

Mortality and cancer morbidity in Cornish tin miners (MR248)

Reasons for the study
A lung cancer excess in a group of Saxony miners first reported in 1878 was attributed in the mid-1920s to ionising radiation. Cornish tin miners, who worked prior to 1939 in an atmosphere rich in radon and its daughter products, have also been found to experience a lung cancer excess. A study was set up to measure the residual hazard in a group of miners employed after ventilation had been enhanced.

Study population
2,104 non-clerical workers at the remaining two active Cornish tin mines employed for at least 1 year between 1945 and 1980 for whom there were adequate identification and work histories.

Nature of the study
Subjects were traced and flagged at NHSCR and coded death certificates were provided. Causes of death were related to job history, total exposure, time since last exposure and arsenic exposure. Lifetime risk estimates were calculated using the BEIR model based on estimated ionizing radiation exposure rates. National rates were used for calculating expected numbers of cases and standard mortality ratios.

Results
A 97.6 per cent rate was achieved. There was a statistically significant excess of 'all cancers' ($p<0.001$) and of cancers of lung, trachea, bronchus and pleura ($p<0.01$) which increased with duration of exposure below ground.

Publications
1. Hodgson, J. T. (1989). 'Mortality in tin miners 1941-1986'. In: *Proceedings of the Seventh International Symposium on Epidemiology in Occupational Health*. Tokyo, Japan 11-13 October 1989. Elsevier Science Publishers B.V.

2. Hodgson, J. T. and Jones, R. D. (1990). 'Mortality of a cohort of Cornish tin miners'. *Br J Ind Med*, 47: 665-676.

Mortality and cancer incidence of workers making phenoxy herbicide (MR250)

Reasons for the study
The wide use of phenoxy herbicides in agriculture, suspicions about their carcinogenicity and that of compounds used as their raw material and produced in their synthesis, and the inconclusive nature of the studies designed to determine their human health hazard, prompted the International Agency for Research on Cancer (WHO) to organize an interna-

tional collaborative study. The cohort from MR187 was contributed to this study with extended follow-up, and workers from four further factories were included.

Study population
Workers from four factories where a range of phenoxy herbicides and chlorophenols were produced and formulated.

Nature of the study
The individual cohorts were defined variously according to the availability of data at the factories. Exposures were determined from job histories. Subjects were traced and flagged at NHSCR and coded copies of causes of death and cancer registrations were provided by OPCS. Standard mortality ratios were derived from national death rates by the person-years method.

Results
A 97.8 per cent trace was achieved. 'All cause' deaths totalled 152 (E=136.22, SMR=112, 95% CI=93-131). The excess was largely attributable to high rates of circulatory disease (SMR=116, 95% CI=91-146) and injury and poisoning (SMR=155, 95% CI=93-242). There was no significant excess of cancer overall. There were 19 deaths from lung cancer with 14.15 expected. There were no cases of soft tissue sarcoma (E=0.18) or of Hodgkin's disease (E=0.43), but there were two cases of non-Hodgkin's lymphoma (E=0.87, SMR=229, 95% CI=28-827). The excess of circulatory disease was confined to one factory but a nested case control study did not suggest underlying occupational causes. Analysis of the international collaborative study showed a statistically non-significant excess of soft tissue sarcoma. (0=4, SMR=196, 95% CI=52-502). Overall cancer mortality was close to expectation.

Publications

1. Coggon, D., Pannett, B. and Winter, P. (1991) 'Mortality and incidence of cancer at four factories making phenoxy herbicides'. *Br J Ind Med*, 48: 173-178.

2. Saracci, R., Kogerinas, M., Bertazzi, P. A. et al (1991). 'Cancer mortality in workers exposed to chlorophenoxy herbicides and chlorophenols'. *Lancet*, 338: 1027-32.

Long-term health effects of working in a building containing friable structural asbestos (MR251)

Reasons for the study
Asbestos arising from deteriorating construction material had been found in the environment of a building. At the time it was considered important to follow up the health of exposed persons.

Study population
60-70 persons were considered to have been exposed.

Nature of the study
Subjects have been traced and flagged at NHSCR and coded death certificates and cancer registrations will be provided.

Morbidity and mortality in silica sand quarry workers (MR256)

Reasons for the study
The extent of silicosis and of non-malignant respiratory diseases in a group of sand quarry workers had not been evaluated. Nor had the hypothesis of an association between heavy dust exposure and the development of bronchial and gastric cancer been investigated.

Study population
Quarrymen employed at a number of silica sand quarries.

Nature of the study
The nominal roll will be traced and flagged at NHSCR and death certificates provided. Predicted values for comparison will be based on National and Regional rates. Any excesses of disease will be related to qualitative and quantitative measures of dust exposure.

Results
There was a non-significant excess of lung cancer which was not positively related to any of the work related variables studied.

Publications

1. Benn, R., Hutchings, S., Thomas, P., Elliott, R. et al. (1993) 'Lung cancer in a population exposed to silica'. *Thorax*, 48:436.

A study of fertiliser workers (MR260)

Reasons for the study
Nitrosamines are powerful animal carcinogens. Consequently they are regarded as potential human carcinogens. There is concern that endogenous nitrosamine production enhanced by an increased intake of nitrate in food and water, might have caused cancer excesses in man. A study of workers producing and formulating inorganic nitrate for fertiliser was mounted to test this hypothesis.

Study population
Nitrate fertiliser workers employed by a large chemical company on or after 1st January 1946, who had been exposed for at least one year, and for whom there were adequate data in company records.

Nature of the study
The population was traced by NHSCR and National Insurance records and flagged. Copies of death certificates were provided for those deceased before 1st March 1981. Crude measures of exposure were derived from the individual's job, the process involved and the period of exposure. Comparisons of the mortality experience were made with predictions derived from National rates, where comparison with those for the region where the factory was situated was not possible.

Results
Of the 1,327 men listed by the company, 99.8 per cent were successfully traced, and of 304 persons known to have died, details were available for 302. The total of deaths overall was less than expected in all exposure groups. For most broad categories of causes, the numbers of observed deaths were fewer than expected. The ratio of observed to expected overall was 1.05, for heavily exposed persons it was 1.15 and for all other men in was 0.90. Of the four cancers associated with experimental exposure to nitrosamine (stomach, oesophagus, liver, urinary bladder), and for lung cancer, observed mortality was slightly higher than expected for oesophagus and for 'cancers at other sites' for the whole cohort. Excesses varied between exposure groups, but no single excess approached statistical significance. A small excess of lung cancer was noted more than 20 years after first exposure in men heavily exposed for more than 10 years. It was concluded that the study did not support the hypothesis that exposure to environmental nitrate leads to the formation of material amounts of carcinogen.

Publications

1. Al-Dabbagh, S., Forman, D., Bryson, D., Stratton, I. and Doll, R. (1986). 'Mortality of nitrate fertiliser workers'. *Br J Ind Med*, 43: 507-515.

2. Forman, D. (1991). 'Nitrate exposure and human cancer'. In: *Nitrate contamination: exposure, consequences and control*. Ed, Bogardi, I. Springer-Verlag, Berlin.

Stress and Health study (civil servants) (MR262)

Reasons for the study
In a previous major study of civil servants, mortality from all causes, cancer and coronary artery disease were related to employment grade. It was decided to follow up the study to determine whether these mortality differences are reflected in differences in cancer morbidity, and to explore reasons.

Study population
Some 11,000 civil servants being screened for coronary artery disease.

Nature of the study
Tracing and flagging was carried out by NHSCR and coded copies of death certificates and cancer registrations provided. Outcomes will be related to sickness absence and psychosocial factors: this will involve studying two subsets of cases and controls. For subjects lost to the civil service, NHSCR data were provided to the researchers to permit tracing through Family Practitioner Committees and their practitioners. Birth certificates were also provided by OPCS for the study of parental social class.

It is proposed to analyse the group after 5 years, but it is appreciated that a 10-15 year follow up may be needed to study the cancer problem.

Mortality in Liverpool University (MR273)

Reasons for the study
The awareness of several deaths from cancer in members of a particular department prompted a request from staff representatives for an epidemiological enquiry.

Study population
Some 150 persons who had worked in the department since 1960.

Nature of the study
Subjects were traced and flagged at NHSCR and coded death certificates provided. Comparisons were made with predicted values derived from national rates.

Results
'All cancer' mortality was greater than predicted but no individual site was outstanding.

Publications
An initial report was presented to the University Health and Safety Committee in 1989.

Mortality of welders (MR274)

Reasons for the study
Welders inhale fumes that contain traces of carcino-

genic agents and irritants. Studies to evaluate the presence of an excess of respiratory cancer have produced inconsistent results. To provide a large population with adequate statistical power for study, an international study was co-ordinated by IARC.

Study population
393 welders who had been employed for a minimum of 5 years in various branches of welding.

Nature of the study
Subjects were traced and flagged at NHSCR and coded death certificates provided. Expected values for all cause and specific cause mortality were calculated from national rates adjusted for man-years of exposure. Subjects were categorised by type of metal or welding process and work sector. Findings were tabulated by time since first exposed, duration of exposure and cumulative fume exposure.

Results
A 95 per cent trace was achieved for the UK group. Analysis of combined data showed all malignancies to be in excess (0=303, E=268.63: SMR=113 95% CI=100-126). These were largely accounted for by 'lung cancer' (0=116, E=86.81: SMR=134 95% CI=110-160). Mild steel and stainless steel welders seemed specially at risk from lung cancer. Risk was not related to duration of exposure but increased with the period since first employed. Cigarette smoking was considered unlikely to account for the cancer excess, but the observation of five cases of malignant mesothelioma suggested a role for asbestos. [This would occur from exposure in shipyard work or from the use of crocidolite wrapped welding rods].

Publications

1. Simonato, L., Fletcher, A. C., Andersen, A. et al (1991). 'A historical prospective study of European stainless-steel, mild steel and shipyard welders'. *Br J Ind Med*, 48: 145-154.

A 50-year mortality follow-up study of brewery workers (MR286)

Reasons for the study
The development of certain tumours has been reported as being associated with substantial alcohol intake. The liberal provision of beer to brewery workers, has led to their study to test the causal hypothesis. As the results of studies in three countries, have not been consistent, a UK study was initiated to test the hypothesis of alcohol carcinogenicity.

Study population
701 men still alive on 1st January 1948, who had been identified from the 1939 national register, were found in the brewery town of Tadmore, together with job descriptions relating to brewing.

Nature of the study
Subjects were traced and flagged at NHSCR and anonymised coded copies of draft death registrations provided. The numbers of deaths expected were calculated from National and where available from local rates, using the person years method, and compared with numbers observed.

Results
In a trial analysis there was a 99.7 per cent trace apparent, with 401 deaths. Overall mortality was similar to that for England and Wales generally. There were however, deficits for 'all cancers' (0=94, E=119) and 'heart disease' (0=130, E=162) and a markedly low death rate for 'digestive diseases' (0=5, E=12). There were no reported deaths from liver cirrhosis. [There were suggestions of a lower death rate in recent years, which might correspond to alterations in the production process.]

For specific sites reported in other studies (oesophagus, larynx, rectum, liver, pancreas), there was no support for an increased cancer risk. For lung cancer, with 28 deaths observed against 43.8 expected, the 95 confidence interval observed to expected ratio was 0.43-0.93 (E and W) and 0.52-1.14 (local rates).

Retrospective study of employees who have worked in a pharmaceutical development experimental plant (MR288)

Reasons for the study
After the closure of a small-scale plant handling biologically active species for pharmaceutical products and pesticides, it was realized that there had been potential exposure to a range of toxic and putative carcinogenic compounds. Although exposures were complex it was considered desirable and feasible to evaluate the health risks of employment in this group.

Study population
All 108 workers employed in the department during the last 14 years of the plant's existence.

Nature of the study
Subjects will be traced and flagged at NHSCR and coded death certificates provided. Specific cause mortality will be compared with rates for England and Wales.

Results
As yet very few deaths have been recorded and it is too soon to draw conclusions, though they do not suggest any unusual pattern.

Mortality of persons exposed to mustard gas (MR297)

Reasons for the study
Mustard gas is suspected to be carcinogenic, but the evidence for this in man is inconclusive. The therapeutic uses of nitrogen mustard make the question a significant one even in peacetime. This prompted a long term study of the mortality of a group of persons heavily exposed accidentally.

Study population
177 men identified from pension records affected by mustard gas exposure following an explosion during the Second World War.

Nature of the study
Subjects have been traced and flagged at NHSCR and causes of death are being assembled. Specific cause mortality will be compared with predicted values from national rates.

Mortality of cooks (MR299)

Reasons for the study
It has been suggested in more than one study that cooks have an elevated mortality from lung cancer. A study of a group of cooks was mounted to investigate the observation.

Study population
1,805 soldiers who had served as cooks in the Army Catering Corps and were receiving a pension between 1974 and 1984.

Nature of the study
Subjects were traced and flagged at NHSCR and death certificates provided to the researchers. Mortality will be compared with that for pensioners from the Royal Army Pay Corps as well as the rates for England and Wales.

1961 census based study of butchers (MR302)

Reasons for the study
Early Registrars' General reported that contrary to popular myth, the occupation of butchers was not a healthy one. Latterly, their occupational mortality supplements have reported butchers to suffer an excess mortality from lung cancer.

Study population
Men aged between 45 and 64 in 1961, whose jobs were identified in the census data as related to butchering (e.g. 'meat cutter', 'slaughterman', 'retail/wholesale butcher').

Nature of the study
Subjects were traced and flagged at NHSCR. Anonymised copies of draft death registrations were provided to the researchers. Their mortality pattern will be compared with National rates and analysed according to job code.

Childhood cancer in West Berkshire and North Hampshire (MR305)

Reasons for the study
Coinciding with the observations of an excess of cancer in the young at Sellafield, a haematologist in Berkshire had the impression of an increase in childhood leukaemia locally. The presence of two nuclear establishments in the catchment area of his clinic and two in the neighbourhood, prompted study of the association between cancer in children under five and parental employment at the two local nuclear establishments.

Study population
699 children under the age of five resident locally, first diagnosed and registered as having cancer 1972-89 in the study area.

Nature of the study
A case control study of leukaemia and non-Hodgkin's lymphoma was designed, with each case matched by 4 controls obtained from the local midwifery labour books, and two from NHSCR. Detailed occupational and residential histories, medical and socio-economic data were obtained by questionnaire.

Results
Of 54 cases, 5 (9 per cent) had fathers and/or mothers employed in the nuclear industry, of 324 controls, 14 (4 per cent) had parents employed in the nuclear industry (relative risk 2.2, 95% C.I = 0.6 - 6.9). Three fathers of cases and 2 fathers of controls were monitored for external radiation prior to conception.

If the relationship is real it might be due to internal contamination by radioactive isotope or some other agent. The local increase in childhood leukaemia was not entirely accounted for by these findings.

Publications

1. Roman, E., Watson, A., Beral, V., Buckle, S., Bull, D., Baker, K., Ryder, H. E. Barton, C. (1993) 'Case-control study of leukaemia and non-Hodgkin's lymphoma among children

aged 0-4 years living in West Berkshire and North Hampshire health districts' *Br Med J*, 306: 615-62

Mortality study of a cohort of workers exposed to cyanide (MR310)

Reasons for the study
While it had been suggested that chronic exposure to cyanides might damage health, no study has been carried out of their long term health effects.

Study population
All males employed in a company's cyanide plants for a minimum of 12 months between 1st January 1948 and 31st December 1975.

Nature of the study
Subjects have been traced and flagged at NHSCR and coded death certificates will be provided. Initially an exploratory investigation of the patterns of all cause and specific cause mortality will be conducted, using National and local rates for comparison. The findings will determine the nature of further investigations.

Publications
1. Leeser, J. (1992). 'Cross-sectional study on a cohort of workers exposed to cyanide'. Presentation at the Medichem Conference, London, 1992.

Mortality and cancer morbidity of Royal Navy submariners (MR314)

Reason for the study
The closed environment of submarine operation raises questions of the long term effects of multiple low level toxic exposures.

This investigation was promoted by reports of the possibility of adverse health effects of exposure to low doses of ionizing radiation in nuclear submarines.

Study population
Submariners who have served in the Royal Navy.

Nature of the study
Subjects were flagged at the NHSCR and copies of death certificates and cancer registrations were provided to the researchers. Comparisons will be made with expected values for the relevant outcomes, based on National and possibly other Royal Navy personnel controls.

Follow up of persons who became ill after eating contaminated bread (Epping jaundice) (MR315)

Reasons for the study
A small group of persons developed jaundice in 1965 that was traced to eating bread made from flour contaminated in transit with methylene diamine. The material had a number of industrial applications, so when it was found to be an animal carcinogen, it was decided that despite the small size of the affected group it would be worthwhile studying their tumour experience.

Study population
This was obtained from the list compiled by an Essex doctor involved in the investigation of the outbreak.

Nature of the study
Subjects were traced and flagged at NHSCR and coded death and cancer registrations provided to the researchers. Using the Family Practitioner Committees as intermediaries, survivors were approached in 1988 to obtain details of their subsequent experience including nutritional, smoking and alcohol histories, and leisure activities. Mortality and cancer morbidity patterns will be compared with those in a control population.

Results
No carcinogenic effect was observed, though it was noted that there were no measures of dose and that follow up requires to be long term for known animal carcinogens.

Publication
1. Hall, A.J., Harrington, J.M and Waterhouse, J.A.H, 'The Epping Jaundice Outbreak'. A 24 year follow up.' *J Epidemiol Community Health*, 1992;46: 327-328.

Mortality and cancer morbidity of production workers in the UK flexible polyurethane foam industry (MR316)

Reasons for the study
The isocyanates and other compounds involved in the manufacture of the polyurethanes, have a long history of use. There had been some suspicion about certain of the constituents being carcinogenic, but there had been no definitive epidemiological or animal experimental studies to test this. The International Trade Association therefore sponsored a study of cause specific mortality and site specific cancer morbidity among workers employed in polyurethane foam factories.

Study population
Some 8,288 workers who had been employed for a minimum of 6 months in the period 1958-79 at 11 UK factories in England and Wales where adequate records were available.

Nature of the study
Subjects were traced and flagged at NHSCR and coded copies of death and cancer registrations at 31st December 1988 were provided to the researchers. Numbers of cases of individual cancers were compared with expected figures derived from National rates adjusted for person years at risk, and the relation of any excesses related to duration and intensity of isocyanate exposure. A nested case control study using internal controls was conducted to evaluate occupational exposures other than to isocyanate.

Results
Statistically significant excess mortality from lung cancer (0=16 E=9.1) and carcinoma of the pancreas (0=6, E=2.2) was seen in women, but not in men. Overall cancer incidence was below expectation, though for women statistically significant excesses were found for cancers of laynx (0=3) and kidney (0=4). Risk was not found to be related to isocyanate exposure.

The case control study found no significant association with 9 other agents. The authors considered cigarette smoking to explain part of the excess, and chance and non-occupational factors possibly the remainder.

Publication
1. Sorahan T. E. and Pope, D (1993) 'Mortality and cancer morbidity of production workers in the United Kingdom flexible polyurethane foam industry'. *Br J Ind Med*, 50: 528-536

Mortality of coal miners exposed to radon and thoron daughters (MR319)

Reasons for the study
Miners exposed to relatively high levels of ionising radiation underground have been known to experience an excess of lung cancer. As measurements in coal mines indicate that miners are exposed underground to levels of radioactivity corresponding to the highest doses met with in the home, they were studied to evaluate the hazard at these low doses.

Study population
Some 15,000 miners employed at 10 fields involved in the long term coal industry field health study.

Nature of the study
Subjects were traced and flagged at NHSCR and coded death entries provided to the researchers. Occupational and domestic radiation exposures were estimated for individuals, and cumulative life-time doses were related to mortality, in particular for malignancies. Due allowance was made for dust exposure and smoking habit. Expected values for individual cause mortality were based on regional rates adjusting for person years at risk. For lung cancer and for stomach cancer, and for a number of other cancers, case-referents analyses were also performed. For lung cancer, comparison was made with values predicted from published models of radiation risk.

Results
A pilot analysis showed there to be an association between crude lung cancer death rates and mean exposures, but the relationship was of the order of 100 times that predicted from uranium miner studies.

The formal analysis showed there to be a significant deficit of lung cancer. Person year analysis showed no relationship between lung cancer and lagged cumulative exposure to radon daughter dose. Case referent analysis showed there to be an association between lung cancer and dose. However, there was an implausible negative association in heavy smokers, with radon daughter dose. A similar anomaly was seen for oesophageal cancer. No relationship was found for stomach cancer and certain other neoplasms and radiation exposures.

Publications
1. Maclaren, W. M. (1992). *Coalminers' mortality in relation to low-level exposures to Radon and Thoron daughter*. Institute of Occupational Medicine, Edinburgh. Report No. TM/92/06.

Institute of Petroleum Epidemiology study (MR321)

Reasons for the study
Petroleum and its products include experimental carcinogens, but studies of exposed workers have produced inconsistent results. The UK industry decided to evaluate the health of employees.

Study population
Males employed at 8 refineries for at least 1 year between 1st January 1950 and 31st December 1975 who had not worked much abroad and for whom adequate records were available.

Similar studies looked at workers at oil distribution centres and bus garage engine maintenance engineers.

Nature of the study
Subjects were primarily traced through Social Security records and failing that through NHSCR with whom the population was flagged. Coded death certificate were provided. Death was analysed by cause, company, job and years of exposure, comparing numbers of cases observed with expected values based on man/years and National rates. Nested case control studies were carried out to relate the risk of death from leukaemia to benzene exposure, and the risk of kidney cancer to total hydrocarbons.

Results
A 99.8 per cent trace was achieved overall. For oil distribution workers there was an overall excess of deaths from myelofibrosis (0=5, E=1.86, p=0.04): numbers of deaths from individual neoplasms in this group were too small to exclude chance effects. The case control study suggested a leukaemia risk for men with high or medium exposure to benzene compared with the low risk group (p=0.05), not increased by length of service, and affecting only a small proportion of workers in refineries.

Garage maintenance men overall, experienced a deficit of lung cancer: raised mortality occurred in certain sub-groups, but numbers were small and follow up short. For oil refinery workers, statistically significant excesses were found for deaths from cancer of the nasal cavity and sinuses, and for melanoma. An inconsistent pattern of mortality was found between the refineries for a number of tumours.

The follow-up analysis to the end of 1989 confirmed the cancer excess for labourers previously observed. In distribution and refinery workers, all cause mortality continued to be reduced as was mortality for many of the major non-malignant causes of death: both groups had raised mortality for arterial disease, especially aortic aneurysm. 'All neoplasm' mortality was reduced, largely due to a deficit of lung cancer. Melanoma continued to be in excess in refinery workers, and workers in distribution and in particular drivers, had a raised mortality from kidney neoplasms. Leukaemia was high in one company and in drivers overall.

Publications

1. Rushton, L. and Alderson, M. R. (1980). 'The influence of occupation on health-some results from a study in the UK oil industry'. *Carcinogenesis*, 1: 739-744.

2. Rushton, L. and Alderson, M. R. (1981). 'A case-control study to investigate the association between exposure to benzene and deaths from leukaemia in oil refinery workers'. *Br J Cancer*, 43: 77-84.

3. Rushton, L. and Alderson, M. R. (1981). 'An epidemiological survey of eight oil refineries in Britain'. *Brit J Ind Med*, 38: 225-234.

4. Alderson, M. R. and Rushton, L. (1982). 'Mortality patterns in eight UK oil refineries'. *Annals NY Acad Sci*, 381: 139-145.

5. Rushton, L. and Alderson, M. R. (1983). 'Epidemiological survey of oil distribution centres in Britain'. *Br J Ind Med*, 40: 330-339.

6. Alderson, M. R. and Rushton, L. (1983). 'Epidemiological studies in the UK petroleum industry'. In: *Health hazards in a changing oil scene*. Wiley and Sons.

7. Rushton, L. and Alderson, M. R. (1983). 'An epidemiological survey of maintenance workers in London Transport Executive bus garages and Chiswick Works'. *Br J Ind Med*, 40: 340-345.

8. Rushton, L. (1991) *The Institute of Petroleum epidemiological study. Distribution centre study principal results 1951-1989*. Institute of Petroleum, London.

9. Rushton, L. (1991) *The Institute of Petroleum epidemiological study. Refinery study principal results 1951-1989*. Institute of Petroleum, London.

10. Rushton, L. (1992) 'UK oil industry and distribution studies'. *Occupational Health Review Issue* 38: 10-12.

11. Rushton, L. (1993) 'Further follow-up of mortality in a UK oil refinery cohort'. *Br J Ind Med*, 50: 549-560

12. Rushton, L. (1993) 'Further follow-up of mortality in a UK oil distribution centre cohort'. *Br J Ind Med*, 50: 561-569.

13. Rushton, L. (1993) 'A 39 year follow-up of the UK oil refinery and distribution centre studies: results for kidney cancer and leukaemia'. *Environmental Health Perspective*.

[Data have also been incorporated with US data in metanalysis to investigate the role of benzene exposure in the causation of leukaemia.]

Mortality in a group of workers exposed to CS_2 (MR324)

Reasons for the study
1. To compare the effects of CS_2 on arterial disease in a group of workers exposed to lower levels than those involved in study MR157.

2. To investigate whether the local high leukaemia rates were associated with exposure to CS_2.

Study population
All persons identified from company records as having been employed during its 40 year history.

Nature of the study

Subjects will be traced and flagged at NHSCR and coded copies of draft death registrations provided. Job descriptions will be used as measures of exposure. Adjusting for person years of exposure, expected deaths will be calculated by cause from National and local rates.

Mortality of workers exposed to Mercaptobenzothiazole (MBT) (MR326)

Reason for the study
Experimental studies showed MBT, which had been used in rubber compounding, to be carcinogenic to rats. A group of workers was identified who had been engaged in the manufacture of MBT and related compounds over a number of years.

Study population
All male and female shop floor employees engaged for a minimum of 6 months who had started employment between 1906 and 1984 with a period of employment prior to 1955.

Nature of the study
Subjects are traced and flagged at NHSCR and coded death certificates provided. Expected values will be derived from national rates and internal comparisons will be made to deal with potential biases.

Cumulative indices of exposure will be derived from years of exposure to individual compounds and to cumulative measures of levels of exposure.

Results
A 98 per cent trace was achieved and the causes of death were ascertained for 98 per cent of those known to have died. The overall mortality experience both from malignant and non-malignant diseases was close to or below expectation. This also applied to exposure sub-groups. Overall 9 deaths from bladder cancer were observed (E=8.6). In the MBT group, 3 deaths from bladder cancer were observed (E=1.1).

Publication

1. Sorahan, T. E and Pope, D. (1993) 'Mortality study of workers employed at a plant manufacturing chemicals for the rubber industry: 1955-86'. *Br J Ind Med*, 50: 998-1002.

Mortality of lead oxide workers (MR344)

Reasons for the study
A suggestion of excesses of specific cancers arose out of certain lead worker studies. The European Trade Association sponsored a study at independent scientific institutions in the UK and Italy to test these findings.

Study population
Workers for whom adequate data existed at three lead oxide works in England. Though numbers were relatively small, the records went back as far as 1918.

Nature of the study
Subjects will be traced and flagged at NHSCR and coded causes of death provided. Cancer mortality will be related to occupational factors in the UK factories and comparisons will be made with National and Regional rates. Comparison will be made with the findings in the Italian study.

Indigo worker mortality study (MR352)

Reasons for the study
There had been little information on the toxicity of indigo or its reaction products, but a study of workers involved in the manufacture of synthetic dyes including indigo, had found a non-significant excess of gastrointestinal cancers.

Study population
Workers employed in an indigo synthesizing plant between 1940 and 1986.

Nature of the study
The vital status of subjects was primarily determined from company records but NHSCR traced those lost from surveillance and flagged the population, providing death certificates. Predicted values for deaths will be derived from Regional rates and compared with observed numbers. An index of exposure will be determined for studying associations.

Publications
An initial proportional mortality ratio analysis has been produced for an academic dissertation. Subse-

quently a full analysis using standard mortality ratios and relating outcomes to exposure levels will form the basis of an MD thesis. It is planned to publish the results in a scientific paper in due course. In the interim, progress reports will be presented to workers' representatives.

Brain tumours in radar research workers (MR354)

Reasons for the study
Seven cases of brain tumour were reported in a group of workers at a research establishment. A study was set up to determine whether this constituted a significant excess. Other studies of cancer in persons exposed to various forms of electromagnetic energy have produced inconsistent results. In this study the putative agent was radio frequency over a particular range of wave lengths.

Study population
Some 4,500 employees at a radar research establishment.

Nature of the study
Subjects will be traced and flagged at NHSCR and copies of coded death and cancer registrations provided to the researchers. Comparisons will be made with patterns of cancer mortality and morbidity in National and Regional data bases. Due allowance will be made for levels and durations of exposure to radio frequency, and to ionizing radiation exposure potential and chemical exposures.

Mortality in men exposed to cutting fluids (MR361)

Reasons for the study
Several studies had suggested that metal machinists might experience excesses of gastrointestinal and respiratory cancers. Animal, vegetable and mineral oils and oil emulsions, together with a range of additives have been widely used in metal machining. During the process, oil commonly contaminates the skin and is inhaled as a mist. Even when the oil is manufactured to produce a low carcinogenic content, subsequent microbiological activity and heating during the cutting process, may lead to the production of carcinogens.

Study population
All men who had been employed at two factories between 1963 and 1988 and who had been hired prior to 1st January 1985.

Nature of the study
Subjects have been traced and flagged at NHSCR and copies of death registrations coded for multiple causes provided. In addition copies of cancer registrations have been provided. Expected values for deaths will be calculated from National and local rates. Mortality will be related to levels and durations of exposure and smoking habits will be taken into consideration. An internal control group will be used for comparison purposes.

Publication
1. Data on a smoking survey of the workforce were presented at the Annual meeting of the Society for Social Medicine, September 1991.

Cancer mortality in the British Rubber Industry (MR363)

Reasons for the study
Antioxidants that had been incriminated as powerful urothelial carcinogens were withdrawn from use in the UK about 1950. With reports of the observation of excesses of other tumours in rubber workers, the UK industry decided to review the situation.

Study population
Some 34,000 men were identified who had been employed for a minimum of 1 year at any one of 13 factories between 1st January and 31st December 1960 and who had survived a further minimum of 9 years before death or loss to follow up on 31st December 1975.

Nature of the study
Subjects were traced and flagged at NHSCR and coded copies of death registrations provided to the researchers. Deaths observed in the study population were compared with predicted values derived from National rates.

Results
An overall cancer excess was observed that was accounted for largely by the excess of bronchial and stomach cancers. A small but statistically significant excess of oesophageal cancer was found in various groups in certain industry sectors. A bladder cancer excess was no longer perceptible in workers first employed subsequent to 1950.

Publication
1. Parkes, H. G., Veys, C. A., Waterhouse, J. A. H. and Peters, A. (1982). 'Cancer mortality in the rubber industry'. *Brit J Ind Med*, 39: 209-220.

Colorectal cancer among polypropylene workers (MR370)

Reasons for the study
After two reports indicating excesses of colorectal cancers in polypropylene workers, a further study was conducted to confirm the findings.

Study population
All persons employed for a minimum of 12 months at two polypropylene plants that had commenced operation in 1972 and 1973.

Nature of the study
All subjects will be traced and flagged at NHSCR and coded copies of draft death registrations and cancer registrations will be provided. Observed mortality and cancer morbidity patterns will be compared with those expected from England and Wales rates and where possible with local rates.

A mortality study of aniline workers (MR373)

Reasons for the study
Aniline had been used as a starting material for synthetic dyestuffs. When exposure to dyes was found to cause urothelial cancers, aniline was suspected of being carcinogenic. No evidence was found at the time but a recent study suggested that it might contribute to the causation of lung cancer. Animal studies have shown it to be carcinogenic, primarily to the spleen.

Study population
All workers employed at an aniline plant from its start up in 1964 to the end of 1985.

Nature of the study
Subjects will be traced and flagged by NHSCR and coded copies of death certificates provided. Numbers of deaths from specific causes will be compared with predicted values derived from national and local regional rates. Mortality will be related to exposure level, duration and years of exposure.

A cohort study of employees in perspex plants (MR384)

Reasons for the study
Acrylates and methacrylates were found to induce nasal epithelial metaplasia experimentally. Subsequently a study of an occupationally exposed group of men failed to detect an excess of respiratory cancer but suggested that there might be an increased risk of colorectal cancer, though support for a causal association was not found. Though further 'non-positive' results were reported, it was decided that a more definitive study was required.

Study population
All workers employed at two perspex factories (established 1948 and 1949), for whom there were records.

Nature of the study
All subjects are to be traced and flagged at NHSCR and coded drafts of death registrations provided. Categories of exposure will be derived for individuals based on job and duration of employment. Four possible comparisons have been considered: England and Wales mortality rates; local mortality rates; a local control population; within group subsets in different exposure categories.

A study of workers exposed to mineral acids (MR396)

Reasons for the study
Four studies over a ten year period had suggested an excess risk of laryngeal cancer in workers exposed to mineral acid mists. It was decided to study a further group of men to seek confirmation of the respiratory cancer hazard in such occupations and to obtain further information about the relation of risk to levels of exposure.

Study population
Employees exposed to mineral acid mists in the manufacture of lead batteries, and in metal treatment.

Nature of the study
Subjects will be traced and flagged at NHSCR and coded drafts of death registrations and cancer registrations will be provided.

Mortality patterns will be compared with those expected from national rates, and mortality and morbidity from laryngeal cancer will be related to duration, timing and level of acid exposures in a nested case-control study.

A cohort study of employees exposed to methylene chloride in the manufacture of triacetate film base (MR 406)

Reasons for the study
Methylene chloride is a volatile organic solvent to which many workers are exposed. Some laboratory carcinogenicity studies have proved positive in certain species. Carbon monoxide is a metabolite which has been suspected to be involved in atherogenesis. The halogenated hydrocarbon is cardiotoxic, compounding the potential hazard.

Study population
Records were found for some 2,500 persons em-

ployed between 1948 and 1988 at a plant where methylene chloride was used in the production of film.

Nature of the study
Subjects will be flagged at NHSCR and draft death entries provided to the researchers. Exposure categories will be allocated to each worker, and mortality experience will be related to these exposure categories and compared with England and Wales and local rates.

Follow up study of photographic chemical workers (MR 409)

Reasons for the study
Five deaths from bladder cancer, where only 0.4 was expected, were noted in the preliminary analysis of mortality of workers at a photographic factory, although known urothelial carcinogens were not employed in the manufacturing processes. As the records were incomplete, better data were sought to confirm the observation.

Study population
All persons employed for at least one month in the particular department of the factory where the original observation was made.

Nature of the study
Subjects will be flagged at NHSCR and coded draft death entries and cancer registrations will be provided to the researchers. All cause mortality and cause specific mortality will be studied, using national reference rates for comparison, and standardised incidence rates will be calculated for cancer registrations. Analysis by time of employment and job category will be used to investigate aetiological factors.

International study of cancer risks in biology research workers (MR 416)

Over the years, excesses of various malignancies and other diseases of putative viral origin have been reported in association with animal husbandry and butchery. The finding of a small cluster of a rare malignancy among workers at the Pasteur Institute in Paris and a separate cluster of malignancy at its counterpart in Rome, prompted the call for a confirmatory study of satisfactory power.

Study population
An estimated total of 8,000 workers potentially exposed to biological material in research laboratories, and 4,000 non-exposed workers in the same establishments, situated in various European countries.

Nature of the study
Within the UK, subjects will be flagged at NHSCR and copies of draft death entries and cancer registrations will be provided to the researchers. Mortality patterns will be studied by occupational history as will cancer incidence. The availability of cancer registration data puts a premium on the UK contribution. In the event of excesses being observed, nested case-control studies will be conducted to determine what agents might be responsible. National studies will be analysed individually, and the total data will be assembled and analysed by a WHO agency to provide an analysis with greater power.

An analysis of asbestos-induced lung disease (MR 419)

Reasons for the study
Within a wider programme of biological research, a mortality component was included to determine the accuracy of expert opinions on life expectancy of plaintiffs in asbestosis litigation, and to study the certified causes of death in these cases.

Study population
Some 1,700 cases, all of whom had been considered to have had lung disease due to asbestos and had reached medico-legal settlement. These had been collected by an expert witness over the course of 40 years.

Nature of the study
Subjects will be traced and flagged at NHSCR and copies of draft entries in the death register provided to the researchers. Survival times will be matched with expert predictions and the modes of death studied.

A cohort study of rubber adhesive workers (MR 426)

Reasons for the study
Seven cases of cerebral tumour were noted at a medical products factory. This constituted a significant excess, even after two cases were excluded on the grounds of tumour type and short latency. Of the remaining cases, 4 had worked in an area where adhesives were mixed and spread. It was decided to study the pattern of death in a comparable population and to relate any abnormal mortality pattern to occupational exposures.

Study population
Some 2,000 workers who had worked at least one year in the adhesives production areas at the index factory, and a further 6,000 are to be targeted at factories where similar work is conducted.

Nature of the study
The populations will be traced and flagged at NHSCR and coded causes of death stated on draft registrations will be provided. Cancer registrations will also be provided as a check for the completeness of death records. Subjects will be categorised by job and by exposure potential to solvents and other chemical agents. Overall mortality patterns and specific cancer rates will be studied and compared with expected values derived from the appropriate rates. An attempt will be made to deal with potential confounding factors by means of a nested case-control study.

Mortality of asbestos workers in a London factory (MR852)

Reasons for the study
To evaluate the health effects of various grades of exposure to asbestos.

Study populations
All male workers employed for a minimum of 30 days between 1st April and 31st March 1964, were included in a general mortality study.

A subset of outside contract laggers was also established. In a separate study of the effects of smoking, 1,834 men born between 1900 and 1930 and who were alive on 1st January 1960 were identified.

Nature of the studies
1. Subject were traced at NHSCR and with assistance from Social Security, and were flagged. Coded death certificates were provided. Jobs were classified according to the severity of dust exposure, and the periods of study related to the date of implementation of regulations. Expected values for numbers of deaths were calculated from national rates, adjusting for man years, age, sex, calendar years of the study.

2. The reanalysis to 31st December 1975 separated the laggers.

3. Smoking histories were sought for a subset dying between 1st January and 31st March 1970.

Results
1. After an interval of 16 years since first exposure, men with heavy levels of exposure had a significant excess of deaths from cancer of the lung and pleura and other sites, whether total exposures were greater or less than 2 years. Excess mortality from non-malignant respiratory disease was only found with long service and heavy exposure.

2. Among the non-laggers, of the 545 deaths, 46 were from malignant mesothelioma. Lung cancer excess was most marked for those most heavily exposed with the greatest duration of exposure. Only 20 had been followed up for 30 years or more. For women, the lung cancer excess was greater for those highly exposed for long durations.

3. Although not conclusive, the lung cancer risk from cigarette smoking and from asbestos exposure was consistent with a multiplicative effect.

Publications
1. Newhouse, M. L. (1969). 'A study of mortality of workers in an asbestos factory'. *Br J Ind Med*, 26: 294-301.

2. Berry, G., Newhouse, M. L. and Turok, M. (1972). 'Combined effects of asbestos exposure and smoking on mortality from lung cancer in factory workers'. *Lancet*, (ii): 476-479.

3. Newhouse, M. L., Berry, G., Wagner, J. C. and Turok, M. E. (1972). 'A study of the mortality of female asbestos workers'. *Br J Ind Med*, 29: 134-141.

4. Newhouse, M. L. (1973). 'Asbestos in the workplace and in the community'. *Ann Occup Hyg*, 16: 97-107.

5. Newhouse, M. L. and Berry, G. (1979) 'Patterns of mortality in asbestos factory workers in London'. *Ann NY Academy of Sciences*, 330: 53-60.

6. Newhouse, M. L., Berry, G. and Wagner, J. C. (1985). 'Mortality of factory workers in East London 1933-80'. *Br J Ind Med*, 42: 4-11.

UK radium luminizer worker study (MR859)

Reasons for the study
Earlier awareness of the adverse effects of ionizing radiation was based on heavy exposures to sealed and to unsealed sources. As the International Commission on Radiation Protection (ICRP) predictions had been challenged, it was decided to evaluate them by studying a group of luminizers.

Study population
Some 1,900 women luminizers for whom there were Ministry of Labour records, together with a further 204 women identified as luminizers. Of these, 1,100 were identified as having been employed on or after 1939 and available for interview in 1961.

Nature of the study
Initially a mortality study was carried out on the population between 1st January 1961 and 31st December 1977. They were traced and flagged at NHSCR and death certificates provided. This was enlarged subsequently to include cancer morbidity, and measures of external and internal dose of radiation were estimated. England and Wales rates were used, adjusted for person years, for deriving expected values.

Results
For those workers who had worked 2 or more years, a 100 per cent trace was achieved. This fell to 90 per cent for those who had worked 1-2 years and to 40 per cent for those working less than 1 year.

For those under 30 when starting work who were exposed to an estimated cumulative absorbed dose of external radiation greater than 0.2Gy to breast tissue, there was initially a statistically significant increase in death from breast cancer (p=0.05) at a level consistent with the ICRP estimate. By 1986, this excess was no longer significant.

The data may be combined with the US dial painter population study.

Publications

1. Baverstock, K. F., Papworth, D. G. and Vennart, J. (1981). 'Risks of radiation at low dose rates'. *Lancet*, i, 430-433.

2. Baverstock, K. F. and Vennart, J. (1983). 'A note on radium body content and breast cancer in the UK radium luminizers'. *Health Physics*, 41: (Supplement No 1) 575-577.

3. Baverstock, K. F. and Papworth, D. G. (1986). 'The UK radium luminizer survey: significance of a lack of excess leukaemia'. In: *The radiobiology of radium and thorotrast*. Eds. Gossner, W., Gerber, G. B., Hagen, U. and Luz, A. Strahlentherapie, 80: 22-25 (Supplement).

4. Baverstock, K. F. and Papworth, D. G. (1989). 'The UK radium luminizer survey'. *Brit J Radiol*, Report 21: 72-76.

Mortality of chrome platers in the West Riding (MR863)

Reasons for the study
Certain chrome salts had been confirmed experimentally to be carcinogens. It was considered important to evaluate the health effects of working as a chrome plater.

Study population
Chrome platers employed at some 50 plants in Yorkshire, together with a matched population from a local industry.

Nature of the study
Subjects and controls were flagged at NHSCR and death certificates were provided to the researcher. Initially, comparisons were made of mortality between the two populations. Subsequently, comparisons were made with England and Wales mortality rates.

A subsidiary study investigated occupational and smoking histories.

Results
An initial analysis indicated a significant excess of deaths from neoplastic disease in the plater group (Platers=39 v Controls 21). These were due not only to a lung cancer excess (Platers=17 v Controls=10).

Subsequent analyses using England and Wales rates for expected values, showed an excess of overall mortality in platers (0=42, E=23.19). Excesses of lung cancer were seen for platers (0=23, E=9.59) and for controls (0=15, E=9.58): the differences were not significant. The excess mortality from neoplasms of colon and rectum was greater in platers. A constant trend with smoking was not observed.

Publications

1. Royle, H. (1975). 'Toxicity of chromic acid in the chrome plating industry'. Part 1 *Environ Res*, 10: 39-53.

2. Royle, H. (1975). 'Toxicity of chromic acid in the chrome plating industry'. Part 2. *Environ Res*, 10: 141-163.

Survey of occupational cancer in the rubber and cable making industries (MR864)

Reasons for the study
The observation of an excess of bladder cancer in rubber workers and cable makers was soon identified as being associated with the use of certain antioxidants. When they were removed from compounding, confirmation of safety of the new process was sought.

Study populations
41,981 persons were studied who were identified in a census of men in employment in the rubber and cable making industries on 1st February 1967, who were aged 35-65 and for whom adequate data were available.

Nature of the study
Subjects were traced at NHSCR and flagged, and coded copies of death certificates were provided to the researchers. Three exposure groups were identified, those employed in factories that had used known carcinogens who started work before 1st January 1950, those employed in these factories after 31st December 1949, and those employed in factories that had not used the specific carcinogenic agents. Mortality rates were compared with those for England and Wales. Analyses were made by industry sector and by jobs within these sectors.

Results
A 97.35 per cent trace was achieved overall. A bladder cancer excess was observed for those persons employed prior to 1949, but there was no evidence of a continued bladder cancer hazard for those joining subsequently. However there was an excess of cancer overall, with a bronchial carcinoma excess in certain sectors. A number of confounding factors were identified but were not considered to account entirely for the excesses observed. The findings of the study prompted a further industry study to investigate in greater detail the lung cancer excess.

Publications

1. Fox, A. J., Lindars, D. C. and Owen, R. (1974). 'A survey of occupational cancer in the rubber and cable making industries'. *Br J Ind Med*, 31: 140-151.

2. Fox, A. J. and Collier, P. F. (1976). 'A survey of occupational cancer in the rubber and cable making industries; analysis of deaths occurring in 1972-1974'. *Br J Ind Med*, 33: 249-264.

3. Baxter, P. J. and Werner, J. B. (1980). *Mortality in the British rubber industries 1967-1976*. London: Health and Safety Executive. HMSO.

Survey of workers manufacturing di-amino-diphenyl methane (DADPM) (MR869)

Reason for the study
DADPM had been used for compounding in the rubber and plastics industries. Although chemically related to certain carcinogens, there had been concern only about its hepatotoxic properties, until an animal study showed it to be a carcinogen.

Study population
All 59 men who were currently or who had ever been employed in the production of DADPM at a plant.

Nature of the study
Subjects have been traced and flagged at NHSCR and coded copies of death registrations are sent to the researcher. A running surveillance will be kept on the causes of death and the findings particularly of deaths from cancer and liver diseases, will be compared with values predicted from National and local rates as available.

Mortality of textile workers exposed to carbon disulphide

Reasons for the study
Reports that exposure to CS_2 might lead to an excess of atherosclerotic heart disease and other diseases prompted a further study.

Study populations

A. All persons employed in a Cumbrian factory since its start up in 1947.

B. All persons employed at a Lancashire factory since 1st January 1945.

Nature of the study
All subjects were traced and flagged by NHSCR and coded death certificates were provided. Indices of CS_2 exposure were available.

Results
The analyses were not able to detect an abnormal mortality pattern. The development of practical ambulatory cardiac monitoring devices provided a more appropriate method for studying disturbance of cardiac function as determined by intermittent arrhythmia.

Publications

1. HM Chief Inspector of Factories. (1972). *Survey into conditions in factories using CS_2*. Annual report 1971. HMSO, London (Cmd 5098) p26.

Pattern of mortality of parents of patients dying of malignant pleural or peritoneal mesothelioma

Reasons for the study
To evaluate a study that reported that parents of persons dying of mesothelial malignancy had a higher cancer death rate that expected. The confirmation of the finding was considered to have importance on scientific grounds.

Study population
A group of persons dying of malignant mesotheliomas together with two controls for each member, matched for sex and age group and with parents matched for age, to be obtained, one forward and one backward in the register.

Nature of the study
Causes of death will be compared between parents of 'cases' and 'controls' and occupations will be taken into consideration in the interpretation of the results. The study is being carried out in parallel with an American study.

A study of British Airways pilots

Reasons for the study
Although a high level of fitness is demanded of pilots who are subjected to periodic medical examination, it was thought that there were certain aspects of occupation, environment and lifestyle that might affect life expectancy and mortality pattern in working and retired pilots.

Study population
All deaths of serving and retired pilots occurring in the UK that were identified between 1 May 1966 and 31 December 1989.

Nature of the study
Death data were derived from a variety of sources including, pension records, a death register, searches of old retiree lists and obituaries in staff newspapers, personnel records and car license information. Causes of death were coded by OPCS and PMRs were calculated using National rates for comparison.

Results
Of a total of 446 deaths, only 411 for which the cause was known, appeared to have occurred in the UK. There was a disproportion of 'all malignancies' and 'accidents'. The specific contributing malignancies included, colon, brain and CNS, and malignant melanoma.

Publication

1. Irvine, D. and Davies, D. M. (1992). 'Mortality of British Airways pilots 1966-1989'. *Aviation, Space and Environ Med*, 63: 276-9

Chapter 16 International comparisons

Morris Greenberg

Health Statistics, OPCS

16.1 Introduction

Effective training and supervision, properly designed work practices and equipment, and if necessary efficient personal protective equipment are required to eradicate or reduce the burden of excess mortality and morbidity associated with occupation. All these activities to protect workers may constitute a charge on production costs, though they need not render processes non-competitive. Model international codes and directives for conditions of work are produced by such bodies as ILO and the European Commission, primarily with a humanitarian end. International agreements on trade also have interests in health and safety at work: partners may be required to shoulder production costs equitably, which include wages, social benefits and the imposts of health and safety measures.

The purpose of analysing routinely collected good quality national statistics on occupational mortality, disease incidence, sickness absence and injuries, would be to provide useful measures of the effectiveness of health and safety legislation, and of its enforcement. Studies of international statistics would also indicate what targets for health and safety it has been possible to achieve with current technology and economic constraints. They could provide economists policing trade agreements with another measure of compliance. The study of international health statistics might also assist in the determination of the aetiology of excess mortality.

16.2 International standards in occupational health

With the widespread use of the ICD, wherever comprehensive death and cancer registration operate and where reliable national censuses are conducted and labour force statistics are collected, valid occupational mortality and cancer morbidity monitoring can be conducted by industry, occupational group or by job at national level. This would also allow international comparisons to be made.

Work accidents and the consequential occupational injuries were defined at the Thirteenth International Conference of Labour Statisticians,[1] to permit valid international comparisons. They remain to be generally adopted. No criticism is implied of any international organisation in drawing attention to delay in its widespread use. From the first proposal to bring the principles of modern medicine to the classification of causes of death to the production of an agreed ICD code, took half a century, and there was further delay before wide adoption.

An international classification of occupations optimum for health studies that takes account of specific agents or groups of agents to which there is exposure, has yet to be designed and adopted. This results in part from the priorities of current national industrial classifications having primarily to do with economic and labour force considerations. Development of an occupational classification with a health slant, would enhance national and international studies.

These are generally categorised as conditions associated with occupation that require to be notified to authorities by the diagnosing doctor or by the employer, and those conditions attracting special sick pay or death benefit. The recognition of a condition as attributable to occupation and meriting special social security consideration, varies over time within a single country, and there is as yet no uniformity of classification between countries. The development by the European Community of a common set of occupational diseases, will aid comparisons.

16.3 Problems arising in comparing national occupational injury analyses

For ten years, the ILO has assembled occupational injury data from a number of countries in its Year Books of Labour Statistics.[2] In keeping with the pragmatism of earlier epidemiologists, it is possible, with circumspection, to analyse these imperfect data in the absence of better, to promote the immediate improvement in the lot of employed persons while awaiting the production of better data.

Numerators provided by national authorities are various, including cases reported, with all the vagaries of completeness of reporting, or cases compensated, the criteria for which vary between countries.

Denominators used for calculating rates, similarly refer either to: numbers insured; the total employed; or, man–hours worked.

Both numerators and denominators may vary in comprehensiveness according to the inclusion or exclusion of the self-employed and the uninsured. To add to the problems of data comparison analyses of national statistics are presented variously as:

rates per 1000 man-years of 300 days each;
rates per 1000 workers exposed to risk;
rates per 1000 employed; or
rates per 1,000,000 man-hours worked.

Provided with the raw data, more ready comparisons could be made. The study of time trends in any one country or between countries, can be frustrated by changes in criteria or even the sources of information. As a consequence, conclusions based on international comparisons of occupational injuries, fatal and non-fatal require to be hedged at present with a number of caveats.

16.4 Some comparisons of fatal occupational accidents

Non-fatal accidents heavily outnumber deaths, and their impact, whether in terms of pain and suffering or loss of an economic resource, is considerable. However, on the grounds of space considerations and the limitations of the data, only fatal accidents will be looked at. Table 16.1 presents the statistics for a group of developed countries for 1986, who present their fatal injuries as rate per 1000 employed.[2] Within the constraints of the variations of national criteria, there would appear to be marked disparities when overall rates were looked at and when the three industry groups generally considered to be most hazardous were compared. Cultural factors as well as reporting bias, may account for gross disparities in experience between countries. Thus in the United States, the commonest work injury for men overall, was from highway vehicle accidents which accounted for 18 per cent of all fatal occupational accidents, with homicide at second place for fatal accidents, accounting for 17 per cent.[3] Finding funds to satisfy the drug habit, and the vulnerability of the lone person at the check out contribute to making homicide the most common "industrial accident" for women in the US.

16.4.1 Industry overall

The fatal injury rate for the UK at 0.017 per thousand workers, is the lowest rate observed in the group, a sixth of that for Poland. Chapter 11 presents in greater detail those factors that contributed to this favourable finding.

16.4.2 Construction

The social and economic characteristics of the workforce in the construction industry, can vary markedly between the specialities and between countries. "Guest workers" who contribute heavily to the workforce in many countries largely in a labouring capacity, may compound occupational hazard with unfavourable lifestyle and they may not be officially registered. Nationals operating within a limited radius from home, may be expected to differ favourably from those committed to an irregular existence, for example those perennially engaged on major constructions and civil engineering contracts required to live away from home and to work and live under difficult conditions.

East Germany appears to have the most favourable fatal accident rate, which does not differ much from that for Denmark, which is substantially lower than those for France, Canada and Austria. This may relate to the scale of construction, the nature of the records, or to the extreme care. (Table 16.1).

It may be that Denmark has a less hazardous building programme, and a stable population of construction workers

Table 16.1 Fatal occupational injuries: comparable international statistics 1986. (Rates per 1000 persons employed)

		All industry	Mining and quarrying	Agriculture, forestry and fisheries	Construction
Austria	(R)	0.079	0.278	0.451	0.291
Canada	(C)	0.069	0.590	-	0.270
Czech.	(R)	0.078	0.251	0.117	0.160
Denmark	(R)	0.030	-	0.090	0.070
France	(C)	0.074	-	-	0.231
Germany E	(C)	-	-	-	0.060
Germany W	(R)	0.080	0.290	0.190	0.190
Poland	(R)	0.109	0.261	0.147	0.134
UK	(R)	0.017	0.161	0.081	0.106

R = reported
C = compensated
See text for caveats
Source: *ILO Year Book of Labour Statistics 1993*. 52nd issue. ILO, Geneva.

that is well trained and supervised and socially advantaged. Under the circumstances, the figures reported for other countries might represent considerable achievement, though not necessarily cause for complacency, considering all matters. Despite having a substantial roving sector in its construction work force, with all the social and health consequences of such a composition, the UK comes between the extremes for fatal accidents in this group of countries. Nevertheless, the rate of 0.106 per 1000, which is 7 times the overall UK all industry rate, still demonstrates the scope for improvement.

16.4.3 Mining and quarrying

Geological conditions, the proportion of open cast mining and quarrying, and the stability of strata in mines may vary between countries, as will the climatic conditions under which workers are required to operate. Further, when these occupations are included under Extractive Industry, the extent to which oil industry is involved and the siting of wells, will determine the nature and order of the hazards. The rate of 0.161 per 1000 fatal injuries reported for the UK, though nine times its all industry rate, is still the lowest in this group with Canada the highest and Germany, Austria, Poland and Czechoslovakia mid-way and with greater and lesser disparities with their overall industry rates. Judgement as to relative safety achievement requires more information on the conditions that prevail.

16.4.4 Agriculture, forestry and fisheries

Each of the three components of this group can present a variety of conditions. If agriculture is restricted to cereal production and to meat production, the extent of mechanisation in the former and the amount of animal contact involved in the latter will make important contributions to risks. Forestry accidents may make a small contribution to the overall industry figure, but in relation to the numbers of foresters involved, could represent a higher rate of injury than the rest of the group. Fishery, when restricted to fish farming is relatively safe, but when coastal and deep sea fishing are involved, is associated with a higher risk. In so far as deaths at sea may not be entered in national records, overall mortality of fishermen can be understated. Without knowing more about the mix of activities in this group therefore, it is even more difficult to evaluate the health data and make useful comparisons between countries.

The UK appears to have the lowest fatal injury rate in this group at 0.081 per 1000, only five times the overall rate. Denmark's mortality rate was similar and only twice its all industry rate. The rate for Austria at the other extreme, was five times greater (also five times its all industry rate).

16.5 Mortality by occupational group: international comparisons

In this section, the term occupational mortality embraces the mortality experiences, derived from national death register data, associated with a history of employment in occupational orders and units. This is distinct from the limited number of deaths due to notifiable or compensatable diseases and injuries.

16.5.1 Problems

Routine periodic occupational mortality analyses on a national basis are not widely available. Attempts to make comparisons between countries that do publish analyses, are complicated by differences in the occupational classifications employed. The Nordic countries, Iceland, Finland, Norway, Sweden, Denmark, have a long history of collaboration, (though some national idiosyncrasy remains) to increase the power of their analyses with the larger cumulative total. They have reported their joint findings for the years 1971-1980.[13] Individual Scandinavian reports are also published, for example by Norway.[14] The prospect of collaborative studies between the UK and Scandinavia (and more widely) have been considered and European comparisons attempted.[11] The United States of America, presumably as a consequence of the difficulties inherent in their administrative system, has not conducted a national study since 1950.[9] Subsequently there have been published analyses of occupational mortality for individual States, and most recently for a group of 12 States.[10] Japan has a long tradition of publishing mortality rates by occupation, though rarely in translation. Constraints of linguistics and time preclude including a comparison with excellent data from other countries.[14,16] In a collaborative study of differences in mortality between Japan and England and Wales,[12] clues were sought for aetiological differences between the two countries by comparing death rates in occupation based social classes.

Further problems in making valid comparisons result when the published analyses are restricted to the working age (15-64), or include subjects aged 65-74. Presentation of results variously as PMRs or SMRs, with and without measures of significance, constitutes a further obstacle to the process of making useful comparisons.

Table 16.2, derived from the Occupational Mortality Analyses published for the US (1984)[10] and for the UK (1979-83)[18] agricultural industries, presents some of the problems and opportunities of international comparisons. (It would have been instructive also to look at 'construction' and 'extractive' industries, as was done for occupational accidents, but it was not possible to disassemble them from the published US data. It may be that the basic US and UK tapes could be made compatible for detailed comparisons.)

It is apparent that health experiences in the US and UK industries were disparate. For 'chronic obstructive pulmonary disease', (COPD), 'chronic bronchitis and emphysema', significant deficits in the US were matched by significant excesses in the UK, whereas for 'ischaemic heart disease' and 'acute myocardial infarction' there were significant excesses in the US but significant deficits in the UK. For 'other ischaemic heart disease' there was significant excess in the UK and deficit in the US, and for 'atherosclerosis', there was a significant excess in the US.

Occupational factors and lifestyle/general environmental factors contribute to cardiovascular diseases and to respiratory diseases. Exposures in the bulk handling and storage of cereals may be to airway irritants (oxides of nitrogen, non-specific dusts) and to dusts with pharmacological and/or immunological properties, as well as to toxic fumigant and fungicide/herbicide residues. Confounding the effects of these agents may be those of cigarette smoke. For the cardiovascular diseases, in the agricultural industry context, occupation is not likely to be an important factor, as compared with the potential adverse effects of cigarette smoke, and diet.

The US industry excess of 'ischaemic heart disease' and 'acute myocardial infarction', in the light of the cancer deficits and the deficits of 'COPD' is unlikely to be due to tobacco habit, and nutrition would need to be excluded as the cause before investigating further. In the UK, the deficits of 'ischaemic heart disease' and 'acute myocardial infarction' do not support the hypothesis that the excess of deaths from 'chronic bronchitis and emphysema' is entirely due to smoking habit. The contributions of occupational dusts and fumes need to be investigated. The agricultural industry is far from a homogenous whole. Hunting, fishing, forestry, farming (arable and animal rearing) present different types of hazard that are reflected in part in their mortality analyses.

In the US analyses the reason that the mortality for 'farm and other agricultural workers' (column 7) closely resembles that for the total industry (column 6) is that the former contribute 21,778 deaths to the 22,788 total. 'Forestry, fishing and hunting' with only 1,010 deaths present different hazards, but apart from significant excesses of deaths from accidents - mainly industrial PMR 361 ($p=<0.001$), 'accidents' (E800-949) PMR=155 ($p<0.01$) - numbers were too small to reach statistical significance for most other causes. Of the UK industry death total of 5,443 (Table GD39[18]), 'farm workers' contributed 2,421, 'forestry workers' 243, 'horticultural, gardeners and groundsmen' 2,673, 'other farm workers' 207, 'agricultural machinery operators' 212 and 'fishermen' 487.

The UK subsets differed in their health experience. Columns 2 and 3 show farm workers and horticultural workers to have had excesses of 'chronic bronchitis and emphysema'. Although cigarette smoke is a powerful cause, there is no support for such an aetiology in these cases from the deaths observed for bronchogenic carcinoma and for ischaemic heart disease. Fishermen had a PMR of 145 ($p=<0.01$) for

Table 16.2 The agriculture, forestry and fishery industries: a comparison of PMRs calculated for a limited range of diseases in the UK study 1979-83[18] and the US 12 state study of 1984[10]

	UK study[18] - all males 20-64					US study[10] - all males 20 years and over		
	1 Farming, fishing and related	2 Farm workers	3 Horticultural gardeners, groundsmen	4 Fishing	5 Forestry	6 Farming, forestry, fishing	7 Farm and other agricultural	8 Forestry, fishing and hunting
All cancers (140-208)	103					89**	88**	103
Ca oral cavity (140-149)	130					63**	64**	51
Ca oesophagus (150)	102					79**	75**	138
Ca stomach (151)	108					82**	80**	112
Ca colon (153)	112	†330	†108	†309	†385	84**	85**	75
Ca rectum (154)	110	121	91	59	153	81**	80	83
Ca larynx (161)	83					63**	61**	108
Ca bronchus (162)	102	101	95	145**	101	82**	80**	120
Malignant melanoma skin (172)	80		†222		†809	61**	60**	75
Leukaemia (204-208)	111					114**	116*	50
Hypertensive heart disease (402)	122					81**	80*	99
Ischaemic heart disease (410-414)	88**					103**	104**	101
Ac myocardial infarction (410)	87**					114**	115**	104
Other ischaemic heart disease (411-414)	124*	94	92	64*	74	90**	89**	96
Cerebro-vascular disease (430-438)	103					111**	112**	99
Atherosclerosis (440)	101					110*	109	134
COPD / Bronchitis and emphysema (490-496)	124**	†148**	128*	†87	†153	91**	91**	99
Gastric and duodenal ulcers (531-533)	102					80**	81	59
Chronic liver disease (571)	51**					81**	79**	100
Nephritis and nephrosis (580-589)	111					110*	110	127
Suicide (E950-E959)	117*					106	108	88

* $p<0.05$
** $p<0.001$
† several diagnoses or sites are involved. The highest PMR is given.

'carcinoma of the bronchus'. An occupational factor is unlikely. The 1979-90 analysis reports significant excesses of deaths from oral, laryngeal and bronchial carcinomas, which can be smoking related.

16.6 Occupational diseases

The sources of information on the numbers of cases of occupational diseases in the UK, is provided in Chapter10, and is discussed at length in the Health and Safety Commission's annual report for 1992/93.[8] As with other occupational health data, there are wide variations in practice between countries. The conditions that are notifiable or compensatable and the completeness of their collection vary, as do the details that are published. For example, there were limited data available for the Netherlands[4] that no longer continue to be available. Whereas extensive lists are published by the UK[8] (some 53 categories plus a number of sub-categories) and by Germany[5] (some 55 categories). As an illustration of the variation in the range of reporting, an arbitrary, less than comprehensive list of conditions is presented in Table 16.3, indicating variations in the range of conditions that are recognised and the numbers reported in five European countries.

Constraints of time and resources have limited data collection and analysis for the purpose of this review, but it is sufficient to indicate what drastic remedies would be required to permit valid comparisons. The position in 1986 was selected as suitable for comparison, because of availability and because it corresponded with a period of vigorous economic activity. To varying extents, the numerators are incomplete. An obvious example is provided by the number of cases of malignant mesothelioma given for the UK. The figure of 305 cases of malignant mesothelioma compensated is a fraction of the total of cases known to have been diagnosed and likely to be attributable to asbestos exposure (see Chapter 10).

Though the denominators of persons at risk are not readily ascertainable, and the precise nature of the exposures responsible are not determined, the study of occupational diseases data does provide a broad brush view of the situation, and alert one to continuing risks that one had mistakenly consigned to the history books. For example such agents as lead, arsenic, carbon monoxide, aromatic amines, halogenated hydrocarbons, benzene and silica, are still being identified as causing disease even in advanced economies. In the 19th century, Ankylostoma infestation was recognised as a hazard of wet mines, and it continues to be listed today in a number of countries, though reported with a zero return. Yet as recently as 1984, 1985 and 1986, there were in West Germany prevalences of 4, 1 and 1 cases respectively.

West Germany in 1986 also had a prevalence of 1,515 cases of infectious diseases with 327 new cases in a group of workers including health care workers and laboratory workers. Infectious diseases acquired from animals had a prevalence of 1,344 and 15 new cases.

West Germany had a prevalence of occupational disease of 44,706 including 3,779 new cases. The presentation of the prevalence as well as the more conventional numbers of new cases, provides a fuller appreciation of the health burden associated with occupational diseases and is a matter to be considered when reviewing occupational health. This is true for a number of diseases and for a number of injuries. For example, extrinsic allergic asthma caused by exposure to an

Table 16.3 Comparisons of some of the data published on occupational diseases in Holland, Germany, Austria, Spain and the UK

	Holland[4]	Germany[5]	Austria[6]	Spain[7]	UK[8]
Radiation effects		2	6	1	
Heat cataract			1	3	2
Decompression disease		2			
All orthopaedic stress conditions		396			446
Cramp, forearm/hand					13
Beat hand					14
Beat knee			3		37
Beat elbow					6
Tenosynovitis (hand/forearm)	52				376
Occupational deafness		992	1087	16	1202
Vibration white finger		9	40	222	1366
All infectious diseases		358	101		
Leptospirosis					1
Tuberculosis				1[a]	13
Brucellosis				265	2
Viral hepatitis				5	5
Strep. suis infection					3
Poisoning by					
Lead or its compounds		5	22	186	3
Arsenic or its compounds		17		1	
Mercury or its compounds		1			3
Chromium or its compounds	12	11		1	
Cadmium or its compounds		2			3
Carbon disulphide	b				
Benzene or homologues		12	6	1	3
Nitro, amino, chloro benzene/homologues			2		
Carbon monoxide		12			
Hydrogen sulphide	b	3			
Nitric acid, NO_x, Ammonia	28				
New growth of skin		6			4
Squamous cell Ca of skin					4
Papilloma of the bladder					21
Occupational vitiligo					2
Inflammation/ulceration of upper respiratory tract or mouth					36
Dermatitis	125	462	759	732	464
Nasal cavity carcinoma					2
Beryllium poisoning					2
Ca bronchus in nickel workers					3
Pneumoconiosis: all causes		958	99	53	836
Coal mining				16	357
Asbestosis		165	17		329
Asbestosis and lung cancer					38
Bilateral pleural thickening					111
Byssinosis		1	3		27
Diffuse mesothelioma		172			305
Occupational asthma		166	95	66	166
Farmers' lung		71	58		11
Chronic obstructive airways disease		215			
All reports of occupational disease	441	3779	2474	2199	3583

This list is not comprehensive. The numbers and natures of the conditions recognized vary widely between countries.

a Bovine

b Sulphurous and sulphuric acids, carbon disulphide and hydrogen sulphide were grouped together in Holland and totalled 42 cases.

allergen, and acute inflammation of the airways from exposure to an irritant may terminate shortly after cessation of exposure: however a proportion of affected persons will continue to be troubled for many years after cessation of exposure. Fractured bones that knit apparently uneventfully after a few weeks, may be the precursor to premature disabling degenerative joint disease.

The differences observed in Table 16.3 may be due to greater or lesser hazard, or to the peculiarities of national recognition and reporting. A remarkable comparison is with Brucellosis which was reported in only 2 cases in the UK among workers with live stock and in veterinarians, whereas in Spain 265 cases were reported, constituting 10 per cent of all their occupational diseases. While it is conceivable that the Brucella organism is more widespread in Spain, there is always the possibility that case finding may differ. With UK policy for the control of brucellosis in cattle, it is unlikely that hundreds of cases are going missing. Until the conditions for reporting occupational diseases and for their recognition are standardised, we shall be able to do little more than note that certain diseases continue to be reported in the developed world, and remain on the alert.

16.7 Conclusions

The case is as strong today as in the time of Farr, on humanitarian and on economic grounds, for the analysis of vital statistics to monitor the health burden of occupation in terms of excess mortality. To this must now be added the analysis of cancer registration data to identify occupations where there are unsuspected excess risks of cancer, a measure that is more important for those tumours with better prognosis. The UK with its Central Cancer Register, and Scandinavian countries[15] with their comparable systems are well endowed for this activity. To obtain a more comprehensive measure of occupational health and safety, it is necessary to conduct periodic analyses of routinely collected occupational accident data, occupational disease data, and patterns of sickness absence. Today, at the international level, the same considerations apply to make a strong case for the assembly of a range of compatible data, to permit valid comparisons between countries, as well as to allow the aggregation of populations to provide sufficient power for the confident assessment of risks of the lower order now required to be achieved.

Equally importantly, different findings between countries might provide clues as to the aetiology of certain disease excesses. Considerable effort has been expended by a number of countries to assemble their accident data, and by ILO to publish them in its Year Book of Labour Statistics. If only criteria for certain data were agreed and adopted and the preliminary analyses presented in a standard form, then they would be suitable for comparative study, and for collaborative analyses of accidents, diseases, mortality and cancer incidence. Experience teaches that the development and then adoption of standards at the international level, takes place over decades. It would be unrealistic therefore, to expect that by the next decennial supplement on occupational health, it will be possible to make valid quantitative comparisons of occupational health and safety throughout the world, even if a programme for international standardisation of comprehensive occupation and health data were initiated immediately. With the expression of a joint humanitarian and economic will, and the provision of the appropriate resources, an international collaborative scheme could be established, and with modern data handling facilities it would not take statisticians and epidemiologists long to establish an economic system of international monitoring of occupational safety and health.

References

1 13th Annual Conference of Labour Statisticians.

2 *Yearbook of Labour Statistics*. 50th issue. Geneva: ILO, 1991.

3 Jack TA and Zak MJ. *Results from the first national census of fatal occupational injuries, 1992*. Compensation and working conditions. US Department of Labor, Bureau of Labor Statistics. December 1993.

4 Centraal Bureau voor de Statistiek. *Statistiek der bedrifjfsongevallen 1984*. 'S-gravenhage, staatsuitgeverij/CBS-Publikaties 1985.

5 Arbeitssicherheit '87. Bericht der Bundesregierung über den Stand der Unfallverhütung und das Unfallgeschehen in der Bundesrepublik Deutschland in Jahre 1986. Der Bundesminister für Arbeit und Sozialordnung.

6 Statistische daten aus der Sozialversicherung. Unfallversicherung Jahresstatistik 1986. Hauptverband der östereichischen Sozialversicherungstüger.

7 Direccion General de Informatica y Estadistica. *Estadistica de Accidentes de Trabajo y Enfermedades Professionales. Ano 1986*. Ministerio de Trabajo y Securidad Social.

8 Health and Safety Commission. *Annual report 1992/93: Statistical Supplement*. HSE Books.

9 Guralnick. L. *Vital statistics - special reports*. Vol 53, numbers 2, 3, 4 and 5. Washington: National Center for Health Statistics, 1962-3.

10 Rosenberg HM, Burnett C, Maurer J and Spiratas R. *Mortality by occupation, industry, and cause of death: 12 reporting States, 1984*. Monthly Vital Statistics Report; Vol 42, number 4. suppl. Hyattsville, Maryland: National Center for Health Statistics, 1993.

11 Lynge E. *Socio-economic differences in mortality in Europe*. Population studies No 9, Strasbourg: Council of Europe, 1984.

12 Kagamimori S, Iibuchi Y and Fox J. *A comparison of socio-economic differences in mortality between Japan and England and Wales*. World Health Statistics Quarterly, 1983; 36:119-128.

13 *Occupational mortality in the Nordic Countries 1971-1980*. Statistical reports of the Nordic countries 49. Copenhagen: Nordic Statistical Secretariat,1988 (In English).

14 Borgan J-K and Kristofersen LB. *Mortality by occupation and socio-economic group in Norway 1970-1980*. (In Norwegian with a list of Contents in English).

15 Lynge E and Thygesen L. Occupational cancer in Denmark: Cancer incidence in the 1970 census population. *Scandinavian Journal of Work, Environment and Health* 1990; 16 (suppl 2): 1-35.

16 Central Statistical Office, Budapest. *Studies in mortality differentials 4. Socio-economic and occupational mortality differential 1980*. Budapest: Central Statistical Office, 1988. (Test in Hungarian, with an English list of Contents).

17 Fox AJ, Leon DA. Review of the use of routine statistical sources for the detection of occupational health surveillance with particular reference to the UK and Scandanavia. *Community Health Studies* VIII (supplement) 1984;3: 28.5-37.5.

18 Office of Population Censuses and Surveys. *Occupational Mortality. The Registrar General's Decennial Supplement for Great Britain, 1979-80, 1982-83*. Series DS No 6. London: HMSO 1986.

Appendices

Appendix 1 Classification of causes of death

Causes of death used in Chapters 4, 5 and 6, were classified according to the ninth revision of the International Classification of Diseases (ICD), and for the analyses presented in this commentary ICD categories were then aggregated into 'diagnostic groups' as defined in the table below. This brought together diagnoses that are known or likely to share the same aetiology. For example, malignant neoplasms of the renal pelvis and ureter (ICD categories 189.1 and 189.2) share many of the known causes of cancer of the bladder (ICD 188), and appear to differ aetiologically from cancer of the renal parenchyma (ICD 189.0). One diagnostic group was therefore formed for urothelial cancer (ICD 188, 189.1-189.8) and another for other cancer of the kidney (ICD 189.0).

Definition of diagnostic groups

Diagnostic group	ICD codes (ninth revision)
Intestinal infectious diseases	001 - 009
Tuberculosis	010 - 018, 137
Zoonotic bacterial diseases	020 - 027
Meningococcal infection	036
Septicaemia	038
Viral hepatitis	070
Sarcoidosis	135
Cancer of the oral cavity	141, 143, 144, 145
Cancer of the salivary glands	142
Cancer of the pharynx (specified)	146 - 148
Cancer of the oesophagus	150
Cancer of the stomach	151
Cancer of the small intestine	152
Cancer of the colon	153
Cancer of the rectum	154
Cancer of the liver	155
Cancer of the gall bladder	156
Cancer of the pancreas	157
Cancer of the retroperitoneum	158.0,
Cancer of the peritoneum	158.8, 158.9
Cancer of the nose and nasal sinuses	160
Cancer of the larynx	161
Cancer of the bronchus	162
Cancer of the pleura	163
Cancer of the thymus	164.0
Cancer of other mediastinum	164.1 - 164.9
Cancer of bone	170
Cancer of soft tissue	171
Melanoma of skin	172
Other cancer of skin	173
Cancer of the female breast	174
Cancer of the male breast	175
Cancer of the uterus, part unspecified	179
Cancer of the cervix	180
Cancer of the placenta	181
Cancer of the body of uterus	182
Cancer of the ovary	183
Cancer of the prostate	185
Cancer of the testis	186
Cancer of the penis	187.1 - 187.4
Urothelial cancer	188, 189.1 - 189.8
Cancer of the kidney (except pelvis)	189.0
Cancer of the eye	190
Cancer of the brain	191
Meningeal tumour	192.1, 192.3, 225.2, 225.4
Cancer of the thyroid	193
Cancer of the suprarenal	194.0
Cancer of other endocrine organs	194.1 - 194.9
Non-Hodgkin's lymphoma	200, 202
Hodgkin's disease	201
Myeloma	203
Acute lymphatic leukaemia	204.0
Chronic lymphatic leukaemia	204.1
Acute myeloid leukaemia	205.0
Chronic myeloid leukaemia	205.1
Acute monocytic leukaemia	206.0
Other leukaemia	207, 208
Adrenal tumour	227.0
Pituitary tumour	227.3
Thyrotoxicosis	242
Hypothyroidism	244
Diabetes	250
Cushing's disease	255.0
Addison's disease	255.4
Amyloidosis	277.3
Immunodeficiency	279.1
Haemolytic anaemia	283
Aplastic anaemia	284
Defibrination syndrome	286.6
Purpura	287
Agranulocytosis	288.0
Dementia	290, 331.0, 331.1
Other alcohol-related diseases	303, 305.0, 425.5, 535.3, 571.0-571.3, 860.0, 860.1
Drug dependence	304
Anorexia nervosa	307.1
Bacterial and unspecified meningitis	320, 322
Encephalitis	323
Intracranial/spinal abscess	324
Parkinson's disease	332

Appendix 1 Classification of causes of death

Diagnostic group	ICD codes (ninth revision)
Motor neurone disease	335.2
Multiple sclerosis	340
Epilepsy	345
Guillain Barré syndrome	357.0
Myasthenia	358
Chronic rheumatic heart disease	394-398
Hypertensive disease	401-405
Ischaemic heart disease	410-414
Pulmonary embolism and phlebitis	415.1, 451, 453
Pulmonary hypertension	416.0
Cor pulmonale	416.9
Pericarditis	420
Endocarditis	421
Acute myocarditis	422
Mitral valve disorders	424.0
Aortic valve disorders	424.1
Other cardiomyopathies	425.0-425.4, 425.6-425.9
Atrial fibrillation	427.3
Chronic and unspecified myocarditis	429.0
Sub-arachnoid haemorrhage	430
Other cerebrovascular disease	431-438
Aortic aneurysm	441
Other aneurysm	442
Peripheral vascular disease	443, 557
Arterial embolism and thrombosis	444
Polyarteritis nodosa	446.0
Wegener's granulomatosis	446.4
Arteritis	447.6
Portal vein thrombosis	452
Oesophageal varices	456.0-456.2
Acute upper respiratory infection	460-465
Acute bronchitis	466
Viral pneumonia	480
Pneumococcal and unspecified lobar pneumonia	481
Other bacterial pneumonia	482
Other specified pneumonia	483
Bronchopneumonia	485
Unspecified pneumonia	486
Influenza	487
Chronic bronchitis and emphysema	491, 492, 496
Asthma	493
Bronchiectasis	494
Farmers' lung disease	495.0
Bird fanciers' lung	495.2
Other and unspecified allergic pneumonitis	495.1, 495.3 - 495.9
Coal workers' pneumoconiosis	500
Asbestosis	501
Silicosis	502
Other pneumoconiosis	503, 505
Byssinosis	504
Respiratory conditions from chemical fumes	506
Pleurisy	511
Pneumothorax	512
Pulmonary fibrosis	515
Fibrosing alvolitis	516.3
Oesophageal disease	530
Gastric ulcer	531
Duodenal ulcer	532
Peptic ulcer	533
Gastrojejunal ulcer	534
Non-alcoholic gastritis and duodenitis	535.0-535.2, 535.4-535.6
Appendicitis	540-542
Inguinal hernia	550
Other hernia	551-553
Crohn's disease	555
Ulcerative colitis	556
Other colitis	558
Volvulus	560.2
Diverticular disease	562
Peritonitis	567
Hepatitis	571.4, 573.3
Cirrhosis (not specified as biliary)	571.5
Biliary cirrhosis	571.6
Cholelithiasis and cholecystitis	574, 575, 576.1-576.4
Pancreatitis	577.0, 577.1
Glomerulonephritis	580-583
Renal failure	584-586
Urinary infections	590, 595, 599.0
Hydronephrosis	591
Renal stones	592
Hyperplasia of prostate	600
Pelvic inflammatory disease	614
Ovarian cysts	620.0-620.2
Ectopic pregnancy	633
Complications of pregnancy	640-648
Complications of delivery	650-669
Complications of puerperium	670-676
Infections of skin, joints and bone	680-686, 711, 730
Erythematous conditions	695
Systemic lupus erythematosus	710.0
Systemic sclerosis	710.1
Myositis	710.3, 710.4
Rheumatoid arthritis	714
Ankylosing spondylitis	720
Osteoporosis	733.0
Railway accidents	E800-E807
Motor vehicle traffic accidents	E810-E819
Off-road motor vehicle traffic accidents	E820-E825
Pedal cycle accidents	E826
Animal transport accidents	E827-E828
Water transport accidents	E830-E838

Appendix 1 Classification of causes of death

Diagnostic group	ICD codes (ninth revision)	Diagnostic group	ICD codes (ninth revision)
Air transport accidents	E840-E845	Cold injury	E901
Other vehicle accidents	E846-E848	Injury by high or low air pressure	E902
Accidental poisoning by drugs	E850-E858	Injury by animals and plants	E905-E906
Methanol poisoning	E860.2	Injury by lightning	E907
		Non-recreational drowning	E910.3
Poisoning by cleansing agents	E861		
Poisoning by solvents	E862	Injury by falling object	E916
Pesticide poisoning	E863	Injury by being caught between objects	E918
Poisoning by corrosive and caustics	E864	Injury by machinery	E919
Other poisoning	E866	Injury by cutting and piercing instruments or objects	E920
Poisoning by gas and other domestic fuels	E867, E868.1, E868.3	Injury by explosion of pressure vessel	E921
Poisoning by liquified petroleum gas	E868.0	Injury by firearms	E922
Poisoning by motor vehicle exhaust	E868.2	Injury by explosive material	E923
Poisoning by carbon monoxide from other sources	E868.8	Injury by hot substances	E924
		Injury by electric current	E925
Poisoning by other gases	E869	Overexertion	E927
Fall on stairs	E880	Other accidents	E928
Fall from ladder or scaffolding	E881	Suicide	E950-E959
Fall from building	E882	Homicide	E960-E969
Fall into hole	E883	Injury undetermined as accidental or purposeful	E980-E989
Other fall	E884	War	E990-E999
Slipping and tripping	E885		
Fracture unspecified	E887		
Fall unspecified	E888		
Injured by fire	E890-E899		
Heat injury	E900		

Appendix 2 Definition of the Southampton classification of job groups

David Coggon
Hazel Inskip
Paul Winter
Brian Pannett

MRC Environmental Epidemiology Unit, University of Southampton

Occupations were coded to the CO80 classification. This was used in the last decennial supplement on occupational mortality. It defines a total of 547 occupational units, and differs from previous classifications in placing supervisors and employees from the same job in different categories. The distinction is important from a socio-economic point of view, but is less relevant to studies of occupational health where supervisors and employees tend to share the same hazards. Furthermore, the numbers of deaths among supervisors are relatively few, limiting the statistical power of analyses which look at them separately. Therefore, the occupational units were aggregated into larger 'job groups' as shown in this appendix.

As well as combining supervisors and employees in the same groups, this revised classification also amalgamates other units in which any occupational hazards that are likely to affect mortality are similar. For example, a large number of clerical workers are included in a single job group.

Definition of job groups

Job group	Occupational units	
001 Lawyers	001	Judges, barristers, advocates, solicitors
002 Accountants	002.1	Chartered and certified accountants
	002.2	Cost and works accountants
	002.3	Estimators
	002.4	Valuers, claims assessors
	002.5	Financial managers
	002.6	Underwriters, brokers, investment analysts
	002.7	Taxation experts
003 Personnel managers etc.	003.1	Personnel and industrial relations officers
	003.2	O and M, work study and OR officers
004 Economists and statisticians	004.1	Economists, statisticians, actuaries
005 Computer programmers	004.2	Systems analysts, computer programmers
006 Sales managers etc	005.1	Marketing and sales managers and executives
	005.2	Advertising and PR executives
	005.3	Buyers (retail trade)
	005.4	Buyers and purchasing officers (not retail)
007 Government inspectors	006.1	Environmental health officers
	006.2	Building inspectors
	006.3	Inspectors (statutory and similar)
008 Government administrators	007.1	General administrators - national government (Assistant Secretary level and above)
	007.2	General administrators - national government (HEO to Senior Principal level)
	008	Local government officers (administrative and executive functions)
009 Other administrators	009.1	Company secretaries
	009.2	Officials of trade associations, trade unions, professional bodies and charities
	009.3	Property and estate managers
	009.4	Librarians, information officers
	009.5	Legal service and related occupations
	009.6	Management consultants
	009.7	Managers' personal assistants
	009.8	Professional workers and related supporting management and administration nec
010 Teachers in higher education	010.1	University academic staff
	010.2	Teachers in establishments for further and higher education
011 Teachers nec	011	Teachers nec
012 Vocational trainers, social scientists etc.	012.1	Vocational and industrial trainers
	012.2	Education officers, school inspectors
	012.3	Social and behavioural scientists

Appendix 2 Definition of the Southampton classification of job groups

Job group	Occupational units	
013 Welfare workers	013.1	Matrons, houseparents
	013.2	Playgroup leaders
	013.3	Welfare occupations nec
014 Clergy	014	Clergy, ministers of religion
015 Doctors	015.1	Medical practitioners
016 Dentists	015.2	Dental practitioners
017 Nurses	016	Nurse administrators, nurses
018 Pharmacists	017.1	Pharmacists
019 Medical radiographers	017.2	Medical radiographers
020 Physiotherapists	017.4	Physiotherapists
021 Health professions nec	017.3	Ophthalmic and dispensing opticians
	017.5	Chiropodists
	017.6	Therapists nec
	018.1	Medical technicians, dental auxiliaries
022 Veterinarians	018.2	Veterinarians
023 Driving instructors	018.3	Driving instructors (not HGV)
024 Literary and artistic occupations	019	Authors, writers, journalists
	020.1	Artists, commercial artists
	020.2	Industrial designers (not clothing)
	020.3	Clothing designers
	020.4	Window dressers
	021.1	Actors, entertainers, singers, stage managers
	021.2	Musicians
	022.1	Photographers, cameramen
	022.2	Sound and vision equipment operators
025 Persons involved in sport	023.1	Professional sportsmen, sports officials
026 Biological scientists	024.1	Biological scientists, biochemists
027 Chemical engineers and scientists	024.2	Chemical scientists
	028.1	Chemical engineers
028 Physical scientists and mathematicians	024.3	Physical and geological scientists, mathematicians
	028.5	Metallurgists
029 Electrical and electronic engineers (professional)	027.1	Electrical engineers
	027.2	Electronic engineers

Job group	Occupational units	
030 Professional engineers nec	025	Civil, structural, municipal, mining and quarrying engineers
	026.1	Mechanical and aeronautical engineers
	026.2	Design and development engineers (mechanical)
	028.2	Production engineers
	028.3	Planning and quality control engineers
	028.4	Engineers nec
	028.6	Technologists nec
031 Draughtspersons	029	Draughtsmen
032 Laboratory technicians	030.1	Laboratory technicians
	030.2	Engineering technicians, technician engineers
	136.8	Foremen - laboratory assistants
	138.1	Laboratory assistants
033 Architects and surveyors	031.1	Architects, town planners
	031.2	Quantity surveyors
	031.3	Building, land and mining surveyors
034 Aircraft flight deck officers	032.1	Aircraft flight deck officers
035 Air traffic controllers	032.2	Air traffic planners and controllers
036 Seafarers	032.3	Deck, engineering and radio officers and pilots, ship
	147	Foremen - ships, lighters and other vessels
	148	Deck, engine-room hands, bargemen, lightermen, boatmen
037 Technicians nec	033.1	Architectural and town planning technicians
	033.2	Building and civil engineering technicians
	033.3	Technical and related workers nec
038 Production and maintenance managers	034	Production, works and maintenance managers, works foremen
039 Managers in construction	035.1	Managers in building and contracting
	035.2	Clerks of works
040 Managers in transport, utilities and mining	036.1	Managers in mining and public utilities
	036.2	Transport managers
	036.3	Stores controllers
	036.4	Managers in warehousing and materials handling nec

Appendix 2 Definition of the Southampton classification of job groups

Job group	Occupational units	
041 Office managers	037.1	Credit controllers
	037.2	Office managers nec
042 Butchers	038.2	Butchers (managers and proprietors)
	090.2	Foremen - butchers, meat cutters
	092.1	Butchers, meat cutters
043 Fishmongers and poultry dressers	038.3	Fishmongers (managers and proprietors)
	090.3	Foremen - fishmongers, poultry dressers
	092.2	Fishmongers, poultry dressers
044 Retailers and dealers	038.1	Garage proprietors
	038.4	Other proprietors and managers (sales)
	054.1	Supervisors of shop salesmen and assistants
	055.1	Shop salesmen and assistants
	055.2	Shelf fillers
	057.2	Market and street traders and assistants
	057.3	Scrap dealers, general dealers, rag and bone merchants
	075.5	Bookmakers, betting shop managers
045 Publicans and bar staff	039.1	Hotel and residential club managers
	039.2	Publicans
	039.4	Club stewards
	063.3	Supervisors of barmen, barmaids
	065.2	Barmen, barmaids
046 Caterers	039.3	Restaurateurs
	063.2	Supervisors of waiters, waitresses
	063.4	Supervisors of counter hands, assistants
	065.1	Waiters, waitresses
	066.1	Counter hands, assistants
047 Farmers	040	Farmers, horticulturists, farm managers
	076.1	Farm foremen
	076.2	Horticultural foremen
	076.3	Foremen gardeners and groundsmen
	076.4	Agricultural machinery foremen
	076.6	Other foremen in farming and related
	077	Farm workers
	078.1	Horticultural workers
	078.2	Gardeners, groundsmen
	079	Agricultural machinery drivers, operators
	083	All other in farming and related
048 Armed forces	041	Officers, UK armed forces
	042	Officers, foreign and Commonwealth armed forces
	058	NCOs and other ranks, UK armed forces
	059	NCOs and other ranks, foreign and Commonwealth armed forces
049 Police	043.2	Police officers (inspectors and above)
	060.1	Police sergeants
	061.1	Policemen (below sergeant)
050 Fire service personnel	043.3	Fire service officers
	060.2	Fire service supervisors
	061.2	Firemen
051 Launderers and dry cleaners	044.2	Managers of laundry and dry cleaning receiving shops
	075.3	Launderers, dry cleaners, pressers
052 Hairdressers	044.3	Hairdressers' and barbers' managers and proprietors
	073	Hairdressing supervisors
	074	Hairdressers, barbers
053 Office workers and cashiers	045.1	Civil service executive officers
	045.2	Supervisors of stores and despatch clerks
	045.3	Supervisors of tracers, drawing office assistants
	045.4	Supervisors of other clerks and cashiers (not retail)
	045.5	Supervisors of retail shop cashiers, check-out and cash and wrap operators
	046.1	Stores and despatch clerks
	046.2	Tracers, drawing office assistants
	046.3	Other clerks and cashiers (not retail)
	047	Retail shop cashiers, check-out and cash and wrap operators
	048.1	Supervisors of typists, shorthand writers, secretaries
	048.2	Supervisors of office machine operators
	048.3	Supervisors of telephone operators
	048.4	Supervisors of radio and telegraph operators
	049.1	Receptionists
	049.2	Typists, shorthand writers, secretaries
	050	Office machine operators
	051.1	Telephonist receptionists
	051.2	Telephone operators
	051.3	Radio and telegraph operators

Appendix 2 Definition of the Southampton classification of job groups

Job group	Occupational units	
054 Postal workers	052.1	Supervisors of postmen, mail sorters
	052.2	Supervisors of messengers
	053.1	Postmen, mail sorters
	053.2	Messengers
055 Petrol pump attendants	054.2	Supervisors of petrol pump, forecourt attendants
	055.3	Petrol pump, forecourt attendants
056 Van sales persons	054.3	Supervisors of roundsmen, van salesmen
	056	Roundsmen, van salesmen
057 Sales representatives	057.1	Importers, exporters, commodity brokers
	057.4	Credit agents, collector salesmen
	057.5	Sales representatives
	057.6	Sales representatives (property and services), other agents
058 Security workers	043.1	Prison officers (chief officers and above)
	060.3	Prison service principal officers
	060.4	Supervisors of security guards and officers, patrolmen, watchmen
	060.5	Supervisors of traffic wardens
	060.6	Supervisors of security and protective service workers nec
	061.3	Prison officers (below principal officer)
	062.1	Security guards and officers, patrolmen, watchmen
	062.2	Traffic wardens
	062.3	Security and protective service workers nec
059 Cooks and kitchen porters	063.1	Supervisors of chefs, cooks
	064	Chefs, cooks
	066.2	Kitchen porters, hands
060 Other service personnel	044.1	Proprietors and managers, service flats, holiday flats, caravan sites, etc
	067.1	Housekeepers (non-domestic)
	067.2	Supervisors of other domestic and school helpers
	067.3	Supervisors of travel stewards and attendants
	067.5	Supervisors of hotel porters
	068.1	Domestic housekeepers
	068.2	Nursery nurses
	068.3	Other domestic and school helpers
	069.1	Travel stewards and attendants
	069.3	Hotel porters
	071.1	Supervisors of caretakers
	071.2	Supervisors of cleaners, window cleaners, chimney sweeps, road sweepers
	071.4	Supervisors of lift and car park attendants
	072.1	Caretakers
	072.2	Cleaners, window cleaners, chimney sweeps, road sweepers
	075.2	Lift and car park attendants
061 Hospital porters and ward orderlies	067.4	Supervisors of hospital porters
	067.7	Supervisors of hospital, ward orderlies
	069.2	Hospital porters
	070.2	Hospital, ward orderlies
062 Ambulance workers	067.6	Supervisors of ambulancemen
	070.1	Ambulancemen
063 Railway station workers	071.3	Supervisors of railway stationmen
	075.1	Railway stationmen
	149.4	Foremen - other foremen rail transport
064 Undertakers	075.4	Undertakers
065 Foresters	076.5	Forestry foremen
	080	Forestry workers
066 Fishing and related workers	081	Supervisors, mates - fishing
	082	Fishermen
067 Tannery workers	084.1	Foremen - tannery production workers
	085.1	Tannery production workers

Appendix 2 Definition of the Southampton classification of job groups

Job group	Occupational units	
068 Leather and shoe workers	084.2	Foremen - shoe repairers
	084.3	Foremen - leather cutters and sewers, footwear lasters, makers, finishers
	084.4	Foremen - other making and repairing, leather
	085.2	Shoe repairers
	085.3	Leather cutters and sewers, footwear lasters, makers, finishers
	107.7	Other making and repairing - leather
069 Preparatory fibre processors	086.1	Foremen - preparatory fibre processors
	087.1	Preparatory fibre processors
070 Spinners and winders	086.2	Foremen - spinners, doublers, twisters
	086.3	Foremen - winders, reelers
	087.2	Spinners, doublers, twisters
	087.3	Winders, reelers
071 Warp preparers and weavers	086.4	Foremen - warp preparers
	086.5	Foremen - weavers
	087.4	Warp preparers
	087.5	Weavers
072 Knitters	086.6	Foremen - knitters
	087.6	Knitters
073 Bleachers, dyers and finishers	086.7	Foremen - bleachers, dyers, finishers
	087.7	Bleachers, dyers, finishers
074 Other textile workers	086.8	Foremen - menders, darners
	086.9	Foremen - other material processing - textiles
	087.8	Menders, darners
	098.4	Other material processing - textiles
	136.2	Foremen inspectors, viewers, examiners - textiles
	138.2	Inspectors, viewers, examiners - textiles
	159.1	Foremen - labourers and unskilled workers nec - textiles (not textile goods)
	160.1	Labourers and unskilled workers nec - textiles (not textile goods)
075 Chemical workers	088	Foremen - chemical processing
	089	Chemical, gas and petroleum process plant operators
	159.2	Foremen - labourers and unskilled workers nec - chemicals and allied trades
	160.2	Labourers and unskilled workers nec - chemicals and allied trades
076 Bakers	090.1	Foremen - bakers, flour confectioners
	091	Bakers, flour confectioners
077 Brewery workers	090.4	Foremen - brewery and vinery process workers
	098.2	Brewery and vinery process workers
078 Food processors	090.5	Foremen - other material processing - bakery and confectionery workers
	090.6	Foremen - other material processing - food and drink nec
	098.5	Other material processing - bakery and confectionery workers
	098.7	Other material processing - food and drink nec
	136.3	Foremen inspectors, viewers, examiners - food
	138.3	Inspectors, viewers, examiners - food
079 Paper manufacturers	093.1	Foremen - paper, paperboard and leatherboard workers
	094.1	Paper, paperboard and leatherboard workers
080 Bookbinders	093.2	Foremen - bookbinders and finishers
	094.2	Bookbinders and finishers
081 Paper cutters	093.3	Foremen - cutting and slitting machine operators (paper and paper products making)
	094.3	Cutting and slitting machine operators (paper and paper products making)
082 Glass and ceramics furnace workers	095.1	Foremen - glass and ceramics furnacemen, kilnsetters
	096.1	Glass and ceramics furnacemen, kilnsetters
083 Glass formers and decorators	095.2	Foremen - glass formers and shapers, finishers, decorators
	096.2	Glass formers and shapers, finishers, decorators
084 Ceramics casters	095.3	Foremen - casters and other pottery makers
	096.3	Casters and other pottery makers
085 Rubber manufacturers	095.4	Foremen - rubber process workers, moulding machine operators, tyre builders
	097.1	Rubber process workers, moulding machine operators, tyre builders

Appendix 2 Definition of the Southampton classification of job groups

Job group	Occupational units	
086 Plastics workers	095.5	Foremen - calender and extruding machine operators, moulders (plastics)
	097.2	Calender and extruding machine operators, moulders (plastics)
087 Man-made fibre makers	095.6	Foremen - man-made fibre makers
	098.1	Man-made fibre makers
088 Other coal miners	095.7	Foremen - washers, screeners and crushers in mines and quarries
	098.3	Washers, screeners and crushers in mines and quarries
	159.7	Foremen - labourers and unskilled workers nec - coal mines
	160.7	Labourers and unskilled workers nec - coal mines
089 Tobacco workers	098.6	Other material processing - tobacco
090 Other wood and paper processors	093.4	Foremen - other material processing, wood and paper
	098.8	Other material processing - wood and paper
	134.4	Foremen assemblers - paper production, processing and printing
	136.9	Foremen - inspectors, sorters in paper production, processing and printing
	138.7	Inspectors, sorters in paper production, processing and printing
	138.8	Assemblers in paper production, processing and printing
091 Other occupations - glass and ceramics	095.8	Foremen - other making and repairing - glass and ceramics
	107.5	Other making and repairing - glass and ceramics
	159.4	Foremen - labourers and unskilled workers nec - glass and ceramics
	160.4	Labourers and unskilled workers nec - glass and ceramics
092 Rubber goods makers	095.9	Foremen - other making and repairing - rubber
	107.10	Other making and repairing - rubber
	136.4	Foremen inspectors, viewers, examiners - rubber goods
	138.4	Inspectors, viewers, examiners - rubber goods
093 Plastic goods makers	095.10	Foremen - other making and repairing - plastics
	107.11	Other making and repairing - plastics
	134.5	Foremen assemblers - plastics goods
	136.5	Foremen inspectors, viewers, examiners - plastics goods
	138.5	Inspectors, viewers, examiners - plastics goods
	138.9	Assemblers (plastic goods)
094 Compositors	099.1	Foremen - compositors
	100.1	Compositors
095 Printing plate preparers	099.2	Foremen - electrotypers, stereotypers, printing plate and cylinder preparers
	100.2	Electrotypers, stereotypers, printing plate and cylinder preparers
096 Printing machine minders	099.3	Foremen - printing machine minders and assistants
	100.3	Printing machine minders and assistants
097 Printers (so described)	099.4	Foremen - screen and block printers
	099.5	Foremen - printers (so described)
	100.4	Screen and block printers
	100.5	Printers (so described)
098 Tailors and dressmakers	101.1	Foremen - tailors, tailoresses, dressmakers
	102.1	Tailors, tailoresses, dressmakers
099 Clothing cutters	101.2	Foremen - clothing cutters, milliners, furriers
	102.2	Clothing cutters, milliners, furriers
100 Sewers and embroiderers	101.3	Foremen - sewers, embroiderers
	102.3	Sewers, embroiderers
101 Upholsterers	101.4	Foremen - coach trimmers, upholsterers, mattress makers
	103	Coach trimmers, upholsterers, mattress makers
102 Carpet fitters	101.5	Foremen - carpet fitters
	107.3	Carpet fitters
103 Other workers with fabrics	101.6	Foremen - other making and repairing, clothing and related products
	107.8	Other making and repairing - clothing and related products

Appendix 2 Definition of the Southampton classification of job groups

Job group	Occupational units	
104 Carpenters	104.1	Foremen - carpenters, joiners
	105.1	Carpenters, joiners
105 Cabinet makers	104.2	Foremen - cabinet makers
	105.2	Cabinet makers
106 Case and box makers	104.3	Foremen - case and box makers
	105.3	Case and box makers
107 Pattern makers	104.4	Foremen - pattern makers (moulds)
	105.4	Pattern makers (moulds)
108 Woodworking machinists	104.5	Foremen - sawyers, veneer cutters, woodworking machinists
	106	Sawyers, veneer cutters, woodworking machinists
109 Other woodworkers	104.6	Foremen - other making and repairing, wood
	107.1	Labourers and mates to woodworking craftsmen
	107.6	Other making and repairing - wood
	136.6	Foremen inspectors, viewers, examiners - woodwork
	138.6	Inspectors, viewers, examiners - woodwork
110 Dental technicians	107.2	Dental technicians
111 Other makers of paper goods	093.5	Foremen - other making and repairing, paper goods and printing
	107.9	Other making and repairing - paper goods and printing
112 Furnace operatives (metal)	108.1	Foremen - furnace operating occupations (metal)
	109.1	Furnace operating occupations (metal)
113 Rollers (metal)	108.2	Foremen - rollermen
	109.2	Rollermen
114 Smiths and forge workers	108.3	Foremen - smiths, forgemen
	109.3	Smiths, forgemen
115 Metal drawers	108.4	Foremen - metal drawers
	110.1	Metal drawers
116 Moulders and coremakers (metal)	108.5	Foremen - moulders, coremakers, die casters
	110.2	Moulders, coremakers, die casters
117 Electroplaters	108.6	Foremen - electroplaters
	110.3	Electroplaters
118 Annealers, hardeners, temperers (metal)	108.7	Foremen - annealers, hardeners, temperers (metal)
	110.4	Annealers, hardeners, temperers (metal)
119 Galvanizers and tin platers	108.8	Foremen - galvanizers, tin platers, dip platers
	131.2	Galvanizers, tin platers, dip platers
120 Other metal manufacturers	108.9	Foremen - metal making and treating workers nec
	131.3	Metal making and treating workers nec
	159.5	Foremen - labourers and unskilled workers nec - foundries in engineering and allied trades
	160.5	Labourers and unskilled workers nec - foundries in engineering and allied trades
121 Press and machine tool setters	111.1	Foremen - press and machine tool setters
	112.1	Press and machine tool setters
122 Centre lathe turners	111.2	Foremen - centre lathe turners
	112.2	Centre lathe turners
123 Machine tool setter operators	111.3	Foremen - machine tool setter operators
	112.3	Machine tool setter operators
124 Machine tool operators	111.4	Foremen - machine tool operators
	112.4	Machine tool operators
125 Press and automatic machine operators	111.5	Foremen - press, stamping and automatic operators
	113.1	Press, stamping and automatic machine operators
126 Metal polishers	111.6	Foremen - metal polishers
	113.2	Metal polishers
127 Fettlers and dressers (metal)	111.7	Foremen - fettlers, dressers
	113.3	Fettlers, dressers
128 Shot blasters	111.8	Foremen - shot blasters
	131.8	Shot blasters
129 Toolmakers	114.1	Foremen - toolmakers, tool fitters, markers out
	115	Toolmakers, tool fitters, markers-out
130 Precision instrument makers	114.2	Foremen - precision instrument makers and repairers
	116.1	Precision instrument makers and repairers

Appendix 2 Definition of the Southampton classification of job groups

Job group	Occupational units	
131 Watch and clock makers	114.3	Foremen - watch and chronometer makers and repairers
	116.2	Watch and chronometer makers and repairers
132 Production fitters	114.4	Foremen - metal working production fitters and fitter/ machinists
	117	Metal working production fitters and fitter/machinists
133 Motor mechanics	114.5	Foremen - motor mechanics, auto engineers
	118.1	Motor mechanics, auto engineers
134 Aircraft engine fitters	114.6	Foremen - maintenance fitters (aircraft engines)
	118.2	Maintenance fitters (aircraft engines)
135 Office machinery mechanics	114.7	Foremen - office machinery mechanics
	119	Office machinery mechanics
136 Electrical and electronic production fitters	120.1	Foremen - production fitters (electrical, electronic)
	121.1	Production fitters (electrical, electronic)
137 Electricians	120.2	Foremen - electricians, electrical maintenance fitters
	121.2	Electricians, electrical maintenance fitters
138 Electrical plant operators	120.3	Foremen - plant operators and attendants nec
	121.4	Plant operators and attendants nec
139 Telephone fitters	120.4	Foremen - telephone fitters
	122.1	Telephone fitters
140 Electric cable and line workers	120.5	Foremen - cable jointers, linesmen
	122.2	Cable jointers, linesmen
141 Radio and TV mechanics	120.6	Foremen - radio and TV mechanics
	123.1	Radio and TV mechanics
142 Other electronic maintenance engineers	120.7	Foremen - other electronic maintenance engineers
	123.2	Other electronic maintenance engineers
143 Electrical engineers (so described)	121.3	Electrical engineers (so described)
144 Plumbers and gas fitters	124.1	Foremen - plumbers, heating and ventilating fitters, gas fitters
	125	Plumbers, heating and ventilating fitters, gas fitters
145 Sheet metal workers	124.2	Foremen - sheet metal workers
	126.1	Sheet metal workers
146 Metal plate workers	124.3	Foremen - metal plate workers, shipwrights, riveters
	126.2	Metal plate workers, shipwrights, riveters
147 Steel erectors	124.4	Foremen - steel erectors, benders, fixers
	127.1	Steel erectors, benders, fixers
148 Scaffolders	124.5	Foremen - scaffolders, stagers
	127.2	Scaffolders, stagers
149 Welders	124.6	Foremen - welders
	128	Welders
150 Riggers	124.7	Foremen - riggers
	131.5	Riggers
151 Jewellery workers	129.1	Foremen - goldsmiths, silversmiths, precious stone workers
	130.1	Goldsmiths, silversmiths, precious stone workers
152 Engravers and etchers (printing)	129.2	Foremen - engravers, etchers (printing)
	130.2	Engravers, etchers (printing)
153 Vehicle body builders	129.3	Foremen - coach and vehicle body builders
	131.1	Coach and vehicle body builders
154 Oilers and greasers	129.4	Foremen - oilers, greasers, lubricators
	131.4	Oilers, greasers, lubricators
155 Electronics wire workers	129.5	Foremen - electronics wiremen
	131.6	Electronics wiremen
156 Coil winders	129.6	Foremen - coil winders
	131.7	Coil winders
157 Pottery decorators	132.1	Foremen - pottery decorators
	133.1	Pottery decorators
158 Coach painters	132.2	Foremen - coach painters (so described)
	133.2	Coach painters (so described)

Appendix 2 Definition of the Southampton classification of job groups

Job group	Occupational units	
159 Other spray painters	132.3	Foremen - other spray painters
	133.3	Other spray painters
160 Painters and decorators nec	132.4	Foremen - painters and decorators nec, french polishers
	133.4	Painters and decorators nec, french polishers
161 Electrical, electronic assemblers	134.1	Foremen assemblers - electrical, electronic
	135.1	Assemblers (electrical, electronic)
162 Instrument assemblers	134.2	Foremen assemblers - instruments
	135.2	Instrument assemblers
163 Assemblers (vehicles and other metal goods)	134.3	Foremen assemblers - (vehicles and other metal goods)
	135.3	Assemblers (vehicles and other metal goods)
164 Packers and sorters	136.7	Foremen - packers, bottlers, canners, fillers
	136.10	Foremen - weighers
	136.11	Foremen - graders, sorters, selectors nec
	137.2	Packers, bottlers, canners, fillers
	138.10	Weighers
	138.11	Graders, sorters, selectors nec
165 Bricklayers and tilesetters	139.1	Foremen - bricklayers, tile setters
	140.1	Bricklayers, tile setters
166 Masons and stonecutters	139.2	Foremen - masons, stone cutters
	140.2	Masons, stone cutters
167 Plasterers	139.3	Foremen - plasterers
	140.3	Plasterers
168 Roofers and glaziers	139.4	Foremen - roofers, glaziers
	140.4	Roofers, glaziers
169 Builders etc.	139.5	Foremen - handymen, general building workers
	140.5	Handymen, general building workers
	140.6	Builders (so described)
170 Rail track workers	139.6	Foremen - railway lengthmen
	141.1	Railway lengthmen
171 Road construction workers and paviors	139.7	Foremen - road surfacers, concreters
	139.8	Foremen - roadmen
	139.9	Foremen - paviors, kerb layers
	141.2	Road surfacers, concreters
	141.3	Roadmen
	141.4	Paviors, kerb layers
172 Sewage plant attendants	139.10	Foremen - sewage plant attendants
	142.1	Sewage plant attendants
173 Mains and service layers	139.11	Foremen - mains and service layers, pipe jointers
	142.2	Mains and service layers, pipe jointers
174 Construction workers nec	139.12	Foremen - construction workers nec
	143.1	Craftsmen's mates
	143.2	Building and civil engineering labourers
	146.2	Construction workers nec
175 Face trained coalminers	144	Foremen/deputies - coalmining
	145	Face-trained coalmining workers
176 Miners (not coal) and quarry workers	146.1	Miners (not coal), quarrymen, well drillers
177 Railway guards	149.1	Foremen - railway guards
	150.2	Railway guards
178 Railway signal workers	149.2	Foremen - signalmen and crossing keepers, railway
	150.3	Signalmen and crossing keepers, railway
179 Shunters and points operators	149.3	Foremen - shunters, pointsmen
	150.4	Shunters, pointsmen
180 Railway engine drivers	150.1	Drivers, motormen, secondmen, railway engines
181 Road transport inspectors	151.1	Bus inspectors
	151.3	Foremen - other foremen road transport
182 Bus and coach drivers	152.1	Bus and coach drivers
183 Lorry drivers	151.2	Foremen - drivers of road goods vehicles
	152.2	Drivers of road goods vehicles
184 Other motor drivers	152.3	Other motor drivers
185 Bus conductors and drivers' mates	153.1	Bus conductors
	153.2	Drivers' mates

Appendix 2 Definition of the Southampton classification of job groups

Job group	Occupational units	
186 Mechanical plant drivers	154.1	Foremen - mechanical plant drivers, operators (earth moving and civil engineering)
	155.1	Mechanical plant drivers, operators (earth moving and civil engineering)
187 Crane drivers	154.2	Foremen - crane drivers, operators
	155.2	Crane drivers, operators
188 Fork lift truck drivers	154.3	Foremen - fork lift, mechanical truck drivers
	155.3	Fork lift, mechanical truck drivers
189 Slingers	154.4	Foremen - slingers
	158.1	Slingers
190 Storekeepers	156.1	Foremen - storekeepers, warehousemen
	157.1	Storekeepers, warehousemen
191 Dockers and goods porters	156.2	Foremen - stevedores, dockers
	156.3	Foremen - goods porters
	157.2	Stevedores, dockers
	157.3	Goods porters
192 Refuse collectors	156.4	Foremen - refuse collectors, dustmen
	157.4	Refuse collectors, dustmen
193 Labourers in coke ovens	159.3	Foremen - labourers and unskilled workers nec - coke ovens and gas works
	160.3	Labourers and unskilled workers nec - coke ovens and gas works
194 Boiler operators	159.9	Foremen - boiler operators
	161.1	Boiler operators

Appendix 3 Description of job groups of the Southampton classification

Brian Pannett
David Coggon
Hazel Inskip
Paul Winter

MRC Environmental Epidemiology Unit, University of Southampton

This appendix provides a brief description of the 194 job groups of the Southampton classification. It sets out what each occupation entails in rather more detail than may be apparent from the job title alone, and indicates the social classes to which the job groups belong. Job groups which are made up of several occupational units may include people from more than one social class.

Also listed are some of the more important hazards that are known or suspected to occur in these occupations. The list is not intended to be exhaustive, but gives emphasis to more serious hazards and especially those which might have an appreciable effect on mortality or cancer incidence. Some established hazards are not mentioned where they apply to only a minority of workers from a job group.

Occupation: 001 - LAWYERS
Judges, barristers, advocates, solicitors
Job description: Presiding over judicial proceedings; preparing and conducting court cases; provision of legal advice. These are indoor, sedentary occupations.
Social class: I
Hazards: None of note

Occupation: 002 - ACCOUNTANTS
Accountants, valuers, finance specialists
Job description: Provision of auditing and accounting services; advising on financial matters and taxation; dealing in stocks and shares. These are indoor, sedentary occupations.
Social class: I/II
Hazards: None of note

Occupation: 003 - PERSONNEL MANAGERS ETC.
Personnel and industrial relations managers; 0 and M, work study and operational research officers
Job description: Advising on and ensuring effective use of manpower and materials; recruitment, training and appraisal of staff; management of industrial relations and personnel policies. These are largely indoor, sedentary occupations.
Social class: II
Hazards: None of note

Occupation: 004 - ECONOMISTS AND STATISTICIANS
Economists, statisticians, actuaries
Job description: Application of economic, statistical and actuarial principles and techniques in the formulation of financial and investment policies. These are indoor, sedentary occupations.
Social class: I
Hazards: None of note

Occupation: 005 - COMPUTER PROGRAMMERS
Systems analysts, computer programmers
Job description: Planning computing operations; writing computer programs.
Social class: II
Hazards: Computer programmers sit in front of visual display screens for a large part of their working day. It has been suggested that magnetic fields from visual display screens might be carcinogenic, although there is no consistent evidence for this. Prolonged use of keyboards has been associated with various musculoskeletal disorders of the upper limb, but these would not be expected to affect mortality.

Occupation: 006 - SALES MANAGERS ETC
Marketing, sales, advertising, public relations and purchasing managers
Job description: Development and implementation of sales policies and advertising strategies; purchasing merchandise for retail or wholesale.
Social class: II
Hazards: None of note

Occupation: 007 - GOVERNMENT INSPECTORS
Statutory and other inspectors
Job description: Conducting inspections and investigations to ensure compliance with legislation relating to buildings, environmental hygiene, public health, ships, schools, factories, weights and measures, etc.
Social class: II
Hazards: These jobs may entail intermittent short-term exposure to various chemical, physical and biological hazards, but there is no major hazard to which a large proportion of the group is consistently exposed.

Occupation: 008 - GOVERNMENT ADMINISTRATORS
General administrators - national government, local government officers (administrative and executive functions)
Job description: Formulation and administration of national and local government policy. These are indoor, sedentary occupations.
Social class: I/II
Hazards: None of note

Occupation: 009 - OTHER ADMINISTRATORS
All other professional and related supporting management and administration
Job description: A range of activities including advising and supporting directors and senior managers; administration of professional associations, unions, charities and libraries; management of estates and property.
Social class: I/II/IIIN
Hazards: None of note

Appendix 3 Description of job groups of the Southampton classification

Occupation: 010 - TEACHERS IN HIGHER EDUCATION
Teachers in higher education
Job description: Teaching in universities and colleges; undertaking research, and writing papers and books for publication.
Social class: I/II
Hazards: Contact with large numbers of young people may lead to unusually frequent exposure to infections. Some members of the group (for example university lecturers who carry out scientific research in laboratories) are exposed to hazardous chemical, physical or biological agents at work. However, only a small proportion come into contact with any single hazard.

Occupation: 011 - TEACHERS NEC
Teachers nec
Job description: Teaching in nursery, primary and secondary schools.
Social class: II
Hazards: School teachers come into frequent contact with childhood infections. Science and technology teachers are intermittently exposed to hazardous materials in laboratories, but not at a level which would be expected to have a discernible effect on the overall mortality of the profession.

Occupation: 012 - VOCATIONAL TRAINERS, SOCIAL SCIENTISTS, ETC.
Vocational and industrial trainers, education officers, social and behavioural scientists
Job description: A range of activities including training adults in vocational skills; planning, organising and directing educational resources of local authorities; studying human social interrelationships.
Social class: I/II
Hazards: None of note

Occupation: 013 - WELFARE WORKERS
Welfare workers
Job description: Organising and managing social and welfare facilities serving children of nursery and pre-school age, young offenders, and elderly people in residential accommodation; provision of other community services.
Social class: II/IIIN
Hazards: Some of these occupations (eg playgroup leaders) entail unusually frequent exposure to childhood infections.

Occupation: 014 - CLERGY
Clergy, ministers of religion
Job description: Conducting religious worship; pastoral care in the community.
Social class: I
Hazards: None of note

Occupation: 015 - DOCTORS
Medical practitioners
Job description: Diagnosis, treatment and prevention of disease and injury.
Social class: I
Hazards: Frequent exposure to infections; toxic chemicals such as anaesthetic gases, cytotoxic drugs, formaldehyde and glutaraldehyde; ionising radiation (especially in radiologists and radiotherapists).

Occupation: 016 - DENTISTS
Dental practitioners
Job description: Diagnosis, treatment and prevention of oral and dental injuries and disease.
Social class: I
Hazards: Infections (especially blood-borne), ionising radiation, anaesthetic gases, mercury.

Occupation: 017 - NURSES
Nurse administrators, nurses
Job description: Provision and management of nursing and midwifery services in hospitals and in the community.
Social class: II
Hazards: Infections; physical injury when lifting and moving patients; toxic chemicals such as anaesthetic gases, cytotoxic drugs and glutaraldehyde; ionising radiation.

Occupation: 018 - PHARMACISTS
Pharmacists
Job description: Compounding and dispensing medicines; management of retail chemist's shops; some pharmacists work in the development of new drugs.
Social class: I
Hazards: Various toxic chemicals including cytotoxic drugs and organic solvents.

Occupation: 019 - MEDICAL RADIOGRAPHERS
Medical radiographers
Job description: Operation of x-ray and similar monitoring equipment under the direction of a radiologist or other medical practitioner.
Social class: II
Hazards: Ionising radiation; photographic processing chemicals.

Occupation: 020 - PHYSIOTHERAPISTS
Physiotherapists
Job description: Planning and provision of massage, exercise and hydro- or electrotherapy in the treatment of injuries, diseases and disabilities.
Social class: II
Hazards: Physical injury from lifting and moving patients; infections.

Occupation: 021 - HEALTH PROFESSIONS NEC
Ophthalmic and dispensing opticians, chiropodists, therapists nec, medical technicians, dental auxiliaries.
Job description: Various activities including testing of eyesight and supplying and fitting of spectacles and contact lenses; diagnosis and treatment of disorders of the feet; provision of occupational and speech therapy; conduct of medical investigations such as cardiography and encephalography; cleaning teeth and assisting dentists.
Social class: II
Hazards: Several of these occupations entail exposure to infectious hazards. Dental auxiliaries may be exposed to mercury.

Occupation: 022 - VETERINARIANS
Veterinarians
Job description: Diagnosis, treatment and prevention of diseases in animals.
Social class: I
Hazards: Physical injury by animals; zoonotic infections; ionising radiation; toxic pharmaceuticals.

Occupation: 023 - DRIVING INSTRUCTORS
Driving instructors (not HGV)
Job description: Teaching people to drive cars, motor cycles and light commercial vehicles.
Social class: II/IIIN
Hazards: Physical injury from road traffic accidents; traffic fumes.

Appendix 3 Description of job groups of the Southampton classification

Occupation: 024 - LITERARY AND ARTISTIC OCCUPATIONS
Authors, writers, journalists, artists, designers, window dressers, actors, musicians, entertainers, stage managers, photographers, camera operators, sound and vision equipment operators.
Job description: A range of activities including writing books and articles for newspapers; art and design work; design and display of clothes; provision of theatrical entertainment; musical performance to live audiences and on record; operation of still, cine and television cameras; and sound recording.
Social class: II/IIIN
Hazards: Several of these occupations may entail exposure to hazardous materials (eg organic solvents in painters, metal fume in sculptors, darkroom chemicals in photographers). Performance on some musical instruments has been associated with musculo-skeletal disorders of the upper limb, but these would not be expected to affect mortality.

Occupation: 025 - PERSONS INVOLVED IN SPORT
Professional athletes, sports officials
Job description: Controlling and participating in sports events; training other sportspeople.
Social class: II/IIIN
Hazards: Physical injury from accidents while training and competing.

Occupation: 026 - BIOLOGICAL SCIENTISTS
Biological scientists, biochemists
Job description: Research and development in the study of living organisms.
Social class: I
Hazards: These occupations entail exposure to a wide range of chemical, physical and biological hazards, but there is no major hazard to which a large majority of biological scientists are exposed.

Occupation: 027 - CHEMICAL ENGINEERS AND SCIENTISTS
Chemical scientists, chemical engineers
Job description: Research into chemical structure and reactions; development, operation and control of commercial, chemical and manufacturing processes.
Social class: I
Hazards: These occupations entail exposure to a wide range of chemicals although there is no major hazard to which the large majority are exposed. Exposure to asbestos may have occurred in chemical engineers supervising the construction, maintenance or demolition of chemical plant, especially in the past.

Occupation: 028 - PHYSICAL SCIENTISTS AND MATHEMATICIANS
Physical and geological scientists, mathematicians, metallurgists
Job description: Research and development in the fields of mathematics, physics (including the physics of living organisms), geology and metallurgy.
Social class: I
Hazards: Some of these occupations entail exposure to ionising and non-ionising radiation. In addition, many different hazardous chemicals may be handled. However, no major chemical hazard applies to the majority of people from the group.

Occupation: 029 - ELECTRICAL AND ELECTRONIC ENGINEERS (PROFESSIONAL)
Electrical engineers, electronic engineers
Job description: Production, design repair and maintenance of electrical and electronic equipment by persons qualified to university degree level.
Social class: I
Hazards: Solvents; solder fumes, (mainly in electronic engineers). "Electrical" occupations have been linked in some studies with high rates of leukaemia and brain cancer, possibly related to magnetic fields. However, this is not a firmly established hazard.

Occupation: 030 - PROFESSIONAL ENGINEERS NEC
Civil, structural, municipal, mining and quarrying engineers, mechanical and aeronautical engineers, production engineers, planning and quality control engineers, engineers nec, technologists nec.
Job description: Planning, organising and supervising the construction and maintenance of roads, bridges, mines, quarries, engines, machines, ships; planning schedules and work processes in production; planning, directing and undertaking research and development in the production of textiles, rubber and other products.
Social class: I
Hazards: Various chemical and physical hazards apply to specific occupations within the group, but there is no major hazard shared by the majority.

Occupation: 031 - DRAUGHTSPERSONS
Draughtspersons
Job description: Preparation of technical drawings, maps, plans, charts and technical illustrations.
Social class: II/IIIN
Hazards: None of note

Occupation: 032 - LABORATORY TECHNICIANS
Laboratory and engineering technicians, technician engineers, supervisor - laboratory assistants, laboratory assistants.
Job description: Carrying out routine laboratory tests and performing technical support functions according to established procedures; assisting scientists and engineers with research, development, testing and analysis.
Social class: II/IIIM
Hazards: These occupations entail exposure to many different chemical, physical and biological hazards, but no major hazard affects the majority of workers from the group.

Occupation: 033 - ARCHITECTS AND SURVEYORS
Architects, town planners, quantity, building and land surveyors.
Job description: Design and planning of buildings and building schemes; provision of financial and contractual advice in relation to building projects; surveying of buildings, land and mines to provide data for valuations, map making, etc.
Social class: I/II
Hazards: There is some possibility of exposure to asbestos in the surveying of old buildings.

Occupation: 034 - AIRCRAFT FLIGHT DECK OFFICERS
Aircraft flight deck officers
Job description: Flying and navigating aircraft; servicing aircraft in flight.
Social class: II
Hazards: Injury from air traffic accidents. Crew members of aircraft flying at high altitude have high exposure to cosmic radiation which in theory might pose a cancer hazard, particularly from leukaemia.

Appendix 3 Description of job groups of the Southampton classification

Occupation: 035 - AIR TRAFFIC CONTROLLERS
Air traffic planners and controllers
Job description: Preparation of flight plans, authorisation of flight departures and arrivals, maintenance of radio/radar/visual contact with aircraft to ensure the safe movement of air traffic.
Social class: II
Hazards: This is a stressful occupation demanding high levels of concentration.

Occupation: 036 - SEAFARERS
Deck, engineering and radio officers and pilots - ship; supervisors - ships, lighters and other vessels; deck, engine-room hands.
Job description: Commanding, navigating, piloting and working on seagoing ships, and on craft operating in coastal and inland waters (including deck and engine room duties).
Social class: II/IIIM/IV
Hazards: Injury from accidents and especially accidental drowning.

Occupation: 037 - TECHNICIANS NEC
Professional and related in science, engineering and other technologies and similar fields nec.
Job description: A range of activities including provision of technical assistance to architects, town planners, and civil and building engineers
Social class: II
Hazards: None of note

Occupation: 038 - PRODUCTION AND MAINTENANCE MANAGERS
Production, works and maintenance managers, works supervisors
Job description: Managing the work and resources necessary for industrial production and processing, and for maintenance of engineering items, equipment and machinery.
Social class: II/IIIN
Hazards: There are no major hazards in this group overall, although individual workers may be exposed to risks relating specifically to the industry in which they are employed.

Occupation: 039 - MANAGERS IN CONSTRUCTION
Site and other managers, agents and clerks of works, general supervisors (building and civil engineering)
Job description: Managing the construction and maintenance of civil and structural engineering work; checking for compliance with design specifications and standards in construction projects.
Social class: II/IIIN
Hazards: Asbestos exposure may occur in a number of these jobs.

Occupation: 040 - MANAGERS IN TRANSPORT, UTILITIES AND MINING
Managers in transport, warehousing, public utilities and mining
Job description: A range of managerial activities including management of mines, quarries, electricity, water and gas services, transport of passengers and freight, and storage of goods and materials.
Social class: II
Hazards: No dominant hazard applies across the majority of these occupations.

Occupation: 041 - OFFICE MANAGERS
Office managers
Job description: Managerial tasks in various offices and businesses.
Social class: II
Hazards: None of note

Occupation: 042 - BUTCHERS
Butchers (managers and proprietors); supervisors - butchers, meat cutters; butchers, meat cutters.
Job description: Slaughtering animals; preparing, processing, storing and selling meat.
Social class: II/IIIN/IIIM
Hazards: Meat handlers are known to have a high prevalence of cutaneous warts caused by the papillomavirus, HPV 7. It is possible that papillomavirus infection could underlie a high rate of lung cancer observed in several previous decennial supplements as well as in other studies. Other potential hazards include polycyclic aromatic hydrocarbons, nitrites, fumes from pyrolysis of plastics used to package meat, and injury by knives used to cut meat.

Occupation: 043 - FISHMONGERS AND POULTRY DRESSERS
Fishmongers (managers and proprietors); Supervisors-fishmongers, poultry dressers; Fishmongers, poultry dressers.
Job description: Cleaning cutting, preparing and selling fish and poultry carcasses.
Social class: II/IIIN/IIIM
Hazards: Poultry dressing has been associated with a high prevalence of musculoskeletal disorders of the upper limb, but these would be unlikely to affect mortality. Fish filleting can cause occupational asthma.

Occupation: 044 - RETAILERS AND DEALERS
Garage proprietors; other proprietors and managers (sales); supervisors of shop salespeople and assistants; shop salespeople and assistants; shelf fillers; market and street traders and assistants; scrap dealers, general dealers, rag and bone merchants; bookmakers, betting shop managers.
Job description: Buying and selling goods and services; management of sales operations.
Social class: II/IIIN/IIIM/IV
Hazards: A number of occupations within the group are associated with specific hazards - eg scrap dealers may be exposed to lead. Also, some of the jobs entail frequent manual handling with a possible risk of back injury. However, this would not be expected to affect mortality patterns.

Occupation: 045 - PUBLICANS AND BAR STAFF
Hotel and residential club managers; publicans; club stewards; supervisors of barpersons ; barpersons.
Job description: Owning and managing hotels, public houses and clubs. Selling and serving alcoholic and non-alcoholic drinks.
Social class: II/IIIN/IIIM/IV
Hazards: Most of these jobs are associated with easy access to alcoholic drinks. Work in bars may entail high levels of passive smoking.

Occupation: 046 - CATERERS
Restaurateurs; supervisors of waiters, waitresses; supervisors of counterhands, assistants; waiters, waitresses; counter hands, assistants.
Job description: Serving food and beverages at the table in restaurants and from counters in eating establishments; managing and supervising such activities; some restaurateurs cook and prepare food.
Social class: II/IIIM/IV
Hazards: Some of these occupations entail easy access to alcoholic drinks. Cooking fumes have been linked in some studies with an increased risk of lung cancer.

Appendix 3 Description of job groups of the Southampton classification

Occupation: 047 - FARMERS
Farmers, horticulturalists, farm managers; farm foremen; horticultural foremen; foremen gardeners and groundsmen; agricultural machinery foremen; other foremen in farming and related; farm workers; horticultural workers; gardeners, groundsmen: agricultural machinery drivers, operators; all others in farming and related.
Job description: Owning, managing and working on farms and in horticultural establishments. This may entail cultivation of arable crops and rearing of various animals. Most jobs are outdoor and physically demanding.
Social class: II/IIIM/IV
Hazards: Accidental injury from many causes including machinery, farm animals and poisoning by agrochemicals; allergic lung disorders; risk of musculo-skeletal disease from heavy manual handling; zoonotic infections such as leptospirosis; occupational stress may be a problem in the self-employed.

Occupation: 048 - ARMED FORCES
Officers, UK armed forces; officers, Foreign and Commonwealth Armed Forces; NCOs and other ranks, UK armed forces; NCOs and other ranks, Foreign and Commonwealth Armed Forces.
Job description: Full-time service in the Army, Air Force and Navy. Many members of the armed services have specific trades (eg pilots, engineers, drivers, cooks, clerks).
Social class: -
Hazards: Injury through military action and in training. A wide range of hazards related to specific trades. Service overseas may entail a risk of unusual infections.

Occupation: 049 - POLICE
Police officers (inspectors and above); police sergeants; police officers below sergeant.
Job description: Investigation and prevention of crime; enforcement of law and order (including traffic control).
Social class: II/IIIN
Hazards: Criminal injury; road traffic accidents.

Occupation: 050 - FIRE SERVICE PERSONNEL
Fire service officers; fire service supervisors.
Job description: Fighting fires, advising on fire hazards and fire prevention, salvaging goods during or after fires. Practising for fire fighting.
Social class: II/IIIN
Hazards: Injury by fire, toxic fumes and falls. Several studies have suggested increased rates of certain cancers in firefighters, but the findings have not been consistent.

Occupation: 051 - LAUNDERERS AND DRY CLEANERS
Managers of laundry and dry cleaning receiving shops; launderers, dry cleaners, pressers.
Job description: Washing, drying cleaning, ironing and pressing garments and other textile articles; serving customers in dry cleaning shops.
Social class: IIIN/IV
Hazards: Organic solvents used in dry cleaning fluids; acids and bleaches. Detergents can cause dermatitis.

Occupation: 052 - HAIRDRESSERS
Hairdressers' and barbers' managers and proprietors; hairdressing supervisors; hairdressers, barbers.
Job description: Cutting, styling, washing and dyeing hair.
Social class: II/IIIN/IIIM
Hazards: Shampoos, dyes and other products used to treat hair can cause dermatitis, but this would not be expected to affect mortality.

Occupation: 053 - OFFICE WORKERS AND CASHIERS
Supervisors of clerks, civil service executive officers, clerks, retail shop cashiers, check-out and cash and wrap operators; supervisors of typists, office machine operators, telephonists etc; secretaries, shorthand typists, receptionists; office machine operators; telephonists, radio and telegraph operators.
Job description: A wide range of clerical and administrative activities, mainly in offices.
Social class: II/IIIN/IV
Hazards: A number of these occupations entail repetitive movements of the hands and arms, and may carry an increased risk of upper limb musculo-skeletal disorders. However, these would not be expected to affect mortality.

Occupation: 054 - POSTAL WORKERS
Supervisors of postal workers, mail sorters, messengers; postal workers, mail sorters, messengers.
Job description: Collecting, sorting and delivering mail.
Social class: IIIM/IV/V
Hazards: Some occupations in this group entail frequent manual handling. (eg lifting and moving mail bags).

Occupation: 055 - PETROL PUMP ATTENDANTS
Supervisors of petrol pump, forecourt attendants; petrol pump, forecourt attendants.
Job description: Most workers today are employed mainly as cashiers, but they may be involved in filling vehicles with petrol or diesel fuel.
Social class: IIIN
Hazards: Petroleum and diesel liquid and vapour. Petroleum contains low levels of benzene which is an established cause of leukaemia.

Occupation: 056 - VAN SALES PERSONS
Supervisors of roundspersons, van salespersons; roundspersons, van salespersons.
Job description: Delivery and sale of food, drink and other goods, either door-to-door or from a mobile shop or van; Collection and delivery of laundry.
Social class: IIIN/IIIM
Hazards: Some of these jobs may entail heavy lifting, but not at a level that would not be expected to affect mortality.

Occupation: 057 - SALES REPRESENTATIVES
Importers, exporters, commodity brokers; credit agents, collector salesmen; sales representatives; sales representatives (property and services), other agents.
Job description: Buying and selling a wide range of goods and services. Some jobs require extensive travel.
Social class: II/IIIN/IV
Hazards: Transport-related accidents may be a hazard in some occupations in the group.

Occupation: 058 - SECURITY WORKERS
Prison officers (chief officers and above); prison service principal officers; supervisors of security guards and officers, patrolmen, watchmen; supervisors of traffic wardens; supervisors of security and protective service workers nec; prison officers (below principal officer); other security and protective service workers.
Job description: Guarding inmates of prisons and detention centres; carrying out investigations or giving protection to persons, premises and property; assisting the police in traffic control.
Social class: II/IIIN/IV
Hazards: Criminal injury.

Appendix 3 Description of job groups of the Southampton classification

Occupation:	059 - COOKS AND KITCHEN PORTERS Supervisors of chefs, cooks; chefs, cooks; kitchen porters, hands.
Job description:	Preparing and cooking food; cleaning; fetching and carrying in kitchens.
Social class:	II/IIIM/V
Hazards:	Some studies have linked cooking fumes with increased risk of lung cancer. Employees in restaurants and hotels may have easy access to alcoholic drinks.

Occupation:	060 - OTHER SERVICE PERSONNEL Proprietors and managers, service flats, holiday flats, caravan sites; housekeepers (non-domestic); supervisors of other domestic and school helpers; supervisors of travel stewards and attendants; supervisors of hotel porters; domestic staff and school helpers; travel stewards and attendants; hotel porters; supervisors of caretakers; supervisors of cleaners, window cleaners, chimney sweeps, road sweepers; supervisors of lift and car park attendants; caretakers, road sweepers and other cleaners; lift and car park attendants.
Job description:	Ownership, management, care and maintenance of buildings and caravan sites; care and guidance of tourists and hotel guests; caring for children in nurseries and assisting with their primary education, outside cleaning tasks such as window cleaning, road sweeping and washing vehicles; attending in lifts and controlling car parks.
Social class:	II/IIIN/IIIM/IV/V
Hazards:	Window cleaners are at increased risk of accidental falls at work. Chimney sweeps are exposed to polycyclic aromatic hydrocarbons in soot. People working in nurseries have high exposure to childhood infections.

Occupation:	061 - HOSPITAL PORTERS AND WARD ORDERLIES Supervisors of hospital porters; supervisors of hospital, ward orderlies, hospital porters; hospital, ward orderlies.
Job description:	Transport of patients, specimens and equipment in hospitals; assisting in the nursing care of patients; helping with cleaning and serving of food in hospitals and nursing homes.
Social class:	IIIM/IV
Hazards:	Exposure to infection; manual handling.

Occupation:	062 - AMBULANCE WORKERS Supervisors of ambulance workers; ambulance workers.
Job description:	Transporting sick, injured and convalescent persons by ambulance; giving first aid treatment.
Social class:	IIIM
Hazards:	Exposure to infections; manual handling.

Occupation:	063 - RAILWAY STATION WORKERS Supervisors of railway station staff, railway station staff; other supervisors rail transport.
Job description:	Performance of various duties at railway stations including ticket collecting and portering.
Social class:	IIIM/V
Hazards:	People working near electrified track may have high exposure to magnetic fields - a hypothesised cause of leukaemia and brain cancer.

Occupation:	064 - UNDERTAKERS Undertakers
Job description:	Making funeral arrangements, supervising and helping to conduct funerals; transporting dead bodies; embalming.
Social class:	II/IIIM
Hazards:	Formaldehyde is used in embalming but long-term exposures are likely to be low.

Occupation:	065 - FORESTERS Forestry supervisors; forestry workers
Job description:	Planting, cultivation and felling of trees.
Social class:	IIIM/IV
Hazards:	Accidental injury associated with tree felling and transport of logs; pesticide poisoning; manual handling; hand-arm vibration syndrome from use of chain saws.

Occupation:	066 - FISHING AND RELATED WORKERS Supervisors, mates - fishing; fishermen
Job description:	Catching fish, mainly by net, in inshore and offshore waters.
Social class:	IIIM/IV
Hazards:	Injury from accidents, especially at sea. High exposure to ultraviolet radiation in sunlight has been associated with increased rates of skin cancer in fishermen.

Occupation:	067 - TANNERY WORKERS Supervisors - tannery production workers; tannery production workers
Job description:	Treating hides, skins and pelts in preparation for the manufacture of leather, skin and fur products.
Social class:	II/IIIM
Hazards:	Various toxic chemicals including hexavalent chromium, chlorophenols and formaldehyde.

Occupation:	068 - LEATHER AND SHOE WORKERS Supervisors - shoe repairers; supervisors - leather cutters and sewers, footwear lasters, makers, finishers; supervisors - other making and repairing, leather; shoe repairers; leather cutters and sewers, footwear lasters, makers, finishers; other making and repairing leather.
Job description:	Repair of worn and damaged footwear manufacture of shoes; Manufacture of leather and artificial leather goods.
Social class:	II/IIIM
Hazards:	Various organic solvents (including benzene and n-hexane in the past). Dust from vegetable tanned leather used in the manufacture of some shoes is a cause of nasal cancer.

Occupation:	069 - PREPARATORY FIBRE PROCESSORS Supervisors - preparatory fibre processors; preparatory fibre processors.
Job description:	Preparation of natural, synthetic and reclaimed fibres for spinning into yarn and making into non-woven fabric (includes binding, combing, carding, hacking, drawing, cutting, flagging and severing).
Social class:	IIIM/IV
Hazards:	Injury from machinery; cotton and certain other natural textile dusts cause byssinosis, some textile workers may be exposed to asbestos.

Occupation:	070 - SPINNERS AND WINDERS Supervisors - spinners, doublers, twisters; supervisors - winders, reelers; spinners, doublers, twister; winders, reelers.
Job description:	Operating textile machinery to spin, double and twist fibres into thread or yarn and wind this from one package to another package, card or spool.
Social class:	IIIM/IV
Hazards:	Injury from machinery; cotton and certain other natural textile dusts cause byssinosis; some textile workers may be exposed to asbestos.

Appendix 3 Description of job groups of the Southampton classification

Occupation: 071 - WARP PREPARERS AND WEAVERS
Supervisors - warp preparers; supervisors - weavers; warp preparers; weavers
Job description: Winding warp yarn onto packages and beams ready for dyeing, knitting or weaving; weaving natural and synthetic fibres into fabrics and carpets.
Social class: II/IIIM
Hazards: Injury by machinery; cotton and certain other natural textile dusts cause byssinosis, but dust levels are low in these occupations as compared with fibre processors; some textile workers may be exposed to asbestos; noise is a particular hazard in weavers.

Occupation: 072 - KNITTERS
Supervisors - knitters; knitters
Job description: Setting up, operating and attending to machines to knit yarn into fabric and articles made from fabric.
Social class: II/IIIM
Hazards: Injury from machinery; cotton and certain other natural textile dusts cause byssinosis, but dust levels tend to be lower in knitters than in preparatory fibre processors.

Occupation: 073 - BLEACHERS, DYERS AND FINISHERS
Supervisors - bleachers, dyers, finishers; bleachers, dyers, finishers
Job description: Bleaching, dyeing and otherwise treating textile fabrics, yarn, cloth and goods.
Social class: II/IIIM
Hazards: Injury by scalding; various irritant and corrosive substances; reactive dyes cause occupational asthma. Manufacture of certain dyestuffs used in the past has been associated with increased rates of bladder cancer, but it is not clear whether this hazard extended to the use of these dyes.

Occupation: 074 - OTHER TEXTILE WORKERS
Supervisors - menders, darners; supervisors - other material processing, textiles; menders, darners; other material processing - textiles; supervisors, inspectors, viewers, examiners - textiles; supervisors - labourers and unskilled workers nec - textiles (not textile goods); labourers and unskilled workers nec - textiles (not textile goods).
Job description: Various tasks including inspecting and testing textiles for defects arising during manufacture; repairing such defects; washing, rinsing, drying and other treatment of fibres to remove impurities; labouring and unskilled work in the textile industry.
Social class: II/IIIM/IV/V
Hazards: Injury by machinery may be a hazard in some of these occupations. Cotton and certain other natural fibre dusts cause byssinosis.

Occupation: 075 - CHEMICAL WORKERS
Supervisors - chemical processing; chemical, gas and petroleum process plant operators; supervisors - labourers and unskilled workers nec chemicals and allied trades; labourers and unskilled workers nec - chemicals and allied trades.
Job description: Operation of plant to process chemicals by crushing, milling, mixing and separating, and by chemical heat and other treatment.
Social class: IIIM/IV/V
Hazards: Exposures to many chemical hazards may occur but these are specific to the processes being carried out. Asbestos exposure has been quite common in chemical works.

Occupation: 076 - BAKERS
Supervisors - bakers, flour confectioners; bakers, flour confectioners
Job description: Baking bread, flour and confectionery and finishing confectionery products by hand.
Social class: II/IIIM
Hazards: Exposure to flour dust is a cause of occupational asthma.

Occupation: 077 - BREWERY WORKERS
Supervisors - brewery and vinery process workers; brewery and vinery process workers
Job description: Crushing, mixing, malting, cooking and fermenting grains and fruits to produce beer, wine, malt liquors, vinegar, yeast and related products.
Social class: II/IIIM
Hazards: These occupations often entail easy access to alcoholic drinks. Some studies have suggested high rates of rectal cancer in brewers.

Occupation: 078 - FOOD PROCESSORS
Supervisors - other material processing - bakery and confectionery workers; supervisors - other material processing - food and drink nec; other material processing - bakery and confectionery workers; supervisors inspectors, viewers, examiners - food; inspectors viewers, examiners - food.
Job description: A variety of food and drink processing operations including curing, freezing and cooling foodstuffs; making glucose, sugar and chocolate, processing milk, cream and other dairy products.
Social class: IIIM/IV
Hazards: No major hazard affects a majority of people from these occupations.

Occupation: 079 - PAPER MANUFACTURERS
Supervisors - paper, paperboard and leatherboard workers; paper, paperboard and leatherboard workers
Job description: Operation of machinery to beat and mix fluid pulps, and to dry, finish, impart a glaze and wind/rewind in the production of paper, paperboard and leatherboard.
Social class: IIIM
Hazards: Injury by machinery; various toxic chemicals including sulphur dioxide, chlorine, formaldehyde, chlorophenols and dioxins.

Occupation: 080 - BOOKBINDERS
Supervisors - bookbinders and finishers; bookbinders and finishers.
Job description: Binding and finishing printed products by hand or machine.
Social class: II/IIIM
Hazards: Adhesives and solvents.

Occupation: 081 - PAPER CUTTERS
Supervisors - cutting and slitting machine operators (paper and paper products making); cutting and slitting machine operators (paper and paper products making).
Job description: Operation of machines to cut paper and paper patterns, paperboard, abrasive cloth, photographic material and other similar material.
Social class: IIIM
Hazards: Injury by machinery.

Appendix 3 Description of job groups of the Southampton classification

Occupation:	082 - GLASS AND CERAMICS FURNACE WORKERS Supervisors - glass and ceramics furnace workers, kilnsetters; glass and ceramics furnace workers, kilnsetters.
Job description:	Operation of furnaces and kilns in the production of glass and ceramics; positioning articles ready for firing in kilns.
Social class:	IIIM
Hazards:	Exposure to heat and inorganic dusts (in the past these dusts included crystalline silica).

Occupation:	083 - GLASS FORMERS AND DECORATORS Supervisors - glass formers and shapers, finishers, decorators; glass formers and shapers, finishers, decorators.
Job description:	Forming glass products by blowing, moulding and pressing; finishing glassware by removing surplus glass, grinding and polishing; decorating glassware by cutting, acid etching and sand blasting.
Social class:	II/IIIM
Hazards:	Infra-red radiation causes heat cataract; exposure to various irritant and corrosive materials may occur.

Occupation:	084 - CERAMICS CASTERS Supervisors - casters and other pottery makers; casters and other pottery makers.
Job description:	Forming ceramic ware (including porcelain, pottery, stoneware and refractory goods) by casting, extruding, moulding, pressing and shaping.
Social class:	II/IIIM
Hazards:	Clay dust which contains crystalline silica.

Occupation:	085 - RUBBER MANUFACTURERS Supervisors - rubber process workers, moulding machine operators, tyre builders; rubber process workers, moulding machine operators, tyre builders.
Job description:	Operation of masticating, calendering, mixing, forming, shaping, moulding, extruding, cutting, trimming and winding machines to make and repair rubber goods.
Social class:	II/IIIM
Hazards:	Rubber fumes; organic solvents; carbon black and numerous other chemical additives. In the past 2-naphthylamine, a potent bladder carcinogen, was used as an antioxidant, but this application was withdrawn in the 1950s.

Occupation:	086 - PLASTICS WORKERS Supervisors - calender and extruding machine operators, moulders (plastics); calender and extruding machine operators, moulders (plastics).
Job description:	Operation of moulding, extruding, thermoforming, calendering, covering, cutting and other machines in the processing of plastics.
Social class:	IIIM/IV
Hazards:	Various fumes from heat-degraded plastics.

Occupation:	087 - MAN-MADE FIBRE MAKERS Supervisors - man-made fibre makers; man-made fibre makers.
Job description:	Operation of plant in which a fibre-forming melt or mix is extruded into air or liquid and wound into packages of continuous filament or chopped into short (staple) lengths.
Social class:	IIIM/IV
Hazards:	Use of carbon-disulphide in the manufacture of rayon has been associated with ischaemic heart disease and neuropsychiatric disorders. Some studies have linked high exposure to rock wool with lung cancer.

Occupation:	088 - OTHER COAL MINERS Supervisors - washers, screeners and crushers in mines and quarries; washers, screeners and crushers in mines and quarries; supervisors - labourers and unskilled workers nec - coal mines; labourers and unskilled workers nec - coal mines.
Job description:	Operation of mechanical crushing, grinding, milling, washing and sieving of coal, stone and other ores; unskilled labourers in coal mines. Many workers are or have been employed below ground in coal or other mines.
Social class:	IIIM/IV
Hazards:	Coal dust, silicaceous dust; injury by machinery and other accidents.

Occupation:	089 - TOBACCO WORKERS Other material processing - tobacco
Job description:	Processing tobacco leaf to make cigarettes, cigars and pipe tobacco.
Social class:	IIIM/IV
Hazards:	Tobacco dust; tobacco workers have easy access to tobacco products.

Occupation:	090 - OTHER WOOD AND PAPER PROCESSORS Supervisors - other material processing, wood and paper; other material processing - wood and paper; supervisors assemblers - paper production, processing and printing; supervisors - inspectors, sorters in paper production, processing and printing; inspectors, sorters in paper production, processing and printing; assemblers in paper production, processing and printing.
Job description:	Various activities including seasoning wood; operating wood mills; assembling and printing paper products.
Social class:	IIIM
Hazards:	No major hazards affect the majority of workers in these occupations.

Occupation:	091 - OTHER OCCUPATIONS - GLASS AND CERAMICS Supervisors - other making and repairing - glass and ceramics; other making and repairing - glass and ceramics; supervisors-labourers and unskilled workers nec - glass and ceramics; labourers and unskilled workers nec - glass and ceramics.
Job description:	Various activities including manufacture of bricks and crucibles; casting plaster; unskilled labouring in the production of glass and ceramics.
Social class:	IIIM/V
Hazards:	Some of these jobs entail exposure to silicaceous dusts.

Occupation:	092 - RUBBER GOODS MAKERS Supervisors - other making and repairing - rubber; other making and repairing - rubber; supervisors inspectors, viewers, examiners - rubber goods; inspectors, viewers, examiners - rubber goods.
Job description:	Making and repairing articles of rubber; sheathing cables and wires with rubber; lining and covering plant and articles with rubber; repairing and restoring tyres; examining and inspecting rubber goods for defects in manufacture or processing.
Social class:	II/IIIM
Hazards:	Organic solvents; various other chemicals. In the past the bladder carcinogen, 2-naphthylamine, was used as an antioxidant in rubber manufacture.

Occupation: 093 - PLASTIC GOODS MAKERS
Supervisors - other making and repairing - plastics; other making and repairing - plastics; supervisors assemblers - plastic goods; supervisors inspectors, viewers, examiners - plastic goods; inspectors, viewers, examiners - plastic goods; assemblers (plastic goods).
Job description: Making and repairing articles of plastic material including glass reinforced plastics; sheathing cables and wires with plastic; inspecting for defects in the manufacture of plastic goods or materials; routine assembly of plastic goods and components.
Social class: IIIM/IV
Hazards: Styrene, used as a solvent in the manufacture of glass-reinforced plastics, has been linked in some studies with increased rates of lymphoma, but this finding has not been consistent; other organic solvents.

Occupation: 094 - COMPOSITORS
Supervisors - compositors; compositors
Job description: Setting up printing type by hand or machine; laying out pages for printing.
Social class: II/IIIM
Hazards: Lead fume and dust; organic solvents.

Occupation: 095 - PRINTING PLATE PREPARERS
Supervisors - electrotypers, stereotypers, printing plate and cylinder preparers; electrotypers, stereotypers, printing plate and cylinder preparers.
Job description: Preparation of printing plates and cylinders.
Social class: II/IIIM
Hazards: Various corrosive chemicals and organic solvents.

Occupation: 096 - PRINTING MACHINE MINDERS
Supervisors - printing machine minders and assistants; printing machine minders and assistants.
Job description: Setting and operating letterpress, lithographic, photogravure and web-offset printing machines.
Social class: IIIM
Hazards: Mineral oil mists; organic solvents.

Occupation: 097 - PRINTERS (SO DESCRIBED)
Supervisors - screen and block printers; supervisors - printers (so described); screen and block printers; printers (so described).
Job description: Setting and operating screen printing machines to print letters and designs on metal, glass, plastic, paper and other materials; operation of printing presses, preparation of printing plates and composition and assembly of type and printing block.
Social class: II/IIIM
Hazards: Organic solvents; mineral oil mists.

Occupation: 098 - TAILORS AND DRESSMAKERS
Supervisors - tailors, tailoresses, dressmakers; tailors, tailoresses, dressmakers.
Job description: Making, fitting and altering tailored garments, dresses and other articles of light clothing.
Social class: II/IIIM
Hazards: None of note

Occupation: 099 - CLOTHING CUTTERS
Supervisors - clothing cutters, milliners, furriers; clothing cutters, milliners, furriers.
Job description: Cutting out, shaping and trimming material (including fur) for the manufacture of garments.
Social class: II/IIIM
Hazards: None of note

Occupation: 100 - SEWERS AND EMBROIDERERS
Supervisors - sewers, embroiderers; sewers, embroiderers
Job description: Sewing and embroidering, by hand or machine, garments and other products made from textile fabric, fur and leather.
Social class: IIIM/IV
Hazards: None of note

Occupation: 101 - UPHOLSTERERS
Supervisors - coach trimmers, upholsterers, mattress makers; coach trimmers, upholsterers, mattress makers
Job description: Upholstering furniture and vehicle seating; making mattresses.
Social class: II/IIIM
Hazards: None of note

Occupation: 102 - CARPET FITTERS
Supervisors - carpet fitters; carpet fitters
Job description: Laying carpets at customers' premises.
Social class: IIIM
Hazards: Pre-patellar bursitis is caused by prolonged kneeling and possibly use of knee kickers, but would not be expected to affect mortality patterns.

Occupation: 103 - OTHER WORKERS WITH FABRICS
Supervisors - other making and preparing, clothing and related products; other making and repairing - clothing and related products.
Job description: Making patterns for the manufacture of garments and upholstery; marking and cutting out parts for articles other than garments; making canvas goods and helping to make fabric bags and sacks.
Social class: II/IIIM
Hazards: None of note

Occupation: 104 - CARPENTERS
Supervisors - carpenters, joiners; carpenters and joiners
Job description: Construction, erection, installation and repair of wooden structures and fittings used in internal and external frameworks.
Social class: II/IIIM
Hazards: Injury by machinery and by falls; hardwood dust (and probably also softwood dust) causes nasal cancer; dust from certain woods (eg cedar) causes asthma; some carpenters work with asbestos board; various adhesives, surface coatings and solvents.

Occupation: 105 - CABINET MAKERS
Supervisors - cabinet makers; cabinet makers
Job description: Making and repairing wooden furniture and piano and cabinet cases.
Social class: II/IIIM
Hazards: Exposure to hardwood dust (and probably also to softwood dust) causes nasal cancer; certain wood dusts (eg cedar) cause asthma; injury by machinery; noise from machinery; various glues, surface coatings and solvents.

Occupation: 106 - CASE AND BOX MAKERS
Supervisors - case and box makers; case and box makers
Job description: Cutting and assembling wood to make packing cases, crates or similar articles.
Social class: II/IIIM
Hazards: Dust exposure is generally less than in cabinet makers and carpenters.

Appendix 3 Description of job groups of the Southampton classification

Occupation: 107 - PATTERN MAKERS
Supervisors - pattern makers (moulds); pattern makers (moulds)
Job description: Making patterns from wood, metal, plaster and plastic to serve as a template for moulds used to cast metal.
Social class: II/IIIM
Hazards: None of note

Occupation: 108 - WOODWORKING MACHINISTS
Supervisors - sawyers, veneer cutters, woodworking machinists; sawyers, veneer cutters, woodworking machinists
Job description: Setting and/or operating wood-cutting and woodworking machinery.
Social class: II/IIIM
Hazards: Injury by machinery; noise from machinery; hardwood dust (and probably softwood dust) causes nasal cancer; some wood dusts (eg cedar) cause asthma.

Occupation: 109 - OTHER WOODWORKERS
Supervisors - the making and repairing, wood; labourers and mates to woodworking craftsmen; other making and repairing - wood; foremen inspectors, viewers, examiners - woodwork; inspectors, viewers, examiners - woodwork.
Job description: Carving and bending wood; laying hardwood strip and woodblock floors; matching, joining and repairing veneers; making other wooden products such as ladders and picture frames; assisting woodworking craftsmen; inspecting and examining wood or wood products for processing or manufacturing defects.
Social class: II/IIIM
Hazards: Some of these occupations entail exposure to wood dust -hardwood dust (and probably also softwood dust) causes nasal cancer, and some wood dusts (eg cedar) cause asthma; exposure to various glues, surface coatings and solvents may also occur in some of these jobs.

Occupation: 110 - DENTAL TECHNICIANS
Dental technicians
Job description: Making and repairing dentures according to individual requirements.
Social class: II/IIIM
Hazards: Solvents; methyl methacrylate; metal dust. Some studies have suggested high rates of lung cancer in dental technicians.

Occupation: 111 - OTHER MAKERS OF PAPER GOODS
Supervisors - other making and repairing, paper goods and printing; other making and repairing - paper goods and printing.
Job description: Various activities including setting and operating die stamping and embossing machines; processing, printing and finishing photographic films; printing textiles and wall paper; operating offset duplicating machines; operating paper pattern cutting machines.
Social class: II/IIIM
Hazards: Exposures vary according to specific jobs within the group. For example, photographic processors are exposed to a range of dark room chemicals and wallpaper printers are exposed to organic solvents.

Occupation: 112 - FURNACE OPERATIVES (METAL)
Supervisors - furnace operating occupations (metal); furnace operating occupations (metal).
Job description: Operation of furnaces to smelt ores, refine metal, melt metal for casting and reheat metal and metal articles for further working.
Social class: IIIM
Hazards: Metal fume; polycyclic aromatic hydrocarbons (especially in aluminium smelters); formaldehyde; phenols; isocyanates; heat.

Occupation: 113 - ROLLERS (METAL)
Supervisors - rollers (metal); rollers (metal)
Job description: Operation of hot and cold rolling mills to roll slabs, bloom, billets, bars, rod, strip, sheet, plates, sections or tubes of metal.
Social class: II/IIIM
Hazards: Oil mist and fumes; heat.

Occupation: 114 - SMITHS AND FORGE WORKERS
Supervisors - smiths, forge workers: smiths, forge workers
Job description: Shaping heated metal by hand hammering on an anvil or by power hammer or forging press.
Social class: II/IIIM
Hazards: Polycyclic aromatic hydrocarbons; oil mists noise; heat.

Occupation: 115 - METAL DRAWERS
Supervisors - metal drawers; metal drawers
Job description: Operation of machines to draw metal tubes, bars, rods and wire through dies to reduce their diameter.
Social class: II/IIIM
Hazards: Mineral oils.

Occupation: 116 - MOULDERS AND COREMAKERS (METAL)
Supervisors - moulders, coremakers, die casters; moulders, coremakers, die casters
Job description: Making sand, loam and plaster moulds and cores, by hand or machine, for use in casting molten metal. Pouring molten metal into dies by hand or machine.
Social class: IIIM
Hazards: Metal fume; formaldehyde; silicaceous dust.

Occupation: 117 - ELECTROPLATERS
Supervisors - electroplaters; electroplaters
Job description: Coating metal articles or parts electrolytically, in tanks or vats, with chromium, copper, nickel or other non-ferrous metal.
Social class: II/IIIM
Hazards: Chromic and other acid mists are known to cause dermatitis and perforation of the nasal septum in these occupations. In addition a number of studies have suggested high rates of lung cancer.

Occupation: 118 - ANNEALERS, HARDENERS, TEMPERERS (METAL)
Supervisors - annealers, hardeners, temperers (metal); annealers, hardeners, temperers (metal).
Job description: Operation of equipment to heat, cool and quench metal and metal articles in order to harden, reduce brittleness or stress and restore ductility.
Social class: IIIM/IV
Hazards: None of note

Occupation: 119 - GALVANISERS AND TIN PLATERS
Supervisors - galvanisers, tin platers, dip platers; galvanisers, tin platers, dip platers.
Job description: Coating metal articles, sheet, wire etc with another metal.
Social class: IIIM
Hazards: Metal fume.

Appendix 3 Description of job groups of the Southampton classification

Occupation: 120 - OTHER METAL MANUFACTURERS
Supervisors - metal making and treating workers nec; metal making and treating workers nec; supervisors - labourers and unskilled workers nec - foundries in engineering and allied trades; labourers and unskilled workers nec - foundries in engineering and allied trades.
Job description: Various activities in the making and treating of metal including assisting in rolling mills and forges and ladling molten metal.
Social class: IIIM/V
Hazards: Metal fume; silicaceous dust; polycyclic aromatic hydrocarbons; heat; noise.

Occupation: 121 - PRESS AND MACHINE TOOL SETTERS
Supervisors - press and machine tool setters; press and machine tool setters.
Job description: Setting up press and machine tools and metal working machines.
Social class: II/IIIM
Hazards: Cutting oils and synthetic cutting fluids can cause dermatitis and asthma; noise; injury by machinery and metal shavings.

Occupation: 122 - CENTRE LATHE TURNERS
Supervisors - centre lathe turners; centre lathe turners
Job description: Setting up and operating centre lathes to shape metal workpieces.
Social class: II/IIIM
Hazards: Cutting oils and synthetic cutting fluids can cause dermatitis and asthma; injury by machinery. Some studies have suggested high rates of bladder cancer in turners, but the finding has not been consistent.

Occupation: 123 - MACHINE TOOL SETTER OPERATORS
Supervisors - machine tool setter operations; machine tool setter operators.
Job description: Setting up and operating boring, drilling, grinding and milling machines.
Social class: II/IIIM
Hazards: Cutting fluids; injury by machinery and metal shavings.

Occupation: 124 - MACHINE TOOL OPERATORS
Supervisors - machine tool operators; machine tool operators
Job description: Operation of previously set up machine tools to cut, shape and otherwise machine metal workpieces.
Social class: IIIM/IV
Hazards: Cutting oils and synthetic cutting fluids can cause dermatitis and asthma; injury by machinery and metal shavings.

Occupation: 125 - PRESS AND AUTOMATIC MACHINE OPERATORS
Supervisors - press, stamping and automatic machine operators; press, stamping and automatic machine operators.
Job description: Tending pre-set automatic machines, operating presses and drop hammers equipped with formers and dies to shape metal articles and form hollows and relief in metal.
Social class: IIIM/IV
Hazards: Noise; injury by machinery.

Occupation: 126 - METAL POLISHERS
Supervisors - metal polishers; metal polishers
Job description: Manipulation of abrasive materials, polishing heads and other tools to clean and polish metal articles and parts.
Social class: II/IIIM
Hazards: Metal dust and dust from abrasive materials.

Occupation: 127 - FETTLERS AND DRESSERS (METAL)
Supervisors - fettlers, dressers; fettlers, dressers
Job description: Removing surplus metal and rough surface from castings, forgings or components with hand and power tools.
Social class: IIIM/IV
Hazards: Silicaceous dust; noise.

Occupation: 128 - SHOT BLASTERS
Supervisors - shot blasters; shot blasters.
Job description: Cleaning and smoothing metal parts and articles using a jet of vapour or compressed air and abrasive material.
Social class: IIIM/IV
Hazards: Silicaceous dust.

Occupation: 129 - TOOLMAKERS
Supervisors - toolmakers, tool fitters, markers out; toolmakers, tool fitters markers out.
Job description: Making, maintaining and repairing tools, dies, gauges, jigs and fixtures; marking out metal for machining.
Social class: II/IIIM
Hazards: None of note

Occupation: 130 - PRECISION INSTRUMENT MAKERS
Supervisors - precision instrument makers and repairers; precision instrument makers and repairers.
Job description: Making, calibrating and repairing precision and optical instruments such as cameras, compasses, barometers and calibrators.
Social class: II/IIIM
Hazards: Exposure to mercury vapour may occur in some of these jobs (eg repairing barometers); organic solvents.

Occupation: 131 - WATCH AND CLOCK MAKERS
Supervisors - watch and chronometer makers and repairers; watch and chronometer makers and repairers.
Job description: Making, repairing, cleaning and adjusting watches, clocks and chronometers.
Social class: II/IIIM
Hazards: None of note

Occupation: 132 - PRODUCTION FITTERS
Supervisors - metal working production fitters and fitter/machinists; metal working production fitters and fitter /machinists.
Job description: Erecting, installing, repairing and servicing engineering and mechanical plant and machinery.
Social class: II/IIIM
Hazards: Oils; greases; organic solvents; asbestos; injury by machinery. In addition, there may be hazards related specifically to the industry in which the job is carried out (eg fitters in the chemical industry may be exposed to process chemicals).

Occupation: 133 - MOTOR MECHANICS
Supervisors - motor mechanics, auto engineers; motor mechanics, auto engineers
Job description: Repairing, servicing, oiling and greasing the mechanical parts of motor vehicles.
Social class: II/IIIM
Hazards: Oils; greases; organic solvents; petroleum; diesel fuel; epoxy resins; motor exhaust fumes; asbestos dust from brake linings.

Occupation: 134 - AIRCRAFT ENGINE FITTERS
Supervisors - maintenance fitters (aircraft engines); maintenance fitters (aircraft engines)
Job description: Repairing and servicing aircraft engines and frames.
Social class: IIIM
Hazards: Oils and greases; organic solvents; motor fuels.

Appendix 3 Description of job groups of the Southampton classification

Occupation: 135 - OFFICE MACHINERY MECHANICS
Supervisors - office machinery mechanics; office machinery mechanics
Job description: Repairing and servicing office machinery, cash tills and registers.
Social class: II/IIIM
Hazards: None of note

Occupation: 136 - ELECTRICAL AND ELECTRONIC PRODUCTION FITTERS
Supervisors - production fitters (electrical, electronic); production fitters (electrical, electronic).
Job description: Fitting, assembling, testing and repairing parts and sub-assemblies in the manufacture of electrical and electronic equipment.
Social class: II/IIIM
Hazards: Solder fumes; organic solvents. 'Electrical' occupations have been linked in some studies with brain cancer and leukaemia, possibly through exposure to magnetic fields. However, the hazard is not firmly established.

Occupation: 137 - ELECTRICIANS
Supervisors - electricians, electrical maintenance fitters; electricians, electrical maintenance fitters.
Job description: Installation, repair and maintenance of electrical wiring, fittings, plant, machinery and other equipment in commercial or domestic premises.
Social class: II/IIIM
Hazards: Electrical injury; asbestos. 'Electrical' occupations have been linked with brain cancer and leukaemia in some studies, but the finding has not been uniformly consistent and the hazard is not proven.

Occupation: 138 - ELECTRICAL PLANT OPERATORS
Supervisors - plant operators and attendants nec; plant operators and attendants nec.
Job description: Operating and attending plant and machinery such as compressors, electrical substations and switchboards, turbines and nuclear generating stations.
Social class: IIIM
Hazards: Asbestos; ionising radiation in nuclear power stations. Some studies have linked 'electrical' occupations with brain cancer and leukaemia, but the hazard is not firmly established.

Occupation: 139 - TELEPHONE FITTERS
Supervisors - telephone fitters; telephone fitters
Job description: Installation, maintenance and repair of public and private telephone systems.
Social class: IIIM
Hazards: Exposure to solder fume may occur. Some studies have linked 'electrical' occupations with brain cancer and leukaemia, but the hazard is not well established.

Occupation: 140 - ELECTRIC CABLE AND LINE WORKERS
Cable jointers, lines repairers
Job description: Installation, maintenance, and repair of overhead electricity supply lines, electric traction and telecommunication lines, submarine and surface power and telecommunication cables.
Social class: IIIM
Hazards: Injury by falls and electricity. Some studies have linked 'electrical' occupations with brain cancer and leukaemia, but the hazard is not firmly established.

Occupation: 141 - RADIO AND TV MECHANICS
Supervisors - radio and TV mechanics; radio and TV mechanics
Job description: Repair and maintenance of domestic radio and television receivers.
Social class: II/IIIM
Hazards: Electrical injury; solder fumes (more in the past than now). Some studies have linked 'electrical' occupations with brain cancer and leukaemia, but the hazard is not well established.

Occupation: 142 - OTHER ELECTRONIC MAINTENANCE ENGINEERS
Supervisors - other electronic maintenance engineers; other electronic maintenance engineers.
Job description: Installation, repair and maintenance of various items of electrical and electronic equipment including computer systems.
Social class: II/IIIM
Hazards: Solder fumes; electrical injury. Some studies have linked 'electrical' occupations with brain cancer and leukaemia, but the hazard is not firmly established.

Occupation: 143 - ELECTRICAL ENGINEERS (SO DESCRIBED)
Electrical engineers (so described)
Job description: Installation, maintenance and repair of electrical plant and machinery, and electrical wiring, fixtures and appliances.
Social class: II/IIIM
Hazards: Electrical injury; asbestos.

Occupation: 144 - PLUMBERS AND GAS FITTERS
Supervisors - plumbers, heating and ventilating fitters, gas fitters; plumbers, heating and ventilating fitters, gas fitters.
Job description: Installation, maintenance and repair of plumbing fixtures, heating and ventilation systems, pipe systems and lead, lead-lined and lead covered plant in domestic, commercial and industrial premises.
Social class: II/IIIM
Hazards: Asbestos; lead; accidental injury from various causes including electricity and falls.

Occupation: 145 - SHEET METAL WORKERS
Supervisors - sheet metal workers; sheet metal workers
Job description: Marking out, cutting, shaping and joining sheet metal by hand or machine.
Social class: II/IIIM
Hazards: Metal fume (from cutting metal with flame jet or lasers); welding fume.

Occupation: 146 - METAL PLATE WORKERS
Supervisors - metal plate workers, shipwrights, riveters; metal plate workers, shipwrights, riveters.
Job description: Marking off, drilling, bending positioning and riveting metal plates and girders; sealing joints in metal plate (includes shipwrights and shipyard steel workers).
Social class: II/IIIM
Hazards: Asbestos (especially in shipyards); noise.

Occupation: 147 - STEEL ERECTORS
Supervisors - steel erectors, benders, fixers; steel erectors, benders, fixers.
Job description: Fitting and erecting structural metal work and making up metal framework by hand to form reinforcing core for concrete structures and products.
Social class: II/IIIM
Hazards: Injury by falls and machinery.

Appendix 3 Description of job groups of the Southampton classification

Occupation: 148 - SCAFFOLDERS
Supervisors - scaffolders, stagers; scaffolders, stagers
Job description: Erecting and dismantling scaffolding and temporary working platforms.
Social class: IIIM/IV
Hazards: Injury by falling.

Occupation: 149 - WELDERS
Supervisors - welders; welders
Job description: Joining metal parts and fabrications by welding, brazing and soldering; cutting metal to specification and removing defects by means of gas, flame and oxygen jet, electric arc or laser beam.
Social class: II/IIIM
Hazards: Welding, brazing and soldering fumes and gases; asbestos (especially in welders working in shipyards); ultraviolet radiation from electric arc welding; noise.

Occupation: 150 - RIGGERS
Supervisors - riggers; riggers
Job description: Preparing, installing, repairing and splicing ropes, wires and cables; setting up lifting equipment.
Social class: II/IIIM
Hazards: Injury by falls.

Occupation: 151 - JEWELLERY WORKERS
Supervisors - goldsmiths, silversmiths, precious stone workers; goldsmiths, silversmiths, precious stone workers.
Job description: Making and repairing jewellery and precious metalware; setting, cutting and polishing precious and semi-precious stones; decoration of metalware and making master patterns for articles of jewellery.
Social class: II/IIIM
Hazards: Solder fume; metal dust; silica; epoxy resins; organic solvents.

Occupation: 152 - ENGRAVERS AND ETCHERS (PRINTING)
Supervisors - engravers, etchers (printing); engravers, etchers (printing)
Job description: Engraving and etching metal printing cylinders, plates and rollers; engraving metal dies and punches.
Social class: IIIM
Hazards: Various corrosive chemicals.

Occupation: 153 - VEHICLE BODY BUILDERS
Supervisors - coach and vehicle body builders; coach and vehicle body builders
Job description: Construction and repair of bodies for road vehicles and railway coaches; fixing of interior and exterior fittings to vehicle and aircraft bodies.
Social class: II/IIIM
Hazards: Asbestos (especially in the construction of railway carriages); styrene; glass fibre; wood dust.

Occupation: 154 - OILERS AND GREASERS
Supervisors - oilers, greasers, lubricators; oilers, greasers, lubricators.
Job description: Lubrication of the moving parts of stationary engines, rolling stock and other machinery.
Social class: IIIM
Hazards: Oils and greases.

Occupation: 155 - ELECTRONICS WIRE WORKERS
Electrical and electronic assemblers/lineworkers
Job description: Wiring up prepared parts or sub-assemblies in the manufacture of various types of electronic equipment.
Social class: IIIM
Hazards: Solder fume.

Occupation: 156 - COIL WINDERS
Supervisors - coil winders; coil winders
Job description: Making coils and wiring harnesses for electrical and electronic equipment (includes cable formers and neon sign assemblers).
Social class: IIIM
Hazards: None of note

Occupation: 157 - POTTERY DECORATORS
Supervisors - pottery decorators; pottery decorators
Job description: Dipping and spraying ceramics with glaze; painting ceramics and applying transfers.
Social class: IIIM
Hazards: Lead; metallic oxides; organic solvents; epoxy resins.

Occupation: 158 - COACH PAINTERS
Supervisors - coach painters (so described); coach painters (so described).
Job description: Application of paint, cellulose and other protective and decorative materials to the bodywork of motor vehicles or railway coaches and wagons using brushes or spray equipment.
Social class: II/IIIM
Hazards: Organic solvents; resins; isocyanates. A number of studies have indicated high rates of lung cancer in painters, but the causes are unclear.

Occupation: 159 - OTHER SPRAY PAINTERS
Supervisors - other spray painters; other spray painters
Job description: Spraying paint, cellulose lacquer and other materials onto furniture, leather goods, musical instruments etc.
Social class: IIIM/IV
Hazards: Organic solvents; resins; isocyanates. A number of studies have indicated high rates of lung cancer in painters, but the causes are unclear.

Occupation: 160 - PAINTERS AND DECORATORS NEC
Supervisors - painters and decorators nec, french polishers; painters and decorators nec, french polishers.
Job description: Application of paint, varnish and other protective and decorative materials to the interior and exterior of buildings, painting designs and lettering; staining, waxing and french polishing wood surfaces by hand.
Social class: II/IIIM
Hazards: Organic solvents; resins; isocyanates; dyes and stains; various dusts associated with surface preparation; injury by falling.

Occupation: 161 - ELECTRICAL, ELECTRONIC ASSEMBLERS
Supervisors assemblers, - electrical, electronic; assemblers (electrical, electronic)
Job description: Semi-skilled or repetitive assembly of previously prepared parts in the batch or mass production of electrical and electronic goods and components.
Social class: IIIM/IV
Hazards: Solder fumes.

Occupation: 162 - INSTRUMENT ASSEMBLERS
Supervisors assemblers - instruments; instrument assemblers
Job description: Routine assembly of precision instruments (gauges, meters, clocks and watches etc).
Social class: IIIM
Hazards: Some of these jobs may entail exposure to mercury.

Appendix 3 Description of job groups of the Southampton classification

Occupation: 163 - ASSEMBLERS (VEHICLES AND OTHER METAL GOODS)
Supervisors assemblers - vehicles and other metal goods; assemblers (vehicles and other metal goods)
Job description: Supervising and undertaking routine assembly of vehicles and other metal goods or components such as frames, axles, wire brushes and wheels.
Social class: IIIM/IV
Hazards: None of note

Occupation: 164 - PACKERS AND SORTERS
Supervisors - packers, bottlers, canners, fillers; supervisors - weighers; supervisors - graders, sorters, selectors nec; packers, bottlers, canners, fillers; weighers; graders, sorters, selectors nec.
Job description: Filling, sealing labelling, wrapping and packing containers by hand or machine; weighing and measuring materials, goods and products; sorting and grading materials and products.
Social class: IIIM/IV
Hazards: Some of these occupations entail repetitive movements of the hand and arm which may cause various musculo-skeletal disorders of the upper limb. However, these would not be expected to affect mortality.

Occupation: 165 - BRICKLAYERS AND TILESETTERS
Supervisors - bricklayers, tile setters; bricklayers, tile setters
Job description: Building and repairing brick structures; setting wall and floor tiles and mosaics.
Social class: II/IIIM
Hazards: Injury by falls; cement dust is a cause of dermatitis but this would not be expected to affect mortality.

Occupation: 166 - MASONS AND STONECUTTERS
Supervisors - masons, stone cutters; masons, stone cutters
Job description: Building and repairing stone structures; facing brick, cement or steel backing with stone; cutting, shaping and finishing granite, marble, slate and other stone.
Social class: II/IIIM
Hazards: Silicaceous dust; injury by falling.

Occupation: 167 - PLASTERERS
Supervisors - plasterers; plasterers
Job description: Applying plaster and cement mixtures to walls and ceilings; casting and fixing ornamental plaster work and carrying out other plastering tasks.
Social class: IIIM/IV
Hazards: None of note

Occupation: 168 - ROOFERS AND GLAZIERS
Supervisors - roofers, glaziers; roofers, glaziers
Job description: Covering roofs with felt, slates, tiles and thatch; cutting, fitting and setting glass in windows, doors etc.
Social class: IIIM/IV
Hazards: Injury by falling; bitumen fumes (in felting of roofs).

Occupation: 169 - BUILDERS ETC.
Supervisors - maintenance hand, general building workers; maintenance hand, general building workers; builders (so described)
Job description: Various activities in the construction, alteration, maintenance and repair of buildings.
Social class: II/IIIM/IV
Hazards: Accidental injury from various causes, especially falls; some exposure to asbestos may occur.

Occupation: 170 - RAIL TRACK WORKERS
Workers in this unit group lay, re-lay, repair and examine railway track
Job description: Laying and maintaining railway tracks.
Social class: IIIM/IV
Hazards: Injury by trains.

Occupation: 171 - ROAD CONSTRUCTION WORKERS AND PAVIORS
Supervisors - road surfacers, concreters; supervisors - road worker; supervisors - paviors, kerb layers; road surfacers, concreters, road worker; paviors, kerb layers.
Job description: Spreading, sealing and smoothing newly laid asphalt on roads and similar surfaces; pouring and levelling concrete in the construction of roads and buildings; laying paving slabs and kerbstones on prepared foundations to form pavements and street gutters.
Social class: IIIM/IV/V
Hazards: Asphalt fumes; electrical injury from accidentally drilling into underground power lines.

Occupation: 172 - SEWAGE PLANT ATTENDANTS
Supervisors - sewage plant attendants; sewage plant attendants
Job description: Attending plant in which sewage is treated and maintaining the main sewerage system.
Social class: IIIM/V
Hazards: Some studies have suggested an increased risk of hepatitis A infection in sewage workers, but this would not be expected to have any discernible effect on mortality.

Occupation: 173 - MAINS AND SERVICE LAYERS
Supervisors - mains and service layers, pipe jointers; mains and service layers, pipe jointers.
Job description: Laying of jointing pipes for drainage, gas, water and similar piping systems.
Social class: IIIM/IV
Hazards: Injury from various causes including drilling into underground electrical cables.

Occupation: 174 - CONSTRUCTION WORKERS NEC
Supervisors - construction workers nec; craftsmen's mates; building and civil engineering labourers; construction workers nec.
Job description: Various activities in building and civil engineering including cleaning the exterior of buildings; laying composition floors and linoleum; steeplejack work; insulating and lagging; demolition work; and gravedigging.
Social class: IIIM/IV/V
Hazards: Asbestos (especially in laggers and in demolition work); accidental injury from various causes, especially falls.

Occupation: 175 - FACE TRAINED COALMINERS
Supervisors/deputies - coal mining; face trained coalmining workers.
Job description: Drilling holes, setting and detonating charges to loosen rock and coal, stowing waste, removing rock to enlarge road and airways, extracting coal, and building and dismantling roof and wall supports in underground coal workings.
Social class: IIIM
Hazards: Coal dust; silicaceous dust; injury by machinery.

Appendix 3 Description of job groups of the Southampton classification

Occupation: 176 - MINERS (NOT COAL) AND QUARRY WORKERS
Miners (not coal), quarry workers, well drillers
Job description: Extraction of minerals other than coal from underground workings or quarries; drilling holes for blasting; erecting supports; setting and detonating explosives to loosen rock; setting up and operating drilling equipment (including on offshore oil rigs).
Social class: IIIM/IV
Hazards: Silicaceous dusts in mines and quarries; injury from various causes including accidents with machinery. Some underground miners are exposed to high levels of diesel fume.

Occupation: 177 - RAILWAY GUARDS
Supervisors - railway guards; railway guards
Job description: Taking charge of passenger and goods trains on surface and underground railways.
Social class: IIIM
Hazards: Injury from railway accidents.

Occupation: 178 - RAILWAY SIGNAL WORKERS
Supervisors - signalmen and crossing keepers, railway; signalmen and crossing keepers, railway.
Job description: Supervising and operating signals; opening and closing gates and barriers at level crossings to control the movement of rail traffic and to safeguard passengers, pedestrians and road users.
Social class: IIIM
Hazards: None of note

Occupation: 179 - SHUNTERS AND POINTS OPERATORS
Supervisors - shunters, points operatives; shunters, point operatives.
Job description: Changing points and guiding wagons and coaches in railway sidings and marshalling yards to form complete trains.
Social class: IIIM
Hazards: Injury by trains.

Occupation: 180 - RAILWAY ENGINE DRIVERS
Drivers, train drivers, second train drivers, railway engine drivers assistants.
Job description: Driving railway engines; driving or acting as an assistant to the driver on all types of locomotives on surface and underground railways.
Social class: IIIM
Hazards: Injury in railway accidents.

Occupation: 181 - ROAD TRANSPORT INSPECTORS
Bus inspectors; supervisors - other supervisors road transport
Job description: Supervising the activities of drivers and conductors of public service vehicles; supervising persons employed in the operation of road transport other than vehicle maintenance workers.
Social class: IIIM
Hazards: None of note

Occupation: 182 - BUS AND COACH DRIVERS
Bus and coach drivers
Job description: Driving road passenger-carrying vehicles such as buses, coaches, mini-buses, trolley buses and trams.
Social class: II/IIIM
Hazards: Injury in road traffic accidents; traffic fume.

Occupation: 183 - LORRY DRIVERS
Supervisors - drivers of road goods vehicles; drivers of road goods vehicles
Job description: Driving heavy goods vehicles (over three tonnes) and light goods vehicles to transport goods and animals.
Social class: II/IIIM
Hazards: Injury in road traffic accidents and from accidents relating to specific cargoes, traffic fume. Some studies have suggested high rates of bladder cancer in lorry drivers, but this has not been a consistent finding.

Occupation: 184 - OTHER MOTOR DRIVERS
Other motor drivers
Job description: Driving taxis; driving mechanical road sweepers; acting as chauffeurs; patrolling roads for motoring organisations.
Social class: II/IIIM
Hazards: Injury in road traffic accidents; traffic fume.

Occupation: 185 - BUS CONDUCTORS AND DRIVERS' MATES
Bus conductors; drivers' mates
Job description: Collecting fares, issuing tickets and controlling passengers on buses; accompanying drivers of road goods vehicles and assisting with the loading and unloading of vehicles.
Social class: IV/V
Hazards: Injury in road traffic accidents; traffic fume; hazards specific to cargoes being transported.

Occupation: 186 - MECHANICAL PLANT DRIVERS
Supervisors - mechanical plant drivers, operators (earth moving and civil engineering); mechanical plant drivers, operators (earth moving and civil engineering).
Job description: Operation of machines to excavate, grade, level and compact sand, gravel, earth and similar materials; driving piles into the ground and laying surfaces of asphalt, concrete and chippings.
Social class: II/IIIM
Hazards: Injury by machinery; asphalt fumes.

Occupation: 187 - CRANE DRIVERS
Supervisors - crane drivers, operators; crane drivers, operators
Job description: Operating cranes, power driven hoisting machinery and stationery engines to raise and lower mine and other cages and lift and move equipment, materials and machinery.
Social class: IIIM
Hazards: Injury by machinery; hazards specific to the industry in which the job is carried out (eg fume in a steelworks).

Occupation: 188 - FORK LIFT TRUCK DRIVERS
Supervisors - fork lift, mechanical truck drivers, fork lift, mechanical truck drivers
Job description: Driving and operating fork lift and similar trucks in factories, warehouses and on industrial and commercial premises to transfer goods and materials.
Social class: IIIM
Hazards: Injury by machinery; hazards specific to the industry in which the job is carried out.

Occupation: 189 - SLINGERS
Supervisors - slingers; slingers
Job description: Placing wire, chains or rope around objects to be lifted and signalling instructions to crane operators.
Social class: IIIM
Hazards: Injury by machinery; hazards specific to the industry in which the job is carried out.

Appendix 3 Description of job groups of the Southampton classification

Occupation: 190 - STOREKEEPERS
Supervisors - storekeepers, warehouse workers; storekeepers, warehouse workers
Job description: Storing, selecting and issuing materials, goods etc; maintaining stock records.
Social class: IIIM/IV
Hazards: Manual handling; there may be hazards from specific materials being stored.

Occupation: 191 - DOCKERS AND GOODS PORTERS
Supervisors - stevedores, dockers; supervisors - goods porters; stevedores, dockers; goods porters
Job description: Loading and unloading cargoes and goods onto and from ships, barges, vehicles and aircraft; loading, unloading and moving goods, materials and equipment about warehouses, stores, goods depots, shops, markets etc; moving furniture; carrying baggage at docks and airports.
Social class: IIIM/V
Hazards: Manual handling; hazards relating to specific cargoes (eg asbestos in dockers); injury by machinery.

Occupation: 192 - REFUSE COLLECTORS
Supervisors - refuse collectors, dustmen/women; refuse collectors, dustmen/women
Job description: Collection of refuse from household, commercial and industrial premises and loading it into refuse vehicles.
Social class: IIIM/V
Hazards: Manual handling; hazards relating to specific waste.

Occupation: 193 - LABOURERS IN COKE OVENS
Supervisors - labourers and unskilled workers nec - coke ovens and gas works; labourers and unskilled workers nec - coke ovens and gas works.
Job description: Unskilled manual work in coke ovens and gas works
Social class: IIIM/V
Hazards: Polycyclic aromatic hydrocarbons; benzene; various other toxic chemicals.

Occupation: 194 - BOILER OPERATORS
Supervisors - boiler operators; boiler operators
Job description: Operation of boilers to produce steam or hot water for central heating systems and industrial use.
Social class: IIIM/IV
Hazards: Asbestos; noise.

Appendix 4 Significant PMRs for each job group

Paul Winter
Hazel Inskip
David Coggon
Brian Pannett

MRC Environmental Epidemiology Unit, University of Southampton

This appendix gives the PMRs, by job group and sex, which differ from 100 at the 5% level of significance. The list comprises all significantly raised PMRs based on three or more observed deaths and all significantly low PMRs where three or more deaths were expected. The PMRs are adjusted for age and social class. The numbers of deaths from all causes in men and women in each job group are also given. Commentaries on mortality by job group are given in Chapters 4 and 5, for men and women respectively.

Job group	Deaths	PMR	95% CI
001 - Lawyers			
Number of deaths in men aged 20-74 during 1979-80 and 1982-90: 3093			
PMRs for men			
Cause of death (ICD)			
Cancer of the stomach (151)	36	69	48-95
Cancer of the liver (155)	32	212	145-300
Cancer of the pancreas (157)	66	129	100-164
Cancer of trachea, bronchus and lung (162)	172	77	66-90
Cancer of the pleura (163)	0	0	0-62
Immunodeficiency (279.1)	12	216	111-377
Other alcohol related diseases (303,305.0,425.5,535.3,571.0-571.3, E860.0,E860.1)	35	173	120-240
Ischaemic heart disease (410-414)	987	90	84-96
Endocarditis (421)	8	274	118-540
Homicide (E960-E969)	5	314	102-733
Injury undetermined whether accidentally purposely inflicted (E980 - E989)	32	165	113-233
Number of deaths in women aged 20-74 during 1979-80 and 1982-90: 170			
			No significant PMRs
002 - Accountants			
Number of deaths in men aged 20-74 during 1979-80 and 1982-90: 12947			
PMRs for men			
Cause of death (ICD)			
Tuberculosis (010-018,137)	27	156	103-227
Melanoma of skin (172)	86	142	114-176
Other cancer of skin (173)	4	39	11-99
Cancer of the penis (187.1-187.4)	11	207	103-370
Cancer of the brain (191)	194	116	100-134
Haemolytic anaemia (283)	4	409	112-1048
Peripheral vascular disease (443,557)	23	65	41-97
Hepatitis (571.4,573.3)	1	17	0-94
Railway accidents (E800-E807)	11	218	109-389
Injury by machinery (E919)	0	0	0-71
Number of deaths in women aged 20-74 during 1979-80 and 1982-90: 560			
PMRs for women			
Cause of death (ICD)			
Myeloma (203)	9	318	145-604
Suicide (E950-E959)	12	56	29-98

Job group	Deaths	PMR	95% CI
003 - Personnel managers etc.			
Number of deaths in men aged 20-74 during 1979-80 and 1982-90: 4157			
PMRs for men			
Cause of death (ICD)			
Other leukaemia (207,208)	7	303	122-623
Fibrosing alveolitis (516.3)	14	210	115-352
Hyperplasia of prostate (600)	8	237	102-467
Suicide (E950-E959)	48	74	54-98
Number of deaths in women aged 20-74 during 1979-80 and 1982-90: 642			
PMRs for women			
Cause of death (ICD)			
Cancer of the ovary (183)	38	142	100-195
Other cerebrovascular disease (431-438)	27	68	45-99
Renal failure (584-586)	6	318	117-691
Suicide (E950-E959)	6	44	16-97
004 - Economists and statisticians			
Number of deaths in men aged 20-74 during 1979-80 and 1982-90: 467			
PMRs for men			
Cause of death (ICD)			
Cancer of the oesophagus (150)	1	17	0-97
Melanoma of skin (172)	8	291	126-574
Number of deaths in women aged 20-74 during 1979-80 and 1982-90: 94			
			No significant PMRs
005 - Computer programmers			
Number of deaths in men aged 20-74 during 1979-80 and 1982-90: 1469			
PMRs for men			
Cause of death (ICD)			
Cancer of bone (170)	11	247	123-442
Cancer of soft tissue (171)	11	203	101-363
Cancer of the brain (191)	43	145	105-195
Other alcohol related diseases (303,305.0,425.5,535.3,571.0-571.3, E860.0,E860.1)	8	41	18-80
Multiple sclerosis (340)	12	202	105-354
Air transport accidents (E840-E845)	0	0	0-76
Number of deaths in women aged 20-74 during 1979-80 and 1982-90: 189			
PMRs for women			
Cause of death (ICD)			
Non-Hodgkin's lymphoma (200,202)	8	298	129-588
006 - Sales managers etc.			
Number of deaths in men aged 20-74 during 1979-80 and 1982-90: 15247			
PMRs for men			
Cause of death (ICD)			
Cancer of the stomach (151)	288	88	79-99
Cancer of the retroperitoneum (158.0)	12	199	103-347

Appendix 4 Significant PMRs for each job group

Job group	Deaths	PMR	95% CI
Melanoma of skin (172)	88	132	105-162
Urothelial cancer (188,189.1-189.8)	206	122	106-140
Acute lymphatic leukaemia (204.0)	21	161	99-246
Chronic lymphatic leukaemia (204.1)	17	60	35-96
Volvulus (560.2)	0	0	0-82
Motor vehicle traffic accidents (E810-E819)	279	127	112-142
Water transport accidents (E830-E838)	1	14	0-77
Fall from building (E882)	1	17	0-93
Injury by machinery (E919)	0	0	0-46
Suicide (E950-E959)	215	75	65-86
Injury undetermined whether accidentally or purposely inflicted (E980-E989)	55	75	57-98

Number of deaths in women aged 20-74 during 1979-80 and 1982-90: 1290

PMRs for women

Cause of death (ICD)	Deaths	PMR	95% CI
Cancer of the ovary (183)	73	146	115-184
Ischaemic heart disease (410-414)	159	75	64-87
Sub-arachnoid haemorrhage (430)	38	147	104-202
Motor vehicle traffic accidents (E810-E819)	46	158	116-211

007 - Government inspectors

Number of deaths in men aged 20-74 during 1979-80 and 1982-90: 3666

PMRs for men

Cause of death (ICD)	Deaths	PMR	95% CI
Cancer of the small intestine (152)	7	274	110-565
Aortic valve disorders (424.1)	8	47	20-92
Cirrhosis of the liver (571.5)	8	50	22-99

Number of deaths in women aged 20-74 during 1979-80 and 1982-90: 104

PMRs for women

Cause of death (ICD)	Deaths	PMR	95% CI
Suicide (E950-E959)	7	397	160-819

008 - Government administrators

Number of deaths in men aged 20-74 during 1979-80 and 1982-90: 5012

PMRs for men

Cause of death (ICD)	Deaths	PMR	95% CI
Cancer of the colon (153)	151	120	102-141
Cancer of trachea, bronchus and lung (162)	403	84	76-93
Cancer of the brain (191)	70	128	100-162
Other alcohol related diseases (303,305.0,425.5,535.3,571.0-571.3, E860.0,E860.1)	18	60	35-94
Aortic aneurysm (441)	143	133	112-156
Chronic bronchitis and emphysema (491,492,496)	135	78	66-93
Motor vehicle traffic accidents (E810-E819)	17	39	23-62
Suicide (E950-E959)	29	44	30-64
Injury undetermined whether accidentally or purposely inflicted (E980-989)	6	36	13-78

Number of deaths in women aged 20-74 during 1979-80 and 1982-90: 693

PMRs for women

Cause of death (ICD)	Deaths	PMR	95% CI
Cancer of the female breast (174)	117	129	106-154
Suicide (E950-E959)	2	20	2-73

009 - Other administrators

Number of deaths in men aged 20-74 during 1979-80 and 1982-90: 8856

PMRs for men

Cause of death (ICD)	Deaths	PMR	95% CI
Cancer of the colon (153)	270	125	111-141
Cancer of trachea, bronchus and lung (162)	743	88	82-95
Immunodeficiency (279.1)	23	162	102-243
Aortic aneurysm (441)	222	118	103-134
Chronic bronchitis and emphysema (491,492,496)	271	87	77-98
Hepatitis (571.4,573.3)	8	245	106-482
Glomerulonephritis (580-583)	16	205	117-333

Number of deaths in women aged 20-74 during 1979-80 and 1982-90: 2881

PMRs for women

Cause of death (ICD)	Deaths	PMR	95% CI
Melanoma of skin (172)	30	155	104-221
Cancer of the female breast (174)	436	111	101-122
Cancer of the ovary (183)	143	120	101-142
Diabetes mellitus (250)	10	40	19-73
Dementia (290,331.0,331.1)	26	186	121-273
Chronic rheumatic heart disease (394-398)	11	52	26-93
Ischaemic heart disease (410-414)	425	86	78-95
Chronic bronchitis and emphysema (491,492,496)	76	129	102-162
Suicide (E950-E959)	34	70	49-98

010 - Teachers in higher education

Number of deaths in men aged 20-74 during 1979-80 and 1982-90: 4521

PMRs for men

Cause of death (ICD)	Deaths	PMR	95% CI
Cancer of the larynx (161)	4	38	10-97
Cancer of trachea, bronchus and lung (162)	284	71	63-80
Cancer of the prostate (185)	150	167	142-196
Cancer of the brain (191)	116	185	153-221
Non-Hodgkin's lymphoma (200,202)	69	146	114-185
Myeloma (203)	38	148	105-204
Chronic lymphatic leukaemia (204.1)	16	184	105-298
Immunodeficiency (279.1)	16	177	101-287
Parkinson's disease (332)	21	170	105-261
Motor neurone disease (335.2)	30	170	115-244
Hypertensive disease (401-405)	20	61	37-95
Viral pneumonia (480)	5	386	125-902
Chronic bronchitis and emphysema (491,492,496)	77	61	48-76
Motor vehicle traffic accidents (E810-E819)	40	73	52-99

Number of deaths in women aged 20-74 during 1979-80 and 1982-90: 830

PMRs for women

Cause of death (ICD)	Deaths	PMR	95% CI
Cancer of the rectum (154)	18	171	101-270
Cancer of trachea, bronchus and lung (162)	30	62	42-89
Cancer of the female breast (174)	167	133	113-154
Ischaemic heart disease (410-414)	91	71	57-87
Pulmonary embolism and phlebitis (415.1,451,453)	2	27	3-98
Chronic bronchitis and emphysema (491,492,496)	5	36	12-83

011 - Teachers nec

Number of deaths in men aged 20-74 during 1979-80 and 1982-90: 15862

PMRs for men

Cause of death (ICD)	Deaths	PMR	95% CI
Cancer of the larynx (161)	19	51	31-80
Cancer of trachea, bronchus and lung (162)	904	61	57-66
Melanoma of skin (172)	105	154	126-186
Cancer of the prostate (185)	409	119	108-131
Cancer of the brain (191)	257	140	124-159
Cancer of the thyroid (193)	21	199	123-305
Non-Hodgkin's lymphoma (200,202)	201	135	117-155
Myeloma (203)	107	133	109-160
Acute myeloid leukaemia (205.0)	85	130	104-161
Chronic myeloid leukaemia (205.1)	40	143	102-194
Immunodeficiency (279.1)	53	138	103-180
Aplastic anaemia (284)	18	266	158-421
Dementia (290,331.0,331.1)	71	138	108-174
Other alcohol related diseases (303,305.0,425.5,535.3,571.0-571.3, E860.0,E860.1)	77	69	54-86
Parkinson's disease (332)	90	186	150-229
Multiple sclerosis (340)	51	152	113-200
Epilepsy (345)	30	154	104-221
Myasthenia (358)	4	432	118-1107
Ischaemic heart disease (410-414)	6038	104	101-107
Pulmonary heart disease unspecified (416.9)	7	42	17-88

Appendix 4 Significant PMRs for each job group

Job group	Deaths	PMR	95% CI
Bronchopneumonia (485)	150	76	65-90
Chronic bronchitis and emphysema (491,492,496)	348	66	59-73
Cirrhosis of the liver (571.5)	54	74	55-96
Osteoporosis (733.0)	7	274	110-565
Pedal cycle accidents (E826)	9	388	178-737
Fall from ladder or scaffolding (E881)	12	235	121-410
Other fall (E884)	19	357	215-557
Injury by falling object (E916)	0	0	0-87
Injury by machinery (E919)	0	0	0-46
Suicide (E950-E959)	380	130	118-144
Injury undetermined whether accidentally or purposely inflicted (E980-E989)	102	135	110-164

Number of deaths in women aged 20-74 during 1979-80 and 1982-90: 14106

PMRs for women

Cause of death (ICD)	Deaths	PMR	95% CI
Tuberculosis (010-018,137)	5	39	13-90
Cancer of the colon (153)	572	118	109-128
Cancer of the gall bladder (156)	57	137	104-178
Cancer of the larynx (161)	5	35	11-81
Cancer of trachea, bronchus and lung (162)	557	62	57-68
Melanoma of skin (172)	134	143	120-169
Cancer of the female breast (174)	2404	128	123-133
Cancer of the cervix (180)	125	53	44-63
Cancer of the body of uterus (182)	140	128	108-152
Cancer of the ovary (183)	740	130	121-140
Cancer of the brain (191)	252	123	108-139
Non-Hodgkin's lymphoma (200,202)	180	118	101-137
Hodgkin's disease (201)	45	142	104-191
Diabetes mellitus (250)	100	79	64-96
Other alcohol related diseases (303,305.0,425.5,535.3,571.0-571.3, E860.0,E860.1)	75	69	54-87
Multiple sclerosis (340)	105	152	125-185
Myasthenia (358)	4	369	101-946
Ischaemic heart disease (410-414)	2174	88	84-92
Other aneurysm (442)	0	0	0-85
Chronic bronchitis and emphysema (491,492,496)	164	56	48-66
Motor vehicle traffic accidents (E810-E819)	218	118	102-134

012 - Vocational trainers, social scientists etc.

Number of deaths in men aged 20-74 during 1979-80 and 1982-90: 4028

PMRs for men

Cause of death (ICD)	Deaths	PMR	95% CI
Cancer of the pleura (163)	13	200	106-342
Non-Hodgkin's lymphoma (200,202)	54	134	101-175
Other alcohol related diseases (303,305.0,425.5,535.3,571.0-571.3, E860.0,E860.1)	16	55	31-89
Hypertensive disease (401-405)	18	61	36-97
Ischaemic heart disease (410-414)	1600	106	101-111
Other cerebrovascular disease (431-438)	203	85	73-97
Oesophageal disease (530)	6	306	112-667
Injury undetermined whether accidentally or purposely inflicted (E980-E989)	8	44	19-86

Number of deaths in women aged 20-74 during 1979-80 and 1982-90: 531

PMRs for women

Cause of death (ICD)	Deaths	PMR	95% CI
Chronic myeloid leukaemia (205.1)	6	350	129-762
Peripheral vascular disease (443,557)	5	366	119-854

013 - Welfare workers

Number of deaths in men aged 20-74 during 1979-80 and 1982-90: 4791

PMRs for men

Cause of death (ICD)	Deaths	PMR	95% CI
Cancer of the oral cavity (141,143,144,145)	3	27	6-79
Cancer of the stomach (151)	147	146	123-171
Cancer of bone (170)	10	225	108-413
Pulmonary hypertension (416.0)	4	373	102-956
Gastric ulcer (531)	2	24	3-88
Injured by fire (E890-E899)	0	0	0-93

Number of deaths in women aged 20-74 during 1979-80 and 1982-90: 4319

PMRs for women

Cause of death (ICD)	Deaths	PMR	95% CI
Cancer of the gall bladder (156)	23	177	112-266
Cancer of trachea, bronchus and lung (162)	335	119	107-133
Cancer of the uterus, part unspecified (179)	19	169	102-264
Cancer of the brain (191)	48	73	54-97
Other alcohol related diseases (303,305.0,425.5,535.3,571.0-571.3, E860.0,E860.1)	11	32	16-57
Multiple sclerosis (340)	9	41	19-78
Ischaemic heart disease (410-414)	828	114	106-122
Accidental poisoning by drugs (E850-E858)	1	11	0-63

014 - Clergy

Number of deaths in men aged 20-74 during 1979-80 and 1982-90: 4008

PMRs for men

Cause of death (ICD)	Deaths	PMR	95% CI
Cancer of the oesophagus (150)	33	64	44-90
Cancer of trachea, bronchus and lung (162)	181	59	51-68
Urothelial cancer (188,189.1-189.8)	31	65	44-92
Diabetes mellitus (250)	59	148	113-191
Other alcohol related diseases (303,305.0,425.5,535.3,571.0-571.3, E860.0,E860.1)	5	24	8-55
Ischaemic heart disease (410-414)	1696	115	109-120
Other cerebrovascular disease (431-438)	336	116	104-129
Aortic aneurysm (441)	76	79	62-99
Chronic bronchitis and emphysema (491,492,496)	71	64	50-81
Duodenal ulcer (532)	3	23	5-66
Motor vehicle traffic accidents (E810-E819)	50	140	104-184
Suicide (E950-E959)	37	70	50-97

Number of deaths in women aged 20-74 during 1979-80 and 1982-90: 834

PMRs for women

Cause of death (ICD)	Deaths	PMR	95% CI
Cancer of the oesophagus (150)	16	187	107-303
Cancer of the pancreas (157)	8	43	19-85
Cancer of trachea, bronchus and lung (162)	13	39	21-66
Other alcohol related diseases (303,305.0,425.5,535.3,571.0-571.3, E860.0,E860.1)	0	0	0-73
Chronic bronchitis and emphysema (491,492,496)	4	38	10-98
Motor vehicle traffic accidents (E810-E819)	21	171	106-262
Suicide (E950-E959)	5	35	11-82

015 - Doctors

Number of deaths in men aged 20-74 during 1979-80 and 1982-90: 4322

PMRs for men

Cause of death (ICD)	Deaths	PMR	95% CI
Viral hepatitis (070)	14	435	238-729
Cancer of the liver (155)	40	190	136-259
Cancer of trachea, bronchus and lung (162)	230	72	63-82
Cancer of the pleura (163)	1	11	0-64
Chronic rheumatic heart disease (394-398)	26	161	105-236
Acute myocarditis (422)	5	385	125-898
Other cardiomyopathies (425.0-425.4, 425.6-425.9)	32	160	109-226
Chronic bronchitis and emphysema (491,492,496)	80	75	60-94
Cirrhosis of the liver (571.5)	49	203	150-268
Accidental poisoning by drugs (E850-E858)	9	272	125-517
Fall on stairs (E880)	10	226	109-416
Suicide (E950-E959)	141	162	136-191
Injury undetermined whether accidentally or purposely inflicted (E980-E989)	41	169	121-229

Appendix 4 Significant PMRs for each job group

Job group	Deaths	PMR	95% CI
Number of deaths in women aged 20-74 during 1979-80 and 1982-90: 699			
PMRs for women			
Cause of death (ICD)			
Cancer of the pancreas (157)	23	160	101-241
Accidental poisoning by drugs (E850-E858)	7	334	134-688
Suicide (E950-E959)	50	193	143-254

016 - Dentists

Number of deaths in men aged 20-74 during 1979-80 and 1982-90: 972

PMRs for men	Deaths	PMR	95% CI
Cause of death (ICD)			
Cancer of the prostate (185)	37	159	112-220
Ischaemic heart disease (410-414)	300	85	76-95
Suicide (E950-E959)	38	194	137-266

Number of deaths in women aged 20-74 during 1979-80 and 1982-90: 79

No significant PMRs

017 - Nurses

Number of deaths in men aged 20-74 during 1979-80 and 1982-90: 4698

PMRs for men	Deaths	PMR	95% CI
Cause of death (ICD)			
Cancer of the colon (153)	76	69	55-87
Cancer of the prostate (185)	73	73	57-92
Cancer of the brain (191)	32	58	40-82
Acute lymphatic leukaemia (204.0)	0	0	0-68
Bronchopneumonia (485)	83	144	115-179
Chronic bronchitis and emphysema (491,492,496)	196	128	110-147
Peptic ulcer (533)	6	352	129-766
Other hernia (551-553)	7	367	148-757
Other colitis (558)	4	419	114-1074
Poisoning by gas and other domestic fuels (E867,E868.1,E868.3)	5	386	125-900
Suicide (E950-E959)	135	127	106-150

Number of deaths in women aged 20-74 during 1979-80 and 1982-90: 20734

PMRs for women	Deaths	PMR	95% CI
Cause of death (ICD)			
Meningococcal infection (036)	1	15	0-83
Sarcoidosis (135)	19	171	103-267
Cancer of the oral cavity (141,143,144,145)	21	54	33-82
Cancer of the colon (153)	635	90	83-97
Melanoma of skin (172)	89	68	55-84
Other cancer of skin (173)	6	43	16-94
Cancer of the female breast (174)	2245	85	82-89
Cancer of the ovary (183)	674	83	76-89
Diabetes mellitus (250)	224	119	104-135
Parkinson's disease (332)	72	128	100-161
Ischaemic heart disease (410-414)	3881	105	102-108
Other aneurysm (442)	15	238	133-392
Chronic bronchitis and emphysema (491,492,496)	482	110	100-120
Accidental poisoning by drugs (E850-E858)	64	152	117-195
Fall unspecified (E888)	49	150	111-199
Suicide (E950-E959)	475	138	126-152
Injury undetermined whether accidentally or purposely inflicted (E980-E989)	201	137	119-157

018 - Pharmacists

Number of deaths in men aged 20-74 during 1979-80 and 1982-90: 1936

PMRs for men	Deaths	PMR	95% CI
Cause of death (ICD)			
Cancer of the stomach (151)	21	64	40-98
Non-Hodgkin's lymphoma (200,202)	10	50	24-92
Chronic and unspecified myocarditis (429.0)	6	295	108-643
Chronic bronchitis and emphysema (491,492,496)	73	129	101-163
Fibrosing alveolitis (516.3)	0	0	0-88
Accidental poisoning by drugs (E850-E858)	4	436	119-1116
Suicide (E950-E959)	46	170	124-227

Number of deaths in women aged 20-74 during 1979-80 and 1982-90: 283

PMRs for women	Deaths	PMR	95% CI
Cause of death (ICD)			
Cancer of the cervix (180)	8	244	105-481

019 - Medical radiographers

Number of deaths in men aged 20-74 during 1979-80 and 1982-90: 157

No significant PMRs

Number of deaths in women aged 20-74 during 1979-80 and 1982-90: 196

No significant PMRs

020 - Physiotherapists

Number of deaths in men aged 20-74 during 1979-80 and 1982-90: 204

No significant PMRs

Number of deaths in women aged 20-74 during 1979-80 and 1982-90: 426

PMRs for women	Deaths	PMR	95% CI
Cause of death (ICD)			
Cancer of the ovary (183)	28	161	107-234
Chronic myeloid leukaemia (205.1)	4	399	109-1021

021 - Health professions nec

Number of deaths in men aged 20-74 during 1979-80 and 1982-90: 1837

PMRs for men	Deaths	PMR	95% CI
Cause of death (ICD)			
Cancer of the testis (186)	6	368	135-801
Atrial fibrillation (427.3)	6	324	119-705
Peripheral vascular disease (443,557)	11	201	100-359
Bronchopneumonia (485)	41	173	124-234

Number of deaths in women aged 20-74 during 1979-80 and 1982-90: 832

PMRs for women	Deaths	PMR	95% CI
Cause of death (ICD)			
Melanoma of skin (172)	12	207	107-362
Suicide (E950-E959)	28	158	105-229

022 - Veterinarians

Number of deaths in men aged 20-74 during 1979-80 and 1982-90: 383

PMRs for men	Deaths	PMR	95% CI
Cause of death (ICD)			
Cancer of trachea, bronchus and lung (162)	17	62	36-100
Suicide (E950-E959)	35	361	252-503

Number of deaths in women aged 20-74 during 1979-80 and 1982-90: 30

PMRs for women	Deaths	PMR	95% CI
Cause of death (ICD)			
Suicide (E950-E959)	7	414	166-853

023 - Driving instructors (not HGV)

Number of deaths in men aged 20-74 during 1979-80 and 1982-90: 1974

PMRs for men	Deaths	PMR	95% CI
Cause of death (ICD)			
Other alcohol related diseases (303,305.0,425.5,535.3,571.0-571.3, E860.0,E860.1)	4	31	9-80
Asthma (493)	14	186	102-312
Suicide (E950-E959)	54	142	107-185

Number of deaths in women aged 20-74 during 1979-80 and 1982-90: 60

PMRs for women

No significant PMRs

Appendix 4 Significant PMRs for each job group

Job group	Deaths	PMR	95% CI

024 - Literary and artistic occupations

Number of deaths in men aged 20-74 during 1979-80 and 1982-90: 11968

PMRs for men

Cause of death (ICD)	Deaths	PMR	95% CI
Tuberculosis (010-018,137)	29	165	111-238
Viral hepatitis (070)	15	242	135-399
Cancer of the oral cavity (141,143,144,145)	42	154	111-209
Cancer of the pharynx (specified) (146-148)	33	166	114-234
Cancer of the stomach (151)	164	68	58-79
Cancer of the liver (155)	60	134	102-172
Other cancer of skin (173)	19	182	110-285
Urothelial cancer (188,189.1-189.8)	102	81	66-98
Immunodeficiency (279.1)	140	341	287-403
Other alcohol related diseases (303,305.0,425.5,535.3,571.0-571.3, E860.0,E860.1)	158	175	149-205
Drug dependence (304)	24	352	225-525
Ischaemic heart disease (410-414)	3679	87	84-90
Acute bronchitis (466)	13	198	105-339
Pneumococcal and unspecified lobar pneumonia (481)	50	155	115-204
Bronchopneumonia (485)	200	135	117-156
Unspecified pneumonia (486)	28	172	114-249
Cirrhosis of the liver (571.5)	82	151	120-187
Motor vehicle traffic accidents (E810-E819)	238	87	76-98
Accidental poisoning by drugs (E850-E858)	44	263	191-353
Fall unspecified (E888)	21	161	100-247
Injured by fire (E890-E899)	26	215	140-315
Injury by machinery (E919)	1	13	0-73
Suicide (E950-E959)	349	114	102-126
Homicide (E960-E969)	29	229	153-329
Injury undetermined whether accidentally or purposely inflicted (E980-E989)	119	142	118-170

Number of deaths in women aged 20-74 during 1979-80 and 1982-90: 2356

PMRs for women

Cause of death (ICD)	Deaths	PMR	95% CI
Meningococcal infection (036)	4	370	101-947
Cancer of the oral cavity (141,143,144,145)	14	336	184-564
Cancer of the oesophagus (150)	31	146	99-208
Cancer of soft tissue (171)	13	218	116-373
Other alcohol related diseases (303,305.0,425.5,535.3,571.0-571.3, E860.0,E860.1)	40	217	155-296
Drug dependence (304)	4	448	122-1148
Anorexia nervosa (307.1)	5	383	124-893
Ischaemic heart disease (410-414)	289	75	67-84
Sub-arachnoid haemorrhage (430)	26	54	36-80
Bronchopneumonia (485)	43	143	103-193
Cirrhosis of the liver (571.5)	16	177	101-288
Motor vehicle traffic accidents (E810-E819)	41	72	51-97
Accidental poisoning by drugs (E850-E858)	15	210	117-346
Injured by fire (E890-E899)	11	317	158-568
Suicide (E950-E959)	90	165	132-202
Injury undetermined whether accidentally or purposely inflicted (E980-E989)	35	143	99-199

025 - Persons involved in sport

Number of deaths in men aged 20-74 during 1979-80 and 1982-90: 515

PMRs for men

Cause of death (ICD)	Deaths	PMR	95% CI
Cancer of the oral cavity (141,143,144,145)	4	394	107-1009
Cancer of the testis (186)	5	358	116-835
Dementia (290,331.0,331.1)	5	316	103-739
Bronchopneumonia (485)	18	222	132-351
Off-road vehicle accidents (E820-E825)	11	3005	1500-5378

Number of deaths in women aged 20-74 during 1979-80 and 1982-90: 74

PMRs for women

Cause of death (ICD)	Deaths	PMR	95% CI
Animal transport accidents (E827-E828)	6	8000	2936-17413

Job group	Deaths	PMR	95% CI

026 - Biological scientists

Number of deaths in men aged 20-74 during 1979-80 and 1982-90: 970

PMRs for men

Cause of death (ICD)	Deaths	PMR	95% CI
Cancer of the testis (186)	6	492	180-1070

Number of deaths in women aged 20-74 during 1979-80 and 1982-90: 162

No significant PMRs

027 - Chemical engineers and scientists

Number of deaths in men aged 20-74 during 1979-80 and 1982-90: 2712

PMRs for men

Cause of death (ICD)	Deaths	PMR	95% CI
Cancer of the pleura (163)	13	238	127-406
Chronic rheumatic heart disease (394-398)	20	198	121-306
Hypertensive disease (401-405)	10	50	24-92
Injury undetermined whether accidentally or purposely inflicted (E980-E989)	29	183	122-263

Number of deaths in women aged 20-74 during 1979-80 and 1982-90: 74

No significant PMRs

028 - Physical scientists and mathematicians

Number of deaths in men aged 20-74 during 1979-80 and 1982-90: 2357

PMRs for men

Cause of death (ICD)	Deaths	PMR	95% CI
Cancer of the pancreas (157)	26	66	43-97
Cancer of other mediastinum (164.1-164.9)	4	787	215-2016
Cancer of the brain (191)	65	169	130-215
Immunodeficiency (279.1)	0	0	0-84
Peripheral vascular disease (443,557)	13	237	126-405

Number of deaths in women aged 20-74 during 1979-80 and 1982-90: 137

No significant PMRs

029 - Electrical and electronic engineers (professional)

Number of deaths in men aged 20-74 during 1979-80 and 1982-90: 3789

PMRs for men

Cause of death (ICD)	Deaths	PMR	95% CI
Cancer of the rectum (154)	69	131	102-166
Cancer of the liver (155)	8	43	19-85
Cancer of the prostate (185)	56	76	57-99
Urothelial cancer (188,189.1-189.8)	58	152	115-196
Acute myeloid leukaemia (205.0)	11	44	22-79
Other cerebrovascular disease (431-438)	177	85	73-98
Cirrhosis of the liver (571.5)	12	57	30-100
Injury undetermined whether accidentally or purposely inflicted (E980-E989)	22	64	40-97

Number of deaths in women aged 20-74 during 1979-80 and 1982-90: 32

No significant PMRs

030 - Professional engineers nec

Number of deaths in men aged 20-74 during 1979-80 and 1982-90: 17780

PMRs for men

Cause of death (ICD)	Deaths	PMR	95% CI
Viral hepatitis (070)	4	30	8-77
Cancer of the pharynx (specified) (146-148)	20	60	37-93
Cancer of the stomach (151)	350	114	102-127
Cancer of the liver (155)	68	78	61-99
Cancer of trachea, bronchus and lung (162)	1616	123	117-129
Cancer of the pleura (163)	53	139	104-182
Myeloma (203)	90	77	62-95
Diabetes mellitus (250)	136	81	68-96

Appendix 4 Significant PMRs for each job group

Job group	Deaths	PMR	95% CI
Immunodeficiency (279.1)	10	41	20-76
Other alcohol related diseases (303,305.0,425.5,535.3,571.0-571.3, E860.0,E860.1)	90	75	60-92
Chronic bronchitis and emphysema (491,492,496)	484	117	107-128
Cirrhosis of the liver (571.5)	79	77	61-96
Suicide (E950-E959)	262	74	65-83
Homicide (E960-E969)	2	23	3-82
Injury undetermined whether accidentally or purposely inflicted (E980-E989)	61	61	47-79

Number of deaths in women aged 20-74 during 1979-80 and 1982-90: 73

No significant PMRs

031 - Draughtspersons

Number of deaths in men aged 20-74 during 1979-80 and 1982-90: 6397

PMRs for men

Cause of death (ICD)	Deaths	PMR	95% CI
Cancer of the larynx (161)	6	45	17-98
Cancer of trachea, bronchus and lung (162)	472	79	72-86
Cancer of the pleura (163)	18	236	140-373
Cancer of the brain (191)	89	133	107-164
Purpura (287)	4	404	110-1033
Other alcohol related diseases (303,305.0,425.5,535.3,571.0-571.3, E860.0,E860.1)	18	48	28-76
Multiple sclerosis (340)	37	190	134-262
Ischaemic heart disease (410-414)	2502	106	102-110
Chronic bronchitis and emphysema (491,492,496)	216	87	75-99
Poisoning by gas and other domestic fuels (E867,E868.1,E868.3)	4	367	100-940

Number of deaths in women aged 20-74 during 1979-80 and 1982-90: 221

PMRs for women

Cause of death (ICD)	Deaths	PMR	95% CI
Cancer of the female breast (174)	44	147	107-197
Cancer of the ovary (183)	18	198	118-313

032 - Laboratory technicians

Number of deaths in men aged 20-74 during 1979-80 and 1982-90: 6117

PMRs for men

Cause of death (ICD)	Deaths	PMR	95% CI
Cancer of the stomach (151)	167	124	106-145
Dementia (290,331.0,331.1)	29	163	109-235
Hypertensive disease (401-405)	25	57	37-84
Pulmonary hypertension (416.0)	5	373	121-871
Non-alcoholic gastritis and duodenitis (535.0-535.2,535.4-535.6)	6	663	243-1443
Inguinal hernia (550)	5	347	113-811
Injury by machinery (E919)	0	0	0-86

Number of deaths in women aged 20-74 during 1979-80 and 1982-90: 1133

PMRs for women

Cause of death (ICD)	Deaths	PMR	95% CI
Cancer of the kidney (except pelvis) (189.0)	14	215	117-360
Hypertensive disease (401-405)	2	25	3-90
Bronchiectasis (494)	5	346	112-807

033 - Architects and surveyors

Number of deaths in men aged 20-74 during 1979-80 and 1982-90: 7463

PMRs for men

Cause of death (ICD)	Deaths	PMR	95% CI
Septicaemia (038)	12	204	106-357
Multiple sclerosis (340)	31	149	101-211

Number of deaths in women aged 20-74 during 1979-80 and 1982-90: 116

PMRs for women

Cause of death (ICD)	Deaths	PMR	95% CI
Pulmonary embolism and phlebitis (415.1,451,453)	4	382	104-978

034 - Aircraft flight deck officers

Number of deaths in men aged 20-74 during 1979-80 and 1982-90: 661

PMRs for men

Cause of death (ICD)	Deaths	PMR	95% CI
Cancer of the colon (153)	26	162	106-237
Cancer of the pleura (163)	4	388	106-994
Melanoma of skin (172)	8	244	105-481
Cancer of the prostate (185)	25	208	134-307
Dementia (290,331.0,331.1)	7	440	177-907
Ischaemic heart disease (410-414)	168	69	59-80
Air transport accidents (E840-E845)	44	7760	5637-10425

Number of deaths in women aged 20-74 during 1979-80 and 1982-90: 4

No significant PMRs

035 - Air traffic controllers

Number of deaths in men aged 20-74 during 1979-80 and 1982-90: 183

PMRs for men

Cause of death (ICD)	Deaths	PMR	95% CI
Melanoma of skin (172)	4	495	135-1268
Other alcohol related diseases (303,305.0,425.5,535.3,571.0-571.3, E860.0,E860.1)	8	597	258-1175

Number of deaths in women aged 20-74 during 1979-80 and 1982-90: 6

No significant PMRs

036 - Seafarers

Number of deaths in men aged 20-74 during 1979-80 and 1982-90: 9184

PMRs for men

Cause of death (ICD)	Deaths	PMR	95% CI
Tuberculosis (010-018,137)	32	185	126-261
Cancer of the oral cavity (141,143,144,145)	56	273	207-355
Cancer of the pharynx (specified) (146-148)	45	290	212-388
Cancer of the liver (155)	46	154	113-206
Cancer of the larynx (161)	67	242	188-307
Cancer of trachea, bronchus and lung (162)	1143	110	104-117
Urothelial cancer (188,189.1-189.8)	78	78	62-97
Cancer of the brain (191)	47	62	46-82
Non-Hodgkin's lymphoma (200,202)	48	72	53-96
Other alcohol related diseases (303,305.0,425.5,535.3,571.0-571.3, E860.0,E860.1)	154	309	262-362
Ischaemic heart disease (410-414)	2609	81	78-84
Other cerebrovascular disease (431-438)	659	110	102-119
Pneumococcal and unspecified lobar pneumonia (481)	46	150	110-200
Bronchopneumonia (485)	222	144	126-165
Unspecified pneumonia (486)	22	175	110-265
Chronic bronchitis and emphysema (491,492,496)	525	114	104-124
Coal workers' pneumoconiosis (500)	0	0	0-34
Gastric ulcer (531)	37	184	129-253
Cirrhosis of the liver (571.5)	81	256	204-319
Pancreatitis (577.0,577.1)	27	170	112-248
Osteoporosis (733.0)	5	355	115-828
Motor vehicle traffic accidents (E810-E819)	107	81	66-97
Water transport accidents (E830-E838)	104	2088	1706-2531
Cold injury (E901)	8	321	139-633
Suicide (E950-E959)	126	83	69-98
Injury undetermined whether accidentally or purposely inflicted (E980-E989)	93	184	149-226

Appendix 4 Significant PMRs for each job group

Job group	Deaths	PMR	95% CI
Number of deaths in women aged 20-74 during 1979-80 and 1982-90: 6			
No significant PMRs			

037 - Technicians nec

Number of deaths in men aged 20-74 during 1979-80 and 1982-90: 4907

PMRs for men

Cause of death (ICD)	Deaths	PMR	95% CI
Cancer of the oral cavity (141,143,144,145)	4	34	9-88
Immunodeficiency (279.1)	4	34	9-86
Other alcohol related diseases (303,305.0,425.5,535.3,571.0-571.3, E860.0,E860.1)	24	65	41-96
Ischaemic heart disease (410-414)	1880	105	100-110
Atrial fibrillation (427.3)	0	0	0-99
Other cerebrovascular disease (431-438)	243	88	77-99
Cirrhosis of the liver (571.5)	13	55	29-95
Railway accidents (E800-E807)	7	329	132-677

Number of deaths in women aged 20-74 during 1979-80 and 1982-90: 223

PMRs for women

Cause of death (ICD)	Deaths	PMR	95% CI
Cancer of the rectum (154)	12	415	215-725
Cancer of the body of uterus (182)	5	311	101-726
Cancer of the kidney (except pelvis) (189.0)	5	357	116-833
Other cerebrovascular disease (431-438)	4	37	10-94

038 - Production and maintenance managers

Number of deaths in men aged 20-74 during 1979-80 and 1982-90: 29048

PMRs for men

Cause of death (ICD)	Deaths	PMR	95% CI
Cancer of the stomach (151)	699	110	102-119
Cancer of the colon (153)	764	108	101-116
Cancer of the rectum (154)	469	112	102-123
Cancer of the gall bladder (156)	64	137	105-175
Cancer of the pancreas (157)	493	110	101-121
Cancer of the peritoneum (158.8,158.9)	12	198	102-345
Cancer of trachea, bronchus and lung (162)	2985	107	103-111
Cancer of the pleura (163)	64	158	121-201
Cancer of the prostate (185)	694	108	100-117
Urothelial cancer (188,189.1-189.8)	378	113	102-125
Cancer of the kidney (except pelvis) (189.0)	265	114	100-128
Immunodeficiency (279.1)	22	47	30-72
Other alcohol related diseases (303,305.0,425.5,535.3,571.0-571.3, E860.0,E860.1)	133	70	59-83
Epilepsy (345)	14	54	29-90
Chronic rheumatic heart disease (394-398)	79	78	62-98
Ischaemic heart disease (410-414)	11432	105	103-107
Pulmonary embolism and phlebitis (415.1,451,453)	127	74	62-88
Endocarditis (421)	9	47	21-89
Sub-arachnoid haemorrhage (430)	219	115	100-131
Other cerebrovascular disease (431-438)	1676	91	87-96
Aortic aneurysms (441)	666	111	103-120
Bronchopneumonia (485)	288	79	71-89
Chronic bronchitis and emphysema (491,492,496)	874	87	81-93
Farmers' lung disease (495.0)	0	0	0-82
Duodenal ulcer (532)	62	75	58-96
Inguinal hernia (550)	2	28	3-100
Pancreatitis (577.0,577.1)	68	136	105-172
Renal stones (592)	0	0	0-69
Motor vehicle traffic accidents (E810-E819)	247	84	74-95
Pedal cycle accidents (E826)	0	0	0-96
Fall on stairs (E880)	12	47	24-81
Fall unspecified (E888)	15	56	31-92
Injured by fire (E890-E899)	9	49	22-93
Suicide (E950-E959)	224	53	46-60
Homicide (E960-E969)	3	18	4-53
Injury undetermined whether accidentally or purposely inflicted (E980-E989)	51	48	36-63

Number of deaths in women aged 20-74 during 1979-80 and 1982-90: 1135

PMRs for women

Cause of death (ICD)	Deaths	PMR	95% CI
Cancer of the stomach (151)	28	156	103-225
Cancer of trachea, bronchus and lung (162)	96	128	103-156
Other cerebrovascular disease (431-438)	59	73	56-94
Suicide (E950-E959)	4	26	7-65

039 - Managers in construction

Number of deaths in men aged 20-74 during 1979-80 and 1982-90: 7404

PMRs for men

Cause of death (ICD)	Deaths	PMR	95% CI
Cancer of the peritoneum (158.8,158.9)	6	426	156-928
Cancer of trachea, bronchus and lung (162)	867	121	113-129
Cancer of the pleura (163)	32	319	218-450
Cancer of the kidney (except pelvis) (189.0)	83	143	114-177
Cancer of the eye (190)	9	242	111-459
Immunodeficiency (279.1)	1	10	0-54
Other alcohol related diseases (303,305.0,425.5,535.3,571.0-571.3, E860.0,E860.1)	29	64	43-92
Other cerebrovascular disease (431-438)	414	88	80-97
Bronchopneumonia (485)	75	75	59-94
Fibrosing alveolitis (516.3)	23	195	124-294
Injured by fire (E890-E899)	0	0	0-79
Suicide (E950-E959)	48	47	35-63
Injury undetermined whether accidentally or purposely inflicted (E980-E989)	9	35	16-67

Number of deaths in women aged 20-74 during 1979-80 and 1982-90: 135

No significant PMRs

040 - Managers in transport, utilities and mining

Number of deaths in men aged 20-74 during 1979-80 and 1982-90: 17068

PMRs for men

Cause of death (ICD)	Deaths	PMR	95% CI
Cancer of the pharynx (specified) (146-148)	12	43	22-76
Cancer of the stomach (151)	428	115	104-126
Cancer of trachea, bronchus and lung (162)	1898	115	110-120
Cancer of the prostate (185)	343	89	80-99
Cancer of the testis (186)	21	174	107-266
Cancer of the penis (187.1-187.4)	13	193	103-330
Immunodeficiency (279.1)	9	35	16-66
Other alcohol related diseases (303,305.0,425.5,535.3,571.0-571.3, E860.0,E860.1)	66	61	47-78
Parkinson's disease (332)	37	70	49-96
Motor neurone disease (335.2)	42	69	50-94
Multiple sclerosis (340)	20	62	38-95
Ischaemic heart disease (410-414)	6853	107	105-110
Other cerebrovascular disease (431-438)	949	87	82-93
Arterial embolism and thrombosis (444)	16	183	105-297
Chronic bronchitis and emphysema (491,492,496)	649	109	101-118
Coal workers' pneumoconiosis (500)	9	901	412-1710
Duodenal ulcer (532)	64	132	102-169
Cirrhosis of the liver (571.5)	53	69	52-91
Air transport accidents (E840-E845)	2	25	3-90
Fall on stairs (E880)	5	34	11-79
Fall unspecified (E888)	5	33	11-76
Injury by falling object (E916)	10	271	130-498
Suicide (E950-E959)	135	55	46-65
Homicide (E960-E969)	3	32	7-92
Injury undetermined whether accidentally or purposely inflicted (E980-E989)	27	44	29-64

Number of deaths in women aged 20-74 during 1979-80 and 1982-90: 466

PMRs for women

Cause of death (ICD)	Deaths	PMR	95% CI
Ischaemic heart disease (410-414)	107	137	112-165
Gastric ulcer (531)	4	389	106-997
Motor vehicle traffic accidents (E810-E819)	2	28	3-100

Appendix 4 Significant PMRs for each job group

041 - Office managers

Number of deaths in men aged 20-74 during 1979-80 and 1982-90: 15987

PMRs for men

Cause of death (ICD)	Deaths	PMR	95% CI
Cancer of the stomach (151)	304	88	78-98
Cancer of the colon (153)	440	113	103-125
Cancer of trachea, bronchus and lung (162)	1423	93	89-98
Melanoma of skin (172)	80	129	102-160
Cancer of the brain (191)	208	118	102-135
Other alcohol related diseases (303,305.0,425.5,535.3,571.0-571.3, E860.0,E860.1)	71	67	53-85
Ischaemic heart disease (410-414)	6261	105	103-108
Other cerebrovascular disease (431-438)	913	90	84-96
Bronchopneumonia (485)	168	85	72-98
Cirrhosis of the liver (571.5)	55	76	57-99
Injured by fire (E890-E899)	3	28	6-82
Injury by machinery (E919)	1	15	0-81
Suicide (E950-E959)	195	78	68-90
Homicide (E960-E969)	3	30	6-88
Injury undetermined whether accidentally or purposely inflicted (E980-E989)	39	62	44-85

Number of deaths in women aged 20-74 during 1979-80 and 1982-90: 2081

PMRs for women

Cause of death (ICD)	Deaths	PMR	95% CI
Urothelial cancer (188,189.1-189.8)	21	178	110-273
Diabetes mellitus (250)	9	51	23-96
Suicide (E950-E959)	22	56	35-85

042 - Butchers

Number of deaths in men aged 20-74 during 1979-80 and 1982-90: 10312

PMRs for men

Cause of death (ICD)	Deaths	PMR	95% CI
Tuberculosis (010-018,137)	7	42	17-87
Cancer of the colon (153)	154	78	66-91
Cancer of trachea, bronchus and lung (162)	1268	108	102-114
Cancer of the pleura (163)	2	11	1-40
Immunodeficiency (279.1)	1	18	0-98
Pulmonary hypertension (416.0)	6	354	130-771
Chronic and unspecified myocarditis (429.0)	23	160	101-241
Chronic bronchitis and emphysema (491,492,496)	611	111	103-120
Renal failure (584-586)	49	135	100-179
Fall from building (E882)	0	0	0-74
Injury by machinery (E919)	1	17	0-96
Injury by cutting and piercing instruments or objects (E920)	4	735	200-1883

Number of deaths in women aged 20-74 during 1979-80 and 1982-90: 256

PMRs for women

Cause of death (ICD)	Deaths	PMR	95% CI
Cancer of the ovary (183)	2	27	3-96

043 - Fishmongers and poultry dressers

Number of deaths in men aged 20-74 during 1979-80 and 1982-90: 1651

PMRs for men

Cause of death (ICD)	Deaths	PMR	95% CI
Cancer of the colon (153)	15	46	26-76
Cancer of the larynx (161)	10	242	116-444
Cancer of trachea, bronchus and lung (162)	255	139	122-157
Dementia (290,331.0,331.1)	13	216	115-369
Other hernia (551-553)	4	580	158-1484

Number of deaths in women aged 20-74 during 1979-80 and 1982-90: 134

PMRs for women

Cause of death (ICD)	Deaths	PMR	95% CI
Cancer of the body of uterus (182)	4	401	109-1027
Diabetes mellitus (250)	5	349	113-814
Other alcohol related diseases (303,305.0,425.5,535.3,571.0-571.3, E860.0,E860.1)	4	683	186-1748

044 - Retailers and dealers

Number of deaths in men aged 20-74 during 1979-80 and 1982-90: 67972

PMRs for men

Cause of death (ICD)	Deaths	PMR	95% CI
Tuberculosis (010-018,137)	139	140	118-165
Cancer of the colon (153)	1318	84	80-89
Cancer of the gall bladder (156)	82	78	62-97
Cancer of the pancreas (157)	904	92	86-98
Cancer of trachea, bronchus and lung (162)	7013	108	105-110
Cancer of the pleura (163)	60	69	53-89
Melanoma of skin (172)	146	63	53-74
Cancer of the prostate (185)	1320	87	83-92
Cancer of the kidney (except pelvis) (189.0)	428	86	78-94
Cancer of the brain (191)	600	88	81-96
Cancer of the thyroid (193)	22	54	34-82
Non-Hodgkin's lymphoma (200,202)	494	86	78-94
Diabetes mellitus (250)	763	116	108-124
Parkinson's disease (332)	192	83	72-96
Motor neurone disease (335.2)	175	79	67-91
Epilepsy (345)	112	121	100-146
Ischaemic heart disease (410-414)	24489	99	97-100
Pulmonary heart disease unspecified (416.9)	97	124	101-152
Other cerebrovascular disease (431-438)	4593	103	100-106
Bronchopneumonia (485)	1139	119	112-126
Chronic bronchitis and emphysema (491,492,496)	2973	115	111-119
Farmers' lung disease (495.0)	0	0	0-43
Coal workers' pneumoconiosis (500)	1	13	0-74
Gastric ulcer (531)	151	118	100-139
Urinary infections (590,595,599.0)	87	126	101-155
Railway accidents (E800-E807)	7	34	14-70
Water transport accidents (E830-E838)	13	54	29-93
Air transport accidents (E840-E845)	14	47	25-78
Injured by fire (E890-E899)	71	141	110-178
Injury by machinery (E919)	11	44	22-78
Suicide (E950-E959)	1190	109	103-116
Homicide (E960-E969)	78	181	143-226
Injury undetermined whether accidentally or purposely inflicted (E980-E989)	353	118	106-131

Number of deaths in women aged 20-74 during 1979-80 and 1982-90: 31493

PMRs for women

Cause of death (ICD)	Deaths	PMR	95% CI
Cancer of the colon (153)	923	89	83-94
Cancer of trachea, bronchus and lung (162)	2220	105	101-110
Cancer of the female breast (174)	2995	84	81-87
Cancer of the cervix (180)	500	117	107-128
Cancer of the ovary (183)	1059	92	86-97
Cancer of the kidney (except pelvis) (189.0)	143	83	70-97
Cancer of the brain (191)	317	88	79-99
Acute lymphatic leukaemia (204.0)	16	59	34-96
Multiple sclerosis (340)	88	65	52-80
Chronic rheumatic heart disease (394-398)	330	117	105-131
Hypertensive disease (401-405)	258	114	100-128
Ischaemic heart disease (410-414)	6997	110	108-113
Acute myocarditis (422)	17	181	106-290
Sub-arachnoid haemorrhage (430)	607	110	101-119
Other cerebrovascular disease (431-438)	2625	106	102-110
Other aneurysm (442)	3	33	7-97
Bronchopneumonia (485)	573	112	103-121
Chronic bronchitis and emphysema (491,492,496)	862	111	103-118
Cirrhosis of the liver (571.5)	82	77	61-95
Glomerulonephritis (580-583)	18	62	37-98
Systemic lupus erythematosus (710.0)	17	61	36-98
Motor vehicle traffic accidents (E810-E819)	279	86	76-97

045 - Publicans and bar staff

Number of deaths in men aged 20-74 during 1979-80 and 1982-90: 18529

PMRs for men

Cause of death (ICD)	Deaths	PMR	95% CI
Cancer of the oral cavity (141,143,144,145)	117	275	227-330

Appendix 4 Significant PMRs for each job group

Job group	Deaths	PMR	95% CI
Cancer of the pharynx (specified) (146-148)	71	230	180-291
Cancer of the oesophagus (150)	261	116	102-131
Cancer of the stomach (151)	333	83	74-92
Cancer of the small intestine (152)	5	39	13-91
Cancer of the colon (153)	307	72	64-80
Cancer of the liver (155)	112	162	134-195
Cancer of the larynx (161)	119	262	217-314
Cancer of trachea, bronchus and lung (162)	2300	129	123-134
Melanoma of skin (172)	40	57	41-78
Cancer of the prostate (185)	326	87	78-97
Cancer of the penis (187.1-187.4)	14	195	106-327
Cancer of the kidney (except pelvis) (189.0)	99	70	57-85
Cancer of the brain (191)	128	63	53-75
Non-Hodgkin's lymphoma (200,202)	116	71	58-85
Hodgkin's disease (201)	19	55	33-85
Myeloma (203)	43	48	35-64
Acute myeloid leukaemia (205.0)	41	58	42-79
Diabetes mellitus (250)	207	118	103-136
Other alcohol related diseases (303,305.0,425.5,535.3,571.0-571.3, E860.0,E860.1)	458	365	333-400
Parkinson's disease (332)	29	55	37-79
Motor neurone disease (335.2)	40	64	45-87
Multiple sclerosis (340)	22	55	35-84
Ischaemic heart disease (410-414)	5443	80	78-82
Other cardiomyopathies (425.0-425.4, 425.6-425.9)	109	131	107-158
Other cerebrovascular disease (431-438)	1372	121	114-127
Aortic aneurysm (441)	302	85	75-95
Peripheral vascular disease (443,557)	83	149	119-185
Oesophageal varices (456.0-456.2)	9	271	124-515
Other bacterial pneumonia (482)	14	222	121-372
Bronchopneumonia (485)	291	122	109-137
Chronic bronchitis and emphysema (491,492,496)	966	146	137-156
Other pneumoconiosis (503,505)	4	425	116-1087
Oesophageal disease (530)	17	186	109-299
Gastric ulcer (531)	64	188	145-240
Duodenal ulcer (532)	71	130	102-164
Cirrhosis of the liver (571.5)	243	301	264-341
Motor vehicle traffic accidents (E810-E819)	171	63	54-73
Fall on stairs (E880)	43	229	166-309
Fall from ladder or scaffolding (E881)	2	27	3-96
Fall unspecified (E888)	37	193	136-267
Injured by fire (E890-E899)	33	202	139-284
Injury by machinery (E919)	3	28	6-80
Injury by electric current (E925)	1	17	0-96
Injury undetermined whether accidentally or purposely inflicted (E980-E989)	125	130	108-155

Number of deaths in women aged 20-74 during 1979-80 and 1982-90: 6092

PMRs for women

Cause of death (ICD)	Deaths	PMR	95% CI
Cancer of the oral cavity (141,143,144,145)	30	271	183-387
Cancer of the pharynx (specified) (146-148)	22	251	157-380
Cancer of trachea, bronchus and lung (162)	544	128	118-139
Melanoma of skin (172)	19	63	38-98
Cancer of the female breast (174)	481	73	67-80
Cancer of the cervix (180)	159	158	134-185
Cancer of the ovary (183)	149	70	59-82
Cancer of the brain (191)	49	70	52-93
Non-Hodgkin's lymphoma (200,202)	40	69	49-94
Hodgkin's disease (201)	3	28	6-83
Myeloma (203)	22	63	39-95
Acute lymphatic leukaemia (204.0)	0	0	0-77
Chronic myeloid leukaemia (205.1)	3	27	6-79
Other alcohol related diseases (303,305.0,425.5,535.3,571.0-571.3, E860.0,E860.1)	101	291	237-354
Multiple sclerosis (340)	13	56	30-96
Oesophageal varices (456.0-456.2)	5	526	171-1227
Chronic bronchitis and emphysema (491,492,496)	239	152	133-172
Asthma (493)	56	138	104-179
Cirrhosis of the liver (571.5)	73	336	264-423

Job group	Deaths	PMR	95% CI
Accidental poisoning by drugs (E850-E858)	19	181	109-283
Suicide (E950-E959)	48	65	48-86
Homicide (E960-E969)	28	315	209-456

046 - Caterers

Number of deaths in men aged 20-74 during 1979-80 and 1982-90: 8915

PMRs for men

Cause of death (ICD)	Deaths	PMR	95% CI
Viral hepatitis (070)	14	352	192-591
Cancer of the pharynx (specified) (146-148)	36	243	170-336
Cancer of the colon (153)	133	74	62-88
Cancer of the liver (155)	56	193	146-251
Melanoma of skin (172)	18	63	37-99
Cancer of the testis (186)	4	37	10-94
Urothelial cancer (188,189.1-189.8)	117	123	102-148
Diabetes mellitus (250)	105	128	104-155
Immunodeficiency (279.1)	62	404	310-518
Dementia (290,331.0,331.1)	14	54	30-91
Other alcohol related diseases (303,305.0,425.5,535.3,571.0-571.3, E860.0,E860.1)	97	177	143-216
Multiple sclerosis (340)	10	47	23-86
Ischaemic heart disease (410-414)	2992	95	92-98
Influenza (487)	0	0	0-86
Coal workers' pneumoconiosis (500)	0	0	0-79
Duodenal ulcer (532)	54	181	136-236
Non-alcoholic gastritis and duodenitis (535.0-535.2,535.4-535.6)	7	627	252-1292
Cirrhosis of the liver (571.5)	52	157	117-206
Injured by fire (E890-E899)	18	170	101-269
Injury by machinery (E919)	0	0	0-59
Injury undetermined whether accidentally or purposely inflicted (E980-E989)	90	148	119-182

Number of deaths in women aged 20-74 during 1979-80 and 1982-90: 17428

PMRs for women

Cause of death (ICD)	Deaths	PMR	95% CI
Cancer of the stomach (151)	376	116	104-128
Cancer of trachea, bronchus and lung (162)	1564	114	108-119
Cancer of the female breast (174)	1389	93	88-98
Multiple sclerosis (340)	26	60	39-88
Guillain Barre syndrome (357.0)	5	363	118-846
Chronic rheumatic heart disease (394-398)	122	75	62-89
Aortic aneurysm (441)	178	117	100-135
Bronchopneumonia (485)	315	88	78-98
Motor vehicle traffic accidents (E810-E819)	168	118	100-137

047 - Farmers

Number of deaths in men aged 20-74 during 1979-80 and 1982-90: 60268

PMRs for men

Cause of death (ICD)	Deaths	PMR	95% CI
Tuberculosis (010-018,137)	76	72	57-90
Cancer of the liver (155)	125	65	54-78
Cancer of the gall bladder (156)	58	67	51-87
Cancer of the peritoneum (158.8,158.9)	6	39	14-85
Cancer of the larynx (161)	125	74	61-88
Cancer of trachea, bronchus and lung (162)	5930	88	86-91
Cancer of the pleura (163)	20	25	15-39
Cancer of the prostate (185)	1361	112	106-118
Urothelial cancer (188,189.1-189.8)	598	86	79-93
Cancer of the kidney (except pelvis) (189.0)	305	84	75-94
Diabetes mellitus (250)	672	119	110-128
Immunodeficiency (279.1)	22	52	33-80
Haemolytic anaemia (283)	10	268	128-492
Other alcohol related diseases (303,305.0,425.5,535.3,571.0-571.3, E860.0,E860.1)	121	47	39-56
Parkinson's disease (332)	250	125	110-142
Multiple sclerosis (340)	110	132	108-159
Epilepsy (345)	148	176	149-207
Hypertensive disease (401-405)	587	126	116-137
Ischaemic heart disease (410-414)	20478	97	96-99

Appendix 4 Significant PMRs for each job group

Job group	Deaths	PMR	95% CI
Pulmonary embolism and phlebitis (415.1,451,453)	492	125	115-137
Mitral valve disorders (424.0)	68	158	123-201
Aortic valve disorders (424.1)	346	127	114-141
Other cardiomyopathies (425.0-425.4,425.6-425.9)	233	114	100-130
Atrial fibrillation (427.3)	81	129	102-160
Other cerebrovascular disease (431-438)	4912	115	111-118
Aortic aneurysm (441)	789	72	67-77
Peripheral vascular disease (443,557)	200	86	75-99
Bronchopneumonia (485)	1232	111	105-117
Influenza (487)	46	163	120-218
Chronic bronchitis and emphysema (491,492,496)	2879	90	86-93
Farmers' lung disease (495.0)	56	1089	823-1416
Other and unspecified allergic pneumonitis (495.1,495.3-495.9)	7	787	317-1622
Coal workers' pneumoconiosis (500)	0	0	0-5
Asbestosis (501)	0	0	0-32
Silicosis (502)	2	21	2-74
Other pneumoconiosis (503,505)	3	30	6-88
Inguinal hernia (550)	41	191	137-259
Other hernia (551-553)	41	149	107-202
Diverticular disease (562)	43	71	51-95
Cirrhosis of the liver (571.5)	110	57	47-69
Urinary infections (590,595,599.0)	88	128	102-157
Hyperplasia of prostate (600)	94	132	106-161
Infections of skin, joints and bone (680-686,711,730)	36	181	127-251
Motor vehicle traffic accidents (E810-E819)	843	111	103-118
Off-road motor vehicle accidents (E820-E825)	38	255	180-350
Pedal cycle accidents (E826)	15	190	106-314
Animal transport accidents (E827-E828)	15	468	262-773
Water transport accidents (E830-E838)	5	21	7-49
Other vehicle accidents (E846-E848)	0	0	0-100
Accidental poisoning by drugs (E850-E858)	27	59	39-87
Pesticide poisoning (E863)	4	1455	396-3724
Poisoning by other gases (E869)	7	417	168-859
Fall on stairs (E880)	39	66	47-90
Slipping and tripping (E885)	17	193	112-309
Cold injury (E901)	22	162	101-246
Injury by animals and plants (E905-E906)	21	775	479-1186
Injury by falling object (E916)	35	156	109-217
Injury by machinery (E919)	147	457	386-538
Injury by firearms (E922)	23	670	424-1006
Injury by electric current (925)	29	213	143-307
Suicide (E950-E959)	1215	156	147-165

Number of deaths in women aged 20-74 during 1979-80 and 1982-90: 2512

PMRs for women

Cause of death (ICD)	Deaths	PMR	95% CI
Other cancer of skin (173)	7	419	169-864
Cancer of the female breast (174)	194	85	74-98
Cancer of the ovary (183)	94	123	100-151
Sub-arachnoid haemorrhage (430)	26	62	40-91
Acute bronchitis (466)	7	288	116-593
Chronic bronchitis and emphysema (491,492,496)	54	72	54-93
Fibrosing alveolitis (516.3)	7	261	105-538
Motor vehicle traffic accidents (E810-E819)	39	146	104-200
Injury by machinery (E919)	4	2632	717-6738
Suicide (E950-E959)	38	145	103-199

048 - Armed forces

Number of deaths in men aged 20-74 during 1979-80 and 1982-90: 10733

No significant PMRs

Number of deaths in women aged 20-74 during 1979-80 and 1982-90: 231

No significant PMRs

049 - Police

Number of deaths in men aged 20-74 during 1979-80 and 1982-90: 7688

PMRs for men

Cause of death (ICD)	Deaths	PMR	95% CI
Tuberculosis (010-018,137)	3	22	4-64
Cancer of the pharynx (specified) (146-148)	5	40	13-94
Cancer of the oesophagus (150)	129	150	125-178
Cancer of the colon (153)	210	128	111-147
Cancer of the rectum (154)	124	125	104-150
Cancer of the pancreas (157)	130	129	108-153
Cancer of the nose and nasal sinuses (160)	9	235	107-446
Melanoma of skin (172)	40	144	103-197
Cancer of the kidney (except pelvis) (189.0)	73	134	105-169
Cancer of the brain (191)	104	127	104-154
Myeloma (203)	47	140	103-186
Immunodeficiency (279.1)	3	17	3-50
Motor neurone disease (335.2)	44	186	135-250
Epilepsy (345)	7	39	16-81
Chronic rheumatic heart disease (394-398)	18	58	35-92
Endocarditis (421)	0	0	0-72
Other cerebrovascular disease (431-438)	413	89	80-98
Aortic aneurysm (441)	188	126	108-145
Peripheral vascular disease (443,557)	11	46	23-82
Bronchopneumonia (485)	80	68	54-84
Chronic bronchitis and emphysema (491,492,496)	190	64	55-74
Asthma (493)	19	61	37-95
Motor vehicle traffic accidents (E810-E819)	226	165	144-188
Accidental poisoning by drugs (E850-E858)	1	11	0-60
Injured by fire (E890-E899)	1	13	0-74
Non recreational drowning (E910.3)	5	1008	327-2352
Suicide (E950-E959)	132	78	65-92
Homicide (E960-E969)	16	263	150-427
Injury undetermined whether accidentally or purposely inflicted (E980-E989)	27	55	36-81

Number of deaths in women aged 20-74 during 1979-80 and 1982-90: 150

PMRs for women

Cause of death (ICD)	Deaths	PMR	95% CI
Melanoma of skin (172)	6	421	154-916

050 - Fire service personnel

Number of deaths in men aged 20-74 during 1979-80 and 1982-90: 2968

PMRs for men

Cause of death (ICD)	Deaths	PMR	95% CI
Cancer of the oesophagus (150)	46	135	99-180
Cancer of the stomach (151)	91	150	120-184
Cancer of the gall bladder (156)	10	224	108-413
Cancer of the larynx (161)	1	16	0-87
Immunodeficiency (279.1)	0	0	0-63
Other fall (E884)	5	455	148-1063
Injured by fire (E890-E899)	8	311	134-612

Number of deaths in women aged 20-74 during 1979-80 and 1982-90: 16

No significant PMRs

051 - Launderers and dry cleaners

Number of deaths in men aged 20-74 during 1979-80 and 1982-90: 1695

PMRs for men

Cause of death (ICD)	Deaths	PMR	95% CI
Cancer of soft tissue (171)	7	415	167-855
Chronic bronchitis and emphysema (491,492,496)	76	75	59-94
Pancreatitis (577.0,577.1)	7	249	100-513
Renal failure (584-586)	12	211	109-368

Number of deaths in women aged 20-74 during 1979-80 and 1982-90: 3660

PMRs for women

Cause of death (ICD)	Deaths	PMR	95% CI
Mitral valve disorders (424.0)	0	0	0-79

Appendix 4 Significant PMRs for each job group

Job group	Deaths	PMR	95% CI
Polyarteritis nodosa (446.0)	4	462	126-1184
Bronchopneumonia (485)	110	127	105-154
Cholelithiasis and cholecystitis (574,575,576.1-576.4)	10	209	100-384
Renal failure (584-586)	23	165	104-247

052 - Hairdressers

Number of deaths in men aged 20-74 during 1979-80 and 1982-90: 3273

PMRs for men

Cause of death (ICD)	Deaths	PMR	95% CI
Viral hepatitis (070)	4	388	106-992
Cancer of the colon (153)	47	73	54-98
Cancer of soft tissue (171)	0	0	0-94
Immunodeficiency (279.1)	31	931	632-1324
Chronic rheumatic heart disease (394-398)	21	185	114-282
Sub-arachnoid haemorrhage (430)	9	50	23-95
Bronchopneumonia (485)	83	134	107-166
Inguinal hernia (550)	5	405	131-945
Motor vehicle traffic accidents (E810-E819)	30	54	36-77
Fall unspecified (E888)	9	223	102-424

Number of deaths in women aged 20-74 during 1979-80 and 1982-90: 2203

PMRs for women

Cause of death (ICD)	Deaths	PMR	95% CI
Cancer of the oral cavity (141,143,144,145)	8	293	127-578
Cancer of the gall bladder (156)	1	17	0-95
Cancer of the female breast (174)	249	114	100-129
Diabetes mellitus (250)	11	48	24-86
Other alcohol related diseases (303,305.0,425.5,535.3,571.0-571.3, E860.0,E860.1)	26	211	138-310
Multiple sclerosis (340)	21	200	123-306
Aortic aneurysm (441)	9	50	23-95
Duodenal ulcer (532)	12	202	104-353
Cirrhosis of the liver (571.5)	14	210	115-352

053 - Office workers and cashiers

Number of deaths in men aged 20-74 during 1979-80 and 1982-90: 107835

PMRs for men

Cause of death (ICD)	Deaths	PMR	95% CI
Cancer of the stomach (151)	2144	95	91-99
Cancer of trachea, bronchus and lung (162)	10304	96	94-98
Cancer of the pleura (163)	92	76	61-94
Immunodeficiency (279.1)	157	129	110-151
Multiple sclerosis (340)	289	113	100-127
Pulmonary embolism and phlebitis (415.1,451,453)	767	108	100-116
Bronchopneumonia (485)	1925	105	100-110
Coal workers' pneumoconiosis (500)	2	23	3-84
Cirrhosis of the liver (571.5)	316	89	79-99
Railway accidents (E800-E807)	45	160	117-215
Motor vehicle traffic accidents (E810-E819)	1034	84	79-89
Water transport accidents (E830-E838)	11	51	25-91
Suicide (E950-E959)	1371	93	88-98
Homicide (E960-E969)	32	64	44-90

Number of deaths in women aged 20-74 during 1979-80 and 1982-90: 80873

PMRs for women

Cause of death (ICD)	Deaths	PMR	95% CI
Cancer of the colon (153)	2780	105	101-109
Cancer of the female breast (174)	10522	106	104-108
Cancer of the brain (191)	1069	107	101-114
Multiple sclerosis (340)	463	111	101-122
Ischaemic heart disease (410-414)	14535	96	95-98

054 - Postal workers

Number of deaths in men aged 20-74 during 1979-80 and 1982-90: 21135

PMRs for men

Cause of death (ICD)	Deaths	PMR	95% CI
Tuberculosis (010-018,137)	30	65	43-92
Viral hepatitis (070)	1	17	0-96
Cancer of the stomach (151)	613	110	101-119
Cancer of the small intestine (152)	3	32	7-93
Cancer of the pancreas (157)	304	118	105-132
Cancer of the prostate (185)	377	116	104-128
Urothelial cancer (188,189.1-189.8)	303	131	116-146
Cancer of the brain (191)	147	123	104-145
Non-Hodgkin's lymphoma (200,202)	158	133	113-155
Hodgkin's disease (201)	36	145	101-200
Immunodeficiency (279.1)	12	208	108-364
Drug dependence (304)	2	25	3-91
Parkinson's disease (332)	77	128	101-160
Motor neurone disease (335.2)	69	149	116-189
Chronic rheumatic heart disease (394-398)	53	74	55-96
Ischaemic heart disease (410-414)	7376	103	101-106
Pulmonary heart disease unspecified (416.9)	23	58	37-87
Other cerebrovascular disease (431-438)	1330	88	83-93
Aortic aneurysm (441)	391	131	119-145
Acute upper respiratory infection (460-465)	4	376	103-963
Pneumococcal and unspecified lobar pneumonia (481)	63	75	58-97
Bronchopneumonia (485)	361	78	71-87
Influenza (487)	4	39	11-99
Chronic bronchitis and emphysema (491,492,496)	1068	79	74-83
Coal workers' pneumoconiosis (500)	1	3	0-17
Fibrosing alveolitis (516.3)	42	157	113-212
Crohn's disease (555)	10	247	118-454
Renal failure (584-586)	49	68	50-90
Systemic lupus erythematosus (710.0)	5	316	103-737
Fall from building (E882)	7	46	18-94
Injured by fire (E890-E899)	10	39	19-72
Injury by falling object (E916)	0	0	0-42
Injury by machinery (E919)	0	0	0-33
Injury undetermined whether accidentally or purposely inflicted (E980-E989)	80	78	62-98

Number of deaths in women aged 20-74 during 1979-80 and 1982-90: 1222

PMRs for women

Cause of death (ICD)	Deaths	PMR	95% CI
Cancer of the female breast (174)	110	122	100-147
Aortic valve disorders (424.1)	0	0	0-99
Other cerebrovascular disease (431-438)	76	74	58-92

055 - Petrol pump attendants

Number of deaths in men aged 20-74 during 1979-80 and 1982-90: 1081

PMRs for men

Cause of death (ICD)	Deaths	PMR	95% CI
Cancer of the colon (153)	13	57	30-97
Cancer of trachea, bronchus and lung (162)	131	122	102-145
Other cardiomyopathies (425.0-425.4, 425.6-425.9)	0	0	0-86
Other cerebrovascular disease (431-438)	90	128	103-158

Number of deaths in women aged 20-74 during 1979-80 and 1982-90: 169

PMRs for women

Cause of death (ICD)	Deaths	PMR	95% CI
Peripheral vascular disease (443,557)	4	828	226-2120
Chronic bronchitis and emphysema (491,492,496)	8	240	103-472

056 - Van sales persons

Number of deaths in men aged 20-74 during 1979-80 and 1982-90: 4670

PMRs for men

Cause of death (ICD)	Deaths	PMR	95% CI
Cancer of the pleura (163)	3	28	6-81

Appendix 4 Significant PMRs for each job group

Job group	Deaths	PMR	95% CI
Ischaemic heart disease (410-414)	1502	93	88-98
Other cardiomyopathies (425.0-425.4, 425.6-425.9)	31	173	117-246
Bronchopneumonia (485)	103	136	111-165
Bronchiectasis (494)	12	197	102-344
Injury by machinery (E919)	0	0	0-85
Suicide (E950-E959)	105	132	108-160

Number of deaths in women aged 20-74 during 1979-80 and 1982-90: 141

No significant PMRs

057 - Sales representatives

Number of deaths in men aged 20-74 during 1979-80 and 1982-90: 25580

PMRs for men

Cause of death (ICD)	Deaths	PMR	95% CI
Tuberculosis (010-018,137)	23	51	33-77
Cancer of the colon (153)	632	114	105-123
Cancer of the rectum (154)	399	119	107-131
Cancer of the male breast (175)	3	30	6-89
Cancer of the kidney (except pelvis) (189.0)	208	115	100-132
Diabetes mellitus (250)	211	86	75-98
Immunodeficiency (279.1)	26	59	39-87
Haemolytic anaemia (283)	5	421	137-981
Agranulocytosis (288.0)	5	476	155-1111
Epilepsy (345)	22	51	32-78
Ischaemic heart disease (410-414)	9765	103	101-105
Aortic valve disorders (424.1)	84	78	63-97
Other cerebrovascular disease (431-438)	1533	95	90-100
Aortic aneurysm (441)	604	117	107-126
Wegener's granulomatosis (446.4)	8	233	101-460
Bronchopneumonia (485)	331	82	73-91
Chronic bronchitis and emphysema (491,492,496)	863	83	78-89
Gastric ulcer (531)	37	71	50-99
Motor vehicle traffic accidents (E810-E819)	460	133	121-145
Fall on stairs (E880)	15	58	32-95
Fall unspecified (E888)	15	47	26-78
Suicide (E950-E959)	512	115	105-126
Injury undetermined whether accidentally or purposely inflicted (E980-E989)	93	74	60-91

Number of deaths in women aged 20-74 during 1979-80 and 1982-90: 1100

PMRs for women

Cause of death (ICD)	Deaths	PMR	95% CI
Pneumococcal and unspecified lobar pneumonia (481)	8	234	101-460
Motor vehicle traffic accidents (E810-E819)	42	206	149-279

058 - Security workers

Number of deaths in men aged 20-74 during 1979-80 and 1982-90: 24116

PMRs for men

Cause of death (ICD)	Deaths	PMR	95% CI
Cancer of the oral cavity (141,143,144,145)	69	129	101-164
Cancer of the peritoneum (158.8,158.9)	1	12	0-66
Non-Hodgkin's lymphoma (200,202)	124	83	69-99
Multiple sclerosis (340)	20	54	33-84
Ischaemic heart disease (410-414)	8847	104	102-106
Other cerebrovascular disease (431-438)	1462	90	85-94
Aortic aneurysm (441)	424	119	108-131
Bronchopneumonia (485)	393	85	77-94
Chronic bronchitis and emphysema (491,492,496)	1324	93	88-98
Coal workers' pneumoconiosis (500)	2	5	1-17
Silicosis (502)	0	0	0-64
Pulmonary fibrosis (515)	15	58	33-96
Fibrosing alveolitis (516.3)	19	58	35-91
Poisoning by liquified petroleum gas (E868.0)	4	660	180-1690
Fall from ladder or scaffolding (E881)	4	38	10-97
Injury by machinery (E919)	3	24	5-70

Job group	Deaths	PMR	95% CI

Number of deaths in women aged 20-74 during 1979-80 and 1982-90: 737

PMRs for women

Cause of death (ICD)	Deaths	PMR	95% CI
Cancer of the stomach (151)	3	22	5-64
Cancer of the colon (153)	35	167	116-233
Motor vehicle traffic accidents (E810-E819)	14	204	112-343

059 - Cooks and kitchen porters

Number of deaths in men aged 20-74 during 1979-80 and 1982-90: 8595

PMRs for men

Cause of death (ICD)	Deaths	PMR	95% CI
Tuberculosis (010-018,137)	34	206	142-288
Viral hepatitis (070)	15	492	275-812
Cancer of the oral cavity (141,143,144,145)	31	152	103-217
Cancer of the pharynx (specified) (146-148)	54	334	251-435
Cancer of the oesophagus (150)	76	79	62-98
Cancer of the stomach (151)	149	72	61-85
Cancer of the rectum (154)	84	78	62-97
Cancer of the liver (155)	70	273	213-345
Cancer of the pleura (163)	4	23	6-58
Cancer of the prostate (185)	93	80	64-98
Cancer of the brain (191)	53	75	56-98
Diabetes mellitus (250)	106	143	117-173
Immunodeficiency (279.1)	31	496	336-704
Other alcohol related diseases (303,305.0,425.5,535.3,571.0-571.3, E860.0,E860.1)	114	220	182-265
Multiple sclerosis (340)	5	35	11-81
Epilepsy (345)	39	172	122-236
Ischaemic heart disease (410-414)	2498	89	85-92
Other cardiomyopathies (425.0-425.4, 425.6-425.9)	52	143	107-187
Other cerebrovascular disease (431-438)	586	111	102-121
Oesophageal varices (456.0-456.2)	4	384	104-982
Bronchopneumonia (485)	182	127	109-147
Gastric ulcer (531)	32	155	106-219
Duodenal ulcer (532)	56	170	129-221
Non-alcoholic gastritis and duodenitis (535.0-535.2,535.4-535.6)	5	431	140-1005
Cirrhosis of the liver (571.5)	41	168	121-229
Renal failure (584-586)	39	147	104-201
Accidental poisoning by drugs (E850-E858)	41	204	147-277
Fall from ladder or scaffolding (E881)	1	14	0-78
Fall unspecified (E888)	21	167	103-256
Injury by falling object (E916)	0	0	0-50
Injury by machinery (E919)	0	0	0-32
Homicide (E960-E969)	27	183	121-267
Injury undetermined whether accidentally or purposely inflicted (E980-E989)	110	140	115-169

Number of deaths in women aged 20-74 during 1979-80 and 1982-90: 11249

PMRs for women

Cause of death (ICD)	Deaths	PMR	95% CI
Cancer of the stomach (151)	249	115	101-130
Cancer of the pancreas (157)	203	118	102-135
Diabetes mellitus (250)	176	126	108-146
Pulmonary heart disease unspecified (416.9)	7	40	16-83
Other cardiomyopathies (425.0-425.4, 425.6-425.9)	39	168	119-230
Other colitis (558)	1	18	0-98
Suicide (E950-E959)	55	73	55-95

060 - Other service personnel

Number of deaths in men aged 20-74 during 1979-80 and 1982-90: 38871

PMRs for men

Cause of death (ICD)	Deaths	PMR	95% CI
Cancer of the prostate (185)	616	109	101-118
Cancer of the thyroid (193)	24	158	101-236
Immunodeficiency (279.1)	41	450	323-611
Agranulocytosis (288.0)	6	484	178-1053
Multiple sclerosis (340)	30	64	43-92
Unspecified pneumonia (486)	76	134	106-168

Appendix 4 Significant PMRs for each job group

Job group	Deaths	PMR	95% CI
Coal workers' pneumoconiosis (500)	3	8	2-23
Silicosis (502)	2	28	3-99
Other pneumoconiosis (503,505)	2	20	2-73
Cirrhosis of the liver (571.5)	130	119	100-142
Rheumatoid arthritis (714)	24	64	41-96
Fall from ladder or scaffolding (E881)	51	259	193-341
Fall from building (E882)	42	162	117-219
Injured by fire (E890-E899)	39	72	51-98
Injury by falling object (E916)	2	13	2-47
Injury by machinery (E919)	8	39	17-77

Number of deaths in women aged 20-74 during 1979-80 and 1982-90: 42482

PMRs for women

Cause of death (ICD)	Deaths	PMR	95% CI
Multiple sclerosis (340)	54	73	54-95
Chronic rheumatic heart disease (394-398)	332	84	75-93
Ischaemic heart disease (410-414)	9805	97	95-99
Chronic bronchitis and emphysema (491,492,496)	1351	87	82-92
Asbestosis (501)	0	0	0-78
Byssinosis (504)	1	6	0-31

061 - Hospital porters and ward orderlies

Number of deaths in men aged 20-74 during 1979-80 and 1982-90: 6889

PMRs for men

Cause of death (ICD)	Deaths	PMR	95% CI
Viral hepatitis (070)	7	334	134-688
Cancer of the nose and nasal sinuses (160)	8	256	111-505
Chronic rheumatic heart disease (394-398)	36	149	104-207
Acute myocarditis (422)	4	388	106-994
Coal workers' pneumoconiosis (500)	3	23	5-67
Accidental poisoning by drugs (E850-E858)	12	218	113-381
Injury by machinery (E919)	0	0	0-88

Number of deaths in women aged 20-74 during 1979-80 and 1982-90: 4211

PMRs for women

Cause of death (ICD)	Deaths	PMR	95% CI
Cancer of trachea, bronchus and lung (162)	399	114	103-126
Diabetes mellitus (250)	65	136	105-173
Chronic rheumatic heart disease (394-398)	26	66	43-97
Bronchopneumonia (485)	58	74	56-95
Hepatitis (571.4,573.3)	8	261	113-515
Slipping and tripping (E885)	6	448	165-976
Suicide (E950-E959)	62	132	101-169

062 - Ambulance workers

Number of deaths in men aged 20-74 during 1979-80 and 1982-90: 1966

PMRs for men

Cause of death (ICD)	Deaths	PMR	95% CI
Cancer of the pancreas (157)	37	143	101-198
Cancer of the brain (191)	24	154	99-229
Ischaemic heart disease (410-414)	771	111	103-119

Number of deaths in women aged 20-74 during 1979-80 and 1982-90: 62

PMRs for women

Cause of death (ICD)	Deaths	PMR	95% CI
Cancer of trachea, bronchus and lung (162)	10	227	109-417
Suicide (E950-E959)	5	520	169-1213

063 - Railway station workers

Number of deaths in men aged 20-74 during 1979-80 and 1982-90: 8293

PMRs for men

Cause of death (ICD)	Deaths	PMR	95% CI
Cancer of the pharynx (specified) (146-148)	7	47	19-96
Cancer of trachea, bronchus and lung (162)	916	86	80-91
Cancer of the eye (190)	7	333	134-686
Non-Hodgkin's lymphoma (200,202)	60	142	108-183
Ischaemic heart disease (410-414)	2886	105	101-109
Bronchopneumonia (485)	179	84	72-97
Chronic bronchitis and emphysema (491,492,496)	507	89	81-97
Renal failure (584-586)	44	147	107-198
Railway accidents (E800-E807)	25	1248	806-1844
Fracture unspecified (E887)	5	331	107-772

Number of deaths in women aged 20-74 during 1979-80 and 1982-90: 166

No significant PMRs

064 - Undertakers

Number of deaths in men aged 20-74 during 1979-80 and 1982-90: 821

PMRs for men

Cause of death (ICD)	Deaths	PMR	95% CI
Cancer of trachea, bronchus and lung (162)	110	123	101-148

Number of deaths in women aged 20-74 during 1979-80 and 1982-90: 26

No significant PMRs

065 - Foresters

Number of deaths in men aged 20-74 during 1979-80 and 1982-90: 1543

PMRs for men

Cause of death (ICD)	Deaths	PMR	95% CI
Cancer of trachea, bronchus and lung (162)	154	83	70-97
Parkinson's disease (332)	10	225	108-413
Hypertensive disease (401-405)	20	170	104-263
Chronic bronchitis and emphysema (491,492,496)	67	71	55-91
Injury by falling object (E916)	14	1566	856-2627
Injury by machinery (E919)	7	585	235-1206
Suicide (E950-E959)	38	159	112-218

Number of deaths in women aged 20-74 during 1979-80 and 1982-90: 12

No significant PMRs

066 - Fishing and related workers

Number of deaths in men aged 20-74 during 1979-80 and 1982-90: 1564

PMRs for men

Cause of death (ICD)	Deaths	PMR	95% CI
Cancer of the oral cavity (141,143,144,145)	9	265	121-503
Cancer of the pancreas (157)	10	53	25-97
Cancer of the larynx (161)	14	279	152-467
Cancer of trachea, bronchus and lung (162)	232	127	111-145
Ischaemic heart disease (410-414)	430	83	76-92
Atrial fibrillation (427.3)	6	439	161-955
Aortic aneurysm (441)	11	54	27-96
Hyperplasia of prostate (600)	6	374	137-814
Motor vehicle traffic accidents (E810-E819)	17	49	28-78
Water transport accidents (E830-E838)	48	3376	2488-4478
Injury undetermined whether accidentally or purposely inflicted (E980-E989)	29	220	147-317

Number of deaths in women aged 20-74 during 1979-80 and 1982-90: 3

No significant PMRs

067 - Tannery workers

Number of deaths in men aged 20-74 during 1979-80 and 1982-90: 786

PMRs for men

Cause of death (ICD)	Deaths	PMR	95% CI
Other cardiomyopathies (425.0-425.4, 425.6-425.9)	8	350	151-690

Number of deaths in women aged 20-74 during 1979-80 and 1982-90: 63

No significant PMRs

Appendix 4 Significant PMRs for each job group

Job group	Deaths	PMR	95% CI
068 - Leather and shoe workers			
Number of deaths in men aged 20-74 during 1979-80 and 1982-90: 5704			
PMRs for men			
Cause of death (ICD)			
Tuberculosis (010-018,137)	17	190	111-304
Cancer of the colon (153)	82	80	63-99
Cancer of trachea, bronchus and lung (162)	611	84	78-91
Cancer of the pleura (163)	5	40	13-93
Epilepsy (345)	16	353	202-573
Chronic rheumatic heart disease (394-398)	29	159	107-229
Bronchopneumonia (485)	145	134	113-158
Renal failure (584-586)	35	176	123-246
Other accidents (E928)	4	416	113-1065
Number of deaths in women aged 20-74 during 1979-80 and 1982-90: 2406			
PMRs for women			
Cause of death (ICD)			
Acute bronchitis (466)	7	277	111-571
Urinary infections (590,595,599.0)	1	16	0-88
Motor vehicle traffic accidents (E810-E819)	5	34	11-80
069 - Preparatory fibre processors			
Number of deaths in men aged 20-74 during 1979-80 and 1982-90: 862			
PMRs for men			
Cause of death (ICD)			
Cancer of trachea, bronchus and lung (162)	84	77	62-96
Suicide (E950-E959)	1	12	0-69
Number of deaths in women aged 20-74 during 1979-80 and 1982-90: 524			
PMRs for women			
Cause of death (ICD)			
Cancer of the female breast (174)	13	42	22-72
Dementia (290,331.0,331.1)	0	0	0-97
Byssinosis (504)	5	7576	2460-17679
Renal failure (584-586)	6	275	101-599
070 - Spinners and winders			
Number of deaths in men aged 20-74 during 1979-80 and 1982-90: 1387			
PMRs for men			
Cause of death (ICD)			
Tuberculosis (010-018,137)	7	255	103-526
Cancer of trachea, bronchus and lung (162)	127	75	62-89
Cancer of the brain (191)	1	13	0-71
Diabetes mellitus (250)	23	182	115-273
Ischaemic heart disease (410-414)	540	114	105-124
Pulmonary heart disease unspecified (416.9)	7	277	111-571
Byssinosis (504)	4	8000	2180-20483
Duodenal ulcer (532)	1	18	0-98
Injury undetermined whether accidentally or purposely inflicted (E980-E989)	1	18	0-100
Number of deaths in women aged 20-74 during 1979-80 and 1982-90: 2562			
PMRs for women			
Cause of death (ICD)			
Cancer of the retroperitoneum (158.0)	5	567	184-1323
Cancer of trachea, bronchus and lung (162)	158	78	66-91
Cancer of the female breast (174)	102	64	52-78
Diabetes mellitus (250)	21	61	38-94
Ischaemic heart disease (410-414)	805	124	115-133
Other cerebrovascular disease (431-438)	289	113	101-127
Chronic bronchitis and emphysema (491,492,496)	127	123	102-146
Byssinosis (504)	5	1511	490-3525
071 - Warp preparers and weavers			
Number of deaths in men aged 20-74 during 1979-80 and 1982-90: 2331			
PMRs for men			
Cause of death (ICD)			
Cancer of trachea, bronchus and lung (162)	200	67	58-77
Cancer of the brain (191)	5	34	11-79
Bacterial and unspecified meningitis (320,322)	4	681	186-1745
Ischaemic heart disease (410-414)	912	112	104-119
Other cerebrovascular disease (431-438)	195	116	100-134
Number of deaths in women aged 20-74 during 1979-80 and 1982-90: 3084			
PMRs for women			
Cause of death (ICD)			
Cancer of trachea, bronchus and lung (162)	164	79	67-92
Cancer of the female breast (174)	136	68	57-80
Myeloma (203)	7	47	19-96
Dementia (290,331.0,331.1)	12	56	29-98
Ischaemic heart disease (410-414)	888	110	103-118
Other cardiomyopathies (425.0-425.4,425.6-425.9)	1	18	0-99
Other cerebrovascular disease (431-438)	374	116	104-128
Bronchopneumonia (485)	117	140	115-167
Chronic bronchitis and emphysema (491,492,496)	152	135	114-158
Byssinosis (504)	4	1338	365-3425
072 - Knitters			
Number of deaths in men aged 20-74 during 1979-80 and 1982-90: 1243			
PMRs for men			
Cause of death (ICD)			
Cancer of trachea, bronchus and lung (162)	104	67	55-82
Bronchopneumonia (485)	33	152	105-214
Bronchiectasis (494)	5	309	100-720
Number of deaths in women aged 20-74 during 1979-80 and 1982-90: 977			
PMRs for women			
Cause of death (ICD)			
Bacterial and unspecified meningitis (320,322)	4	743	203-1904
Aortic aneurysm (441)	15	181	101-298
073 - Bleachers, dyers and finishers			
Number of deaths in men aged 20-74 during 1979-80 and 1982-90: 2080			
PMRs for men			
Cause of death (ICD)			
Tuberculosis (010-018,137)	9	271	124-515
Diabetes mellitus (250)	28	153	102-222
Other cerebrovascular disease (431-438)	187	128	110-147
Bronchopneumonia (485)	57	150	114-195
Unspecified pneumonia (486)	7	252	101-520
Gastric ulcer (531)	12	262	136-458
Number of deaths in women aged 20-74 during 1979-80 and 1982-90: 255			
PMRs for women			
Cause of death (ICD)			
Cancer of the colon (153)	1	13	0-75
074 - Other textile workers			
Number of deaths in men aged 20-74 during 1979-80 and 1982-90: 8534			
PMRs for men			
Cause of death (ICD)			
Viral hepatitis (070)	8	359	155-707
Cancer of the oral cavity (141,143,144,145)	11	51	25-91
Cancer of the oesophagus (150)	73	76	59-95
Cancer of the rectum (154)	80	74	59-92
Cancer of trachea, bronchus and lung (162)	847	79	73-84
Diabetes mellitus (250)	102	127	104-155
Other alcohol related diseases (303,305.0,425.5,535.3,571.0-571.3, E860.0,E860.1)	25	62	40-91
Chronic rheumatic heart disease (394-398)	49	175	129-231
Ischaemic heart disease (410-414)	3305	117	113-121
Other cardiomyopathies (425.0-425.4, 425.6-425.9)	11	41	20-73
Other cerebrovascular disease (431-438)	717	113	105-122
Byssinosis (504)	14	2414	1320-4050
Hepatitis (571.4,573.3)	6	288	106-627

Appendix 4 Significant PMRs for each job group

Job group	Deaths	PMR	95% CI
Motor vehicle traffic accidents (E810-E819)	73	78	61-98
Accidental poisoning by drugs (E850-E858)	2	24	3-86
Fall from building (E882)	1	17	0-97
Suicide (E950-E959)	74	73	57-91
Injury undetermined whether accidentally or purposely inflicted (E980-E989)	26	57	37-83

Number of deaths in women aged 20-74 during 1979-80 and 1982-90: 5189

PMRs for women

Cause of death (ICD)	Deaths	PMR	95% CI
Cancer of the oesophagus (150)	31	68	46-97
Cancer of trachea, bronchus and lung (162)	300	74	66-83
Other alcohol related diseases (303,305.0,425.5,535.3,571.0-571.3, E860.0,E860.1)	6	44	16-95
Ischaemic heart disease (410-414)	1481	113	108-119
Pulmonary heart disease unspecified (416.9)	16	184	105-299
Mitral valve disorders (424.0)	0	0	0-56
Arterial embolism and thrombosis (444)	0	0	0-89
Chronic bronchitis and emphysema (491,492,496)	227	115	100-131
Byssinosis (504)	25	955	617-1411
Systemic lupus erythematosus (710.0)	7	312	125-642

075 - Chemical workers

Number of deaths in men aged 20-74 during 1979-80 and 1982-90: 15480

PMRs for men

Cause of death (ICD)	Deaths	PMR	95% CI
Cancer of the stomach (151)	448	111	101-122
Cancer of the pleura (163)	39	159	113-217
Epilepsy (345)	18	59	35-93
Ischaemic heart disease (410-414)	5477	104	101-107
Pulmonary embolism and phlebitis (415.1,451,453)	81	77	61-96
Chronic and unspecified myocarditis (429.0)	8	30	13-60
Sub-arachnoid haemorrhage (430)	104	122	100-148
Bronchopneumonia (485)	257	83	73-94
Chronic bronchitis and emphysema (491,492,496)	815	87	81-93
Asthma (493)	35	67	47-94
Coal workers' pneumoconiosis (500)	2	9	1-32
Cirrhosis of the liver (571.5)	30	68	46-98
Poisoning by other gases (869)	4	1093	298-2798
Injury by falling object (E916)	2	23	3-85
Injury by machinery (E919)	22	199	125-302
Injury by explosive material (E923)	12	797	412-1392
Injury by hot substances (E924)	5	559	182-1305
Injury undetermined whether accidentally or purposely inflicted (E980-E989)	55	64	48-83

Number of deaths in women aged 20-74 during 1979-80 and 1982-90: 1180

PMRs for women

Cause of death (ICD)	Deaths	PMR	95% CI
Cancer of the pleura (163)	4	498	136-1275
Urothelial cancer (188,189.1-189.8)	0	0	0-46
Aortic aneurysm (441)	3	32	7-93

076 - Bakers

Number of deaths in men aged 20-74 during 1979-80 and 1982-90: 5697

PMRs for men

Cause of death (ICD)	Deaths	PMR	95% CI
Cancer of the pleura (163)	3	24	5-71
Melanoma of skin (172)	3	31	6-91
Cancer of the penis (187.1-187.4)	6	312	115-680
Epilepsy (345)	13	237	126-405
Other cerebrovascular disease (431-438)	460	113	103-124
Influenza (487)	10	390	187-716
Asthma (493)	26	151	98-221
Other hernia (551-553)	6	282	104-615
Crohn's disease (555)	7	570	229-1174

Number of deaths in women aged 20-74 during 1979-80 and 1982-90: 734

PMRs for women

Cause of death (ICD)	Deaths	PMR	95% CI
Ischaemic heart disease (410-414)	209	117	102-134

077 - Brewery workers

Number of deaths in men aged 20-74 during 1979-80 and 1982-90: 1742

PMRs for men

Cause of death (ICD)	Deaths	PMR	95% CI
Other alcohol related diseases (303,305.0,425.5,535.3,571.0-571.3, E860.0,E860.1)	14	232	127-388
Epilepsy (345)	5	354	115-826
Chronic rheumatic heart disease (394-398)	1	17	0-96
Bronchopneumonia (485)	45	148	108-198
Peptic ulcer (533)	4	501	137-1283
Fall unspecified (E888)	7	404	162-832

Number of deaths in women aged 20-74 during 1979-80 and 1982-90: 117

No significant PMRs

078 - Food processors

Number of deaths in men aged 20-74 during 1979-80 and 1982-90: 8748

PMRs for men

Cause of death (ICD)	Deaths	PMR	95% CI
Cancer of the oral cavity (141,143,144,145)	9	48	22-92
Cancer of the pleura (163)	7	45	18-93
Chronic myeloid leukaemia (205.1)	3	33	7-95
Ischaemic heart disease (410-414)	3160	105	101-108
Coal workers' pneumoconiosis (500)	1	8	0-43
Motor vehicle traffic accidents (E810-E819)	84	72	58-90
Accidental poisoning by drugs (E850-E858)	2	27	3-98
Injured by fire (E890-E899)	3	33	7-97
Injury by falling object (E916)	0	0	0-83
Suicide (E950-E959)	81	68	54-85
Injury undetermined whether accidentally or purposely inflicted (E980-E989)	24	57	36-84

Number of deaths in women aged 20-74 during 1979-80 and 1982-90: 3132

PMRs for women

Cause of death (ICD)	Deaths	PMR	95% CI
Cancer of the thyroid (193)	0	0	0-96
Chronic rheumatic heart disease (394-398)	17	57	33-91
Ischaemic heart disease (410-414)	790	110	102-118
Other cerebrovascular disease (431-438)	306	114	102-127
Gastric ulcer (531)	17	179	104-287
Other hernia (551-553)	10	292	140-536
Motor vehicle traffic accidents (E810-E819)	13	55	29-93
Suicide (E950-E959)	11	46	23-83

079 - Paper manufacturers

Number of deaths in men aged 20-74 during 1979-80 and 1982-90: 1088

PMRs for men

Cause of death (ICD)	Deaths	PMR	95% CI
Cancer of the prostate (185)	11	55	28-99
Non-Hodgkin's lymphoma (200,202)	14	203	111-341

Number of deaths in women aged 20-74 during 1979-80 and 1982-90: 75

PMRs for women

Cause of death (ICD)	Deaths	PMR	95% CI
Cancer of the female breast (174)	13	238	127-407

080 - Bookbinders

Number of deaths in men aged 20-74 during 1979-80 and 1982-90: 744

PMRs for men

Cause of death (ICD)	Deaths	PMR	95% CI
Cancer of trachea, bronchus and lung (162)	66	72	56-92
Multiple sclerosis (340)	4	380	103-972
Epilepsy (345)	8	955	412-1881

Appendix 4 Significant PMRs for each job group

Job group	Deaths	PMR	95% CI

Number of deaths in women aged 20-74 during 1979-80 and 1982-90: 749
PMRs for women

Cause of death (ICD)			
Cancer of the female breast (174)	82	126	100-156

081 - Paper cutters

Number of deaths in men aged 20-74 during 1979-80 and 1982-90: 452
PMRs for men

Cause of death (ICD)			
Aplastic anaemia (284)	4	2667	727-6828

Number of deaths in women aged 20-74 during 1979-80 and 1982-90: 35

No significant PMRs

082 - Glass and ceramics furnacemen

Number of deaths in men aged 20-74 during 1979-80 and 1982-90: 1168
PMRs for men

Cause of death (ICD)			
Cancer of the prostate (185)	12	56	29-98
Chronic bronchitis and emphysema (491,492,496)	93	135	109-166

Number of deaths in women aged 20-74 during 1979-80 and 1982-90: 26

No significant PMRs

083 - Glass formers and decorators

Number of deaths in men aged 20-74 during 1979-80 and 1982-90: 1317

No significant PMRs

Number of deaths in women aged 20-74 during 1979-80 and 1982-90: 162
PMRs for women

Cause of death (ICD)			
Sub-arachnoid haemorrhage (430)	8	302	130-594

084 - Ceramics casters

Number of deaths in men aged 20-74 during 1979-80 and 1982-90: 722
PMRs for men

Cause of death (ICD)			
Other pneumoconiosis (503,505)	4	3846	1048-9848

Number of deaths in women aged 20-74 during 1979-80 and 1982-90: 253
PMRs for women

Cause of death (ICD)			
Ischaemic heart disease (410-414)	74	127	99-159
Fall unspecified (E888)	5	1185	385-2765

085 - Rubber manufacturers

Number of deaths in men aged 20-74 during 1979-80 and 1982-90: 1937
PMRs for men

Cause of death (ICD)			
Cancer of the testis (186)	4	381	104-975
Motor neurone disease (335.2)	12	244	126-426
Other cerebrovascular disease (431-438)	161	121	103-142
Suicide (E950-E959)	9	46	21-87

Number of deaths in women aged 20-74 during 1979-80 and 1982-90: 143
PMRs for women

Cause of death (ICD)			
Other cerebrovascular disease (431-438)	21	161	99-246

086 - Plastics workers

Number of deaths in men aged 20-74 during 1979-80 and 1982-90: 1188
PMRs for men

Cause of death (ICD)			
Cancer of the rectum (154)	7	45	18-94
Cancer of trachea, bronchus and lung (162)	190	128	111-148
Fibrosing alveolitis (516.3)	6	384	141-837
Suicide (E950-E959)	8	46	20-92

Number of deaths in women aged 20-74 during 1979-80 and 1982-90: 102

No significant PMRs

087 - Man-made fibre makers

Number of deaths in men aged 20-74 during 1979-80 and 1982-90: 440
PMRs for men

Cause of death (ICD)			
Chronic and unspecified myocarditis (429.0)	8	1040	449-2050

Number of deaths in women aged 20-74 during 1979-80 and 1982-90: 13

No significant PMRs

088 - Other coal miners

Number of deaths in men aged 20-74 during 1979-80 and 1982-90: 39844
PMRs for men

Cause of death (ICD)			
Viral hepatitis (070)	2	20	2-74
Cancer of the oesophagus (150)	374	85	77-94
Cancer of the colon (153)	606	89	82-96
Cancer of the rectum (154)	578	113	104-123
Cancer of trachea, bronchus and lung (162)	4610	92	89-94
Cancer of the pleura (163)	11	20	10-35
Cancer of the prostate (185)	519	77	70-83
Urothelial cancer (188,189.1-189.8)	387	84	76-93
Cancer of the brain (191)	176	84	72-97
Acute myeloid leukaemia (205.0)	73	77	60-97
Diabetes mellitus (250)	238	66	58-75
Immunodeficiency (279.1)	1	12	0-68
Other alcohol related diseases (303,305.0,425.5,535.3,571.0-571.3, E860.0,E860.1)	100	81	66-98
Parkinson's disease (332)	92	76	61-93
Ischaemic heart disease (410-414)	13213	97	95-98
Pulmonary heart disease unspecified (416.9)	110	146	120-176
Endocarditis (421)	7	44	18-91
Mitral valve disorders (424.0)	16	61	35-100
Other cardiomyopathies (425.0-425.4,425.6-425.9)	80	72	57-90
Chronic and unspecified myocarditis (429.0)	157	226	192-264
Sub-arachnoid haemorrhage (430)	126	74	62-88
Other cerebrovascular disease (431-438)	3118	105	102-109
Aortic aneurysm (441)	368	60	54-67
Acute bronchitis (466)	53	161	121-211
Influenza (487)	9	50	23-95
Chronic bronchitis and emphysema (491,492,496)	3799	142	137-146
Asthma (493)	91	81	65-99
Coal workers' pneumoconiosis (500)	693	770	713-829
Asbestosis (501)	0	0	0-28
Silicosis (502)	21	179	110-274
Pulmonary fibrosis (515)	66	143	111-182
Crohn's disease (555)	2	26	3-94
Cholelithiasis and cholecystitis (574,575,576.1-576.4)	57	136	103-176
Glomerulonephritis (580-583)	24	65	42-97
Water transport accidents (E830-E838)	2	21	3-75
Other vehicle accidents (E846-E848)	32	1261	862-1782
Fall on stairs (E880)	73	175	137-221
Fall from building (E882)	9	44	20-83
Injured by fire (E890-E899)	57	165	125-214
Injury by falling object (E916)	33	275	189-386
Injury by machinery (E919)	34	223	154-312
Suicide (E950-E959)	410	113	102-124

Number of deaths in women aged 20-74 during 1979-80 and 1982-90: 10

No significant PMRs

Appendix 4 Significant PMRs for each job group

Job group	Deaths	PMR	95% CI
089 - Tobacco workers			
Number of deaths in men aged 20-74 during 1979-80 and 1982-90: 491			
PMRs for men			
Cause of death (ICD)			
Aortic aneurysm (441)	15	207	116-341
Number of deaths in women aged 20-74 during 1979-80 and 1982-90: 393			
PMRs for women			
Cause of death (ICD)			
Bronchopneumonia (485)	3	32	7-94
090 - Other wood and paper processors			
Number of deaths in men aged 20-74 during 1979-80 and 1982-90: 1429			
PMRs for men			
Cause of death (ICD)			
Parkinson's disease (332)	11	280	140-501
Ischaemic heart disease (410-414)	559	110	101-120
Renal failure (584-586)	10	217	104-399
Number of deaths in women aged 20-74 during 1979-80 and 1982-90: 470			
PMRs for women			
Cause of death (ICD)			
Dementia (290,331.0,331.1)	7	285	114-587
091 - Other occupations - glass and ceramics			
Number of deaths in men aged 20-74 during 1979-80 and 1982-90: 4924			
PMRs for men			
Cause of death (ICD)			
Cancer of the oesophagus (150)	36	63	44-87
Cancer of the pleura (163)	2	22	3-78
Chronic and unspecified myocarditis (429.0)	17	237	138-380
Aortic aneurysm (441)	46	64	47-85
Chronic bronchitis and emphysema (491,492,496)	375	125	112-138
Other pneumoconiosis (503,505)	6	593	218-1290
Homicide (E960-E969)	0	0	0-89
Number of deaths in women aged 20-74 during 1979-80 and 1982-90: 1024			
PMRs for women			
Cause of death (ICD)			
Tuberculosis (010-018,137)	4	368	100-941
Cancer of the oesophagus (150)	15	182	102-300
Cancer of the ovary (183)	17	62	36-98
Other cardiomyopathies (425.0-425.4,425.6-425.9)	6	298	109-648
Peripheral vascular disease (443,557)	12	244	126-425
Unspecified pneumonia (486)	5	408	133-953
Chronic bronchitis and emphysema (491,492,496)	61	174	133-223
092 - Rubber goods makers			
Number of deaths in men aged 20-74 during 1979-80 and 1982-90: 1546			
PMRs for men			
Cause of death (ICD)			
Bronchopneumonia (485)	37	142	100-195
Oesophageal disease (530)	4	462	126-1184
Number of deaths in women aged 20-74 during 1979-80 and 1982-90: 288			
PMRs for women			
Cause of death (ICD)			
Urothelial cancer (188,189.1-189.8)	9	457	209-867
093 - Plastic goods makers			
Number of deaths in men aged 20-74 during 1979-80 and 1982-90: 1942			
PMRs for men			
Cause of death (ICD)			
Cancer of soft tissue (171)	6	275	101-599
Acute lymphatic leukaemia (204.0)	6	574	211-1250
Diabetes mellitus (250)	27	158	104-230

Job group	Deaths	PMR	95% CI
Other cerebrovascular disease (431-438)	149	121	103-143
Suicide (E950-E959)	20	58	35-89
Injury undetermined whether accidentally or purposely inflicted (E980-E989)	3	24	5-70
Number of deaths in women aged 20-74 during 1979-80 and 1982-90: 550			
PMRs for women			
Cause of death (ICD)			
Chronic and unspecified myocarditis (429.0)	4	560	153-1434
Suicide (E950-E959)	0	0	0-83
094 - Compositors			
Number of deaths in men aged 20-74 during 1979-80 and 1982-90: 2054			
PMRs for men			
Cause of death (ICD)			
Cancer of the stomach (151)	39	71	51-98
Cancer of trachea, bronchus and lung (162)	195	76	65-87
Parkinson's disease (332)	13	190	101-325
Chronic rheumatic heart disease (394-398)	1	15	0-85
Ischaemic heart disease (410-414)	817	114	106-122
Chronic bronchitis and emphysema (491,492,496)	77	63	50-79
Injury undetermined whether accidentally or purposely inflicted (E980-E989)	0	0	0-60
Number of deaths in women aged 20-74 during 1979-80 and 1982-90: 48			
PMRs for women			
Cause of death (ICD)			
Cancer of the female breast (174)	12	214	111-374
095 - Printing plate preparers			
Number of deaths in men aged 20-74 during 1979-80 and 1982-90: 561			
PMRs for men			
Cause of death (ICD)			
Acute myeloid leukaemia (205.0)	7	426	171-878
Multiple sclerosis (340)	5	626	203-1460
Chronic bronchitis and emphysema (491,492,496)	18	57	34-91
Fibrosing alveolitis (516.3)	4	523	142-1339
Suicide (E950-E959)	0	0	0-52
Number of deaths in women aged 20-74 during 1979-80 and 1982-90: 19			
No significant PMRs			
096 - Printing machine minders			
Number of deaths in men aged 20-74 during 1979-80 and 1982-90: 2766			
PMRs for men			
Cause of death (ICD)			
Motor neurone disease (335.2)	15	208	116-343
Arterial embolism and thrombosis (444)	6	362	133-788
Number of deaths in women aged 20-74 during 1979-80 and 1982-90: 553			
PMRs for women			
Cause of death (ICD)			
Cancer of the rectum (154)	1	14	0-79
097 - Printers (so described)			
Number of deaths in men aged 20-74 during 1979-80 and 1982-90: 8250			
PMRs for men			
Cause of death (ICD)			
Cancer of the stomach (151)	164	77	65-89
Cancer of the colon (153)	173	117	100-136
Cancer of trachea, bronchus and lung (162)	933	92	86-98
Cancer of the pleura (163)	3	16	3-46
Cancer of the prostate (185)	167	118	100-137
Cancer of the brain (191)	78	126	100-158
Acute myeloid leukaemia (205.0)	15	60	34-99
Dementia (290,331.0,331.1)	38	144	102-198
Multiple sclerosis (340)	23	189	119-283
Hypertensive disease (401-405)	91	148	119-182

Appendix 4 Significant PMRs for each job group

Job group	Deaths	PMR	95% CI
Aortic aneurysm (441)	168	122	104-142
Chronic bronchitis and emphysema (491,492,496)	405	89	81-98
Oesophageal disease (530)	0	0	0-81
Injury by falling object (E916)	0	0	0-88
Suicide (E950-E959)	151	127	107-149

Number of deaths in women aged 20-74 during 1979-80 and 1982-90: 395
PMRs for women
Cause of death (ICD)

Cause of death (ICD)	Deaths	PMR	95% CI
Cancer of trachea, bronchus and lung (162)	47	161	118-214

098 - Tailors and dressmakers

Number of deaths in men aged 20-74 during 1979-80 and 1982-90: 2552
PMRs for men
Cause of death (ICD)

Cause of death (ICD)	Deaths	PMR	95% CI
Tuberculosis (010-018,137)	16	402	230-653
Cancer of the larynx (161)	1	13	0-74
Cancer of trachea, bronchus and lung (162)	203	62	54-72
Chronic lymphatic leukaemia (204.1)	9	234	107-445
Diabetes mellitus (250)	49	215	159-284
Immunodeficiency (279.1)	5	1057	343-2467
Other alcohol related diseases (303,305.0,425.5,535.3,571.0-571.3, E860.0,E860.1)	14	197	108-330
Other cardiomyopathies (425.0-425.4,425.6-425.9)	18	239	142-378
Chronic bronchitis and emphysema (491,492,496)	101	65	53-78

Number of deaths in women aged 20-74 during 1979-80 and 1982-90: 3761
PMRs for women
Cause of death (ICD)

Cause of death (ICD)	Deaths	PMR	95% CI
Cancer of trachea, bronchus and lung (162)	214	81	71-93
Cancer of the pleura (163)	12	420	217-734
Cancer of the female breast (174)	326	120	107-133

099 - Clothing cutters

Number of deaths in men aged 20-74 during 1979-80 and 1982-90: 1250
PMRs for men
Cause of death (ICD)

Cause of death (ICD)	Deaths	PMR	95% CI
Tuberculosis (010-018,137)	6	308	113-671
Cancer of trachea, bronchus and lung (162)	130	83	69-99
Epilepsy (345)	7	599	241-1235
Urinary infections (590,595,599.0)	5	347	113-810

Number of deaths in women aged 20-74 during 1979-80 and 1982-90: 567
PMRs for women
Cause of death (ICD)

Cause of death (ICD)	Deaths	PMR	95% CI
Cancer of the gall bladder (156)	6	343	126-747

100 - Sewers and embroiderers

Number of deaths in men aged 20-74 during 1979-80 and 1982-90: 704
PMRs for men
Cause of death (ICD)

Cause of death (ICD)	Deaths	PMR	95% CI
Multiple sclerosis (340)	4	375	102-960
Motor vehicle traffic accidents (E810-E819)	4	38	10-97
Injury undetermined whether accidentally or purposely inflicted (E980-E989)	0	0	0-87

Number of deaths in women aged 20-74 during 1979-80 and 1982-90: 10002
PMRs for women
Cause of death (ICD)

Cause of death (ICD)	Deaths	PMR	95% CI
Cancer of the oral cavity (141,143,144,145)	9	50	23-95
Cancer of trachea, bronchus and lung (162)	738	92	85-99
Cancer of the female breast (174)	835	109	101-116
Amyloidosis (277.3)	7	252	101-520
Chronic rheumatic heart disease (394-398)	138	145	122-172
Pulmonary hypertension (416.0)	9	228	104-433
Rheumatoid arthritis (714)	43	138	100-186
Suicide (E950-E959)	109	122	100-147
Homicide (E960-E969)	8	50	22-99

101 - Upholsterers

Number of deaths in men aged 20-74 during 1979-80 and 1982-90: 2763
PMRs for men
Cause of death (ICD)

Cause of death (ICD)	Deaths	PMR	95% CI
Cancer of the pleura (163)	16	264	151-429
Cancer of the prostate (185)	76	149	117-186
Meningeal tumour (192.1,192.3,225.2,225.4)	5	563	183-1314
Chronic lymphatic leukaemia (204.1)	9	222	102-422
Bronchopneumonia (485)	67	130	101-165

Number of deaths in women aged 20-74 during 1979-80 and 1982-90: 385

No significant PMRs

102 - Carpet fitters

Number of deaths in men aged 20-74 during 1979-80 and 1982-90: 577
PMRs for men
Cause of death (ICD)

Cause of death (ICD)	Deaths	PMR	95% CI
Cancer of the rectum 154)	13	188	100-322

Number of deaths in women aged 20-74 during 1979-80 and 1982-90: 13

No significant PMRs

103 - Other workers with fabrics

Number of deaths in men aged 20-74 during 1979-80 and 1982-90: 867
PMRs for men
Cause of death (ICD)

Cause of death (ICD)	Deaths	PMR	95% CI
Diabetes mellitus (250)	14	184	100-308
Cirrhosis of the liver (571.5)	7	324	130-668

Number of deaths in women aged 20-74 during 1979-80 and 1982-90: 691

No significant PMRs

104 - Carpenters

Number of deaths in men aged 20-74 during 1979-80 and 1982-90: 27425
PMRs for men
Cause of death (ICD)

Cause of death (ICD)	Deaths	PMR	95% CI
Tuberculosis (010-018,137)	69	157	122-199
Cancer of the oral cavity (141,143,144,145)	37	68	48-94
Cancer of the stomach (151)	651	91	85-99
Cancer of the small intestine (152)	8	49	21-97
Cancer of the rectum (154)	314	86	77-96
Cancer of trachea, bronchus and lung (162)	3192	94	91-98
Cancer of the pleura (163)	167	262	224-305
Cancer of the prostate (185)	564	121	111-131
Cancer of the eye (190)	19	208	125-325
Myeloma (203)	125	120	100-143
Epilepsy (345)	45	138	101-185
Hypertensive disease (401-405)	247	121	106-137
Pulmonary embolism and phlebitis (415.1,451,453)	196	116	101-134
Endocarditis (421)	27	179	118-260
Arteritis (447.6)	9	234	107-444
Chronic bronchitis and emphysema (491,492,496)	1269	84	80-89
Coal workers' pneumoconiosis (500)	0	0	0-43
Appendicitis (540-542)	15	209	117-345
Off-road motor vehicle accidents (E820-E825)	2	19	2-67
Pedal cycle accidents (E826)	8	261	113-515
Fall from ladder or scaffolding (E881)	31	182	123-258
Fall from building (E882)	40	227	162-309
Injury by machinery (E919)	3	14	3-41
Suicide (E950-E959)	461	117	107-128
Injury undetermined whether accidentally or purposely inflicted (E980-E989)	174	140	120-163

Appendix 4 Significant PMRs for each job group

Job group	Deaths	PMR	95% CI
Number of deaths in women aged 20-74 during 1979-80 and 1982-90: 37			
No significant PMRs			
105 - Cabinet makers			
Number of deaths in men aged 20-74 during 1979-80 and 1982-90: 3006			
PMRs for men			
Cause of death (ICD)			
Cancer of trachea, bronchus and lung (162)	316	85	76-95
Cancer of the eye (190)	4	416	113-1066
Myeloma (203)	23	204	129-307
Immunodeficiency (279.1)	5	530	172-1237
Pulmonary embolism and phlebitis (415.1,451,453)	30	159	107-227
Hyperplasia of prostate (600)	10	300	144-551
Cancer of the nose and nasal sinuses (160)	8	522	225-1029
Number of deaths in women aged 20-74 during 1979-80 and 1982-90: 45			
PMRs for women			
Cause of death (ICD)			
Chronic bronchitis and emphysema (491,492,496)	5	371	121-867
106 - Case and box makers			
Number of deaths in men aged 20-74 during 1979-80 and 1982-90: 749			
PMRs for men			
Cause of death (ICD)			
Cancer of the stomach (151)	30	152	103-217
Bronchiectasis (494)	4	415	113-1061
Motor vehicle traffic accidents (E810-E819)	3	30	6-86
Number of deaths in women aged 20-74 during 1979-80 and 1982-90: 42			
No significant PMRs			
107 - Pattern makers			
Number of deaths in men aged 20-74 during 1979-80 and 1982-90: 1223			
PMRs for men			
Cause of death (ICD)			
Cancer of the colon (153)	32	144	99-204
Cancer of trachea, bronchus and lung (162)	103	66	54-81
Parkinson's disease (332)	10	269	129-495
Ischaemic heart disease (410-414)	516	119	109-130
Pneumococcal and unspecified lobar pneumonia (481)	0	0	0-92
Chronic bronchitis and emphysema (491,492,496)	39	56	40-76
Number of deaths in women aged 20-74 during 1979-80 and 1982-90: 7			
No significant PMRs			
108 - Woodworking machinists			
Number of deaths in men aged 20-74 during 1979-80 and 1982-90: 4216			
PMRs for men			
Cause of death (ICD)			
Cancer of bone (170)	9	344	157-653
Chronic lymphatic leukaemia (204.1)	1	16	0-90
Cancer of the nose and nasal sinuses (160)	8	357	154-704
Other cardiomyopathies (425.0-425.4, 425.6-425.9)	4	28	8-72
Acute bronchitis (466)	9	303	139-575
Motor vehicle traffic accidents (E810-E819)	45	73	54-98
Injury by falling object (E916)	6	325	119-707
Number of deaths in women aged 20-74 during 1979-80 and 1982-90: 125			
No significant PMRs			

Job group	Deaths	PMR	95% CI
109 - Other woodworkers			
Number of deaths in men aged 20-74 during 1979-80 and 1982-90: 1236			
PMRs for men			
Cause of death (ICD)			
Melanoma of skin (172)	7	275	110-566
Bronchiectasis (494)	5	320	104-746
Number of deaths in women aged 20-74 during 1979-80 and 1982-90: 69			
PMRs for women			
Cause of death (ICD)			
Cancer of trachea, bronchus and lung (162)	10	209	100-384
110 - Dental technicians			
Number of deaths in men aged 20-74 during 1979-80 and 1982-90: 675			
PMRs for men			
Cause of death (ICD)			
Cancer of the prostate (185)	24	204	131-305
Hypertensive disease (401-405)	0	0	0-74
Number of deaths in women aged 20-74 during 1979-80 and 1982-90: 22			
No significant PMRs			
111 - Other makers of paper goods			
Number of deaths in men aged 20-74 during 1979-80 and 1982-90: 1462			
PMRs for men			
Cause of death (ICD)			
Suicide (E950-E959)	13	55	29-93
Number of deaths in women aged 20-74 during 1979-80 and 1982-90: 998			
PMRs for women			
Cause of death (ICD)			
Cancer of the colon (153)	18	61	36-97
Cancer of the brain (191)	16	180	103-292
Rheumatoid arthritis (714)	9	243	111-460
112 - Furnace operatives (metal)			
Number of deaths in men aged 20-74 during 1979-80 and 1982-90: 3309			
PMRs for men			
Cause of death (ICD)			
Multiple sclerosis (340)	0	0	0-87
Aortic aneurysm (441)	37	63	44-87
Bronchopneumonia (485)	81	133	106-166
Chronic bronchitis and emphysema (491,492,496)	248	126	111-143
Gastric ulcer (531)	14	192	105-322
Peptic ulcer (533)	5	320	104-746
Volvulus (560.2)	4	421	115-1077
Suicide (E950-E959)	18	62	37-98
Number of deaths in women aged 20-74 during 1979-80 and 1982-90: 9			
No significant PMRs			
113 - Rollers (metal)			
Number of deaths in men aged 20-74 during 1979-80 and 1982-90: 606			
PMRs for men			
Cause of death (ICD)			
Ischaemic heart disease (410-414)	184	85	74-99
Bronchopneumonia (485)	21	189	117-289
Chronic bronchitis and emphysema (491,492,496)	50	140	104-184
Cirrhosis of the liver (571.5)	6	407	149-887
Number of deaths in women aged 20-74 during 1979-80 and 1982-90: 6			
No significant PMRs			

Appendix 4 Significant PMRs for each job group

114 - Smiths and forge workers

Number of deaths in men aged 20-74 during 1979-80 and 1982-90: 2770

PMRs for men

Cause of death (ICD)	Deaths	PMR	95% CI
Ischaemic heart disease (410-414)	1039	106	100-113
Aortic aneurysm (441)	34	69	48-96
Duodenal ulcer (532)	21	202	125-309

Number of deaths in women aged 20-74 during 1979-80 and 1982-90: 9

No significant PMRs

115 - Metal drawers

Number of deaths in men aged 20-74 during 1979-80 and 1982-90: 843

PMRs for men

Cause of death (ICD)	Deaths	PMR	95% CI
Cancer of the pancreas (157)	3	27	6-80
Other cancer of skin (173)	4	660	180-1690
Cancer of the brain (191)	1	17	0-96
Gastric ulcer (531)	8	433	187-853

Number of deaths in women aged 20-74 during 1979-80 and 1982-90: 40

No significant PMRs

116 - Moulders and coremakers (metal)

Number of deaths in men aged 20-74 during 1979-80 and 1982-90: 4843

PMRs for men

Cause of death (ICD)	Deaths	PMR	95% CI
Cancer of the pancreas (157)	47	74	54-98
Cancer of trachea, bronchus and lung (162)	736	118	110-127
Cancer of the prostate (185)	56	65	49-84
Urothelial cancer (188,189.1-189.8)	41	72	51-97
Diabetes mellitus (250)	28	67	44-96
Other alcohol related diseases (303,305.0,425.5,535.3,571.0-571.3, E860.0,E860.1)	27	170	112-247
Hypertensive disease (401-405)	58	158	120-204
Ischaemic heart disease (410-414)	1565	91	87-96
Other cerebrovascular disease (431-438)	370	111	100-123
Aortic aneurysm (441)	47	56	41-75
Pneumococcal and unspecified lobar pneumonia (481)	32	201	138-284
Chronic bronchitis and emphysema (491,492,496)	409	146	132-161
Other pneumoconiosis (503,505)	9	1240	567-2353
Osteoporosis (733.0)	4	484	132-1240

Number of deaths in women aged 20-74 during 1979-80 and 1982-90: 189

No significant PMRs

117 - Electroplaters

Number of deaths in men aged 20-74 during 1979-80 and 1982-90: 850

PMRs for men

Cause of death (ICD)	Deaths	PMR	95% CI
Cancer of trachea, bronchus and lung (162)	134	126	105-149
Diabetes mellitus (250)	14	192	105-323
Ischaemic heart disease (410-414)	255	86	76-98
Aortic valve disorders (424.1)	0	0	0-98
Chronic bronchitis and emphysema (491,492,496)	68	144	112-182

Number of deaths in women aged 20-74 during 1979-80 and 1982-90: 31

PMRs for women

Cause of death (ICD)	Deaths	PMR	95% CI
Cancer of trachea, bronchus and lung (162)	6	273	100-594

118 - Annealers, hardeners, temperers (metal)

Number of deaths in men aged 20-74 during 1979-80 and 1982-90: 749

PMRs for men

Cause of death (ICD)	Deaths	PMR	95% CI
Pneumococcal and unspecified lobar pneumonia (481)	8	288	125-568
Suicide (E950-E959)	1	15	0-83

Number of deaths in women aged 20-74 during 1979-80 and 1982-90: 11

No significant PMRs

119 - Galvanisers and tin platers

Number of deaths in men aged 20-74 during 1979-80 and 1982-90: 497

No significant PMRs

Number of deaths in women aged 20-74 during 1979-80 and 1982-90: 26

No significant PMRs

120 - Other metal manufacturers

Number of deaths in men aged 20-74 during 1979-80 and 1982-90: 7388

PMRs for men

Cause of death (ICD)	Deaths	PMR	95% CI
Cancer of the pancreas (157)	59	68	52-88
Cancer of the larynx (161)	40	153	109-208
Cancer of trachea, bronchus and lung (162)	1054	112	105-119
Viral pneumonia (480)	7	309	124-637
Chronic bronchitis and emphysema (491,492,496)	524	110	101-120
Silicosis (502)	6	740	272-1610
Other pneumoconiosis (503,505)	12	677	350-1182
Gastric ulcer (531)	33	144	99-203
Cirrhosis of the liver (571.5)	31	156	106-222
Glomerulonephritis (580-583)	13	198	105-339
Systemic lupus erythematosus (710.0)	4	515	140-1320
Motor vehicle traffic accidents (E810-E819)	60	75	57-97
Accidental poisoning by drugs (E850-E858)	2	28	3-99
Injury by machinery (E919)	9	230	105-437
Suicide (E950-E959)	67	77	60-98

Number of deaths in women aged 20-74 during 1979-80 and 1982-90: 126

PMRs for women

Cause of death (ICD)	Deaths	PMR	95% CI
Cancer of the female breast (174)	3	31	6-91
Other cerebrovascular disease (431-438)	20	181	111-280

121 - Press and machine tool setters

Number of deaths in men aged 20-74 during 1979-80 and 1982-90: 5719

PMRs for men

Cause of death (ICD)	Deaths	PMR	95% CI
Cancer of the colon (153)	125	120	100-143
Asthma (493)	28	149	99-215
Urinary infections (590,595,599.0)	1	17	0-95

Number of deaths in women aged 20-74 during 1979-80 and 1982-90: 35

No significant PMRs

122 - Centre lathe turners

Number of deaths in men aged 20-74 during 1979-80 and 1982-90: 5150

PMRs for men

Cause of death (ICD)	Deaths	PMR	95% CI
Cancer of the rectum (154)	89	126	101-155
Cancer of trachea, bronchus and lung (162)	598	90	83-97
Ischaemic heart disease (410-414)	1959	106	102-111
Sub-arachnoid haemorrhage (430)	17	60	35-95
Other cerebrovascular disease (431-438)	385	114	103-126
Peripheral vascular disease (443,557)	33	167	115-235
Suicide (E950-E959)	32	60	41-84

Job group	Deaths	PMR	95% CI
Number of deaths in women aged 20-74 during 1979-80 and 1982-90: 48			
No significant PMRs			
123 - Machine tool setter operators			
Number of deaths in men aged 20-74 during 1979-80 and 1982-90: 1099			
PMRs for men			
Cause of death (ICD)			
Ischaemic heart disease (410-414)	435	110	100-121
Bronchiectasis (494)	5	349	113-815
Motor vehicle traffic accidents (E810-E819)	5	42	14-98
Suicide (E950-E959)	3	23	5-67
Number of deaths in women aged 20-74 during 1979-80 and 1982-90: 10			
No significant PMRs			
124 - Machine tool operators			
Number of deaths in men aged 20-74 during 1979-80 and 1982-90: 50854			
PMRs for men			
Cause of death (ICD)			
Tuberculosis (010-018,137)	82	76	61-95
Viral hepatitis (070)	28	175	116-253
Cancer of the stomach (151)	1221	92	86-97
Cancer of the colon (153)	937	109	102-116
Cancer of the nose and nasal sinuses (160)	34	146	101-204
Cancer of the pleura (163)	116	143	119-172
Cancer of soft tissue (171)	65	129	99-164
Melanoma of skin (172)	101	123	101-150
Cancer of the prostate (185)	851	111	103-118
Cancer of the brain (191)	373	114	103-126
Myeloma (203)	204	117	102-134
Acute lymphatic leukaemia (204.0)	33	144	99-203
Acute myeloid leukaemia (205.0)	167	125	107-146
Diabetes mellitus (250)	504	112	102-122
Immunodeficiency (279.1)	8	36	16-71
Multiple sclerosis (340)	91	127	102-156
Epilepsy (345)	63	67	51-86
Ischaemic heart disease (410-414)	17863	103	101-104
Pulmonary embolism and phlebitis (415.1,451,453)	295	85	76-95
Aortic aneurysm (441)	884	122	114-130
Acute bronchitis (466)	27	67	44-98
Bronchopneumonia (485)	769	78	73-84
Chronic bronchitis and emphysema (491,492,496)	2646	86	83-90
Coal workers' pneumoconiosis (500)	2	2	0-8
Silicosis (502)	3	24	5-71
Pleurisy (511)	11	206	103-368
Fibrosing alveolitis (516.3)	84	126	100-156
Gastric ulcer (531)	99	80	65-97
Duodenal ulcer (532)	163	79	67-92
Peritonitis (567)	26	163	106-239
Biliary cirrhosis (571.6)	0	0	0-65
Motor vehicle traffic accidents (E810-E819)	850	123	115-132
Other vehicle accidents (E846-E848)	1	17	0-97
Fall from building (E882)	19	40	24-63
Injury by falling object (E916)	16	58	33-95
Suicide (E950-E959)	824	113	105-121
Number of deaths in women aged 20-74 during 1979-80 and 1982-90: 3523			
PMRs for women			
Cause of death (ICD)			
Cancer of the gall bladder (156)	4	34	9-87
Cancer of trachea, bronchus and lung (162)	363	122	109-135
Cancer of the female breast (174)	223	83	72-95
Motor neurone disease (335.2)	4	37	10-95
Peripheral vascular disease (443,557)	26	167	109-245
Chronic bronchitis and emphysema (491,492,496)	183	140	121-162
Motor vehicle traffic accidents (E810-E819)	12	53	27-92

Job group	Deaths	PMR	95% CI
125 - Press and automatic machine operators			
Number of deaths in men aged 20-74 during 1979-80 and 1982-90: 2363			
PMRs for men			
Cause of death (ICD)			
Cancer of trachea, bronchus and lung (162)	354	122	109-135
Other alcohol related diseases (303,305.0,425.5,535.3,571.0-571.3, E860.0,E860.1)	4	36	10-93
Coal workers' pneumoconiosis (500)	0	0	0-89
Fibrosing alveolitis (516.3)	8	260	112-513
Number of deaths in women aged 20-74 during 1979-80 and 1982-90: 1403			
PMRs for women			
Cause of death (ICD)			
Cancer of bone (170)	4	594	162-1522
Aortic aneurysm (441)	3	25	5-72
Chronic bronchitis and emphysema (491,492,496)	76	145	114-181
126 - Metal polishers			
Number of deaths in men aged 20-74 during 1979-80 and 1982-90: 2139			
PMRs for men			
Cause of death (ICD)			
Cancer of the stomach (151)	75	131	103-165
Cancer of the colon (153)	20	52	32-80
Cancer of trachea, bronchus and lung (162)	332	122	109-136
Ischaemic heart disease (410-414)	668	89	82-96
Chronic bronchitis and emphysema (491,492,496)	168	136	116-158
Number of deaths in women aged 20-74 during 1979-80 and 1982-90: 249			
No significant PMRs			
127 - Fettlers and dressers (metal)			
Number of deaths in men aged 20-74 during 1979-80 and 1982-90: 1608			
PMRs for men			
Cause of death (ICD)			
Tuberculosis (010-018,137)	11	319	159-570
Cancer of trachea, bronchus and lung (162)	259	127	112-143
Chronic rheumatic heart disease (394-398)	1	18	0-98
Ischaemic heart disease (410-414)	506	90	83-99
Sub-arachnoid haemorrhage (430)	2	25	3-92
Chronic bronchitis and emphysema (491,492,496)	131	128	107-152
Silicosis (502)	4	911	248-2333
Other pneumoconiosis (503,505)	15	3650	2043-6020
Number of deaths in women aged 20-74 during 1979-80 and 1982-90: 96			
No significant PMRs			
128 - Shot blasters			
Number of deaths in men aged 20-74 during 1979-80 and 1982-90: 575			
PMRs for men			
Cause of death (ICD)			
Pneumococcal and unspecified lobar pneumonia (481)	6	284	104-619
Number of deaths in women aged 20-74 during 1979-80 and 1982-90: 2			
No significant PMRs			
129 - Toolmakers			
Number of deaths in men aged 20-74 during 1979-80 and 1982-90: 9341			
PMRs for men			
Cause of death (ICD)			
Cancer of the small intestine (152)	12	221	114-387

Appendix 4 Significant PMRs for each job group

Job group	Deaths	PMR	95% CI
Cancer of trachea, bronchus and lung (162)	1059	89	83-94
Cancer of the pleura (163)	10	45	22-83
Cancer of the testis (186)	12	203	105-354
Chronic myeloid leukaemia (205.1)	20	181	110-279
Thyrotoxicosis (242)	4	470	128-1203
Multiple sclerosis (340)	24	182	116-271
Chronic rheumatic heart disease (394-398)	46	148	109-198
Ischaemic heart disease (410-414)	3487	106	102-109
Aortic aneurysm (441)	199	126	110-145
Chronic bronchitis and emphysema (491,492,496)	429	82	74-90
Renal stones (592)	5	375	122-876
Pedal cycle accidents (E826)	4	449	122-1149
Fall from ladder or scaffolding (E881)	0	0	0-71

Number of deaths in women aged 20-74 during 1979-80 and 1982-90: 18

No significant PMRs

130 - Precision instrument makers

Number of deaths in men aged 20-74 during 1979-80 and 1982-90: 2910

PMRs for men

Cause of death (ICD)			
Cancer of the colon (153)	70	133	104-168
Cancer of trachea, bronchus and lung (162)	278	75	67-85
Cancer of the prostate (185)	65	134	104-171
Diabetes mellitus (250)	13	53	28-90
Multiple sclerosis (340)	11	248	124-444
Bronchopneumonia (485)	32	68	46-96
Chronic bronchitis and emphysema (491,492,496)	116	74	61-88
Infections of skin, joints and bone (680-686,711,730)	4	496	135-1271

Number of deaths in women aged 20-74 during 1979-80 and 1982-90: 63

No significant PMRs

131 - Watch and clock makers

Number of deaths in men aged 20-74 during 1979-80 and 1982-90: 1111

PMRs for men

Cause of death (ICD)			
Tuberculosis (010-018,137)	8	445	192-876
Cancer of the stomach (151)	15	50	28-83
Cancer of trachea, bronchus and lung (162)	101	71	58-86
Chronic rheumatic heart disease (394-398)	9	244	112-464
Aortic valve disorders (424.1)	11	221	110-395
Other cerebrovascular disease (431-438)	94	123	100-151
Cholelithiasis and cholecystitis (574,575,576.1-576.4)	5	486	158-1135

Number of deaths in women aged 20-74 during 1979-80 and 1982-90: 11

No significant PMRs

132 - Production fitters

Number of deaths in men aged 20-74 during 1979-80 and 1982-90: 52566

PMRs for men

Cause of death (ICD)			
Cancer of the salivary glands (142)	31	153	104-218
Cancer of the stomach (151)	1289	94	89-99
Cancer of the colon (153)	1017	108	101-114
Cancer of the pleura (163)	192	151	131-174
Urothelial cancer (188,189.1-189.8)	661	111	103-120
Cancer of the eye (190)	9	51	23-97
Immunodeficiency (279.1)	8	38	16-74
Parkinson's disease (332)	171	121	104-141
Ischaemic heart disease (410-414)	18789	102	100-103
Aortic aneurysm (441)	949	112	105-119
Chronic bronchitis and emphysema (491,492,496)	2568	91	88-95
Asthma (493)	146	82	69-96
Byssinosis (504)	4	629	171-1610

Job group	Deaths	PMR	95% CI
Renal stones (592)	1	14	0-77
Hyperplasia of prostate (600)	32	69	47-97
Motor vehicle traffic accidents (E810-E819)	831	107	100-115
Air transport accidents (E840-E845)	17	234	137-375
Injured by fire (E890-E899)	29	68	46-98
Injury by machinery (E919)	107	271	222-327
Injury by explosions (E921)	5	424	138-989
Injury by hot substances (E924)	10	310	149-570
Homicide (E960-E969)	17	48	28-77

Number of deaths in women aged 20-74 during 1979-80 and 1982-90: 121

No significant PMRs

133 - Motor mechanics

Number of deaths in men aged 20-74 during 1979-80 and 1982-90: 15435

PMRs for men

Cause of death (ICD)			
Cancer of the larynx (161)	30	67	45-96
Cancer of the pleura (163)	12	33	17-58
Urothelial cancer (188,189.1-189.8)	202	120	104-138
Cancer of the brain (191)	153	121	103-142
Immunodeficiency (279.1)	3	32	7-93
Motor neurone disease (335.2)	55	139	105-181
Hypertensive disease (401-405)	90	80	64-98
Other cardiomyopathies (425.0-425.4, 425.6-425.9)	82	138	109-171
Sub-arachnoid haemorrhage (430)	129	122	102-145
Bronchopneumonia (485)	218	87	76-100
Chronic bronchitis and emphysema (491,492,496)	736	92	85-99
Fibrosing alveolitis (516.3)	10	49	24-91
Off-road motor vehicle accidents (E820-E825)	17	201	117-322
Poisoning by motor vehicle exhaust (E868.2)	11	245	122-438
Fall from building (E882)	3	23	5-68
Injury by falling object (E916)	21	202	125-309
Injury by machinery (E919)	2	12	1-44
Injury undetermined whether accidentally or purposely inflicted (E980-E989)	69	74	58-94

Number of deaths in women aged 20-74 during 1979-80 and 1982-90: 44

No significant PMRs

134 - Aircraft engine fitters

Number of deaths in men aged 20-74 during 1979-80 and 1982-90: 165

PMRs for men

Cause of death (ICD)			
Other cerebrovascular disease (431-438)	20	179	110-277
Aortic aneurysm (441)	7	251	101-517

Number of deaths in women aged 20-74 during 1979-80 and 1982-90: 0

No significant PMRs

135 - Office machinery mechanics

Number of deaths in men aged 20-74 during 1979-80 and 1982-90: 325

PMRs for men

Cause of death (ICD)			
Cancer of trachea, bronchus and lung (162)	24	60	38-89
Non-Hodgkin's lymphoma (200,202)	7	299	120-616
Aortic valve disorders (424.1)	5	359	117-839

Number of deaths in women aged 20-74 during 1979-80 and 1982-90: 1

No significant PMRs

136 - Electrical and electronic production fitters

Number of deaths in men aged 20-74 during 1979-80 and 1982-90: 1426

PMRs for men

Cause of death (ICD)			
Epilepsy (345)	8	534	231-1052

Job group	Deaths	PMR	95% CI
Endocarditis (421)	4	501	136-1282
Cholelithiasis and cholecystitis (574,575,576.1-576.4)	7	552	222-1137
Fall on stairs (E880)	5	382	124-892

Number of deaths in women aged 20-74 during 1979-80 and 1982-90: 75

No significant PMRs

137 - Electricians

Number of deaths in men aged 20-74 during 1979-80 and 1982-90: 20517

PMRs for men

Cause of death (ICD)	Deaths	PMR	95% CI
Cancer of the liver (155)	46	72	53-97
Cancer of the larynx (161)	41	67	48-91
Cancer of trachea, bronchus and lung (162)	2154	87	83-91
Cancer of the pleura (163)	127	254	212-302
Cancer of bone (170)	25	167	108-247
Melanoma of skin (172)	79	162	128-202
Cancer of the brain (191)	204	117	101-134
Meningeal tumour (192.1,192.3,225.2,225.4)	1	14	0-78
Cancer of the thyroid (193)	18	183	109-290
Myeloma (203)	97	125	102-153
Amyloidosis (277.3)	14	259	142-435
Immunodeficiency (279.1)	3	25	5-73
Other alcohol related diseases (303,305.0,425.5,535.3,571.0-571.3, E860.0,E860.1)	118	122	101-146
Multiple sclerosis (340)	70	197	153-249
Aortic valve disorders (424.1)	65	74	57-95
Sub-arachnoid haemorrhage (430)	184	126	109-146
Aortic aneurysm (441)	361	115	104-128
Bronchopneumonia (485)	266	85	75-95
Chronic bronchitis and emphysema (491,492,496)	841	81	76-87
Fibrosing alveolitis (516.3)	39	144	102-196
Peritonitis (567)	13	198	106-339
Pancreatitis (577.0,577.1)	24	66	42-98
Injury by explosive material (E923)	10	317	152-583
Injury by electric current (E925)	42	385	277-521
Suicide (E950-E959)	433	116	105-127

Number of deaths in women aged 20-74 during 1979-80 and 1982-90: 196

No significant PMRs

138 - Electrical plant operators

Number of deaths in men aged 20-74 during 1979-80 and 1982-90: 2681

PMRs for men

Cause of death (ICD)	Deaths	PMR	95% CI
Cancer of the oral cavity (141,143,144,145)	13	243	130-416
Cancer of the oesophagus (150)	52	161	120-211
Cancer of the pleura (163)	14	219	119-367
Epilepsy (345)	7	336	135-693
Ischaemic heart disease (410-414)	1020	107	100-113

Number of deaths in women aged 20-74 during 1979-80 and 1982-90: 9

No significant PMRs

139 - Telephone fitters

Number of deaths in men aged 20-74 during 1979-80 and 1982-90: 7323

PMRs for men

Cause of death (ICD)	Deaths	PMR	95% CI
Cancer of the stomach (151)	225	118	103-135
Cancer of the colon (153)	169	128	110-149
Cancer of the pancreas (157)	120	127	105-151
Cancer of trachea, bronchus and lung (162)	780	86	80-92
Cancer of the pleura (163)	9	51	24-98
Cancer of the prostate (185)	159	132	112-154
Cancer of the brain (191)	86	149	119-184
Non-Hodgkin's lymphoma (200,202)	73	145	113-182
Multiple sclerosis (340)	26	222	145-326
Chronic rheumatic heart disease (394-398)	7	29	12-59
Ischaemic heart disease (410-414)	2730	107	103-111
Other cerebrovascular disease (431-438)	406	85	77-94
Aortic aneurysm (441)	146	124	104-145
Chronic bronchitis and emphysema (491,492,496)	256	65	58-74
Injury by machinery (E919)	1	17	0-94
Homicide (E960-E969)	0	0	0-69

Number of deaths in women aged 20-74 during 1979-80 and 1982-90: 63

No significant PMRs

140 - Electric cable and line workers

Number of deaths in men aged 20-74 during 1979-80 and 1982-90: 1845

PMRs for men

Cause of death (ICD)	Deaths	PMR	95% CI
Cancer of trachea, ronchus and lung (162)	188	81	70-94
Fall from building (E882)	7	693	279-1428
Injury by electric current (E925)	12	2178	1125-3804
Homicide (E960-E969)	4	391	107-1001

Number of deaths in women aged 20-74 during 1979-80 and 1982-90: 10

No significant PMRs

141 - Radio and TV mechanics

Number of deaths in men aged 20-74 during 1979-80 and 1982-90: 2163

PMRs for men

Cause of death (ICD)	Deaths	PMR	95% CI
Cancer of trachea, bronchus and lung (162)	223	86	75-98
Motor neurone disease (335.2)	14	240	131-402
Multiple sclerosis (340)	12	290	150-507
Pancreatitis (577.0,577.1)	10	246	118-452
Injury by electric current (E925)	5	387	126-903
Suicide (E950-E959)	62	142	109-182

Number of deaths in women aged 20-74 during 1979-80 and 1982-90: 22

No significant PMRs

142 - Other electronic maintenance engineers

Number of deaths in men aged 20-74 during 1979-80 and 1982-90: 2535

PMRs for men

Cause of death (ICD)	Deaths	PMR	95% CI
Cancer of the colon (153)	71	161	126-203
Cancer of trachea, bronchus and lung (162)	221	76	66-86
Cancer of the pleura (163)	1	16	0-88
Cancer of the prostate (185)	54	159	119-208
Cancer of the brain (191)	39	156	111-214
Non-Hodgkin's lymphoma (200,202)	32	161	110-228
Acute myeloid leukaemia (205.0)	17	179	104-287
Chronic myeloid leukaemia (205.1)	10	242	116-445
Dementia (290,331.0,331.1)	14	235	129-395
Chronic and unspecified myocarditis (429.0)	0	0	0-95
Aortic aneurysm (441)	47	135	99-179
Bronchopneumonia (485)	21	61	38-94
Chronic bronchitis and emphysema (491,492,496)	69	61	47-77
Injury by machinery (E919)	0	0	0-98

Number of deaths in women aged 20-74 during 1979-80 and 1982-90: 53

PMRs for women

Cause of death (ICD)	Deaths	PMR	95% CI
Ischaemic heart disease (410-414)	5	43	14-99

Appendix 4 Significant PMRs for each job group

143 - Electrical engineers (so described)

Number of deaths in men aged 20-74 during 1979-80 and 1982-90: 7930

PMRs for men

Cause of death (ICD)	Deaths	PMR	95% CI
Intestinal infectious diseases (001-009)	5	399	130-931
Cancer of the colon (153)	208	133	115-152
Cancer of the pancreas (157)	136	126	105-149
Cancer of the larynx (161)	13	57	31-98
Cancer of trachea, bronchus and lung (162)	798	85	79-91
Cancer of the pleura (163)	31	186	126-264
Cancer of the thyroid (193)	11	273	136-489
Myeloma (203)	47	141	103-187
Motor neurone disease (335.2)	36	161	113-223
Multiple sclerosis (340)	23	185	117-278
Aortic aneurysm (441)	197	138	120-159
Oesophageal varices (456.0-456.2)	4	373	102-954
Chronic bronchitis and emphysema (491,492,496)	275	69	61-78
Hyperplasia of prostate (600)	2	27	3-96
Air transport accidents (E840-E845)	7	488	196-1006
Injury by explosive material (E923)	4	536	146-1373

Number of deaths in women aged 20-74 during 1979-80 and 1982-90: 18

No significant PMRs

144 - Plumbers and gas fitters

Number of deaths in men aged 20-74 during 1979-80 and 1982-90: 16941

PMRs for men

Cause of death (ICD)	Deaths	PMR	95% CI
Tuberculosis (010-018,137)	17	62	36-99
Cancer of the peritoneum (158.8,158.9)	11	309	154-553
Cancer of trachea, bronchus and lung (162)	2235	107	103-112
Cancer of the pleura (163)	134	327	274-387
Cancer of the brain (191)	165	120	103-140
Cancer of the suprarenal (194.0)	7	294	118-607
Encephalitis (323)	5	344	112-802
Bronchopneumonia (485)	240	88	77-100
Chronic bronchitis and emphysema (491,492,496)	810	91	85-98
Coal workers' pneumoconiosis (500)	0	0	0-74
Asbestosis (501)	13	411	219-703
Cholelithiasis and cholecystitis (574,575,576.1-576.4)	5	33	11-77
Motor vehicle traffic accidents (E810-E819)	224	84	73-96
Water transport accidents (E830-E838)	13	307	163-525
Poisoning by other gases (E869)	4	600	163-1535
Fall on stairs (E880)	7	42	17-86
Fall from ladder or scaffolding (E881)	21	184	114-282
Fall from building (E882)	21	179	111-275
Injury by falling object (E916)	3	31	6-92
Injury by machinery (E919)	5	34	11-80

Number of deaths in women aged 20-74 during 1979-80 and 1982-90: 12

No significant PMRs

145 - Sheet metal workers

Number of deaths in men aged 20-74 during 1979-80 and 1982-90: 7961

PMRs for men

Cause of death (ICD)	Deaths	PMR	95% CI
Intestinal infectious diseases (001-009)	5	388	126-905
Cancer of the pharynx (specified) (146-148)	22	167	105-254
Cancer of trachea, bronchus and lung (162)	1072	109	102-115
Ischaemic heart disease (410-414)	2628	95	92-99
Other cardiomyopathies (425.0-425.4, 425.6-425.9)	18	63	37-99
Pneumococcal and unspecified lobar pneumonia (481)	38	147	104-203
Poisoning by motor vehicle exhaust (E868.2)	7	394	158-811

Number of deaths in women aged 20-74 during 1979-80 and 1982-90: 42

No significant PMRs

146 - Metal plate workers

Number of deaths in men aged 20-74 during 1979-80 and 1982-90: 5867

PMRs for men

Cause of death (ICD)	Deaths	PMR	95% CI
Cancer of the rectum (154)	101	127	103-154
Cancer of the liver (155)	29	159	106-228
Cancer of the larynx (161)	37	205	144-283
Cancer of trachea, bronchus and lung (162)	894	120	112-128
Cancer of the pleura (163)	73	515	404-648
Dementia (290,331.0,331.1)	30	167	113-239
Ischaemic heart disease (410-414)	1879	90	86-94
Aortic aneurysm (441)	60	62	47-80
Bronchopneumonia (485)	123	127	105-151
Pancreatitis (577.0,577.1)	17	182	106-291

Number of deaths in women aged 20-74 during 1979-80 and 1982-90: 33

No significant PMRs

147 - Steel erectors

Number of deaths in men aged 20-74 during 1979-80 and 1982-90: 3720

PMRs for men

Cause of death (ICD)	Deaths	PMR	95% CI
Cancer of the oral cavity (141,143,144,145)	20	247	151-381
Cancer of the stomach (151)	76	79	62-99
Cancer of the larynx (161)	21	182	112-278
Cancer of trachea, bronchus and lung (162)	567	123	113-133
Dementia (290,331.0,331.1)	17	172	100-276
Other alcohol related diseases (303,305.0,425.5,535.3,571.0-571.3, E860.0,E860.1)	29	165	110-237
Ischaemic heart disease (410-414)	1132	86	81-91
Pulmonary embolism and phlebitis (415.1,451,453)	12	54	28-94
Chronic bronchitis and emphysema (491,492,496)	227	123	107-140
Fall from ladder or scaffolding (E881)	20	754	461-1165
Fall from building (E882)	30	1136	766-1623
Other fall (E884)	8	563	243-1110
Injury by falling object (E916)	9	409	187-777
Injury by machinery (E919)	8	243	105-480

Number of deaths in women aged 20-74 during 1979-80 and 1982-90: 2

No significant PMRs

148 - Scaffolders

Number of deaths in men aged 20-74 during 1979-80 and 1982-90: 1737

PMRs for men

Cause of death (ICD)	Deaths	PMR	95% CI
Cancer of the stomach (151)	59	143	109-185
Cancer of trachea, bronchus and lung (162)	221	115	101-132
Acute myeloid leukaemia (205.0)	12	214	110-373
Drug dependence (304)	7	340	137-701
Chronic rheumatic heart disease (394-398)	1	15	0-86
Ischaemic heart disease (410-414)	495	88	80-96
Chronic bronchitis and emphysema (491,492,496)	100	123	100-149
Fall from ladder or scaffolding (E881)	38	2175	1539-2988
Fall from building (E882)	9	228	104-433
Other fall (E884)	4	370	101-948
Homicide (E960-E969)	9	271	124-515

Number of deaths in women aged 20-74 during 1979-80 and 1982-90: 1

No significant PMRs

Appendix 4 Significant PMRs for each job group

149 - Welders

Number of deaths in men aged 20-74 during 1979-80 and 1982-90: 12467

PMRs for men

Cause of death (ICD)	Deaths	PMR	95% CI
Cancer of trachea, bronchus and lung (162)	1658	109	103-114
Cancer of the pleura (163)	56	179	136-233
Ischaemic heart disease (410-414)	4217	97	94-100
Pneumococcal and unspecified lobar pneumonia (481)	75	186	147-234
Chronic bronchitis and emphysema (491,492,496)	678	109	101-118

Number of deaths in women aged 20-74 during 1979-80 and 1982-90: 361

PMRs for women

Cause of death (ICD)	Deaths	PMR	95% CI
Cancer of the stomach (151)	14	203	111-340

150 - Riggers

Number of deaths in men aged 20-74 during 1979-80 and 1982-90: 1756

PMRs for men

Cause of death (ICD)	Deaths	PMR	95% CI
Cancer of the oesophagus (150)	38	177	125-243
Cancer of the larynx (161)	12	219	113-383
Cancer of soft tissue (171)	6	282	104-615
Hypertensive disease (401-405)	5	38	12-89
Pulmonary heart disease unspecified (416.9)	8	315	136-621
Fall from ladder or scaffolding (E881)	4	369	100-944
Fall from building (E882)	10	1025	491-1884

Number of deaths in women aged 20-74 during 1979-80 and 1982-90: 3

No significant PMRs

151 - Jewellery workers

Number of deaths in men aged 20-74 during 1979-80 and 1982-90: 495

PMRs for men

Cause of death (ICD)	Deaths	PMR	95% CI
Cancer of trachea, bronchus and lung (162)	39	65	46-88
Bronchopneumonia (485)	21	245	151-375

Number of deaths in women aged 20-74 during 1979-80 and 1982-90: 84

PMRs for women

Cause of death (ICD)	Deaths	PMR	95% CI
Diabetes mellitus (250)	4	388	106-993

152 - Engravers and etchers (printing)

Number of deaths in men aged 20-74 during 1979-80 and 1982-90: 385

PMRs for men

Cause of death (ICD)	Deaths	PMR	95% CI
Cancer of trachea, bronchus and lung (162)	31	65	44-92
Diabetes mellitus (250)	8	242	104-476
Other cerebrovascular disease (431-438)	38	148	105-203

Number of deaths in women aged 20-74 during 1979-80 and 1982-90: 53

No significant PMRs

153 - Vehicle body builders

Number of deaths in men aged 20-74 during 1979-80 and 1982-90: 2256

PMRs for men

Cause of death (ICD)	Deaths	PMR	95% CI
Cancer of the peritoneum (158.8,158.9)	4	957	261-2450
Cancer of the larynx (161)	1	15	0-83
Cancer of the pleura (163)	24	470	301-700
Chronic bronchitis and emphysema (491,492,496)	104	80	65-97
Asbestosis (501)	5	1139	370-2658

Number of deaths in women aged 20-74 during 1979-80 and 1982-90: 6

No significant PMRs

154 - Oilers and greasers

Number of deaths in men aged 20-74 during 1979-80 and 1982-90: 585

No significant PMRs

Number of deaths in women aged 20-74 during 1979-80 and 1982-90: 8

No significant PMRs

155 - Electronics wire workers

Number of deaths in men aged 20-74 during 1979-80 and 1982-90: 432

No significant PMRs

Number of deaths in women aged 20-74 during 1979-80 and 1982-90: 160

No significant PMRs

156 - Coil winders

Number of deaths in men aged 20-74 during 1979-80 and 1982-90: 369

PMRs for men

Cause of death (ICD)	Deaths	PMR	95% CI
Cancer of the brain (191)	10	364	174-669
Aortic aneurysm (441)	12	198	102-346

Number of deaths in women aged 20-74 during 1979-80 and 1982-90: 341

No significant PMRs

157 - Pottery decorators

Number of deaths in men aged 20-74 during 1979-80 and 1982-90: 157

PMRs for men

Cause of death (ICD)	Deaths	PMR	95% CI
Suicide (E950-E959)	9	371	170-704

Number of deaths in women aged 20-74 during 1979-80 and 1982-90: 484

PMRs for women

Cause of death (ICD)	Deaths	PMR	95% CI
Bronchopneumonia (485)	4	35	10-90
Gastric ulcer (531)	5	331	107-772
Fall unspecified (E888)	5	579	188-1351
Suicide (E950-E959)	10	338	162-621

158 - Coach painters

Number of deaths in men aged 20-74 during 1979-80 and 1982-90: 615

PMRs for men

Cause of death (ICD)	Deaths	PMR	95% CI
Hypertensive disease (401-405)	10	211	101-387

Number of deaths in women aged 20-74 during 1979-80 and 1982-90: 5

No significant PMRs

159 - Other spray painters

Number of deaths in men aged 20-74 during 1979-80 and 1982-90: 3705

PMRs for men

Cause of death (ICD)	Deaths	PMR	95% CI
Cancer of the oesophagus (150)	26	63	41-93
Cancer of the liver (155)	19	176	106-275
Cancer of trachea, bronchus and lung (162)	557	126	116-137
Cancer of the testis (186)	10	315	151-579
Other alcohol related diseases (303,305.0,425.5,535.3,571.0-571.3, E860.0,E860.1)	7	39	15-79
Sub-arachnoid haemorrhage (430)	33	146	100-205
Coal workers' pneumoconiosis (500)	0	0	0-56
Inguinal hernia (550)	5	348	113-812
Homicide (E960-E969)	0	0	0-90

Appendix 4 Significant PMRs for each job group

Job group	Deaths	PMR	95% CI
Number of deaths in women aged 20-74 during 1979-80 and 1982-90: 141			
			No significant PMRs

160 - Painters and decorators nec

Job group	Deaths	PMR	95% CI
Number of deaths in men aged 20-74 during 1979-80 and 1982-90: 29689			
PMRs for men			
Cause of death (ICD)			
Tuberculosis (010-018,137)	63	133	102-170
Cancer of the oral cavity (141,143,144,145)	75	127	100-159
Cancer of trachea, bronchus and lung (162)	4110	112	109-116
Cancer of the kidney (except pelvis) (189.0)	130	77	64-91
Diabetes mellitus (250)	218	85	74-97
Other alcohol related diseases (303,305.0,425.5,535.3,571.0-571.3, E860.0,E860.1)	138	119	100-140
Drug dependence (304)	38	297	210-408
Ischaemic heart disease (410-414)	9677	94	92-96
Bronchopneumonia (485)	565	111	102-121
Chronic bronchitis and emphysema (491,492,496)	1869	114	109-120
Coal workers' pneumoconiosis (500)	1	11	0-59
Motor vehicle traffic accidents (E810-E819)	381	85	77-94
Off-road motor vehicle accidents (E820-E825)	4	35	10-90
Accidental poisoning by drugs (E850-E858)	76	251	198-315
Fall on stairs (E880)	44	158	115-213
Fall from ladder or scaffolding (E881)	64	350	269-447
Other fall (E884)	18	175	104-277
Injury by falling object (E916)	1	7	0-37
Injury by machinery (E919)	7	31	12-64
Suicide (E950-E959)	498	119	109-130
Homicide (E960-E969)	42	203	146-274
Injury undetermined whether accidentally or purposely inflicted (E980-E989)	214	163	141-186
Number of deaths in women aged 20-74 during 1979-80 and 1982-90: 155			
PMRs for women			
Cause of death (ICD)			
Urinary infections (590,595,599.0)	5	1166	378-2720

161 - Electrical, electronic assemblers

Job group	Deaths	PMR	95% CI
Number of deaths in men aged 20-74 during 1979-80 and 1982-90: 1092			
PMRs for men			
Cause of death (ICD)			
Sub-arachnoid haemorrhage (430)	15	221	124-364
Asthma (493)	10	239	114-439
Number of deaths in women aged 20-74 during 1979-80 and 1982-90: 2069			
PMRs for women			
Cause of death (ICD)			
Chronic myeloid leukaemia (205.1)	7	269	108-555
Diabetes mellitus (250)	40	162	116-221
Pulmonary embolism and phlebitis (415.1,451,453)	11	48	24-86
Pneumococcal and unspecified lobar pneumonia (481)	2	23	3-85
Bronchopneumonia (485)	27	64	42-93
Asthma (493)	23	169	107-254
Motor vehicle traffic accidents (E810-E819)	8	44	19-88
Suicide (E950-E959)	9	50	23-96
Injury undetermined whether accidentally or purposely inflicted (E980-E989)	2	17	2-63

162 - Instrument assemblers

Job group	Deaths	PMR	95% CI
Number of deaths in men aged 20-74 during 1979-80 and 1982-90: 72			
			No significant PMRs
Number of deaths in women aged 20-74 during 1979-80 and 1982-90: 74			
PMRs for women			
Cause of death (ICD)			
Chronic rheumatic heart disease (394-398)	5	729	237-1701

163 - Assemblers (vehicles and other metal goods)

Job group	Deaths	PMR	95% CI
Number of deaths in men aged 20-74 during 1979-80 and 1982-90: 5297			
PMRs for men			
Cause of death (ICD)			
Cancer of the liver (155)	25	157	102-232
Non-Hodgkin's lymphoma (200,202)	49	148	109-195
Other cardiomyopathies (425.0-425.4, 425.6-425.9)	28	150	99-217
Coal workers' pneumoconiosis (500)	1	11	0-64
Injury by machinery (E919)	0	0	0-95
Injury undetermined whether accidentally or purposely inflicted (E980-E989)	17	60	35-96
Number of deaths in women aged 20-74 during 1979-80 and 1982-90: 1214			
PMRs for women			
Cause of death (ICD)			
Aortic aneurysm (441)	2	20	2-72
Motor vehicle traffic accidents (E810-E819)	0	0	0-45

164 - Packers and sorters

Job group	Deaths	PMR	95% CI
Number of deaths in men aged 20-74 during 1979-80 and 1982-90: 8337			
PMRs for men			
Cause of death (ICD)			
Cancer of the stomach (151)	250	114	100-129
Other alcohol related diseases (303,305.0,425.5,535.3,571.0-571.3, E860.0,E860.1)	18	54	32-86
Epilepsy (345)	33	230	158-323
Bronchopneumonia (485)	204	122	106-140
Coal workers' pneumoconiosis (500)	1	6	0-35
Pulmonary fibrosis (515)	18	197	117-311
Water transport accidents (E830-E838)	0	0	0-99
Number of deaths in women aged 20-74 during 1979-80 and 1982-90: 7143			
PMRs for women			
Cause of death (ICD)			
Epilepsy (345)	24	153	98-228
Chronic rheumatic heart disease (394-398)	86	126	101-155
Mitral valve disorders (424.0)	16	182	104-295
Other cardiomyopathies (425.0-425.4, 425.6-425.9)	31	168	114-239
Peripheral vascular disease (443,557)	15	49	28-81
Appendicitis (540-542)	6	317	116-690
Cholelithiasis and cholecystitis (574,575,576.1-576.4)	16	177	101-287
Glomerulonephritis (580-583)	1	17	0-93
Motor vehicle traffic accidents (E810-E819)	46	71	52-95
Homicide (E960-E969)	4	37	10-94

165 - Bricklayers and tilesetters

Job group	Deaths	PMR	95% CI
Number of deaths in men aged 20-74 during 1979-80 and 1982-90: 15970			
PMRs for men			
Cause of death (ICD)			
Viral hepatitis (070)	0	0	0-100
Cancer of the rectum (154)	253	119	104-134
Cancer of trachea, bronchus and lung (162)	2233	112	107-117
Cancer of the thyroid (193)	1	14	0-77
Diabetes mellitus (250)	104	75	61-91
Immunodeficiency (279.1)	1	17	0-97
Chronic rheumatic heart disease (394-398)	68	131	102-166
Hypertensive disease (401-405)	153	128	108-150
Ischaemic heart disease (410-414)	5207	93	91-96
Pulmonary heart disease unspecified (416.9)	36	150	105-208
Aortic aneurysm (441)	227	84	73-95
Bronchopneumonia (485)	325	116	103-129
Chronic bronchitis and emphysema (491,492,496)	964	107	100-114
Bronchiectasis (494)	33	159	109-224
Duodenal ulcer (532)	75	128	101-161
Other hernia (551-553)	1	17	0-96

Job group	Deaths	PMR	95% CI
Accidental poisoning by drugs (E850-E858)	27	195	129-285
Fall on stairs (E880)	25	175	113-259
Fall from ladder or scaffolding (E881)	19	208	125-325
Injury undetermined whether accidentally or purposely inflicted (E980-E989)	92	147	119-181

Number of deaths in women aged 20-74 during 1979-80 and 1982-90: 2

No significant PMRs

166 - Masons and stonecutters

Number of deaths in men aged 20-74 during 1979-80 and 1982-90: 2388

PMRs for men

Cause of death (ICD)	Deaths	PMR	95% CI
Cancer of the oesophagus (150)	40	145	103-197
Pulmonary heart disease unspecified (416.9)	0	0	0-99
Chronic bronchitis and emphysema (491,492,496)	117	83	68-99
Silicosis (502)	11	6471	3230-11578
Other pneumoconiosis (503,505)	5	1351	439-3154
Injured by fire (E890-E899)	5	314	102-734

Number of deaths in women aged 20-74 during 1979-80 and 1982-90: 0

No significant PMRs

167 - Plasterers

Number of deaths in men aged 20-74 during 1979-80 and 1982-90: 5198

PMRs for men

Cause of death (ICD)	Deaths	PMR	95% CI
Cancer of trachea, bronchus and lung (162)	805	127	119-136
Diabetes mellitus (250)	29	65	43-93
Ischaemic heart disease (410-414)	1582	88	84-93
Pulmonary embolism and phlebitis (415.1,451,453)	44	139	101-186
Chronic bronchitis and emphysema (491,492,496)	343	122	109-136
Accidental poisoning by drugs (E850-E858)	12	196	101-342
Injury by machinery (E919)	0	0	0-82
Homicide (E960-E969)	10	240	115-442
Injury undetermined whether accidentally or purposely inflicted (E980-E989)	41	159	114-216
Cancer of the nose and nasal sinuses (160)	7	249	100-513

Number of deaths in women aged 20-74 during 1979-80 and 1982-90: 1

No significant PMRs

168 - Roofers and glaziers

Number of deaths in men aged 20-74 during 1979-80 and 1982-90: 3444

PMRs for men

Cause of death (ICD)	Deaths	PMR	95% CI
Tuberculosis (010-018,137)	2	27	3-96
Cancer of the gall bladder (156)	0	0	0-92
Cancer of trachea, bronchus and lung (162)	439	121	110-133
Cancer of the prostate (185)	53	136	102-178
Immunodeficiency	0	0	0-79
Other alcohol related diseases (303,305.0,425.5,535.3,571.0-571.3, E860.0,E860.1)	38	154	109-211
Drug dependence (304)	13	238	127-408
Epilepsy (345)	6	35	13-76
Ischaemic heart disease (410-414)	950	89	83-95
Other cardiomyopathies (425.0-425.4, 425.6-425.9)	27	164	108-239
Coal workers' pneumoconiosis (500)	0	0	0-78
Fall from ladder or scaffolding (E881)	27	725	477-1056
Fall from building (E882)	116	1238	1023-1485
Injury by machinery (E919)	1	15	0-83

Number of deaths in women aged 20-74 during 1979-80 and 1982-90: 4

No significant PMRs

169 - Builders etc.

Number of deaths in men aged 20-74 during 1979-80 and 1982-90: 22433

PMRs for men

Cause of death (ICD)	Deaths	PMR	95% CI
Cancer of the oesophagus (150)	298	114	102-128
Cancer of the gall bladder (156)	43	141	102-191
Cancer of trachea, bronchus and lung (162)	2896	106	103-110
Cancer of the pleura (163)	65	136	105-174
Cancer of the male breast (175)	16	197	112-320
Cancer of the brain (191)	215	124	108-142
Non-Hodgkin's lymphoma (200,202)	183	119	103-138
Acute myeloid leukaemia (205.0)	85	125	100-155
Diabetes mellitus (250)	159	81	69-95
Chronic rheumatic heart disease (394-398)	54	71	53-93
Ischaemic heart disease (410-414)	7374	95	92-97
Other cardiomyopathies (425.0-425.4, 425.6-425.9)	112	135	111-163
Asthma (493)	60	76	58-98
Coal workers' pneumoconiosis (500)	3	18	4-52
Asbestosis (501)	0	0	0-74
Fall from ladder or scaffolding (E881)	53	378	283-495
Fall from building (E882)	35	211	147-293
Fall into hole (E883)	6	332	122-723
Injury by firearms (E922)	5	470	153-1098
Suicide (E950-E959)	441	128	116-140
Injury undetermined whether accidentally or purposely inflicted (E980-E989)	135	121	101-143

Number of deaths in women aged 20-74 during 1979-80 and 1982-90: 23

No significant PMRs

170 - Railway track workers

Number of deaths in men aged 20-74 during 1979-80 and 1982-90: 3209

PMRs for men

Cause of death (ICD)	Deaths	PMR	95% CI
Cancer of the eye (190)	4	430	117-1101
Aortic aneurysm (441)	33	66	46-93
Bronchopneumonia (485)	90	132	106-162
Coal workers' pneumoconiosis (500)	0	0	0-65
Railway accidents (E800-E807)	55	4754	3580-6191

Number of deaths in women aged 20-74 during 1979-80 and 1982-90: 1

No significant PMRs

171 - Road construction workers and paviors

Number of deaths in men aged 20-74 during 1979-80 and 1982-90: 5207

PMRs for men

Cause of death (ICD)	Deaths	PMR	95% CI
Cancer of the oesophagus (150)	77	132	104-165
Acute lymphatic leukaemia (204.0)	6	323	119-703
Chronic myeloid leukaemia (205.1)	11	211	105-377
Parkinson's disease (332)	25	155	100-229
Ischaemic heart disease (410-414)	1812	105	100-110
Crohn's disease (555)	4	408	111-1045
Motor vehicle traffic accidents (E810-E819)	89	140	113-173
Injury by electric current (E925)	6	494	181-1076

Number of deaths in women aged 20-74 during 1979-80 and 1982-90: 2

No significant PMRs

172 - Sewage plant attendants

Number of deaths in men aged 20-74 during 1979-80 and 1982-90: 1710

PMRs for men

Cause of death (ICD)	Deaths	PMR	95% CI
Cancer of the small intestine (152)	4	440	120-1127
Cancer of the prostate (185)	38	154	109-212
Non-Hodgkin's lymphoma (200,202)	16	182	104-295
Dementia (290,331.0,331.1)	1	16	0-88

Appendix 4 Significant PMRs for each job group

Job group	Deaths	PMR	95% CI
Number of deaths in women aged 20-74 during 1979-80 and 1982-90: 3			
			No significant PMRs

173 - Mains and service layers

Number of deaths in men aged 20-74 during 1979-80 and 1982-90: 2172

PMRs for men

Cause of death (ICD)	Deaths	PMR	95% CI
Cancer of the rectum (154)	53	187	140-245
Cancer of the pancreas (157)	39	141	100-193
Cancer of the larynx (161)	16	224	128-364
Other cancer of skin (173)	5	321	104-749
Cancer of the prostate (185)	50	155	115-205
Pneumococcal and unspecified lobar pneumonia (481)	2	26	3-92
Injury by falling object (E916)	7	537	216-1106
Suicide (E950-E959)	20	61	37-94

Number of deaths in women aged 20-74 during 1979-80 and 1982-90: 0

No significant PMRs

174 - Construction workers nec

Number of deaths in men aged 20-74 during 1979-80 and 1982-90: 26886

PMRs for men

Cause of death (ICD)	Deaths	PMR	95% CI
Cancer of the peritoneum (158.8,158.9)	64	956	736-1221
Cancer of the nose and nasal sinuses (160)	22	157	98-238
Cancer of trachea, bronchus and lung (162)	3679	114	110-118
Cancer of the pleura (163)	77	191	151-239
Cancer of other endocrine organs (194.1-194.9)	6	295	108-641
Other leukaemia (207,208)	19	184	111-288
Diabetes mellitus (250)	164	66	57-77
Immunodeficiency (279.1)	4	33	9-85
Parkinson's disease (332)	50	73	54-96
Multiple sclerosis (340)	20	53	32-81
Epilepsy (345)	66	73	57-93
Ischaemic heart disease (410-414)	8354	96	94-98
Wegener's granulomatosis (446.4)	8	290	125-572
Bronchopneumonia (485)	516	92	84-100
Asthma (493)	67	69	54-88
Coal workers' pneumoconiosis (500)	1	7	0-41
Asbestosis (501)	71	1274	995-1608
Other colitis (558)	1	17	0-95
Urinary infections (590,595,599.0)	18	51	30-80
Railway accidents (E800-E807)	3	18	4-54
Water transport accidents (E830-E838)	3	26	5-76
Fall from ladder or scaffolding (E881)	37	184	129-253
Fall from building (E882)	87	259	208-320
Fall into hole (E883)	13	379	202-649
Slipping and tripping (E885)	0	0	0-93
Injury by falling object (E916)	56	289	218-375
Injury by machinery (E919)	42	170	122-229
Suicide (E950-E959)	394	73	66-81
Injury undetermined whether accidentally or purposely inflicted (E980-E989)	209	83	72-95

Number of deaths in women aged 20-74 during 1979-80 and 1982-90: 19

No significant PMRs

175 - Face trained coalminers

Number of deaths in men aged 20-74 during 1979-80 and 1982-90: 9816

PMRs for men

Cause of death (ICD)	Deaths	PMR	95% CI
Cancer of the pharynx (specified) (146-148)	7	46	19-95
Cancer of the oesophagus (150)	82	70	56-88
Cancer of the stomach (151)	326	123	110-137
Cancer of the rectum (154)	171	128	109-149
Cancer of the pancreas (157)	152	118	100-139
Cancer of trachea, bronchus and lung (162)	1137	90	84-95
Cancer of the pleura (163)	4	18	5-46
Cancer of the prostate (185)	143	78	65-91
Urothelial cancer (188,189.1-189.8)	82	69	55-86
Non-Hodgkin's lymphoma (200,202)	44	73	53-98
Diabetes mellitus (250)	43	50	36-67
Motor neurone disease (335.2)	12	49	25-86
Pulmonary embolism and phlebitis (415.1,451,453)	88	141	113-174
Aortic valve disorders (424.1)	68	152	118-193
Other cardiomyopathies (425.0-425.4, 425.6-425.9)	12	40	21-70
Chronic and unspecified myocarditis (429.0)	41	288	207-391
Aortic aneurysm (441)	94	53	43-65
Acute bronchitis (466)	17	246	143-394
Chronic bronchitis and emphysema (491,492,496)	920	156	146-166
Coal workers' pneumoconiosis (500)	131	3771	3153-4476
Silicosis (502)	5	695	226-1623
Pneumothorax (512)	5	336	109-785
Renal failure (584-586)	20	59	36-91
Motor vehicle traffic accidents (E810-E819)	28	43	28-62
Other vehicle accidents (E846-E848)	17	4749	2766-7603
Fall on stairs (E880)	1	13	0-72
Injury by falling object (E916)	21	816	504-1249
Injury by machinery (E919)	16	458	262-744
Injury by explosive material (E923)	7	1102	443-2271
Injury undetermined whether accidentally or purposely inflicted (E980-E989)	13	56	30-96

Number of deaths in women aged 20-74 during 1979-80 and 1982-90: 1

No significant PMRs

176 - Miners (not coal) and quarry workers

Number of deaths in men aged 20-74 during 1979-80 and 1982-90: 2717

PMRs for men

Cause of death (ICD)	Deaths	PMR	95% CI
Tuberculosis (010-018,137)	24	431	276-643
Other alcohol related diseases (303,305.0,425.5,535.3,571.0-571.3, E860.0,E860.1)	3	30	6-87
Chronic rheumatic heart disease (394-398)	3	32	7-95
Other cerebrovascular disease (431-438)	164	84	72-98
Coal workers' pneumoconiosis (500)	0	0	0-68
Silicosis (502)	58	8123	6167-10506
Other pneumoconiosis (503,505)	7	992	399-2043
Pulmonary fibrosis (515)	11	362	181-648
Injury by falling object (E916)	8	646	279-1272
Injury by machinery (E919)	7	445	179-917

Number of deaths in women aged 20-74 during 1979-80 and 1982-90: 2

No significant PMRs

177 - Railway guards

Number of deaths in men aged 20-74 during 1979-80 and 1982-90: 2029

PMRs for men

Cause of death (ICD)	Deaths	PMR	95% CI
Diabetes mellitus (250)	28	158	105-229
Sub-arachnoid haemorrhage (430)	4	37	10-95
Pulmonary fibrosis (515)	7	344	138-708
Rheumatoid arthritis (714)	6	275	101-599
Railway accidents (E800-E807)	17	2751	1602-4404
Suicide (E950-E959)	10	43	21-80

Number of deaths in women aged 20-74 during 1979-80 and 1982-90: 5

No significant PMRs

178 - Railway workers

Number of deaths in men aged 20-74 during 1979-80 and 1982-90: 2185

PMRs for men

Cause of death (ICD)	Deaths	PMR	95% CI
Cancer of trachea, bronchus and lung (162)	219	78	68-89

Appendix 4 Significant PMRs for each job group

Job group	Deaths	PMR	95% CI
Diabetes mellitus (250)	29	150	101-216
Other cerebrovascular disease (431-438)	185	118	102-136
Chronic bronchitis and emphysema (491,492,496)	106	80	66-97

Number of deaths in women aged 20-74 during 1979-80 and 1982-90: 54

PMRs for women

Cause of death (ICD)	Deaths	PMR	95% CI
Ischaemic heart disease (410-414)	21	164	101-251

179 - Shunters and points operators

Number of deaths in men aged 20-74 during 1979-80 and 1982-90: 984

PMRs for men

Cause of death (ICD)	Deaths	PMR	95% CI
Railway accidents (E800-E807)	13	4887	2602-8357

Number of deaths in women aged 20-74 during 1979-80 and 1982-90: 2

No significant PMRs

180 - Railway engine drivers

Number of deaths in men aged 20-74 during 1979-80 and 1982-90: 4418

PMRs for men

Cause of death (ICD)	Deaths	PMR	95% CI
Cancer of trachea, bronchus and lung (162)	483	84	76-91
Motor neurone disease (335.2)	5	43	14-100
Ischaemic heart disease (410-414)	1716	107	102-112
Pulmonary embolism and phlebitis (415.1,451,453)	17	62	36-100
Pulmonary heart disease unspecified (416.9)	14	216	118-362
Chronic bronchitis and emphysema (491,492,496)	206	86	75-99
Other colitis (558)	4	449	122-1151
Rheumatoid arthritis (714)	12	248	128-434
Railway accidents (800-807)	28	2244	1490-3246
Suicide (E950-E959)	23	53	33-79

Number of deaths in women aged 20-74 during 1979-80 and 1982-90: 0

No significant PMRs

181 - Road transport inspectors

Number of deaths in men aged 20-74 during 1979-80 and 1982-90: 2715

PMRs for men

Cause of death (ICD)	Deaths	PMR	95% CI
Cancer of the oral cavity (141,143,144,145)	0	0	0-70
Melanoma of skin (172)	10	221	106-407
Ischaemic heart disease (410-414)	1121	115	109-122
Aortic valve disorders (424.1)	5	41	13-95
Aortic aneurysm (441)	62	130	100-167
Bronchopneumonia (485)	30	62	41-88
Chronic bronchitis and emphysema (491,492,496)	125	78	65-93
Bronchiectasis (494)	11	310	155-554
Other colitis (558)	4	675	184-1727
Pancreatitis (577.0,577.1)	0	0	0-90

Number of deaths in women aged 20-74 during 1979-80 and 1982-90: 9

No significant PMRs

182 - Bus and coach drivers

Number of deaths in men aged 20-74 during 1979-80 and 1982-90: 11782

PMRs for men

Cause of death (ICD)	Deaths	PMR	95% CI
Cancer of the oesophagus (150)	166	118	101-138
Cancer of the rectum (154)	125	79	66-94
Cancer of the pleura (163)	7	25	10-51
Cancer of bone (170)	1	15	0-82
Epilepsy (345)	5	40	13-94
Hypertensive disease (401-405)	63	71	55-91
Ischaemic heart disease (410-414)	4436	106	103-110
Atrial fibrillation (427.3)	4	39	11-100
Other cerebrovascular disease (431-438)	721	93	86-100
Chronic bronchitis and emphysema (491,492,496)	762	120	112-129
Hydronephrosis (591)	4	402	110-1029
Railway accidents (800-807)	0	0	0-91
Water transport accidents (E830-E838)	8	357	154-704
Accidental poisoning by drugs (E850-E858)	3	31	6-89
Injured by fire (E890-E899)	2	22	3-79
Injury by machinery (E919)	1	12	0-66
Injury by electric current (E925)	0	0	0-93
Suicide (E950-E959)	125	80	67-96
Homicide (E960-E969)	1	14	0-78

Number of deaths in women aged 20-74 during 1979-80 and 1982-90: 64

No significant PMRs

183 - Lorry drivers

Number of deaths in men aged 20-74 during 1979-80 and 1982-90: 61558

PMRs for men

Cause of death (ICD)	Deaths	PMR	95% CI
Cancer of the colon (153)	980	88	83-94
Cancer of the larynx (161)	249	132	116-150
Cancer of trachea, bronchus and lung (162)	8294	109	107-112
Cancer of the pleura (163)	55	35	27-46
Melanoma of skin (172)	113	79	65-95
Cancer of the prostate (185)	884	93	87-99
Cancer of the kidney (except pelvis) (189.0)	318	85	76-94
Cancer of the brain (191)	447	86	78-94
Cancer of the thyroid (193)	19	63	38-99
Non-Hodgkin's lymphoma (200,202)	390	88	80-98
Acute monocytic leukaemia (206.0)	15	203	113-334
Immunodeficiency (279.1)	12	39	20-69
Other alcohol related diseases (303,305.0,425.5,535.3,571.0-571.3, E860.0,E860.1)	219	76	66-87
Drug dependence (304)	15	60	34-99
Motor neurone disease (335.2)	123	74	61-88
Multiple sclerosis (340)	64	59	46-76
Epilepsy (345)	38	50	35-68
Ischaemic heart disease (410-414)	21486	99	97-100
Pulmonary embolism and phlebitis (415.1,451,453)	308	83	74-93
Sub-arachnoid haemorrhage (430)	381	89	80-98
Pneumococcal and unspecified lobar pneumonia (481)	166	83	71-96
Chronic bronchitis and emphysema (491,492,496)	3740	120	116-124
Coal workers' pneumoconiosis (500)	3	18	4-52
Asbestosis (501)	4	35	9-89
Oesophageal disease (530)	21	62	38-95
Cirrhosis of the liver (571.5)	129	75	63-89
Railway accidents (800-807)	15	58	32-95
Motor vehicle traffic accidents (E810-E819)	1409	155	147-163
Off-road motor vehicle accidents (E820-E825)	75	291	229-364
Poisoning by liquified petroleum gas (E868.0)	10	253	121-465
Fall from ladder or scaffolding (E881)	7	16	6-33
Fall from building (E882)	13	30	16-52
Other fall (E884)	37	159	112-220
Fall unspecified (E888)	47	70	52-94
Injury by falling object (E916)	65	181	140-231
Injury undetermined whether accidentally or purposely inflicted (E980-E989)	243	82	72-93

Number of deaths in women aged 20-74 during 1979-80 and 1982-90: 373

PMRs for women

Cause of death (ICD)	Deaths	PMR	95% CI
Motor vehicle traffic accidents (E810-E819)	19	271	163-423

Appendix 4 Significant PMRs for each job group

184 - Other motor drivers

Number of deaths in men aged 20-74 during 1979-80 and 1982-90: 9569

PMRs for men

Cause of death (ICD)	Deaths	PMR	95% CI
Viral hepatitis (070)	8	278	120-548
Cancer of the pharynx (specified) (146-148)	6	34	13-75
Cancer of the larynx (161)	15	52	29-86
Cancer of the pleura (163)	10	41	20-76
Parkinson's disease (332)	10	43	21-80
Ischaemic heart disease (410-414)	3558	105	102-109
Other cerebrovascular disease (431-438)	505	86	79-94
Aortic aneurysm (441)	169	117	100-136
Pneumococcal and unspecified lobar pneumonia (481)	13	42	22-71
Gastric ulcer (531)	11	55	28-99
Peptic ulcer (533)	10	256	123-471
Railway accidents (E800-E807)	0	0	0-83
Fall from building (E882)	0	0	0-50
Injury by machinery (E919)	1	11	0-60
Suicide (E950-E959)	227	136	119-155
Homicide (E960-E969)	15	183	103-302

Number of deaths in women aged 20-74 during 1979-80 and 1982-90: 146

PMRs for women

Cause of death (ICD)	Deaths	PMR	95% CI
Cancer of trachea, bronchus and lung (162)	20	197	120-304
Melanoma of skin (172)	4	488	133-1249
Other cerebrovascular disease (431-438)	4	39	11-100

185 - Bus conductors and drivers' mates

Number of deaths in men aged 20-74 during 1979-80 and 1982-90: 3628

PMRs for men

Cause of death (ICD)	Deaths	PMR	95% CI
Cancer of trachea, bronchus and lung (162)	388	87	79-96
Cancer of the kidney (except pelvis) (189.0)	30	172	116-246
Non-Hodgkin's lymphoma (200,202)	10	51	24-93
Myeloma (203)	24	201	129-300
Diabetes mellitus (250)	51	154	115-202
Pneumococcal and unspecified lobar pneumonia (481)	22	157	98-238
Influenza (487)	6	341	125-743
Chronic bronchitis and emphysema (491,492,496)	291	119	106-134
Bronchiectasis (494)	15	275	154-454
Coal workers' pneumoconiosis (500)	0	0	0-49
Suicide (E950-E959)	28	65	43-94

Number of deaths in women aged 20-74 during 1979-80 and 1982-90: 918

PMRs for women

Cause of death (ICD)	Deaths	PMR	95% CI
Cancer of trachea, bronchus and lung (162)	103	131	107-159
Pulmonary heart disease unspecified (416.9)	5	337	110-787
Chronic bronchitis and emphysema (491,492,496)	51	145	108-191

186 - Mechanical plant drivers

Number of deaths in men aged 20-74 during 1979-80 and 1982-90: 2937

PMRs for men

Cause of death (ICD)	Deaths	PMR	95% CI
Cancer of the pleura (163)	1	13	0-73
Melanoma of skin (172)	1	14	0-76
Acute bronchitis (466)	6	305	112-664
Chronic bronchitis and emphysema (491,492,496)	168	118	101-137
Injury by machinery (E919)	22	768	481-1164

Number of deaths in women aged 20-74 during 1979-80 and 1982-90: 1

No significant PMRs

187 - Crane drivers

Number of deaths in men aged 20-74 during 1979-80 and 1982-90: 8562

PMRs for men

Cause of death (ICD)	Deaths	PMR	95% CI
Cancer of soft tissue (171)	3	31	6-91
Cancer of the prostate (185)	89	60	48-74
Cancer of the brain (191)	41	66	47-90
Endocarditis (421)	0	0	0-82
Other cerebrovascular disease (431-438)	655	114	105-123
Aortic aneurysm (441)	115	80	66-96
Chronic bronchitis and emphysema (491,492,496)	573	119	110-130
Cirrhosis of the liver (571.5)	32	145	99-204
Off-road motor vehicle accidents (E820-E825)	7	332	134-684
Fall from ladder or scaffolding (E881)	0	0	0-78
Injury by machinery (E919)	24	549	352-819

Number of deaths in women aged 20-74 during 1979-80 and 1982-90: 70

PMRs for women

Cause of death (ICD)	Deaths	PMR	95% CI
Chronic bronchitis and emphysema (491,492,496)	7	273	110-562

188 - Fork lift truck drivers

Number of deaths in men aged 20-74 during 1979-80 and 1982-90: 6697

PMRs for men

Cause of death (ICD)	Deaths	PMR	95% CI
Cancer of the pleura (163)	5	29	9-67
Cancer of the brain (191)	45	72	52-96
Multiple sclerosis (340)	6	46	17-100
Atrial fibrillation (427.3)	10	211	101-387
Chronic bronchitis and emphysema (491,492,496)	373	120	109-133
Fall unspecified (E888)	16	212	121-344
Injury by machinery (E919)	27	351	231-511
Injury by electric current (E925)	0	0	0-94

Number of deaths in women aged 20-74 during 1979-80 and 1982-90: 28

No significant PMRs

189 - Slingers

Number of deaths in men aged 20-74 during 1979-80 and 1982-90: 1497

PMRs for men

Cause of death (ICD)	Deaths	PMR	95% CI
Hypertensive disease (401-405)	21	185	114-283
Pulmonary heart disease unspecified (416.9)	10	441	211-811
Injury by machinery (E919)	6	836	307-1819

Number of deaths in women aged 20-74 during 1979-80 and 1982-90: 0

No significant PMRs

190 - Storekeepers

Number of deaths in men aged 20-74 during 1979-80 and 1982-90: 49431

PMRs for men

Cause of death (ICD)	Deaths	PMR	95% CI
Cancer of the oral cavity (141,143,144,145)	71	69	54-88
Cancer of the colon (153)	917	109	102-116
Cancer of the pleura (163)	58	74	56-96
Dementia (290,331.0,331.1)	130	83	70-99
Other alcohol related diseases (303,305.0,425.5,535.3,571.0-571.3, E860.0,E860.1)	161	85	73-100
Epilepsy (345)	108	140	115-169
Ischaemic heart disease (410-414)	17676	104	103-106
Atrial fibrillation (427.3)	63	131	100-167
Chronic and unspecified myocarditis (429.0)	66	77	59-98
Aortic aneurysm (441)	782	108	101-116
Bronchopneumonia (485)	1061	107	101-114
Chronic bronchitis and emphysema (491,492,496)	2862	93	90-97

Appendix 4 Significant PMRs for each job group

Job group	Deaths	PMR	95% CI
Asthma (493)	184	117	101-136
Coal workers' pneumoconiosis (500)	2	2	0-8
Silicosis (502)	2	16	2-59
Ulcerative colitis (556)	20	167	102-258
Cholelithiasis and cholecystitis (574,575,576.1-576.4)	35	70	49-98
Urinary infections (590,595,599.0)	40	68	49-93
Erythematous conditions (695)	5	387	126-903
Railway accidents (E800-E807)	8	43	19-85
Motor vehicle traffic accidents (E810-E819)	481	82	75-90
Water transport accidents (E830-E838)	5	26	8-61
Other vehicle accidents (E846-E848)	0	0	0-79
Fall from ladder or scaffolding (E881)	5	22	7-50
Fall from building (E882)	14	37	20-62
Injured by fire (E890-E899)	35	69	48-96
Injury by machinery (E919)	15	51	29-84
Injury by electric current (E925)	4	39	11-99
Suicide (E950-E959)	477	77	71-85
Injury undetermined whether accidentally or purposely inflicted (E980-E989)	160	71	60-83

Number of deaths in women aged 20-74 during 1979-80 and 1982-90: 2021

PMRs for women

Cause of death (ICD)	Deaths	PMR	95% CI
Cancer of the gall bladder (156)	14	206	113-346
Cancer of the female breast (174)	195	117	101-134
Cancer of the body of uterus (182)	23	173	109-260
Hodgkin's disease (201)	7	317	127-653
Chronic myeloid leukaemia (205.1)	7	268	108-552
Dementia (290,331.0,331.1)	4	39	11-99
Pulmonary embolism and phlebitis (415.1,451,453)	33	148	102-208
Peritonitis (567)	5	475	154-1108

191 - Dockers and goods porters

Number of deaths in men aged 20-74 during 1979-80 and 1982-90: 13603

PMRs for men

Cause of death (ICD)	Deaths	PMR	95% CI
Cancer of the oral cavity (141,143,144,145)	53	136	102-178
Cancer of the liver (155)	25	65	42-96
Cancer of the retroperitoneum (158.0)	7	254	102-524
Cancer of the peritoneum (158.8,158.9)	7	293	118-603
Cancer of trachea, bronchus and lung (162)	1936	112	107-117
Cancer of the pleura (163)	36	214	150-297
Diabetes mellitus (250)	98	75	61-91
Aplastic anaemia (284)	0	0	0-94
Other alcohol related diseases (303,305.0,425.5,535.3,571.0-571.3, E860.0,E860.1)	102	138	112-167
Ischaemic heart disease (410-414)	4160	95	92-98
Acute myocarditis (422)	5	482	156-1124
Other cerebrovascular disease (431-438)	923	88	83-94
Influenza (487)	3	33	7-96
Cirrhosis of the liver (571.5)	61	163	124-209
Water transport accidents (E830-E838)	12	683	353-1192
Injured by fire (E890-E899)	12	48	25-84

Number of deaths in women aged 20-74 during 1979-80 and 1982-90: 49

No significant PMRs

192 - Refuse collectors

Number of deaths in men aged 20-74 during 1979-80 and 1982-90: 4610

PMRs for men

Cause of death (ICD)	Deaths	PMR	95% CI
Cancer of trachea, bronchus and lung (162)	629	109	101-118
Cancer of bone (170)	6	287	106-626
Urothelial cancer (188,189.1-189.8)	63	141	108-180
Hypertensive disease (401-405)	25	64	41-94
Fall from building (E882)	0	0	0-96

Number of deaths in women aged 20-74 during 1979-80 and 1982-90: 26

No significant PMRs

193 - Labourers in coke ovens

Number of deaths in men aged 20-74 during 1979-80 and 1982-90: 1039

PMRs for men

Cause of death (ICD)	Deaths	PMR	95% CI
Cancer of trachea, bronchus and lung (162)	156	118	100-138
Cancer of the prostate (185)	5	33	11-77
Bronchopneumonia (485)	44	140	102-189
Asthma (493)	7	269	108-555

Number of deaths in women aged 20-74 during 1979-80 and 1982-90: 6

No significant PMRs

194 - Boiler operators

Number of deaths in men aged 20-74 during 1979-80 and 1982-90: 6212

PMRs for men

Cause of death (ICD)	Deaths	PMR	95% CI
Cancer of the larynx (161)	34	166	115-232
Cancer of trachea, bronchus and lung (162)	914	115	108-123
Cancer of the pleura (163)	24	269	172-401
Urothelial cancer (188,189.1-189.8)	99	136	111-166
Epilepsy (345)	1	18	0-99
Chronic bronchitis and emphysema (491,492,496)	464	110	100-120
Coal workers' pneumoconiosis (500)	6	44	16-95
Motor vehicle traffic accidents (E810-E819)	22	59	37-89

Number of deaths in women aged 20-74 during 1979-80 and 1982-90: 1

No significant PMRs

Appendix 5 Cancer incidence in men by job group: cancers with statistically significant PRRs

Eve Roman
Cancer Epidemiology Unit, Imperial Cancer Research Fund

Lucy Carpenter
Department of Public Health and Primary Care, University of Oxford

This appendix presents the findings by job group for men aged 20-74 years registered with cancer in England 1981-87. Results are listed for each cancer for which the PRR is significantly greater than 100 at the two-sided 5 per cent level of statistical significance, based on three or more registrations. PRRs which are significantly less than 100 are listed, regardless of the number of cases on which they are based. PRRs are adjusted for age (in five-year age groups), social class (six classes) and region of registration. The total number of cancers registered in each occupational group is given in section 7.8, which provides a commentary on cancer incidence by job group and gives results for certain specific cancers where an association is reported in the mortality analyses.

Job group	Registrations	PRR	95% CI
001 Lawyers			
Lip (140)	3	763	158-2231
Colon (153)	51	143	107-189
Melanoma of skin (172)	22	168	105-255
Brain (191)	27	152	100-222
Lung (162)	67	75	58-96
002 Accountants			
Liver (155.0-155.1)	5	39	13-93
003 Personnel managers etc			
Pleura (163)	8	336	146-664
005 Computer programmers			
Liver (155.0-155.1)	5	387	126-904
006 Sales managers etc			
Other cancer of skin (173)	253	118	104-134
Bladder (188, 189.1-189.9)	168	117	100-137
Chronic lymphatic leukaemia (204.1)	9	52	24-99
007 Government inspectors			
Chronic lymphatic leukaemia (204.1)	9	234	107-445
008 Government administrators			
Colon (153)	35	155	108-216
Brain (191)	16	181	104-295
Lung (162)	63	76	59-98
009 Other administrators			
Eye (190)	6	281	103-612
Non-Hodgkin's lymphoma (200,202)	48	142	105-189

Job group	Registrations	PRR	95% CI
010 Teachers in higher education			
Lung (162)	83	78	63-98
011 Teachers nec			
Colon (153)	255	114	101-130
Melanoma of skin (172)	93	151	123-186
Other cancer of skin (173)	512	163	150-178
Prostate (185)	255	129	114-146
Testis (186)	117	128	106-154
Brain (191)	144	126	107-149
Myeloma (203)	60	144	111-187
Chronic myeloid leukaemia (205)	25	155	100-229
Mouth and pharynx (140-149)	43	70	51-96
Pancreas (157)	33	67	47-95
Lung (162)	425	56	51-62
Ill-defined + secondary (195-199)	125	75	63-90
012 Vocational trainers, social scientists etc.			
Retroperitoneum (158.0)	3	577	119-1689
Lung (162)	135	121	102-144
Other cancer of skin (173)	33	66	46-94
013 Welfare workers			
Other cancer of skin (173)	86	129	104-160
014 Clergy			
Other cancer of skin (173)	101	133	109-163
Lung (162)	69	54	43-69
015 Doctors			
Mouth and pharynx (140-149)	37	178	125-246
Oral cavity (141, 143-145)	24	258	166-385
Other parts of mouth (145)	8	342	148-674
Liver (155.0-155.1)	18	178	106-282
Other cancer of skin (173)	161	148	126-173
Non-Hodgkin's lymphoma (200,202)	50	139	104-184
Stomach (151)	34	70	49-99
Lung (162)	127	73	62-88
Pleura (163)	0	0	0-73
Ill-defined + secondary (195-199)	26	56	37-83
017 Nurses			
Rectum (154)	52	141	105-185
Oesophagus (150)	10	53	26-99

Appendix 5 Cancer incidence in men by job group: cancers with statistically significant PRRs

Job group	Registrations	PRR	95% CI
018 Pharmacists			
Bladder (188, 189.1-189.9)	35	146	102-203
020 Physiotherapists			
Pancreas (157)	3	767	158-2243
022 Veterinarians			
Lung (162)	4	33	9-85
023 Driving instructors			
Bladder (188, 189.1-189.9)	36	161	113-224
024 Literary and artistic occupations			
Mouth and pharynx (140-149)	45	139	102-186
Soft tissue (171)	20	170	104-263
025 Persons involved in sport			
Prostate (185)	16	185	106-302
027 Chemical engineers and scientists			
Chronic lymphatic leukaemia (204.1)	7	270	109-557
028 Physical scientists and mathematicians			
Other cancer of skin (173)	60	153	117-198
029 Electrical and electronic engineers (professional)			
Stomach (151)	35	169	118-235
Lung (162)	86	124	100-154
030 Professional engineers nec			
Lung (162)	549	125	115-136
Melanoma of skin (172)	36	68	48-94
Other cancer of skin (173)	221	77	68-88
031 Draughtspersons			
Pleura (163)	9	234	108-445
Brain (191)	50	168	125-222
032 Laboratory technicians			
Stomach (151)	68	133	104-169
Colon (153)	32	62	43-88
Melanoma of skin (172)	5	40	13-95
033 Architects and surveyors			
Small intestine (152)	7	278	112-574
Prostate (185)	101	128	104-156
Mouth and pharynx (140-149)	13	53	29-92
Stomach (151)	40	72	52-98
034 Aircraft flight deck officers			
Other cancer of skin (173)	25	207	134-306

Job group	Registrations	PRR	95% CI
036 Seafarers			
Mouth and pharynx (140-149)	58	240	183-311
Oral Cavity (141, 143-145)	34	300	208-420
Floor of mouth (144)	12	388	201-678
Pharynx (146-148)	18	214	127-339
Oropharynx (146)	8	242	105-479
Hypopharynx (148)	9	299	137-569
Larynx (161)	33	173	119-244
Other cancer of skin (173)	142	150	127-178
Rectum (154)	41	67	49-92
Bladder (188, 189.1-189.9)	58	75	58-98
Hodgkin's disease (201)	4	37	10-96
Myeloma (203)	4	29	8-77
038 Production and maintenance managers			
Lung (162)	632	108	101-118
Pleura (163)	23	227	144-341
Soft tissue (171)	20	173	106-268
Mouth and pharynx (140-149)	25	63	41-94
Oral Cavity (141, 143-145)	10	50	24-92
Prostate (185)	118	79	66-96
039 Managers in construction			
Other cancer of skin (173)	81	127	101-158
040 Managers in transport, utilities and mining			
Lung (162)	560	120	111-131
Other endocrine organs (194.1,194.7-194.9)	3	541	112-1582
Ill-defined + secondary (195-199)	120	124	103-149
Other cancer of skin (173)	135	83	70-99
Prostate (185)	99	78	64-95
Myeloma (203)	13	54	29-93
041 Office managers			
Hypopharynx (148)	9	229	105-435
042 Butchers			
Chronic lymphatic leukaemia (204.1)	18	175	104-278
Colon (153)	65	75	58-96
Pleura (163)	3	27	6-80
043 Fishmongers and poultry dressers			
Larynx (161)	9	219	100-417
044 Retailers and dealers			
Pharynx (146-148)	62	131	101-169
Nasopharynx (147)	20	184	112-284
Lung (162)	2536	114	110-119
Colon (153)	531	90	83-99
Other cancer of skin (173)	624	80	74-87
Non-Hodgkin's lymphoma (200, 202)	219	86	75-98

Appendix 5 Cancer incidence in men by job group: cancers with statistically significant PRRs

Job group	Registrations	PRR	95% CI
045 Publicans and bar staff			
Mouth and pharynx (140-149)	138	300	253-355
Oral cavity (141,143-145)	70	316	247-400
Floor of mouth (144)	28	482	321-697
Pharynx (146-148)	48	334	247-444
Oropharynx (146)	25	431	279-636
Hypopharynx (148)	19	356	215-556
Ill defined mouth and pharynx (149)	14	613	336-1030
Liver (155.0-155.1)	38	202	143-278
Larynx (161)	75	194	153-244
Lung (162)	816	122	115-131
Ill-defined + secondary (195-199)	160	118	101-139
Colon (153)	136	78	66-93
Melanoma of skin (172)	17	55	32-89
Other cancer of skin (173)	166	70	60-82
Prostate (185)	122	72	60-87
Testis (186)	18	60	36-95
Bladder (188,189.1-189.9)	140	84	71-100
Brain (191)	45	63	47-86
Non-Hodgkin's lymphoma (200, 202)	50	67	50-89
Myeloma (203)	17	51	30-83
All leukaemia (204-208)	37	58	41-81
Acute myeloid leukaemia (205.0)	7	35	14-73
046 Caterers			
Mouth and pharynx (140-149)	40	212	152-289
Oral cavity (141, 143-145)	20	234	143-361
Floor of mouth (144)	6	295	108-643
Other parts of mouth (145)	10	513	246-945
Pharynx (146-148)	15	235	132-389
Nasopharynx (147)	9	520	238-988
Liver (155.0-155.1)	17	227	133-364
Colon (153)	45	72	53-97
Other cancer of skin (173)	55	61	46-80
047 Farmers			
Lip (140)	47	288	212-383
Other cancer of skin (173)	745	118	110-127
Prostate (185)	641	117	108-127
Myeloma (203)	120	126	105-151
All leukaemia (204-208)	213	117	102-134
Chronic lymphatic leukaemia (204.1)	74	127	100-160
Oral cavity (141, 143-145)	41	68	50-94
Gallbladder (156)	29	69	47-100
Larynx (161)	88	80	65-99
Lung (162)	2258	92	89-97
Pleura (163)	10	28	14-53
050 Fire service personnel			
Oropharynx (146)	4	469	128-1203
Liver (155.0-155.1)	0	0	0-97
051 Launderers and dry cleaners			
Larynx (161)	8	233	101-461
Other leukaemia (207, 208)	3	675	139-1974
Mouth and pharynx (140-149)	11	250	125-448
Hypopharynx (148)	3	580	120-1695
052 Hairdressers			
Prostate (185)	45	146	107-196
Oesophagus (150)	3	25	5-76
Lung (162)	133	83	70-99
054 Postal workers			
Bladder (188, 189.1-189.9)	222	114	100-131
Oesophagus (150)	61	72	55-93
Lung (162)	1010	92	87-98
Pleura (163)	12	52	27-91
055 Petrol pump attendants			
Larynx (161)	6	368	135-801
056 Van sales persons			
Lung (162)	233	115	101-131
Mouth and pharynx (140-149)	5	41	13-97
057 Sales representatives			
Lung (162)	794	89	84-96
Mouth and pharynx (140-149)	5	41	13-97
058 Security workers			
Oral cavity (141, 143-145)	45	140	103-189
Other parts of mouth (145)	14	205	113-346
Other endocrine organ (194.1, 194.7-194.9)	5	433	141-1012
059 Cooks and kitchen porters			
Mouth and pharynx (140-149)	67	256	199-325
Oral Cavity (141, 143-145)	22	185	116-281
Floor of mouth (144)	8	248	107-490
Pharynx (146-148)	37	397	280-548
Oropharynx (146)	9	229	105-437
Nasopharynx (147)	18	847	502-1340
Hypopharynx (148)	10	306	147-563
Liver (155.0-155.1)	17	215	126-345
Penis (187.1-187.5)	57	264	106-544
Chronic lymphatic leukaemia (204.1)	13	203	108-348
Colon (153)	43	71	52-96
Other cancer of skin (173)	69	74	58-94
061 Hospital porters and ward orderlies			
Pleura (163)	0	0	0-81
062 Ambulance workers			
Other cancer of skin (173)	43	147	107-199
063 Railway station workers			
Other cancer of skin (173)	108	126	104-152
Bladder (188,189.1-189.9)	90	126	102-156
Eye (190)	6	441	162-960
Pharynx (146-148)	3	32	7-96
Lung (162)	416	86	78-95
064 Undertakers			
Lung (162)	49	136	101-180
065 Foresters			
Myeloma (203)	7	322	130-664

Appendix 5 Cancer incidence in men by job group: cancers with statistically significant PRRs

Job group	Registrations	PRR	95% CI
066 Fishing & related workers			
Mouth and pharynx (140-149)	12	227	118-397
Lip (140)	3	504	104-1474
Other cancer of skin (173)	28	151	101-219
Bladder (188,189.1-189.9)	6	36	13-79
068 Leather and shoe workers			
Gum (143)	3	714	147-2088
Rectum (154)	48	136	101-181
Soft tissue (171)	7	273	110-564
All leukaemia (204-208)	28	159	106-230
Acute myeloid leukaemia (205.0)	13	222	119-381
069 Preparatory fibre processors			
Other cancer of skin (173)	3	30	6-88
070 Spinners and winders			
Ill-defined + secondary (195-199)	21	167	104-256
071 Warp preparers and weavers			
Colon (153)	32	155	106-220
Lung (162)	111	81	67-98
072 Knitters			
Gallbladder (156)	5	426	139-996
073 Bleachers, dyers and finishers			
Gallbladder (156)	6	326	120-712
074 Other textile workers			
Thyroid (193)	7	397	160-819
Other cancer of skin (173)	52	69	52-91
075 Chemical workers			
Pleura (163)	30	169	114-242
076 Bakers			
Nasopharynx (147)	4	366	100-939
077 Brewery workers			
Pharynx (146-148)	6	311	114-678
Rectum (154)	24	165	106-246
Thyroid (193)	3	562	116-1643
Colon (153)	7	44	18-93
083 Glass formers and decorators			
Liver (155.0-155.1)	4	372	101-953
085 Rubber manufacturers			
Mouth and pharynx (140-149)	14	184	101-310
Bladder (188,189.1-189.9)	58	226	172-293
087 Man-made fibre makers			
Ill-defined + secondary (195-199)	5	386	126-902
Non-Hodgkin's lymphoma (200, 202)	3	616	127-1802

Job group	Registrations	PRR	95% CI
088 Other coal miners			
Rectum (154)	388	116	106-129
Lung (162)	2337	104	100-109
Oesophagus (150)	118	77	64-92
Melanoma of skin (172)	19	62	38-98
089 Tobacco workers			
Colon (153)	7	305	123-630
091 Other occupations - glass and ceramics			
Acute myeloid leukaemia (205.0)	0	0	0-87
093 Plastic goods makers			
Bladder (188,189.1-189.9)	19	187	113-293
094 Compositors			
Kidney (except pelvis) (189.0)	11	238	119-426
097 Printers (so described)			
Other cancer of skin (173)	146	121	103-143
Prostate (185)	107	122	101-149
Bladder (188,189.1-189.9)	115	123	102-148
Myeloma (203)	6	38	14-84
098 Tailors and dressmakers			
Stomach (151)	46	163	120-218
Melanoma of skin (172)	6	275	101-600
Lung (162)	94	74	60-91
100 Sewers and embroiderers			
Other cancer of skin (173)	3	31	7-93
101 Upholsterers			
Pleura (163)	7	257	104-531
Prostate (185)	39	151	108-207
Oesophagus (150)	4	37	10-97
103 Other workers with fabrics			
Colon (153)	15	231	129-382
104 Carpenters			
Pleura (163)	70	206	161-261
Other cancer of skin (173)	357	112	101-125
Stomach (151)	278	87	78-98
Lung (162)	1314	91	87-97
Chronic myeloid leukaemia (205.1)	7	48	19-99
105 Cabinet makers			
Nose and nasal sinuses (160)	9	803	367-1525
Testis (186)	10	237	114-436
107 Pattern makers			
Hodgkin's disease (201)	4	399	109-1024
108 Woodworking machinists			
Oesophagus (150)	24	165	106-246
Nose and nasal sinuses (160)	10	710	341-1307

Appendix 5 Cancer incidence in men by job group: cancers with statistically significant PRRs

Job group	Registrations	PRR	95% CI
109 Other woodworkers			
Nose and nasal sinuses (160)	5	676	220-1580
110 Dental technicians			
Prostate (185)	15	249	140- 412
112 Furnace operatives (metal)			
Ill-defined + secondary (195-199)	40	143	103-196
115 Metal drawers			
Lung (162)	64	130	100-166
116 Moulders and coremakers (metal)			
Lung (162)	329	123	111-138
Other cancer of skin (173)	27	57	38-84
Prostate (185)	24	54	35-80
117 Electroplaters			
Oesophagus (150)	0	0	0-84
118 Annealers, hardeners, temperers (metal)			
Liver (155.0-155.1)	3	649	134-1898
119 Galvanisers and tin platers			
Prostate (185)	0	0	0-96
120 Other metal manufacturers			
Larynx (161)	27	154	102-225
Lung (162)	507	111	102-122
Other cancer of skin (173)	48	71	53-95
121 Press and machine tool setters			
Scrotal skin (187.7)	3	676	140-1979
Brain (191)	12	56	29-100
122 Centre lathe turners			
Salivary (142)	4	395	108-1014
Ill-defined mouth and pharynx (149)	5	777	253-1815
Scrotal skin (187.7)	4	1132	309-2901
124 Machine tool operators			
Colon (153)	654	110	102- 119
Soft tissue (171)	60	134	103- 173
Melanoma of skin (172)	102	139	114- 170
Other cancer of skin (173)	905	111	105- 119
Prostate (185)	682	117	109- 127
Testis (186)	141	119	100- 141
Scrotal skin (187.7)	18	261	155- 413
Bladder (188,189.1-189.9)	735	112	105- 121
Kidney (except pelvis) (189.0)	210	115	101- 132
Mouth and pharynx (140-149)	181	83	72-96
Pharynx (146-148)	48	67	50-89
Oropharynx (146)	16	52	30-85
Stomach (151)	717	88	83-96
Liver (155.0-155.1)	52	72	54-95
Lung (162)	3319	90	87-94
126 Metal polishers			
Lung (162)	157	120	102-141
Colon (153)	11	54	27-97
Other cancer of skin (173)	12	50	26-88
127 Fettlers and dressers (metal)			
Lung (162)	102	124	101-151
128 Shot blasters			
Prostate (185)	0	0	0-83
129 Toolmakers			
Colon (153)	94	129	104-158
Lung (162)	397	86	79-96
Pleura (163)	0	0	0-45
130 Precision instrument makers			
Myeloma (203)	9	284	130-540
All leukaemia (204-208)	13	190	101-325
132 Production fitters			
Other parts of mouth (145)	24	169	109-253
Pleura (163)	121	181	150-217
Eye (190)	21	175	108-268
Nose and nasal sinuses (160)	9	48	22-93
133 Motor mechanics			
Gallbladder (156)	17	187	109-300
Pleura(163)	3	23	5-68
Other cancer of skin (173)	95	74	61-92
Ill-defined + secondary (195-199)	70	73	57-93
137 Electricians			
Pleura (163)	61	250	192-322
Meningeal (192.1-192.3, 225.2-225.4)	17	214	125-343
Myeloma (203)	43	143	104-193
Liver (155.0-155.1)	8	43	19-86
Lung (162)	822	85	80-91
138 Electrical plant operators			
Small intestine (152)	5	566	184-1322
Peritoneum (158.8-158.9)	3	990	204-2894
Pleura (163)	16	382	219- 622
Lung (162)	145	83	71-99
139 Telephone fitters			
Other cancer of skin (173)	107	128	106-156
Kidney (except pelvis) (189.0)	29	153	103-221
Lung (162)	283	83	75-94
140 Electric cable and line workers			
Stomach (151)	28	157	105-228
Testis (186)	8	385	167-760
Colon (153)	4	33	9-85

Appendix 5 Cancer incidence in men by job group: cancers with statistically significant PRRs

Job group	Registrations	PRR	95% CI
141 Radio and TV mechanics			
Hodgkin's disease (201)	9	240	110-457
142 Other electronic maintenance engineers			
All leukaemia (204-208)	19	277	167-434
Chronic lymphatic leukaemia (204.1)	5	408	133-954
Acute myeloid leukaemia (205.0)	8	293	127-579
Lung (162)	51	72	54-95
143 Electrical engineers (so described)			
Pleura (163)	18	207	123-328
Prostate (185)	82	132	106-165
Larynx (161)	7	43	17-89
Lung (162)	277	75	67-85
144 Plumbers and gas fitters			
Pleura (163)	50	220	164-291
Myeloma (203)	17	61	36-99
145 Sheet metal workers			
Pleura (163)	25	266	173-394
146 Metal plate workers			
Pleura (163)	41	378	272-513
Non-Hodgkin's lymphoma (200, 202)	11	55	28-99
Myeloma (203)	3	30	6-90
147 Steel erectors			
Lip (140)	6	311	114-678
Oral cavity (141, 143-145)	11	237	118-424
Floor of mouth (144)	5	380	124-888
148 Scaffolders			
Oesophagus (150)	14	187	102-314
149 Welders			
Pleura (163)	34	196	136-274
150 Riggers			
Chronic myeloid leukaemia (205.1)	4	484	132-1239
Prostate (185)	4	34	10-89
151 Jewellery workers			
Lung (162)	10	51	25-95
153 Vehicle body builders			
Pleura (163)	10	385	185-709
158 Coach painters			
Larynx (161)	5	365	119-853
Bladder (188,189.1-189.9)	12	198	103-347
159 Other spray painters			
Other leukaemia (207, 208)	4	404	110-1035
Prostate (185)	17	58	34-93
160 Painters and decorators nec			
Hypopharynx (148)	18	174	103-275
Lung (162)	1664	108	103-114
Pleura (163)	20	57	35-89
Melanoma of skin (172)	20	62	38-97
Prostate (185)	213	81	72-94
161 Electrical, electronic assemblers			
Pharynx (146-148)	3	556	115-1628
163 Assemblers (vehicles and other metal goods)			
Acute lymphatic leukaemia (204. 0)	3	520	108-1522
164 Packers and sorters			
Nose and nasal sinuses (160)	6	312	115-681
Lung (162)	421	111	101-123
Colon (153)	37	62	44-86
Other cancer of skin (173)	55	71	54-94
165 Bricklayers and tilesetters			
Lung (162)	1059	107	101-114
Other cancer of skin (173)	256	126	112-144
Testis (186)	37	142	100-196
Pharynx (146-148)	8	47	20-93
Colon (153)	111	72	60-88
Bladder (188,189.1-189.9)	125	74	62-89
166 Masons and stonecutters			
Bladder (188,189.1-189.9)	23	164	104-247
167 Plasterers			
Other parts of the mouth (145)	5	396	129-926
Lung (162)	306	115	103-129
Gallbladder (156)	0	0	0-88
168 Roofers and glaziers			
Pancreas (157)	16	208	119-339
169 Builders etc.			
Other cancer of skin (173)	346	119	108-133
Brain (191)	94	123	100-151
171 Road construction workers and paviors			
Ill-defined + secondary (195-199)	24	64	41-96
173 Mains and service layers			
Oesophagus (150)	16	194	111-317
Liver (155.0-155.1)	6	289	106-631
Pancreas (157)	10	222	107-410
Chronic myeloid leukaemia (205.1)	5	450	146-1052
174 Construction workers nec			
Peritoneum (158.8-158.9)	14	449	246-754
Lung (162)	1144	114	109-122
Pleura (163)	42	245	177-331
Prostate (185)	111	77	64-93
Bladder (188,189.1-189.9)	134	78	66-93
Myeloma (203)	17	57	33-91

Appendix 5 Cancer incidence in men by job group: cancers with statistically significant PRRs

Job group	Registrations	PRR	95% CI
175 Face trained coalminers			
Kidney (except pelvis) (189.0)	54	146	110-191
Ill-defined + secondary (195-199)	203	116	102-134
Pleura (163)	6	22	8-49
Other cancer of skin (173)	83	72	58-90
177 Railway guards			
Chronic lymphatic leukaemia (205.1)	5	319	104-745
179 Shunters and points operators			
Pancreas (157)	8	383	166-756
Lung (162)	65	141	109-181
180 Railway engine drivers			
Thyroid (193)	5	382	124-892
Acute myeloid leukaemia (205.0)	9	223	102-425
Pharynx (146-148)	0	0	0-98
181 Road transport inspectors			
Small intestine (152)	3	553	114-1618
182 Bus and coach drivers			
Oesophagus (150)	59	132	101-171
Colon (153)	114	125	103-150
Retroperitoneum (158.0)	5	336	109-785
Mouth and pharynx (140-149)	18	58	35-93
Pharynx (146-148)	4	35	10-92
Pleura (163)	3	21	5-63
183 Lorry drivers			
Ill-defined mouth and pharynx (149)	17	176	103-283
Lung (162)	3666	107	104-111
Colon (153)	505	91	84-100
Pleura (163)	26	30	20-45
Bone (170)	10	48	23-90
Male breast (175)	7	48	20-100
184 Other motor drivers			
Bladder (188,189.1-189.9)	112	125	103-151
Mouth and pharynx (140-149)	17	60	35-96
185 Bus conductors and drivers' mates			
Stomach (151)	50	144	107-190
Gallbladder (156)	7	315	127-650
Myeloma (203)	10	220	106-405
186 Mechanical plant drivers			
Male breast (175)	4	829	226-2125
187 Crane drivers			
Lung (162)	526	109	100-119
Male breast (175)	5	336	109-784
Prostate (185)	53	73	55-97
190 Storekeepers			
Pancreas (157)	129	125	105-149
Other parts of mouth (145)	4	33	9-85
Pleura (163)	18	52	31-83
Soft tissue (171)	12	49	26-87
Meningeal (192.1, 192.3, 225.2, 225.4)	4	36	10-94
Chronic myeloid leukaemia (204.1)	11	55	28-99
191 Dockers and goods porters			
Non-Hodgkin's lymphoma (200, 202)	29	65	44-95
194 Boiler operators			
Lung (162)	387	114	104-127
Pleura (163)	11	231	115-414

Appendix 6 Cancer incidence in women by job group: cancers with statistically significant PRRs

Eve Roman
Cancer Epidemiology Unit, Imperial Cancer Research Fund

Lucy Carpenter
Department of Public Health and Primary Care, University of Oxford

This appendix presents the findings by job group for women aged 20-74 years registered with cancer in England 1981-87. Results are listed for each cancer for which the PRR is significantly greater than 100 at the two-sided 5 per cent level of statistical significance, based on three or more registrations. PRRs which are significantly less than 100 are listed, regardless of the number of cases on which they are based. PRRs are adjusted for age (in five-year age-groups), social class (six classes) and region of registration. The total number of cancers registered in each occupational group is given in section 7.8, which provides a commentary on cancer incidence by job group and gives results for certain specific cancers where an association is reported in the mortality analyses.

Job group	Registrations	PRR	95% CI
003 Personnel managers etc.			
Myeloma (203)	5	473	154-1105
006 Sales managers etc			
Non-Hodgkin's lymphoma (200,202)	4	37	10-96
007 Government inspectors			
Ca in Situ Cervix (233.1)	0	0	0-96
009 Other administrators			
Ovary (183)	54	135	102-177
010 Teachers in higher education			
Lung (162)	6	43	16-95
011 Teachers nec			
Colon (153)	315	116	104-130
Melanoma of skin (172)	223	145	128-166
Other cancer of skin (173)	400	117	107-130
Female breast (174)	1942	118	113-123
Uterus (179, 181, 182)	244	114	101-130
Pharynx (146-148)	7	46	18-95
Stomach (151)	72	68	54-86
Larynx (161)	6	42	16-93
Lung (162)	231	51	46-59
Cervix (180)	177	64	55-75
Ca in Situ Cervix (223.1)	714	81	76-87
012 Vocational trainers, social scientists etc.			
Lung (162)	15	180	101-297
Non-Hodgkin's lymphoma (200,202)	8	271	117-535
Ill-defined + secondary (195-199)	12	274	142-480

Job group	Registrations	PRR	95% CI
013 Welfare workers			
Meningeal (192.1, 192.3, 225.2, 225.4)	15	183	103-303
Brain (191)	17	61	36-98
014 Clergy			
Uterus (179, 181, 182)	23	165	105-248
Lung (162)	7	39	16-81
015 Doctors			
Other cancer of skin (173)	43	144	104-194
Cervix (180)	4	36	10-94
Ovary (183)	11	52	26-94
017 Nurses			
Nasopharynx (147)	13	204	109-350
Other cancer of skin (173)	460	110	101-122
Bladder (188, 189.1-189.9)	138	120	101-142
019 Medical radiographers			
Melanoma of skin (172)	10	272	131-501
020 Physiotherapists			
Female breast (174)	71	128	100-162
Stomach (151)	0	0	0-99
021 Health professions nec			
Meningeal (192.1, 192.3, 225.2, 225.4)	5	322	105-752
026 Biological scientists			
Acute myeloid leukaemia (205.0)	4	1244	339-3188
030 Professional engineers nec			
Ca in Situ Cervix (233.1)	19	187	113-292
032 Laboratory technicians			
Kidney(except pelvis) (189.0)	9	255	117-486
Uterus (179, 181, 182)	7	46	19-96

Appendix 6 Cancer incidence in women by job group: cancers with statistically significant PRRs

Job group	Registrations	PRR	95% CI
037 Technicians nec			
Hodgkin's disease (201)	4	428	117-1098
038 Production and maintenance managers			
Stomach (151)	9	353	162- 671
All leukaemia (204-208)	6	290	107- 632
Acute myeloid leukaemia (205.0)	5	500	162-1168
039 Managers in construction			
Colon (153)	3	629	130-1840
040 Managers in transport, utilities and mining			
Mouth and pharynx (140-149)	5	544	177-1270
042 Butchers			
Oesophagus (150)	5	402	131- 940
044 Retailers and dealers			
Lung (162)	1036	112	106-120
Cervix (180)	547	125	115-136
Ca in Situ Cervix (233.1)	1536	107	103-113
Melanoma of skin (172)	162	75	65-88
Female breast (174)	2471	95	92-99
045 Publicans and bar staff			
Mouth and pharynx (140-149)	36	181	127-252
Oral Cavity (141, 143-145)	17	192	112-309
Gum (143)	6	554	203-1206
Pharynx (146-148))	12	201	104-351
Oropharynx (146)	8	524	227-1034
Ill-defined mouth and pharynx (149)	5	490	159-1145
Lung (162)	293	138	124-156
Mediastinum (164.1-164.9)	7	691	278-1426
Cervix (180)	198	140	122-162
Ca in Situ Cervix (233.1)	640	116	108-126
Melanoma of skin (172)	26	59	39-87
Other cancer of skin (173)	80	62	50-78
Female breast (174)	416	80	73-88
Uterus (179,181,182)	45	65	48-88
Kidney(except pelvis) (189.0)	9	46	21-89
All leukaemia (204-208)	17	55	32-89
046 Caterers			
Myeloma (203)	19	54	33-85
047 Farmers			
Non-Hodgkin's lymphoma (200,202)	25	166	108-246
051 Launderers and dry cleaners			
Ovary (183)	48	139	103-185
Brain (191)	3	34	7-100
052 Hairdressers			
Stomach (151)	13	57	31-98
053 Office workers and cashiers			
Cervix (180)	1194	92	87-98
057 Sales representatives			
Oesophagus (150)	9	228	104-434
Lung (162)	54	165	124-216
Ca in Situ Cervix (233.1)	140	123	104-146
Female breast (174)	86	71	58-89
058 Security workers			
Uterus (179,181,182)	20	165	101-255
Stomach (151)	3	28	6-83
060 Other service personnel			
Nose and nasal sinuses (160)	29	153	103-220
Female breast (174)	3003	103	100-108
Lung (162)	1782	94	90-99
Pleura (163)	15	57	32-95
Ill-defined + secondary (195-199)	577	92	85-100
061 Hospital porters and ward orderlies			
Thyroid (193)	13	195	104-335
069 Preparatory fibre processors			
Gallbladder (156.0)	4	625	171-1602
Other cancer of skin (173)	17	172	101- 277
Ill-defined + secondary (195-199)	17	209	122- 336
070 Spinners and winders			
Stomach (151)	46	154	113- 207
Female breast (174)	73	76	60- 96
072 Knitters			
Bladder (188,189.1 189.9)	14	193	106-324
Lung (162)	20	58	36-91
075 Chemical workers			
Stomach (151)	31	185	126-263
All leukaemia (204-208)	18	305	181-483
Acute myeloid leukaemia (205.0)	10	382	183-703
Other leukaemia (207, 208)	3	558	115-1632
Female breast (174)	69	73	58-94
077 Brewery workers			
Bladder (188,189.1-189.9)	3	589	122-1723
085 Rubber manufacturers			
Bladder (188,189.1-189.9)	7	350	141- 723
086 Plastics workers			
Lung (162)	7	306	123-631
089 Tobacco workers			
Cervix (180)	6	281	103-602
091 Other occupations - glass and ceramics			
Other cancer of skin (173)	3	23	5-70

Appendix 6 Cancer incidence in women by job group: cancers with statistically significant PRRs

Job group	Registrations	PRR	95% CI
092 Rubber goods makers			
Ill-defined + secondary (195-199)	6	362	133-790
096 Printing machine minders			
Ovary (183)	10	233	112-429
098 Tailors and dressmakers			
Cervix (180)	42	142	103-193
100 Sewers and embroiderers			
Meningeal (192.1, 192.3, 225.2, 225.4)	22	162	102-246
Other leukaemia (207, 208)	0	0	0-82
104 Carpenters			
Lung (162)	20	171	105-266
Pleura (163)	3	1596	329-4665
124 Machine tool operators			
Larynx (161)	5	325	106-759
Lung (162)	94	129	105-159
Female breast (174)	81	75	60-94
125 Press and automatic machine operators			
Lung (162)	69	127	100-162
Female breast (174)	64	76	59-98
126 Metal polishers			
Lung (162)	16	191	109-310
132 Production fitters			
Gallbladder (156)	5	619	201-1446
Female breast (174)	18	58	35-93
137 Electricians			
Cervix (180)	3	625	129-1828
143 Electrical engineers (so described)			
Mouth and pharynx (140-149)	3	609	126-1780
Brain (191)	3	630	130-1843
159 Other spray painters			
Stomach (151)	5	371	121- 867
161 Electrical, electronic assemblers			
Brain (191)	7	311	125- 642
All leukaemia (204-208)	8	320	138- 632
174 Construction workers nec			
Lung (162)	16	269	154- 437
180 Railway engine drivers			
Other cancer of skin (173)	5	354	115- 828
182 Bus and coach drivers			
Acute myeloid leukaemia (205.0)	4	584	159-1497
185 Bus conductors and drivers' mates			
Pleura (163)	3	753	156-2203
Female breast (174)	20	62	38- 96
190 Storekeepers			
Cervix (180)	47	138	102- 184

Appendix 7 Mortality of men aged 20 to 64 by cause of deaths and job group, England and Wales, 1979-80, 1982-90 (PMR - all men with classifiable occupations = 100)

Job group		I Infections and parasitic diseases (001-139)	II Neoplasms (140-239)	III Endocrine nutritional and metabolic diseases and immunity disorders (240-279)	IV Diseases of blood and blood-forming organs (280-289)	V Mental disorders (290-319)	VI Diseases of the nervous system and sense organs (320-389)	VII Diseases of the circulatory system (390-459)	VIII Diseases of the respiratory system (460-519)
001 Lawyers	Deaths	19	444	31	6	5	30	524	47
	PMR	163	103	145	130	104	122	87	110
002 Accountants	Deaths	46	1962	101	18	18	101	2931	226
	PMR	106	101	102	109	85	103	102	96
003 Personnel managers etc	Deaths	11	640	31	3	4	33	1082	88
	PMR	93	97	102	65	68	118	107	102
004 Economists and statisticians	Deaths	0	71	3	0	1	5	90	8
	PMR	0	103	87	0	129	126	96	115
005 Computer programmers	Deaths	8	369	34	6	2	40	406	51
	PMR	58	108	104	145	23	150	94	105
006 Sales managers etc	Deaths	31	2522	105	18	19	96	3815	313
	PMR	63	103	85	101	76	86	103	99
007 Government inspectors	Deaths	3	509	10	4	3	18	799	64
	PMR	34	102	43	114	65	85	105	96
008 Government administrators	Deaths	19	746	25	4	4	28	1158	83
	PMR	154	105	80	81	67	93	106	90
009 Other administrators	Deaths	22	1128	65	8	11	51	1774	143
	PMR	98	97	117	94	98	98	100	95
010 Teachers in higher education	Deaths	12	912	43	6	15	45	1262	71
	PMR	72	109	112	97	198	118	99	71
011 Teachers nec	Deaths	51	2238	118	30	26	179	3398	234
	PMR	105	97	97	175	107	166	98	78
012 Vocational trainers, social scientists etc	Deaths	11	728	39	6	5	29	1112	79
	PMR	83	103	121	115	77	92	103	90
013 Welfare workers	Deaths	12	798	35	4	10	46	1232	118
	PMR	72	100	84	66	112	123	103	112
014 Clergy	Deaths	14	398	26	4	4	22	723	35
	PMR	149	90	136	102	91	94	110	75
015 Doctors	Deaths	26	567	31	3	15	34	892	58
	PMR	170	89	105	47	221	98	97	88
016 Dentists	Deaths	4	147	7	0	4	4	179	9
	PMR	118	103	107	0	268	51	88	62
017 Nurses	Deaths	17	549	47	7	15	37	983	115
	PMR	107	83	119	127	168	107	102	131
018 Pharmacists	Deaths	7	113	9	2	3	12	219	18
	PMR	182	72	122	118	175	139	99	112
019 Medical radiographers	Deaths	1	19	2	0	1	1	28	9
	PMR	199	85	161	0	309	87	85	299
020 Physiotherapists	Deaths	3	19	3	0	0	3	29	2
	PMR	792	90	307	0	0	340	90	69
021 Health professionals nec	Deaths	3	211	11	3	5	9	366	30
	PMR	63	91	98	161	206	81	106	103
022 Veterinarians	Deaths	2	53	2	0	0	2	72	6
	PMR	125	87	67	0	0	59	85	97
023 Driving instructors	Deaths	5	338	18	3	2	9	579	50
	PMR	69	105	100	126	57	46	105	91
024 Literary and artistic occupations	Deaths	92	1772	210	12	60	122	2390	336
	PMR	196	96	181	76	232	116	87	129
025 Persons involved in sport	Deaths	1	55	3	1	1	6	75	16
	PMR	50	97	57	132	68	112	89	159
026 Biological scientists	Deaths	4	151	8	4	1	4	187	17
	PMR	107	106	115	241	62	50	95	116
027 Chemical engineers and scientists	Deaths	8	417	18	3	5	28	556	37
	PMR	82	103	95	73	116	125	95	89
028 Physical scientists and mathematicians	Deaths	10	455	12	4	7	21	568	44
	PMR	98	110	62	93	158	92	97	104
029 Electrical and electronic engineers (professional)	Deaths	16	707	31	9	7	41	974	61
	PMR	82	100	86	108	86	99	102	86
030 Professional engineers nec	Deaths	48	3040	100	24	17	140	4526	324
	PMR	72	102	76	86	56	88	104	106

IX Diseases of the digestive system (520-579)	X Diseases of the genito-urinary system (520-629)	XII Diseases of the skin subcutaneous tissue (680-709)	XIII Diseases of the musculo-skeletal system and connective tissue (710-739)	XIV Congenital anomalies (740-759)	XVI Symptoms signs and ill-defined conditions (780-799)	EXVII External causes of injury and poisoning (E800-E999)		Job group	
66	7	0	3	8	3	170	Deaths	Lawyers	001
147	103	0	82	142	159	109	PMR		
173	28	2	20	27	10	586	Deaths	Accountants	002
88	94	95	123	116	119	93	PMR		
49	9	0	5	7	3	113	Deaths	Personnel managers etc	003
76	91	0	91	112	138	77	PMR		
8	2	0	1	0	2	29	Deaths	Economists and statisticians	004
115	182	0	170	0	602	99	PMR		
26	8	0	3	11	3	328	Deaths	Computer programmers	005
63	140	0	102	109	74	102	PMR		
219	37	4	20	25	6	613	Deaths	Sales managers etc	006
88	99	153	98	92	61	90	PMR		
38	9	0	5	7	0	103	Deaths	Government inspectors	007
80	118	0	120	147	0	90	PMR		
58	7	0	7	9	1	64	Deaths	Government administrators	008
86	65	0	117	149	48	47	PMR		
124	25	2	7	13	6	306	Deaths	Other administrators	009
108	140	158	72	110	144	104	PMR		
72	14	2	6	3	2	157	Deaths	Teachers in higher education	010
86	112	225	87	37	69	81	PMR		
187	47	5	29	35	15	820	Deaths	Teachers nec	011
79	133	202	151	129	147	121	PMR		
54	13	1	8	10	0	134	Deaths	Vocational trainers,	012
78	122	133	137	147	0	81	PMR	social scientists etc	
55	12	1	4	8	4	243	Deaths	Welfare workers	013
69	98	120	60	83	114	96	PMR		
30	6	0	4	4	3	99	Deaths	Clergy	014
68	91	0	105	98	231	98	PMR		
79	14	1	12	4	2	262	Deaths	Doctors	015
123	142	136	217	57	82	134	PMR		
21	2	0	0	1	0	68	Deaths	Dentists	016
147	91	0	0	62	0	154	PMR		
92	16	1	6	9	9	293	Deaths	Nurses	017
134	155	138	109	91	244	106	PMR		
26	6	0	3	2	1	74	Deaths	Pharmacists	018
167	249	0	223	105	149	131	PMR		
4	0	0	0	1	0	9	Deaths	Medical radiographers	019
182	0	0	0	297	0	89	PMR		
5	0	0	0	0	0	2	Deaths	Physiotherapists	020
254	0	0	0	0	0	44	PMR		
22	5	0	1	6	0	67	Deaths	Health professionals nec	021
98	142	0	51	234	0	96	PMR		
7	0	0	2	0	0	47	Deaths	Veterinarians	022
114	0	0	382	0	0	205	PMR		
23	6	0	1	2	0	81	Deaths	Driving instructors	023
65	91	0	33	43	0	95	PMR		
275	35	1	10	20	27	900	Deaths	Literary and artistic	024
139	113	50	63	66	242	113	PMR	occupations	
6	0	0	0	2	0	62	Deaths	Persons involved in sport	025
93	0	0	0	103	0	119	PMR		
12	2	0	2	2	1	61	Deaths	Biological scientists	026
85	89	0	165	101	145	103	PMR		
46	5	0	2	3	2	148	Deaths	Chemical engineers and	027
112	81	0	58	63	128	113	PMR	scientists	
34	4	2	3	7	3	128	Deaths	Physical scientists and	028
82	63	418	86	139	175	90	PMR	mathematicians	
70	16	0	6	14	5	320	Deaths	Electrical and electronic	029
97	143	0	102	132	136	98	PMR	engineers (professional)	
242	39	3	24	33	2	695	Deaths	Professional engineers nec	030
81	88	87	96	110	19	86	PMR		

Appendix 7 Mortality of men aged 20-64 by cause of deaths and job-group, England and Wales, 1979-80, 1982-90

Job group		I Infections and parasitic diseases (001-139)	II Neoplasms (140-239)	III Endocrine nutritional and metabolic diseases and immunity disorders (240-279)	IV Diseases of blood and blood-forming organs (280-289)	V Mental disorders (290-319)	VI Diseases of the nervous system and and sense organs (320-389)	VII Diseases of the circulatory system (390-459)	VIII Diseases of the respiratory system (460-519)
031 Draughtspersons	Deaths	15	914	50	7	11	85	1642	154
	PMR	66	97	87	90	90	137	104	92
032 Laboratory technicians	Deaths	13	1000	45	6	16	47	1498	144
	PMR	65	101	91	79	133	101	102	102
033 Architects and surveyors	Deaths	27	1144	53	18	15	87	1635	132
	PMR	99	99	95	162	119	141	98	105
034 Aircraft flight deck officers	Deaths	5	142	3	0	3	6	133	14
	PMR	191	111	45	0	225	101	69	83
035 Air traffic controllers	Deaths	0	36	1	0	1	2	42	3
	PMR	0	115	64	0	309	140	89	73
036 Seafarers	Deaths	40	1345	47	5	35	53	1700	294
	PMR	158	101	82	58	210	84	84	116
037 Technicians nec	Deaths	12	859	29	5	11	49	1376	111
	PMR	69	98	65	75	113	120	104	96
038 Production and maintenance managers	Deaths	68	4633	156	25	21	154	6905	462
	PMR	88	109	79	85	56	84	104	83
039 Managers in construction	Deaths	11	1232	32	6	7	32	1709	114
	PMR	57	115	65	83	76	69	101	77
040 Managers in transport, utilities and mining	Deaths	33	2441	91	20	13	72	4056	314
	PMR	76	100	82	118	60	70	108	98
041 Office managers	Deaths	50	2368	110	21	13	102	3710	279
	PMR	115	102	99	128	60	101	104	92
042 Butchers	Deaths	15	1096	48	6	12	43	1559	225
	PMR	81	104	111	84	82	87	97	117
043 Fishmongers and poultry dressers	Deaths	1	175	6	1	0	7	250	34
	PMR	34	107	85	91	0	89	97	114
044 Retailers and dealers	Deaths	186	8059	485	65	114	390	13190	1344
	PMR	107	95	110	100	120	92	101	110
045 Publicans and bar staff	Deaths	66	2813	155	12	51	88	3817	614
	PMR	118	100	113	60	158	65	87	150
046 Caterers	Deaths	55	1269	135	6	23	59	1891	225
	PMR	186	99	196	62	118	77	93	99
047 Farmers	Deaths	118	6353	288	47	80	387	9189	1231
	PMR	100	100	106	111	98	131	95	103
048 Armed forces	Deaths	44	1261	33	9	41	108	1569	243
	PMR	100	100	100	100	100	100	100	100
049 Police	Deaths	13	1191	37	6	10	72	1652	113
	PMR	49	116	55	68	69	102	98	65
050 Fire service personnel	Deaths	3	432	10	4	8	20	581	50
	PMR	34	116	45	132	169	86	96	83
051 Launderers and dry cleaners	Deaths	3	220	14	1	3	14	335	46
	PMR	73	101	156	74	101	132	100	96
052 Hairdressers	Deaths	13	258	46	2	14	20	418	69
	PMR	205	92	308	89	264	118	96	135
053 Office workers and cashiers	Deaths	296	12181	777	112	157	805	21684	2449
	PMR	112	96	115	117	110	109	101	105
054 Postal workers	Deaths	34	2847	94	13	35	131	3876	464
	PMR	76	111	96	88	94	115	99	76
055 Petrol pump attendants	Deaths	1	116	9	0	1	4	218	30
	PMR	39	93	134	0	68	55	103	128
056 Vans sales persons	Deaths	13	710	26	3	8	24	920	141
	PMR	117	105	102	70	85	82	92	116
057 Sales representatives	Deaths	67	3441	145	23	22	177	5419	443
	PMR	89	104	77	89	56	89	100	80
058 Security workers	Deaths	51	3457	127	23	30	124	5543	647
	PMR	86	100	96	113	84	84	103	84
059 Cooks and kitchen porters	Deaths	55	1237	100	10	42	79	1814	294
	PMR	208	92	177	112	136	112	92	107
060 Other service personnel	Deaths	90	5312	203	29	64	206	7944	1273
	PMR	97	102	102	101	75	87	100	98
061 Hospital porters and ward orderlies	Deaths	18	898	41	7	14	46	1369	191
	PMR	115	100	120	130	128	118	100	93
062 Ambulance workers	Deaths	4	317	6	0	1	15	461	48
	PMR	89	105	57	0	30	127	103	86
063 Railway station workers	Deaths	12	824	33	8	5	53	1438	193
	PMR	81	93	102	173	37	141	107	86
064 Undertakers	Deaths	0	100	3	1	1	4	166	18
	PMR	0	97	74	152	118	100	106	118
065 Foresters	Deaths	3	158	5	1	3	7	256	26
	PMR	89	91	70	90	109	80	98	70

IX Diseases of the digestive system (520-579)	X Diseases of the genito-urinary system (520-629)	XII Diseases of the skin subcutaneous tissue (680-709)	XIII Diseases of the musculo-skeletal system and connective tissue (710-739)	XIV Congenital anomalies (740-759)	XVI Symptoms signs and ill-defined conditions (780-799)	EXVII External causes of injury and poisoning (E800-E999)		Job group	
62	15	2	10	15	9	340	Deaths	Draughtspersons	031
61	76	209	112	94	165	102	PMR		
70	15	1	4	13	9	328	Deaths	Laboratory technicians	032
75	98	98	50	109	213	96	PMR		
125	16	2	3	9	10	381	Deaths	Architects and surveyors	033
107	91	152	31	65	204	98	PMR		
17	2	1	0	0	0	84	Deaths	Aircraft flight deck officers	034
132	102	763	0	0	0	227	PMR		
5	0	0	1	1	0	8	Deaths	Air traffic controllers	035
163	0	0	391	287	0	88	PMR		
259	28	1	8	7	15	538	Deaths	Seafarers	036
203	118	75	81	49	214	131	PMR		
82	10	1	7	13	1	276	Deaths	Technicians nec	037
95	75	107	96	127	29	99	PMR		
312	43	0	35	39	6	604	Deaths	Production and maintenance managers	038
74	66	0	98	97	43	65	PMR		
73	8	0	7	8	0	160	Deaths	Managers in construction	039
70	47	0	78	81	0	73	PMR		
189	30	2	22	21	3	372	Deaths	Managers in transport, utilities and mining	040
80	81	78	109	92	39	69	PMR		
190	32	2	17	30	5	437	Deaths	Office managers	041
83	90	82	88	129	61	78	PMR		
109	17	2	6	7	3	355	Deaths	Butchers	042
119	93	215	73	62	68	96	PMR		
18	3	0	1	2	0	42	Deaths	Fishmongers and poultry dressers	043
123	101	0	75	119	0	93	PMR		
910	167	9	71	111	35	2682	Deaths	Retailers and dealers	044
107	119	102	99	105	94	103	PMR		
723	39	3	10	22	15	791	Deaths	Publicans and bar staff	045
254	84	100	43	68	121	94	PMR		
213	21	2	12	15	13	546	Deaths	Caterers	046
157	85	143	112	77	164	103	PMR		
456	134	9	74	66	28	2526	Deaths	Farmers	047
77	119	144	153	97	92	125	PMR		
190	24	3	5	12	18	1157	Deaths	Armed forces	048
100	100	100	100	100	100	100	PMR		
95	12	0	9	10	3	444	Deaths	Police	049
82	56	0	92	51	47	107	PMR		
36	4	0	4	2	1	135	Deaths	Fire service personnel	050
88	55	0	116	32	51	103	PMR		
28	4	1	0	4	1	57	Deaths	Launderers and dry cleaners	051
137	96	483	0	171	80	81	PMR		
29	3	0	2	4	1	119	Deaths	Hairdressers	052
103	54	0	84	89	61	84	PMR		
1311	299	13	132	201	72	3062	Deaths	Office workers and cashiers	053
104	114	116	112	120	117	90	PMR		
235	48	1	19	20	13	620	Deaths	Postal workers	054
98	95	41	106	92	89	89	PMR		
14	3	0	0	5	3	29	Deaths	Petrol pump attendants	055
115	115	0	0	298	497	81	PMR		
63	15	1	8	9	3	247	Deaths	Vans sales persons	056
113	137	168	165	136	110	104	PMR		
321	56	3	27	46	11	1158	Deaths	Sales representatives	057
91	86	87	89	93	63	112	PMR		
332	54	0	18	21	23	728	Deaths	Security workers	058
106	82	0	71	72	136	98	PMR		
210	46	6	7	24	19	743	Deaths	Cooks and kitchen porters	059
162	188	454	75	149	213	105	PMR		
525	94	5	23	39	38	1372	Deaths	Other service personnel	060
101	89	95	64	100	116	102	PMR		
86	16	0	5	11	8	222	Deaths	Hospital porters and ward orderlies	061
108	95	0	79	133	179	91	PMR		
11	6	1	2	2	2	70	Deaths	Ambulance workers	062
46	125	370	93	86	200	98	PMR		
96	15	0	5	3	8	164	Deaths	Railway station workers	063
109	83	0	83	56	152	93	PMR		
8	0	0	1	1	0	14	Deaths	Undertakers	064
88	0	0	123	125	0	76	PMR		
13	4	0	1	1	2	109	Deaths	Foresters	065
81	123	0	81	48	196	155	PMR		

Appendix 7 Mortality of men aged 20-64 by cause of deaths and job-group, England and Wales, 1979-80, 1982-90

Job group		I Infections and parasitic diseases (001-139)	II Neoplasms (140-239)	III Endocrine nutritional and metabolic diseases and immunity disorders (240-279)	IV Diseases of blood and blood-forming organs (280-289)	V Mental disorders (290-319)	VI Diseases of the nervous system and and sense organs (320-389)	VII Diseases of the circulatory system (390-459)	VIII Diseases of the respiratory system (460-519)
066 Fishing and related workers	Deaths	5	214	6	0	5	8	296	46
	PMR	104	97	60	0	119	63	88	93
067 Tannery workers	Deaths	2	82	3	0	2	7	113	19
	PMR	189	103	120	0	287	252	97	123
068 Leather and shoe workers	Deaths	18	550	21	4	6	38	916	136
	PMR	208	90	103	113	93	167	102	117
069 Preparatory fibre processors	Deaths	3	85	6	0	0	7	173	29
	PMR	184	83	163	0	0	174	111	129
070 Spinners and winders	Deaths	3	109	16	1	1	6	254	32
	PMR	118	76	287	117	59	92	115	102
071 Warp preparers and weavers	Deaths	5	188	14	3	2	12	403	52
	PMR	147	78	174	217	82	134	113	114
072 Knitters	Deaths	0	120	7	2	5	9	259	31
	PMR	0	78	128	215	266	144	114	110
073 Bleachers, dyers and finishers	Deaths	8	234	10	2	0	10	351	49
	PMR	237	96	126	145	0	113	98	106
074 Other textile workers	Deaths	31	850	66	1	13	54	1805	267
	PMR	162	83	164	18	69	110	115	102
075 Chemical workers	Deaths	29	2249	62	15	23	87	3242	431
	PMR	75	106	74	119	77	88	100	89
076 Bakers	Deaths	7	516	30	8	10	36	887	119
	PMR	78	87	147	218	134	154	103	106
077 Brewery workers	Deaths	4	227	15	1	1	9	328	42
	PMR	125	99	199	78	47	107	96	98
078 Food processors	Deaths	22	1197	59	6	5	57	1941	291
	PMR	107	99	130	83	34	109	106	112
079 Paper manufacturers	Deaths	3	130	1	0	1	6	209	27
	PMR	167	100	24	0	83	127	108	110
080 Bookbinders	Deaths	2	72	0	0	0	20	148	21
	PMR	140	79	0	0	0	535	110	125
081 Paper cutters	Deaths	0	57	2	2	0	2	90	8
	PMR	0	102	115	656	0	104	109	75
082 Glass and ceramics furnace workers	Deaths	3	147	4	2	0	2	197	38
	PMR	158	105	89	259	0	40	95	144
083 Glass formers and decorators	Deaths	2	183	11	3	2	5	276	37
	PMR	79	101	186	301	123	75	103	111
084 Ceramics casters	Deaths	2	89	4	1	0	5	156	23
	PMR	135	88	115	168	0	128	105	121
085 Rubber manufacturers	Deaths	5	255	8	1	2	10	369	58
	PMR	144	103	97	71	82	109	101	124
086 Plastics workers	Deaths	3	186	6	0	0	7	272	34
	PMR	99	109	91	0	0	92	104	89
087 Man-made fibre makers	Deaths	1	67	5	0	0	4	107	11
	PMR	96	104	215	0	0	159	107	77
088 Other coal workers	Deaths	66	3700	101	27	39	148	6297	1307
	PMR	98	90	68	114	93	90	101	136
089 Tobacco workers	Deaths	0	76	2	1	1	3	112	16
	PMR	0	107	79	248	142	108	104	101
090 Other wood and paper processors	Deaths	3	194	13	1	0	7	332	40
	PMR	101	93	186	84	0	89	108	103
091 Other occupations - glass and ceramics	Deaths	14	572	14	2	10	28	971	184
	PMR	133	91	60	57	97	100	103	131
092 Rubber goods makers	Deaths	1	200	3	2	2	7	359	48
	PMR	32	90	41	160	95	86	109	114
093 Plastic goods makers	Deaths	5	291	14	0	2	17	522	79
	PMR	87	94	112	0	47	114	110	120
094 Compositors	Deaths	6	194	8	1	4	8	353	37
	PMR	187	90	108	81	177	96	110	93
095 Printing plate preparers	Deaths	0	69	3	0	0	5	105	15
	PMR	0	101	123	0	0	181	104	122
096 Printing machine minders	Deaths	4	416	5	4	1	20	584	62
	PMR	71	108	38	185	26	136	102	89
097 Printers (so described)	Deaths	11	963	44	9	19	63	1553	185
	PMR	66	92	116	140	136	145	101	98
098 Tailors and dressmakers	Deaths	12	184	29	3	1	14	401	43
	PMR	348	73	355	214	44	154	107	91
099 Clothing cutters	Deaths	1	103	4	2	3	19	194	27
	PMR	52	81	91	253	188	378	105	110
100 Sewers and embroiderers	Deaths	4	99	8	0	1	10	183	20
	PMR	204	94	189	0	73	202	113	86

IX Diseases of the digestive system (520-579)	X Diseases of the genito-urinary system (520-629)	XII Diseases of the skin subcutaneous tissue (680-709)	XIII Diseases of the musculo-skeletal system and connective tissue (710-739)	XIV Congenital anomalies (740-759)	XVI Symptoms signs and ill-defined conditions (780-799)	EXVII External causes of injury and poisoning (E800-E999)		Job group	
25	3	0	1	2	2	168	Deaths	Fishing and related workers	066
113	68	0	64	63	124	157	PMR		
6	0	0	0	0	0	7	Deaths	Tannery workers	067
103	0	0	0	0	0	52	PMR		
48	17	1	8	8	2	120	Deaths	Leather and shoe workers	068
104	175	199	183	182	110	88	PMR		
7	2	0	0	2	0	9	Deaths	Preparatory fibre processors	069
79	109	0	0	265	0	48	PMR		
11	9	1	0	0	1	26	Deaths	Spinners and winders	070
83	338	735	0	0	130	67	PMR		
11	3	0	4	3	0	43	Deaths	Warp preparers and weavers	071
60	79	0	233	176	0	86	PMR		
13	1	0	3	1	1	39	Deaths	Knitters	072
106	41	0	273	76	182	85	PMR		
30	4	0	1	6	2	39	Deaths	Bleachers, dyers and finishers	073
164	105	0	58	370	288	83	PMR		
97	33	1	12	13	5	183	Deaths	Other textile workers	074
90	154	91	173	162	69	63	PMR		
170	34	0	12	21	8	641	Deaths	Chemical workers	075
85	83	0	81	101	65	100	PMR		
61	9	1	4	7	3	176	Deaths	Bakers	076
134	95	203	95	138	149	96	PMR		
27	7	0	2	1	2	38	Deaths	Brewery workers	077
153	194	0	122	65	303	88	PMR		
105	16	0	8	10	1	220	Deaths	Food processors	078
99	73	0	93	90	17	65	PMR		
7	0	0	0	1	0	14	Deaths	Paper manufacturers	079
71	0	0	0	116	0	58	PMR		
7	3	0	0	2	0	18	Deaths	Bookbinders	080
96	204	0	0	246	0	62	PMR		
3	0	0	1	0	0	3	Deaths	Paper cutters	081
73	0	0	252	0	0	37	PMR		
8	6	0	1	1	0	17	Deaths	Glass and ceramics furnace workers	082
76	275	0	100	114	0	72	PMR		
9	4	0	0	0	0	22	Deaths	Glass formers and decorators	083
65	143	0	0	0	0	64	PMR		
8	4	0	2	1	0	20	Deaths	Ceramics casters	084
101	248	0	277	128	0	85	PMR		
17	6	0	1	1	0	30	Deaths	Rubber manufacturers	085
90	154	0	57	59	0	60	PMR		
12	2	0	1	3	0	36	Deaths	Plastics workers	086
78	62	0	83	181	0	73	PMR		
5	1	0	0	0	0	4	Deaths	Man-made fibre makers	087
88	85	0	0	0	0	35	PMR		
357	62	6	26	27	13	976	Deaths	Other coal workers	088
100	80	166	89	85	69	114	PMR		
3	3	0	0	1	0	6	Deaths	Tobacco workers	089
50	232	0	0	189	0	44	PMR		
13	5	0	4	1	0	28	Deaths	Other wood and paper processors	090
81	153	0	271	68	0	62	PMR		
66	14	0	7	9	5	149	Deaths	Other occupations - glass and ceramics	091
112	120	0	162	187	144	85	PMR		
22	6	0	1	4	1	26	Deaths	Rubber goods makers	092
130	171	0	63	268	159	61	PMR		
22	9	0	4	8	1	58	Deaths	Plastic goods makers	093
76	156	0	180	242	56	57	PMR		
15	5	0	2	3	1	34	Deaths	Compositors	094
87	147	0	130	186	142	69	PMR		
8	0	0	0	2	0	10	Deaths	Printing plate preparers	095
146	0	0	0	341	0	49	PMR		
25	7	0	7	3	2	52	Deaths	Printing machine minders	096
83	117	0	257	107	163	61	PMR		
110	19	2	5	19	4	350	Deaths	Printers (so described)	097
130	114	222	67	198	101	101	PMR		
24	5	0	6	3	1	44	Deaths	Tailors and dressmakers	098
126	127	0	333	187	144	101	PMR		
10	4	0	1	2	0	32	Deaths	Clothing cutters	099
101	196	0	110	187	0	86	PMR		
11	2	0	0	4	0	9	Deaths	Sewers and embroiderers	100
112	100	0	0	365	0	27	PMR		

Appendix 7 Mortality of men aged 20-64 by cause of deaths and job-group, England and Wales, 1979-80, 1982-90

Job group		I Infections and parasitic diseases (001-139)	II Neoplasms (140-239)	III Endocrine nutritional and metabolic diseases and immunity disorders (240-279)	IV Diseases of blood and blood-forming organs (280-289)	V Mental disorders (290-319)	VI Diseases of the nervous system and and sense organs (320-389)	VII Diseases of the circulatory system (390-459)	VIII Diseases of the respiratory system (460-519)
101 Upholsterers	Deaths	2	291	8	2	2	15	409	64
	PMR	45	100	79	113	56	129	96	119
102 Carpet fitters	Deaths	1	98	3	1	3	6	104	14
	PMR	53	113	73	140	129	120	85	94
103 Other workers with fabrics	Deaths	4	78	5	1	2	3	147	19
	PMR	262	83	145	178	168	75	105	112
104 Carpenters	Deaths	74	3722	132	28	46	165	5171	625
	PMR	130	102	101	126	98	110	97	94
105 Cabinet makers	Deaths	7	313	18	2	7	14	387	48
	PMR	152	108	173	108	175	116	92	89
106 Case and box makers	Deaths	1	84	1	0	1	8	128	17
	PMR	79	101	35	0	98	243	105	110
107 Pattern makers	Deaths	1	143	4	1	4	4	250	18
	PMR	46	93	78	114	262	69	110	63
108 Woodworking machinists	Deaths	12	516	21	3	2	29	759	127
	PMR	151	98	115	94	31	140	99	129
109 Other woodworkers	Deaths	2	155	9	2	4	11	216	25
	PMR	76	99	152	189	156	159	96	87
110 Dental technicians	Deaths	1	73	6	3	1	4	141	15
	PMR	76	84	197	566	91	115	111	93
111 Other makers of paper goods	Deaths	5	231	8	1	3	10	335	28
	PMR	145	107	103	75	106	111	105	72
112 Furnace operatives (metal)	Deaths	6	390	7	0	4	7	566	96
	PMR	113	100	56	0	112	50	99	130
113 Rollers (metal)	Deaths	0	68	3	0	1	4	104	19
	PMR	0	94	132	0	166	157	97	141
114 Smiths and forge workers	Deaths	1	302	7	0	3	8	500	48
	PMR	23	95	69	0	103	71	108	79
115 Metal drawers	Deaths	2	104	7	1	0	3	155	20
	PMR	140	103	209	178	0	80	103	106
116 Moulders and coremakers	Deaths	4	601	12	4	7	17	885	161
(metal)	PMR	47	98	60	116	122	77	98	138
117 Electroplaters	Deaths	1	129	3	0	4	2	136	32
	PMR	60	117	78	0	282	46	85	154
118 Annealers, hardeners, temperers	Deaths	0	94	4	1	0	1	132	20
(metal)	PMR	0	110	131	207	0	30	101	103
119 Galvanisers and tin platers	Deaths	1	73	5	0	1	4	89	17
	PMR	88	107	195	0	102	136	89	138
120 Other metal manufacturers	Deaths	16	890	43	8	18	36	1420	248
	PMR	98	97	124	163	108	85	102	111
121 Press and machine tool	Deaths	13	922	33	4	9	31	1308	159
setters	PMR	105	105	114	80	104	95	101	96
122 Centre lathe turners	Deaths	2	789	22	4	9	23	1223	140
	PMR	19	100	87	91	125	82	106	93
123 Machine tool setter operators	Deaths	3	183	8	0	0	9	287	40
	PMR	119	101	134	0	0	135	107	117
124 Machine tool operators	Deaths	135	6815	297	37	83	302	10653	1327
	PMR	109	99	111	89	93	96	102	86
125 Press and automatic machine	Deaths	7	377	18	3	4	13	486	78
operators	PMR	112	110	132	145	91	82	92	102
126 Metal polishers	Deaths	1	271	10	2	2	10	358	65
	PMR	27	105	117	134	76	104	94	134
127 Fettlers and dressers (metal)	Deaths	5	219	7	1	1	7	310	63
	PMR	146	105	92	84	50	84	97	131
128 Shot blasters	Deaths	6	80	5	0	1	1	146	22
	PMR	337	88	132	0	75	22	104	108
129 Toolmakers	Deaths	17	1247	39	11	5	65	2026	206
	PMR	93	96	91	146	37	136	107	83
130 Precision instrument makers	Deaths	7	455	9	2	4	23	654	60
	PMR	111	104	61	78	84	139	102	73
131 Watch and clock makers	Deaths	8	114	6	3	0	9	225	26
	PMR	402	79	127	374	0	171	105	96
132 Production fitters	Deaths	89	7890	231	42	90	295	11543	1378
	PMR	76	101	86	90	99	96	101	95
133 Motor mechanics	Deaths	37	2163	88	15	30	104	3256	383
	PMR	100	99	105	104	86	107	103	97
134 Aircraft engine fitters	Deaths	0	18	1	0	0	0	39	5
	PMR	0	85	137	0	0	0	125	126
135 Office machinery mechanics	Deaths	1	35	1	0	0	2	100	15
	PMR	120	64	52	0	0	91	124	151

IX Diseases of the digestive system (520-579)	X Diseases of the genito-urinary system (520-629)	XII Diseases of the skin subcutaneous tissue (680-709)	XIII Diseases of the musculo-skeletal system and connective tissue (710-739)	XIV Congenital anomalies (740-759)	XVI Symptoms signs and ill-defined conditions (780-799)	EXVII External causes of injury and poisoning (E800-E999)		Job group	
26	9	0	5	4	1	84	Deaths	Upholsterers	101
114	195	0	242	160	98	95	PMR		
9	3	0	2	2	0	73	Deaths	Carpet fitters	102
112	200	0	325	131	0	106	PMR		
11	2	0	1	4	0	26	Deaths	Other workers with fabrics	103
139	133	0	148	481	0	87	PMR		
265	57	4	35	33	16	1227	Deaths	Carpenters	104
91	98	129	136	102	118	108	PMR		
35	9	1	2	2	1	89	Deaths	Cabinet makers	105
154	194	410	97	72	92	86	PMR		
5	3	0	1	3	0	12	Deaths	Case and box makers	106
77	228	0	169	433	0	48	PMR		
10	4	0	4	0	2	29	Deaths	Pattern makers	107
85	167	0	369	0	428	86	PMR		
44	10	0	1	6	0	131	Deaths	Woodworking machinists	108
108	120	0	27	137	0	85	PMR		
12	3	0	2	1	2	75	Deaths	Other woodworkers	109
95	118	0	179	58	299	108	PMR		
2	4	0	2	0	0	26	Deaths	Dental technicians	110
30	290	0	323	0	0	93	PMR		
10	4	0	2	3	0	57	Deaths	Other makers of paper goods	111
58	117	0	131	151	0	77	PMR		
30	9	0	1	4	0	66	Deaths	Furnace operatives (metal)	112
103	148	0	36	158	0	94	PMR		
9	2	0	0	0	0	9	Deaths	Rollers (metal)	113
168	180	0	0	0	0	73	PMR		
38	5	0	0	2	0	49	Deaths	Smiths and forge workers	114
163	101	0	0	98	0	85	PMR		
5	1	0	0	0	0	13	Deaths	Metal drawers	115
64	63	0	0	0	0	65	PMR		
54	14	0	6	4	1	112	Deaths	Moulders and coremakers	116
117	145	0	137	99	58	97	PMR	(metal)	
13	3	0	0	1	0	26	Deaths	Electroplaters	117
153	171	0	0	105	0	76	PMR		
6	1	0	1	1	0	10	Deaths	Annealers, hardeners, temperers	118
81	63	0	162	163	0	64	PMR	(metal)	
3	0	0	0	2	0	27	Deaths	Galvanisers and tin platers	119
53	0	0	0	305	0	110	PMR		
107	16	3	10	9	2	191	Deaths	Other metal manufacturers	120
114	87	310	160	135	33	78	PMR		
59	11	1	5	11	4	142	Deaths	Press and machine tool	121
88	80	136	80	180	155	78	PMR	setters	
59	11	1	4	2	0	102	Deaths	Centre lathe turners	122
101	89	154	72	40	0	73	PMR		
6	0	0	2	1	0	17	Deaths	Machine tool setter operators	123
44	0	0	155	80	0	48	PMR		
608	154	10	54	66	39	2153	Deaths	Machine tool operators	124
97	118	156	111	97	105	104	PMR		
33	10	1	8	6	2	94	Deaths	Press and automatic machine	125
103	153	308	326	172	102	90	PMR	operators	
21	5	0	2	2	1	48	Deaths	Metal polishers	126
107	124	0	109	108	130	81	PMR		
19	4	0	4	1	0	26	Deaths	Fettlers and dressers (metal)	127
103	102	0	268	63	0	66	PMR		
6	1	0	0	1	0	42	Deaths	Shot blasters	128
68	56	0	0	95	0	126	PMR		
84	23	2	9	8	3	259	Deaths	Toolmakers	129
86	112	186	97	87	79	90	PMR		
31	8	0	5	5	0	94	Deaths	Precision instrument makers	130
93	116	0	161	150	0	88	PMR		
14	2	0	4	1	1	29	Deaths	Watch and clock makers	131
128	88	0	390	104	245	109	PMR		
609	101	9	40	50	25	2142	Deaths	Production fitters	132
100	82	137	72	80	96	102	PMR		
188	53	2	12	14	9	845	Deaths	Motor mechanics	133
104	149	105	77	60	95	93	PMR		
0	0	0	0	1	0	3	Deaths	Aircraft engine fitters	134
0	0	0	0	588	0	49	PMR		
2	0	0	1	0	0	16	Deaths	Office machinery mechanics	135
46	0	0	256	0	0	106	PMR		

Appendix 7 Mortality of men aged 20-64 by cause of deaths and job-group, England and Wales, 1979-80, 1982-90

Job group		I Infections and parasitic diseases (001-139)	II Neoplasms (140-239)	III Endocrine nutritional and metabolic diseases and immunity disorders (240-279)	IV Diseases of blood and blood-forming organs (280-289)	V Mental disorders (290-319)	VI Diseases of the nervous system and and sense organs (320-389)	VII Diseases of the circulatory system (390-459)	VIII Diseases of the respiratory system (460-519)
136 Electrical and electronic production fitters	Deaths	5	234	7	1	0	14	324	33
	PMR	157	106	95	79	0	167	100	81
137 Electricians	Deaths	72	3223	119	22	45	167	4750	473
	PMR	137	100	99	109	99	121	100	81
138 Electrical plant operators	Deaths	2	359	11	0	0	21	556	55
	PMR	40	99	94	0	0	161	103	80
139 Telephone fitters	Deaths	19	1074	35	11	10	53	1574	137
	PMR	118	103	94	175	76	124	103	72
140 Electric cable and line workers	Deaths	3	252	9	2	1	7	341	49
	PMR	82	102	106	141	38	73	93	109
141 Radio and TV mechanics	Deaths	6	336	12	2	2	31	527	77
	PMR	99	93	86	88	37	194	99	118
142 Other electronic maintenance engineers	Deaths	8	481	19	5	11	23	702	43
	PMR	98	104	103	163	140	108	104	52
143 Electrical engineers (so described)	Deaths	17	1063	44	11	12	65	1615	133
	PMR	105	99	113	170	108	154	101	72
144 Plumbers and gas fitters	Deaths	35	2715	82	13	34	112	3638	374
	PMR	89	108	91	86	108	108	98	82
145 Sheet metal workers	Deaths	18	1189	36	7	8	49	1511	228
	PMR	106	109	93	103	56	110	95	111
146 Metal plate workers	Deaths	17	1013	22	4	12	22	1119	160
	PMR	141	118	78	81	138	69	89	100
147 Steel erectors	Deaths	11	674	19	6	14	27	852	141
	PMR	112	102	83	157	191	104	87	118
148 Scaffolders	Deaths	4	323	12	4	10	16	406	67
	PMR	56	106	82	187	154	83	87	103
149 Welders	Deaths	33	2133	87	11	34	90	3005	461
	PMR	100	101	114	85	123	104	97	119
150 Riggers	Deaths	1	307	5	0	3	4	406	58
	PMR	25	107	53	0	108	38	96	109
151 Jewellery workers	Deaths	2	52	6	2	2	1	85	15
	PMR	196	84	261	495	219	37	94	131
152 Engravers and etchers (printing)	Deaths	0	29	3	0	0	3	107	5
	PMR	0	56	167	0	0	145	140	52
153 Vehicle body builders	Deaths	4	297	7	3	2	13	395	36
	PMR	101	111	78	188	68	127	100	72
154 Oilers and greasers	Deaths	1	74	2	0	0	1	118	19
	PMR	98	96	84	0	0	38	104	131
155 Electronics wire workers	Deaths	2	71	8	0	1	7	121	11
	PMR	142	88	247	0	72	187	102	77
156 Coil winders	Deaths	2	55	4	1	0	1	79	12
	PMR	245	99	214	292	0	47	98	113
157 Pottery decorators	Deaths	2	23	0	0	0	0	34	1
	PMR	556	101	0	0	0	0	100	25
158 Coach painters	Deaths	0	63	0	0	1	2	85	9
	PMR	0	110	0	0	230	103	100	82
159 Other spray painters	Deaths	5	558	27	3	4	24	761	104
	PMR	48	109	124	91	48	89	97	91
160 Painters and decorators nec	Deaths	58	3946	129	13	90	172	5381	800
	PMR	95	101	92	55	181	108	94	112
161 Electrical/ electronic assemblers	Deaths	8	150	9	0	1	10	259	44
	PMR	249	94	134	0	36	120	107	125
162 Instrument assemblers	Deaths	0	7	0	0	0	0	21	2
	PMR	0	68	0	0	0	0	138	101
163 Assemblers (vehicles and other metal goods)	Deaths	14	818	21	6	13	28	1280	173
	PMR	99	103	68	126	134	78	105	98
164 Packers and sorters	Deaths	19	1025	49	5	8	62	1695	266
	PMR	99	95	118	77	59	129	103	109
165 Bricklayers and tilesetters	Deaths	39	2048	52	10	47	87	2771	397
	PMR	130	103	75	84	199	110	95	108
166 Masons and stonecutters	Deaths	5	264	7	0	5	12	405	43
	PMR	130	99	78	0	170	119	104	85
167 Plasterers	Deaths	10	779	19	4	12	21	913	123
	PMR	90	114	75	93	126	72	91	100
168 Roofers and glaziers	Deaths	8	604	15	1	30	20	742	104
	PMR	55	109	52	24	203	50	89	87
169 Builders etc.	Deaths	55	3343	94	21	47	131	4583	526
	PMR	104	105	78	109	115	95	96	87
170 Rail track workers	Deaths	4	356	9	0	4	9	518	66
	PMR	66	102	68	0	90	58	98	86

IX Diseases of the digestive system (520-579)	X Diseases of the genito-urinary system (520-629)	XII Diseases of the skin subcutaneous tissue (680-709)	XIII Diseases of the musculo-skeletal system and connective tissue (710-739)	XIV Congenital anomalies (740-759)	XVI Symptoms signs and ill-defined conditions (780-799)	EXVII External causes of injury and poisoning (E800-E999)		Job group	
14	4	0	2	1	1	44	Deaths	Electrical and electronic production fitters	136
83	116	0	128	61	145	84	PMR		
266	60	8	31	30	14	1196	Deaths	Electricians	137
100	115	286	135	96	108	104	PMR		
34	1	0	1	4	2	63	Deaths	Electrical plant operators	138
124	18	0	39	172	202	99	PMR		
84	15	0	5	13	2	271	Deaths	Telephone fitters	139
100	90	0	68	144	52	88	PMR		
16	4	0	5	1	1	80	Deaths	Electric cable and line workers	140
82	103	0	285	53	124	134	PMR		
38	2	0	5	4	1	135	Deaths	Radio and TV mechanics	141
124	34	0	193	111	65	105	PMR		
46	4	0	4	10	3	178	Deaths	Other electronic maintenance engineers	142
117	53	0	122	193	138	88	PMR		
79	17	1	9	8	10	256	Deaths	Electrical engineers (so described)	143
90	102	104	114	95	295	103	PMR		
182	41	0	16	13	12	736	Deaths	Plumbers and gas fitters	144
89	102	0	90	59	128	98	PMR		
73	22	1	9	9	4	330	Deaths	Sheet metal workers	145
85	126	107	115	93	101	95	PMR		
77	12	1	4	7	1	172	Deaths	Metal plate workers	146
119	89	142	66	115	39	91	PMR		
58	11	1	4	2	4	240	Deaths	Steel erectors	147
111	106	180	86	38	179	144	PMR		
34	2	1	2	2	2	221	Deaths	Scaffolders	148
106	33	299	93	40	79	130	PMR		
161	31	1	14	14	6	666	Deaths	Welders	149
95	92	55	93	74	76	100	PMR		
29	3	0	1	2	0	63	Deaths	Riggers	150
132	67	0	49	101	0	110	PMR		
8	0	0	1	4	0	25	Deaths	Jewellery workers	151
159	0	0	227	634	0	103	PMR		
2	3	0	1	1	0	11	Deaths	Engravers and etchers (printing)	152
49	363	0	271	226	0	72	PMR		
20	1	0	4	1	0	57	Deaths	Vehicle body builders	153
97	24	0	210	48	0	81	PMR		
1	1	0	0	2	0	14	Deaths	Oilers and greasers	154
18	84	0	0	442	0	113	PMR		
8	3	0	1	1	0	35	Deaths	Electronics wire workers	155
117	226	0	174	108	0	100	PMR		
9	1	0	0	0	0	10	Deaths	Coil winders	156
216	113	0	0	0	0	65	PMR		
1	0	0	0	0	0	12	Deaths	Pottery decorators	157
54	0	0	0	0	0	175	PMR		
9	0	0	0	0	0	3	Deaths	Coach painters	158
215	0	0	0	0	0	38	PMR		
40	9	0	8	4	1	218	Deaths	Other spray painters	159
81	90	0	221	61	29	101	PMR		
329	43	3	28	32	30	1395	Deaths	Painters and decorators nec	160
105	69	89	101	93	209	116	PMR		
16	2	0	3	5	1	42	Deaths	Electrical/ electronic assemblers	161
107	65	0	264	242	100	60	PMR		
0	1	0	0	0	0	1	Deaths	Instrument assemblers	162
0	617	0	0	0	0	43	PMR		
67	19	0	4	7	1	165	Deaths	Assemblers (vehicles and other metal goods)	163
92	127	0	70	93	23	75	PMR		
91	23	0	13	21	2	269	Deaths	Packers and sorters	164
94	113	0	170	201	35	85	PMR		
152	34	3	18	14	5	590	Deaths	Bricklayers and tilesetters	165
97	108	177	127	85	74	106	PMR		
10	5	0	3	4	2	65	Deaths	Masons and stonecutters	166
49	119	0	158	198	241	98	PMR		
54	13	2	4	6	4	253	Deaths	Plasterers	167
97	118	334	82	91	147	105	PMR		
55	5	1	1	4	3	524	Deaths	Roofers and glaziers	168
91	43	158	26	36	57	128	PMR		
263	46	1	14	21	18	1153	Deaths	Builders etc.	169
96	86	35	61	71	131	118	PMR		
27	5	1	1	2	0	136	Deaths	Rail track workers	170
88	77	310	41	60	0	135	PMR		

Appendix 7 Mortality of men aged 20-64 by cause of deaths and job-group, England and Wales, 1979-80, 1982-90

Job group		I Infections and parasitic diseases (001-139)	II Neoplasms (140-239)	III Endocrine nutritional and metabolic diseases and immunity disorders (240-279)	IV Diseases of blood and blood-forming organs (280-289)	V Mental disorders (290-319)	VI Diseases of the nervous system and and sense organs (320-389)	VII Diseases of the circulatory system (390-459)	VIII Diseases of the respiratory system (460-519)
171 Road construction workers and paviors	Deaths	6	638	20	4	10	39	970	135
	PMR	52	103	81	113	95	130	102	91
172 Sewage plant attendants	Deaths	2	239	10	1	0	8	336	36
	PMR	52	113	125	93	0	82	104	66
173 Mains and service layers	Deaths	4	357	8	2	6	17	476	58
	PMR	70	111	63	104	150	114	97	85
174 Construction workers nec	Deaths	67	4499	99	16	103	182	5825	993
	PMR	79	114	58	71	100	78	96	102
175 Face-trained coal miners	Deaths	11	1039	18	3	8	28	1693	306
	PMR	74	93	51	49	85	72	103	142
176 Miners (not coal) and quarry workers	Deaths	6	294	7	6	2	14	462	79
	PMR	110	96	59	323	51	100	100	115
177 Railway guards	Deaths	5	233	13	2	2	11	369	34
	PMR	138	95	155	137	71	116	103	74
178 Railway signal workers	Deaths	4	202	13	2	1	10	396	51
	PMR	121	81	166	145	47	115	108	107
179 Shunters and points operators	Deaths	4	109	4	0	4	5	165	25
	PMR	239	92	102	0	337	114	94	112
180 Railway engine drivers	Deaths	8	696	19	6	5	16	1174	123
	PMR	82	94	82	148	81	62	107	87
181 Road transport inspectors	Deaths	4	348	13	1	0	10	546	47
	PMR	89	101	121	54	0	84	107	71
182 Bus and coach drivers	Deaths	27	1584	55	9	15	69	2609	349
	PMR	106	95	93	94	82	103	105	117
183 Lorry drivers	Deaths	129	10066	328	52	82	277	14905	1954
	PMR	83	100	91	88	70	68	99	108
184 Other motor drivers	Deaths	20	1533	64	12	18	45	2502	243
	PMR	78	95	107	125	89	67	104	85
185 Bus conductors and drivers' mates	Deaths	6	300	20	1	3	27	547	111
	PMR	91	86	144	49	48	156	102	132
186 Mechanical plant drivers	Deaths	8	488	18	2	6	18	797	115
	PMR	99	94	96	66	96	85	103	123
187 Crane drivers	Deaths	16	1111	39	4	16	34	1830	273
	PMR	95	91	99	59	147	77	101	120
188 Fork lift truck drivers	Deaths	22	1212	39	7	10	47	1939	256
	PMR	109	96	84	91	60	89	104	112
189 Slingers	Deaths	3	209	10	1	0	5	286	38
	PMR	109	104	155	90	0	69	96	101
190 Storekeepers	Deaths	101	6406	239	37	64	303	10047	1403
	PMR	93	101	100	98	85	111	104	98
191 Dockers and goods porters	Deaths	24	1787	55	10	34	74	2367	408
	PMR	75	112	85	122	95	89	96	92
192 Refuse collectors	Deaths	8	734	26	4	18	37	914	149
	PMR	60	118	98	126	112	103	94	87
193 Labourers in coke ovens	Deaths	0	99	0	1	1	4	159	31
	PMR	0	101	0	204	53	86	105	110
194 Boiler operators	Deaths	5	680	18	1	3	11	972	154
	PMR	51	106	81	28	55	47	100	104

IX Diseases of the digestive system (520-579)	X Diseases of the genito-urinary system (520-629)	XII Diseases of the skin subcutaneous tissue (680-709)	XIII Diseases of the musculo-skeletal system and connective tissue (710-739)	XIV Congenital anomalies (740-759)	XVI Symptoms signs and ill-defined conditions (780-799)	EXVII External causes of injury and poisoning (E800-E999)		Job group	
58	7	0	2	1	2	188	Deaths	Road construction workers and paviors	171
93	56	0	46	18	49	99	PMR		
22	3	0	0	2	0	37	Deaths	Sewage plant attendants	172
98	68	0	0	143	0	74	PMR		
21	6	0	2	0	1	96	Deaths	Mains and service layers	173
71	102	0	87	0	58	107	PMR		
415	56	3	15	31	28	1674	Deaths	Construction workers nec	174
92	68	66	56	74	78	96	PMR		
83	8	0	6	6	0	170	Deaths	Face-trained coal miners	175
101	46	0	75	90	0	99	PMR		
18	4	0	0	4	0	109	Deaths	Miners (not coal) and quarry workers	176
65	69	0	0	134	0	121	PMR		
19	2	0	3	2	0	69	Deaths	Railway guards	177
100	52	0	172	105	0	110	PMR		
22	2	0	2	1	1	46	Deaths	Railway signal workers	178
120	52	0	113	66	158	113	PMR		
14	1	0	0	0	0	36	Deaths	Shunters and points operators	179
155	54	0	0	0	0	139	PMR		
66	4	0	6	6	1	112	Deaths	Railway engine drivers	180
121	35	0	114	135	53	98	PMR		
24	5	0	2	2	0	31	Deaths	Road transport inspectors	181
94	93	0	82	103	0	67	PMR		
120	17	1	15	9	9	337	Deaths	Bus and coach drivers	182
88	65	68	127	68	153	82	PMR		
695	158	8	65	81	28	3099	Deaths	Lorry drivers	183
85	99	91	91	98	78	114	PMR		
133	27	2	11	14	5	512	Deaths	Other motor drivers	184
99	105	140	96	99	80	109	PMR		
47	6	0	5	4	7	100	Deaths	Bus conductors and drivers' mates	185
137	85	0	204	123	299	85	PMR		
40	13	1	0	4	0	143	Deaths	Mechanical plant drivers	186
94	158	221	0	91	0	97	PMR		
110	19	1	7	6	2	248	Deaths	Crane drivers	187
119	100	98	81	77	59	114	PMR		
100	21	1	8	15	6	380	Deaths	Fork lift truck drivers	188
97	104	92	89	130	122	93	PMR		
17	3	1	2	1	0	37	Deaths	Slingers	189
112	96	606	140	77	0	100	PMR		
528	111	7	46	67	27	1302	Deaths	Storekeepers	190
95	93	123	102	117	87	77	PMR		
220	35	2	14	7	26	416	Deaths	Dockers and goods porters	191
117	97	105	134	61	188	88	PMR		
72	15	1	2	10	5	189	Deaths	Refuse collectors	192
93	108	131	50	194	83	88	PMR		
13	2	0	1	0	1	17	Deaths	Labourers in coke ovens	193
114	88	0	157	0	125	77	PMR		
54	11	0	5	6	2	80	Deaths	Boiler operators	194
100	93	0	111	145	79	83	PMR		

Appendix 8 Mortality of women aged 20 to 59 by cause of deaths and job group, England and Wales, 1979-80, 1982-90 (PMR - all women with classifiable occupations = 100)

Job group		I Infections and parasitic diseases (001-139)	II Neoplasms (140-239)	III Endocrine nutritional and metabolic diseases and immunity disorders (240-279)	IV Diseases of blood and blood-forming organs (280-289)	V Mental disorders (290-319)	VI Diseases of the nervous system and and sense organs (320-389)	VII Diseases of the circulatory system (390-459)	VIII Diseases of the respiratory system (460-519)
001 Lawyers	Deaths	1	53	0	1	2	0	12	3
	PMR	89	100	0	188	203	0	111	126
002 Accountants	Deaths	8	148	1	2	1	4	40	12
	PMR	289	104	27	179	48	63	105	144
003 Personnel managers etc	Deaths	2	175	4	1	0	7	54	4
	PMR	80	108	133	102	0	113	113	41
004 Economists and statisticians	Deaths	0	22	2	0	0	0	6	0
	PMR	0	113	377	0	0	0	146	0
005 Computer programmers	Deaths	3	70	3	1	1	5	21	2
	PMR	149	98	155	148	73	122	109	39
006 Sales managers etc	Deaths	3	305	7	1	2	10	79	13
	PMR	61	104	121	56	65	84	93	73
007 Government inspectors	Deaths	0	15	0	0	1	2	3	3
	PMR	0	67	0	0	505	238	42	223
008 Government administrators	Deaths	0	153	0	1	0	3	46	5
	PMR	0	114	0	131	0	65	109	64
009 Other administrators	Deaths	6	710	12	0	5	22	176	43
	PMR	68	108	106	0	100	96	89	114
010 Teachers in higher education	Deaths	2	262	4	0	1	9	44	9
	PMR	71	112	92	0	55	112	68	76
011 Teachers nec	Deaths	42	3497	58	14	22	135	724	140
	PMR	102	110	108	90	95	124	76	79
012 Vocational trainers, social scientists etc	Deaths	4	161	7	0	0	4	34	6
	PMR	202	110	232	0	0	77	86	81
013 Welfare workers	Deaths	12	1057	18	1	4	25	358	61
	PMR	91	101	101	19	53	69	111	100
014 Clergy	Deaths	0	83	6	2	0	7	17	1
	PMR	0	99	307	209	0	265	88	37
015 Doctors	Deaths	1	148	4	1	4	5	33	8
	PMR	45	90	82	48	204	85	90	135
016 Dentists	Deaths	1	28	0	0	0	1	2	1
	PMR	306	105	0	0	0	98	38	104
017 Nurses	Deaths	58	3818	81	32	49	164	1394	284
	PMR	97	90	108	142	140	108	109	114
018 Pharmacists	Deaths	0	72	2	0	1	0	11	2
	PMR	0	101	108	0	121	0	70	76
019 Medical radiographers	Deaths	2	66	4	1	1	3	7	2
	PMR	177	105	315	243	143	116	38	50
020 Physiotherapists	Deaths	0	106	1	1	0	6	16	2
	PMR	0	111	61	214	0	182	56	37
021 Health professionals nec	Deaths	4	201	1	1	0	6	53	13
	PMR	128	101	26	80	0	80	89	107
022 Veterinarians	Deaths	0	4	0	0	0	1	1	0
	PMR	0	53	0	0	0	307	54	0
023 Driving instructors	Deaths	0	21	0	0	0	0	6	2
	PMR	0	112	0	0	0	0	90	158
024 Literary and artistic occupations	Deaths	19	507	12	2	20	25	107	39
	PMR	193	95	104	54	318	107	68	115
025 Persons involved in sport	Deaths	0	20	1	1	0	0	2	0
	PMR	0	112	121	412	0	0	32	0
026 Biological scientists	Deaths	1	49	1	0	0	4	16	1
	PMR	136	102	67	0	0	196	156	55
027 Chemical engineers and scientists	Deaths	0	21	0	1	0	2	6	0
	PMR	0	103	0	422	0	271	131	0
028 Physical scientists and mathematicians	Deaths	3	52	2	0	1	0	12	0
	PMR	341	118	135	0	134	0	133	0
029 Electrical and electronic engineers (professional)	Deaths	0	12	0	0	0	1	0	1
	PMR	0	124	0	0	0	210	0	325
030 Professional engineers nec	Deaths	0	21	2	1	0	1	9	1
	PMR	0	92	265	332	0	125	197	135

IX Diseases of the digestive system (520-579)	X Diseases of the genito-urinary system (520-629)	XI Complications of pregnancy childbirth and the puerperium (630-679)	XII Diseases of the skin subcutaneous tissue (680-709)	XIII Diseases of the musculo-skeletal system and connective tissue (710-739)	XIV Congenital anomalies (740-759)	XVI Symptoms signs and ill-defined conditions (780-799)	EXVII External causes of injury and poisoning (E800-E999)		Job group	
2	0	0	0	1	1	0	25	Deaths	Lawyers	001
70	0	0	0	109	110	0	107	PMR		
13	2	1	0	2	1	0	43	Deaths	Accountants	002
161	157	77	0	97	48	0	75	PMR		
9	1	1	0	0	0	0	27	Deaths	Personnel managers etc	003
96	62	112	0	0	0	0	76	PMR		
0	0	0	0	0	0	0	3	Deaths	Economists and statisticians	004
0	0	0	0	0	0	0	57	PMR		
4	0	1	0	1	1	0	37	Deaths	Computer programmers	005
94	0	89	0	90	106	0	105	PMR		
16	3	1	0	2	2	0	87	Deaths	Sales managers etc	006
94	99	46	0	55	80	0	111	PMR		
3	0	0	0	0	1	0	11	Deaths	Government inspectors	007
234	0	0	0	0	714	0	258	PMR		
8	2	1	0	3	0	1	5	Deaths	Government administrators	008
103	163	258	0	204	0	297	24	PMR		
36	8	0	0	4	5	2	91	Deaths	Other administrators	009
95	127	0	0	55	118	114	78	PMR		
15	1	0	1	3	1	0	40	Deaths	Teachers in higher education	010
114	52	0	621	113	67	0	94	PMR		
145	30	10	1	41	21	8	496	Deaths	Teachers nec	011
79	99	91	55	118	107	97	92	PMR		
9	0	0	0	2	0	1	26	Deaths	Vocational trainers, social scientists etc	012
109	0	0	0	122	0	170	76	PMR		
49	9	2	1	9	5	0	170	Deaths	Welfare workers	013
82	90	60	173	78	78	0	99	PMR		
3	1	0	0	0	1	0	17	Deaths	Clergy	014
63	714	0	0	0	188	0	93	PMR		
10	0	1	1	2	2	1	66	Deaths	Doctors	015
112	0	235	386	110	135	125	138	PMR		
0	0	1	0	1	1	1	9	Deaths	Dentists	016
0	0	1099	0	275	457	637	111	PMR		
243	54	22	3	55	30	25	1028	Deaths	Nurses	017
100	132	121	127	115	105	209	120	PMR		
3	0	0	0	3	0	0	30	Deaths	Pharmacists	018
75	0	0	0	348	0	0	140	PMR		
9	0	1	0	1	1	0	18	Deaths	Medical radiographers	019
246	0	235	0	131	189	0	100	PMR		
5	1	2	0	0	1	0	22	Deaths	Physiotherapists	020
91	106	519	0	0	161	0	129	PMR		
11	1	1	0	4	2	0	58	Deaths	Health professionals nec	021
95	52	98	0	175	131	0	118	PMR		
0	0	0	0	0	0	0	10	Deaths	Veterinarians	022
0	0	0	0	0	0	0	231	PMR		
1	0	0	0	0	0	0	3	Deaths	Driving instructors	023
97	0	0	0	0	0	0	105	PMR		
45	8	0	1	5	4	4	193	Deaths	Literary and artistic occupations	024
143	144	0	324	72	77	187	125	PMR		
0	0	1	0	0	0	0	15	Deaths	Persons involved in sport	025
0	0	395	0	0	0	0	178	PMR		
1	1	0	0	0	1	0	14	Deaths	Biological scientists	026
38	1075	0	0	0	146	0	75	PMR		
0	0	0	0	0	1	0	5	Deaths	Chemical engineers and scientists	027
0	0	0	0	0	441	0	78	PMR		
0	0	0	0	0	2	1	14	Deaths	Physical scientists and mathematicians	028
0	0	0	0	0	219	467	64	PMR		
0	0	1	0	1	0	0	2	Deaths	Electrical and electronic engineers (professional)	029
0	0	1316	0	549	0	0	48	PMR		
1	0	0	0	1	0	0	3	Deaths	Professional engineers nec	030
83	0	0	0	365	0	0	40	PMR		

Appendix 8 Mortality of women aged 20-64 by cause of deaths and job group, England and Wales, 1979-80, 1982-90

Job group		I Infections and parasitic diseases (001-139)	II Neoplasms (140-239)	III Endocrine nutritional and metabolic diseases and immunity disorders (240-279)	IV Diseases of blood and blood-forming organs (280-289)	V Mental disorders (290-319)	VI Diseases of the nervous system and sense organs (320-389)	VII Diseases of the circulatory system (390-459)	VIII Diseases of the respiratory system (460-519)
031 Draughtspersons	Deaths	2	67	4	1	0	3	11	1
	PMR	213	111	230	204	0	92	55	23
032 Laboratory technicians	Deaths	5	284	5	4	3	11	65	23
	PMR	129	108	84	270	124	108	68	120
033 Architects and surveyors	Deaths	1	42	0	1	1	1	7	1
	PMR	137	107	0	211	169	64	80	57
034 Aircraft flight deck officers	Deaths	0	1	0	0	0	0	0	0
	PMR	0	130	0	0	0	0	0	0
035 Air traffic controllers	Deaths	0	1	0	0	0	0	0	0
	PMR	0	80	0	0	0	0	0	0
036 Seafarers	Deaths	0	2	0	0	0	0	0	0
	PMR	0	126	0	0	0	0	0	0
037 Technicians nec	Deaths	1	78	2	0	0	1	15	5
	PMR	79	107	138	0	0	34	70	110
038 Production and maintenance managers	Deaths	4	266	5	3	0	6	68	14
	PMR	142	111	131	277	0	78	94	106
039 Managers in construction	Deaths	1	38	0	0	0	3	11	2
	PMR	208	99	0	0	0	231	93	91
040 Managers in transport, utilities and mining	Deaths	2	105	0	0	0	2	48	4
	PMR	136	100	0	0	0	54	152	65
041 Office managers	Deaths	7	546	5	5	0	22	154	26
	PMR	98	109	56	188	0	121	103	89
042 Butchers	Deaths	0	24	0	0	0	0	8	3
	PMR	0	99	0	0	0	0	75	151
043 Fishmongers and poultry dressers	Deaths	0	23	1	0	1	1	5	1
	PMR	0	116	215	0	667	137	59	63
044 Retailers and dealers	Deaths	56	4580	95	26	32	166	1940	362
	PMR	92	95	93	92	87	90	117	112
045 Publicans and bar staff	Deaths	10	860	14	3	16	29	395	100
	PMR	76	85	69	49	193	76	104	140
046 Caterers	Deaths	21	2362	50	17	16	68	1151	206
	PMR	75	99	91	109	85	79	102	96
047 Farmers	Deaths	0	338	10	2	5	20	123	13
	PMR	0	101	128	80	154	141	86	46
048 Armed forces	Deaths	1	30	0	0	0	6	9	1
	PMR	100	100	0	0	0	100	100	100
049 Police	Deaths	2	46	2	0	0	1	9	5
	PMR	233	117	121	0	0	37	66	140
050 Fire service personnel	Deaths	0	2	0	0	0	0	0	0
	PMR	0	107	0	0	0	0	0	0
051 Launderers and dry cleaners	Deaths	5	341	10	2	4	14	237	39
	PMR	126	91	114	82	144	110	123	109
052 Hairdressers	Deaths	6	412	8	3	6	27	125	39
	PMR	84	104	58	73	119	114	81	114
053 Office workers and cashiers	Deaths	210	16511	398	104	139	710	5668	1120
	PMR	99	101	102	97	101	103	97	96
054 Postal workers	Deaths	3	184	2	0	1	3	76	12
	PMR	172	116	54	0	85	57	86	73
055 Petrol pump attendants	Deaths	1	40	0	0	0	0	16	4
	PMR	169	87	0	0	0	0	103	126
056 Vans sales persons	Deaths	1	17	1	0	0	0	14	1
	PMR	433	83	184	0	0	0	149	58
057 Sales representatives	Deaths	8	278	6	1	0	6	91	20
	PMR	199	98	87	51	0	46	95	101
058 Security workers	Deaths	1	133	4	0	1	3	68	18
	PMR	58	91	120	0	87	57	94	134
059 Cooks and kitchen porters	Deaths	16	1381	36	9	6	35	721	107
	PMR	113	103	108	148	67	80	104	85
060 Other service personnel	Deaths	57	5503	123	29	38	161	2971	507
	PMR	102	101	99	94	100	91	100	92
061 Hospital porters and ward orderlies	Deaths	5	861	16	4	5	37	401	66
	PMR	48	105	79	65	68	114	96	83
062 Ambulance workers	Deaths	0	15	0	0	0	0	5	2
	PMR	0	90	0	0	0	0	78	168
063 Railway station workers	Deaths	0	16	1	0	0	0	13	1
	PMR	0	87	262	0	0	0	118	50
064 Undertakers	Deaths	0	5	0	0	0	0	1	0
	PMR	0	156	0	0	0	0	90	0
065 Foresters	Deaths	0	0	0	0	0	0	0	0
	PMR	0	0	0	0	0	0	0	0

IX Diseases of the digestive system (520-579)	X Diseases of the genito-urinary system (520-629)	XI Complications of pregnancy childbirth and the puerperium (630-679)	XII Diseases of the skin subcutaneous tissue (680-709)	XIII Diseases of the musculo-skeletal system and connective tissue (710-739)	XIV Congenital anomalies (740-759)	XVI Symptoms signs and ill-defined conditions (780-799)	EXVII External causes of injury and poisoning (E800-E999)		Job group	
6	0	0	0	0	1	0	18	Deaths	Draughtspersons	031
166	0	0	0	0	95	0	118	PMR		
15	3	4	0	1	3	3	61	Deaths	Laboratory technicians	032
99	104	281	0	32	129	333	101	PMR		
2	0	0	0	0	0	0	18	Deaths	Architects and surveyors	033
91	0	0	0	0	0	0	115	PMR		
0	0	0	0	0	0	0	1	Deaths	Aircraft flight deck officers	034
0	0	0	0	0	0	0	142	PMR		
1	0	0	0	0	0	0	0	Deaths	Air traffic controllers	035
1563	0	0	0	0	0	0	0	PMR		
0	0	0	0	0	0	0	1	Deaths	Seafarers	036
0	0	0	0	0	0	0	391	PMR		
7	1	1	0	0	0	0	21	Deaths	Technicians nec	037
169	133	214	0	0	0	0	111	PMR		
15	1	0	0	1	0	0	13	Deaths	Production and maintenance managers	038
110	45	0	0	40	0	0	39	PMR		
2	1	1	0	1	0	0	5	Deaths	Managers in construction	039
93	274	862	0	236	0	0	80	PMR		
4	2	0	0	2	2	0	9	Deaths	Managers in transport, utilities and mining	040
68	193	0	0	172	307	0	46	PMR		
21	4	4	1	6	2	0	66	Deaths	Office managers	041
73	82	168	353	104	58	0	67	PMR		
2	1	0	0	0	0	0	7	Deaths	Butchers	042
149	365	0	0	0	0	0	187	PMR		
2	0	0	0	1	0	0	1	Deaths	Fishmongers and poultry dressers	043
183	0	0	0	452	0	0	37	PMR		
260	43	19	3	41	49	12	736	Deaths	Retailers and dealers	044
99	89	130	145	79	105	94	94	PMR		
139	9	0	0	9	9	0	228	Deaths	Publicans and bar staff	045
237	75	0	0	83	98	0	128	PMR		
145	32	9	1	31	15	4	404	Deaths	Caterers	046
107	104	150	43	130	70	62	109	PMR		
18	8	0	0	3	5	1	98	Deaths	Farmers	047
91	174	0	0	83	136	91	139	PMR		
3	0	0	0	0	0	0	26	Deaths	Armed forces	048
100	0	0	0	0	0	0	100	PMR		
2	0	0	0	0	1	0	16	Deaths	Police	049
78	0	0	0	0	95	0	102	PMR		
0	0	0	0	0	0	0	1	Deaths	Fire service personnel	050
0	0	0	0	0	0	0	380	PMR		
17	6	0	0	4	1	1	38	Deaths	Launderers and dry cleaners	051
80	120	0	0	110	30	101	77	PMR		
33	4	4	0	12	3	3	145	Deaths	Hairdressers	052
130	77	115	0	184	42	138	102	PMR		
891	173	56	7	200	204	48	2796	Deaths	Office workers and cashiers	053
100	100	94	114	110	103	99	98	PMR		
8	1	0	0	1	1	0	23	Deaths	Postal workers	054
89	44	0	0	71	70	0	97	PMR		
7	0	1	0	0	0	0	13	Deaths	Petrol pump attendants	055
272	0	602	0	0	0	0	150	PMR		
0	0	0	0	0	1	0	4	Deaths	Vans sales persons	056
0	0	0	0	0	526	0	117	PMR		
12	2	2	0	2	1	0	88	Deaths	Sales representatives	057
72	63	140	0	58	26	0	147	PMR		
11	2	0	0	1	0	1	37	Deaths	Security workers	058
132	101	0	0	70	0	253	170	PMR		
69	13	1	1	6	9	4	160	Deaths	Cooks and kitchen porters	059
93	76	48	110	45	86	99	87	PMR		
315	60	8	5	33	41	16	832	Deaths	Other service personnel	060
101	81	88	93	67	87	95	104	PMR		
49	7	3	0	6	4	3	158	Deaths	Hospital porters and ward orderlies	061
102	58	119	0	72	44	117	108	PMR		
2	0	0	0	0	0	0	6	Deaths	Ambulance workers	062
196	0	0	0	0	0	0	227	PMR		
2	1	0	0	0	0	0	2	Deaths	Railway station workers	063
199	429	0	0	0	0	0	102	PMR		
0	0	0	0	0	0	0	0	Deaths	Undertakers	064
0	0	0	0	0	0	0	0	PMR		
0	0	0	0	0	0	0	1	Deaths	Foresters	065
0	0	0	0	0	0	0	270	PMR		

Appendix 8 Mortality of women aged 20-64 by cause of deaths and job group, England and Wales, 1979-80, 1982-90

Job group		I Infections and parasitic diseases (001-139)	II Neoplasms (140-239)	III Endocrine nutritional and metabolic diseases and immunity disorders (240-279)	IV Diseases of blood and blood-forming organs (280-289)	V Mental disorders (290-319)	VI Diseases of the nervous system and sense organs (320-389)	VII Diseases of the circulatory system (390-459)	VIII Diseases of the respiratory system (460-519)
066 Fishing and related workers	Deaths	0	0	0	0	0	0	0	0
	PMR	0	0	0	0	0	0	0	0
067 Tannery workers	Deaths	0	9	0	0	0	0	5	1
	PMR	0	95	0	0	0	0	108	119
068 Leather and shoe workers	Deaths	4	261	10	1	1	7	130	35
	PMR	149	98	147	95	59	80	99	149
069 Preparatory fibre processors	Deaths	0	23	1	0	0	2	18	8
	PMR	0	80	157	0	0	243	118	281
070 Spinners and winders	Deaths	2	123	2	0	2	4	101	34
	PMR	146	81	57	0	209	89	124	221
071 Warp preparers and weavers	Deaths	1	123	1	2	0	3	70	19
	PMR	74	90	28	372	0	67	103	155
072 Knitters	Deaths	1	120	5	2	2	8	69	13
	PMR	70	96	147	312	218	169	116	118
073 Bleachers, dyers and finishers	Deaths	1	30	2	0	0	2	13	2
	PMR	305	107	252	0	0	184	94	78
074 Other textile workers	Deaths	4	402	8	3	5	18	262	52
	PMR	84	93	77	134	165	128	112	122
075 Chemical workers	Deaths	4	165	7	3	2	4	77	16
	PMR	156	99	157	208	116	54	90	95
076 Bakers	Deaths	0	60	2	0	1	4	39	6
	PMR	0	88	102	0	174	155	118	96
077 Brewery workers	Deaths	0	13	0	0	0	0	6	1
	PMR	0	112	0	0	0	0	102	94
078 Food processors	Deaths	4	454	15	4	2	9	216	46
	PMR	83	105	142	151	62	59	99	113
079 Paper manufacturers	Deaths	0	7	1	0	0	0	2	1
	PMR	0	108	568	0	0	0	58	164
080 Bookbinders	Deaths	2	106	3	0	0	2	56	9
	PMR	178	101	111	0	0	55	112	100
081 Paper cutters	Deaths	0	6	0	0	0	0	3	1
	PMR	0	111	0	0	0	0	107	202
082 Glass and ceramics furnace workers	Deaths	0	3	0	0	0	0	3	0
	PMR	0	84	0	0	0	0	167	0
083 Glass formers and decorators	Deaths	0	17	0	0	0	0	11	5
	PMR	0	84	0	0	0	0	112	279
084 Ceramics casters	Deaths	1	35	0	0	0	1	21	3
	PMR	260	95	0	0	0	76	115	90
085 Rubber manufacturers	Deaths	0	15	1	0	0	2	11	2
	PMR	0	87	223	0	0	323	131	134
086 Plastics workers	Deaths	0	22	0	0	0	1	9	1
	PMR	0	112	0	0	0	153	94	59
087 Man-made fibre makers	Deaths	0	1	0	0	0	0	1	1
	PMR	0	63	0	0	0	0	113	637
088 Other coal workers	Deaths	0	0	0	0	0	0	0	0
	PMR	0	0	0	0	0	0	0	0
089 Tobacco workers	Deaths	0	47	1	0	0	1	22	3
	PMR	0	110	100	0	0	69	100	72
090 Other wood and paper processors	Deaths	0	58	1	0	0	3	24	11
	PMR	0	102	66	0	0	150	86	217
091 Other occupations - glass and ceramics	Deaths	0	90	1	0	0	4	70	13
	PMR	0	84	36	0	0	113	132	137
092 Rubber goods makers	Deaths	0	25	2	0	0	0	25	3
	PMR	0	82	247	0	0	0	161	107
093 Plastic goods makers	Deaths	3	106	4	0	1	2	36	4
	PMR	326	125	201	0	154	68	83	50
094 Compositors	Deaths	1	15	0	0	0	2	2	0
	PMR	585	115	0	0	0	321	39	0
095 Printing plate preparers	Deaths	0	5	0	0	0	0	1	0
	PMR	0	106	0	0	0	0	52	0
096 Printing machine minders	Deaths	2	70	4	0	0	4	33	13
	PMR	258	93	207	0	0	153	92	202
097 Printers (so described)	Deaths	1	59	2	0	0	5	22	3
	PMR	155	106	126	0	0	222	82	60
098 Tailors and dressmakers	Deaths	1	269	7	3	1	13	146	25
	PMR	37	96	100	306	61	146	105	102
099 Clothing cutters	Deaths	0	45	2	1	0	1	24	2
	PMR	0	101	171	524	0	62	113	52
100 Sewers and embroiderers	Deaths	25	1337	31	14	14	64	693	132
	PMR	138	99	91	132	109	115	101	100

IX Diseases of the digestive system (520-579)	X Diseases of the genito-urinary system (520-629)	XI Complications of pregnancy childbirth and the puerperium (630-679)	XII Diseases of the skin subcutaneous tissue (680-709)	XIII Diseases of the musculo-skeletal system and connective tissue (710-739)	XIV Congenital anomalies (740-759)	XVI Symptoms signs and ill-defined conditions (780-799)	EXVII External causes of injury and poisoning (E800-E999)		Job group	
0	0	0	0	0	0	0	1	Deaths	Fishing and related workers	066
0	0	0	0	0	0	0	3448	PMR		
1	1	0	0	0	1	0	0	Deaths	Tannery workers	067
183	952	0	0	0	1163	0	0	PMR		
20	5	1	0	2	2	0	16	Deaths	Leather and shoe workers	068
139	164	299	0	72	106	0	56	PMR		
1	0	0	0	0	0	0	1	Deaths	Preparatory fibre processors	069
62	0	0	0	0	0	0	37	PMR		
5	2	0	0	2	1	0	10	Deaths	Spinners and winders	070
59	103	0	0	146	96	0	66	PMR		
11	2	0	0	4	2	0	16	Deaths	Warp preparers and weavers	071
153	125	0	0	289	230	0	113	PMR		
2	2	0	0	0	0	1	14	Deaths	Knitters	072
29	131	0	0	0	0	248	71	PMR		
1	0	0	0	0	0	0	3	Deaths	Bleachers, dyers and finishers	073
64	0	0	0	0	0	0	72	PMR		
25	6	1	1	10	3	0	47	Deaths	Other textile workers	074
102	103	149	289	242	79	0	75	PMR		
6	2	0	1	2	1	0	54	Deaths	Chemical workers	075
60	74	0	538	116	43	0	139	PMR		
4	0	0	0	0	2	0	14	Deaths	Bakers	076
103	0	0	0	0	288	0	119	PMR		
0	0	0	0	0	0	0	2	Deaths	Brewery workers	077
0	0	0	0	0	0	0	145	PMR		
22	4	1	0	5	6	0	42	Deaths	Food processors	078
90	70	104	0	113	156	0	68	PMR		
1	0	0	0	0	0	0	0	Deaths	Paper manufacturers	079
289	0	0	0	0	0	0	0	PMR		
4	1	0	0	0	1	1	10	Deaths	Bookbinders	080
69	83	0	0	0	118	336	75	PMR		
0	0	0	0	0	0	0	0	Deaths	Paper cutters	081
0	0	0	0	0	0	0	0	PMR		
1	0	0	0	0	0	0	0	Deaths	Glass and ceramics furnace workers	082
472	0	0	0	0	0	0	0	PMR		
0	0	0	0	0	0	0	5	Deaths	Glass formers and decorators	083
0	0	0	0	0	0	0	192	PMR		
1	1	0	0	0	0	0	7	Deaths	Ceramics casters	084
50	220	0	0	0	0	0	136	PMR		
1	0	0	0	0	0	0	0	Deaths	Rubber manufacturers	085
106	0	0	0	0	0	0	0	PMR		
0	1	0	0	0	0	0	3	Deaths	Plastics workers	086
0	398	0	0	0	0	0	117	PMR		
0	0	0	0	0	0	0	0	Deaths	Man-made fibre makers	087
0	0	0	0	0	0	0	0	PMR		
0	0	0	0	0	0	0	0	Deaths	Other coal workers	088
0	0	0	0	0	0	0	0	PMR		
2	1	0	0	0	0	0	5	Deaths	Tobacco workers	089
83	173	0	0	0	0	0	87	PMR		
5	0	0	0	0	0	0	4	Deaths	Other wood and paper processors	090
163	0	0	0	0	0	0	61	PMR		
9	1	0	1	1	1	0	9	Deaths	Other occupations - glass and ceramics	091
153	80	0	1493	90	121	0	72	PMR		
0	0	0	0	0	0	0	2	Deaths	Rubber goods makers	092
0	0	0	0	0	0	0	64	PMR		
3	1	0	0	1	2	0	0	Deaths	Plastic goods makers	093
62	88	0	0	119	257	0	0	PMR		
3	0	0	0	0	0	0	3	Deaths	Compositors	094
354	0	0	0	0	0	0	79	PMR		
1	0	0	0	0	0	0	1	Deaths	Printing plate preparers	095
431	0	0	0	0	0	0	245	PMR		
7	1	0	0	1	1	0	3	Deaths	Printing machine minders	096
167	118	0	0	118	166	0	33	PMR		
4	0	0	0	0	0	1	10	Deaths	Printers (so described)	097
127	0	0	0	0	0	483	112	PMR		
17	1	1	0	6	1	2	23	Deaths	Tailors and dressmakers	098
114	31	373	0	211	58	314	88	PMR		
3	0	0	0	0	0	0	5	Deaths	Clothing cutters	099
124	0	0	0	0	0	0	85	PMR		
71	21	8	1	16	17	8	257	Deaths	Sewers and embroiderers	100
89	103	168	67	114	106	180	99	PMR		

Appendix 8 Mortality of women aged 20-64 by cause of deaths and job group, England and Wales, 1979-80, 1982-90

Job group		I Infections and parasitic diseases (001-139)	II Neoplasms (140-239)	III Endocrine nutritional and metabolic diseases and immunity disorders (240-279)	IV Diseases of blood and blood-forming organs (280-289)	V Mental disorders (290-319)	VI Diseases of the nervous system and and sense organs (320-389)	VII Diseases of the circulatory system (390-459)	VIII Diseases of the respiratory system (460-519)
101 Upholsterers	Deaths	1	48	0	0	1	0	12	3
	PMR	253	131	0	0	386	0	68	92
102 Carpet fitters	Deaths	0	0	0	0	0	0	1	0
	PMR	0	0	0	0	0	0	317	0
103 Other workers with fabrics	Deaths	2	96	3	0	0	2	35	6
	PMR	228	115	139	0	0	70	87	82
104 Carpenters	Deaths	0	3	0	0	0	0	2	0
	PMR	0	84	0	0	0	0	116	0
105 Cabinet makers	Deaths	0	7	0	0	0	0	4	1
	PMR	0	95	0	0	0	0	118	159
106 Case and box makers	Deaths	0	6	0	0	0	0	1	0
	PMR	0	142	0	0	0	0	50	0
107 Pattern makers	Deaths	0	1	0	0	0	0	0	0
	PMR	0	89	0	0	0	0	0	0
108 Woodworking machinists	Deaths	0	11	0	0	1	1	9	1
	PMR	0	84	0	0	909	204	126	76
109 Other woodworkers	Deaths	0	11	0	0	0	0	2	0
	PMR	0	126	0	0	0	0	52	0
110 Dental technicians	Deaths	0	5	0	0	0	0	3	0
	PMR	0	117	0	0	0	0	155	0
111 Other makers of paper goods	Deaths	1	144	2	0	0	5	76	14
	PMR	66	99	52	0	0	98	108	109
112 Furnace operatives (metal)	Deaths	0	2	0	0	0	0	1	1
	PMR	0	102	0	0	0	0	118	521
113 Rollers (metal)	Deaths	0	0	0	0	0	0	1	0
	PMR	0	0	0	0	0	0	317	0
114 Smiths and forge workers	Deaths	0	2	0	0	0	0	0	0
	PMR	0	117	0	0	0	0	0	0
115 Metal drawers	Deaths	0	1	0	0	0	0	1	1
	PMR	0	45	0	0	0	0	105	595
116 Moulders and coremakers (metal)	Deaths	0	14	0	0	0	0	16	1
	PMR	0	73	0	0	0	0	166	59
117 Electroplaters	Deaths	0	3	0	0	0	1	3	0
	PMR	0	88	0	0	0	709	158	0
118 Annealers, hardeners, temperers (metal)	Deaths	0	1	0	0	0	0	2	0
	PMR	0	70	0	0	0	0	268	0
119 Galvanisers and tin platers	Deaths	0	1	0	0	0	0	1	0
	PMR	0	86	0	0	0	0	189	0
120 Other metal manufacturers	Deaths	0	14	1	0	0	0	8	1
	PMR	0	107	321	0	0	0	113	79
121 Press and machine tool setters	Deaths	0	7	0	0	0	1	5	0
	PMR	0	94	0	0	0	356	144	0
122 Centre lathe turners	Deaths	0	2	0	0	1	0	1	0
	PMR	0	92	0	0	7692	0	104	0
123 Machine tool setter operators	Deaths	0	3	0	0	0	0	2	0
	PMR	0	109	0	0	0	0	151	0
124 Machine tool operators	Deaths	4	424	11	7	6	14	253	47
	PMR	88	98	108	244	192	97	112	110
125 Press and automatic machine operators	Deaths	1	152	3	1	0	1	98	21
	PMR	60	95	81	96	0	19	117	134
126 Metal polishers	Deaths	0	12	2	0	0	2	6	0
	PMR	0	83	595	0	0	460	87	0
127 Fettlers and dressers (metal)	Deaths	0	17	1	0	0	1	6	2
	PMR	0	113	311	0	0	228	77	139
128 Shot blasters	Deaths	0	0	0	0	0	0	0	0
	PMR	0	0	0	0	0	0	0	0
129 Toolmakers	Deaths	0	2	0	0	0	0	0	2
	PMR	0	72	0	0	0	0	0	826
130 Precision instrument makers	Deaths	0	7	1	0	0	0	1	0
	PMR	0	149	758	0	0	0	44	0
131 Watch and clock makers	Deaths	1	2	0	0	0	0	0	0
	PMR	3125	129	0	0	0	0	0	0
132 Production fitters	Deaths	1	11	0	1	0	0	8	1
	PMR	685	83	0	1923	0	0	123	86
133 Motor mechanics	Deaths	0	11	2	0	0	0	5	1
	PMR	0	92	660	0	0	0	99	106
134 Aircraft engine fitters	Deaths	0	0	0	0	0	0	0	0
	PMR	0	0	0	0	0	0	0	0
135 Office machinery mechanics	Deaths	0	0	0	0	0	0	0	0
	PMR	0	0	0	0	0	0	0	0

IX Diseases of the digestive system (520-579)	X Diseases of the genito-urinary system (520-629)	XI Complications of pregnancy childbirth and the puerperium (630-679)	XII Diseases of the skin subcutaneous tissue (680-709)	XIII Diseases of the musculo-skeletal system and connective tissue (710-739)	XIV Congenital anomalies (740-759)	XVI Symptoms signs and ill-defined conditions (780-799)	EXVII External causes of injury and poisoning (E800-E999)		Job group	
0	1	0	0	1	0	0	3	Deaths	Upholsterers	101
0	218	0	0	226	0	0	54	PMR		
0	0	0	0	0	0	0	0	Deaths	Carpet fitters	102
0	0	0	0	0	0	0	0	PMR		
4	1	0	0	1	0	0	5	Deaths	Other workers with fabrics	103
89	104	0	0	114	0	0	50	PMR		
1	0	0	0	0	0	0	1	Deaths	Carpenters	104
526	0	0	0	0	0	0	135	PMR		
0	0	0	0	0	1	0	1	Deaths	Cabinet makers	105
0	0	0	0	0	1786	0	85	PMR		
0	0	0	0	0	0	0	1	Deaths	Case and box makers	106
0	0	0	0	0	0	0	157	PMR		
0	0	0	0	0	0	0	1	Deaths	Pattern makers	107
0	0	0	0	0	0	0	552	PMR		
1	0	0	0	0	0	0	2	Deaths	Woodworking machinists	108
138	0	0	0	0	0	0	105	PMR		
0	0	0	0	0	0	0	5	Deaths	Other woodworkers	109
0	0	0	0	0	0	0	199	PMR		
0	0	0	0	0	0	0	1	Deaths	Dental technicians	110
0	0	0	0	0	0	0	70	PMR		
4	5	0	0	3	3	0	15	Deaths	Other makers of paper goods	111
50	300	0	0	188	268	0	82	PMR		
0	0	0	0	0	0	0	0	Deaths	Furnace operatives (metal)	112
0	0	0	0	0	0	0	0	PMR		
0	0	0	0	0	0	0	0	Deaths	Rollers (metal)	113
0	0	0	0	0	0	0	0	PMR		
0	0	0	0	0	0	0	1	Deaths	Smiths and forge workers	114
0	0	0	0	0	0	0	372	PMR		
0	0	0	0	0	0	0	1	Deaths	Metal drawers	115
0	0	0	0	0	0	0	379	PMR		
2	0	0	0	0	0	0	2	Deaths	Moulders and coremakers (metal)	116
199	0	0	0	0	0	0	130	PMR		
0	0	0	0	0	0	0	0	Deaths	Electroplaters	117
0	0	0	0	0	0	0	0	PMR		
0	0	0	0	0	0	0	0	Deaths	Annealers, hardeners, temperers (metal)	118
0	0	0	0	0	0	0	0	PMR		
0	0	0	0	0	0	0	0	Deaths	Galvanisers and tin platers	119
0	0	0	0	0	0	0	0	PMR		
0	0	0	0	0	0	0	1	Deaths	Other metal manufacturers	120
0	0	0	0	0	0	0	71	PMR		
0	0	0	0	0	1	0	0	Deaths	Press and machine tool setters	121
0	0	0	0	0	1961	0	0	PMR		
0	0	0	0	0	0	0	0	Deaths	Centre lathe turners	122
0	0	0	0	0	0	0	0	PMR		
0	0	0	0	0	0	0	0	Deaths	Machine tool setter operators	123
0	0	0	0	0	0	0	0	PMR		
21	9	0	1	6	1	0	31	Deaths	Machine tool operators	124
86	157	0	198	148	28	0	55	PMR		
8	3	0	0	1	1	0	18	Deaths	Press and automatic machine operators	125
87	139	0	0	65	72	0	89	PMR		
0	2	0	0	0	0	0	2	Deaths	Metal polishers	126
0	1282	0	0	0	0	0	163	PMR		
1	0	0	0	0	0	0	0	Deaths	Fettlers and dressers (metal)	127
122	0	0	0	0	0	0	0	PMR		
0	0	0	0	0	0	0	1	Deaths	Shot blasters	128
0	0	0	0	0	0	0	3448	PMR		
0	0	0	0	1	0	0	0	Deaths	Toolmakers	129
0	0	0	0	4348	0	0	0	PMR		
0	0	0	0	0	0	0	0	Deaths	Precision instrument makers	130
0	0	0	0	0	0	0	0	PMR		
1	0	0	0	0	0	0	0	Deaths	Watch and clock makers	131
917	0	0	0	0	0	0	0	PMR		
0	0	0	0	0	1	0	2	Deaths	Production fitters	132
0	0	0	0	0	1000	0	122	PMR		
0	0	0	0	1	0	0	2	Deaths	Motor mechanics	133
0	0	0	0	794	0	0	110	PMR		
0	0	0	0	0	0	0	0	Deaths	Aircraft engine fitters	134
0	0	0	0	0	0	0	0	PMR		
0	0	0	0	0	0	0	0	Deaths	Office machinery mechanics	135
0	0	0	0	0	0	0	0	PMR		

Appendix 8 Mortality of women aged 20-64 by cause of deaths and job group, England and Wales, 1979-80, 1982-90

Job group		I Infections and parasitic diseases (001-139)	II Neoplasms (140-239)	III Endocrine nutritional and metabolic diseases and immunity disorders (240-279)	IV Diseases of blood and blood-forming organs (280-289)	V Mental disorders (290-319)	VI Diseases of the nervous system and and sense organs (320-389)	VII Diseases of the circulatory system (390-459)	VIII Diseases of the respiratory system (460-519)
136 Electrical and electronic	Deaths	1	17	1	0	0	0	1	0
production fitters	PMR	769	141	310	0	0	0	17	0
137 Electricians	Deaths	0	34	0	0	0	0	17	4
	PMR	0	106	0	0	0	0	115	149
138 Electrical plant operators	Deaths	0	1	0	0	0	0	0	0
	PMR	0	190	0	0	0	0	0	0
139 Telephone fitters	Deaths	0	7	0	0	0	1	4	0
	PMR	0	89	0	0	0	258	112	0
140 Electric cable and line workers	Deaths	0	2	0	0	0	0	3	0
	PMR	0	59	0	0	0	0	202	0
141 Radio and TV mechanics	Deaths	0	3	0	0	0	0	1	0
	PMR	0	149	0	0	0	0	108	0
142 Other electronic maintenance	Deaths	0	9	0	0	0	0	6	0
engineers	PMR	0	123	0	0	0	0	157	0
143 Electrical engineers (so	Deaths	0	3	0	0	0	0	1	0
described)	PMR	0	107	0	0	0	0	83	0
144 Plumbers and gas fitters	Deaths	0	4	0	0	0	0	1	0
	PMR	0	138	0	0	0	0	81	0
145 Sheet metal workers	Deaths	1	2	0	0	0	0	3	0
	PMR	1852	69	0	0	0	0	196	0
146 Metal plate workers	Deaths	0	4	0	0	0	0	0	0
	PMR	0	136	0	0	0	0	0	0
147 Steel erectors	Deaths	0	1	0	0	0	0	0	0
	PMR	0	190	0	0	0	0	0	0
148 Scaffolders	Deaths	0	0	0	0	0	0	0	0
	PMR	0	0	0	0	0	0	0	0
149 Welders	Deaths	0	37	0	0	0	1	26	6
	PMR	0	81	0	0	0	60	118	150
150 Riggers	Deaths	0	0	0	0	0	0	0	0
	PMR	0	0	0	0	0	0	0	0
151 Jewellery workers	Deaths	0	13	0	0	1	0	4	0
	PMR	0	123	0	0	917	0	78	0
152 Engravers and etchers (printing)	Deaths	0	4	0	0	0	0	1	1
	PMR	0	110	0	0	0	0	59	294
153 Vehicle body builders	Deaths	0	1	0	0	0	0	0	0
	PMR	0	171	0	0	0	0	0	0
154 Oilers and greasers	Deaths	0	1	0	0	0	0	1	1
	PMR	0	53	0	0	0	0	115	521
155 Electronics wire workers	Deaths	0	40	2	1	0	0	17	2
	PMR	0	111	211	578	0	0	98	63
156 Coil winders	Deaths	1	40	2	0	0	1	17	6
	PMR	242	93	192	0	0	74	83	164
157 Pottery decorators	Deaths	1	48	0	1	2	3	19	1
	PMR	207	106	0	461	667	183	90	26
158 Coach painters	Deaths	0	0	0	0	0	0	0	0
	PMR	0	0	0	0	0	0	0	0
159 Other spray painters	Deaths	0	22	0	0	0	3	4	0
	PMR	0	136	0	0	0	482	49	0
160 Painters and decorators nec	Deaths	0	13	0	0	0	1	6	3
	PMR	0	97	0	0	0	142	92	220
161 Electrical/ electronic assemblers	Deaths	6	334	9	1	1	15	165	34
	PMR	161	107	118	44	38	129	102	110
162 Instrument assemblers	Deaths	1	10	0	0	0	0	7	1
	PMR	943	95	0	0	0	0	138	114
163 Assemblers (vehicles and other	Deaths	3	170	4	0	2	6	97	17
metal goods)	PMR	166	102	105	0	163	105	114	108
164 Packers and sorters	Deaths	15	982	26	5	8	41	623	108
	PMR	116	95	99	63	86	102	116	104
165 Bricklayers and tilesetters	Deaths	0	0	0	0	0	0	1	0
	PMR	0	0	0	0	0	0	226	0
166 Masons and stonecutters	Deaths	0	0	0	0	0	0	0	0
	PMR	0	0	0	0	0	0	0	0
167 Plasterers	Deaths	0	0	0	0	0	0	0	0
	PMR	0	0	0	0	0	0	0	0
168 Roofers and glaziers	Deaths	0	1	0	0	0	0	0	0
	PMR	0	176	0	0	0	0	0	0
169 Builders etc.	Deaths	0	4	0	0	0	0	2	3
	PMR	0	81	0	0	0	0	94	767
170 Rail track workers	Deaths	0	0	0	0	0	0	0	0
	PMR	0	0	0	0	0	0	0	0

IX Diseases of the digestive system (520-579)	X Diseases of the genito-urinary system (520-629)	XI Complications of pregnancy childbirth and the puerperium (630-679)	XII Diseases of the skin subcutaneous tissue (680-709)	XIII Diseases of the musculo-skeletal system and connective tissue (710-739)	XIV Congenital anomalies (740-759)	XVI Symptoms signs and ill-defined conditions (780-799)	EXVII External causes of injury and poisoning (E800-E999)		Job group	
1	0	0	0	0	0	0	2	Deaths	Electrical and electronic	136
155	0	0	0	0	0	0	116	PMR	production fitters	
0	0	0	0	0	0	0	4	Deaths	Electricians	137
0	0	0	0	0	0	0	100	PMR		
0	0	0	0	0	0	0	0	Deaths	Electrical plant operators	138
0	0	0	0	0	0	0	0	PMR		
0	0	0	0	0	1	0	3	Deaths	Telephone fitters	139
0	0	0	0	0	1124	0	145	PMR		
0	0	0	0	0	0	0	1	Deaths	Electric cable and line workers	140
0	0	0	0	0	0	0	301	PMR		
0	0	0	0	0	0	0	0	Deaths	Radio and TV mechanics	141
0	0	0	0	0	0	0	0	PMR		
0	0	0	0	0	0	0	0	Deaths	Other electronic maintenance	142
0	0	0	0	0	0	0	0	PMR	engineers	
0	0	0	0	0	0	0	2	Deaths	Electrical engineers (so	143
0	0	0	0	0	0	0	193	PMR	described)	
0	0	0	0	0	0	0	0	Deaths	Plumbers and gas fitters	144
0	0	0	0	0	0	0	0	PMR		
0	0	1	0	0	0	0	0	Deaths	Sheet metal workers	145
0	0	2703	0	0	0	0	0	PMR		
0	0	0	0	0	0	0	2	Deaths	Metal plate workers	146
0	0	0	0	0	0	0	354	PMR		
0	0	0	0	0	0	0	0	Deaths	Steel erectors	147
0	0	0	0	0	0	0	0	PMR		
0	0	0	0	0	0	0	0	Deaths	Scaffolders	148
0	0	0	0	0	0	0	0	PMR		
2	2	0	0	3	0	0	9	Deaths	Welders	149
78	369	0	0	546	0	0	147	PMR		
0	0	0	0	0	0	0	0	Deaths	Riggers	150
0	0	0	0	0	0	0	0	PMR		
0	0	0	0	0	0	1	2	Deaths	Jewellery workers	151
0	0	0	0	0	0	2128	93	PMR		
0	0	0	0	0	0	0	1	Deaths	Engravers and etchers (printing)	152
0	0	0	0	0	0	0	153	PMR		
0	0	0	0	0	0	0	0	Deaths	Vehicle body builders	153
0	0	0	0	0	0	0	0	PMR		
0	0	0	0	0	0	0	1	Deaths	Oilers and greasers	154
0	0	0	0	0	0	0	173	PMR		
2	0	0	0	0	0	0	4	Deaths	Electronics wire workers	155
102	0	0	0	0	0	0	79	PMR		
2	1	0	0	2	0	0	7	Deaths	Coil winders	156
87	214	0	0	454	0	0	145	PMR		
1	1	0	0	0	0	0	8	Deaths	Pottery decorators	157
40	190	0	0	0	0	0	122	PMR		
0	0	0	0	0	0	0	0	Deaths	Coach painters	158
0	0	0	0	0	0	0	0	PMR		
0	0	0	0	2	0	0	1	Deaths	Other spray painters	159
0	0	0	0	1093	0	0	35	PMR		
0	1	0	0	0	0	0	5	Deaths	Painters and decorators nec	160
0	541	0	0	0	0	0	115	PMR		
11	4	1	1	1	6	0	24	Deaths	Electrical/ electronic assemblers	161
61	91	121	276	33	190	0	47	PMR		
0	0	0	0	0	0	0	0	Deaths	Instrument assemblers	162
0	0	0	0	0	0	0	0	PMR		
7	2	1	0	2	4	0	5	Deaths	Assemblers (vehicles and other	163
73	88	299	0	123	269	0	22	PMR	metal goods)	
52	18	2	0	6	24	4	147	Deaths	Packers and sorters	164
87	120	60	0	58	211	127	80	PMR		
0	0	0	0	0	0	0	1	Deaths	Bricklayers and tilesetters	165
0	0	0	0	0	0	0	234	PMR		
0	0	0	0	0	0	0	0	Deaths	Masons and stonecutters	166
0	0	0	0	0	0	0	0	PMR		
0	0	0	0	0	0	0	0	Deaths	Plasterers	167
0	0	0	0	0	0	0	0	PMR		
0	0	0	0	0	0	0	0	Deaths	Roofers and glaziers	168
0	0	0	0	0	0	0	0	PMR		
0	0	0	0	0	0	0	0	Deaths	Builders etc.	169
0	0	0	0	0	0	0	0	PMR		
0	0	0	0	0	0	0	0	Deaths	Rail track workers	170
0	0	0	0	0	0	0	0	PMR		

Appendix 8 Mortality of women aged 20-64 by cause of deaths and job group, England and Wales, 1979-80, 1982-90

Job group		I Infections and parasitic diseases (001-139)	II Neoplasms (140-239)	III Endocrine nutritional and metabolic diseases and immunity disorders (240-279)	IV Diseases of blood and blood-forming organs (280-289)	V Mental disorders (290-319)	VI Diseases of the nervous system and and sense organs (320-389)	VII Diseases of the circulatory system (390-459)	VIII Diseases of the respiratory system (460-519)
171 Road construction workers and paviors	Deaths	0	1	0	0	0	0	0	0
	PMR	0	176	0	0	0	0	0	0
172 Sewage plant attendants	Deaths	0	0	0	0	0	0	0	0
	PMR	0	0	0	0	0	0	0	0
173 Mains and service layers	Deaths	0	0	0	0	0	0	0	0
	PMR	0	0	0	0	0	0	0	0
174 Construction workers nec	Deaths	0	3	0	0	0	1	2	0
	PMR	0	89	0	0	0	649	123	0
175 Face-trained coal miners	Deaths	0	1	0	0	0	0	0	0
	PMR	0	171	0	0	0	0	0	0
176 Miners (not coal) and quarry workers	Deaths	0	0	0	0	0	0	1	0
	PMR	0	0	0	0	0	0	322	0
177 Railway guards	Deaths	0	2	0	0	0	0	0	0
	PMR	0	122	0	0	0	0	0	0
178 Railway signal workers	Deaths	0	6	1	0	0	0	0	0
	PMR	0	97	735	0	0	0	0	0
179 Shunters and points workers	Deaths	0	0	0	0	0	0	0	0
	PMR	0	0	0	0	0	0	0	0
180 Railway engine drivers	Deaths	0	0	0	0	0	0	0	0
	PMR	0	0	0	0	0	0	0	0
181 Road transport inspectors	Deaths	0	1	0	0	0	0	1	0
	PMR	0	45	0	0	0	0	102	0
182 Bus and coach drivers	Deaths	0	21	0	0	0	0	11	1
	PMR	0	92	0	0	0	0	112	56
183 Lorry drivers	Deaths	0	82	3	0	0	3	27	9
	PMR	0	97	123	0	0	75	73	124
184 Other motor drivers	Deaths	0	36	1	0	1	1	12	0
	PMR	0	104	114	0	361	68	85	0
185 Bus conductors and drivers' mates	Deaths	1	99	0	0	1	4	52	15
	PMR	112	97	0	0	160	133	96	149
186 Mechanical plant drivers	Deaths	0	0	0	0	0	0	0	0
	PMR	0	0	0	0	0	0	0	0
187 Crane drivers	Deaths	0	0	0	0	0	0	6	1
	PMR	0	0	0	0	0	0	272	260
188 Fork lift truck drivers	Deaths	0	7	0	0	0	1	5	0
	PMR	0	87	0	0	0	364	134	0
189 Slingers	Deaths	0	0	0	0	0	0	0	0
	PMR	0	0	0	0	0	0	0	0
190 Storekeepers	Deaths	2	327	8	3	0	7	151	24
	PMR	62	107	111	160	0	67	96	82
191 Dockers and goods porters	Deaths	0	5	0	0	0	0	3	2
	PMR	0	101	0	0	0	0	97	375
192 Refuse collectors	Deaths	0	4	0	0	0	0	0	0
	PMR	0	192	0	0	0	0	0	0
193 Labourers in coke ovens	Deaths	0	2	0	0	0	0	0	0
	PMR	0	245	0	0	0	0	0	0
194 Boiler operators	Deaths	0	0	0	0	0	0	1	0
	PMR	0	0	0	0	0	0	317	0

IX Diseases of the digestive system (520-579)	X Diseases of the genito-urinary system (520-629)	XI Complications of pregnancy childbirth and the puerperium (630-679)	XII Diseases of the skin subcutaneous tissue (680-709)	XIII Diseases of the musculo-skeletal system and connective tissue (710-739)	XIV Congenital anomalies (740-759)	XVI Symptoms signs and ill-defined conditions (780-799)	EXVII External causes of injury and poisoning (E800-E999)		Job group	
0	0	0	0	0	0	0	0	Deaths	Road construction workers and paviors	171
0	0	0	0	0	0	0	0	PMR		
0	0	0	0	0	0	0	0	Deaths	Sewage plant attendants	172
0	0	0	0	0	0	0	0	PMR		
0	0	0	0	0	0	0	0	Deaths	Mains and service layers	173
0	0	0	0	0	0	0	0	PMR		
0	0	0	0	0	0	0	1	Deaths	Construction workers nec	174
0	0	0	0	0	0	0	107	PMR		
0	0	0	0	0	0	0	0	Deaths	Face-trained coal miners	175
0	0	0	0	0	0	0	0	PMR		
0	0	0	0	0	0	0	0	Deaths	Other miners (not coal) and quarry workers	176
0	0	0	0	0	0	0	0	PMR		
0	0	0	0	0	0	1	0	Deaths	Railway guards	177
0	0	0	0	0	0	33333	0	PMR		
0	0	0	0	0	0	0	4	Deaths	Railway signal workers	178
0	0	0	0	0	0	0	653	PMR		
0	0	0	0	0	0	0	0	Deaths	Shunters and points operators	179
0	0	0	0	0	0	0	0	PMR		
0	0	0	0	0	0	0	0	Deaths	Railway engine drivers	180
0	0	0	0	0	0	0	0	PMR		
0	0	0	0	0	0	0	3	Deaths	Road transport inspectors	181
0	0	0	0	0	0	0	332	PMR		
1	1	0	0	1	1	0	5	Deaths	Bus and coach drivers	182
73	386	0	0	377	398	0	152	PMR		
5	2	1	0	1	1	0	32	Deaths	Lorry drivers	183
99	189	261	0	89	99	0	172	PMR		
1	0	0	0	0	0	0	13	Deaths	Other motor drivers	184
47	0	0	0	0	0	0	200	PMR		
9	2	0	0	0	0	0	9	Deaths	Bus conductors and drivers' mates	185
160	156	0	0	0	0	0	95	PMR		
0	0	0	0	0	0	0	0	Deaths	Mechanical plant drivers	186
0	0	0	0	0	0	0	0	PMR		
0	0	0	0	0	0	0	0	Deaths	Crane drivers	187
0	0	0	0	0	0	0	0	PMR		
0	0	0	0	1	0	0	1	Deaths	Fork lift truck drivers	188
0	0	0	0	1087	0	0	84	PMR		
0	0	0	0	0	0	0	0	Deaths	Slingers	189
0	0	0	0	0	0	0	0	PMR		
15	7	0	0	4	1	1	35	Deaths	Storekeepers	190
87	171	0	0	134	40	134	89	PMR		
0	0	0	0	0	0	0	0	Deaths	Dockers and goods porters	191
0	0	0	0	0	0	0	0	PMR		
0	0	0	0	0	0	0	0	Deaths	Refuse collectors	192
0	0	0	0	0	0	0	0	PMR		
0	0	0	0	0	0	0	0	Deaths	Labourers in coke ovens	193
0	0	0	0	0	0	0	0	PMR		
0	0	0	0	0	0	0	0	Deaths	Boiler operators	194
0	0	0	0	0	0	0	0	PMR		

Abbreviations

CI	confidence interval
CO80	OPCS Classification of Occupations (1980)
CODOT	Classification of Occupations and Directory of Occupational Titles (ED)
CSO	Central Statistical Office
EC	European Community
ED	Employment Department
FHSA	Family Health Services Authority
GHS	General Household Survey
HSE	Health and Safety Executive
ICD9	International Classification of Diseases, 9th Revision
II	Industrial injuries
ISCO 88	International Standard Classification of Occupations
LFS	Labour Force Survey
LS	Longitudinal Study
MRC	Medical Research Council
MSGP	Morbidity Statistics from General Practice
NADOR	Notification of Accidents and Dangerous Occurrences Regulations (1981-86)
nec	Not elsewhere classified
NHSCR	National Health Service Central Register
OPCS	Office of Population Censuses and Surveys
RHA	Regional Health Authority
RIDDOR	Reporting of Injuries, Diseases and Dangerous Occupations Regulations (1987-)
SEG	Socio-economic group
SIC(80)	Standard Industrial Classification (revised 1980)
SIC(92)	Standard Industrial Classification (revised 1992)
SOC	Standard Occupational Classification (1990)
WHO	World Health Organisation